5

Springer Series in Chemical Physics

Edited by Fritz Peter Schäfer

Springer Series in Chemical Physics

Editors: V. I. Goldanskii R. Gomer F. P. Schäfer J. P. Toennies

Volume 1 **Atomic Spectra and Radiative Transitions** By I. I. Sobelman

Volume 2 **Surface Crystallography by LEED** Theory, Computation and Structural Results By M. A. Van Hove, S. Y. Tong

Volume 3 **Advances in Laser Chemistry** Editor: A. H. Zewail

Volume 4 **Picosecond Phenomena** Editors: C. V. Shank, E. P. Ippen, S. L. Shapiro

Volume 5 **Laser Spectroscopy** Basic Concepts and Instrumentation By W. Demtröder

Volume 6 **Laser-Induced Processes in Molecules** Physics and Chemistry Editors: K. L. Kompa, S. D. Smith

Volume 7 **Excitation of Atoms and Broadening of Spectral Lines** By I. I. Sobelman, L. A. Vainshtein, E. A. Yukov

Volume 8 **Spin Exchange** Principles and Applications in Chemistry and Biology By Yu. N. Molin, K. M. Salikhov, K. I. Zamaraev

Volume 9 **Secondary Ion Mass Spectrometry SIMS II** Editors: A. Benninghoven, C. A. Evans, Jr., R. A. Powell, R. Shimizu, H. A. Storms

Volume 10 **Lasers and Chemical Change** By A. Ben-Shaul, Y. Haas, K. L. Kompa, R. D. Levine

Volume 11 **Liquid Crystals of One- and Two-Dimensional Order** Editors: W. Helfrich, G. Heppke

Volume 12 **Gasdynamic Laser** By S. A. Losev

Volume 13 **Atomic Many-Body Theory** By I. Lindgren, J. Morrison

Volume 14 **Picosecond Phenomena II** Editors: R. M. Hochstrasser, W. Kaiser, C. V. Shank

Volume 15 **Vibrational Spectroscopy of Adsorbates** Editor: R. F. Willis

Volume 16 **Spectroscopy of Molecular Excitons** By V. L. Broude, E. I. Rashba, E. F. Sheka

Volume 17 **Inelastic Particle-Surface Collisions** Editors: E. Taglauer, W. Heiland

Volume 18 **Modelling of Chemical Reaction Systems** Editors: K. H. Ebert, P. Deuflhard, W. Jäger

Volume 19 **Secondary Ion Mass Spectrometry SIMS III** Editors: A. Benninghoven, J. Giber, J. László, M. Riedel, H. W. Werner

Volume 20 **Chemistry and Physics of Solid Surfaces IV** Editors: R. Vanselow, R. Howe

Volume 21 **Dynamics of Gas-Surface Interaction** Editors: G. Benedek, U. Valbusa

Volume 22 **Nonlinear Laser Chemistry** Multiple - Photon Excitation By V.S. Letokhov

Volume 23 **Picosecond Phenomena III** Editors: K.B. Eisenthal, R.M. Hochstrasser, W. Kaiser, A. Laubereau

Wolfgang Demtröder

Laser Spectroscopy

Basic Concepts and Instrumentation

Second Corrected Printing

With 431 Figures

Springer-Verlag Berlin Heidelberg New York 1982

Professor Dr. Wolfgang Demtröder
Fachbereich Physik, Universität Kaiserslautern
D-6750 Kaiserslautern, Fed. Rep. of Germany

Series Editors

Professor Vitalii I. Goldanskii
Institute of Chemical Physics
Academy of Sciences
Vorobyevskoye Chaussee 2-b
Moscow V-334, USSR

Professor Robert Gomer
The James Franck Institute
The University of Chicago
5640 Ellis Avenue
Chicago, IL 60637, USA

Professor Dr. Fritz Peter Schäfer
Max-Planck-Institut für
Biophysikalische Chemie
D-3400 Göttingen-Nikolausberg
Fed. Rep. of Germany

Professor Dr. J. Peter Toennies
Max-Planck-Institut für Strömungsforschung
Böttingerstraße 6–8
D-3400 Göttingen
Fed. Rep. of Germany

ISBN 3-540-10343-0 Springer-Verlag Berlin Heidelberg New York
ISBN 0-387-10343-0 Springer-Verlag New York Heidelberg Berlin

Library of Congress Cataloging in Publication Data. Demtröder, W. Laser spectroscopy. (Springer series in chemical physics ; v. 5) Includes bibliographical references and index. 1. Laser spectroscopy. I. Title. II. Series. QC454.L3D46 535.5'8 80-24952

Printed in Germany

Offset printing: Beltz Offsetdruck, Hemsbach/Bergstr. Bookbinding: J. Schäffer oHG, Grünstadt.
2153/3130-543210

Preface

The impact of lasers on spectroscopy can hardly be overestimated. Lasers represent intense light sources with spectral energy densities which may exceed those of incoherent sources by several orders of magnitude. Furthermore because of their extremely small bandwidth, single-mode lasers allow a spectral resolution which far exceeds that of conventional spectrometers. Many experiments which could not be done before the application of lasers because of lack of intensity or insufficient resolution are readily performed with lasers.

Now several thousands of laser lines are known which span the whole spectral range from the vacuum-ultraviolet to the far-infrared region. Of particular interest are the continuously tunable lasers which may in many cases replace wavelength-selecting elements, such as spectrometers or interferometers. In combination with optical frequency mixing techniques such continuously tunable monochromatic coherent light sources are available at nearly any desired wavelength above 100 nm.

The high intensity and spectral monochromasy of lasers have opened a new class of spectroscopic techniques which allow investigation of the structure of atoms and molecules in much more detail. Stimulated by the variety of new experimental possibilities that lasers give to spectroscopists, very lively research activities have developed in this field, as manifested by an avalanche of publications. A good survey about recent progress in laser spectroscopy is given by the proceedings of various conferences on laser spectroscopy (see "Springer Series in Optical Sciences"), on picosecond phenomena (see "Springer Series in Chemical Physics"), and by several quasi-monographs on laser spectroscopy published in "Topics in Applied Physics".

For nonspecialists, however, or for people who are just starting in this field, it is often difficult to find out of many articles scattered over many journals a coherent representation of the basic principles of laser spectroscopy. This textbook intends to close this gap between the advanced research papers and the representation of fundamental principles and experimental techniques. It is addressed to physicists and chemists who want to study laser spectroscopy in more detail. Students who have some knowledge in atomic and molecular physics, in electrodynamics, and in optics should be able to follow the presentation.

The fundamental principles of lasers are covered only very briefly because many excellent textbooks on lasers already exist.

On the other hand, those characteristics of the laser, which are important for its applications in spectroscopy are treated in more detail. Examples are the frequency spectrum of different types of lasers, their linewidths, amplitude and frequency stability, tunability, and tuning ranges. The optical components such as mirrors, prisms, and gratings, and the experimental equipment of spectroscopy, for example monochromators, interferometers, photon detectors, etc., are discussed extensively because detailed knowledge of modern spectroscopic equipment may be crucial for the successful performance of an experiment.

Each chapter gives several examples to illustrate the subject discussed. Problems at the end of each chapter may serve as a test of the reader's understanding. The literature cited for each chapter is, of course, not complete but should inspire further studies. Many subjects which could be covered only briefly in this book can be found in the references in a more detailed and often more advanced treatment. The literature selection does not represent any priority list but has didactical purposes and intends to illustrate the subject of each chapter more thoroughly.

The spectroscopic applications of lasers covered in this book are restricted to the spectroscopy of free atoms, molecules, or ions. There exists, of course, a wide range of applications in plasma physics, solid-state physics, or fluid dynamics which are not discussed because they are beyond the scope of this book. It is hoped that this book may be of help to both students and researchers. Although it is meant as an introduction to laser spectroscopy, it may also facilitate the understanding of advanced papers on special subjects in laser spectroscopy. Since laser spectroscopy is a very fascinating field of research, the author would be happy if this book can transfer to the reader some of his excitement and pleasure experienced in the laboratory while looking for new lines or unexpected results.

I want to thank many people who have helped to complete this book. In particular there are the students in my research group who by their experimental work have contributed to many of the examples given for illustration and who have spent their time to read the galley proofs. I am grateful to colleagues from many laboratories who have supplied me with figures from their publications. Special thanks are due to Mrs. Keck and Mrs. Ofiara who typed the manuscript and to Mrs. Wollscheid and Mrs. Ullmer who made the drawings. Last but not least I would like to thank Dr. U. Hebgen, Dr. H. Lotsch, Mr. K.-H. Winter, and other coworkers of Springer-Verlag who showed much patience with a dilatory author and who tried hard to complete the book in a short time.

Kaiserslautern, March 1981 *W. Demtröder*

Contents

1. Introduction

Most of our knowledge about the structure of atoms and molecules is based on spectroscopic investigations. Thus spectroscopy has made an outstanding contribution to the present state of atomic and molecular physics. Information on molecular structure and on the interaction of molecules with their surroundings may be derived in various ways from the absorption or emission spectra generated when electromagnetic radiation interacts with matter.

Wavelength measurements of spectral lines allow the determination of energy levels of the atomic or molecular system. The *line intensity* is proportional to the transition probability which measures how strongly the two levels of a molecular transition are coupled. Since the transition probability depends on the wave functions of both levels, intensity measurements are useful to verify the spatial charge distribution of excited electrons, which can only be roughly calculated from approximate solutions of the Schrödinger equation. The *natural linewidth* of a spectral line may be resolved by special techniques, allowing mean lifetimes of excited molecular states to be determined. Measurements of the *Doppler width* yield the velocity distribution of the emitting or absorbing molecules and with it the temperature of the sample. From *pressure broadening* and *pressure shifts* of spectral lines, information about collision processes and interatomic potentials can be extracted. *Zeemann* and *Stark splittings* by external magnetic or electric fields are important means of measuring magnetic or electric moments and elucidating the coupling of the different angular momenta in atoms or molecules, even with complex electron configurations. The *hyperfine structure* of spectral lines yields information about the interaction between the nuclei and the electron cloud and allows nuclear magnetic dipole moments or electric quadrupole moments to be determined.

These examples represent only a small selection of the many possible ways by which spectroscopy provides tools to explore the microworld of atoms and molecules. However, the amount of information which can be extracted from a spectrum depends essentially on the attainable spectral resolution and on the detection sensitivity that can be achieved.

The application of new technologies to optical instrumentation (for instance, the production of larger and better ruled gratings in spectrographs, the use of highly reflecting dielectric coatings in interferometers, and the development of image intensifiers) has certainly significantly extended the sensitivity limits. Considerable progress was furthermore achieved through the introduction of new spectroscopic techniques such as Fourier spectroscopy, optical pumping, level-crossing techniques, and various kinds of double-resonance methods and molecular beam spectroscopy.

Although these new techniques have proved to be very fruitful, the really stimulating impetus to the whole field of spectroscopy was given by the introduction of lasers. In many cases these new spectroscopic light sources may increase spectral resolution and sensitivity by several orders of magnitude. Combined with new spectroscopic techniques, lasers are able to surpass basic limitations of classical spectroscopy. Many experiments that could not be performed with incoherent light sources are now feasible or have already been successfully completed recently. This book deals with such new techniques of laser spectroscopy and explains the necessary instrumentation.

The book begins with a discussion of the fundamental definitions and concepts of classical spectroscopy, such as thermal radiation, induced and spontaneous emission, radiation power and intensity, transition probabilities and oscillator strengths, linear and nonlinear absorption and dispersion, and coherent and incoherent radiation fields. In order to understand the theoretical limitations of spectral resolution in classical spectroscopy, the next chapter treats the different causes of the broadening of spectral lines. Numerical examples at the end of each section illustrate the order of magnitude of the different effects.

The contents of Chap.4, which covers spectroscopic instrumentation and its application to wavelength and intensity measurements, are essential for the experimental realization of laser spectroscopy. Although spectrographs and monochromators, which played a major rule in classical spectroscopy, may be abandoned for many experiments in laser spectroscopy, there are still numerous applications where these instruments are indispensible. Of major importance for laser spectroscopists are the different kinds of interferometers. They are used not only in laser resonators to realize single mode operation, but also for line-profile measurements of spectral lines and for very precise wavelength measurements. Since the determination of wavelength is a central problem in spectroscopy, a whole section discusses the different modern techniques and their accuracy.

Lack of intensity is one of the major limitations in many spectroscopic investigations. It is therefore often vital for the experimentalist to choose the proper light detector. Section 4.5 surveys several light detectors and sensitive techniques such as photon counting, which is coming more and more into use.

This chapter concludes the first part of the book, which covers fundamental concepts and basic instrumentation of general spectroscopy. The second part discusses in more detail subjects more specific to laser spectroscopy.

Chapter 5 is a short recapitulation of the fundamentals of lasers, such as threshold conditions, optical resonators, and laser modes. Only those laser characteristics that are important in laser spectroscopy are discussed here. For a more detailed treatment the reader is referred to the extensive laser literature, for example [1.1-5].

The next chapter discusses in more detail the basic properties and techniques which make the laser such an attractive spectroscopic light source. The important questions of wavelength stabilization and continuous wavelength tuning are treated, and experimental realizations of single-mode tunable lasers are presented.

Different types of tunable coherent light sources have been developed for the different spectral regions. Chapter 7 gives a brief survey of the various types, discussing their advantages and limitations. This chapter ends the second part of the book devoted to the basic concepts and the instrumentation of laser spectroscopy.

The third part presents various applications of lasers in spectroscopy and discusses the different methods that have been developed recently. Chapter 8 covers the field of laser absorption spectroscopy with its different high-sensitivity techniques such as intracavity absorption, excitation spectroscopy, photoacoustic spectroscopy, laser magnetic resonance, and methods that allow single atoms to be detected. Furthermore several interesting spectroscopic techniques are presented which are based on optical pumping with lasers. The large population densities achievable in excited states by optical pumping with lasers greatly facilitate the spectroscopy of these states.

Of particular interest are the high-lying Rydberg states of atoms or molecules, which can be investigated with high resolution. While the spectral resolution in all examples discussed in this chapter is limited in principle by the Doppler width of the absorbing molecules, several methods which overcome this limitation and which allow essentially "Doppler-free" absorption spectroscopy are explained in Chap.10.

Raman spectroscopy has been revolutionized by the use of lasers. Not only spontaneous Raman spectroscopy with greatly enhanced sensitivity, but also new techniques such as induced Raman spectroscopy or coherent anti-Stokes Raman spectroscopy have contributed greatly to the rapid development of sensitive, high-resolution detection techniques.

Really impressive progress towards higher spectral resolution has been achieved by the development of various "Doppler-free" techniques; these are discussed extensively in Chap.10. They rely either on linear spectroscopy using single-mode tunable lasers or on nonlinear spectroscopy where the high intensity of the laser is essential. Several Doppler-free techniques, such as level crossing spectroscopy or the various double-resonance methods which had been already applied in *atomic* spectroscopy using incoherent atomic resonance lamps, have been extended to molecular spectroscopy with the help of lasers.

Also in the *time domain*, investigations of transient phenomena with short laser pulses can reach a time resolution below 10^{-12} s, hitherto unobtained. This picosecond spectroscopy has unravelled many interesting fast-relaxation phenomena which had been unknown before. Some time resolving techniques, such as quantum beats, photon echoes, and pulsed Fourier transform spectroscopy, are outlined briefly.

Besides spectroscopy, the investigation of atomic and molecular scattering processes is the second most important source of information about atomic structure and interatomic potentials. The introduction of lasers to the studies of collision processes supports a closer cooperation between spectroscopists and "scattering people". The relevance of this new field of laser spectroscopy of collision processes is outlined in Chap.12.

The interesting question of whether there exists a fundamental resolution limit, is discussed in Chap.13. Some recently developed techniques such as optical Ramsay fringes and trapping and cooling of atoms, illustrate how laser spectroscopy approaches this ultimate limit of spectral resolution.

The last chapter illustrates the expansion of laser spectroscopy into other fields by summarizing some of its possible applications. Although the increasing relevance of laser spectroscopy to numerous applications in science and technology can be predicted, the rate of this increase will depend on the technical development and reliability of versatile, low-cost tunable lasers.

This book is intended as an *introduction* to the basic methods and instrumentation of laser spectroscopy. The examples in each chapter illustrate the text and may suggest other possible applications. They are of course not complete, and have been selected from the literature or from our own laboratory work for didactic purposes and may not represent the priorities of publication dates. For a far more extensive survey of the latest publications in the wide field of laser spectroscopy the reader is referred to the proceedings of various conferences on laser spectroscopy [1.6-11] and to several excellent books which cover the recent literature of different fields in laser spectroscopy [1.12-19].

2. Absorption and Emission of Light

This chapter contains basic considerations about absorption, emission, and dispersion of electromagnetic waves (E.M.) interacting with matter. Especially emphasized are those aspects which are important for the spectroscopy of gaseous media. The discussion starts with thermal radiation fields and the concept of cavity modes in order to elucidate differences and connections between spontaneous and induced emission and absorption. This leads to the definition of Einstein coefficients and their mutual relations. The next section explains some definitions used in photometry such as radiation power, intensity, and spectral power density.

It is possible to understand many phenomena in optics and spectroscopy in terms of classical models based on concepts of classical electrodynamics. For example, the absorption and dispersion of electromagnetic waves in matter can be described using the model of damped oscillators for the atomic electrons. This model leads to a complex refractive index and to *dispersion relations* which give the link between absorption and dispersion. It is not too difficult to give a quantum mechanical formulation of the classical results in most cases. The semiclassical approach will be outlined briefly.

The high laser intensities which spectroscopists use, often result in a partial or complete depletion of molecules in the absorbing state. This saturation effect leads to nonlinear absorption phenomena which will be treated in Sect.2.8.

The intensities of spectral lines are proportional to the transition probabilities of the corresponding atomic or molecular transitions. Sections 2.7 and 2.9 cover some experimental and theoretical methods for the determination of transition probabilities and lifetimes of excited states.

Many experiments in laser spectroscopy depend on the coherence properties of the radiation. Some basic ideas of temporal and spatial coherence are therefore discussed at the end of this chapter.

Throughout this book the term "light" is often used for electromagnetic radiation of all spectral regions. Likewise the term "molecule" in general statements includes atoms as well. We shall, however, restrict the discussion and most of the examples to *gaseous* media, which means essentially free atoms or molecules.

For more detailed or more advanced presentations of the subject summarized in this chapter, the reader is referred to the extensive literature on spectroscopy [2.1-6].

2.1 Cavity Modes

Consider a cubic cavity with sides L at a temperature T. The walls of the cavity absorb and emit electromagnetic radiation. At thermal equilibrium the absorbed power $P_a(\omega)$ has to be equal to the emitted power $P_e(\omega)$ for all frequencies ω. Inside the cavity there is a *stationary radiation field* $\underline{E}$ which can be described by a superposition of plane waves with amplitudes $\underline{A}_p$, wave vectors $\underline{k}_p$, and frequencies ω_p,

$$\underline{E} = \sum_p \underline{A}_p \cdot e^{i(\omega_p t - \underline{k}_p \cdot \underline{r})} + \text{compl. conj.} \quad . \tag{2.1}$$

The waves are reflected at the walls of the cavity. For each wave vector $\underline{k} = (k_x, k_y, k_z)$ this leads to 8 possible combinations $\underline{k} = (\pm k_x, \pm k_y, \pm k_z)$ which interfere with each other. A stationary field configuration only occurs if this superpositions result in *standing waves*. This imposes boundary conditions for the wave vector,

$$\underline{k} = \frac{\pi}{L}(n_1, n_2, n_3) \tag{2.2}$$

where n_1, n_2, n_3 are positive integers.

The magnitudes of the wave vectors allowed by the boundary conditions are

$$|\underline{k}| = \frac{\pi}{L} \cdot \sqrt{n_1^2 + n_2^2 + n_3^2} \quad , \tag{2.3}$$

which can be written in terms of the wavelength $\lambda = 2\pi/|\underline{k}|$ or the frequencies $\omega = c \cdot |\underline{k}|$

$$L = \frac{\lambda}{2} \cdot \sqrt{n_1^2 + n_2^2 + n_3^2} \quad \text{or} \quad \omega = \frac{\pi \cdot c}{L} \sqrt{n_1^2 + n_2^2 + n_3^2} \quad . \tag{2.4}$$

These standing waves are called *cavity modes* (Fig.2.1.a).

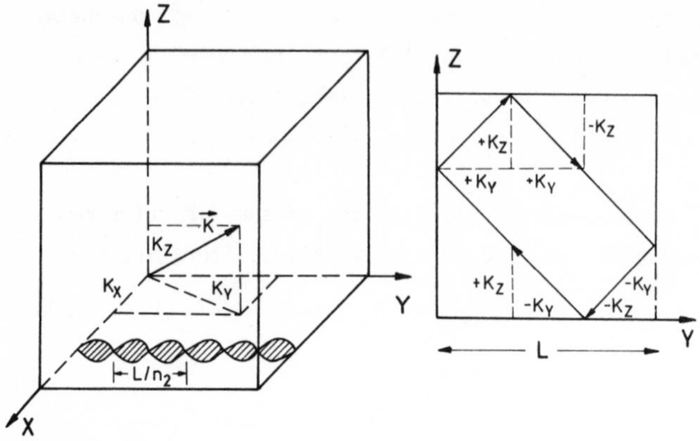

Fig.2.1a,b. Modes of a stationary E.M. field in a cavity. (a) Standing waves in a cubic cavity; (b) superposition of possible k vectors to form standing waves, illustrated in a two-dimensional coordinate system

Since the amplitude vector $\underline{A}$ of a transverse wave $\underline{E}$ is always perpendicular to the wave vector $\underline{k}$, it can be composed of two components a_1 and a_2 with unit vectors $\hat{e}_1$ and $\hat{e}_2$

$$(\hat{e}_1 \cdot \hat{e}_2 = \delta_{12} \quad ; \quad \hat{e}_1, \hat{e}_2 \perp \underline{k}) \quad ,$$

$$\underline{A} = a_1\hat{e}_1 + a_2\hat{e}_2 \quad . \tag{2.5}$$

The complex numbers a_1, a_2 define the polarization of the standing wave. Equation (2.5) states that any arbitrary polarization can be always expressed by a linear combination of two mutually orthogonal linear polarizations. To each cavity mode defined by the wave vector $\underline{k}_p$ therefore belong two possible polarization states. This means

each triple of integers (n_1, n_2, n_3) represents two cavity modes. Any arbitrary stationary field configuration can be expressed as a linear combination of cavity modes.

We will now investigate how many modes with frequencies $\omega \leq \omega_m$ are possible. Because of the boundary condition (2.2) this number is equal to the number of all integer triples (n_1, n_2, n_3) which fulfil the condition $c^2k^2 = \omega^2 \leq \omega_m^2$ (Fig.2.2).

In a coordinate system with the coordinates (π/L) (n_1, n_2, n_3), each triple (n_1, n_2, n_3) represents a lattice point in a three-dimensional lattice with

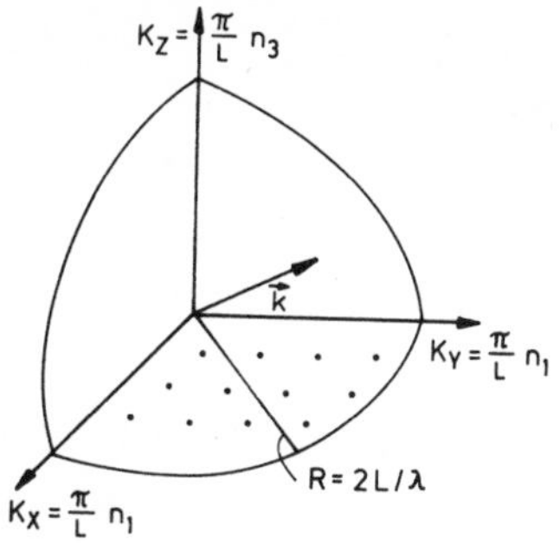

Fig.2.2. Illustration of the calculation of the maximum number of modes in momentum space

a lattice constant of unity. In this system, (2.4) describes a sphere with radius $2L/\lambda$ or $\omega L/(\pi c)$. If this radius is large compared to unity, which means that $2L \gg \lambda_m$, the number of lattice points (n_1, n_2, n_3) with $\omega^2 \leqq \omega_m^2$ is roughly given by the volume of the octant of the sphere shown in Fig.2.2. With the two possible polarization states of each mode, one therefore obtains for the number of allowed modes with frequencies between $\omega = 0$ and ω_m in a cubic cavity of volume L^3 with $L \gg \lambda$

$$N(\omega_m) = 2 \cdot \frac{1}{8} \cdot \frac{4\pi}{3} \left(\frac{L\omega_m}{\pi c}\right)^3 = \frac{1}{3} \cdot \frac{L^3\omega_m^3}{\pi^2 c^3} \quad . \tag{2.6}$$

N/L^3 yields the number of modes per unit volume.

It is often interesting to know the number $n(\omega)d\omega$ of modes per unit volume within a certain frequency interval $d\omega$, for instance within the width of a spectral line. The spectral mode density $n(\omega)$ can be obtained directly from (2.6) by differentiating $N(\omega)/L^3$ with respect to ω. $N(\omega)$ is assumed to be a continuous function of ω, which is strictly speaking only the case for $L \to \infty$. We get

$$n(\omega)d\omega = \frac{\omega^2}{\pi^2 c^3} d\omega \quad . \tag{2.7a}$$

In spectroscopy the frequency $\nu = \omega/2\pi$ is often used instead of the angular frequency ω. The number of modes per unit volume within the frequency interval $d\nu$ is then

$$\boxed{n(\nu)d\nu = \frac{8\pi\nu^2}{c^3} d\nu} \quad . \tag{2.7b}$$

Examples

a) In the visible part of the spectrum (λ = 500 nm; $\nu = 6\times10^{14}\ s^{-1}$), (2.7b) yields for the number of modes per m^3 within the Doppler width of a spectral line ($d\nu = 10^9\ s^{-1}$)

$$n(\nu)d\nu = 3\times10^{14}\ m^{-3} \quad .$$

b) In the microwave region (λ = 1 cm; $\nu = 3\times10^{10}\ s^{-1}$) the number of modes per m^3 within the typical Doppler width $d\nu = 10^5\ s^{-1}$ is only $n(\nu)d\nu = 10^2\ m^{-3}$.

c) In the X-ray region (λ = 1 nm; $\nu = 3\times10^{17}\ s^{-1}$), one finds $n(\nu)d\nu = 8.4\times10^{21}\ m^{-3}$ within a typical natural linewidth $d\nu = 10^{11}\ s^{-1}$ of an X-ray transition.

2.2 Thermal Radiation and Planck's Law

In classical thermodynamics each degree of freedom of a system in thermal equilibrium has the mean energy $\frac{1}{2}kT$. Since classical oscillators have kinetic as well as potential energy their mean energy is kT, where k is the Boltzmann constant. If this classical concept is applied to the electromagnetic field discussed in Sect.2.1, each mode would represent a classical oscillator with mean energy kT. According to (2.7b) the spectral energy density of the radiation field would therefore be

$$\rho(\nu)d\nu = n(\nu)kT\,d\nu = \frac{8\pi\nu^2 k}{c^3}T\,d\nu \quad . \tag{2.8}$$

This "Rayleigh-Jeans" law matches the experimental results fairly well at low frequencies (in the infrared region) but is in strong disagreement with experiment at higher frequencies (in the ultraviolet region).

In order to explain this discrepancy, Planck suggested in 1900 that each mode of the radiation field can only emit or absorb energy in discrete amounts $qh\nu$, which are integer multiples q of a minimum energy quantum $h\nu$. These energy quanta $h\nu$ are called *photons*. Planck's constant h can be determined from experiments. *A mode with* q *photons therefore has energy* $qh\nu$.

In thermal equilibrium the partition of the total energy into the different modes is governed by the Maxwell-Boltzmann distribution, so that the probability $p(q)$ that a mode contains the energy $qh\nu$ is

$$p(q) = (1/Z)e^{-qh\nu/kT} \quad , \tag{2.9}$$

where k is the Boltzmann constant and

$$Z = \sum_q \exp(-q \cdot h\nu/kT) \tag{2.10}$$

is the partition function summed over all modes. Z acts as a normalization factor which makes $\sum_q p(q) = 1$, as can be immediately seen by inserting (2.10) into (2.9).

The mean energy per mode is therefore

$$W = \sum_{q=0}^{\infty} p(q)qh\nu = \frac{1}{Z}\sum_{q=0}^{\infty} qh\nu e^{-qh\nu/kT} \quad . \tag{2.11}$$

The evaluation of the sum yields (see for instance [2.6])

$$W = \frac{h\nu}{e^{h\nu/kT} - 1} \quad . \tag{2.12}$$

The thermal radiation field has an energy density $\rho(\nu)d\nu$ within a frequency interval ν to $\nu + d\nu$ which is equal to the number $n(\nu)d\nu$ of modes in the interval $d\nu$ times the mean energy W per mode. Using (2.7b) and (2.12) one obtains

$$\boxed{\rho(\nu)d\nu = \frac{8\pi\nu^2}{c^3} \cdot \frac{h\nu}{e^{h\nu/kT} - 1} \cdot d\nu} \quad . \tag{2.13}$$

This is Planck's famous radiation law (Fig.2.3) which predicts a spectral energy density of the thermal radiation that is fully consistent with experiments. The expression "thermal radiation" comes from the fact that the spectral energy distribution (2.13) is characteristic of a radiation field which is in thermal equilibrium with its surroundings (in Sect.2.1 the surroundings were the cavity walls).

The thermal radiation field described by its energy density $\rho(\nu)$ is *isotropic*. This means that through any transparent surface element dA of a sphere containing a thermal radiation field, the same energy flux dW/dt per second is emitted into the solid angle $d\Omega$ at an angle ϑ to the surface normal

$$dW/dt = \frac{c}{4\pi} \rho(\nu) \, dA \, d\Omega \, d\nu \cos\vartheta \quad . \tag{2.14}$$

It is therefore possible to determine $\rho(\nu)$ experimentally by measuring the spectral distribution of the radiation penetrating through a small hole in

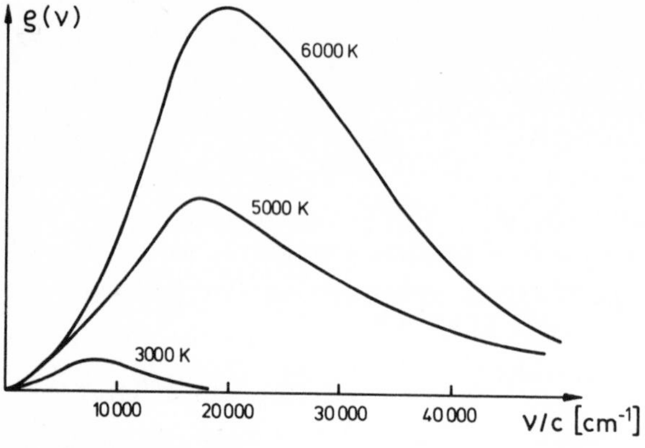

Fig.2.3. Spectral distribution of the energy density $\rho(\nu)$ for different temperatures

the walls of the cavity. If the hole is sufficiently small, the energy loss through this hole is negligibly small and does not disturb the thermal equilibrium inside the cavity.

Examples

a) Examples of real radiation sources with spectral energy distributions close to the Planck distribution (2.13) are the sun, the bright tungsten wire of a light bulb, flash lamps, and high-pressure discharge lamps.

b) Spectral lamps which emit discrete spectra are examples of *nonthermal* radiation sources. In these gas-discharge lamps, the light-emitting atoms or molecules may be in thermal equilibrium with respect to their translational energy, which means that their velocity distribution is Maxwellian. However, the population of the different excited atomic levels may not necessarily follow a Boltzmann distribution and there is generally also no thermal equilibrium between the atoms and the radiation field. The radiation may nevertheless be isotropic.

c) Lasers are examples of nonthermal and anisotropic radiation sources (see Chap.5). The radiation field is concentrated in a few modes, and most of the radiation energy is emitted into a small solid angle. This means that the laser represents an extreme anisotropic radiation source.

2.3 Absorption, Induced and Spontaneous Emission

Assume that molecules with energy levels E_1 and E_2 have been brought into the thermal radiation field of Sect.2.2. If a molecule absorbs a photon of energy $h\nu$ with $h\nu = E_2 - E_1$, it is excited from the lower energy level E_1 into the higher level E_2 (see Fig.2.4). This process is called *induced absorption*. The probability per second that a molecule will absorb a photon, dP_{12}/dt, is proportional to the number of photons of energy $h\nu$ per unit volume, and can be expressed in terms of the spectral energy density $\rho(\nu)$ of the radiation field as

$$\frac{d}{dt} P_{12} = B_{12}\rho(\nu) \quad . \tag{2.15}$$

The constant factor B_{12} is the *Einstein coefficient of induced absorption*. Each absorbed photon of energy $h\nu$ decreases the number of photons in one mode of the radiation field by one.

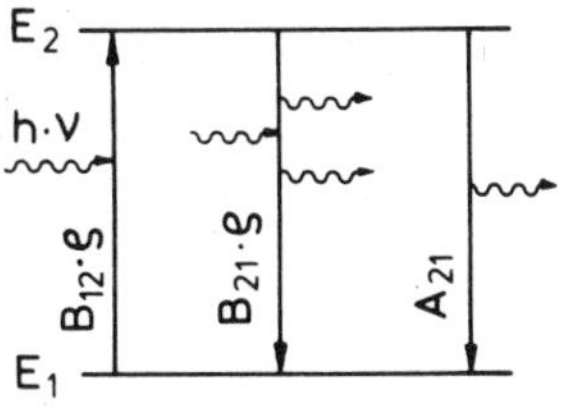

Fig.2.4. Schematic diagram of the interaction of a two-level system with a radiation field

The radiation field can also induce molecules in the excited state E_2 to make a transition to the lower state E_1 with simultaneous emission of a photon of energy $h\nu$. This process is called *induced (or stimulated) emission*. The induced photon of energy $h\nu$ is emitted into the same mode which caused the emission. This means that the number of photons in this mode is increased by one. The probability dP_{21}/dt that one molecule emits one induced photon per second is, analogous to (2.15),

$$\frac{d}{dt} P_{21} = B_{21}\rho(\nu) \quad . \tag{2.16}$$

The constant factor B_{21} is the *Einstein coefficient of induced emission*.

An excited molecule in a state E_2 may also *spontaneously* convert its excitation energy into an emitted photon $h\nu$. This spontaneous radiation can be emitted in an arbitrary direction $\underline{k}$ and increases the number of photons in the mode with frequency ν and wave vector $\underline{k}$ by one. In case of isotropic emission the probability of gaining a spontaneous photon is equal for all modes with the same frequency ν but different directions $\underline{k}$.

The probability $dP_{21}^{spont.}/dt$ per second that a photon $h\nu = E_2 - E_1$ is spontaneously emitted by a molecule, depends on the structure of the molecule and the selected transition $E_2 \rightarrow E_1$ *but is independent of the external radiation field*,

$$\frac{d}{dt} P_{21}^{spont.} = A_{21} \quad . \tag{2.17}$$

A_{21} is the *Einstein coefficient of spontaneous emission* and is often called the *spontaneous transition probability*.

Let us now look for relations between the three Einstein coefficients B_{12}, B_{21}, and A_{21}. The total number N of all molecules per unit volume is distributed among various energy levels E_i of population density N_i such that $\sum_i N_i = N$. In thermal equilibrium the population distribution $N_i(E_i)$ is given by the Boltzmann distribution

$$N_i = (Ng_i/Z) \exp(-E_i/kT) \quad . \tag{2.18}$$

The statistical weight g_i gives the number of degenerate sublevels of the energy level E_i and the partition function

$$Z = \sum_i g_i \exp(-E_i/kT) \quad ,$$

acts again as normalization factor which ensures that $\sum_i N_i = N$.

In a stationary field the total absorption rate $N_1 B_{12} \rho(\nu)$ which gives the number of photons absorbed per unit volume per second, has to equal the total emission rate $N_2 B_{21} \rho(\nu) + N_2 A_{21}$ (otherwise the spectral energy density $\rho(\nu)$ of the radiation field would change).

$$\left[B_{21}\rho(\nu) + A_{21}\right] N_2 = B_{12} N_1 \rho(\nu) \quad . \tag{2.19}$$

Using the relation

$$N_2/N_1 = (g_2/g_1) e^{-(E_2-E_1)/kT} = (g_2/g_1) e^{-h\nu/kT}$$

deduced from (2.18) and solving (2.19) for $\rho(\nu)$ yields

$$\rho(\nu) = \frac{A_{21}/B_{21}}{\frac{g_1}{g_2}\frac{B_{12}}{B_{21}} e^{h\nu/kT} - 1} \quad . \tag{2.20}$$

In Sect.2.2 we derived Planck's law (2.13) for the spectral energy density $\rho(\nu)$ of the thermal radiation field. Since both (2.13) and (2.20) must be valid for arbitrary temperatures T and all frequencies ν, comparison of the constant coefficients yields the relations

$$B_{12} = \frac{g_2}{g_1} B_{21} \quad , \tag{2.21}$$

$$A_{21} = \frac{8\pi h\nu^3}{c^3} B_{21} \quad . \tag{2.22}$$

Equation (2.21) states that for levels E_2, E_1 with equal statistical weights $g_2 = g_1$ *the probability of induced emission is equal to that of induced absorption.*

From (2.22) the following illustrative result can be extracted: The probability of induced emission $B_{21}\rho(\nu)$ is always larger than the spontaneous transition probability if

$$\rho(\nu) > \frac{8\pi h\nu^3}{c^3} = n(\nu)h\nu \quad \text{or} \tag{2.23}$$

$$\rho(\nu)/n(\nu) > h\nu \quad .$$

Since $n(\nu) = 8\pi\nu^2/c^3$ gives the number of modes per unit volume and frequency interval $d\nu = 1\ s^{-1}$ [see (2.7b)], the inequality (2.23) is equivalent to $\rho(\nu) > 1$ photon per mode. This means

the ratio of the induced to the spontaneous emission rates in an arbitrary mode is equal to the number of photons in this mode.

In Fig.2.5 the mean number of photons per mode in a thermal radiation field at different absolute temperatures is plotted as a function of frequency ν. The graphs illustrate that in the visible spectrum this number is small compared to unity at temperatures which can be realized in a laboratory. *This implies that in thermal radiation fields the spontaneous emission per mode exceeds by far the induced emission.* If it is possible, however, to concentrate most of the radiation energy onto a few modes, the number of photons *in these modes* may become exceedingly large and the induced emission in these modes dominates, although the total spontaneous emission into *all* modes may still be larger than the induced rate. Such a selection of a few modes is realized in a laser (see Chap.5).

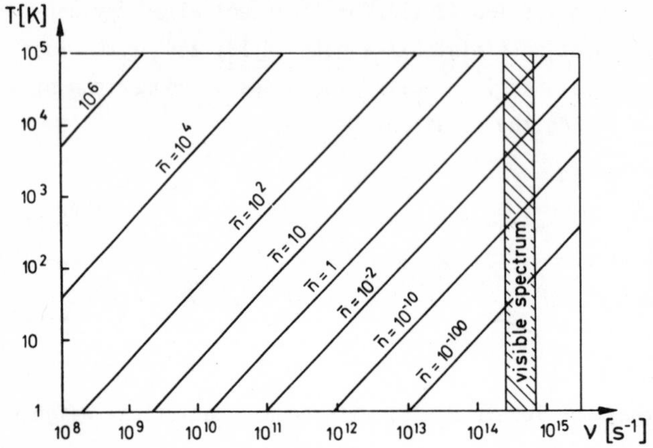

Fig.2.5. Average number of photons per mode in a thermal radiation field as a function of temperature and frequency ν

Comment

Note that the relations (2.21) and (2.22) are valid for all kinds of radiation fields. Although they have been derived for stationary fields at thermal equilibrium, the Einstein coefficients are constants which depend only on the molecular properties and not on external fields as far as these fields do not alter the molecular properties. Equations (2.21,22) therefore hold for arbitrary $\rho(\nu)$. Using the frequency $\omega = 2\pi\nu$ instead of ν the unit frequency interval $d\omega = 1$ corresponds to $d\nu = 1/2\pi$. The energy density $\rho(\omega) = n(\omega)W$ is then, according to (2.7a),

$$\rho(\omega) = \frac{\omega^2}{\pi^2 c^3} \frac{\hbar\omega}{e^{\hbar\omega/kT} - 1} , \qquad (2.13a)$$

where $\hbar$ is Planck's constant divided by 2π. The ratio of the Einstein coefficients

$$A_{21}/B_{21} = \hbar\omega^3/(\pi^2 c^3) \qquad (2.22a)$$

now contains $\hbar$ instead of h and is smaller by a factor of 2π. However, the ratio $A_{21}/[B_{21}\rho(\omega)]$ which gives the ratio of the spontaneous to the induced transition probabilities remains the same.

Examples

a) In the thermal radiation field of a 100 W light bulb, 10 cm away from the tungsten wire, the number of photons per mode at $\lambda = 500$ nm is about 10^{-8}. If a molecular probe is placed in this field, the induced emission is therefore completely negligible.

b) In the center spot of a high-current mercury discharge lamp with very high pressure the number of photons per mode is about 10^{-2} at the center frequency of the strongest emission line at $\lambda = 253.6$ mm. This shows that even in this very bright light source the induced emission only plays a minor role.

c) Inside the cavity of a He-Ne laser (output power 1 mW with mirror transmittance T = 1%) which oscillates in a single mode, the number of photons in this mode is about 10^7. In this example the *spontaneous* emission into this mode is completely negligible. Note, however, that the total spontaneous emission power at $\lambda = 632.2$ nm, which is emitted into all directions, is much larger than the induced emission. This spontaneous emission is more or less uniformly distributed among all modes. Assuming a volume of 1 cm^3 for the gas discharge, the number of modes within the Doppler width of the neon transition is about 10^8, which means that the total spontaneous rate is about 10 times the induced rate.

2.4 Basic Photometric Quantities

In spectroscopic applications of light sources it is very useful to define some characteristic quantities of the emitted and absorbed radiation. This allows a proper comparison of different light sources and detectors and enables one to make an appropriate choice of apparatus for a particular experiment.

The *radiant energy* W (measured in joules) refers to the total amount of energy emitted by a light source, transferred through a surface, or collected by a detector. The *radiant power* or *radiant flux* Φ [W] is the radiant energy per second. The *radiant energy density* ρ [J/m^3] is the radiant energy per unit volume of space.

The above three quantities refer to the total radiation integrated over the whole spectrum. Their spectral versions $W_\nu(\nu)$, $\Phi_\nu(\nu)$, and $\rho_\nu(\nu)$ are called the *spectral densities* and are defined as the amounts of W, Φ, or ρ within the unit frequency interval $d\nu = 1\ s^{-1}$ around the frequency ν.

$$W = \int_{\nu=0}^{\infty} W_\nu(\nu)d\nu \quad ; \quad \Phi = \int_0^\infty \Phi_\nu(\nu)d\nu \quad ; \quad \rho = \int_0^\infty \rho_\nu(\nu)d\nu \quad . \tag{2.24}$$

Consider a unit surface element $dA = 1\ m^2$ of a light source (Fig.2.6). The radiant power emitted from dA into the unit solid angle $d\Omega = 1$ ster is called the *radiance* L [$W\ m^{-2}\ ster^{-1}$] of the source. Its spectral density $L_\nu(\nu, \vartheta)$ [$W\ m^{-2}\ s\ ster^{-1}$] generally depends on the spectral region and on the angle ϑ to the surface normal $\hat{n}$. The radiant flux of the source is

$$\Phi = \oint_A \int_{\nu=0}^{\infty} \int_\Omega L_\nu(\nu,\vartheta)\cos\vartheta\ d\Omega\ d\nu\ dA \quad . \tag{2.25}$$

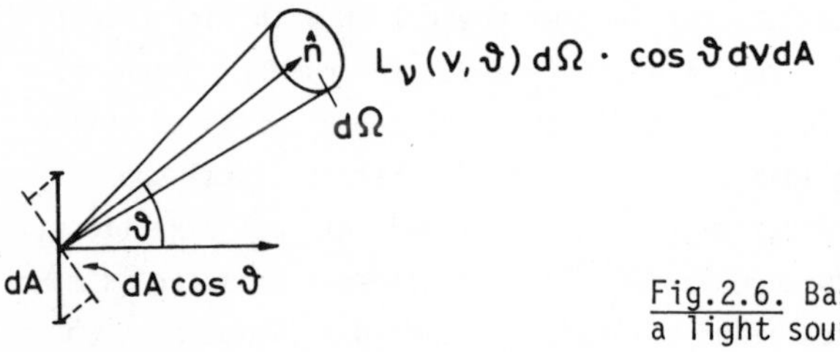

Fig.2.6. Basic radiant quantities of a light source

An important quantity is the *radiant intensity* $I^* = \int L\ dA$ of the source. This represents the flux per unit solid angle from the entire source [$W\ ster^{-1}$]. The total flux is then

$$\Phi = \int_\Omega I\ d\Omega \quad . \tag{2.26}$$

For a uniformly emitting isotropic source, I^* does not depend on the angle ϑ and (2.26) yields

$$\Phi = 4\pi I^* \quad . \tag{2.27}$$

For a thermal radiation source, for example the blackbody radiator of Sect. 2.2 with a spectral energy density ρ_ν, the spectral radiance $L_\nu(\nu)$ is independent of ϑ and can be expressed by

$$L_\nu(\nu) = \rho_\nu(\nu)c/4\pi = \frac{2h\nu^3}{c^2} \frac{1}{e^{h\nu/kT} - 1} \quad . \tag{2.28}$$

A surface element dA' of a detector at distance R from the source element dA covers a solid angle $d\Omega = dA' \cos\vartheta'/R^2$ as seen from the source (Fig.2.7). With $R^2 \gg dA, dA'$ the radiant power received by dA' is

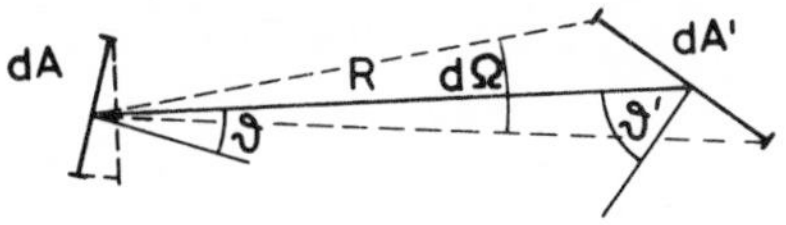

Fig.2.7. Radiance and irradiance of source and detector

$$d\Phi = L(\vartheta)\, dA \cos\vartheta \, d\Omega$$

$$= L(\vartheta) \cos\vartheta \, dA \cos\vartheta' \, dA'/R^2 \quad , \tag{2.29}$$

$dA \cos\vartheta$ is the projection of dA, as seen from dA'. For isotropic sources (2.29) is symmetric with regard to ϑ and ϑ' or dA and dA'. The positions of detector and source may be exchanged without altering (2.29). Because of this reciprocity L may be interpreted either as the *radiance of the source* at the angle ϑ to the surface normal or, equally well, as the *radiance incident onto the detector at the angle* ϑ'.

For isotropic sources, where L is independent of ϑ, (2.29) shows that the radiant flux emitted into the unit solid angle is proportional to $\cos\vartheta$ (*Lambert's Law*). An example for such a source is a hole with area dA in a blackbody radiation cavity.

The radiant flux incident on unit detector area is called *irradiance* or *flux density* [W/m^2] which is proportional to the square of the electric field strength. In spectroscopic literature it is often called *intensity*, which, however, should not be confused with the radiant intensity I^* [W/ster] defined above. The flux density or intensity I [W/m^2] of a plane wave $\underline{E} = \underline{E}_0 \cos(\omega t - kx)$ travelling in vacuum in the x direction is given by

$$I = c \int \rho(\omega) d\omega = c\varepsilon_0 E^2 = c\varepsilon_0 E_0^2 \cos^2(\omega t - kx) \quad . \tag{2.30a}$$

With the complex notation

$$\underline{E} = \underline{A}_0 \, e^{i(\omega t - kx)} + \underline{A}_0^* \, e^{-i(\omega t - kx)} \quad , \text{ where } \quad \underline{A}_0 = \frac{1}{2} \underline{E}_0 \tag{2.31}$$

the intensity becomes

$$I = c\varepsilon_0 E^2 = 4c\varepsilon_0 A_0^2 \cos^2(\omega t - kx) \quad . \tag{2.30b}$$

Most detectors cannot follow the rapid oscillations of the light wave with frequencies $\omega \approx 10^{13} - 10^{15}$ s^{-1} in the visible and near infrared region. They measure at a fixed position x the time-averaged intensity

$$\bar{I} = c\varepsilon_0 E_0^2 \, \overline{\cos^2(\omega t - kx)} = \frac{1}{2} c\varepsilon_0 E_0^2 = 2c\varepsilon_0 A_0^2 \quad . \tag{2.30c}$$

Note

It is often convenient to write the E.M. wave in a shorthand complex notation

$$\underline{E} = \underline{E}_0\, e^{i(\omega t - kx)} \quad , \quad E_0 = |E_0| e^{i\varphi} \quad . \tag{2.31a}$$

The actual field is then the real part of the complex vector E and the intensity becomes

$$I = c\varepsilon_0 E_0^2 \cos^2(\omega t - kx + \varphi) \quad , \tag{2.30d}$$

with the time-averaged intensity (2.30c).

For many spectroscopic applications the spectral density $I_\nu(\nu)$ of the intensity I is important. This is the intensity within a frequency interval $d\nu = 1\ s^{-1}$ [$W\ s/m^2$] and is related to the spectral energy density $\rho_\nu(\nu)$ of a plane wave by $I_\nu = c\rho_\nu$. In case of a spherical wave

$$E = (E_0/r)\cos(\omega t - kr) \quad ,$$

the intensity I of the wave,

$$I = \frac{1}{2} E_0^2 c\varepsilon_0/r^2 \quad , \tag{2.32}$$

decreases as $1/r^2$ with increasing distance from the source.

The radiation flux incident on a surface may be partly reflected, transmitted, or absorbed. The *reflectance* R, the *transmittance* T, and the *absorbtance* A of a surface element are defined, respectively, as the ratios of the reflected, transmitted, and absorbed intensities to the incident intensity I_0,

$$R = I_R/I_0 \quad ; \quad T = I_T/I_0 \quad ; \quad A = I_A/I_0 \quad . \tag{2.33}$$

In case of extended detector areas, the total power received by the detector is obtained by integration over all detector elements dA' (Fig.2.8). The detector receives all the radiation which is emitted from the source element dA within the angles $-u \leq \vartheta \leq +u$. The same radiation passes an imaginary spherical surface in front of the detector. We choose as surface elements of this spherical surface circular rings with $dA' = 2\pi r dr = 2\pi R\, R\sin\vartheta\, d\vartheta$. From (2.29) one obtains for the total flux Φ onto the detector with $\cos\vartheta' = 1$

$$\Phi = \int_0^u L\, dA \cos\vartheta\, 2\pi \sin\vartheta\, d\vartheta \quad . \tag{2.34}$$

If the source is isotropic, L does not depend on ϑ and (2.34) yields

$$\Phi = \pi L \sin^2 u \; dA \quad . \tag{2.35}$$

Note that it is impossible to increase the radiance of a source by any sophisticated imaging optics. This means that the image dA^* of a radiation source dA never has a larger radiance than the source itself. It is true that the flux density can be increased by demagnification. The solid angle, however, into which radiation from the image dA^* is emitted has also been increased by the same factor. Therefore, the radiance *does not* increase. In fact, because of inevitable reflection, scattering, and absorption losses of the imaging optics, the radiance of the image dA^* is in practice always less than that of the source (see Fig.2.9).

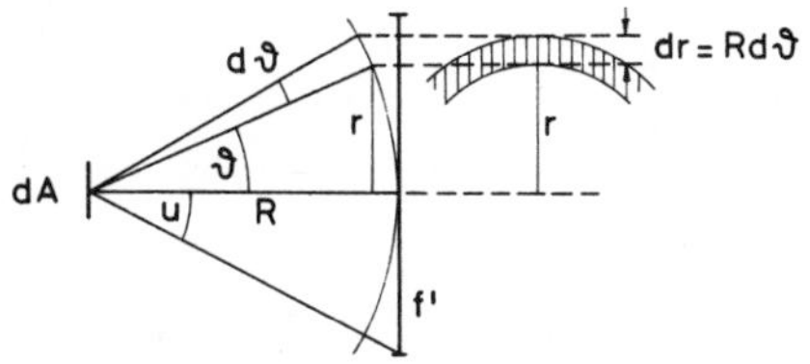

Fig.2.8. Flux densities of detectors with extended area

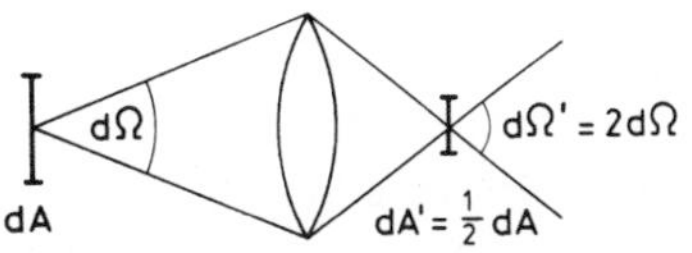

Fig.2.9. The radiance of a source cannot be increased by optical imaging

A strictly parallel light beam would be emitted into a solid angle $d\Omega = 0$. With a finite radiant power this would imply an infinite radiance L, which is impossible. This illustrates that such a light beam cannot be realized. The radiation source for a strictly parallel beam anyway has to be a point source in the focal plane of a lens. Such a point source with zero surface cannot emit any power.

For more extensive treatments of photometry, see [2.7-9].

Examples

a) *Radiance of the sun.* 1 m^2 of the earth's surface receives at normal incidence without reflection or absorption through the atmosphere an incident radiant flux I_e of about 1.35 kW/m^2 (solar constant). Because of the symmetry of (2.29) we may regard dA' as emitter and dA as receiver. The sun is seen from the earth under an angle of 2u = 32 minutes of arc. This yields $\sin u = 4.7 \times 10^{-3}$. Inserting this number into (2.35) one obtains $L_s = 2 \times 10^4$ $kW/(m^2$ ster) for the radiance of the sun's surface. The total radiant power Φ of the sun can be obtained from (2.25) or from the relation $\Phi = 4\pi R^2 I_e$, where $R = 1.5 \times 10^{11}$ m is the distance from the earth to the sun. These numbers give $\Phi = 4 \times 10^{26}$ W.

b) *Radiance of a He-Ne Laser*. We assume that the output power of 1 mW is emitted from 1 mm^2 of the mirror surface into an angle of 4 minutes of arc, which is equivalent to a solid angle of 1×10^{-6} ster. The maximum radiance in the direction of the laser beam is then $L = 10^{-3}/(10^{-6} \times 10^{-6})$ W/(m^2 ster) $= 10^9$ W/(m^2 ster). This is about 50 times larger than the radiance of the sun. For the spectral density of the radiance the comparison is even more dramatic. Since the emission of an unstabilized single-mode laser is restricted to a spectral range of about 1 MHz, the laser has a spectral radiance density $L_\nu = 1 \times 10^3$(W s m^{-2} ster^{-1}), whereas the sun, which emits within a mean spectral range of about 10^{15} s^{-1}, only reaches $L_\nu = 2 \cdot 10^{-8}$ W s m^{-2} ster^{-1}.

c) Looking directly into the sun, the retina receives a radiant flux of 1 mW if the diameter of the iris is 1 mm. This is just the same flux the retina receives staring into the laser beam of Example b). There is, however, a big difference regarding the irradiance of the retina. The image of the sun on the retina is about 100 times as large as the focal area of the laser beam. This means that the power density incident on single retina cells is about 100 times larger in the case of the laser radiation.

2.5 Discrete and Continuous Spectra

The spectral distribution of the radiant flux from a source is called its spectrum. The thermal radiation discussed in Sect.2.2 has a *continuous* spectral distribution described by its spectral energy density (2.13). *Discrete* spectra where the radiant flux has distinct maxima at certain frequencies ν_{ik}, are generated by transitions of atoms or molecules between two bound states, a higher energy state E_k and a lower state E_i with the relation

$$h\nu_{ik} = E_k - E_i \quad . \tag{2.36}$$

In a spectrograph (see Sect.4.1 for a detailed description) the entrance slit is imaged into the focal plane of the camera lens. Because of dispersive elements in the spectrograph, the position of this image depends on the wavelength of the incident radiation. In a discrete spectrum each wavelength λ_{ik} produces a separate line in the imaging plane, provided the spectrograph has sufficiently high resolving power (see Fig.2.10). Discrete spectra are therefore also called *line spectra*, as opposed to *continuous spectra* where the slit images form a continuous band in the focal plane, even with infinite resolving power.

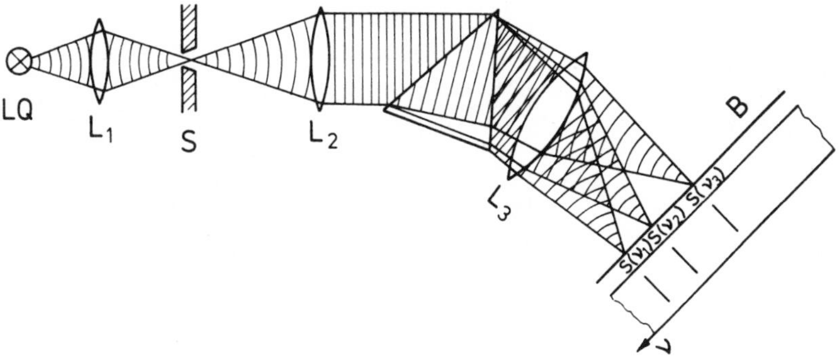

Fig.2.10. Spectral lines in a discrete spectrum as images of the entrance slit of a spectrograph

If radiation with a continuous spectrum passes through a gaseous molecular probe, molecules in the lower state E_i may absorb radiant power at the eigenfrequencies $\nu_{ik} = (E_k - E_i)/h$, which is thus missing in the transmitted power. The difference in the spectral distributions of incident minus transmitted power is the *absorption spectrum* of the probe. The absorbed energy $h\nu_{ik}$ brings a molecule into the higher energy level E_k. If these levels are bound levels, the resulting spectrum is a discrete absorption spectrum. If E_k is above the dissociation limit or above the ionization energy, the absorption spectrum becomes continuous. In Fig.2.11 both cases are schematically illustrated for atoms (a) and molecules (b).

Examples of discrete absorption lines are the Fraunhofer lines in the spectrum of the sun which appear as dark lines in the bright continuous spectrum (see Fig.2.12). They are produced by atoms in the sun's atmosphere which absorb at their specific eigenfrequencies the continuous blackbody radiation from the sun's photosphere.

The absorbed power is proportional to the density N_i of molecules in level E_i. The absorption lines are only measurable if the absorbed power is sufficiently high, which means that the density N_i or else the path length through the probe must be large enough. In a gas at thermal equilibrium, the Boltzmann relation (2.18)

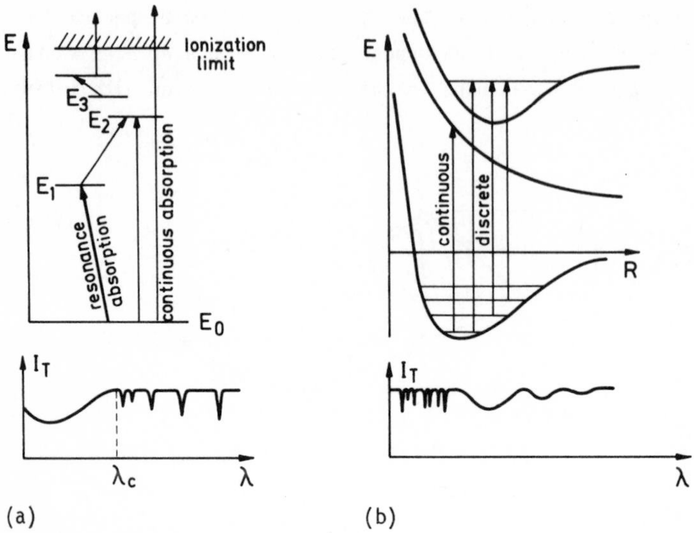

Fig.2.11a,b. Schematic diagram to illustrate the origin of discrete and continuous absorption spectra for (a) atomic and (b) molecular absorption

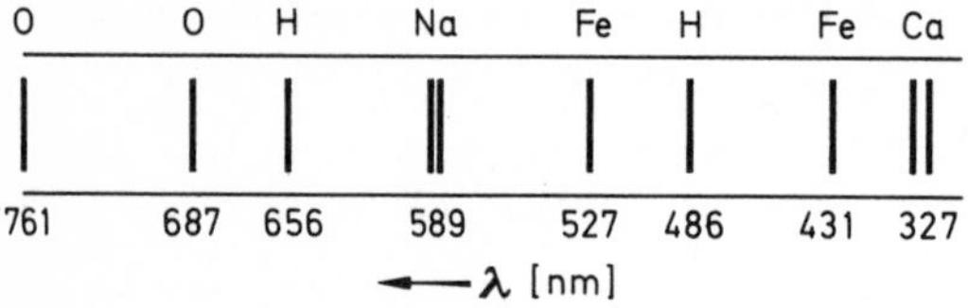

Fig.2.12. Prominent Fraunhofer lines in the spectrum of the sun

$$N_i \propto q_i \exp(-E_i/kT)$$

holds. Absorption lines in gases at thermal equilibrium are therefore only intense for transitions from low-lying energy levels E_i for which E_i is not much larger than kT. It is, however, possible to pump molecules into higher energy states by various excitation mechanisms. This allows measurement of absorption spectra for transitions from these states to even higher molecular levels.

Examples of such pumping mechanisms are absorption of light,

$$M + h\nu \rightarrow M^* \quad \text{(optical pumping)} \quad , \tag{2.37}$$

and collisional excitation by electron impact,

$$M + e^- + E_{kin} \rightarrow M^* + e^- + E'_{kin} \quad , \quad \text{with} \quad E_{kin} - E'_{kin} = E(M^*) - E(M) = \Delta E_{kin} \quad , \tag{2.38}$$

which represent the main contribution to excitation mechanisms in gas discharges. Often collisions between two atoms or molecules play an important role. One or both of the atoms or molecules may be in an excited state. For example,

$$A + B + E_{kin} \rightarrow A^* + B + \Delta E_{kin} \quad ; \tag{2.39}$$

$$A + B^* \rightarrow A^* + B + \Delta E_{kin} \quad . \tag{2.40}$$

The last process (2.40) has a large probability if the collision partners have nearly equal energy levels $E_i(A)$ and $E_k(B)$. This process, for instance, is mainly responsible for achieving population inversion in a He-Ne laser.

The excited molecules release their energy either by spontaneous or induced emission or by collisional deactivation (Fig.2.13a). The spatial distribution of spontaneous emission depends on the orientation of the excited molecules and on the symmetry properties of the excited state. If the molecules are randomly oriented, the spontaneous emission (often called fluorescence) is isotropic.

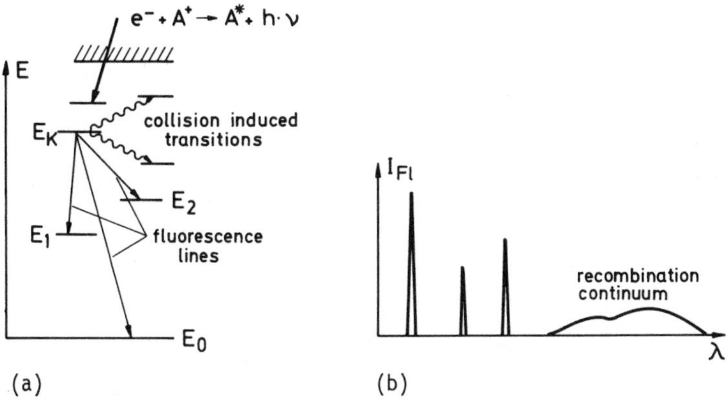

Fig.2.13a,b. Discrete and continuous emission spectra and the corresponding energy level diagrams

The fluorescence spectrum (emission spectrum) emitted from a discrete upper level E_k consists of discrete lines if the terminating lower levels E_i are bound states, while a continuum is emitted if E_i belongs to a repulsive state which dissociates. As an example the fluorescence of the $^3\Pi \rightarrow {}^3\Sigma$ transition spectrum of the NaK molecule is shown in Fig.2.14. It is emitted from a selectively excited level in a bound $^3\Pi$ state to a repulsive $^3\Sigma$ state, which has a shallow van der Waals minimum. Transitions terminating to lower energies E_k above the dissociation energy form the continuous part of the spectrum whereas transitions to bound levels in the van der Waals potential well produce discrete lines. The modulation of the continuum reflects the maxima and nodes of the vibrational wave function in the upper bound level [2.10].

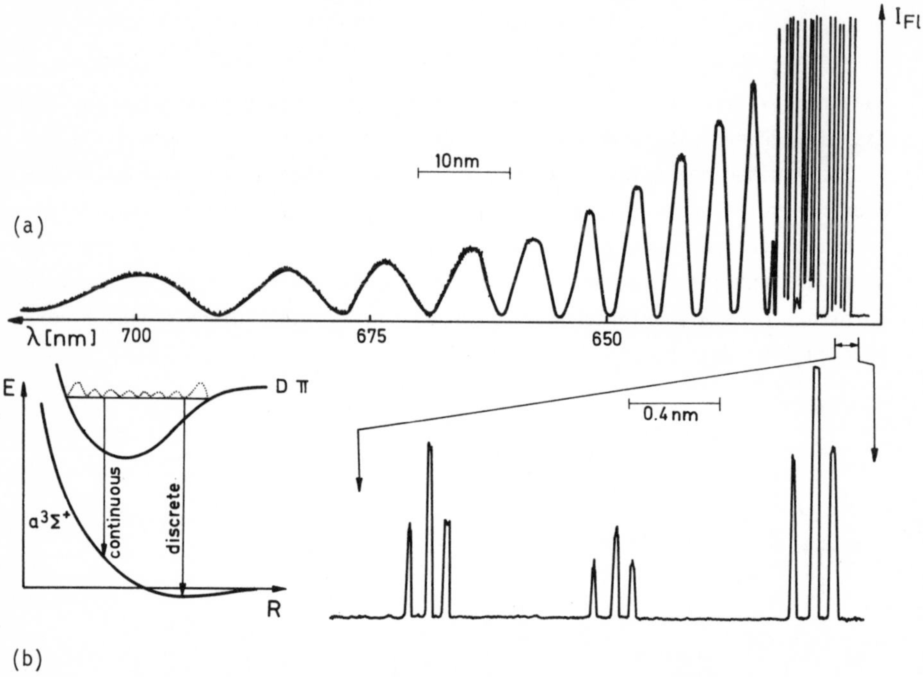

Fig.2.14a,b. Bound-free and bound-bound transitions in the NaK molecule. (a) Part of the spectrum; (b) schematic energy level diagram

2.6 Absorption and Dispersion

When an electromagnetic wave passes through a medium with refractive index n, not only the *wave amplitude* decreases (absorption) but also the *phase velocity* changes from its value c in vacuum to $v = c/n$ (dispersion). The refractive index $n = n(\omega)$ depends on the frequency ω of the E.M. wave.

The classical model describing the atomic electrons as damped harmonic oscillators which are forced to oscillate by the electric field $\underline{E}(\omega)$ of the wave, gives a clear picture of the relation between absorption and dispersion (Kramers-Kronig relation). This model allows the macroscopic refractive index to be related to its microscopic origin, namely the electronic charge distribution of the atoms or molecules and its response to the E.M. wave. The conclusions of this classical model can be transferred to real molecules in a relatively simple way by introducing the concept of *oscillator strength.*

At first we shall discuss this classical model. Although the quantum mechanical treatment will be only briefly outlined, this should still allow the reader to understand more advanced presentations [2.6,11].

2.6.1 Classical Model of the Refractive Index

The forced oscillation of a damped oscillator with charge q, mass m, and damping coefficient b under the influence of the external force $q\underline{E}$ in the x direction $[\underline{E} = (E, 0, 0) = \underline{E}_0 \exp(i\omega t)]$ is described in complex notation by the differential equation

$$m\ddot{x} + b\dot{x} + Dx = qE_0 \exp(i\omega t) \quad . \tag{2.41}$$

Assuming a solution of the form $x = x_0 \exp(i\omega t)$, we find for the amplitude x_0 the expression

$$x_0 = \frac{qE_0}{m(\omega_0^2 - \omega^2 + i\gamma\omega)} \tag{2.42}$$

using the abbreviations $\gamma = b/m$ and $\omega_0^2 = D/m$. The forced oscillation of a charge q generates an induced electric dipole moment

$$p = qx = \frac{q^2 E_0\, e^{i\omega t}}{m(\omega_0^2 - \omega^2 + i\gamma\omega)} \quad . \tag{2.43}$$

In a sample with N oscillators per unit volume the macroscopic polarization P which is the sum of all dipole moments per unit volume is therefore

$$P = Nqx \quad . \tag{2.44}$$

On the other hand, the polarization can be derived in classical electrodynamics from Maxwell's equations using the dielectric constant ε_0 or the susceptibility χ,

$$\underline{P} = \varepsilon_0(\varepsilon - 1)\underline{E} = \varepsilon_0\chi\underline{E} \quad . \tag{2.45}$$

The relative dielectric constant ε is related to the refractive index n by

$$n = \varepsilon^{\frac{1}{2}} \quad . \tag{2.46}$$

This can be easily verified from the relation

$$v = (\varepsilon\varepsilon_0\mu\mu_0)^{-\frac{1}{2}} = c/n \tag{2.47}$$

for the velocity of light which follows from Maxwell's equations in media. Except for ferromagnetic materials the relative permeability $\mu \approx 1$ is close to unity and since $c = (\varepsilon_0\mu_0)^{-\frac{1}{2}}$, (2.46) follows immediately.

Combining (2.42-46) the refractive index n can be written as

$$n^2 = 1 + \frac{Nq^2}{\varepsilon_0 m(\omega_0^2 - \omega^2 + i\gamma\omega)} \quad . \tag{2.48}$$

In gaseous media at sufficiently low pressure the index of refraction is close to unity. (For example in air at atmöspheric pressure, n = 1.00028 for λ = 500 nm). In this case the approximation

$$n^2 - 1 = (n + 1)(n - 1) \approx 2(n - 1)$$

is sufficiently accurate for most purposes, and (2.48) can be reduced to

$$n = 1 + \frac{Nq^2}{2\varepsilon_0 m(\omega_0^2 - \omega^2 + i\gamma\omega)} \quad . \tag{2.49}$$

In order to make clear the physical implication of this complex index of refraction, we separate the real and the imaginary parts and write

$$n = n' - i\varkappa \quad . \tag{2.50}$$

An E.M. wave $E = E_0 \exp[i(\omega t - kz)]$ passing in the z direction through a medium with refractive index n has the same frequency $\omega_n = \omega_0$ as in vacuum but a different wave vector $k_n = k_0 \cdot n$. Inserting (2.50) yields with $|\underline{k}| = 2\pi/\lambda$

$$\begin{aligned} E &= E_0 \exp(-k_0\varkappa z) \exp[i(\omega t - k_0 n' z)] \\ &= E_0 \exp(-2\pi\varkappa z/\lambda_0) \exp[ik_0(ct - n'z)] \quad . \end{aligned} \tag{2.51}$$

Equation (2.51) shows that the imaginary part $\varkappa(\omega)$ of the complex refractive index n describes the *absorption* of the E.M. wave. At a penetration depth of $\Delta z = \lambda_0/(2\pi\varkappa)$ the amplitude $E_0 \exp(-k_0\varkappa z)$ has decreased to 1/e of its value at z = 0. The real part $n'(\omega)$ represents the *dispersion* of the wave, i.e., the dependence of the phase velocity $v(\omega) = c/n'(\omega)$ on the frequency.

Generally the absorption of light passing through a medium is characterized by the absorption coefficient α which describes the attentuation of the *intensity* rather than the *amplitude*. If the intensity of a plane wave is I(z), its attenuation along the distance dz is (see Fig.2.15)

$$dI = -\alpha I\, dz \quad . \tag{2.52}$$

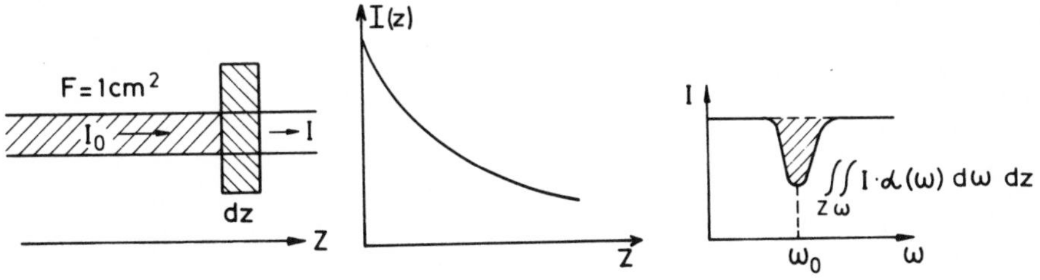

Fig.2.15. Absorption of light in an optically thin layer

The absorption coefficient α represents the fraction dI/I absorbed within the unit interval dz = 1 cm. For constant α (i.e., independent of I) integration of (2.52) yields Beer's law for linear absorption ($dI \propto I$)

$$I(Z) = I_0\, e^{-\alpha Z} \quad , \tag{2.53}$$

where I_0 is the intensity at z = 0. Since the intensity $I \propto AA^*$ is proportional to the square of the amplitude, comparison of (2.53) with (2.51) gives the relation

$$\alpha = 4\pi\varkappa/\lambda_0 = 2\varkappa k_0 \quad . \tag{2.54}$$

The absorption coefficient α *is proportional to the imaginary part* $\varkappa$ *of the complex refractive index* $n = n' - i\varkappa$.

The frequency dependence of κ and n' can be obtained by inserting (2.50) into (2.49). Separating the real and imaginary parts, we get

$$\kappa = \frac{Nq^2}{2\varepsilon_0 m}\,\frac{\gamma\omega}{(\omega_0^2-\omega^2)^2+\gamma^2\omega^2} \tag{2.55a}$$

$$n' = 1 + \frac{Nq^2}{2\varepsilon_0 m}\,\frac{\omega_0^2-\omega^2}{(\omega_0^2-\omega^2)^2+\gamma^2\omega^2} \quad . \tag{2.56a}$$

The equations (2.55a) and (2.56a) are the *Kramers-Kronig dispersion relations*. They relate absorption and dispersion through the complex refractive index $n = n' - i\kappa$.

In the neighborhood of a molecular transition frequency ω_0 where $|\omega_0 - \omega| << \omega_0$, the dispersion relations reduce to

$$\kappa = \frac{Nq^2}{8\varepsilon_0 m\omega_0} \frac{\gamma}{(\omega_0 - \omega)^2 + (\gamma/2)^2} \qquad (2.55b)$$

$$n' = 1 + \frac{Nq^2}{4\varepsilon_0 m\omega_0} \frac{\omega_0 - \omega}{(\omega_0 - \omega)^2 + (\gamma/2)^2} \quad . \qquad (2.56b)$$

Figure 4.16 shows the frequency dependence of $\kappa(\omega)$ and $n'(\omega)$ in the vicinity of the eigenfrequency ω_0 of an atomic transition.

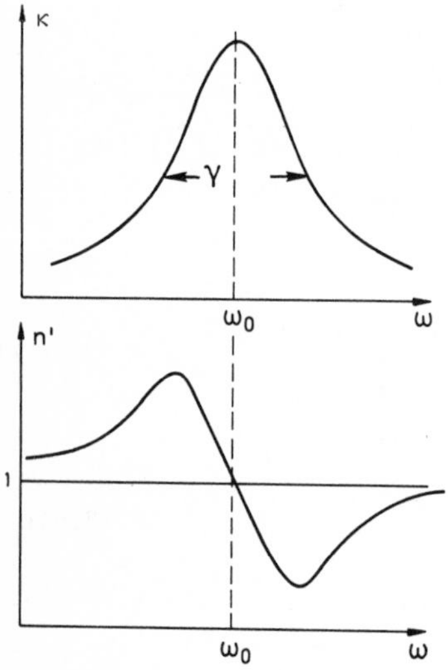

Fig.2.16. Absorption coefficient $\alpha = 2k\kappa(\omega)$ and dispersion $n'(\omega)$ in the vicinity of an atomic transition with center frequency ω_0

Note; the relations derived above are only valid for oscillators at rest in the observer's coordinate system. The thermal motion of real atoms in a gas introduces an additional broadening of the line profiles, the "Doppler broadening" which will be discussed in Sect.3.2. The profiles (2.55,56) can therefore be observed only with Doppler-free techniques (see Chap.10).

2.6.2 Oscillator Strengths and Einstein Coefficients

Because of the large number of possible energy levels, atoms and molecules have not only one but many eigenfrequencies at which they can absorb radiation. We shall now relate the absorption on these lines with the absorption coefficient and its frequency dependence as derived in the previous section from the classical oscillator model. The value of the absorption coefficient depends on the electronic structure and on the symmetry of initial and final states of a molecular transition and can be expressed as a *transition probability* (see Sect.2.3). Although the calculation of these transition probabilities demands quantum mechanical approximations, it is possible to give phenomenological expressions for them by introducing the so-called oscillator strengths. These semiclassical terms have the advantages that they can be used to relate the classical and quantum mechanical treatments and that they can be directly measured (see next section).

The meaning of the oscillator strength is the following.
An atom with one outer-shell electron having excitation energies in the spectral range under consideration can be described with regard to its absorption as a classical oscillator with oscillating charge $q = -e$. The total absorption of an atom in the level E_i, however, is distributed among many transitions $E_i \rightarrow E_k$ to all higher levels E_k which are optically connected with E_i (see Fig.2.17). Each of these transitions contributes only a fraction f_{ik} to the total absorption. This number $f_{ik} < 1$ is called the *oscillator strength* of the transition $E_i \rightarrow E_k$. *The absorption of N atoms on the transition* $E_i \rightarrow E_k$ *is equal to that of* $f_{ik}N$ *classical oscillators.*

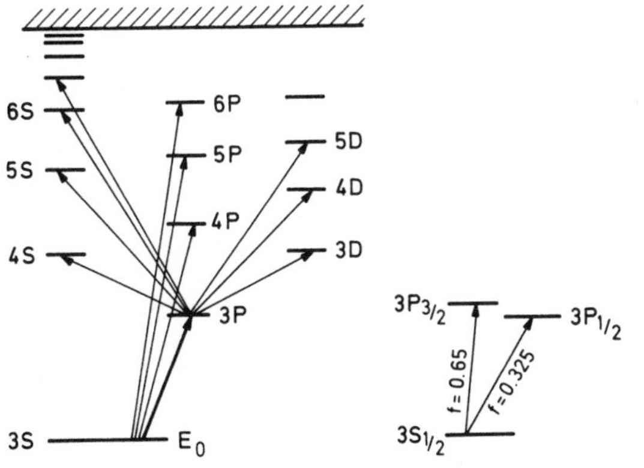

Fig.2.17. Schematic diagram of the Na energy levels. The oscillator strength $f(3s \rightarrow 3p)$ equals to the intensity ratio $I(3s \rightarrow 3p)/(\sum_k I_k(3s \rightarrow kp)$ of the resonance absorption lines

From the definition of the oscillator strength f_{ik} we obtain the relation

$$\sum_k Nf_{ik} = N \quad \text{or} \quad \sum_k f_{ik} = 1 \quad . \tag{2.57a}$$

For atoms with p outer-shell electrons which have excitation energies within the spectral range under consideration, (2.57a) can be generalized to the *sum role of Thomas, Reiche, and Kuhn*

$$\sum_k f_{ik} = p \quad . \tag{2.57b}$$

The summation extends over all levels E_k (including the continuum) which are accessible from level E_i by electric dipole transitions. If E_i is an excited level, induced emission to lower levels may occur, which diminishes the effective absorption. The corresponding oscillator strengths f_{ik} with $E_k < E_i$ are therefore negative.

Examples

1) The f value of the two Na D-lines is

$$f(3S - 3P_{1/2}) = 0.325 \; ; \quad f(3S - 3P_{3/2}) = 0.65 \quad .$$

This implies that the two D-lines carry 98% of the total oscillator strength.

2) The f value of the intense line $\lambda = 3720$ Å in the iron spectrum which belongs to the $^5F \rightarrow {}^5D$ transition is $f = 0.04$. It is important for astrophysicists in determining the iron concentration in the sun's atmosphere. The small f-value shows that only a small fraction of the total oscillator strength $f = 2$ is concentrated in this transition.

Using the concept of oscillator strength, the absorption and dispersion of real atoms or molecules in a level E_i with absorption frequencies ω_{ik} can be described by modifying the classical formulas (2.55,56) to

$$\kappa_i = \frac{N_i e^2}{2\varepsilon_0 m} \sum_k \frac{\omega f_{ik}\gamma_{ik}}{(\omega_{ik}^2 - \omega^2)^2 + \gamma_{ik}^2\omega^2} \tag{2.58}$$

$$n_i' = 1 + \frac{N_i e^2}{2\varepsilon_0 m} \sum_k \frac{(\omega_{ik}^2 - \omega^2) f_{ik}}{(\omega_{ik}^2 - \omega^2)^2 + \gamma_{ik}^2\omega^2} \quad . \tag{2.59}$$

In these equations $\alpha_i = 2k\kappa_i$ and n_i' are the absorption coefficient and the dispersion which are caused by N_i molecules per unit volume in level E_i. Figure 2.18 illustrates absorption and dispersion of sodium vapor in the

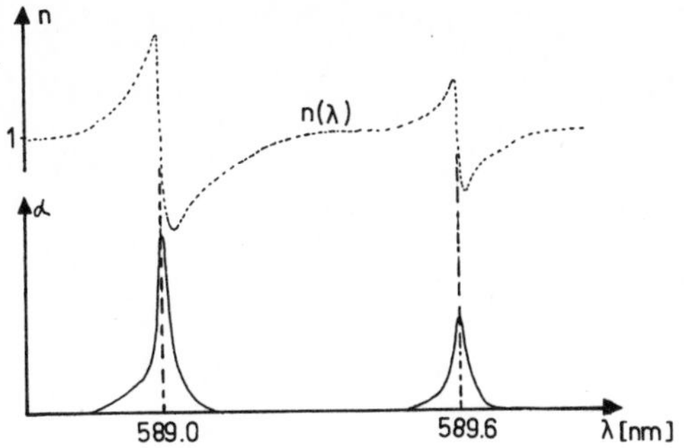

Fig.2.18. Absorption and dispersion of sodium vapor measured with low resolution in the spectral range around the two Na D-lines at λ = 589.0 nm and λ = 589,6 nm

vicinity of the two Na D-lines ($3S \rightarrow 3P_{1/2}$ and $3S \rightarrow 3P_{3/2}$) at low resolution where the hyperfine splitting is not resolved.

According to (2.58,59) the oscillator strength can be determined experimentally from measurements of absorption and dispersion profiles of spectral lines (see Sect.2.7.2). Another, widely used method derives transition probabilities and oscillator strengths from measurements of spontaneous lifetimes of excited levels. This will be discussed in the next section.

We now derive the relations between the oscillator strengths f_{ik} and the Einstein coefficients A_{ik} and B_{ik}. Equation (2.52) shows that the radiation power absorbed per unit volume within the frequency interval $d\omega$ is

$$dW/dt = -\alpha(\omega)I(\omega)d\omega \quad . \tag{2.60}$$

The power absorbed per unit volume on the transition $E_i \rightarrow E_k$ is then

$$dW_{ik}/dt = \int \alpha_{ik}(\omega)I(\omega)d\omega \quad , \tag{2.61}$$

where the integration extends over all frequencies ω which contribute to this transition.

If the incident intensity $I(\omega)$ does not change much within the frequency range of the absorption profile (which means roughly within the absorption halfwidth) $I(\omega) = I(\omega_{ik})$ can be assumed to be constant and (2.61) becomes

$$dW_{ik}/dt = I(\omega_{ik}) \int_0^\infty \alpha_{ik}(\omega)d\omega \quad . \tag{2.62}$$

In Fig.2.15 the absorbed radiation power is marked by the hatched area.

According to (2.15) the probability that one molecule undergoes such an absorbing transition is determined by the Einstein coefficient B_{ik}. For N_i molecules per unit volume the transition rate becomes $N_iB_{ik}\rho(\omega_{ik})$ and the radiation power absorbed per unit volume on the transition $E_i \rightarrow E_k$ is

$$dW_{ik}/dt = N_i B_{ik} \hbar\omega_{ik} \rho(\omega_{ik}) \quad . \tag{2.63}$$

Since the intensity $I(\omega)$ of a plane wave is related to the energy density $\rho(\omega)$ by $I(\omega) = c\rho(\omega)$, (see Sect.2.4), comparison of (2.63) and (2.62) yields

$$\int \alpha_{ik} d\omega = (\hbar\omega_{ik}/c) B_{ik} N_i \quad . \tag{2.64}$$

Equation (2.64) shows that the integral of the absorption coefficient is a constant, independent of the line-broadening process. With $\alpha_{ik} = (4\pi/\lambda)\kappa_{ik}$, the integrand of (2.64) can be replaced by (2.58). This yields

$$\frac{e^2 f_{ik} \gamma_{ik}}{\varepsilon_0 m} \int_0^\infty \frac{\omega^2 d\omega}{(\omega_{ik}^2 - \omega^2)^2 + (\omega\gamma_{ik})^2} = \hbar\omega B_{ik} \quad . \tag{2.65}$$

Near of the line center $(\omega_{ik} - \omega) \ll \omega_{ik}$ the integral can be performed by an elementary procedure and the relation between oscillator strength f_{ik} and the Einstein coefficient B_{ik} comes out to be

$$\boxed{f_{ik} = \frac{2m\varepsilon_0 \hbar\omega_{ik}}{\pi \cdot e^2} B_{ik}} \quad . \tag{2.66}$$

Note

If the upper levels E_k are significantly populated, one has to take into account the *induced emission* which causes an effective *decrease* of absorption. In this case, a term has to be added to (2.64), which, because of (2.21), becomes

$$\int \alpha_{ik}(\omega) d\omega = (\hbar\omega_{ik}/c)(N_i - N_k g_i/g_k) B_{ik} \quad . \tag{2.67}$$

Often it is useful to express the absorption probability by the *absorption cross section* σ_{ik} which specifies the absorption per molecule. It is related to the absorption coefficient α_{ik} by

$$\alpha_{ik} = \sigma_{ik}(N_i - N_k g_i/g_k) \quad . \tag{2.68}$$

In terms of the absorption cross section σ_{ik}, (2.64) can be written in the form

$$\int \sigma_{ik} \, d\nu = (h\nu_{ik}/c) B_{ik} \quad . \tag{2.69}$$

Using (2.22) the total absorption cross section of a transition $E_i \to E_k$ can be expressed by the spontaneous transition probability A_{ik}

$$\sigma_{tot} = \int \sigma_{ik}\, d\nu = (\lambda^2/8\pi)A_{ik} \quad . \tag{2.70}$$

If the halfwidth of the corresponding spectral line is only determined by the natural linewidth $\delta\nu_n$ this converts with $A_{ik} = 2\pi\delta\nu_n$ to

$$\sigma_{tot} = (\lambda/2)^2\delta\nu_n \quad . \tag{2.71}$$

The optical absorption cross section of a spectral line with natural broadening is equal to $(\lambda/2)^2\delta\nu_n$.

In laser spectroscopic experiments the linewidth $\Delta\nu_L$ of the incident radiation is often smaller than the linewidth of the absorbing transition. In this case $I(\nu)$ in (2.61) cannot be taken out of the integral. One has to evaluate the integral $\int \alpha(\nu)I(\nu)d\nu$ by some approximate method. If $\Delta\nu_L \ll \delta\nu_n$ (this is for instance always the case if a single-mode laser is used as the radiation source) the integration yields the integrand at the frequency ν_L. With a laser intensity I_L the radiation power absorbed per unit length is

$$dW_{ik}/dt = \alpha_{ik}I(\nu_L) \quad . \tag{2.72}$$

If the absorption line profile is described by the function

$$\alpha(\nu) = \alpha_0 g(\nu - \nu_0) \quad \text{with} \quad \int g(\nu - \nu_0)d\nu = 1 \quad ,$$

the cross section $\sigma_{ik}(\nu)$ can also be expressed by the Einstein coefficient B_{ik} and the normalized line profile function $g(\nu - \nu_0)$. From (2.67-69) one derives

$$\sigma_{ik}(\nu) = (h\nu/c)g(\nu - \nu_0)B_{ik} \quad . \tag{2.73}$$

Equation (2.73) allows the optical absorption cross section $\sigma_{ik}(\nu)$ to be determined directly as a function of the frequency ν, if the monochromatic incident radiation can be tuned over the absorption line.

2.7 Transition Probabilities

In the previous sections we have seen that the intensities of spectral lines depend not only on the population density of molecules in the absorbing or emitting level but also on the transition probabilities of the corresponding molecular transitions. If these probabilities are known, the population density can be obtained from measurements of line intensities. This is very important, for example, in astrophysics, where spectral lines represent the main source of information from the extraterrestrial world. Intensity measurements of absorption and emission lines allow the concentration of the elements in stellar atmospheres or in interstellar space to be determined. Comparing the intensities of different lines of the same element (e.g., on transitions $E_i \rightarrow E_k$ and $E_e \rightarrow E_k$ from different upper levels E_i, E_e to the same lower level E_k) furthermore enables us to derive the temperature of the radiation source or of the absorbing medium from the relative population densities N_i, N_e in levels E_i and E_e at thermal equilibrium [see (2.18)]. *All these experiments, however, demand a knowledge of the corresponding transition probabilities.*

There is another aspect which makes measurements of transition probabilities very attractive with regard to a more detailed knowledge of molecular structure. Transition probabilities derived from computed wave functions of upper and lower states are much more sensitive to approximation errors in these functions than are the energies of these states. Experimentally determined transition probabilities are therefore well suited to test the validity of calculated approximate wave functions and the comparison with computed probabilities allows theoretical models of electronic charge distribution in excited molecular states to be improved [2.12].

In this section we discuss some experimental methods for the determination of transition probabilities, while in Sect.2.9, we give a survey of the quantum mechanical description of transition probabilities and a brief discussion of some calculational methods.

2.7.1 Lifetimes and Spontaneous Transition Probabilities

The probability P_{ik} that an excited molecule in the level E_i makes a transition to a lower level E_k by spontaneous emission of a fluorescene quantum $h\nu_{ik} = E_i - E_k$ is, according to (2.17), related to the Einstein coefficient A_{ik} by

$$dP_{ik}/dt = A_{ik} \quad .$$

When several transition paths from E_i to different lower levels E_k are possible (see Fig.2.19), the total transition probability is given by

$$A_i = \sum_k A_{ik} \quad . \tag{2.74}$$

The decrease dN_i of the population density N_i during the time interval dt is then

$$dN_i = -A_i N_i dt \quad . \tag{2.75}$$

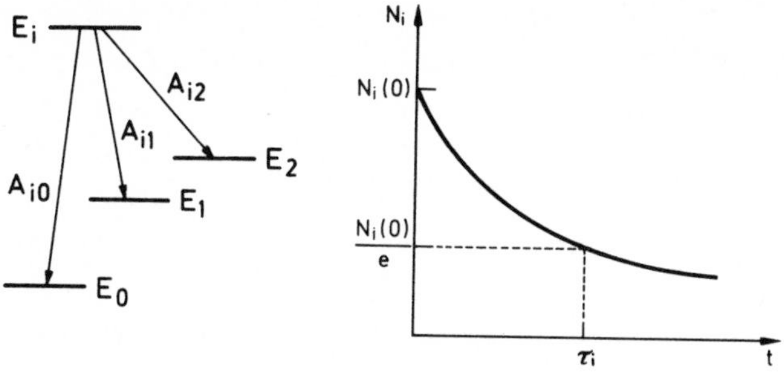

Fig.2.19. Transition probabilities A_{ik} and the exponential decay of level E_i with a mean lifetime τ_i

Integration of (2.75) yields

$$N_i(t) = N_{i0}\, e^{-A_i t} \quad , \tag{2.76}$$

where N_{i0} is the population density at $t = 0$.

After a time $\tau_i = 1/A_i$ the population density N_i has decreased to 1/e of its initial value at $t = 0$. τ_i is the *mean spontaneous lifetime* of the level E_i. This can be seen immediately from the definition of the mean time

$$\bar{t}_i = \int_0^\infty t P_i(t) dt = \int_0^\infty t A_i\, e^{-A_i t}\, dt = 1/A_i = \tau_i \quad , \tag{2.77}$$

where $P_i(t)\,dt$ is the probability that one atom in the level E_i makes a spontaneous transition within the time interval between t and t + dt.

The *radiant power* emitted from N_i molecules on the transition $E_i \to E_k$ is

$$dW_{ik}/dt = N_i h\nu_{ik} A_{ik} \quad . \tag{2.78}$$

If several transitions $E_i \to E_k$ from the same upper level E_i to different lower levels E_k are possible, the radiant powers of the corresponding spectral lines are proportional to the Einstein coefficients A_{ik}. The relative radiant intensities in a certain direction may also depend on the spatial distribution of the fluorescence, which can be different for the different transitions.

A molecular level E_i of a molecule A can be depopulated not only by spontaneous emission but also by collision-induced radiationless transitions (see Fig.2.20). The probability dP_{ik}^{coll}/dt of such a transition depends on the density N_B of the collision partner B, on the mean relative velocity $\bar{v}$ between A and B, and on the collision cross section σ_{ik}^{coll} for an inelastic collision which induces the transition $E_i \to E_k$ in the molecule A.

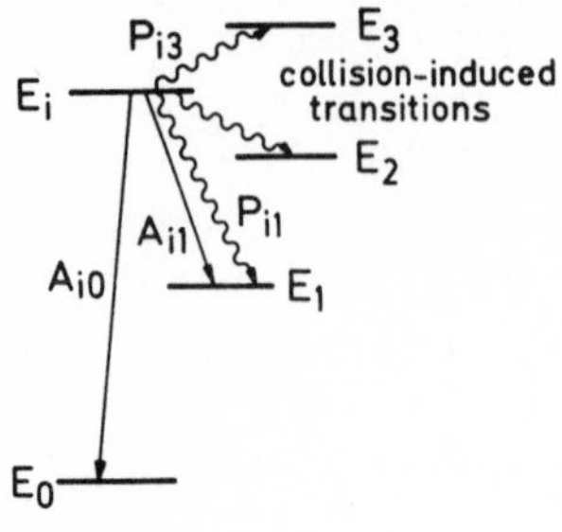

Fig.2.20. Depopulation paths of an excited level

$$dP_{ik}^{coll}/dt = \bar{v}N_B\sigma_{ik}^{coll} \quad . \tag{2.79}$$

At thermal equilibrium the relative velocities follow a Maxwellian distribution with a mean value $\bar{v}$

$$\bar{v} = \sqrt{\frac{8kT}{\pi\mu}} \quad , \tag{2.80}$$

where $\mu = M_AM_B/(M_A + M_B)$ is the reduced mass and M_A, M_B are the masses of the molecules A and B.

When the excited molecule $A(E_k)$ is exposed to an intense radiation field, the *induced emission* may become noticeable. It contributes to the depopulation of level E_i with a probability

$$dP_{ik}^{ind}/dt = \rho(\nu_{ik})B_{ik}[N_i - (g_i/g_k)N_k] \quad . \tag{2.81}$$

The total transition probability which determines the effective lifetime of a level E_i is then the sum of spontaneous, induced and collisional contributions and the mean lifetime τ_i^{eff} becomes

$$1/\tau_i^{eff} = \sum_k \left[A_{ik} + \rho(\nu_{ik})B_{ik}(N_i - N_kg_i/g_k) + N_B\sigma_{ik}\sqrt{\frac{8kT}{\pi\mu}}\right] . \tag{2.82}$$

Measuring the effective lifetime τ_{eff} as a function of exciting radiation intensity and also its dependence on the density N_B of collision partners (Stern-Vollmer plot) allows one to determine the three transition probabilities separately (see Fig.2.21).

2.7.2 Experimental Methods for the Determination of Transition Probabilities

The previous section has shown that the spontaneous lifetime of an excited level E_k is determined by the sum of the transition probabilities A_{km} of all transitions from E_k to lower levels E_m.

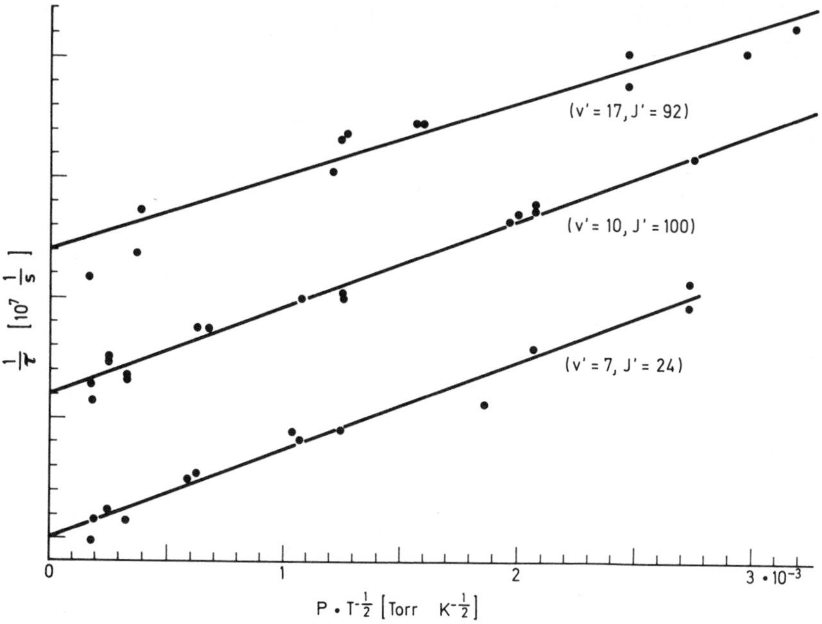

Fig.2.21. Inverse effective lifetimes of some excited (v', J') levels in the $^1D\Pi$ state of the NaK molecule as a function of pressure P of the K-atoms (Stern-Vollmer plot)

$$1/\tau_k = A_k = \sum_{m=0}^{k-1} A_{km} = A_{k0} \sum_{m=0}^{k-1} A_{km}/A_{k0} \quad . \tag{2.83}$$

From measurements of the spontaneous lifetime τ_k together with the determination of the relative intensities I_{km}/I_{k0} of the different transitions $E_k \to E_m$ the *absolute* transition probabilities A_{km} can be obtained.

Since lifetime measurements represent important contributions to the determination of absolute transition probabilities, several experimental techniques were developed before the invention of lasers. For a recent review, see [2.13]. The introduction of pulsed or mode-locked lasers has greatly enhanced the experimental capabilities. Because of their high intensity and the short pulse duration, the sensitivity and the time resolution have been greatly improved and lifetimes in the picosecond range can now be measured accurately. Some commonly used techniques of lifetime measurements with lasers will be discussed in Sect.11.2.

Besides lifetime measurements the experimental investigation of absorption or dispersion of spectral lines can be used to obtain transition probabilities. According to (2.64) the absorption coefficient α_{ik} integrated over the line profile of an absorbing transition $E_i \to E_k$

$$\int \alpha_{ik}\, d\omega = N_i(\hbar\omega_{ik}/c)B_{ik}$$

is related to the Einstein coefficient B_{ik}. If the absorption path is optically thin ($\alpha z \ll 1$), B_{ik} can be directly determined from the absorbed radiation power [see (2.62)], provided the density N_i of absorbing atoms is known. The use of lasers allows measurements of absorption coefficients with greatly enhanced sensitivity. Several techniques will be discussed in Chap.8.

The density N_i can be determined from temperature and pressure measurements in absorption cells at thermal equilibrium. In the case of vaporized samples the vapor pressure has to be known [2.14]. In heat pipes [2.15] the vapor pressure is determined by the pressure of the noble gas which encloses the vapor. Since in heat pipes a definite zone with constant temperature and vapor pressure and with relatively sharp edges can be realized, these devices are very useful in the absorption spectroscopy of vapors [2.16a].

Another method to determine the density N_i uses an atomic beam with known geometrical dimensions in a vacuum chamber [2.16b]. The beam atoms are condensed on a cold collector and the mass deposit per second dM/dt is measured with a sensitive balance. The mass flux of atoms with mass m through the oven orifice (cross section A) is

$$dM_0/dt = N_0\bar{v}Am \quad .$$

The mean velocity $\bar{v}$ can be determined from the oven temperature. The density N_0 at the orifice is related to the density N_i at a distance R from the orifice, where the optical path crosses the atomic beam, by $N_i = N_0/R^2$ if $R^2 \gg A$. The mass deposit on the collector is then

$$dM/dt = b\, dM_0/dt \quad ,$$

where the constant factor b can be readily calculated from the geometry of the beam (see Fig.2.22).

A different technique for the determination of oscillator strengths or transition probabilities is the "hook method", which is based on anomalous dispersion close to a spectrum line. This technique has been extensively used by Russian spectroscopists [2.16c]. According to (2.59) the oscillator strengths f_{ik} can be obtained from measurements of the refractive index n_i' at frequencies ω close to the frequency ω_{ik}. Such measurements have been performed with Mach-Zehnder interferometers (see Sect.4.2) and photographic recording. A possible experimental arrangement is shown in Fig.2.23. When a light continuum is sent through the absorption cell, the anomalous disper-

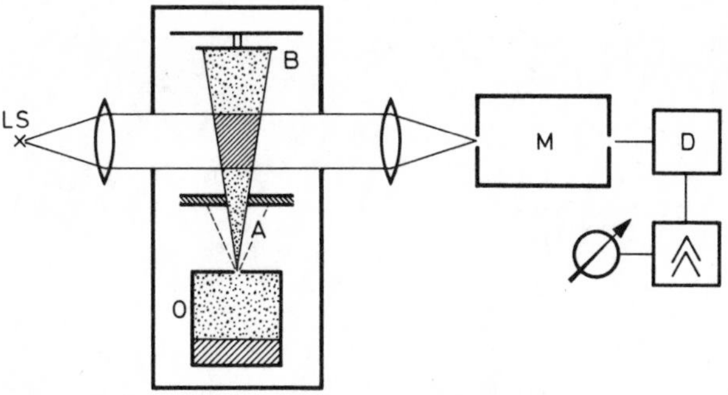

Fig.2.22. Atomic beam apparatus for simultaneous measurements of absorber density and integrated absorption coefficient (A = aperture, B = balance, D = detector, M = monochromator, O = oven)

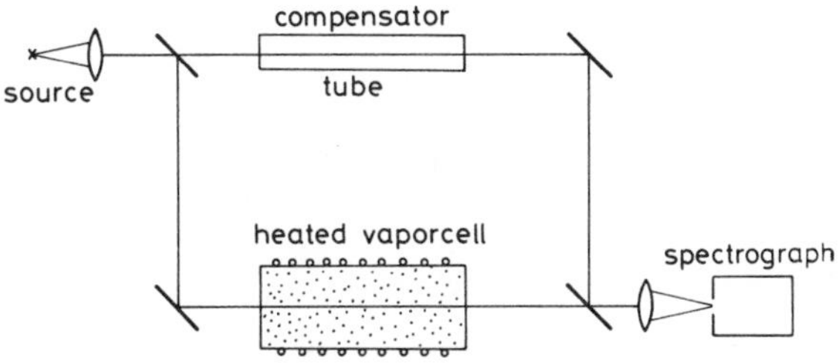

Fig.2.23. Experimental arrangement for the measurement of dispersion in atomic vapors (hook method)

sion close to the absorption lines causes interference patterns behind the Mach-Zehnder interferometer, which can be displayed behind a spectrograph as a function of the wavelength λ. This dispersed pattern has the form of hooks (see Fig.2.24, also Fig.4.33), which gave the method its name. With simultaneous detection of absorption and dispersion the density N_i of absorbing molecules can be determined. This allows the oscillator strength to be derived from one experiment without additional information [2.17]. Such combined techniques have considerably increased the accuracy and relevance of the hook method.

Using tunable lasers the sensitivity and accuracy of dispersion measurements can be further enhanced. Some of these laser methods will be discussed in Sect.10.4.

An extensive class of experimental methods for the determination of transition probabilities is based on direct or indirect measurements of the natural linewidth $\delta\omega_n$ of a transition $E_k \to E_i$. As will be shown in Sect.3.1,

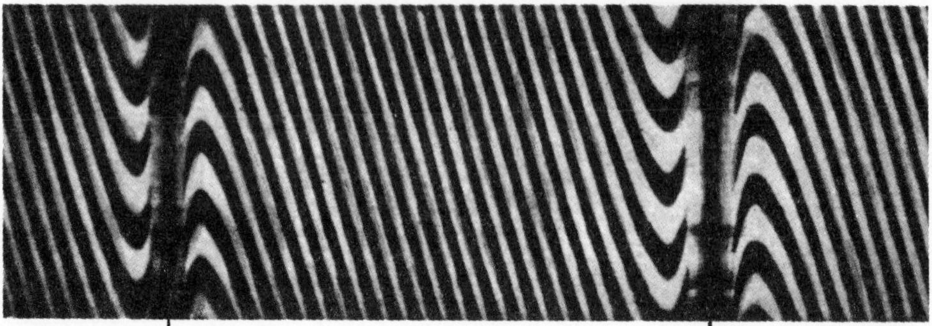

Fig.2.24. Interference pattern (hooks) as a function of wavelength λ in the vicinity of the AlI resonance lines

the spontaneous transition probabilities A_i, A_k from levels E_i, E_k, respectively, are related to $\delta\omega_n$ by

$$\delta\omega_n = A_i + A_k \quad .$$

All methods which allow the natural linewidth of spectral lines to be measured even in the presence of other broadening effects, may be used to obtain f values and Einstein coefficients. In this field, laser spectroscopy has brought a great variety of different Doppler-free techniques which will be discussed in Chap.10. A review of different methods of the prelaser era can be found in [2.18a,b].

2.8 Linear and Nonlinear Absorption

In Sect.2.6 we saw that the intensity decrease dI of a wave with intensity I propagating along the z direction through an absorbing sample is

$$dI = -I\sigma_{ik}[N_i - (g_i/g_k)N_k]dz \quad . \tag{2.84a}$$

As long as the population densities N_i and N_k of the two levels E_i and E_k are not noticeably altered by the interaction with the radiation field (weak-signal approximation of Sect.2.9.2), one can regard them as constant. The *absorbed* intensity is then proportional to the *incident* intensity (*linear absorption*). Integration of (2.84a) over the absorption path z gives Beer's law for linear absorption,

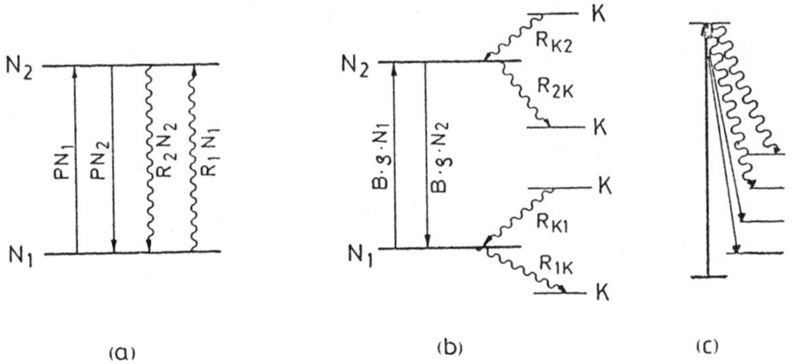

Fig.2.25a-c. Two-level system with relaxation channels. (a) Relaxation only between levels 1 and 2; (b) Additional relaxation from and to other levels E_k (c) Laser induced molecular fluorescence as example for multichannel relaxation

$$I(z) = I_0\, e^{-\sigma_{ik}[N_i - (g_i/g_k)N_k]z}$$

$$= I_0\, e^{-\alpha z} \quad , \qquad \text{with} \tag{2.84b}$$

$$\alpha = \sigma_{ik}[N_i - (g_i/g_k)N_k] \quad .$$

At larger intensities I, the density N_i of the lower state E_i can noticeably decrease while the upper state density N_k increases. This corresponds to the strong field case in Sect.2.9.4 and means that $N_i(I)$ and $N_k(I)$ are functions of I and therefore dI is no longer proportional to I (*nonlinear absorption*).

We shall illustrate this nonlinear absorption by the simple example of a two-level system with population densities N_1 and N_2 and equal statistical weights $g_1 = g_2 = 1$. The total number density $N = N_1 + N_2$ of the two-level system will be constant, if we exclude for the present all decay channels to other levels than these two levels with energies E_1, E_2 and $E_2 > E_1$ (Fig.2.25a).

According to (2.15-17) the time derivations of N_1 and N_2 can be related to the Einstein coefficients $B_{21} = B_{12}$ for induced emission and absorption and A_{21} for spontaneous emission. If we allow additional collision induced transitions with probabilities C_{12} and C_{21} between the two levels we obtain under stationary conditions:

$$dN_1/dt = -dN_2/dt = B_{12}\rho(\omega_{12})\cdot(N_2 - N_1) + (A_{21} + C_{21})N_2 - C_{12}N_1 = 0 \quad . \tag{2.84c}$$

With the abbreviations $\Delta N = N_1 - N_2$; $N = N_1 + N_2 = \text{const.}$, $R_2 = A_{21} + C_{21}$, $R_1 = C_{12}$ the population difference ΔN can be written as

$$\Delta N = \frac{\Delta N_0}{1 + 2B_{12} \cdot \rho(\omega_{12})/(R_1 + R_2)} = \frac{\Delta N_0}{1 + S} \tag{2.84d}$$

where $\Delta N_0 = N \cdot (R_2 - R_1)/(R_2 + R_1)$ is the population difference for $\rho = 0$.

The *saturation parameter*

$$S = \frac{2B_{12}\rho(\omega_{12})}{R_1 + R_2} = \frac{B_{12} \cdot \rho(\omega_{12})}{\bar{R}} \tag{2.84e}$$

represents the ratio of the induced transition probability $B_{12} \cdot \rho(\omega_{12})$ to the mean relaxation probability $\bar{R} = (R_1 + R_2)/2$. If the only relaxation process is spontaneous emission ($R_1 = 0$, $R_2 = A_{21}$), then the saturation parameter S yields the ratio of induced to spontaneous transition rates. For $S = 1$ the population difference ΔN drops to one half of its unsaturated value ΔN. With $I(\omega) = c \cdot \rho(\omega)$ (2.84d) can be written as

$$\Delta N = \frac{\Delta N_0}{1 + \dfrac{B_{12} \cdot I(\omega_{12})}{c \cdot \bar{R}}} = \frac{\Delta N_0}{1 + I/I_s} \tag{2.84f}$$

where the *saturation intensity* $I_S = c \cdot \bar{R}/B_{12}$ stands for that incident intensity which decreases ΔN to $\Delta N_0/2$.

Since the absorption coefficient $\alpha = \sigma \cdot \Delta N$ is proportional to ΔN we obtain the remarkable result, that with increasing incident intensity I the absorption coefficient

$$\alpha = \frac{\alpha_0}{1 + S} = \frac{\alpha_0}{1 + I/I_s} \tag{2.84g}$$

of a two level system approaches zero for $I \to \infty$ (Fig.2.26).

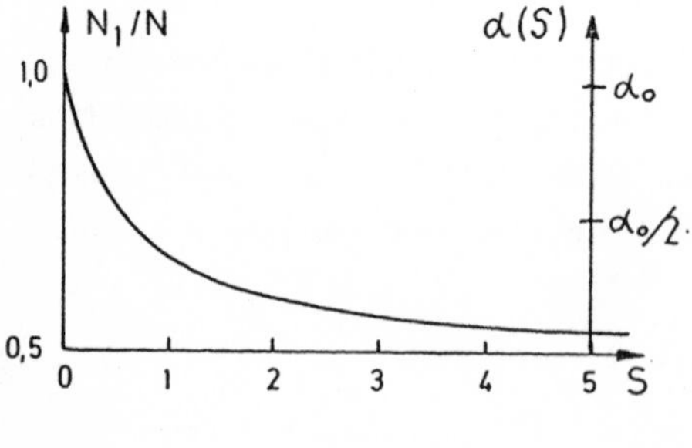

Fig.2.26. Ratio N_1/N in a two level system with $N_1 + N_2 = N$ (left scale) and absorption coefficient α (right scale) as a function of the saturation parameter S

Note

When the radiation field is switched on at time $t = 0$, one may observe at first a transient behavior of the population densities. The probability amplitudes $a(t)$ and $b(t)$ show a damped oscillation (see Sect.2.9.5) and approach their stationary values $|a|^2 = |b|^2 = 0.5$ only after the oscillations have died away (see Fig.2.30).

We now generalize our example by allowing additional relaxation paths to other levels k (Fig.2.26b). Note that now $N_1 + N_2 = N = \text{const.}$ no longer holds since a change ΔN_2 of the population N_2 does not necessarily cause an equal change ΔN_1. This situation is relevant, for example, in the case of molecules where the excited rovibronic level of an upper electronic state may decay into many vibrational levels of the electronic ground state (Fig.2.25c).

The stationary population densities N_1 and N_2 can now be obtained from

$$dN_1/dt = B_{12}\rho(N_2 - N_1) - R_1N_1 + \sum_K R_{K1}N_K = 0$$
$$dN_2/dt = B_{12}\rho(N_1 - N_2) - R_2N_2 + \sum_K R_{K2}N_K = 0 \qquad (2.84h)$$

where R_iN_i represents the total relaxation rate (including spontaneous emission) which depopulates N_i, and $\sum R_{Ki}N_K$ takes care of all relaxation paths from other levels k which contribute to the repopulation of N_i. If this sum is not noticeably changed by the radiation field (ω_{12}), we can set $\sum_K R_{Ki}\, N_K = C_i$ and obtain from (2.84b) for the unsaturated population difference ($\rho = 0$)

$$\Delta N_0 = \frac{C_2R_1 - C_1R_2}{R_1R_2} \qquad (2.84i)$$

and for the saturated population difference ($\rho \neq 0$)

$$\Delta N = \frac{\Delta N_0}{1 + B_{12}\rho(\omega_{12})/\left(\frac{1}{R_1} + \frac{1}{R_2}\right)} \quad . \qquad (2.84j)$$

This can again be expressed with a saturation parameter

$$S = \frac{B_{12}\rho}{R^*} \quad \text{with} \quad \frac{1}{R^*} = \frac{1}{R_1} + \frac{1}{R_2} \quad . \qquad (2.84k)$$

Note that for the second example the "mean relaxation probability" is defined as $R^* = R_1 \cdot R_2/(R_1 + R_2)$, while in the case of the isolated two level system, where no relaxation channels to other levels are open, the "mean relaxation probability" is $\bar{R} = (R_1 + R_2)/2$.

There is another important difference between the two systems: while for the isolated two level system the population N_1 of the lower level is

$$N_1 = N \cdot \frac{1 + S}{1 + 2S} \tag{2.84l}$$

and can never drop below N/2 for $S \to \infty$ (Fig.2.26) it can be nearly completely depopulated for the second example. This can be seen from (2.84h) which yields

$$N_1 = \frac{(C_1 + C_2)B_{12}\rho(\omega_{12}) + R_2C_1}{(R_1 + R_2)B_{12}\rho(\omega_{12}) + R_1R_2} \quad . \tag{2.84m}$$

If the repopulation rates C_1 and C_2 are small compared to the depopulation probabilities R_1 and R_2 the population N_1 may become very small with increasing radiation field density $\rho(\omega_{12})$. This is, for instance, true for molecules in a molecular beam, which are excited by the laser from the ground state level 1 to level 2 in the excited state. Since collisions are negligible, we can set $C_2 = 0$, $C_1 = A_{21}N_2$, $R_2 = A_2$, $R_1 = 0$. From (2.84m) we obtain

$$N_1/N_2 = \frac{A_{21}(B_{12}\rho(\omega_{12}) + A_2)}{A_2B_{12}\rho(\omega_{12})}$$

which approaches the value A_{21}/A_2 for $B\rho \gg A_2$. Since the fluorescence from level 2 terminates on many vibrational levels of the ground state, the total spontaneous transition probability A_2 is generally much larger than the probabilities A_{21} of the transition $2 \to 1$.

The frequency dependence of the absorption coefficient $\alpha_{12}(\omega) = (N_1 - N_2)\sigma_{12}$ in the saturation case is crucially dependent on the broadening mechanism of the absorption line and is completely different for homogeneously broadened lines than for inhomogeneously broadened lines. Nonlinear absorption and its importance for high-resolution spectroscopy is treated in Chap.10 after a discussion of various broadening mechanisms in Chap.3 (see also [1.13]).

2.9 Semiclassical Description

This section briefly outlines the semiclassical treatment of the interaction of E.M. radiation with atoms. A more extensive presentation can be found in [2.6] or in textbooks on quantum mechanics.

The radiation incident upon an atom is described by a classical E.M. plane wave

$$\underline{E} = \underline{A}_0 \cos(\omega t - kz) \quad . \tag{2.85a}$$

The atom, on the other hand, is treated quantum mechanically. In order to simplify the equations we restrict ourselves to a two-level system with eigenstates E_a and E_b(Fig.2.27).

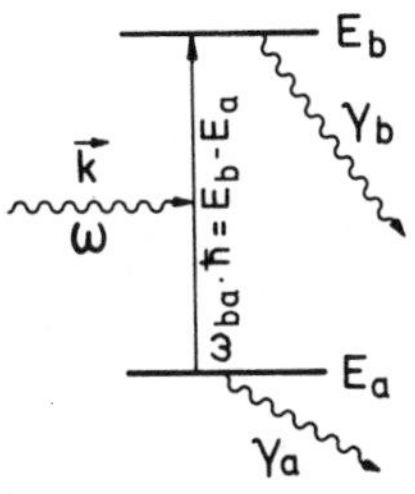

Fig.2.27. Two-level system with decay channels, interacting with a monochromatic field

Laser spectroscopy is performed in spectral regions where the wavelength λ is large compared to the diameter d of an atom (e.g., in the visible spectrum λ is 500 nm, but d only about 0.5 nm). For $\lambda \gg d$ the phase of the E.M. wave does not change much within the volume of an atom because $k \cdot z = (2\pi/\lambda) \cdot z \ll 1$ for $z \leq d$. We can therefore neglect the spatial derivatives of the field amplitude. The interaction term in the Hamiltonian reduces in this *dipole approximation* to the product of the mean atomic dipole moment $\bar{\underline{p}} = e \cdot \bar{\underline{r}}$ with the electric field amplitude $\underline{E}$. In a coordinate system with its origin in the center of the atom we can assume $k \cdot z \approx 0$ within the atomic volume, and write (2.85a) in the form

$$\underline{E} = \underline{A}_0 \cos\omega t = (\underline{A}_0/2)[e^{i\omega t} + e^{-i\omega t}] \quad . \tag{2.85b}$$

2.9.1 Basic Equations

We start with the time-dependent Schrödinger equation [2.6]

$$\tilde{H}\psi = i\hbar \frac{\partial\psi}{\partial t} \quad . \tag{2.86}$$

The Hamiltonian operator

$$\tilde{H} = \tilde{H}_0 + \tilde{V} \quad , \tag{2.87}$$

can be written as a sum of the unperturbed Hamiltonian $\tilde{H}_0$ of the free atom plus the perturbation operator

$$\tilde{V} = \tilde{\underline{p}} \cdot \underline{E} = \tilde{\underline{p}} \cdot \underline{A}_0 \cos\omega t \quad , \tag{2.88}$$

which describes the interaction of the atom with the E.M. field. $\tilde{V}$ is the scalar product of the dipole operator $\tilde{\underline{p}} = -e \cdot \tilde{\underline{r}}$ and the electric field $\underline{E}$.

The general solution $\psi(\underline{r},t)$ of (2.86) can be expressed as a linear superposition

$$\psi(\underline{r},t) = \sum_{n=1}^{\infty} c_n(t)u_n(\underline{r})\exp(-iE_n t/\hbar) \tag{2.89}$$

of the eigenfunctions of the unperturbed atom.

$$\phi_n(r,t) = u_n(\underline{r})\exp(-iE_n t/\hbar) \quad . \tag{2.90}$$

The spatial parts $u_n(\underline{r})$ of these eigenfunctions are solutions of the time-independent Schrödinger equation

$$\tilde{H}_0 u_n(\underline{r}) = E_n u_n(\underline{r}) \quad , \tag{2.91}$$

and satisfy the orthogonality relations

$$\int u_i^* u_k d\tau = \delta_{ik} \quad . \tag{2.92}$$

Note: In (2.89-91) a nondegenerate system has been assumed.

For our two-level system with eigenstates a and b and energies E_a and E_b, the expansion (2.89) reduces to a sum of two terms

$$\psi(\underline{r},t) = a(t)u_a\, e^{-iE_a t/\hbar} + b(t)u_b\, e^{-iE_b t/\hbar} \quad . \tag{2.93}$$

The coefficients a(t), b(t) are the time-dependent *probability amplitudes* of the atomic states E_a, E_b. This means that the value $|a(t)|^2$ gives the probability of finding the system in level a at time t. Obviously the relation $|a(t)|^2 + |b(t)|^2 = 1$ must hold at all times t, if decay into other levels is neglected.

Substituting (2.93) into (2.86) gives

$$i\hbar\dot{a}(t)u_a\, e^{-iE_a t/\hbar} + i\hbar\dot{b}(t)u_b\, e^{-iE_b t/\hbar} = a\tilde{V}u_a\, e^{-iE_a t/\hbar} + b\tilde{V}u_b\, e^{-iE_b t/\hbar} \quad , \qquad (2.94)$$

where the relation $\tilde{H}_0 u_n = E_n u_n$ has been used to cancel equal terms on both sides. Multiplication with u_n^* (n = a,b) and spatial integration results in the following two equations

$$\dot{a}(t) = -(i/\hbar)\left[a(t)V_{aa} + b(t)V_{ab}\, e^{i(E_a - E_b)t/\hbar}\right] \quad , \qquad (2.95a)$$

$$\dot{b}(t) = -(i/\hbar)\left[b(t)V_{bb} + a(t)V_{ba}\, e^{-i(E_a - E_b)t/\hbar}\right] \quad . \qquad (2.95b)$$

The spatial integral

$$V_{ik} = \int u_i^* \tilde{V} u_k d\tau = -e\underline{E} \int u_i^* \underline{r} u_k d\tau$$

$$= -e\underline{E}\underline{R}_{ik} \quad \text{with} \quad \underline{R}_{ik} = \underline{R}_{ki} = \int u_i^* \underline{r} u_k d\tau \qquad (2.96)$$

is the *matrix element* of the interaction operator $\tilde{V} = \tilde{\underline{p}} \cdot \underline{E}$. The integral $e\,\underline{R}_{ik}$ is called the atomic *dipole matrix element*. It depends on the stationary wave functions u_i and u_k of the two states E_i, E_k and is determined by the charge distribution in the atomic states.

Since $\underline{r}$ has odd parity the integrals V_{aa} and V_{bb} vanish when integrating over all coordinates from $-\infty$ to $+\infty$. Using (2.85) for the E.M. field, (2.95), with the abbreviations

$$\omega_{ba} = (E_b - E_a)/\hbar = -\omega_{ab} \quad \text{and} \quad R_{ab} = -e\underline{R}_{ab}\underline{A}_0/\hbar = R_{ba} \qquad (2.97)$$

reduce to

$$\dot{a}(t) = (i/2)R_{ab}\left[e^{i(\omega_{ab} - \omega)t} + e^{i(\omega_{ab} + \omega)t}\right]b(t) \qquad (2.98a)$$

$$\dot{b}(t) = (i/2)R_{ab}\left[e^{-i(\omega_{ab} - \omega)t} + e^{-i(\omega_{ab} + \omega)t}\right]a(t) \quad . \qquad (2.98b)$$

These are the basic equations which must be solved to obtain the probability amplitudes a(t) and b(t).

Suppose that the atoms are in the lower state E_a at time $t = 0$, which implies that $a(0) = 1$ and $b(0) = 0$. We assume the field amplitude A_0 to be sufficiently small that for times $t < T$ the population of E_b remains small compared with that of E_a, i.e, $|b(t < T)|^2 \ll 1$. Under this *weak-field con-*

dition we can solve (2.98) with an iterative procedure starting with a = 1 and b = 0. Using thermal radiation sources the field amplitude A_0 is generally small enough to make the first iteration step already sufficiently accurate.

With these assumptions the first approximation of (2.98) gives

$$\dot{a}(t) = 0 \qquad (2.99a)$$

$$\dot{b}(t) = (i/2)R_{ab}[e^{i(\omega_{ba} - \omega)t} + e^{i(\omega_{ba} + \omega)t}] \ . \qquad (2.99b)$$

With the initial conditions a(0) = 1 and b(0) = 0 integration of (2.99) yields

$$a(t) = a(0) = 1 \qquad (2.100a)$$

$$b(t) = (R_{ab}/2)\left[\frac{e^{i(\omega_{ba} - \omega)t} - 1}{\omega_{ba} - \omega} + \frac{e^{i(\omega_{ba} + \omega)t} - 1}{\omega_{ba} + \omega}\right] \ . \qquad (2.100b)$$

For $E_b > E_a$ the term $\omega_{ba} = (E_b - E_a)/\hbar$ is positive. In a transition $E_a \rightarrow E_b$, the atomic system absorbs energy from the radiation field. Noticeable absorption occurs, however, only if the field frequency ω is close to the eigenfrequency ω_{ba}. In the optical frequency range this implies that $|\omega_{ba} - \omega| \ll \omega_{ba}$. The second term in (2.100b) is then small compared to the first and may be neglected. This is called the *rotating-wave approximation* for only that term is kept in which the atomic wave functions and the field waves with phasors $\exp(-i\omega_{ab}t)$ and $\exp(-i\omega t)$ rotate together.

In the rotating-wave approximation with

$$b(t) = (R_{ab}/2)\ \frac{e^{i(\omega_{ba} - \omega)t} - 1}{\omega_{ba} - \omega} \ , \qquad (2.101)$$

we obtain for the probability $|b(t)|^2$ that the system is at time t in the upper level E_b

$$|b(t)|^2 = (R_{ab}/2)^2\left(\frac{\sin(\omega_{ba} - \omega)t/2}{(\omega_{ba} - \omega)/2}\right)^2 \ . \qquad (2.102)$$

Since we had assumed that the atom was at t = 0 in the lower level E_a, (2.102) *gives the transition probability for the atom to go from* E_a *to* E_b *during the time* t. Figure 2.28a illustrates this transition probability as a function of the detuning $\Delta\omega = \omega - \omega_{ba}$.

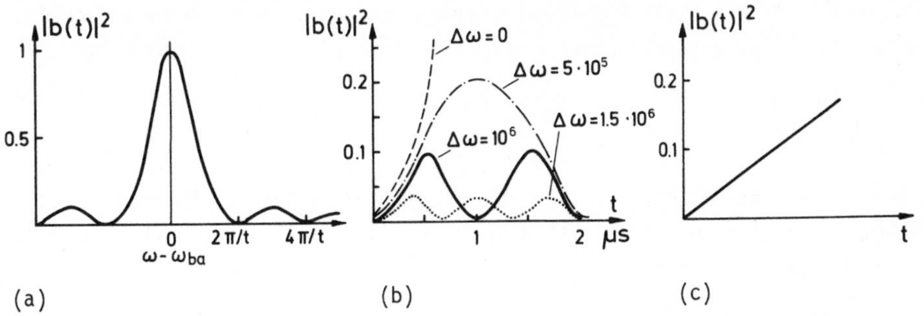

Fig.2.28. (a) Normalized transition probability as a function of the detuning ($\omega - \omega_{ba}$); (b) Probability of a transition to the upper level as a function of time for different detuning; (c) $|b(t)|^2$ under broad-band excitation and weak fields

Equation (2.102) shows that the probability $|b(t)|^2$ depends on the detuning $\Delta\omega = |\omega_{ba} - \omega|$ of the field frequency ω from the eigenfrequency ω_{ba}. When tuning the frequency ω into resonance with the atomic system ($\omega \to \omega_{ba}$), the second factor in (2.102) approaches the value t^2 because $(\sin^2 xt)/x^2 \to t^2$ as $x \to 0$. The transition probability at resonance,

$$|b(t)|^2_{\omega=\omega_{ba}} = (R_{ab}/2)^2 t^2 \quad , \tag{2.103}$$

increases proportional to t^2. The approximation used in deriving (2.102) has, however, anticipated that $|b(t)|^2 \ll 1$. According to (2.103) for the resonance case this assumption is equivalent to

$$R^2_{ab} t^2 \ll 1 \quad \text{or} \quad t \ll T = \hbar/(e\underline{R}_{ab}\underline{A}_0) \ . \tag{2.104}$$

In order for our small-signal approximation to hold, the maximum interaction time T of the field (amplitude A_0) with the atom (matrix element R_{ab}) is therefore restricted. Because the spectral analysis of a wave with finite detection time T gives a spectral width $\Delta\omega \approx 1/T$ (see also Sect.3.4), we cannot assume monochromaticity but have to take into account the frequency distribution of the interaction term.

2.9.2 Transition Probabilities with Broad-Band Excitation

In general, thermal radiation sources have a bandwidth $\delta\omega$ which is much larger than the Fourier limit $\Delta\omega = 1/T$ anyway. Therefore the finite interaction time imposes no extra limitations. This may change, however, when lasers are considered.

Instead of the field amplitude $\underline{A}_0$ (which refers to unit frequency interval) we introduce the spectral energy density $\rho(\omega)$ by the relation [see (2.30c)]

$$\rho(\omega) = \varepsilon_0 A_0^2/2 \quad .$$

We can now generalize (2.102) to include the interaction of broad-band radiation with our two-level system by integrating (2.102) over all frequencies ω of the radiation field. This yields the total transition probability $P_{ab}(t)$ within the time t, if $\underline{R}_{ab} \parallel \underline{A}_0$;

$$P_{ab}(t) = \int |b(t)|^2 d\omega = \frac{(e\underline{R}_{ab})^2}{2\varepsilon_0 \hbar^2} \int \rho(\omega) \left(\frac{\sin(\omega_{ba} - \omega)t/2}{(\omega_{ba} - \omega)/2} \right)^2 d\omega \quad . \tag{2.105}$$

For thermal light sources $\rho(\omega)$ is slowly varying compared with the $(\sin^2 xt)/x^2$ factor, and is essentially constant over the frequency range where the factor $[\sin^2(\omega_{ba} - \omega)t/2]/[(\omega_{ba} - \omega)/2]^2$ is large (see Fig.2.28a). We can therefore replace $\rho(\omega)$ by its resonance value $\rho(\omega_{ab})$. The integration can then be performed and gives

$$\rho(\omega_{ba}) 2\pi t \quad .$$

The transition probability for the time interval between 0 and t is therefore (Fig.2.28c)

$$P_{ab}(t) = \frac{\pi e^2}{\varepsilon_0 \hbar^2} R_{ab}^2 \rho(\omega_{ba}) t \tag{2.106}$$

and the *transition probability per second* becomes

$$d\, P_{ab}/dt = \frac{\pi e^2}{\varepsilon_0 \hbar^2} R_{ab}^2 \rho(\omega_{ba}) \quad . \tag{2.107}$$

To compare this result with the Einstein coefficient B_{ab} derived in Sect.2.3, we must take into account that the blackbody radiation in Sect. 2.3 was isotropic whereas the E.M. wave (2.84) used in the derivation of (2.107) propagates into one direction. For randomly oriented atoms with dipole moment $\underline{p}$, the averaged component in the z direction is $\bar{p}_z = p/3$.

In the case of isotropic radiation the interaction term $R_{ab}^2 \rho(\omega_{ba})$ therefore has to be divided by a factor of 3. The comparison of (2.16) and the modified (2.107) yields

$$d\,P_{ab}/dt = \frac{\pi e^2}{3\varepsilon_0\hbar^2}\,\rho(\omega_{ba})R_{ab}^2 = \rho(\omega_{ba})B_{ab} \quad . \tag{2.108}$$

With the definition (2.96) for the dipole matrix element $\underline{R}_{ik}$, the Einstein coefficient B_{ik} of induced absorption $E_i \rightarrow E_k$ finally becomes

$$\boxed{B_{ik}^{(\omega)} = \frac{\pi e^2}{3\varepsilon_0\hbar^2}\left|\int u_i^*\underline{r}u_k d\tau\right|^2} \quad . \tag{2.109}$$

Equation (2.109) gives the Einstein coefficient for a one-electron system if $\underline{r} = (x, y, z)$ is the vector from the nucleus to the electron and the $u_n(x, y, z)$ are the one-electron wave functions. For atoms with N electrons all of which may contribute to the dipole moment, the matrix element has to be generalized to

$$\underline{R}_{ik} = e\int\ldots\int u_i^*\left(\sum_{n=1}^{N}\right)\underline{r}_n\, u_k d\tau_1 \ldots d\tau_N \quad , \tag{2.110}$$

where the $u_n(\underline{r}_1 \ldots \underline{r}_N)$ are now multielectron wave functions and the integration extends over all 3N coordinates of the N electrons. Such matrix elements can be calculated of course only by using approximation methods.

Note

When using the frequency $\nu = \omega/2\pi$ instead of ω, the spectral energy density $\rho(\nu)$ per unit frequency interval is larger by a factor of 2π because a unit frequency interval $d\omega = 1$ corresponds to $d\nu = 1/2\pi$. The right-hand side of (2.109) has then to be divided by a factor of 2π since $B_{ik}\rho(\nu) = B_{ik}\rho(\omega)$.

2.9.3 Phenomenological Inclusion of Decay Phenomena

So far we have neglected the fact that the levels E_a and E_b are not only coupled by transitions induced by the external field but may also decay by spontaneous emission or by other relaxation processes such as collision-induced transitions. We can include these decay phenomena in our formulas by adding phenomenological decay terms to (2.98) which can be expressed by the decay constants γ_a and γ_b (see Fig.2.27).

In the rotating-wave approximation, where the term with frequency $(\omega_{ba} + \omega)$ is neglected, (2.98) then become

$$\dot{a}(t) = -\frac{1}{2}\gamma_a a + \frac{i}{2}R_{ab}\, e^{-i(\omega_{ba} - \omega)t}\, b(t) \tag{2.111a}$$

$$\dot{b}(t) = -\frac{1}{2}\gamma_b b + \frac{i}{2} R_{ab}\, e^{+i(\omega_{ba} - \omega)t}\, a(t) \quad . \qquad (2.111b)$$

The expectation value D of the dipole moment for our two-level system which interacts with the radiation field is

$$\underline{D} = -e \int \psi^* \underline{r} \psi d\tau \quad . \qquad (2.112)$$

Using the expansion (2.93) and the definition (2.96) this can be expressed by the coefficients a(t) and b(t) and by the matrix element R_{ab} as

$$\underline{D} = -e\,\underline{R}_{ab}(a^* b\, e^{-i\omega_{ba}t} + ab^* e^{+i\omega_{ba}t}) \quad . \qquad (2.113)$$

When the field amplitude A_0 is sufficiently small [see (2.104)] we can use the weak-signal approximation of the previous section. This means that $|a(t)|^2 = 1$ and $|b(t)|^2 \ll 1$ and also $aa^* - bb^* \approx 1$. With this approximation one obtains, after taking the second time derivative of (2.113) and using (2.111), the equation of motion for the dipole moment of the atom under the influence of the radiation field,

$$\ddot{D} + \gamma_{ab}\dot{D} + (\omega_{ba}^2 + \gamma_{ab}^2/4)D = (eR_{ab}^2 A_0/\hbar)\,[(\omega_{ba} + \omega)\cos\omega t + (\gamma_{ab/2})\sin\omega t] \qquad (2.114)$$

where $\gamma_{ab} = \gamma_a + \gamma_b$.

The homogeneous equation

$$\ddot{D} + \gamma_{ab}\dot{D} + (\omega_{ba}^2 + \gamma_{ab}^2/4)D = 0 \qquad (2.115)$$

which describes the atomic dipoles without the driving field ($A_0 = 0$), has for weak damping ($\gamma_{ab} \ll \omega_{ba}$) the solution

$$D(t) = D_0\, e^{-(\gamma ab/2)t} \cos\omega_{ba} t \quad . \qquad (2.116)$$

The inhomogeneous equation (2.114) is the quantum mechanical equivalent of the classical equation (2.41) and *shows that the induced dipole moment of the atom interacting with a monochromatic radiation field behaves like a damped harmonic oscillator with the eigenfrequency* $\omega_{ba} = (E_b - E_a)/\hbar$ *and a damping constant* $\gamma_{ab} = (\gamma_a + \gamma_b)$.

Using the approximations $(\omega_{ba} + \omega) \approx 2\omega$ and $\gamma_{ab} \ll \omega_{ba}$ which means weak damping and close-to-resonance situation, we obtain solutions of the form

$$D = D_1 \cos\omega t + D_2 \sin\omega t \quad , \qquad (2.117)$$

where the factors D_1 and D_2 include the frequency dependence,

$$D_1 = \frac{B_{ab}A_0(\omega_{ba} - \omega)}{(\omega_{ba} - \omega)^2 + (\gamma_{ab}/2)^2} \ , \quad \text{with} \quad B_{ab} = eR_{ab}^2/\hbar \tag{2.118a}$$

$$D_2 = \frac{B_{ab}A_0\gamma_{ab}}{(\omega_{ba} - \omega)^2 + (\gamma_{ab}/2)^2} \ . \tag{2.118b}$$

Comparison with (2.55b,56n) shows that the two equations for D_1 and D_2 describe dispersion and absorption of the E.M. wave. The former is caused by the phase lag between the radiation field and the induced dipole oscillation, and the latter by the atomic transition from the lower level E_a to the upper level E_b and the resultant conversion of field energy into potential energy $(E_b - E_a)$.

The macroscopic polarization P of a sample with N atoms/cm^2 is related to the induced dipole moment by $P = N \cdot D$.

2.9.4 Interaction with Strong Fields

In the previous sections we assumed weak-field conditions where the probability of finding the atom in the initial state was not essentially changed by the interaction with the field. This means that the population in the initial state remained approximately constant during the interaction time. In case of broad-band radiation this approximation resulted in a *time-independent transition probability*. Also the inclusion of weak damping terms with $\gamma_{ab} \ll \omega_{ba}$ did not affect the assumption of constant population in the initial state.

In this section, we consider cases where the radiation field is so intense that such an approximation is no longer valid. This strong-signal theory, developed by Rabi, leads to a time-dependent probability of the atom being in either the upper or lower level. The representation shown below follows that of [1.5].

We consider a monochromatic field of frequency ω and start from the basic equations (2.98) for the probability amplitudes in the rotating wave approximation

$$\dot{a}(t) = \frac{i}{2} R_{ab} e^{-i(\omega_{ba} - \omega)t} \ b(t) \tag{2.119a}$$

$$\dot{b}(t) = \frac{i}{2} R_{ab} e^{+i(\omega_{ba} - \omega)t} \ a(t) \ . \tag{2.119b}$$

Inserting a trial solution

$$a(t) = \exp(i\mu t) \Longrightarrow \dot{a}(t) = i\mu\exp(i\mu t)$$

into (2.119a) yields

$$b(t) = \frac{2\mu}{R_{ab}} e^{i(\omega_{ba} - \omega + \mu)t} \quad . \tag{2.120}$$

Substituting this back into (2.119b) gives the relation

$$2\mu(\omega_{ba} - \omega + \mu) = R_{ab}^2/2 \quad . \tag{2.121}$$

This is a quadratic equation for the unknown quantity μ with the two solutions

$$\mu_{1,2} = -\frac{1}{2}(\omega_{ba} - \omega) \pm \frac{1}{2}[(\omega_{ba} - \omega)^2 + R_{ab}^2]^{\frac{1}{2}} \quad . \tag{2.122}$$

The general solutions for the amplitudes a and b are then

$$a(t) = C_1 e^{i\mu_1 t} + C_2 e^{i\mu_2 t}$$

$$b(t) = (2/R_{ab})e^{i(\omega_{ba} - \omega)t}(C_1\mu_1 e^{i\mu_1 t} + C_2\mu_2 e^{i\mu_2 t}) \quad . \tag{2.123}$$

With the initial conditions $a(0) = 1$ and $b(0) = 0$ we find for the coefficients

$$C_1 + C_2 = 1 \quad \text{and} \quad C_1\mu_1 = -C_2\mu_2$$

$$\Longrightarrow C_1 = -\mu_2/(\mu_1 - \mu_2)$$

$$C_2 = +\mu_1/(\mu_1 - \mu_2) \quad . \tag{2.124}$$

From (2.122) we obtain $\mu_1 \cdot \mu_2 = -R_{ab}^2/4$.

With the abbreviation

$$\mu = \mu_1 - \mu_2$$

we get the probability amplitude

$$b(t) = i(R_{ab}/\mu)e^{i(\omega_{ba} - \omega)t/2} \sin(\mu t/2) \quad . \tag{2.125}$$

The probability $b(t) \cdot b^*(t)$ of finding the system in level E_b is then

$$|b(t)|^2 = R_{ab}^2 \frac{\sin^2(\mu t/2)}{\mu^2} \tag{2.126}$$

where

$$\mu = \sqrt{(\omega_{ba} - \omega)^2 + (e\, \underline{R}_{ab} \cdot \underline{A}_0/\hbar)^2} \tag{2.127}$$

is called the *Rabi - "flopping frequency"*. Equation (2.126) shows that the transition probability is a periodic function of time. Since

$$|a(t)|^2 = 1 - |b(t)|^2 = 1 - (R_{ab}/\mu)^2 \sin^2(\mu t/2) \quad (2.128)$$

the system oscillates with frequency μ between levels E_a and E_b, where the flopping frequency μ depends on the detuning $(\omega_{ba} - \omega)$, on the field amplitude A_0 and the matrix element R_{ab} (see Fig.2.28b).

On resonance $(\omega_{ba} = \omega)$ (2.126 and 2.128) reduce to

$$|b(t)|^2 = \sin^2[e\, \underline{R}_{ab}\underline{A}_0 t/2\hbar]$$

$$|a(t)|^2 = \cos^2[e\, \underline{R}_{ab}\underline{A}_0 t/2\hbar] \quad . \quad (2.129)$$

After a time

$$T = \pi\hbar/(e\, \underline{R}_{ab} \cdot \underline{A}_0) \quad (2.130)$$

the probability $|b(t)|^2$ of finding the system in level E_b becomes unity. This means that the state $|a(0)|^2 = 1$, $|b(0)|^2 = 0$ of the initial system has been inverted to $|a(T)|^2 = 0$, $|b(T)|^2 = 1$.

Radiation with amplitude A_0 which interacts during the time interval $T = \pi\hbar/(e\, \underline{R}_{ab} \cdot \underline{A}_0)$ with the atomic system is called a *π-pulse* because it changes the phases of the probability amplitudes a(t), b(t) by π [see (2.123, 125 for $\omega_{ba} = \omega$].

We now include the damping terms γ_a and γ_b and insert the trial solution

$$a(t) = e^{i\mu t} \quad (2.131)$$

into (2.111a,b). Similar to the procedure used for the undamped case, we obtain a quadratic equation for the parameter μ with the two complex solutions

$$\mu_{1,2} = -\frac{1}{2}\left(\omega_{ba} - \omega - \frac{i}{2}\gamma_{ab}\right) \pm \frac{1}{2}\sqrt{\left(\omega_{ba} - \omega - \frac{i}{2}\gamma\right)^2 + R_{ab}^2} \quad (2.132)$$

where

$$\gamma_{ab} = \gamma_a + \gamma_b \quad \text{and} \quad \gamma = \gamma_a - \gamma_b \quad .$$

From the general solution

$$a(t) = C_1\, e^{i\mu 2t} + C_2\, e^{i\mu 2t} \quad (2.133)$$

we obtain from (2.111a) with the initial conditions $|a(0)|^2 = 1$, $|b(0)|^2 = 0$ the transition probability

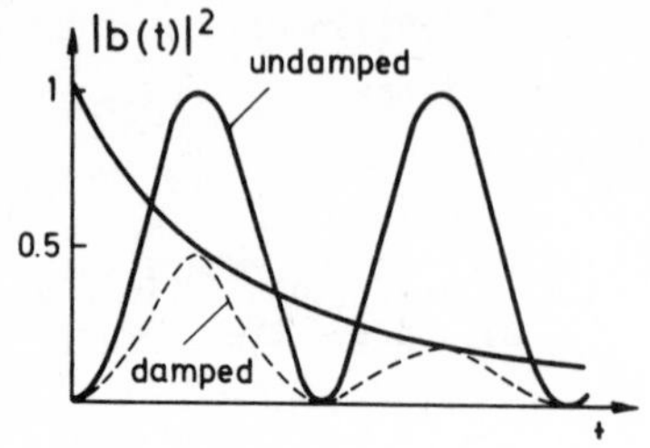

Fig.2.29. Population of levels E_a and E_b altering with the Rabbi flopping frequency due to the interaction with a strong field. The resonant case is shown without damping and with damping due to decay channels into other levels

$$|b(t)|^2 = \frac{R_{ab}^2 \, e^{-(\gamma ab/2)t} \, |\sin(\mu/2)t|^2}{(\omega_{ab} - \omega)^2 + (\gamma/2)^2 + R_{ab}^2} \tag{2.134}$$

with $\mu = \mu_1 - \mu_2 = [(\omega_{ba} - \omega - \frac{i}{2}\gamma)^2 + R_{ab}^2]^{\frac{1}{2}}$.

This is a damped oscillation (see Fig.2.29) with the damping constant $(\gamma_{ab/2}) = (\gamma_a + \gamma_b)/2$. The line profile of the transition probability is Lorentzian (see Sect.3.1) with a halfwidth, depending on $\gamma = \gamma_a - \gamma_b$ *and* on the strength of the interaction. Since $R_{ab}^2 = (\underline{R}_{ab} \cdot \underline{A}_0/\hbar)^2$ is proportional to the intensity of the electromagnetic wave, the linewidth *increases* with increasing intensity (saturation broadening, see Sect.3.6). *Note*, that $|a(t)|^2 + |b(t)|^2 < 1$ for $t > 0$, because the levels a and b decay into other levels.

In some cases the two level system may be regarded as isolated from its environment. The relaxation processes then occur only between the levels E_a and E_b but do not connect the system with other levels. (This implies $|a(t)|^2 + |b(t)|^2 = 1$.) Equations (2.111) have then to be modified as

$$\dot{a}(t) = -\frac{1}{2}\gamma_a a(t) + \frac{1}{2}\gamma_b b(t) + \frac{i}{2} R_{ab} \, e^{-i(\omega_{ba} - \omega)t} \, b(t) \tag{2.135a}$$

$$\dot{b}(t) = -\frac{1}{2}\gamma_b b(t) + \frac{1}{2}\gamma_a a(t) + \frac{i}{2} R_{ab} \, e^{+i(\omega_{ba} - \omega)t} \, a(t) \quad . \tag{2.135b}$$

The trial solution $a = \exp(i\mu t)$ yields for the resonance case $\omega = \omega_{ba}$ the two solutions

$$\mu_1 = \frac{1}{2} R_{ab} + \frac{i}{2}\gamma_{ab}$$

$$\mu_2 = -\frac{1}{2} R_{ab} \tag{2.136}$$

and one obtains with $|a(0)|^2 = 1$, $|b(0)|^2 = 0$ for the transition probability $|b(t)|^2$ a damped oscillation which approches the steady state value

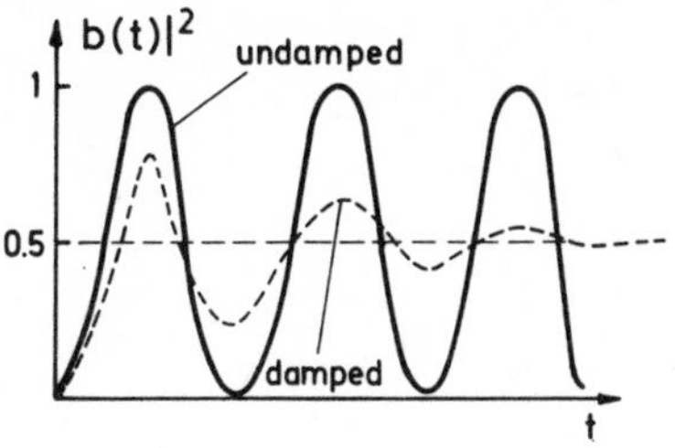

Fig.2.30. Two-level system under the influence of a strong radiation field with relaxation channels connecting only these two levels

$$|b(t=\infty)|^2 = \frac{1}{2}\,\frac{R_{ab}^2+\gamma_a\gamma_b}{R_{ab}^2+(\gamma_{ab/2})^2} \quad . \tag{2.137}$$

This is drawn in Fig.2.30 for the special case $\gamma_a = \gamma_b$ where $|b(\infty)|^2 = 1/2$, which means that the two levels become equally populated.

Density Matrix

It is often convenient to combine the probability amplitudes a(t) and b(t) of equation (2.93) in a matrix

$$\rho(t) = \begin{pmatrix} \rho_{aa} & \rho_{ab} \\ \rho_{ba} & \rho_{bb} \end{pmatrix} = \begin{pmatrix} a(t)\cdot a^*(t) & a(t)\cdot b^*(t) \\ b(t)\cdot a^*(t) & b(t)\cdot b^*(t) \end{pmatrix} . \tag{2.138}$$

The elements of this "density matrix" have the following physical meaning: The diagonal elements $\rho_{aa} = a(t)\cdot a^*(t)$ and $\rho_{bb} = b(t)b^*(t)$ give the probabilities of finding the system in levels E_a, respectively in E_b. The off-diagonal elements $\rho_{ab} = a(t)b^*(t) = \rho_{ba}^*$ are proportional to the complex dipole moment of an electric dipole transition $E_a \rightarrow E_b$ [see (2.113)]. The dipole moment is given by

$$\langle e\underline{r}\rangle = e\underline{r}(\rho_{ab}+\rho_{ba}) \quad . \tag{2.139}$$

2.9.5 Transition Probabilities and Line Strengths

So far we have assumed that the energy levels E_i and E_k are not degenerate and therefore have the statistical weight factor g = 1. In the case of a degenerate level E_k the total transition probability ρB_{ik} of a transition $E_i \rightarrow E_k$ is the sum

$$\rho B_{ik} = \rho \sum_n B_{ik_n} \tag{2.140}$$

over all transitions to the sublevels k_n of E_k. If also level E_i is degenerate, an additional summation over all sublevels i_m is necessary, taking into account that the population of each sublevel i_m is only the fraction N_i/g_i.

The Einstein coefficient B_{ik} for a transition $E_i \rightarrow E_k$ between two degenerate levels E_i, E_k is therefore

$$B_{ik} = \frac{\pi e^2}{3\varepsilon_0 \hbar^2} \frac{1}{g_i} \sum_{m=1}^{g_i} \sum_{n=1}^{g_k} |R_{i_m k_n}|^2 \tag{2.141}$$

The double sum is called the *line strength* S_{ik}. With this abbreviation, the relations between the Einstein coefficients A_{ik}, B_{ik} and the oscillator strength f_{ik} can be summarized

$$A_{ik} = \frac{1}{g_i} \frac{e^2 \omega_{ik}^3}{3\pi\varepsilon_0 \hbar c^3} S_{ik} = \frac{16\pi^3 \nu_{ik}^3 e^2}{3\varepsilon_0 h c^3 q_i} S_{ik} \tag{2.142}$$

$$B_{ik}^{(\omega)} = \frac{1}{g_i} \frac{\pi e^2}{3\varepsilon_0 \hbar^2} S_{ik} \quad \Big| \quad B_{ik}^{(\nu)} = \frac{1}{g_i} \frac{2\pi^2 e^2}{3\varepsilon_0 h^2} S_{ik} \tag{2.143}$$

$$f_{ik} = \frac{2m_e \omega_{ik}}{3g_i \hbar} S_{ik} = \frac{8\pi^2 m_e \nu_{ik}}{3gih} S_{ik} \tag{2.144}$$

For the transformation from ω to ν in (2.142-144) see the note at the end of Sect.2.9.2.

Example

Each atomic level E_k with the total angular momentum $J = \sqrt{j(j+1)}\hbar$ has $g_k = 2j + 1$ sublevels (Zeeman levels) which belong to the different projections J_z of the angular momentum J onto the selected z direction and which are degenerate without an external magnetic field. For a transition $E_i \rightarrow E_k$, the summation in (2.141) extends in this case from $n = -j_k$ to $n = +j_k$ and $m = -j_i$ to $+j_i$ although many of the $R_{i_m k_n}$ in the double sum will vanish because of symmetry selection rules, which only allow transitions with $n = m$ or $n = m \pm 1$ (see Fig.2.31).

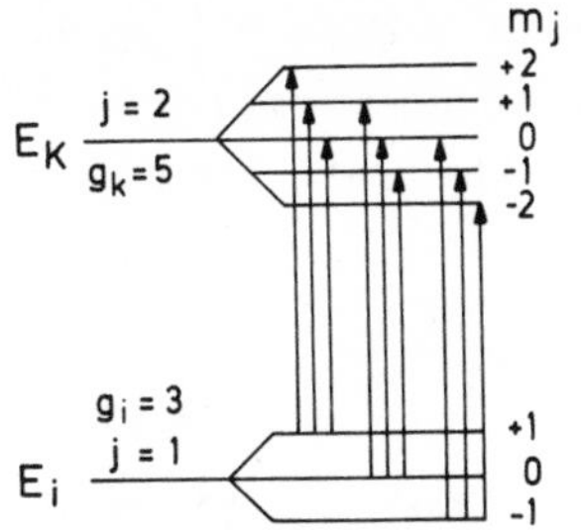

Fig.2.31. Illustration of (2.141)

2.9.6 Intensity and Polarization of Spectral Lines

In the preceding sections we have characterized the levels $E_i(a)$ and $E_k(b)$ of a two-level system by two sets a and b of quantum numbers without further specification of the individual quantum numbers. A closer inspection of the intensity and polarization of emitted or absorbed radiation has to take into account the orientation of the atoms or molecules with respect to a specified direction, called the *quantization axis*. We therefore describe the two levels E_i and E_k by two sets (α, J, M) and (α', J', M') of quantum numbers, where J stands for the total angular momentum and M for its projection on the quantization axis. All other quantum numbers are represented as a whole by α.

The spontaneous transition rate dn/dt for transitions $(\alpha'\ J'\ M') \rightarrow (\alpha\ J\ M)$ is given by

$$dn/dt = N(\alpha'\ J'\ M') \cdot A_{ik}(\alpha'\ J'\ M',\ \alpha\ J\ M) \quad , \tag{2.145}$$

where N is the number of molecules in the level $E_i(\alpha'\ J'\ M')$ and A_{ik} the spontaneous transition probability. According to (2.109) and (2.22a) this can be written for electric dipole transitions as

$$dn/dt = \frac{\omega^3}{3\pi c^3} \frac{e^2}{\varepsilon_0 \hbar} |R_{ik}(\alpha'\ J'\ M',\ \alpha\ J\ M)|^2 N(\alpha'\ J'\ M') \quad . \tag{2.146}$$

The matrix element

$$R_{ik}(\alpha'\ J'\ M',\ \alpha\ J\ M) = \langle \alpha\ J\ M \| \underline{r} \| \alpha'\ J'\ M' \rangle \tag{2.147}$$

can be separated into a product of two factors: a "geometrical factor" which describes the orientation of the molecule and an "intrinsic factor" which depends on the radiative coupling of the two levels. According to the Wigner-Eckart theorem [2.20,21,21a] we can write

$$R_{ik} = (-1)^{J-M} \begin{pmatrix} J & 1 & J' \\ -M & m & M' \end{pmatrix} \langle \alpha\ J \| \underline{r} \| \alpha'\ J' \rangle \quad . \tag{2.148}$$

The first factor, called the 3J-symbol, depends on M, M' and m = M' - M. It can be expressed by the Clebsch-Gordan coefficients which describe the coupling of angular momenta for a system initially in a state (J,M). The photon transfers an angular momentum of $1\hbar$ with projection $m\hbar$ (m = 0, ±1) and brings the system into a state (J'M'). The second factor

$$\langle \alpha \; J \, \| \, \underline{r} \, \| \, \alpha' \; J' \rangle \tag{2.149}$$

which is *independent* of the molecular orientation gives the "physical part" of the transition probability and depends on the molecular wave functions $\psi_i(\alpha' \; J')\psi_k(\alpha, J)$ in the coordinate frame of the molecule. It is often called the *reduced matrix element*.

The intensity of the fluorescence emitted into the direction θ against the quantization axis is obtained as the product of the emission rate dn/dt of individual photons and the appropriate angular distribution of the electric dipole field. In the far field zone at a distance $\rho \gg \lambda$ one obtains

$$I(\theta) = \frac{\omega^4 e^2}{2\varepsilon_0 c^3 4\pi\rho^2} N(\alpha', J' M') |\langle \alpha \; J \, \| \, r \, \| \, \alpha' \; J' \rangle|^2$$

$$\times \left| \begin{pmatrix} J & 1 & J' \\ -M & m & M' \end{pmatrix} \right|^2 \begin{cases} \frac{1}{2}(1 + \cos^2\theta) & \text{for} \quad m = \pm 1 \\ \sin^2\theta & \text{for} \quad m = 0 \; . \end{cases} \tag{2.150}$$

Figure 2.32 shows the angular distribution for $\Delta M = 0$ and $\Delta M = \pm 1$ transitions. The polarization characteristics of the emitted fluorescence are different for the two cases.

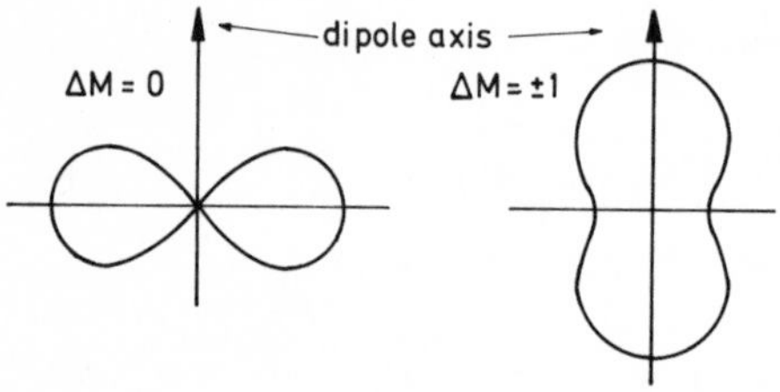

Fig.2.32. Angular distribution of the fluorescence emitted on transitions $\Delta M = 0, \pm 1$ between the Zeeman components $(\alpha' \; J' \; M') \rightarrow (\alpha \; J \; M)$

The Clebsch-Gordan coefficients CG (J, M, J', M') also contain a selection rule on $\Delta J = J' - J$. For electric or magnetic dipole radiation E1 or M1 the evaluation of the CG coefficients yields

$$\Delta J = 0, \; \pm 1, \quad \begin{matrix} \text{but } J' = 0 \not\leftrightarrow J = 0 \\ \text{and } J' = 1/2 \not\leftrightarrow J = 1/2 \end{matrix} \; . \tag{2.151}$$

Unless the Zeeman sublevels M' of the upper state $(\alpha' \; J')$ can be selectively populated one observes the total fluorescence of a line $(\alpha' \; J') \rightarrow (\alpha \; J)$. The observed intensity is then the sume over all components

$$I_{ik} \propto \sum_{M} \sum_{M'} N(\alpha' \; J' \; M') A_{ik}(\alpha' \; J' \; M', \; \alpha \; J \; M) \quad . \tag{2.152}$$

Without external magnetic field the different Zeeman sublevels M are degenerate. Under thermal equilibrium all sublevels are equally populated and the population $N(\alpha' \; J')$ of the level $E(\alpha' \; J')$ is

$$N(\alpha' \; J') = (2J' + 1) N(\alpha' \; J' \; M') \quad . \tag{2.153}$$

The transition probability $A(\alpha' \; J', \; J)$ of the total transition $(\alpha' \; J') \rightarrow (\alpha \; J)$ can then be defined by

$$N(\alpha' \; J') A(\alpha' \; J', \; J) = \sum_{M'} \sum_{M} N(\alpha' \; J' \; M') A(\alpha' \; J' \; M', \; \alpha \; J \; M) \quad . \tag{2.154}$$

From (2.154) and (2.153) we obtain

$$A(\alpha' \; J', \; \alpha J) = \frac{1}{2J' + 1} \sum_{MM'} A(\alpha' \; J' \; M', \; \alpha \; J \; M) \quad . \tag{2.155}$$

The summation over M and M' can be readily performed using the summation properties of the CG coefficients. We obtain the line strength $S(\alpha' \; J', \; \alpha \; J)$ of the total transition [see (2.142)]

$$\sum_{M,M'} A(\alpha' \; J' \; M', \; J \; M) = \frac{\omega^3 e^2}{3\pi\varepsilon_0 c^3 \hbar} S(\alpha' \; J', \; J) \tag{2.156}$$

which can be compared with (2.142) using (2.155).
The general expression for the intensity I of fluorescence with polarization vector $\underline{E}_2$ following excitation by absorption of light with frequency ω and polarization vector $\underline{E}_1$ is given by the Breit formula [2.21]. Let $a = (\alpha'', \; J'', \; M'')$ be the initial state of the atomic system, $b = (\alpha', \; J', \; M')$ the intermediate state and $c = (\alpha, \; J, \; M)$ the final state of the absorption-emission sequence $a \rightarrow b \rightarrow c$. We then obtain

$$I(a,b,\underline{E}_1,\underline{E}_2) = K(M'',\underline{E},\omega) N(a) \rho(\omega) S(a,b) S(b,c) \cdot \omega^3$$
$$\times \sum_{M_1'} \sum_{M_2'} (-1)^{M_1'-M_2'} \, CG(J''J'M_1'M_2') \; CG(J'J,M_1'M_2') \quad . \tag{2.157}$$

The factor K depends on the detector efficiency and geometry and on the polarization characteristics of the incident radiation. M_1' is the sublevel of the state b excited on the transition $a \rightarrow b$, while M_2' is the sublevel in b emitting on the transition $b \rightarrow c$. The evaluation of the corresponding Clebsch-Gordon coefficients CG yields the polarization characteristics of the emitted fluorescence.

If the detector collects the total fluorescence into all polarization states the summation yields the result

$$I(a,b,c) = K(\omega)\rho(\omega)N_a B(a,b)A(b,c) \quad , \tag{2.158}$$

which turns out to be independent of the polarization of the incident light, and can be expressed by the Einstein coefficients B(a,b) for absorption and A(b,c) for spontaneous emission. If the sample molecules are exposed to an external magnetic field, the Zeeman sublevels are no longer degenerate. This causes interference effects in the molecular fluorescence which depend on the Zeeman splittings. These phenomena play an important role in level crossing spectroscopy (see Chap.10) and quantum beats (Chap.11).

2.9.7 Molecular Transitions

For diatomic molecules [2.22] the wave functions of the energy levels depend on the electron coordinates $\underline{r}_e$ as well as on the nuclear coordinates $\underline{R}_n$. If the Born-Oppenheimer approximation can be used [2.23] the total wave function can be separated into a product of electronic, vibrational, and rotational wave functions,

$$\psi(\underline{r}_e,\underline{R}_n) = \psi_{el}(\underline{r}_{el}R)\psi_{vib}(R)\psi_{rot}(\theta,\phi) \quad , \tag{2.159}$$

where R is the internuclear distance. The matrix element for a transition between two different electronic states can then be written as

$$R_{ik} = \int \psi_i{}'\underline{M}\psi_k'' d\tau_{el} d\tau_N \quad , \tag{2.160}$$

where the upper state is indicated by a single prime, the lower by double prime. The integration extends over the electronic coordinates and the nuclear coordinates. The dipole operator

$$\underline{M} = \underline{M}_e + \underline{M}_N \tag{2.161}$$

consists of an electronic part $\underline{M}_e = -e \sum_i \underline{r}_i$ and a nuclear part $M_N = +e \sum_i \underline{R}_i$. Substituting (2.161) and (2.159) into (2.160) we obtain

$$R_{ik} = \int \psi_{el}'^{*}\psi_{vib}'^{*}\psi_{rot}'^{*} \underline{M}_e \psi_{el}''\psi_{vib}''\psi_{rot}'' d\tau_{el} d\tau_N$$

$$+ \int \psi_{el}'^{*}\psi_{vib}'^{*}\psi_{rot}'^{*}\underline{M}_N\psi_{el}''\psi_{vib}''\psi_{rot}'' d\tau_{el} d\tau_N \quad . \tag{2.162}$$

Since M_N does not depend on the electron coordinates, we can write the second integral as

$$\int \psi_{vib}'^{*}\psi_{rot}'^{*}M_N\psi_{vib}''\psi_{rot}'' d\tau_N \int \psi_{el}'^{*}\psi_{el}'' d\tau_{el} = 0 \quad . \tag{2.163}$$

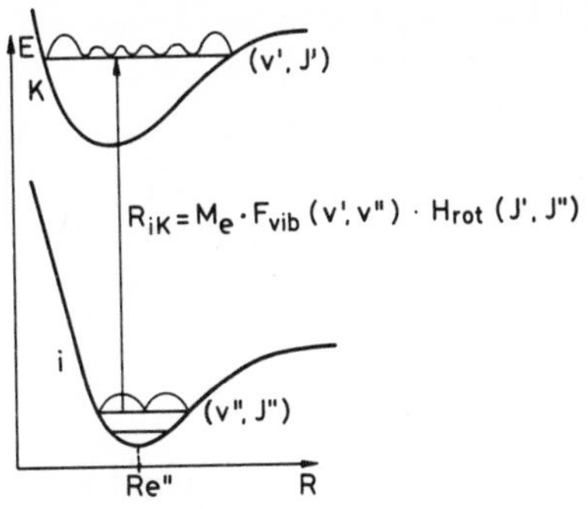

Fig.2.33. Potential energy curves and electronic transitions between two rotational-vibrational levels in different electronic states of a diatomic molecule

The second factor of (2.163) is zero because ψ_e' and ψ_e'' belong to two different electronic states and are therefore orthogonal. The first term in (2.162) may be rearranged and gives with $d\tau_N = d\tau_{vib} d\tau_{rot} = R^2 dR \sin\theta \, d\theta \, d\phi$

$$R_{ik} = \int_R \left[\int_{\underline{r}} \psi_e'^* \underline{M}_e \psi_e'' d\tau_{el}\right] \psi_{vib}'^* \psi_{vib}'' d\tau_{vib} \int_{\theta,\sigma} \psi_{rot}'^* \psi_{rot}'' d\tau_{rot} \quad . \tag{2.164}$$

The last integral is called *Hönl-London factor* H_{rot}. It contains the wave functions of the rigid rotator which are essentially represented by spherical harmonics. The Hönl-London factor can be therefore readily calculated for a given rotational quantum number and the corresponding formulas are found in textbooks on molecular physics [2.24]. The electronic part of the transition moment $\underline{M}_{el}$

$$\underline{M}_{el}(R) = \int_{\underline{r}_e} \psi_e'^* (\underline{r}_{el}, R) M_e \psi'' (\underline{r}_{el}, R) d\tau_{el} \tag{2.165}$$

often depends only slightly on the internuclear separation R and may be therefore expanded into a Taylor series around the equilibrium separation $R = R_e$,

$$\underline{M}_{el}(R) = \underline{M}_{el}(R_e) + \left(\frac{d\underline{M}_{el}}{dR}\right)_{R_e} (R + R_e) + \ldots \quad . \tag{2.166}$$

Neglecting all but the first term of this expansion yields for the total matrix element (2.164) the product form

$$R_{ik} = \underline{M}_{el}(R_e) \, F_{vib} \, H_{rot} \quad , \tag{2.167}$$

where the factor F_{vib} is the Franck-Condon factor

$$F_{vib} = \int \psi_{vib}'^* \psi_{vib}'' d\tau_{vib} \quad , \tag{2.168}$$

which represents the overlap integral of the vibrational wave functions in the two electronic states.

With the notation of Sect.2.9.6 we obtain for the line strength S(a,b) of a transition (n_a, v_a, J_a) → (n_b, v_b, J_b) between two levels with electronic quantum number n, vibrational quantum number v, and rotational quantum number J

$$S(a,b) = |\underline{M}_{el}(a,b)|^2 |F_{vib}(v_a, n_a, v_b, n_b)|^2 |H_{rot}(J_a,J_b)|^2 \quad . \tag{2.169}$$

For a more detailed discussion on the subject of this section see, for instance [2.21-25].

2.10 Coherence

The radiation emitted by an extended source generates a total field amplitude A at the point P which is a superposition of an infinite number of partial waves with amplitudes A_n and phases φ_n emitted from the different surface elements dS (Fig.2.34).

$$A(P) = \sum_n A_n(P)\, e^{i\varphi_n(P)} = \sum_n [A_n(0)/r_n^2]\, e^{i[\varphi_{n_0} + 2\pi r_n/\lambda]} \quad . \tag{2.170}$$

Where $\varphi_{no}(t) = \omega t + \varphi_n(0)$ is the phase of the nth partial wave at the surface element ds of the source. The phases $\varphi_n(r_n,t)$ depend on the distances r_n from the source and on the frequency ω. If the phase differences $\Delta\varphi_n = \varphi_n(P,t_1) - \varphi_n(P,t_2)$ *at a given point P* between two different times t_1, t_2 are nearly the same for all partial waves, the radiation field at P is called *temporally coherent*. The maximum time intervall $\Delta t = t_2 - t_1$ for which the $\Delta\varphi_n$ for all partial waves differ by less than π is called the *coherence time* of the radiation source. The path length $\Delta s_c = c\cdot\Delta t$ travelled by the wave during the coherence time Δt is the *coherence length*.

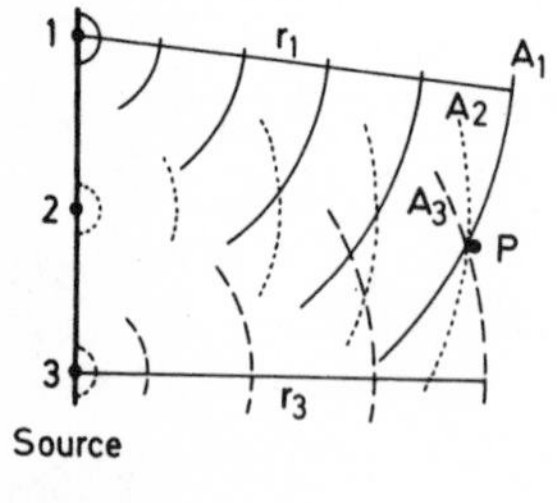

Fig.2.34. The field amplitudes A_n at a point P in a radiation field as superposition of an infinite number of waves from different points of an extended source

If a constant time-independent phase difference $\Delta\varphi = \varphi(P_1) - \varphi(P_2)$ exists for the total amplitudes $A = A_0\, e^{i\varphi}$ at two different points P_1, P_2 the radiation field is called *spatially coherent*. All points P_m, P_n which fulfil the condition that for all times t, $|\varphi(P_m,t) - \varphi(P_n,t)| < \pi$, have nearly the same optical path difference from the source and form the *coherence surface*. The product of coherence surface and coherence length is the *coherence volume*.

The superposition of coherent waves results in interference phenomena which, however, can be directly observed only within the coherence volume. The dimensions of this coherence volume depend on the size of the radiation source, on the spectral width of the radiation and on the distance between source and observation point P.

The following examples illustrate these different expressions for the coherence properties of radiation fields.

2.10.1 Temporal Coherence

Consider a point source in the focal plane of a lens forming a parallel light beam which is divided by a beam splitter S into two partial beams (see Fig.2.35), which are superimposed in the plane of observation B after reflection from the mirrors M_1, M_2. This arrangement is called a *Michelson interferometer* (see Sect.4.2). The two beams with wavelength λ travel different optical path lengths SM_1SB and SM_2SB and their path difference in the plane B is

$$\Delta s = 2(SM_1 - SM_2) \quad .$$

The mirror M_2 is mounted on a carriage and can be moved, resulting in a continuous change of Δs. In the plane B, one obtains maximum intensity when both amplitudes have the same phase, which means $\Delta s = m\lambda$, and minimum intensity if $\Delta s = (2m + 1)\lambda/2$. With increasing Δs, the contrast $(I_{max} - I_{min})/(I_{max} + I_{min})$ decreases and vanishes if Δs becomes larger than the coherence length Δs_c. Experiments show that Δs_c is related to the spectral width $\Delta\omega$ of the incident wave by

$$\boxed{\Delta s_c \approx c/\Delta\omega = c/(2\pi\Delta\nu)} \quad . \tag{2.171}$$

This observation may be explained as follows. A wave emitted from a point source with spectral width $\Delta\omega$ can be regarded as a superposition of many quasi-monochromatic components with frequencies ω_n within the interval $\Delta\omega$. The superposition results in wave trains of finite length $\Delta s_c = c\Delta t = c/\Delta\omega$

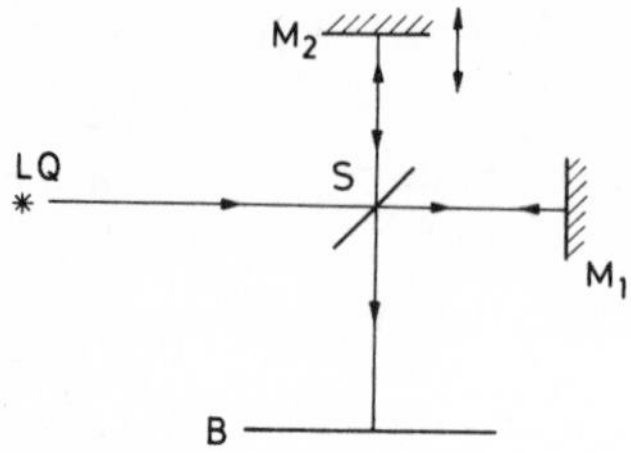

Fig.2.35. Michelson interferometer for measurement of the temporal coherence of radiation from the source S

because the different components with slightly different frequencies ω_n come out of phase during the time interval Δt and interfere destructively causing the total amplitude to decrease (see also Sect.3.1). If the path difference Δs in the Michelson interferometer becomes larger than Δs_c, the split wave trains no longer overlap in the plane B. The coherence lengths Δs_c of a light source therefore becomes larger with decreasing spectral width $\Delta\omega$.

Examples

a) A low-pressure mercury spectral lamp with a spectral filter which only transmits the green line $\lambda = 546$ nm has, because of the Doppler width $\Delta\omega_D = 4\times10^9\ s^{-1}$, a coherence length of $\Delta s_c \approx 8$ cm.

b) A single-mode He-Ne laser with a bandwidth of $\Delta\nu = 1$ MHz has a coherence length of about 50 m.

2.10.2 Spatial Coherence

The radiation from an *extended source* of size b illuminates two slits S_1 and S_2 in the plane A a distance d apart (Young's double-slit interference experiment, Fig.2.36a). The total amplitude and phase at each of the two slits are obtained by superposition of all partial waves emitted from the different surface elements of the source, taking into account the different paths df S_1 and df S_2.

The intensity at the point of observation P in the plane B depends on the path difference $S_1P - S_2P$ and on the phase difference $\Delta\varphi = \varphi(S_1) - \varphi(S_2)$ of the total field amplitudes in S_1 and S_2. If the different surface elements df of the source emit independently with random phases (thermal radiation source) the phases of the total amplitudes in S_1 and S_2 will also fluctuate randomly. However, this would not influence the intensity in P as long as these fluctuations occur in S_1 and S_2 synchronously, because then the phase difference $\Delta\varphi$ would remain constant. In this case, the two slits

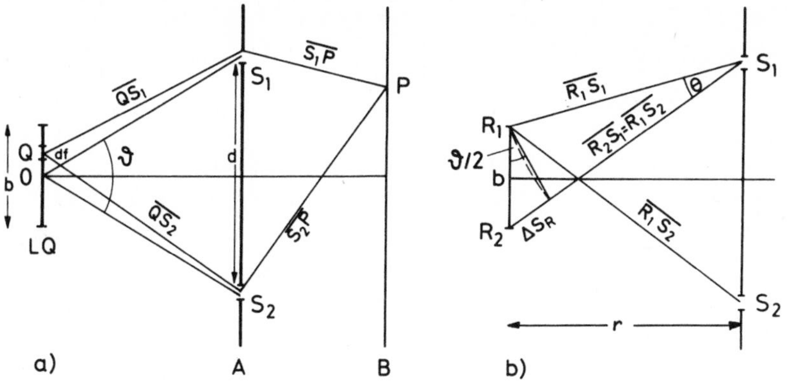

Fig.2.36a,b. Young's double-slit arrangement for measurement of spatial coherence

form two coherent sources which generate an interference pattern in the plane B.

For radiation emitted from the central part 0 of the light source this proves to be true since the paths OS_1 and OS_2 are equal and all phase fluctuations in 0 arrive simultaneously in S_1 and S_2. For all other points Q of the source, however, path differences $\Delta s_Q = QS_1 - QS_2$ exist which are largest for the edges R of the source. From Fig.2.36b one can infer for $b \ll r$ the relation

$$\Delta s_R = R_1S_2 - R_1S_1 = R_2S_1 - R_1S_1 \approx b\sin(\vartheta/2) \quad .$$

For $\Delta s_R > \lambda/2$ the phase difference $\Delta\varphi$ of the partial amplitudes in S_1 and S_2 exceeds π. With random emission from the different surface elements df of the source, the time-averaged interference pattern in the plane B will be washed out. The condition for coherent illumination of S_1 and S_2 from a light source with dimensions b is therefore

$$\Delta s = b\sin(\vartheta/2) < \lambda/2 \quad . \tag{2.172a}$$

With $2\sin(\vartheta/2) = d/r$, this condition can be written as

$$b\,d/r < \lambda \quad . \tag{2.172b}$$

Extension of this coherence condition to two dimensions yields for a source area $A_s = b^2$ the following condition for the maximum surface $A_c = d^2$ which can be illuminated coherently:

$$b^2\,d^2/r^2 \le \lambda^2 \quad . \tag{2.172c}$$

Since $d\Omega = d^2/r^2$ is the solid angle accepted by the illuminated surface $A_c = d^2$ (2.172c) can be formulated as

$$A_s d\Omega \leq \lambda^2 \quad . \tag{2.173}$$

The source surface $A_s = b^2$ determines the maximum solid angle $d\Omega \leq \lambda^2/A_s$ inside which the radiation field shows spatial coherence. Equation (2.173) shows that the radiation from a point source (spherical waves) is spatially coherent within the whole solid angle $d\Omega = 4\pi$. The coherence surfaces are spheres with the source in the center. Likewise, a plane wave produced by a point source in the focus of a lens shows spatial coherence over the whole aperture confining the light beam. For given source dimensions, the coherence surface $A_c = d^2$ increases with the square of the distance from the source. Because of the vast distances of stars, the starlight received by telescopes is spatially coherent across the telescope aperture in spite of the large diameter of the radiation source.

The arguments above may be summarized as follows. The coherence surface S, i.e., that area A_c which can be coherently illuminated at a distance r from an extended quasi-monochromatic light source with area A_s emitting at a wavelength λ is determined by

$$\boxed{S = \lambda^2 r^2/A_s} \quad . \tag{2.173a}$$

2.10.3 Coherence Volume

With the coherence length $\Delta s_c = c/\Delta\omega$ in the propagation direction of the radiation with spectral width $\Delta\omega$ and the coherence surface $S_c = \lambda^2 r^2/A_s$, the coherence volume $V_c = S_c \Delta s_c$ becomes

$$V_c = \lambda^2 r^2 c/(\Delta\omega A_s) \quad . \tag{2.174}$$

A unit surface element of a source with the spectral radiance L_ω(W/m^2 ster) emits within the frequency interval $d\omega = 1\ s^{-1}$ $L_\omega/\hbar\omega)$ photons per second into the unit solid angle 1 ster.

The mean number $\bar{n}$ of photons in the spectral range $\Delta\omega$ within the coherence volume defined by the solid angle $\Delta\Omega = \lambda^2/A_s$ and the coherence length $\Delta s_c = c\Delta t_c$ generated by a source with area A_s is therefore

$$\bar{n} = (L\omega/\hbar\omega) A_s \Delta\Omega \Delta\omega \Delta t_c \quad . \tag{2.175}$$

With $\Delta\Omega = \lambda^2/A_s$ and $\Delta t_c \approx 1/\Delta\omega$ this gives

$$\boxed{\bar{n} = (L\omega/\hbar\omega)\lambda^2} \quad . \tag{2.176}$$

Example

For a thermal radiation source, the spectral radiance for linearly polarized light [(2.28) divided by a factor 2] is for $\cos\vartheta = 1$ and $L_\nu d\nu = L_\omega d\omega$

$$L_\nu = (h\nu^3/c^2)/[\exp(h\nu/kT)-1] \quad .$$

The mean number of photons within the coherence volume is then with $\lambda = c/\nu$

$$\bar{n} = 1/[\exp(h\nu/kT)-1] \quad .$$

This is identical to the mean number of photons per mode of the thermal radiation as derived in Sect.2.2. The mean number $\bar{n}$ of photons per mode is often called the *degeneracy parameter* of the radiation field.

The coherence volume is directly related to the modes of the radiation field. This can be seen as follows. Each mode of the radiation field in a cavity is represented by a plane wave with wave vector $\underline{k}$ indicating the direction of propagation, with frequency $\omega = c|\underline{k}|$ and with an intensity which is determined by the number n of photons in this mode (see Sect.2.1).

If we allow the radiation from all modes with the same direction of $\underline{k}$ to escape through a hole in the cavity wall with area $A_s = b^2$, the wave emitted from A_s will not be strictly parallel but will have a diffraction-limited divergence angle $\theta \approx \lambda/b$ around the direction of $\underline{k}$. This means that the radiation is emitted into a solid angle $d\Omega = \lambda^2/b^2$. This is the same solid angle (2.173) which limits the spatial coherence.

The modes with the same direction of $\underline{k}$ (which we assume to be the z direction) may still differ in $|\underline{k}|$, i.e., they may have different frequencies ω. The coherence length is determined by the spectral width $\Delta\omega$ of the radiation emitted from A_s. With $|\underline{k}| = \omega/c$ the spectral width $\Delta\omega$ corresponds to an interval $\Delta k = \Delta\omega/c$ of the k values.

As is well known from atomic physics the diffraction of light can be explained by Heisenberg's uncertainty relation. Photons passing through a slit of width Δx have an uncertainty Δp_x of the x component p_x of their momentum $\underline{p}$, given by $\Delta p_x \Delta x \geq \hbar$ (see Fig.2.37).

Generalized to three dimensions, the uncertainty principle postulates that the simultaneous measurements of momentum and location of a photon has a minimum uncertainty,

$$\Delta p_x \Delta p_y \Delta p_z \Delta x \Delta y \Delta z \geq \hbar^3 = V_{ph} \quad , \tag{2.177}$$

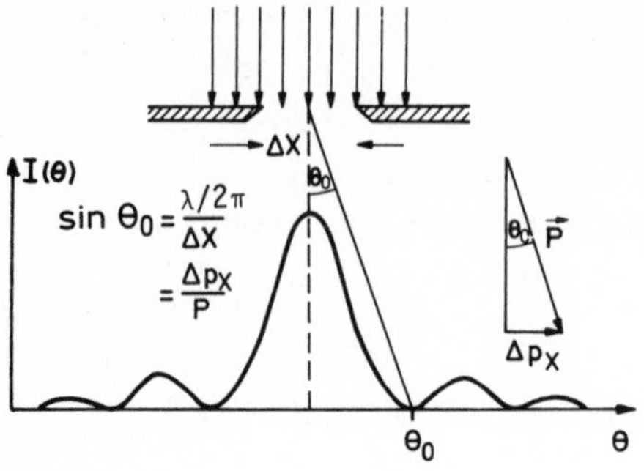

Fig.2.37. The uncertainty principle applied to the diffraction of light by a slit

where $V_{ph} = \hbar^3$ is the volume of one cell in phase space. *Photons within the same cell of phase space are indistinguishable and can be therefore regarded as identical.*

Photons which are emitted from the hole $A_s = b^2$ within the diffraction angle $\theta = \lambda/b$ against the surface normal, which may point into the z direction, have a minimum uncertainty

$$\Delta p_x = \Delta p_y = |\underline{p}|\lambda/(2\pi b) = (\hbar\omega/c)\lambda/(2\pi b) = (\hbar\omega/c)d/(2\pi r) \quad , \tag{2.178}$$

of the momentum components p_x and p_y (the last equality follows from 2.172b).

The uncertainty Δp_z is mainly caused by the spectral width $\Delta\omega$. Since $p = \hbar\omega/c$, we find

$$\Delta p_z = (\hbar/c)\Delta\omega \quad . \tag{2.179}$$

Substituting (2.178,179) into (2.177) we obtain for the spatial components of the phase-space cell

$$\Delta x\Delta y\Delta z = \frac{\lambda^2 r^2 c}{\Delta\omega A_s} = V_c \quad , \tag{2.180}$$

which turns out to be identical with the coherence volume defined by (2.174).

2.10.4 The Mutual Coherence Function and the Degree of Coherence

In the previous sections we have described the coherence properties of radiation fields in a more conspicuous way. We now briefly discuss a more quantitative description which allows partial coherence to be treated and a measure for the degree of coherence to be obtained.

In the cases of both temporal and spatial coherence we are concerned with the correlation between optical fields either at the same point P but at different times $[E(P_0,t_1)$ and $E(P_0,t_2)]$ or at the same time but at two different points $[E(P_1,t)$ and $E(P_2,t)]$. The subsequent description follows the representation in [2.3,26].

Suppose we have an extended source which generates a radiation field with narrow spectral bandwidth $\Delta\omega$ which we shall represent by the complex notation of a plane wave,

$$\underline{E}(\underline{r},t) = \underline{A}_0\, e^{i(\omega t-\underline{k}\underline{r})} \quad . \tag{2.181}$$

The field at two points in space S_1 and S_2 (e.g., the two apertures in Young's experiment) is then $E(S_1,t)$ and $E(S_2,t)$. The two apertures serve as secondary sources (see Fig.2.36) and the resultant field at the point of observation P is

$$E(P,t) = k_1E_1(S_1,t - r_1/c) + k_2E_2(S_2,t - r_2/c) \quad , \tag{2.182}$$

where the imaginary numbers k_1 and k_2 depend on the size of the apertures and on the distances $r_1 = S_1P$ and $r_2 = S_2P$.

The resultant irradiance at P measured over a time interval which is long compared to the coherence time is

$$I_p = \varepsilon_0 c\langle E(P,t)E^*(P,t)\rangle \quad , \tag{2.183}$$

where the brackets $\langle\cdot\cdot\rangle$ indicate the time average. Using (2.182) this becomes

$$\begin{aligned} I_p = c\varepsilon_0[&k_1k_1^*\langle E_1(t - t_1)E_1^*(t - t_1)\rangle + k_2k_2^*\langle E_2(t - t_2)E_2^*(t - t_2)\rangle \\ &+ k_1k_2^*\langle E_1(t - t_1)E_2^*(t - t_2)\rangle + k_1^*k_2\langle E_1^*(t - t_1)E_2(t - t_2)\rangle] \quad . \end{aligned} \tag{2.184}$$

If the field is stationary, the time-averaged values do not depend on time. We can therefore shift the time origin without changing the irradiances (2.183). Accordingly the first two time averages in (2.184) can be transformed to $\langle E_1(t)E_1^*(t)\rangle$ and $\langle E_2(t)E_2^*(t)\rangle$. In the last two terms we shift the time origin by an amount t_2 and write them with $\tau = t_2 - t_1$

$$k_1k_2^*\langle E_1(t_1 + \tau)E_2^*(t)\rangle + k_1^*k_2\langle E_1^*(t + \tau)E_2(t)\rangle \quad . \tag{2.185}$$

The second term in (2.185) is just the complex conjugate of the first term. We can therefore write (2.185) as

$$2\mathrm{Re}\{k_1k_2^*\langle E_1(t + \tau)E_2^*(t)\rangle\} \quad .$$

The term

$$\Gamma_{12}(\tau) = \langle E_1(t + \tau)E_2^*(t)\rangle \qquad (2.186)$$

is called the *mutual coherence function* and describes the cross correlation of the light field at S_1 and S_2. When the amplitudes and phases of E_1 and E_2 fluctuate in time, the time average $\Gamma_{12}(\tau)$ will be zero if these fluctuations of the two fields at two different points and at two different times are completely uncorrelated. If the field at S_1 at time $t + \tau$ were perfectly correlated with the field at S_2, the relative phase would be unaltered despite individual fluctuations. Inserting (2.186) into (2.184) gives for the irradiance at P (note that k_1 and k_2 are pure imaginary numbers for which $2\mathrm{Re}\{k_1k_2\} = 2|k_1||k_2|$)

$$I_P = \varepsilon_0 c[|k_1|^2 I_{S1} + |k_2|^2 I_{S2} + 2|k_1||k_2|\mathrm{Re}\{\Gamma_{12}(\tau)\}] \quad . \qquad (2.187)$$

The first term $I_1 = \varepsilon_0 c|k_1|I_{S1}$ gives the irradiance at P when only the aperture S_1 is open ($k_2 = 0$); the second term $I_2 = \varepsilon_0 c|k_2|I_{S2}$ is that for $k_1 = 0$.

Let us introduce the coherence functions

$$\Gamma_{11}(\tau) = \langle E_1(t + \tau)E_1^*(t)\rangle$$
$$\Gamma_{22}(\tau) = \langle E_2(t + \tau)E_2^*(t)\rangle \quad , \qquad (2.188)$$

which correlate the field amplitudes at the same point but at different times. For $\tau = 0$ the *self-coherence functions*,

$$\Gamma_{11}(0) = \langle E_1(t)E_1^*(t)\rangle = I_1/(\varepsilon_0 c) \quad ,$$
$$\Gamma_{22}(0) = I_2/(\varepsilon_0 c)$$

are proportional to the irradiance I at S_1, S_2, respectively.

With the definition of the normalized form of the mutual coherence function,

$$\gamma_{12}(\tau) = \frac{\Gamma_{12}(\tau)}{\sqrt{\Gamma_{11}(0)\Gamma_{22}(0)}} = \frac{\langle E_1(t + \tau)E_2^*(t)\rangle}{\sqrt{\langle|E_1(t)|^2\rangle\langle|E_2(t)|^2\rangle}} \quad , \qquad (2.189)$$

(2.187) can be written as

$$\boxed{I_P = I_1 + I_2 + 2\sqrt{I_1 I_2}\ \mathrm{Re}\{\gamma_{12}(\tau)\}} \quad . \qquad (2.190)$$

This is the general interference law for partially coherent light. $\gamma_{12}(\tau)$ is called the *complex degree of coherence*. Its meaning will be illustrated by the following: We express the complex quantity $\gamma_{12}(\tau)$ as

$$\gamma_{12}(\tau) = |\gamma_{12}(\tau)|e^{i\phi_{12}(\tau)} \quad , \tag{2.191}$$

where the phase angle $\phi_{12}(\tau)$ is related to the phases of the fields E_1 and E_2 in (2.186).

For a quasi-monochromatic wave $\underline{E} = \underline{E}_0 \exp(i\ t - i\underline{k}\underline{r})$ an optical path difference $(r_2 - r_1)$ causes a corresponding phase difference

$$\varphi = (2\pi/\lambda)(r_2 - r_1) = \omega\tau \quad , \quad \text{since} \quad \tau = (r_2 - r_1)/c \quad . \tag{2.192}$$

If we set $\phi_{12}(\tau) = \alpha_{12}(\tau) + \varphi$ we obtain

$$\mathrm{Re}\{\gamma_{12}(\tau)\} = |\gamma_{12}(\tau)|\cos[\alpha_{12}(\tau) + \varphi] \quad ,$$

and (2.190) can be expressed by

$$I_p = I_1 + I_2 + 2\sqrt{I_1 I_2}|\gamma_{12}(\tau)|\cos[\alpha_{12}(\tau) + \varphi] \quad . \tag{2.193}$$

For $|\gamma_{12}(\tau)| = 1$, (2.193) describes the interference of two completely coherent waves out of phase at S_1 or S_2 by an amount $\alpha_{12}(\tau)$. For $|\gamma_{12}(\tau)| = 0$, the interference term vanishes. The two waves are said to be completely incoherent. For $0<|\gamma_{12}(\tau)|<1$ we have *partial coherence*. $\gamma_{12}(\tau)$ is therefore a measure of the degree of coherence.

We illustrate this by applying the correlation functions to our two examples of Sects.2.10.1,2. In the Michelson interferometer, the incoming nearly parallel light beam is split by S (Fig.2.35) and recombined in the plane B. If both partial beams have the same amplitude $E = E_0 \exp[i\phi(t)]$, the degree of coherence becomes

$$\gamma_{11}(\tau) = \frac{\langle E(t+\tau)E^*(t)\rangle}{|E(t)|^2} = \langle e^{i\phi(t+\tau)}\, e^{-i\phi(t)}\rangle \quad . \tag{2.194}$$

For long averaging times T,

$$\langle e^{i\phi(t+\tau)}\, e^{-i\phi(t)}\rangle = \lim_{T\to\infty} \frac{1}{T}\int_0^T e^{i[\phi(t+\tau)-\phi(t)]}dt \quad , \tag{2.195}$$

and we obtain with $\Delta\phi = \phi(t+\tau) - \phi(t)$,

$$\gamma_{11}(\tau) = \lim_{T\to\infty} \frac{1}{T}\int_0^T (\cos\Delta\phi + i\ \sin\Delta\phi)dt \quad .$$

For a strictly monochromatic wave with infinite coherence length Δs_c, the phase function is $\phi(t) = \omega t - \underline{k} \cdot \underline{r}$ and $\Delta\phi = +\omega\tau$. This yields

$$\gamma_{11}(\tau) = \cos\omega\tau + i\,\sin\omega\tau = e^{i\omega\tau} \quad , \tag{2.196}$$

$$|\gamma_{11}(\tau)| = 1 \quad .$$

For a wave with spectral width $\Delta\omega$ so large that $\tau > \Delta s_c/c = {}^1/\Delta\omega$, the phase differences $\Delta\phi$ will vary randomly between 0 and 2π and the integral averages to zero, giving $|\gamma_{11}(\tau)| = 0$.

Referring to Young's experiment (Fig.2.36) with a narrow bandwidth but extended source, *spatial* coherence effects will predominate. The fringe pattern in the plane B will depend on $\Gamma(S_1,S_2,\tau) = \Gamma_{12}(\tau)$. In the region about the central fringe $(r_2 - r_1) = 0$, $\tau = 0$ and $\Gamma_{12}(0)$ and $\gamma_{12}(0)$ can be determined from the visibility of the interference pattern.

To find the value $\gamma_{12}(\tau)$ for any point P on the screen B in Fig.2.36, the intensity $I = I(P)$ is measured when both slits are open and also the intensities $I_1(P)$ and $I_2(P)$ separately when one of the pinholes is blocked. In terms of these observed quantities the degree of coherence can be determined from (2.190) to be

$$\mathrm{Re}[\gamma_{12}(P)] = \frac{I(P) - I_1(P) - I_2(P)}{2\sqrt{I_1(P)I_2(P)}} \quad . \tag{2.197}$$

This yields the desired information about the spatial coherence of the source which depends on the size of the source. The visibility of the fringes at P is defined to be

$$V(P) = \frac{I_{max} - I_{min}}{I_{max} + I_{min}} = \frac{2\sqrt{I_1(P)}\,\sqrt{I_2(P)}}{I_1(P) + I_2(P)}\,|\gamma_{12}(\tau)| \quad , \tag{2.198}$$

where the last equality follows from (2.190) and (2.197). If $I_1 = I_2$ (equal size pinholes), we see from (2.198) that

$$V(P) = |\gamma_{12}(\tau)| \quad . \tag{2.199}$$

The visibility is then equal to the degree of coherence. Figure 2.38a-c illustrates the interference pattern I(P) caused by the superposition of two waves with $I_1 = I_2$ in the plane of observation B. The total intensity depends on the phase difference $\Delta\varphi = \omega\tau$ (2.192). For completely coherent light ($|\gamma_{12}(\tau)| = 1$) the intensity $I(\tau)$ changes between $4 \cdot I_1$ and zero, whereas for

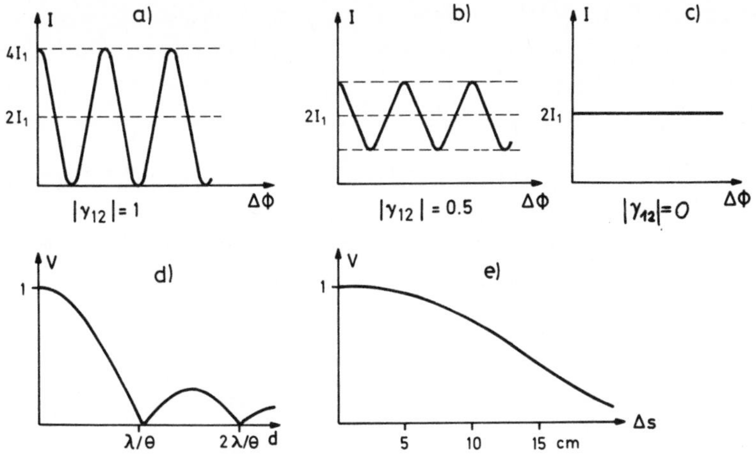

Fig.2.38a-e. Interference pattern of two-beam interference for different degrees of coherence (a-c). Visibility behind the two slits of Fig.2.36, if they are illuminated by a monochromatic extended source (d). Visibility of a Doppler-broadened line behind a Michelson interferometer as a function of path difference Δs (e)

$|\gamma(t)| = 0$ the interference term vanishes and the total intensity does not depend on τ. Figure 2.38d shows the visibility V of the fringe pattern in P as a function of the slit separation d in Fig.2.36 when these slits are illuminated by monochromatic light from an extended uniform source with rectangular size b which appears from S_1 under the angle θ. Figure 2.38e illustrates the visibility as a function of path difference Δs in a Michelson interferometer which is illuminated with the Doppler-broadened line $\lambda = 632.8$ nm from a neon discharge lamp.

For more detailed presentations of coherence see the textbooks [2.5,27, 28].

Problems: Chapter 2

2.1. The angular divergence of the output from a 1 W argon laser is assumed to be 4×10^{-3} rad. Calculate the radiance L and the radiant intensity I^* of the laser beam and the irradiance I (intensity) at a surface 1 m away from the output mirror, when the laser beam diameter at the mirror is 2 mm. What is the spectral power density $\rho(\nu)$ if the laser bandwidth is 1 MHz?

2.2. Unpolarized light of intensity I_0 is transmitted through a dichroitic polarizer with thickness 1 mm. Calculate the transmitted intensity when the absorption coefficients for the two polarizations $\alpha_{\parallel} = 100\ \mathrm{cm}^{-1}$ and $\alpha_{\perp} = 5\ \mathrm{cm}^{-1}$.

2.3. The beam of a monochromatic laser passes through an absorbing atomic vapor with path length $L = 5$ cm. If the laser frequency is tuned to the center of an absorbing transition with oscillator strength $f = 0.1$, the attenuation of the transmitted intensity is 10%. Calculate the atomic density N.

2.4. An excited molecular level E_i is connected with three lower levels E_n by radiative transitions with spontaneous probabilities $A_{i3} = 5\times10^7\ s^{-1}$, $A_{i1} = 3\times10^7\ s^{-1}$, and $A_{i2} = 1\times10^7\ s^{-1}$. Calculate the spontaneous lifetime τ_i and the relative populations N_n/N_i under cw excitation of E_i, when $\tau_3 = 10^{-8}$ s, $\tau_1 = 5\times10^{-7}$ s, and $\tau_2 = 6\times10^{-9}$ s.

2.5. Which pumping rate is necessary on the transition $E_0 \rightarrow E_i$ of problem 2.4 at $\lambda = 5000$ Å from the ground state E_0 to achieve steady-state inversion of $(N_i - N_n)$ with n = 1, 2 or 3?

2.6. The frequency of a monochromatic tunable laser is tuned to the center of an atomic transition $E_k \rightarrow E_i$ with transition probability $A_{ki} = 5\times10^7\ s^{-1}$. Calculate the saturation intensity I_s which depletes the lower level population to 0.5 of its unsaturated value if the only repopulation path is spontaneous emission $E_i \rightarrow E_k$.

2.7. Expansion of a laser beam is accomplished by two lenses with different focal lengths. Why does an

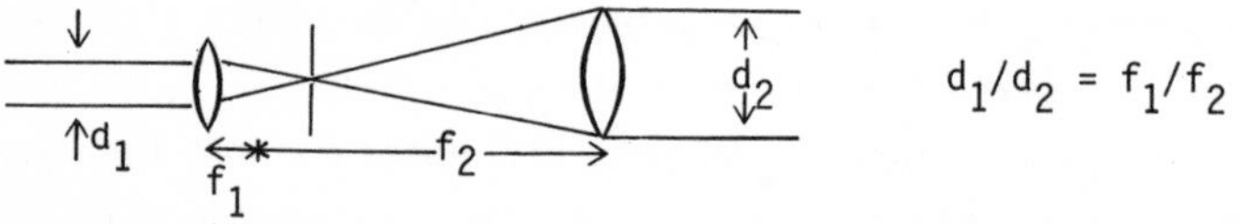

aperture in the focal plane improve the quality of the wavefronts in the expanded beam by eliminating perturbations due to diffraction effects by dust and other imperfections on the lens surfaces?

2.8. Calculate the maximum slit separation in Young's interference experiments which still gives distinct interference fringes, if the two slits are illuminated

a) by incoherent light of $\lambda = 500$ nm from a hole with 1 mm diameter, 1 m away from the slits,

b) by a star with 10^6 km diameter, at a distance of 4 light years,

c) by the expanded beam from an He-Ne laser with diffraction-limited divergence, emitted from a spot size of 1 mm at the output mirror.

3. Widths and Profiles of Spectral Lines

Spectral lines in discrete absorption or emission spectra are never strictly monochromatic. Even with the very high resolution of interferometers one observes a spectral distribution $I(\nu)$ of the absorbed or emitted intensity around the central frequency $\nu_0 = (E_i - E_k)/h$ corresponding to a molecular transition with the energy difference $\Delta E = E_i - E_k$ between upper and lower level. The function $I(\nu)$ in the vicinity of ν_0 is called the *line profile* (see Fig.3.1). The frequency interval $\delta\nu = |\nu_2 - \nu_1|$ between the two frequencies ν_1 and ν_2 for which $I(\nu_1) = I(\nu_2) = I(\nu_0)/2$ is the *full width at half maximum* of the line (FWHM), often shortly called the *linewidth* or *halfwidth* of the spectral line.

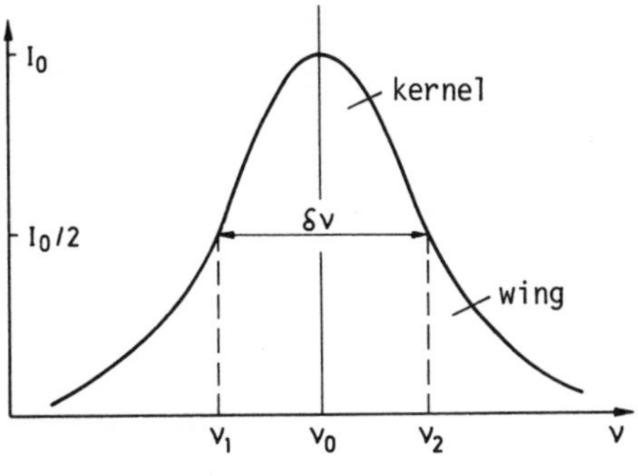

Fig.3.1. Line profile, halfwidth, kernel and wings of a spectral line

The halfwidth is sometimes written in terms of the angular frequency $\omega = 2\pi\nu$ with $\delta\omega = 2\pi\delta\nu$ or in terms of the wavelength λ (in units of nm or Å) with $\delta\lambda = |\lambda_1 - \lambda_2|$. From $\lambda = c/\nu$, it follows that

$$\delta\lambda = (c/\nu^2)\delta\nu \quad . \tag{3.1}$$

The *relative* halfwidths, however, are the same in all three schemes,

$$\left|\frac{\delta\nu}{\nu}\right| = \left|\frac{\delta\omega}{\omega}\right| = \left|\frac{\delta\lambda}{\lambda}\right| \quad . \tag{3.2}$$

The spectral region within the halfwidth is called the kernel of the line, the regions outside ($\nu < \nu_1$ and $\nu > \nu_2$) are the *line wings*.

In the following sections we discuss various causes of the finite linewidth. Several examples illustrate the order of magnitude of the various effects in different spectral regions and their importance for high-resolution spectroscopy [3.1-3]. Following the usual convention we shall often use the angular frequency ω to avoid factors of 2π in the equations.

3.1 Natural Linewidth

An excited atom can emit its excitation energy as spontaneous radiation (see Sect.2.7). In order to investigate the spectral distribution of this spontaneous emission on a transition $E_i \to E_k$ we shall describe the excited atomic electron, as in Sect.2.6, by the classical model of a damped harmonic oscillator with frequency ω, mass m, and restoring force constant D. The radiative energy loss results in a damping of the oscillation described by the damping constant γ. We shall see, however, that for real atoms the damping is extremely small, which means that $\gamma \ll \omega$.

The amplitude x(t) of the oscillation can be obtained by solving the differential equation of motion,

$$\ddot{x} + \gamma\dot{x} + \omega_0^2 x = 0 \quad , \tag{3.3}$$

where $\omega_0^2 = D/m$.

The real solution of (3.3) with the initial values $x(0) = x_0$ and $\dot{x}(0) = 0$ is

$$x(t) = x_0\, e^{-(\gamma/2)t}[\cos\omega t + (\gamma/2\omega)\sin\omega t] \quad . \tag{3.4}$$

The frequency $\omega = \sqrt{(\omega_0^2 - \gamma^2/4)}$ of the damped oscillation is slightly lower than the frequency ω_0 of the undamped case. However, for small damping ($\gamma \ll \omega_0$) we can put $\omega \approx \omega_0$ and we also may neglect the second term in (3.4). With this approximation, which still is very accurate for real atoms, we obtain the solution of (3.3) as

$$x(t) = x_0\, e^{-(\gamma/2)t} \cos\omega_0 t \quad . \tag{3.5}$$

The frequency $\omega_0 = 2\pi\nu_0$ of the oscillator corresponds to the central frequency $\omega_{ik} = (E_i - E_k)/\hbar$ of an atomic transition $E_i \to E_k$.

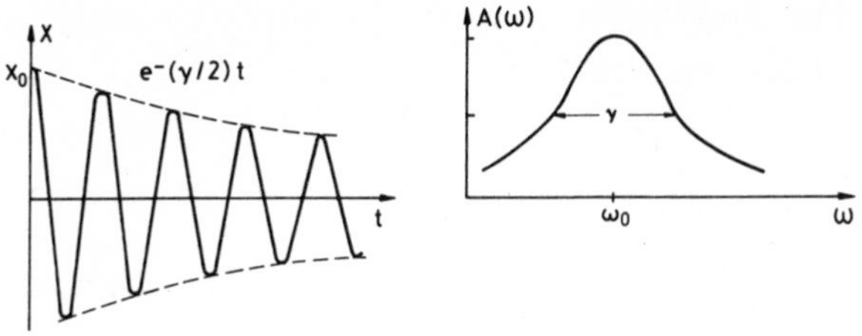

Fig.3.2. Damped oscillation x(t) and its frequency distribution $A(\omega)$ as obtained by a Fourier transformation of x(t)

Because the amplitude x(t) of the oscillation decreases gradually, the frequency of the emitted radiation is no longer monochromatic as it would be for an oscillation with constant amplitude, but shows a frequency distribution related to the function x(t) in (3.5) by a Fourier transformation (see Fig.3.2).

The oscillation x(t) can be described as a superposition of monochromatic oscillations $\exp(i\omega t)$ with amplitudes $A(\omega)$ (see Fig.3.2),

$$x(t) = \frac{1}{\sqrt{2\pi}} \int_0^\infty A(\omega) e^{i\omega t} d\omega \quad . \tag{3.6}$$

The amplitudes $A(\omega)$ can be calculated from (3.6) by the Fourier transformation

$$A(\omega) = \frac{1}{\sqrt{2\pi}} \int_{-\infty}^{+\infty} x(t) e^{-i\omega t} dt = \frac{1}{\sqrt{2\pi}} \int_0^\infty x_0 \, e^{-(\gamma/2)t} \cos\omega_0 t \, e^{-i\omega t} dt \quad . \tag{3.7}$$

The lower integration limit is taken to be zero because $x(t) = 0$ for $t < 0$. Equation (3.7) can be easily integrated to give the complex amplitudes $A(\omega)$:

$$A(\omega) = \frac{x_0}{\sqrt{8\pi}} \left(\frac{1}{i(\omega - \omega_0) + \gamma/2} + \frac{1}{i(\omega + \omega_0) + \gamma/2} \right) \quad . \tag{3.8}$$

The real intensity $I(\omega) \propto A(\omega) \cdot A^*(\omega)$ contains terms with $(\omega - \omega_0)$ and $(\omega + \omega_0)$. In the vicinity of the central frequency ω_0 of an atomic transition where $(\omega - \omega_0)^2 \ll \omega_0^2$, the terms with $(\omega + \omega_0)$ can be neglected and the intensity profile of the spectral line becomes

$$I(\omega - \omega_0) = I_0 \frac{1}{(\omega - \omega_0)^2 + (\gamma/2)^2} \quad . \tag{3.9}$$

For comparison of different line profiles it is useful to define a normalized intensity profile $g(\omega - \omega_0) = CI(\omega)$ such that

$$\int_0^{\infty} g(\omega - \omega_0)d\omega = \int_{-\infty}^{+\infty} g(\omega - \omega_0)d(\omega - \omega_0) = 1 \quad .$$

Performing the integration yields $C = \gamma/(2\pi I_0)$. The normalized line profile of the intensity of a damped oscillator is then

$$\boxed{g(\omega - \omega_0) = \frac{1}{2\pi}\,\frac{\gamma}{(\omega - \omega_0)^2 + (\gamma/2)^2}} \quad . \tag{3.10}$$

$g(\omega - \omega_0)$ is called the normalized *Lorentzian profile*. Its full halfwidth follows from (3.10),

$$\delta\omega_n = \gamma \quad \text{or} \quad \delta\nu_n = \gamma/2\pi \quad . \tag{3.11}$$

The radiant power of the damped oscillator can be obtained from (3.3) if both sides of the equation are multiplied by $m\dot{x}$ which yields after rearranging

$$m\ddot{x}\dot{x} + m\omega_0^2 x\dot{x} = -\gamma m\dot{x}^2 \quad . \tag{3.12}$$

The left-hand side of (3.12) is the time derivative of the total energy W (sum of kinetic energy $m\dot{x}^2/2$ and potential energy $Dx^2/2 = m\omega_0^2 x^2/2$), and can therefore be written as

$$\frac{d}{dt}\left(\frac{m}{2}\dot{x}^2 + \frac{m}{2}\omega_0^2 x^2\right) = \frac{dW}{dt} = -\gamma m\dot{x}^2 \quad . \tag{3.13}$$

Inserting $x(t)$ from (3.5) yields when neglecting terms with γ^2

$$\frac{dW}{dt} = -\gamma m x_0^2 \omega_0^2 e^{-\gamma t} \sin^2\omega_0 t \quad . \tag{3.14}$$

Because $\overline{\sin^2\omega t} = 1/2$, the time average $\overline{dW/dt}$ is

$$\overline{\frac{dW}{dt}} = -\frac{\gamma}{2} m x_0^2 \omega_0^2 \, e^{-\gamma t} \quad . \tag{3.15}$$

Equation (3.15) shows that the time-averaged radiant power $\bar{P} = \overline{dW/dt}$ [and with it the intensity $I(t)$ of the spectral line] has decreased to $1/e$ of its initial value $I(t = 0)$ after the decay time $\tau = 1/\gamma$.

In Sect.2.7 we saw that the mean lifetime τ_i of a molecular level E_i, which decays exponentially by spontaneous emission, is related to the Einstein coefficient A_i by $\tau_i = 1/A_i$. Replacing the classical damping constant

γ by the spontaneous transition probability A_i we can use the classical formulas (3.9-11) as a correct description of the frequency distribution of spontaneous emission and its linewidth. The halfwidth of a spectral line spontaneously emitted from the level E_i is, according to (3.11),

$$\delta\nu_n = A_i/2\pi = 1/(2\pi\tau_i) \quad \text{or} \quad \delta\omega_n = A_i = 1/\tau_i \quad . \tag{3.16}$$

Since the radiant power emitted from N_i excited atoms on a transition $E_i \rightarrow E_k$ is given by

$$dW_{ik}/dt = N_i A_{ik} \hbar\omega_{ik} \quad , \tag{3.17}$$

the spectral distribution of a fluorescence line emitted by atoms at rest is written, with the normalized Lorentzian profile and $\gamma = A_{ik}$, as

$$\frac{dW_{ik}(\omega)}{dt} d\omega = \frac{dW_{ik}}{dt} g(\omega_{ik} - \omega) d\omega = \frac{N_i \cdot A_{ik} \hbar\omega_{ik}/2\pi}{(\omega_{ik} - \omega)^2 + (A_i/2)^2} \quad 3.18)$$

in which $A_i = \sum_m A_{im}$ and the lower state E_k is assumed to be the ground state.

If the emission is isotropic, the irradiance of a detector with area F at a distance r receiving a solid angle $d\Omega = F/r^2$, is

$$I_{ik} = (dW_{ik}/dt) d\Omega/(4\pi) \quad . \tag{3.19}$$

When the instrument resolution $\Delta\omega$ is small compared to the halfwidth $\delta\omega = \gamma$, the line profile can be resolved and the maximum spectral intensity at the line center is measured as

$$I(\omega_{ik})\Delta\omega = [N_i \hbar\omega_{ik} A_{ik}/(2\pi^2 A_i^2)]\Delta\omega d\Omega \tag{3.20}$$

In a similar way the spectral profile of an *absorption* line can be derived, using (2.54), (2.58), and (2.66). For linear absorption and sufficiently weak fields without Doppler broadening or power broadening (see Sect.3.6), one obtains for the transmitted intensity of an incident spectral continuum after passing through an optically thin absorption layer with path length Δz the expression

$$I(\omega) = I_0 \left(1 - \frac{(\omega_{ik}/c) B_{ik} [N_i - (g_i/g_k) N_k] \gamma}{(\omega_{ik} - \omega)^2 + (\gamma/2)^2}\right) \quad . \tag{3.21}$$

Note: Equation (3.16) can be also derived from the uncertainty principle (see Fig.3.3). With a mean lifetime τ_i of the excited level E_i, its energy E_i can be determined only with an uncertainty $\Delta E_i \approx \hbar/\tau_i$. The frequency

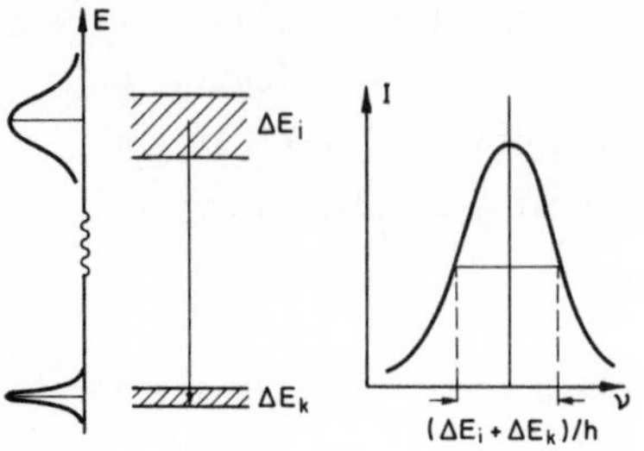

Fig.3.3. Illustration of the uncertainty principle which relates the natural linewidth to the energy uncertainties of upper and lower level

$\omega_{ik} = (E_i - E_k)/\hbar$ of a transition terminating in the stable ground state E_k has therefore an uncertainty

$$\delta\omega = \Delta E_i/\hbar = 1/\tau_i \quad . \tag{3.22}$$

If the lower level E_k is not the ground state but also an excited state with a lifetime τ_k, the uncertainties ΔE_i and ΔE_k of the two levels both contribute to the linewidth. This yields for the total uncertainty

$$\Delta E = \Delta E_i + \Delta E_k \Rightarrow \delta\omega_n = (1/\tau_i + 1/\tau_k) \quad . \tag{3.23}$$

In the general case, where the decay of both levels is caused not only by spontaneous emission but also by nonradiative relaxation, the line profile is determined by the total decay constants γ_i and γ_k, and the normalized line profile becomes

$$g(\omega - \omega_{ik}) = \frac{1}{2\pi} \frac{\gamma_i + \gamma_k}{(\omega_{ik} - \omega)^2 + [(\gamma_i + \gamma_k)/2]^2} \quad , \tag{3.24}$$

where both γ_i and γ_k are the sum $(\gamma = \gamma_R + \gamma_{NR})$ of radiative and nonradiative contributions. With the abbreviation $\gamma = (\gamma_i + \gamma_k)$ (3.24) becomes identical to (3.10).

Note, that in the literature often the normalization factor $\gamma/2\pi$ is omitted, and furthermore the full halfwidth is denoted by 2Γ! In this notation the Lorentzian profile is

$$L(\omega - \omega_{ik}) = \frac{\Gamma^2}{(\omega_{ik} - \omega)^2 + \Gamma^2} \quad . \tag{3.24a}$$

For $\omega = \omega_{ik}$ now $L(\omega_{ik}) = 1$.

With $x = (\omega_{ik} - \omega)/\Gamma$ this can be abbreviated as

$$L(\omega - \omega_{ik}) = \frac{1}{1 + x^2} \quad . \tag{3.24b}$$

Examples

a) The natural linewidth of the sodium D_1 line at $\lambda = 589.1$ nm which corresponds to a transition from the $3P_{3/2}$ level ($\tau = 16$ ns) to the $3S_{1/2}$ ground state is

$$\delta\nu_n = \frac{10^9}{16\times 2\pi} = 10^7\ s^{-1} = 10\ \text{MHz} \quad .$$

Note that with a central frequency $\nu_0 = 5\times 10^{14}\ s^{-1}$ and a lifetime of 16 ns, the damping of the corresponding classical oscillator is extremely small. Only after 8×10^6 periods of oscillation has the amplitude decreased to 1/e of its initial value.

b) The natural linewidth of a molecular transition between two vibrational levels of the electronic ground state with a wavelength in the infrared region is very small because of the long spontaneous lifetimes of vibrational levels. For a typical lifetime of $\tau = 10^{-3}$ s the natural linewidth becomes $\delta\nu_n = 160$ Hz.

c) Even in the visible or ultraviolet range, atomic or molecular electronic transitions with very small transition probabilities exist. In a dipole approximation these are "forbidden" transitions. One example is the $2S \rightarrow 1S$ transition for the hydrogen atom. The upper level 2S cannot decay by electric dipole transition but a two-photon transition to the 1S ground state is possible. The natural lifetime τ is about 1 s and the natural linewidth of such a two-photon line is therefore $\delta\nu_n = 0.15\ s^{-1}$!

3.2 Doppler Width

Generally the Lorentzian line profile with the natural linewidth $\delta\nu_n$, as discussed in the previous section, cannot be observed without special techniques, because it is completely concealed by other broadening effects. One of the major contributions to the spectral linewidth in gases at low pressures is the Doppler width, which is due to the thermal motion of the absorbing or emitting molecules.

Consider an excited molecule with a velocity $\underline{v} = \{v_x, v_y, v_z\}$ relative to the rest frame of the observer. The central frequency of a molecular emission line that is ω_0 in the coordinate system of the molecule, is Doppler-shifted to

$$\omega_e = \omega_0 + \underline{k}\cdot\underline{v} \tag{3.25}$$

for an observer looking towards the emitting molecule (that is, against the direction of the wave vector $\underline{k}$ of the emitted radiation; see Fig.3.4a). The apparent emission frequency ω_e is increased if the molecule moves towards the observer ($\underline{k}\cdot\underline{v} > 0$), and decreased if the molecule moves away ($\underline{k}\cdot\underline{v} < 0$).

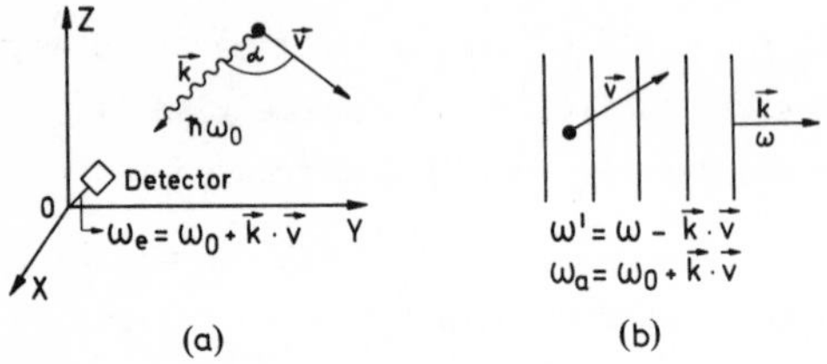

Fig.3.4a,b. Doppler shift of a monochromatic emission line (a) and absorption line (b)

Similarly, one can see that the absorption frequency ω_0 of a molecule moving with the velocity $\underline{v}$ across a plane E.M. wave $\underline{E} = \underline{E}_0 \exp(i\omega t - \underline{k}\cdot\underline{r})$ is shifted. The wave frequency ω in the rest frame appears in the frame of the moving molecule as

$$\omega' = \omega - \underline{k}\cdot\underline{v} \quad . \tag{3.26a}$$

The molecule can only absorb if ω' coincides with its eigenfrequency ω_0. The absorption frequency $\omega = \omega_a$ is then

$$\omega_a = \omega_0 + \underline{k}\cdot\underline{v} \quad . \tag{3.26b}$$

As in the emission case the absorption frequency ω_a is increased for $\underline{k}\cdot\underline{v} > 0$. This happens, for example, if the molecule moves parallel to the wave propagation. It is decreased if $\underline{k}\cdot\underline{v} < 0$; e.g., when the molecule moves against the light propagation. If we choose the +z direction to coincide with the light propagation, (3.26b) becomes with $\underline{k} = \{0, 0, k_z\}$ and $|k| = 2\pi/\lambda$,

$$\omega_a = \omega_0(1 + v_z/c) \quad . \tag{3.26c}$$

At thermal equilibrium, the molecules of a gas follow a Maxwellian velocity distribution. At the temperature T, the number of molecules $n_i(v_z)dv_z$ in the level E_i per unit volume with a velocity component between v_z and $v_z + dv_z$ is

$$n_i(v_z)dv_z = \frac{N_i}{v_p\sqrt{\pi}}\, e^{-(v_z/v_p)^2} dv_z \quad , \tag{3.27}$$

where $N_i = \int n_i(v_z)dv_z$ is the density of all molecules in level E_i, $v_p = (2kT/m)^{\frac{1}{2}}$ is the most probable velocity, m is the mass of a molecule, and k is Boltzmann's constant. Inserting the relation (3.26c) between velocity component and frequency shift into (3.27) gives the number of molecules with absorption frequencies shifted from ω_0 into the interval from ω to $\omega + d\omega$,

$$n_i(\omega)d\omega = \frac{N_i c/\omega_0}{v_p \pi^{\frac{1}{2}}} e^{-[(c/v_p)(\omega-\omega_0)/\omega_0]^2} d\omega \quad . \tag{3.28}$$

Since the emitted or absorbed radiant power $P(\omega)d\omega$ is proportional to the density $n_i(\omega)d\omega$, the intensity profile of a Doppler-broadened spectral line becomes

$$I(\omega) = I_0 \exp\left[-\left(\frac{c(\omega - \omega_0)}{\omega_0 v_p}\right)^2\right] \quad . \tag{3.29}$$

This is a Gaussian profile with a halfwidth $\delta\omega_D = |\omega_1 - \omega_2|$ which can be calculated from (3.29) with $I(\omega_1) = I(\omega_2) = I(\omega_0)/2$,

$$\delta\omega_D = 2\sqrt{\ln 2}\ \omega_0 v_p/c \quad ; \tag{3.30a}$$

or, with $v_p = (2kT/m)^{\frac{1}{2}}$,

$$\boxed{\delta\omega_D = \frac{\omega_0}{c}(8kT\ \ln 2/m)^{\frac{1}{2}}} \quad . \tag{3.30b}$$

This quantity is called the *Doppler width*; (3.29) can be written in terms of it with $1/(4\ \ln 2) = 0.36$ as

$$I(\omega) = I_0\, e^{-[(\omega-\omega_0)^2/0.36\delta\omega_D^2]} \quad . \tag{3.31}$$

Note that $\delta\omega_D$ increases linearly with frequency ω_0 and is proportional to $(T/m)^{\frac{1}{2}}$. The largest Doppler width is thus expected for hydrogen (M = 1) at large frequencies and high temperatures.

Equation (3.30) can be written more conveniently in terms of the Avogadro number N_A (the number of molecules per mole), the mass of a mole, $M = N_A m$, and the gas constant $R = N_A k$. Inserting these relations into (3.30b) gives for the Doppler width

$$\delta\omega_D = (2\omega_0/c)\sqrt{2RT\ \ln 2/M} \quad , \tag{3.30c}$$

or in frequency units, using the values for c and R,

$$\delta\nu_D = 7.16 \times 10^{-7}\nu_0\sqrt{T/M}\quad [s^{-1}] \quad . \tag{3.30d}$$

Examples

a) Vacuum ultraviolet: For the Lyman α line (2P → 1S transition in the H atom) in a discharge with temperature T = 1000 K, M = 1, λ = 1216 Å, $\nu_0 = 2.47 \times 10^{15}\ s^{-1} \Rightarrow \delta\nu_D = 5.6 \times 10^9\ s^{-1}$, $\delta\lambda_D = 2.8 \times 10^{-1}$ Å.

b) Visible spectral region: For the sodium D line (3P → 3S transition of the Na atom) in a sodium-vapor cell at T = 500 K, λ = 5891 Å, $\nu_0 = 5.1 \times 10^{14}\ s^{-1} \Rightarrow \delta\nu_D = 1.7 \times 10^9\ s^{-1}$, $\delta\lambda_D = 1 \times 10^{-2}$ Å.

c) Infrared region: For a vibrational transition between two rovibronic levels (quantum numbers J,v) $(J_i,v_i) \leftrightarrow (J_k,v_k)$ of the CO_2 molecule in a CO_2 cell at room temperature (T = 300 K), λ = 10 µm, $\nu = 3 \times 10^{13}\ s^1$, $M = 44 \Rightarrow \delta\nu_D = 5.6 \times 10^7\ s^{-1}$, $\delta\nu_D$ = 0.19 Å.

These examples illustrate that in the visible and uv regions, the Doppler width exceeds the natural linewidth by about two orders of magnitude. Note, however, that the intensity I approaches zero for large arguments $(\nu - \nu_0)$ much faster for a Gaussian line profile than for a Lorentzian profile (see Fig.3.5). It is therefore possible to obtain information about the Lorentzian profile from the extreme line wings even if the Doppler width is much larger than the natural linewidth (see below).

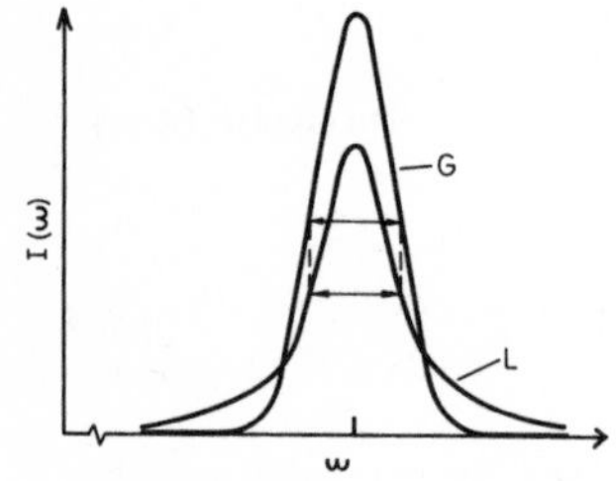

Fig.3.5. Comparison between Lorentzian and Gaussian line profile of equal halfwidths

More detailed consideration shows that a Doppler-broadened spectral line cannot be strictly represented by a pure Gaussian profile as has been assumed in the foregoing discussion. The reason is that not all molecules with a definite velocity component v_z emit or absorb radiation at the same frequency $\omega' = \omega_0(1 - v_z/c)$. Because of the finite lifetimes of the molecular energy levels, the frequency response of these molecules is represented by a Lorentzian profile (see previous section)

$$g(\omega - \omega') = \frac{\gamma/2\pi}{(\omega - \omega')^2 + (\gamma/2)^2} , \tag{3.10}$$

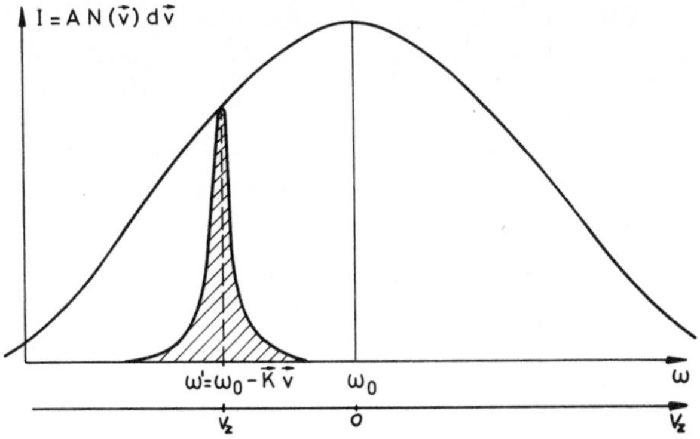

Fig.3.6. Lorentzian profile centered at $\omega' = \omega_0 - v_z/c$ which belongs to molecules with a definite velocity component v_z

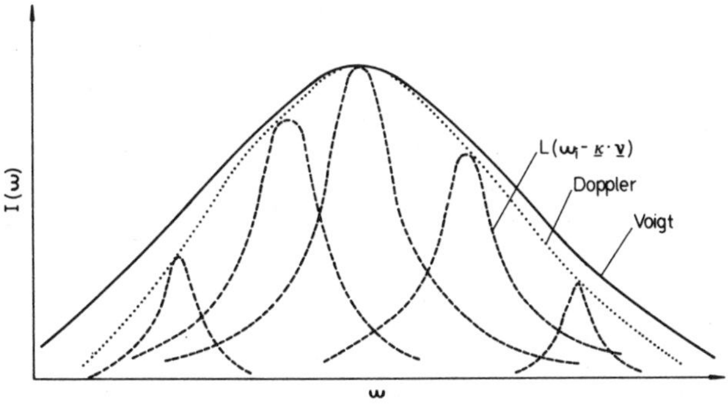

Fig.3.7. Voigt profile

with a central frequency ω' (see Fig.3.6). Let $n(\omega')d\omega' = n(v_z)dv_z$ be the number of molecules per unit volume with velocity components within the interval v_z to $v_z + dv_z$. The spectral intensity distribution $I(\omega)$ of the total absorption or emission of all molecules at the transition $E_i \to E_k$ is then

$$I(\omega) = I_0 \int n(\omega')g(\omega - \omega')d\omega' \quad . \tag{3.32}$$

Inserting (3.10) for $g(\omega - \omega')$ and (3.28) for $n(\omega')$ we obtain

$$I(\omega) = C \int_0^\infty \frac{e^{-[(c/v_p)(\omega_0-\omega')/\omega_0]^2}}{(\omega - \omega')^2 + (\gamma/2)^2} d\omega' \quad \text{with} \quad C \doteq \frac{\gamma N_i c}{2v_p \pi^{3/2} \omega_0} \quad . \tag{3.33}$$

This intensity profile, which is a convolution of Lorentzian and Gaussian profiles, [3.4] is called a *Voigt profile* (see Fig.3.7). Voigt profiles play an important role in the spectroscopy of stellar atmospheres where accurate measurements of line wings allow the contributions of Doppler broadening and natural linewidth or collision line broadening to be separated (see next section and [3.5]). From such measurements the temperature and pressure of the emitting or absorbing layers in the stellar atmospheres may be deduced.

3.3 Collision Broadening of Spectral Lines

When an atom A with energy levels E_i and E_k approaches another atom or molecule B, the energy levels of A are shifted because of the interaction between A and B. This shift depends on the electron configurations of A and B and on the distance R(A,B) between both collision partners, which, for definiteness, we define as the distance between the centers of mass of A and B.

The energy shifts ΔE are in general different for the levels E_i and E_k and they may be positive as well as negative. ΔE is positive if the interaction between A and B is repulsive and negative if it is attractive. When plotting the energy E(R) for the different energy levels as a function of the interatomic distance R the potential curves of Fig.3.8 are obtained.

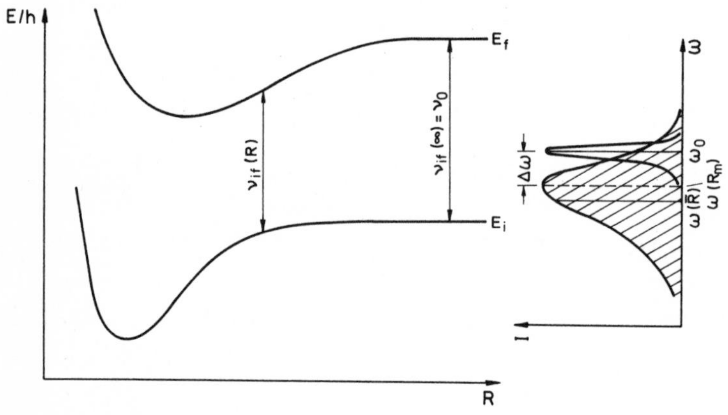

Fig.3.8. Illustration of collisional line broadening explained with the potential curves of the collision pair AB

This mututal interaction of both partners at distances $R \le R_c$ is called a *collision* and $2R_c$ is the *collision diameter*. If no internal energy of the collision partners is transferred during the collision by nonradiative transitions, the collision is called *elastic*. Without additional stabilizing mechanisms (recombination) the partners will separate again after the collision time $\tau_c \approx R_c/v$ which depends on the relative velocity v.

Numerical Example

$v = 5 \times 10^2$ m/s, $R_c = 1$ nm $\Rightarrow \tau_c = 2 \times 10^{-12}$ s. During this time the electronic charge distribution generally follows the perturbation "adiabatically", which justifies the potential curve model of Fig.3.8.

If atom A undergoes a *radiative* transition between levels E_i and E_k during the collision time, the frequency ω_{ik} of absorbed or emitted radiation, satisfying

$$\hbar\omega_{ik} = |E_i(R) - E_k(R)| \quad , \tag{3.34}$$

depends on the distance R(t) at the time t of the transition. We assume that the radiative transition takes place in a time interval short compared to the collision time, so that the distance R does not change during the transition. In Fig.3.8, this assumption leads to vertical radiative transitions.

In a gas mixture of atoms A and B the mutual distances R(A,B) show random fluctuations with a distribution around a mean value $\bar{R}$ which depends on pressure and temperature. According to (3.34) the fluorescence shows a corresponding frequency distribution around a most probable value $\overline{\omega_{ik}(R_m)}$ which may be shifted against the frequency ω_0 of the unperturbed atom A. The shift $\Delta\omega = \omega_0 - \omega_{ik}$ depends on how differently the two energy levels E_i and E_k are shifted at a distance $R_m(A,B)$ where the emission probability has a maximum. The intensity profile $I(\omega)$ of the collision-broadened and shifted emission line can be obtained from

$$I(\omega) \propto \int A_{ik}(R)P_{col}(R)[E_i(R) - E_k(R)]dR \quad , \tag{3.35}$$

where $A_{ik}(R)$ is the spontaneous transition probability which depends on R because the electronic wave functions of the collision pair (AB) depend on R, and $P_{col}(R)$ is the probability per unit time that the distance between A and B lies in the range from R to R + dR.

From (3.35) it can be seen that the intensity profile of the collision-broadened line reflects the difference of the potential curve

$$E_i(R) - E_k(R) = V[A(E_i),B] - V[A(E_k),B] \quad .$$

The line shift is caused by elastic collisions as will be shown below. The energy defect $\Delta E = \hbar\Delta\omega$ is supplied from the kinetic energy of the collision partners. This means that in case of positive shifts ($\Delta\omega > 0$), the kinetic energy is smaller after the collision than before.

Besides elastic collisions, inelastic collisions may also occur in which the excitation energy E_i of atom A is either partly or completely transferred into internal energy of the collision partner B or into translational energy of both partners. Such inelastic collisions are often called *quenching collisions* because they decrease the number of excited atoms in level E_i and therefore quench the fluorescence intensity. According to Sect.2.7.1, the total transition probability A_i for the depopulation of level E_i is a sum of spontaneous and collision-induced probabilities. Using the relation

$$p_B = N_B kT \tag{3.36}$$

between density N_B and pressure p_B of the collision partner B, one obtains from (2.82) with $\rho_{ik} = 0$ the total transition probability

$$A_i^{eff} = \sum_k A_{ik} + ap_B \quad \text{with} \quad a = \sigma_{ik}^{col} \frac{(2kT)^{3/2}}{(\pi\mu)^{1/2}} \quad . \tag{3.37}$$

It is evident from (3.24) that this pressure-dependent transition probability causes a corresponding pressure-dependent linewidth $\delta\omega$ which is proportional to the pressure p of the collision partners and which can be described by a sum of two damping terms

$$\delta\omega = \delta\omega_n + \delta\omega_{col} = \gamma_n + \gamma_{col} = \gamma_n + ap \quad . \tag{3.38}$$

The collision-induced additional line broadening is therefore often called *pressure broadening*.

The preceding discussion has shown that both elastic and inelastic collisions cause spectral line broadening. The elastic collisions may additionally cause a line shift which depends on the potential curves $E_i(R)$ and $E_k(R)$. This can be quantitatively seen from a model introduced by LINDHOLM [3.6], which treats the excited atom A as a damped oscillator which suffers collisions with particles B (atoms or molecules). In this model inelastic collisions damp the *amplitude* of the oscillation. This is described by introducing a damping constant γ_{col} such that the sum of radiative and collisional damping is represented by $\gamma = \gamma_n + \gamma_{col}$. From the derivation in Sect.3.1 one obtains for the line broadened by inelastic collisions a Lorentzian profile with halfwidth (3.38)

$$I(\omega) = \frac{C}{(\omega - \omega_0)^2 + [(\gamma_n + \gamma_{col})/2]^2} \quad .$$

The elastic collisions change not the *amplitude*, but the *phase* of the damped oscillator, due to the frequency shift $\Delta\omega(R)$ during the collisions. They are often called *phase-perturbing collisions*. We now give a short outline of a quantitative description based on this model which gives a better insight into the relation between line profile and collision cross sections [3.7,8].

As in Sect.3.1, we describe the damped oscillatory by its time-dependent amplitude

$$x(t) = x_0\, e^{i\omega_0 t + i\eta(t) - \gamma t/2} \quad , \tag{3.39}$$

in which $\gamma = \gamma_n + \gamma_{col}$ includes both radiative damping and damping by inelastic collisions. The term

$$\eta(t) = \int_0^t [\omega(t) - \omega_0]dt = \int_0^t \Delta\omega(t)dt \tag{3.40}$$

represents the sum of all phase shifts of our oscillator due to all collisions within the time from 0 to t. We shall at first neglected the damping by radiative transitions and inelastic collisions and treat the elastic collisions separately, so that $\gamma = 0$ in (3.39).

When the integration time is long compared to the mean time $T = \Lambda/\bar{v}$ between successive collisions (which depends on the mean free path Λ and the mean relative velocity $\bar{v}$) the total phase shift $\eta(t)$ is the sum of many random phase shifts caused by collisions with different impact parameters (see Fig.3.9). The mean phase shift per collision depends on the correlation between the oscillation before and after the collisions, which can be described by the correlation function

$$\begin{aligned}\varphi(\tau) &= \frac{1}{x_0^2} e^{-i\omega_0\tau} \lim_{t\to\infty} \frac{1}{t} \int_{-t/2}^{+t/2} x^*(t)x(t+\tau)dt \\ &= \lim_{t\to\infty} \frac{1}{t} \int_{-t/2}^{+t/2} e^{i[\eta(t+\tau)-\eta(t)]}dt \quad .\end{aligned} \tag{3.41}$$

During the time interval $d\tau$ the change of $\varphi(\tau)$ depends on the number of collisions and on the mean phase jump per collision. What is the physical meaning of $\varphi(\tau)$?

From (3.41) we obtain

$$x_0^2 \int_{-\tau/2}^{+\tau/2} \varphi(\tau)e^{i(\omega_0-\omega)\tau}d\tau = \lim_{t\to\infty} \frac{1}{t} \int_{-t/2}^{+t/2} x^*(t)e^{i\omega t} \int_{-\tau/2}^{+\tau/2} x(t+\tau)e^{-i\omega(t+\tau)}d\tau \quad . \tag{3.42}$$

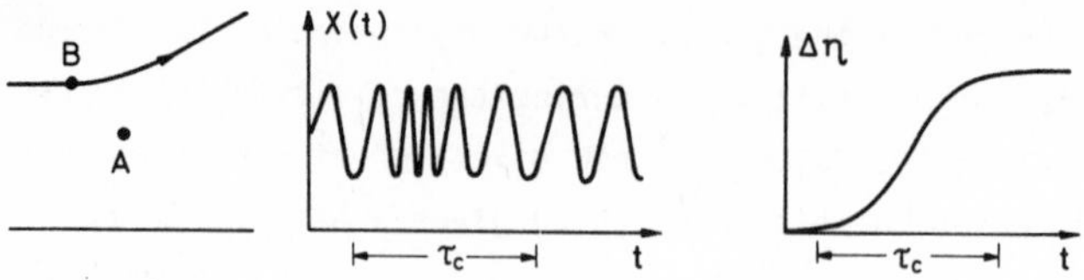

Fig.3.9. Phase perturbations of an oscillator by collisions

The integral over $d\tau$ does not depend on our choice of time origin. We may therefore shift the time scale by an amount τ from $t + \tau$ to t (see the analogous discussion in Sect.2.9.4). Replacing the dummy variable τ on the right-hand side of (3.42) by t we can write

$$x_0^2 \lim_{\tau\to\infty} \frac{1}{2\pi} \int_{-\tau/2}^{+\tau/2} \varphi(\tau) e^{i(\omega_0-\omega)\tau} d\tau = \lim_{t\to\infty} \frac{1}{2\pi t} \int_{-t/2}^{+t/2} x^*(t) e^{i\omega t} dt \int_{-t/2}^{+t/2} x(t) e^{-i\omega t} dt$$

$$= \text{F.T.}(xx^*) = I(\omega) \quad . \tag{3.43}$$

The right-hand side of (3.43) is just the *Fourier transform of the square of the amplitude* $X^*(t)X(t)$ *and represents therefore the intensity profile* $I(\omega)$ *of the spectral line* (see also Sect.3.1). Provided that we know the correlation function $\varphi(\tau)$ we can use (3.43) to determine the line profile produced by the elastic collisions. This can be performed as follows.

With

$$\Delta\eta = \eta(t + \tau) - \eta(t) \quad , \tag{3.44}$$

we can write

$$\varphi(\tau) = \lim_{t\to\infty} \frac{1}{t} \int_{-t/2}^{+t/2} e^{i\Delta\eta} dt = \langle e^{i\Delta\eta(\tau)} \rangle \tag{3.45}$$

the increment $d\varphi(\tau)$ during the time interval $d\tau$ is

$$d\varphi(\tau) = \varphi(\tau + d\tau) - \varphi(\tau) = \langle e^{i\Delta\eta(\tau+\Delta\tau)} \rangle - \langle e^{i\Delta\eta(\tau)} \rangle$$

$$= \langle e^{i\Delta\eta(\tau)} (e^{i\varepsilon} - 1) \rangle \quad , \tag{3.46}$$

where ε stands for the additional phase shift during the time interval $d\tau$. When the phase jump occurring at a collision is independent of the phase η before the collision, we can replace the time average of the product by the product of the averaged factors, which gives

$$d\varphi(\tau) = \langle e^{i\Delta\eta(\tau)} \rangle \langle e^{i\varepsilon} - 1 \rangle = \varphi(\tau) \langle e^{i\varepsilon} - 1 \rangle \quad . \tag{3.47}$$

We can further replace the time average by an ensemble average which is equivalent to an average over all possible collisions [3.9].

In a gas with N particles per unit volume, during the time interval $d\tau$ our oscillator suffers

$$dZ = 2\pi R dR N\bar{v} d\tau \tag{3.48}$$

collisions with impact parameters between R and R + dR, where $\bar{v}$ is the mean relative velocity. The ensemble average $\langle \exp(i\varepsilon) - 1 \rangle$ is therefore

$$\begin{aligned} \langle e^{i\varepsilon} - 1 \rangle &= 2\pi N\bar{v} d\tau \int_0^\infty \langle e^{i\eta(R)} - 1 \rangle R dR \\ &= N\bar{v} d\tau (\sigma_b - i\sigma_s) \quad , \end{aligned} \tag{3.49}$$

with the abbreviations

$$\sigma_b = 2\pi \int_0^\infty [1 - \cos\eta(R)] R dR \quad , \tag{3.50}$$

$$\sigma_s = 2\pi \int_0^\infty [\sin\eta(R)] R dR \quad . \tag{3.51}$$

Integration of (3.46) yields the correlation function

$$\varphi(\tau) = e^{-N\bar{v}\tau(\sigma_b - i\sigma_s)} \quad . \tag{3.52}$$

By substituting (3.52) into (3.43) we finally obtain the line profile

$$\boxed{I(\omega) = I_0 \frac{N\bar{v}\sigma_b}{(\omega - \omega_0 - N\bar{v}\sigma_s)^2 + (N\bar{v}\sigma_b)^2}} \quad , \tag{3.53}$$

which shows that the line profile in the presence of elastic collisions is a Lorentzian with a halfwidth

$$\delta\omega = 2N\bar{v}\sigma_b \quad , \tag{3.54}$$

and a shift of the line center

$$\Delta\omega = N\bar{v}\sigma_s \quad . \tag{3.55}$$

In this model of the phase-perturbed oscillator, both line broadening and line shift are proportional to the density N of collision partners and to the mean relative velocity $\bar{v}$. The line *broadening* is determined by the cross section σ_b (3.50) and the line *shift* by σ_s (3.51).

If we now include additional damping by spontaneous emission of the oscillator and by inelastic collisions, we can consider these effects on the line profiles by introducing into (3.53) the damping constant $\gamma = \gamma_n + \gamma_{col}^{inel}$ and obtain the Lorentzian profile of a damped oscillator which suffers elastic and inelastic collisions.

$$I(\omega) = I_0 \frac{(\gamma_n/2 + \gamma_{col}^{inel}/2 + N\bar{v}\sigma_b)^2}{(\omega - \omega_0 - N\bar{v}\sigma_s)^2 + [(\gamma_n + \gamma_{col}^{inel})/2 + N\bar{v}\sigma_b]^2} \quad . \qquad (3.56)$$

In order to gain more insight into the physical meaning of the cross sections σ_s and σ_b we have to discover the relation between the phase shift $\eta(R)$ and the potential $V(R)$. Assume a potential of the form

$$V_i(R) = C_i/R^n \qquad (3.57)$$

between the atom in level E_i and the perturbing atom B. The frequency shift $\Delta\omega$ for the transition $E_i \rightarrow E_k$ is then

$$\hbar\Delta\omega(r) = \frac{C_i - C_k}{R^n} \quad , \qquad (3.58)$$

and the corresponding phase shift for a collision with impact parameter R_0, where we neglect the scattering of B and assume that the path of B is not deflected but follows a straight line (Fig.3.10), is

$$\eta(R_0) = \int_{-\infty}^{+\infty} \Delta\omega dt = \int \frac{(C_i - C_k)dt}{[R_0^2 + \bar{v}^2(t - t_0)^2]^{n/2}} = \frac{\alpha_n(C_i - C_k)}{\bar{v}R_0^{n-1}} \quad . \qquad (3.59)$$

Equation (3.59) shows the relation between phase shift $\eta(R_0)$ and the interaction potential (3.57), where α_n is a numerical constant depending on the exponent n in (3.57).

Substituting $\eta(R)$ into (3.50) and (3.51) allows the cross sections σ_s and σ_b to be calculated. The calculation for n = 4, for example, yields the result that the main contribution to σ_b comes from collisions with *small* impact parameters whereas σ_s still has large values for *large* impact parameters. This means that *collisions at large distances do not cause noticeable broadening of the line but can still very effectively shift the line center.*

From the preceding discussion it might appear that only the *difference* $V_i(R) - V_k(R)$ of the interaction potentials between B and the atom A in its

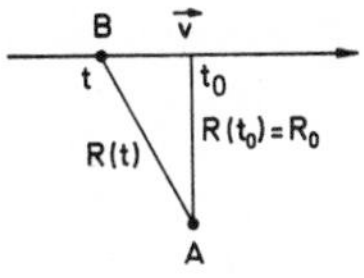

(a)

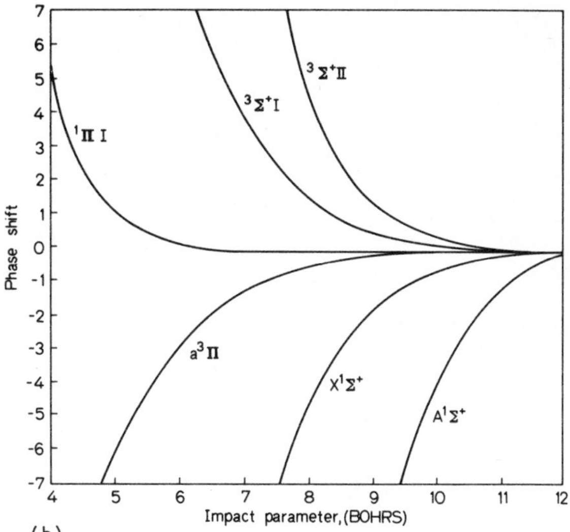

(b)

Fig.3.10. (a) Linear path approximation of a collision between A and B. (b) Phase shift versus impact parameter for Na-H collisions. The various adiabatic molecular states associated with $Na^*(3P)$ are indicated [3.10]

two states E_i and E_k can be obtained from line broadening and line shifts. However, it is also possible to obtain separately the upper and lower potentials as follows:

Assume an interaction potential V(R) between atom A in its ground state and its collision partner B. The probability that B has a distance between R and R + dR is proportional to $4\pi R^2 dR$ and (in thermal equilibrium) to the Boltzmann factor $\exp[-V(R)/kT]$. The density n(R) of collision pairs A, B with distance R is therefore

$$n_{AB}(R)dR = CR^2\, e^{-V(R)/(kT)}dR \quad . \tag{3.60}$$

Because the intensity of an absorption line is proportional to the density of absorbing atoms while they are forming collision pairs, using the relation $\hbar\omega(R) = [V_i(R) - V_k(R)] \Rightarrow \hbar d\omega/dR = d[V_i(R) - V_k(R)]/dR$, the intensity profile of the absorption line can be written as

$$I(\omega)d\omega = C^*\left\{R^2\exp\left(-\frac{V_i(R)}{kT}\right)\frac{d}{dR}\,[V_i(R) - V_k(R)]\right\}dR \quad . \tag{3.61}$$

Measuring the line profile as a function of temperature yields $V_i(R)$ separately. Frequently, different spherical model potentials V(R) are substituted in (3.61) such as the potential (3.57) or the Lennard-Jones potential

$$V(R) = a/R^{12} - b/R^6 \tag{3.62}$$

and the coefficients c_i [for (3.57)] or a, b [for (3.62)] are adjusted for optimum agreement between theory and experiment [3.11,12].

Because of the long-range Coulomb interactions between charged particles (electrons and ions) described by the potential (3.57) with $n = 1$, pressure broadening and shift is particularly large in plasmas and gas discharges. The interaction between charged particles can be described by the linear and quadratic Stark effects. It can be shown that the linear Stark effect causes only line broadening, while the quadratic effect leads also to line shifts. From measurements of line profiles in plasmas, very detailed plasma characteristics, such as electron or ion densities and temperatures, can be determined. Plasma spectroscopy has therefore become an extensive field of research, of interest not only for astrophysics, but also for fusion research in high-temperature plasmas [3.13].

The classical models used to explain collision broadening and line shifts can be improved by using quantum mechanical calculations. These are, however, beyond the scope of this book, and the reader is referred to the literature [3.8-16].

Examples

1) The pressure broadening of the sodium D line $\lambda = 589$ nm by argon is 3×10^{-4} Å/torr, equivalent to 30 MHz/torr. The shift is about minus 1 MHz/torr. The self-broadening of 150 MHz/torr, due to collision between Na atoms, is much larger. However, at pressures of several torr, the pressure broadening is still smaller than the Doppler width.

2) The pressure broadening of molecular vibration-rotation transitions with wavelengths $\lambda \approx 5$ µm is a few MHz/torr. At atmospheric pressure, the collision broadening therefore exceeds the Doppler width. For example, the rotational lines of the ν_2 band of H_2O in air at normal pressure (760 torr) have a Doppler width of 150 MHz but a pressure-broadened linewidth of 930 MHz.

3) The collision broadening of the red neon line at $\lambda = 633$ nm in the low-pressure discharge of a He-Ne laser is about $\delta\nu = 150$ MHz/torr; the pressure shift $\Delta\nu = 20$ MHz/torr. In high-current discharges, such as the argon laser discharge, the degree of ionization is much higher than in the He-Ne laser and the Coulomb interaction between ions and electrons plays a major role. The pressure broadening is therefore much larger: $\delta\nu = 1500$ MHz/torr. Because of the high temperature in the plasma, the Doppler width $\delta\nu_D \approx 5000$ MHz is yet still larger [3.17].

	Helium		Neon		Argon		Krypton		Xenon		Nitrogen		Carbon tetrafluoride	
	Width	Shift	Width	Shift	Width	Shift	Width	Shift	Width	Shift	Width	Shift	Width	Shift
Na 5896 $P_{1/2}$					0.742	-0.196					0.49	-0.214		
5890 $P_{3/2}$					0.689	-0.213					0.49	-0.223		
K 7699 $P_{1/2}$					1.01	-0.42					0.82	-0.36		
7665 $P_{3/2}$					1.01	-0.36					0.82	-0.30		
K 4047 $P_{1/2}$	1.45		0.71		2.20	-9.2					1.65	-0.83		
4044 $P_{3/2}$	2.02		0.82	-0.16	2.58	-8.3					1.89	-1.02		
Rb 7947 $P_{1/2}$	0.595	+0.228			0.627									
7800 $P_{3/2}$	0.735	-0.092			0.855									
Rb 4216 $P_{1/2}$	2.77	+0.93	1.30	+0.22	2.21	-1.2					1.51			
4202 $P_{3/2}$	1.88	+0.43	0.73	-0.16	2.56	-1.0					1.01			
Cs 8943 $P_{1/2}$		+0.128		-0.04	0.30	-0.238	0.28	-0.20	0.49	-0.25				-0.29
8521 $P_{3/2}$		+0.015		-0.08	0.30	-0.215	0.28	-0.20	0.23	-0.25				-0.23
Cs 4593 $P_{1/2}$		-0.51				-0.70	1.1	-0.62		-0.65				
4555 $P_{3/2}$		-0.26				-0.63	3.2	-0.62		-0.65				

(a)

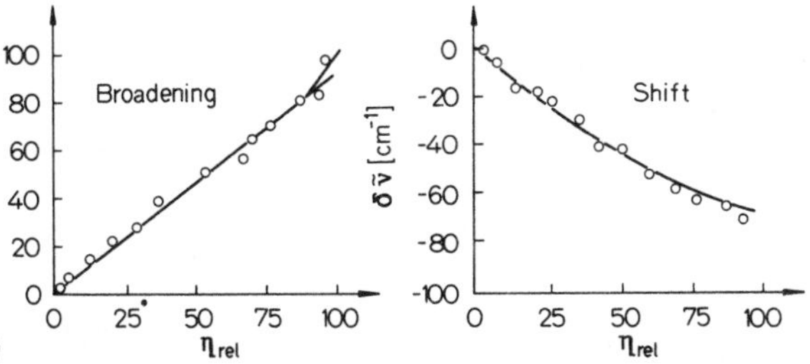

(b)

Fig.3.11. (a) Pressure broadening and shifts of some alkali resonance lines by different gases. The numbers are given in cm^{-1} for standard conditions (1 atm, 15°C). (b) Halfwidth and line shift (in cm^{-1}) of the Cs line λ = 894.3 nm as a function of relative argon densities (density n at pressure p divided by the density at standard conditions) [3.2]

4) Figure 3.11a gives as examples pressure broadening and shifts of some alkali resonance lines perturbed by different gases. The numbers are in cm^{-1} at standard conditions of 1 atm and 15°C.

Figure 3.11b gives the halfwidth and line shift of the Cs line λ = 894.3 nm in cm^{-1} as a function of relative density (i.e., density divided by density at 1 atm [3.2]).

Note

In the infrared and microwave range, collisions may sometimes cause a narrowing of the linewidth instead of a broadening (Dicke narrowing). This can be explained as follows. If the life time of the upper molecular level (e.g., an excited vibrational level in the electronic ground state) is long compared to the mean time between successive collisions, the velocity of the

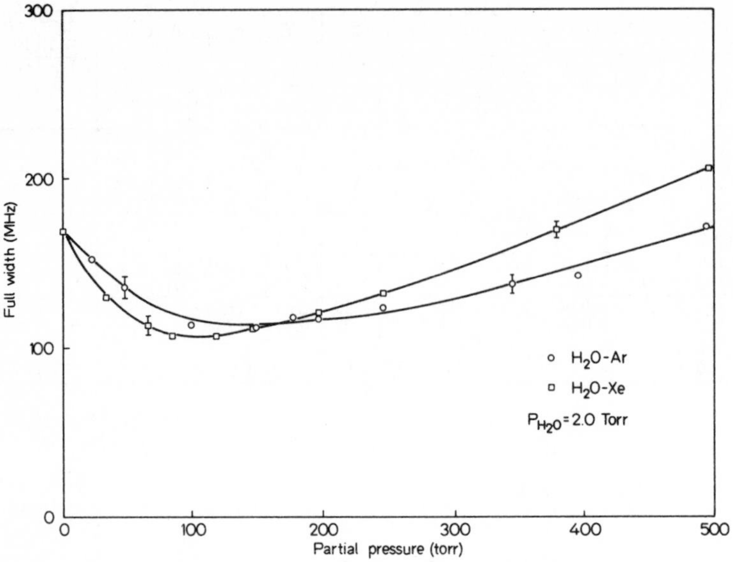

Fig.3.12. Dicke narrowing and pressure broadening of a rotational transition in H_2O at 1871 cm^{-1}, as a function of Ar and Xe pressure [3.18]

oscillator is often altered by elastic collisions and the mean velocity component is smaller than without these collisions, resulting in a smaller Doppler shift. When the Doppler width is larger than the pressure-broadened width, this causes a narrowing of the lines, if the mean free path is smaller than the wavelength of the molecular transition [3.18]. Figure 3.12 illustrates this "Dicke narrowing" for a rotational transition of the H_2O molecule at 1871 cm^{-1}. The linewidth decreases with increasing pressure up to pressures of about 100-150 torr, depending on the collision partner, which determines the mean free path Λ. For higher pressures, the pressure broadening overcompensates the Dicke narrowing, and the linewidth increases again.

3.4 Time-of-Flight Broadening

In many experiments in laser spectroscopy, the interaction time of molecules with the radiation field is small compared with the spontaneous lifetimes of excited levels. Particularly for transitions between rotational-vibrational levels of molecules with spontaneous lifetimes in the millisecond range, the time of flight $T = d/\bar{v}$ of molecules with a mean thermal velocity $\bar{v}$ through a laser beam of diameter d may be smaller than the spontaneous lifetime by several orders of magnitude. With $\bar{v} = 5 \times 10^4$ cm/s and $d = 0.1$ cm, the time of flight becomes $T = 2$ μs. For a beam of fast ions with velocities $\bar{v} = 3 \times 10^8$ cm/s the time required to traverse a laser beam with $d = 0.1$ cm is already below 10^{-9} s, which is shorter than the spontaneous lifetimes of most atomic levels.

In such cases, the linewidth of a Doppler-free molecular transition is no longer limited by the spontaneous transition probabilities (see Sect. 3.1) but by the time of flight through the laser beam which determines the interaction time of the molecule with the radiation field. This can be seen as follows. Consider an undamped oscillator $x = x_0 \cos\omega_0 t$ which oscillates with constant amplitude during the time interval T and which then suddenly stops oscillating. Its frequency spectrum is obtained from the Fourier transform

$$A(\omega) = \frac{1}{\sqrt{2\pi}} \int_0^T X_0 \cos(\omega_0 t) e^{-i\omega t} dt \quad . \tag{3.63}$$

The spectral intensity profile $I(\omega) = A^*A$ is, for $(\omega - \omega_0) \ll \omega_0$,

$$I(\omega) = C \frac{\sin^2[(\omega - \omega_0)T/2]}{(\omega - \omega_0)^2} \quad , \tag{3.64}$$

according to the discussion in Sect.3.1. This is a function of the form $(\sin^2 x)/x^2$ with a halfwidth of $\delta\omega_T = 5.6/T$ (see Fig.3.13).

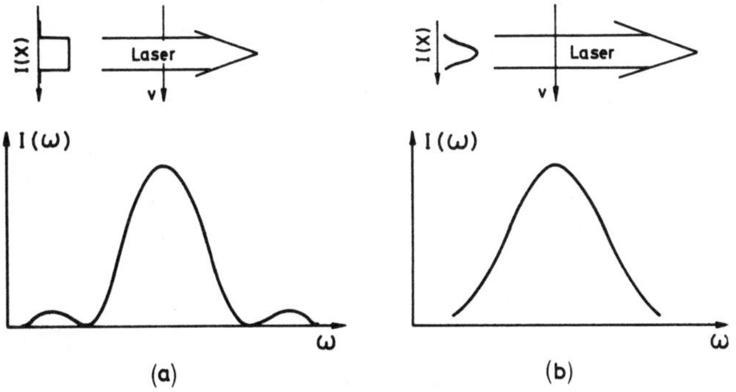

Fig.3.13a,b. Time-of-flight broadening: Frequency distribution of the transition probability of an atom traversing a laser beam. (a) With rectangular intensity profile; (b) with a Gaussian intensity profile

This example can be applied to an atom which traverses a laser beam with a rectangular intensity profile (see Fig.3.13a). The oscillator amplitude $x(t)$ is proportional to the field amplitude $E = E_0(\underline{r})\cos\omega t$. If the interaction time $T = d/v$ is small compared to the damping time $T = 1/\gamma$ the oscillation amplitude can be regarded as constant during the time T. The halfwidth of the line in frequency units is then $\delta\nu \approx v/d$.

The field distribution across a laser beam which oscillates in the fundamental mode is given by (see Sect.5.3)

$$E = E_0 \exp(-r^2/w^2)\cos\omega t \quad ,$$

in which 2w gives the diameter of the beam across the points where $E = E_0/e$. Substituting this with $x = \alpha E$ into (3.63) one obtains instead of (3.64) a Gaussian distribution

$$I(\omega) = C^* e^{-[(\omega-\omega_0)^2 w^2/2v^2]} \quad , \tag{3.65}$$

with a halfwidth

$$\delta\omega = 2(v/w)\sqrt{2\ \ln 2} \approx 2.4\ v/w \quad . \tag{3.66}$$

There are two possible ways of reducing the time-of-flight broadening: one may either enlarge the laser beam diameter w, or one may decrease the temperature, thus reducing the molecular velocity v. Both methods have been verified experimentally and will be discussed in Chap.10.

So far we have assumed that the wave fronts of the radiation field are planes and that the molecules move parallel to these planes. However, as will be shown in Sect.5.11, the phase surfaces of a Gaussian beam are curved except at the beam waist, which is the region around the focus of a Gaussian beam. As Fig.3.14 illustrates, the spatial phase shift $x2\pi/\lambda$ experienced by an atom moving along the r direction perpendicular to the laser beam axis z is with $r^2 = R^2 - (R - x)^2 \Rightarrow x \approx r^2/2R$ for $x \ll R$

$$\Delta\varphi = kr^2/2R = \omega r^2/(2cR) \quad , \tag{3.67}$$

where $k = \omega/c$ is the magnitude of the wave vector and R is the radius of curvature of the wave fronts. This phase shift restricts the effective transit time and causes additional l·ne broadening [1.13]. In order to minimize this effect, the radius of curvature has to be made as large as possible. If $\Delta\varphi \ll \pi$ for a distance $r = d$, the broadening by the wave-front curvature is small compared to the transit-time broadening. This imposes the condition $R \gg d^2/\lambda$ on the radius of curvature. Example: $d = 1$ cm, $\lambda = 1$ µm $\Rightarrow R \gg 10$ m.

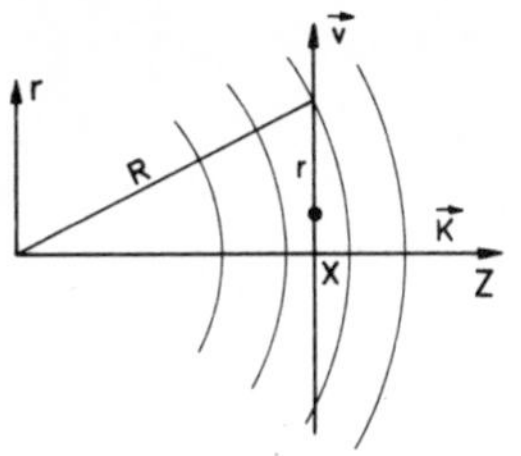

Fig.3.14. Line broadening caused by the curvature of wave fronts

3.5 Homogeneous and Inhomogeneous Line Broadening

If the probability $P_{ik}(\omega)$ of absorption or emission of radiation with frequency ω causing a transition $E_i \rightarrow E_k$ is equal for all the molecules of a sample that are in the same level E_i, we call the spectral line profile of this transition *homogeneously broadened*. Natural line broadening is an example which yields a homogeneous line profile. In this case, the probability for emission of light with frequency ω on a transition $E_i \rightarrow E_k$,

$$P_{ik}(\omega) = A_{ik} g(\omega - \omega_0) \quad ,$$

is equal for all atoms in level E_i. $g(\omega - \omega_0)$ is the normalized Lorentzian profile with central frequency ω_0.

The standard example of inhomogeneous line broadening is the Doppler broadening. In this case, the probability of absorption or emission of monochromatic radiation $E(\omega)$ is not equal for all molecules, but depends on their velocity v (see Sect.3.2). We divide the molecules in level E_i into subgroups such that all molecules with a velocity component within the interval v_z to $v_z + \Delta v_z$ belong to one subgroup. If we choose Δv_z to be $\delta\omega_n/k$ where $\delta\omega_n$ is the natural linewidth, we may consider the frequency interval $\delta\omega_n$ to be homogeneously broadened inside the much larger inhomogeneous Doppler width. That is to say, all molecules in the subgroup can absorb or emit radiation with wave vector $\underline{k}$ and frequency $\omega = \omega_0 - v_z|\underline{k}|$ (see Fig.3.6), because in the coordinate system of the moving molecules, this frequency is within the natural width $\delta\omega_n$ around ω_0 (see Sect.3.2).

In Sect.3.3, we saw that the spectral line profile is altered by two kinds of collisions. Inelastic collisions cause additional damping, resulting in pure broadening of the Lorentzian line profile. This broadening by inelastic collisions brings about a homogeneous Lorentzian line profile. The elastic collisions could be described as phase-perturbing collisions.

The Fourier transform of the oscillation trains with random phase jumps yields again a Lorentzian line profile as derived in Sect.3.3. Summarizing, we can state that elastic and inelastic collisions which only perturb the phase or amplitude of an oscillating atom without changing its velocity cause homogeneous line broadening.

So far, we have neglected the fact that collisions also change the velocity of both collision partners. If the velocity component v_z of a molecule is altered by an amount u_z during the collision, the molecule is transferred from one subgroup $(v_z \pm \Delta v_z)$ within the Doppler profile to another subgroup $(v_z + u_z \pm \Delta v_z)$. This causes a shift of its absorption or emission frequency from ω to $\omega + ku_z$. This shift should not be confused with the line shift caused by phase-perturbing elastic collisions which occurs also when the velocity of the oscillator does not change.

At thermal equilibrium, the changes u_z of v_z by velocity-changing collisions are randomly distributed. Therefore the whole Doppler profile will in general not be affected and the effect of these collisions is washed out in Doppler-limited spectroscopy. In Doppler-free laser spectroscopy, however, the velocity-changing collisions may play a nonnegligible role, and they cause effects which depend on the ratio of the collision time $T = \Lambda/\bar{v}$ to the interaction time τ_c with the radiation field. For $T > \tau_c$, the redistribution of molecules by velocity-changing collisions causes only a small change of the population densities $n_i(v_z)dv_z$ within the different subgroups, without noticeably changing the homogeneous width of this subgroup. If $T \ll \tau_c$, the different subgroups are uniformly mixed. This results in a broadening of the homogeneous linewidth associated with each subgroup. The effective interaction time of the molecules with a monochromatic laser field is shortened because the velocity-changing collisions move a molecule out of resonance with the field. The resultant change of the line shape can be monitored with the technique of saturation spectroscopy (see Sect.10.2).

Under certain conditions, if the mean free path Λ of the molecules is smaller than the wavelength of the radiation field, velocity-changing collisions may also result in a narrowing of a Doppler-broadened line profile (Dicke narrowing; see Sect.3.3).

3.6 Saturation Broadening

In Sect.2.8, we saw that a sufficiently strong radiation field can significantly change the population densities N_1 and N_2 of an atomic system by induced absorption and emission. This saturation of the population densities also causes additional line broadening. The spectral line profiles of such partially saturated transitions are different for homogeneously and for inhomogeneously broadened lines [3.19]. We treat first the homogeneous case.

3.6.1 Homogeneous Saturation

The population difference $\Delta N_0 = N_{10} - N_{20}$ in a stationary two-level system without radiation field decreases to $\Delta N = \Delta N_0/(1 + S)$ [see (2.84h)] in the presence of the radiation field because of the induced emission and absorption processes due to the interaction with the field. The decrease of ΔN depends on the saturation parameter $S = B_{12}\rho(\omega)/R$ which is the ratio of induced transition rate $B_{12}\rho$ to the mean value $\bar{R}$ of the relaxation rates of levels 1 and 2 (Fig.2.25a).

The radiation power absorbed per unit volume is

$$dW_{12}/dt = \hbar\omega B_{12}\rho\Delta N = \frac{\hbar\omega B_{12}\rho\Delta N_0}{1 + S} = \frac{\hbar\omega R\Delta N_0}{1 + 1/S} \quad . \tag{3.68}$$

Since the absorption profile $\alpha(\omega)$ of a homogeneously broadened line is Lorentzian, the induced absorption probability of a monochromatic wave with frequency ω follows a Lorentzian line profile $B_{12}\rho(\omega)L(\omega - \omega_0)$. We can therefore introduce a frequency-dependent saturation parameter for the transition $E_1 \rightarrow E_2$,

$$S_\omega = (B_{12}\rho(\omega)/R)L(\omega - \omega_0) \quad , \qquad \text{and} \qquad S = \bar{S}_\omega = \frac{1}{\Delta\omega}\int S_\omega d\omega \quad , \tag{3.69}$$

where we can assume that the mean relaxation rate $\bar{R}$ is independent of ω within the frequency range of the line profile. With the definition (3.24a) of the Lorentzian profile $L(\omega - \omega_0)$, we obtain for the spectral saturation parameter S_ω

$$S_\omega = S_0\frac{(\gamma/2)^2}{(\omega - \omega_0)^2 + (\gamma/2)^2} \quad \text{with} \quad S_0 = S_\omega(\omega_0) \quad . \tag{3.70}$$

Substituting (3.70) into (3.68) yields the frequency dependence of the absorbed radiation power per unit frequency interval $d\omega = 1$

$$d[W_{12}(\omega)]/dt = \frac{\hbar\omega R\Delta N_0}{1+S_\omega^{-1}} = \frac{\hbar\omega R\Delta N_0 S_0(\gamma/2)^2}{(\omega-\omega_0)^2+(\gamma/2)^2(1+S_0)}$$

$$= \frac{C}{(\omega-\omega_0)^2+(\gamma_s/2)^2} \quad . \qquad (3.71)$$

This is a Lorentzian profile with the increased halfwidth $\gamma_s = \gamma\sqrt{1+S_0}$, or, because $\delta\omega = \gamma$,

$$\boxed{\delta\omega_s = \delta\omega\sqrt{1+S_0}} \quad . \qquad (3.72)$$

The halfwidth $\delta\omega_s$ of the saturation-broadened line increases with the saturation parameter S_0 at the line center ω_0. If the induced transition rate at ω_0 equals the total relaxation rate R, the saturation parameter $S_0 = [B_{12}\rho(\omega_0)]/R$ becomes 1, which increases the linewidth by a factor $\sqrt{2}$, compared to the unsaturated linewidth $\delta\omega$ for weak radiation fields ($\rho \to 0$).

Since the power dW_{12}/dt absorbed per unit volume equals the intensity decrease per cm, $dI = -\alpha_s I$, of an incident wave with intensity I, we can obtain the absorption coefficient α from (3.71) with $I = c\rho$ and $S_\omega = [B_{12}\rho(\omega)]/R$

$$\alpha_s(\omega) = C_s\frac{(\gamma_s/2)^2}{(\omega-\omega_0)^2+(\gamma_s/2)^2} \quad \text{with} \quad C_s = \frac{2(\hbar\omega)\Delta N_0 B_{12}}{\pi c\gamma(1+S_0)} \quad . \qquad (3.73)$$

Comparing this with the unsaturated absorption profile,

$$\alpha(\omega) = \frac{2\hbar\omega B_{12}\Delta N_0}{\pi c\gamma}\,\frac{(\gamma/2)^2}{(\omega-\omega_0)^2+(\gamma/2)^2} \quad , \qquad (3.74)$$

one sees that the saturation decreases the absorption coefficient $\alpha(\omega)$ by a factor $(1+S_\omega)$. At the line center, this factor has its maximum $(1+S_0)$, while it decreases for increasing $(\omega-\omega_0)$ [see (3.69)]. The saturation is

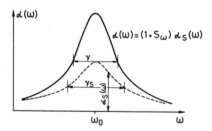

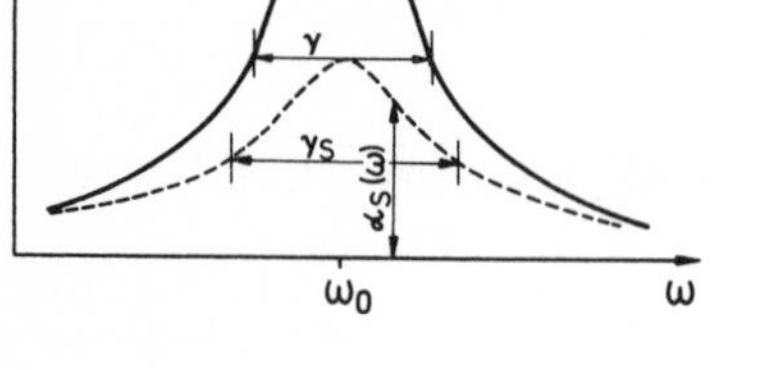

Fig.3.15. Saturation broadening of a homogeneous line profile

therefore strongest at the line center, and approaches zero for $(\omega - \omega_0) \to \infty$ (Fig.3.15). This is the reason why the line broadens.

3.6.2 Saturation of Inhomogeneous Line Profiles

If a monochromatic wave with frequency ω and wave vector $\underline{k}$ is incident on a sample of molecules with a thermal velocity distribution, only those molecules which are "Doppler shifted" into resonance with the field can interact with the radiation field. When the resonance frequency $\omega_{12} = (E_2 - E_1)/\hbar$ of the absorbing molecular transition has a homogeneous linewidth $\delta\omega$ (caused by natural or pressure broadening) the velocities of the absorbing molecules must fall into the interval $\Delta\bar{v}$ defined by

$$\omega - \underline{k} \cdot (\bar{v} \pm \Delta\bar{v}) = \omega_{12} \pm \delta\omega \quad . \tag{3.75a}$$

Assume the wave vector $\underline{k}$ to be parallel to the z direction. Equation (3.75a) then reduces to

$$\omega - k \cdot (v_z \pm \Delta v_z) = \omega_{12} \pm \delta\omega \quad . \tag{3.75b}$$

The velocity distribution $n_1(v_z)dv_z$ of molecules in level E_1 is the Maxwellian distribution (3.27) and the total population of E_1 is $N_1 = \int n_1(v_z)dv_z$. Because of saturation the population density $n_1(v_z)dv_z$ of the absorbing subgroup in the interval $dv_z = \delta\omega/k$ around $v_z = (\omega - \omega_{12})/k$ *decreases* while that of the upper state E_2 increases correspondingly. This causes a dip in the population distribution $n_1(v_z)$ (Bennet hole) and a corresponding peak in the distribution $n_2(v_z)$ of the upper state (see Figs.3.16,17)

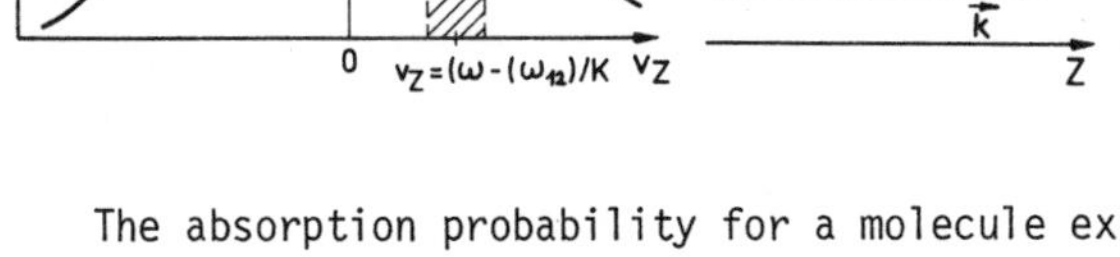

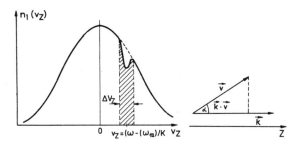

Fig.3.16. Selective absorption of a monochromatic wave by a subgroup within an inhomogeneous line profile

The absorption probability for a molecule expressed by the absorption cross section $\sigma_{12}(\underline{v},\omega)$ therefore depends on its velocity $\underline{v}$ and on the frequency ω and wave vector $\underline{k}$ of the light wave. From the considerations above one obtains [see (3.70)]

$$\sigma_{12}(\underline{v},\omega) = \sigma_0 \frac{(\gamma/2)^2}{(\omega_0 - \omega - \underline{k}\cdot\underline{v})^2 + (\gamma/2)^2} \quad . \tag{3.76}$$

For the population difference $\Delta n(v) = n_1(v) - n_2(v)$ altered by the velocity-selective saturation, we derive from (2.84h) with $\Delta n = \Delta n_0/(1 + S_\omega)$, using the frequency-dependent saturation parameter $S_\omega(v,\omega) \propto \sigma_{12}(v,\omega)$ corresponding to (3.70),

$$\Delta n(v) = \Delta n_0(v)\left[1 + \frac{S_0(\gamma/2)^2}{(\omega_0 - \omega - \underline{k}\cdot\underline{v})^2 + (\gamma/2)^2}\right]^{-1} \quad , \tag{3.77a}$$

which can be transformed to

$$\Delta n(v) = \Delta n_0(v)\left[1 - \frac{S_0(\gamma/2)^2}{(\omega_0 - \omega - \underline{k}\cdot\underline{v})^2 + (\gamma_S/2)^2}\right] . \tag{3.77b}$$

According to (2.68) the absorption coefficient $\alpha(v,\omega)$ is

$$\alpha(v,\omega) = \sigma(v,\omega)\Delta n(v) \quad . \tag{3.78}$$

$\alpha(v,\omega)$ represents the absorption of radiation with frequency ω by molecules with a given velocity component v in the direction of $\underline{k}$, which we choose to be the z direction. The total absorption coefficient $\alpha(\omega)$ caused by all molecules $(N_1 - N_2)$ within the whole velocity distribution is given by

$$\alpha(\omega) = \int_{-\infty}^{+\infty} \Delta n(v_z)\sigma(v_z,\omega)dv_z \quad . \tag{3.79}$$

For weak fields (unsaturated case with $S = 0$), we can replace $\Delta n(v)$ by $\Delta n_0(v)$ [see (3.77)] which represents the Gaussian profile of the Maxwellian velocity distribution (3.27). We obtain for $\alpha_0(\omega)$ the Voigt profile (3.33),

$$\alpha_0(\omega) = \frac{\gamma^2\sigma_0\Delta N_0}{4\sqrt{\pi}\, v_p} \int_{-\infty}^{+\infty} \frac{e^{-(v_z/v_p)^2}dv_z}{(\omega - \omega_0 - kv_z)^2 + (\gamma/2)^2} \quad . \tag{3.80}$$

With strong fields causing saturation we find from (3.79), (3.77a), and (3.76)

$$\alpha_S(\omega) = \frac{\gamma^2\sigma_0\Delta N_0}{4\sqrt{\pi}\, v_p} \int_{-\infty}^{+\infty} \frac{e^{-(v_z/v_p)^2}dv_z}{(\omega - \omega_0 - kv_z)^2 + (\gamma_S/2)^2} \quad . \tag{3.81}$$

This saturated absorption profile is similar to the unsaturated profile (3.80). The only difference is that γ is replaced by $\gamma_S = \gamma(1 + S_0)^{1/2}$ in the denominator of the integrand. The evaluation of the integral yields for $\gamma_s << \Delta\omega_D$

$$\alpha_s(\omega) = \alpha_0(\omega)/\sqrt{1 + S_0} \tag{3.81a}$$

with

$$\alpha_0(\omega) = \alpha_0(\omega_0)\exp\left[-\left(\frac{\omega - \omega_0}{\delta\omega_D}\right)^2\right]$$

where $\delta\omega_D = 2\sqrt{\ln 2}\ \omega_0 v_p/c$ is the Doppler width (see Sect.3.2). The saturated Doppler-broadened line profile therefore becomes

$$\alpha_s(\omega) = \frac{\alpha_0(\omega_0)}{\sqrt{1 + S_0}} \exp\left[-\left(\frac{\omega - \omega_0}{0.36\delta\omega_D}\right)^2\right] . \tag{3.81b}$$

This demonstrates the following remarkable fact. Although the monochromatic wave with frequency ω burns a narrow Bennet hole, centered at $\omega = \omega_0 - kv_z$, into the Doppler-broadened population distribution $n_1(v)$, this hole cannot be detected just by tuning the frequency ω of the saturating wave across the Doppler-broadened absorption profile. $\alpha_s(\omega)$ represents a Voigt profile without any hole. The reason for this is of course that the Bennett hole is burned *at any* frequency ω of the incident wave, which experiences a decreased absorption $\alpha_s(\omega) = \alpha_0(\omega)/f(\omega - \omega_0 - kv)$. The factor $f(\omega - \omega_0 - kv)$ varies slowly over the Gaussian line profile, and has its maximum value $f(0) = (1 + S_0)$ at the center of the line (see Fig.3.17). When tuning a monochromatic laser across the inhomogeneous line profile, saturation therefore manifests itself in a way analogous to the homogeneous case.

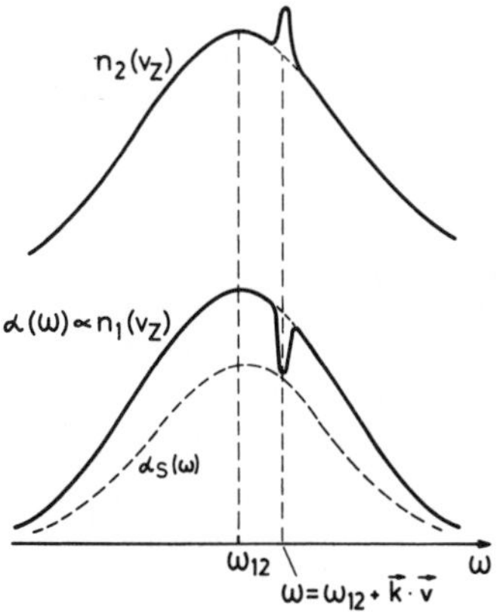

Fig.3.17. Saturation of an inhomogeneously broadened transition. Generation of a Bennet hole in the lower and a population peak in the upper level

This is drastically changed when besides the saturating light wave, a second wave is used which can probe the Bennet hole burned by the saturating wave. We shall assume the probe wave to be sufficiently weak for saturation by the probe to be negligible, and that the saturating wave has a given frequency ω_s. The absorption coefficient $\alpha(\omega)$ for the probe wave is then

$$\alpha(\omega,\omega_s) = \int \sigma_{12}(v,\omega)\Delta n(v,\omega_s)dv \quad , \tag{3.82}$$

and from (3.77b) and (3.78) one obtains

$$\alpha(\omega,\omega_s) = \frac{\gamma^2\sigma_0}{4\sqrt{\pi}v_p} \int_{-\infty}^{+\infty} \left[1 - \frac{S_0(\gamma/2)^2}{(\omega_0 - \omega_s - k_s v_z)^2 + (\gamma_s/2)^2} \cdot \frac{e^{-(v_z/v_p)^2}dv_z}{(\omega_0 - \omega - k_p v_z)^2 + (\gamma/2)^2}\right] dv_z \quad . \tag{3.83}$$

This is a *Voigt profile with a Bennet hole centered at* $\omega = \omega_s$ (see Fig.3.18). The halfwidth $\delta\omega_s$ of this hole is determined by the saturated homogeneous linewidth $\gamma_s = \gamma\sqrt{1 + S_0}$ and the depth $4S_0/\gamma$ is also determined by the intensity of the saturating field and the Einstein coefficient B_{12} of the transition [3.19].

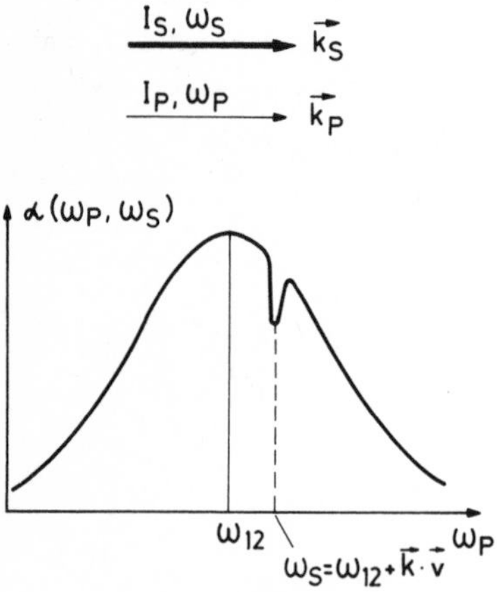

Fig.3.18. Absorption profile of an inhomogeneous line saturated by a monochromatic pump wave at ω_S and probed by a weak probe wave tuned across the line profile

If the absorbing molecules are exposed to a monochromatic *standing* wave with light frequency ω, this field can be composed of two travelling waves propagating in opposite directions.

$$A\cos(\omega t - kz) + A\cos(\omega t + kz) = 2A\cos(kz)\cos(\omega t).$$

Due to saturation, both waves burn a Bennet hole into the population distribution $n_1(v_z)$ which are located symmetrical to $v_z = 0$ at $v_z = \pm(\omega_{12} - \omega)/k$ because of the opposite Doppler shift for both waves (Fig.3.19a). If the laser frequency ω is tuned across the Doppler width of the absorption line, both holes coincide at $\omega = \omega_{12} \rightarrow v_z = 0$. At $\omega = \omega_{12}$ the intensity $I = I_1 + I_2$ of the saturating wave is twice the intensity I_1 or I_2 and the Bennet hole at ω_{12} is therefore deeper than the separated holes at $\omega \neq \omega_{12}$. This implies that the total absorption of the standing wave has a dip at the line center at ω_{12} (Fig.3.19b) which is called the "Lamb dip" after W. Lamb who has described this phenomenon theoretically [10.29]. The absorption coefficient $\alpha(\omega)$ for this standing wave case can be immediately obtained from (3.83), if we put $\omega_s = \omega$ and $\underline{k}_s = \underline{k}_+ + \underline{k}_- = 0$. The width of this Lamb dip is determined by the homogeneous width γ_s, its depth by the intensity of the saturating field, expressed by the saturation parameter S_0. For homogeneous line profiles this Lamb dip does not occur since the whole line profile is homogeneously saturated (Fig.3.19c).

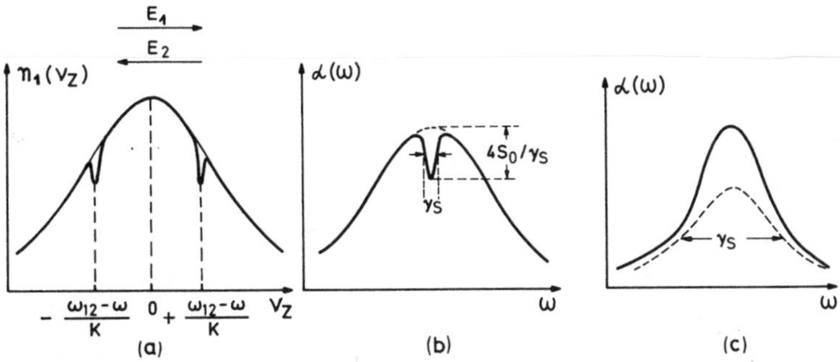

Fig.3.19a-c. Saturation by a monochromatic standing wave. (a) Population distribution $n(v_z)$ for $\omega \neq \omega_0$. (b) Absorption coefficient $\alpha(\omega)$ of an inhomogeneously broadened line, as obtained when tuning the frequency ω of the standing wave across the line profile. (c) Saturation of a homogeneous absorption profile

3.7 Spectral Line Profiles in Liquids and Solids

Many different types of lasers use liquids or solids as amplifying media. Since the spectral characteristics of such lasers play a significant role in applications of laser spectroscopy, we briefly outline the spectral linewidths of optical transitions in liquids and solids. Because of the large densities compared with the gaseous state, the mean relative distances $R(A,B_j)$ between an atom or molecule A and its surrounding partners B_j is very small (typically a few angströms), and the interaction between A and the adjacent partners B is accordingly large.

In general the atoms or molecules used for laser action are diluted to small concentrations in liquids or solids. Examples are the dye laser, where dye molecules are dissolved in organic solutions at concentrations of 10^{-4} - 10^{-3} moles/liter, or the ruby laser, where the concentration of the active Cr^{3+} ions in Al_3O_3 is of the order of 10^{-3}. The optically pumped laser molecules A^* interact with their surrounding host molecules B. The resulting broadening of the excited levels of A^* depends on the total electric field produced at the location of A by all adjacent molecules B_j, and on the dipole moment or the polarizability of A^*. The linewidth $\Delta\omega_{ik}$ of a transition $A^*(E_i) \rightarrow A^*(E_k)$ is determined by the difference of the level shifts $(\delta E_i - \delta E_k)$.

In liquids, the distances $R_j(A^*,B_j)$ show random fluctuations analogous to the situation in a high-pressure gas. The linewidth $\Delta\omega_{ik}$ is therefore determined by the probability distribution $P(R_j)$ of the mutal distances $R_j(A^*,B_j)$ and the correlation between the phase perturbations at A^* caused by elastic collisions during the lifetime of the levels E_i, E_k (see the analogous discussion in Sect.3.3).

Inelastic collisions of A^* with molecules B of the liquid host may cause radiationless transitions from the level E_i populated by optical pumping to lower levels E_n. These radiationless transitions shorten the lifetime of E_i and cause collision line broadening. In liquids the mean time between successive inelastic collisions is of the order of 10^{-11} - 10^{-13} s and the spectral lines $E_i \rightarrow E_k$ are therefore greatly broadened with a *homogeneously broadened profile*. When the line broadening becomes larger than the separation of the different spectral lines, a broad continuum arises. In the case of molecular spectra with their many closely spaced rotational-vibrational lines within an electronic transition, such a continuum inevitably appears since the broadening at liquid densities is always much larger than the line separation.

Examples of such continuous absorption and emission line profiles are the optical dye spectra in organic solvents, such as the spectrum of rhodamine 6G shown in Fig.3.20, together with a schematic level diagram. The optically pumped level E_i is collisionally deactivated by radiationless

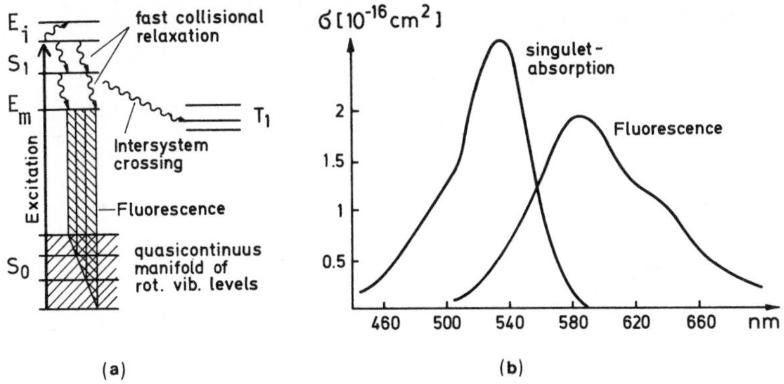

Fig.3.20a,b. Absorption and emission spectrum of rhodamine 6G, dissolved in ethanol (a). Schematic level diagram to illustrate radiative and radiationless transitions (b)

transitions to the lowest vibrational level E_m of the excited electronic state. The fluorescence starts therefore from E_m instead of E_i and ends on various vibrational levels of the electronic ground state (see Fig.3.19b). The emission spectrum is therefore shifted to larger wavelengths compared with the absorption spectrum (see Fig.3.19b).

In crystalline solids the electric field $E(\underline{R})$ at the location $\underline{R}$ of the excited molecule A^* has a symmetry depending on that of the host lattice. Because the lattice atoms perform vibrations with amplitudes depending on the temperature T, the electric field will vary in time and the time average $\langle E(T, t, \underline{R})\rangle$ will depend on temperature and crystal structure [3.20]. Since the oscillation period is short compared with the mean lifetime of $A^*(E_i)$, these vibrations cause a homogeneous line broadening for the emission or absorption of the atom A. If all atoms are placed at completely equivalent lattice points of an ideal lattice, the total emission or absorption of all atoms A on a transition $E_i \rightarrow E_k$ would be homogeneously broadened.

However, in reality it often happens that the different atoms A are placed at nonequivalent lattice points with nonequal electric fields. This is particularly true in amorphous solids or in supercooled liquids such as glass which have no regular lattice structure. In such cases the line centers ω_{0j} of the homogeneously broadened lines from the different atoms A_j are placed at different frequencies. *The total emission or absorption forms an inhomogeneously broadened line profile* which is composed of homogeneous subgroups. This is completely analogous to the gaseous case of

Doppler broadening, although the resultant linewidth in solids may be larger by several orders of magnitude. An example of such inhomogeneous line broadening is the emission of excited neodymium ions in glass which is used in the Nd-glass laser.

Problems to Chapter 3

3.1. Determine the natural linewidth, the Doppler width, pressure broadening and shifts for the neon transition $3s_2 \rightarrow 2p_4$ at $\lambda = 632.8$ nm in an He-Ne discharge at $p_{He} = 1$ torr, $p_{Ne} = 0.2$ torr at a gas temperature of 400 K. The relevant data are: $\tau(3s_2) = 58$ ns, $\tau(2p_4) = 18$ ns, $\sigma_B(Ne - He) = 6 \times 10^{-14}$ cm^2, $\sigma_s(Ne - He) \approx 1 \times 10^{-14}$ cm^2.

3.2. What is the dominant broadening mechanism for absorption lines in the following examples:

a) The output from a CO_2 laser with 50 W at $\lambda = 10$ μm is focussed into an absorbing probe of SF_6 molecules. Beam waist 0.5 mm diameter, T = 300 K, p = 100 torr, $\sigma_B = 5 \times 10^{-14}$ cm^2.

b) Radiation from a star passes through an absorbing atomic hydrogen cloud with N = 100 cm^{-3}, T = 10 K, and a path length of 3×10^9 km. The oscillator strength for the $\lambda = 21$ cm line is f = 5.7×10^{-12}, for the Lyman α line at $\lambda = 121.6$ nm is f = 0.416. Note that the cloud is optically thick for the Lyman α radiation.

c) The expanded beam from an He-Ne laser at $\lambda = 3.39$ μm is sent through a methane cell. T = 300 K, p = 0.1 torr, beam diameter 1 cm. The absorbing CH_4 transition is from the vibrational ground state ($\tau \approx \infty$) to an excited vibrational level with $\tau \approx 20$ ms.

3.3. Calculate the minimum beam diameter which is necessary to bring time of flight broadening in problem 3.2c below the natural linewidth.

3.4. The sodium D-line at $\lambda = 589$ nm has a natural linewidth of 20 MHz. How far away from the line center do the wings of the Lorentzian line profile exceed the Doppler profile at T = 500 K? Calculate the intensity $g(\omega - \omega_0)$ of the Lorentzian at this frequency ω relative to the line center ω_0.

3.5. A cw dye laser is tuned to the center of the Na D-line. At which laser intensity is the power broadening equal to half of the Doppler width at 500 K?

3.6. Estimate the collision-broadened width of the Na D_2-line at λ = 589 nm due to Na-Na collisions (resonance broadening) at 1 torr atomic vapor pressure. The resonance broadening γ_{res} is caused by an r^{-n} potential term with n = 3 and can be calculated as $\gamma_{res} = Ne^2 f_{ik}/(m\omega_{ik})$. The oscillator strength is $f_{ik} = 0.65$.

3.7. A collimated beam of Na atoms which diffuse with thermal velocities out of an orifice at T = 500 K is perpendicularly crossed with a single-mode laser. What is the maximum allowed divergence angle of the beam, if the residual Doppler width shall be smaller than the natural linewidth of 20 MHz? Calculate the power broadening, when the laser power is 10 mW and the laser beam diameter 0.1 mm. How large is the time of flight broadening?

4. Spectroscopic Instrumentation

This chapter is devoted to a discussion of instruments and techniques which are of fundamental importance for measurements of wavelengths and line profiles or for sensitive detection of radiation. The optimum selection of proper equipment or the application of a new technique is often decisive for the success of an experimental investigation. Since the development of spectroscopic instrumentation has shown great progress in recent years it is most important for any spectroscopist to be informed about the state of the art regarding sensitivity, spectral resolving power, and signal-to-noise ratios attainable with modern equipment.

At first we discuss the basic properties of *spectrographs* and *monochromators*. Although for many experiments in laser spectroscopy these instruments can be replaced by monochromatic tunable lasers (see Chaps.6-8) they are still indispensible for the solution of quite a number of problems in spectroscopy.

Probably the most important instrument in laser spectroscopy is the *interferometer*, applicable in its various modifications to numerous problems. We therefore treat these devices in somewhat more detail. Recently new techniques of measuring laser wavelengths with high accuracy have been developed which are mainly based on interferometric devices. Because of their relevance in laser spectroscopy they will be discussed in a separate section.

Great progress has also been achieved in the field of low-level signal detection. Apart from new photomultipliers with an extended spectral sensitivity range and large quantum efficiencies, new detection instruments have been developed such as image intensifiers, infrared detectors, or optical multichannel analyzers, which could escape from classified military research into the open market. For many spectroscopic applications they prove to be extremely useful.

4.1 Spectrographs and Monochromators

Spectrographs were the first instruments for measuring wavelengths and they still hold their position in spectroscopic laboratories, particularly when equipped with modern accessories such as computerized microdensitometers. Spectrographs are optical instruments which form images $S_2(\lambda)$ of an entrance slit S_1 which are laterally separated for different wavelengths λ of the incident radiation (see Fig.2.10). This lateral dispersion is achieved either by spectral dispersion in prisms or by diffraction on plane or concave reflection gratings.

Figure 4.1 shows the schematic arrangement of optical components in a *prism spectrograph*. The light source L illuminates the entrance slit S_1 which is placed in the focal plane of the collimator lens L_1. Behind L_1 the parallel light beam passes through the prism P where it is difracted by an angle $\theta(\lambda)$ depending on the wavelength λ. The camera lens L_2 forms an image $S_2(\lambda)$ of the entrance slit S_1. The position $x(\lambda)$ of this image in the focal plane of L_2 is a function of the wavelength λ. The *linear dispersion* $dx/d\lambda$ of the spectrograph depends on the spectral dispersion $dn/d\lambda$ of the prism material and on the focal length of L_2.

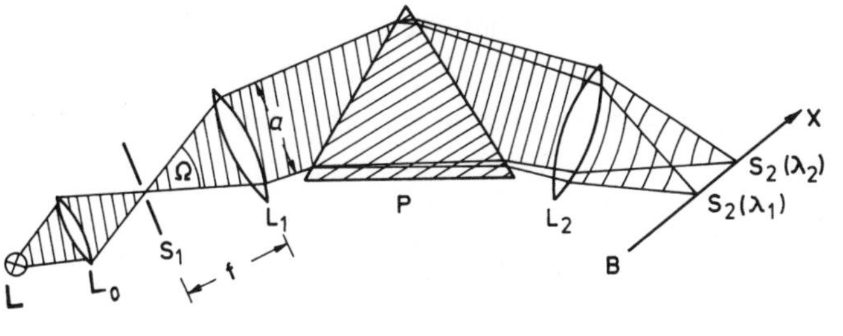

Fig.4.1. Prism spectrograph

When a reflecting diffraction grating is used to separate the spectral lines $S_2(\lambda)$, the two lenses L_1 and L_2 are commonly replaced by two spherical mirrors M_1 and M_2 which image the entrance slit onto the plane of observation (see Fig.4.2). Both systems can use either photographic or photoelectric recording. According to the kind of detection we distinguish between *spectrographs* and *monochromators*.

In spectrographs a photoplate or photographic film is placed in the focal plane of L_2 or M_2. The whole spectral range $\Delta\lambda = \lambda_1(x_1) - \lambda_2(x_2)$ covered by the lateral extension $\Delta x = x_1 - x_2$ of the photoplate can be simultaneously

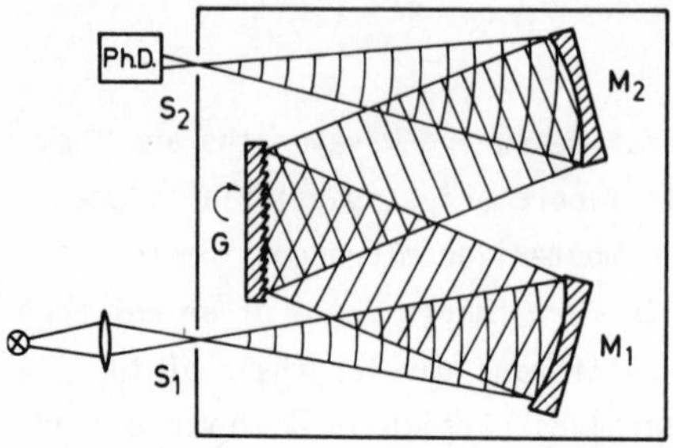

Fig.4.2.
Grating spectrometer

recorded. If the exposure of the plate remains within the linear part of the photographic density range, the density $D_a(x)$ of the developed photoplate at the position $x(\lambda)$

$$D_a(x) = C(\lambda) \int_0^T I(\lambda)dt \tag{4.1}$$

is proportional to the spectral irradiance $I(\lambda)$ as received in the focal plane B, integrated over the exposure time T. The sensitivity factor $C(\lambda)$ depends on the wavelength λ and furthermore on the developing procedure and the history of the photoplate (presensitizing). The photoplate can accumulate the incident radiant power over long periods (up to 50 hours). In astrophysics, for instance, presensitized photographic plates are still the most often used detectors for spectrally dispersed radiation from distant stars or galaxies. Photographic detection can be used for both pulsed and cw light sources. The spectral range is limited by the spectral sensitivity of available photoplates and covers the region between about 200-1000 nm.

Monochromators, on the other hand, use photoelectric recording. An exit slit S_2, selecting an interval Δx_2 in the focal plane B, lets only a limited range $\Delta\lambda$ through to the photoelectric detector. Different spectral ranges can be detected by shifting S_2 in the x direction. A more convenient solution (which is also easier to construct) turns the prism or grating by a gear-box drive, wich allows the different spectral regions to be tuned across the fixed exit slit S_2. Unlike the spectrograph, different spectral regions are detected not simultaneously, but successively. The signal received by the detector is proportional to the product of the exit-slit area $h\Delta x_2$ with the spectral intensity $\int I(\lambda)d\lambda$, where the integration extends over the spectral range dispersed within the width Δx_2 of S_2.

Whereas the spectrograph allows the simultaneous measurement of a large spectral region with moderate time resolution, photoelectric detection allows large time resolution but permits, for a given spectral resolution, only a small wavelength interval $\Delta\lambda$ to be measured at a time. With integration times below one minute, photoelectric recording shows a higher sensitivity, while for longer detection times, photoplates may be more convenient.

In spectroscopic literature [4.1-5] the name *spectrometer* is often used for both types of instruments. The introduction of optical multichannel analyzers (see Sect.4.5.9) as detectors behind spectrographs combines the advantages of high spectral and time resolution with simultaneous detection of extended spectral ranges.

We now discuss the basic properties of spectrometers, relevant for laser spectroscopy. For a more detailed treatment see for instance [4.1-4].

4.1.1 Basic Properties

The selection of the optimum type of spectrometer for a particular experiment is guided by some basic characteristics of spectrometers and their relevance for the particular application. The basic properties which are important not only for spectrographs but for all dispersive optical instruments may be listed as follows:

1) The light-gathering power or *étendue*, often called the "speed" of the instrument. It is determined by the maximum acceptance angle for the incident radiation, measured by the ratio a/f of diameter a to focal length f of the collimating lens L_1 or of the mirror M_1 (Figs.4.1,2).

2) The spectral transmittance $T(\lambda)$ of the instrument, which is limited by the transparency of the lenses and prism in the prism spectrograph or by the reflectivity $R(\lambda)$ of the mirrors and grating in grating spectrographs.

3) The spectral resolving power $\lambda/\Delta\lambda$ which specifies the minimum separation $\Delta\lambda$ of two spectral lines that can just be resolved.

4) The free spectral range of the instrument, i.e., the wavelength range $\delta\lambda$ in which the wavelength λ can be unambigiuously determined from the position $x(\lambda)$.

1) The Speed of a Spectrometer

When the spectral radiant intensity I^*_λ within the solid angle of one steradian is incident on the entrance slit of area A, a spectrometer with an acceptance angle Ω transmits the radiant flux within the spectral interval $d\lambda$,

$$\phi_\lambda d\lambda = I^*_\lambda A\Omega T(\lambda)d\lambda \quad . \tag{4.2}$$

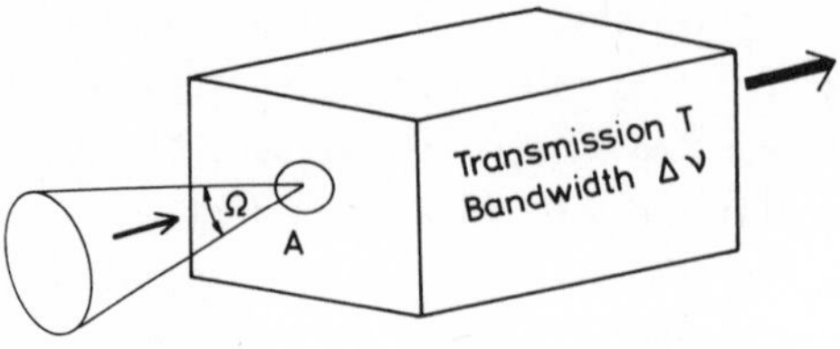

Fig.4.3. Light-gathering power of a spectrometer

The product $U = A\Omega$ is often named *étendue* (Fig.4.3). For the prism spectrograph the maximum solid angle of acceptance, $\Omega = F/f_1^2$, is limited by the effective area $F = h \cdot a$ of the prism which represents the limiting aperture with height h and width a for the light beam (Fig.4.1).

Example

For a prism of height h = 6 cm, a = 6 cm, f_1 = 30 cm ⇒ a/f = 1:5 and Ω = 0.04.

It is advantageous to image the light source onto the entrance slit in such a way that the acceptance angle Ω is fully used (see Fig.4.4). Although more radiant power from an extended source can pass the entrance slit by using a converging lens to reduce the source image on the entrance slit, the divergence is increased and the radiation outside the acceptance angle cannot be detected but may increase the background by scattering from lens holders and spectrometer walls.

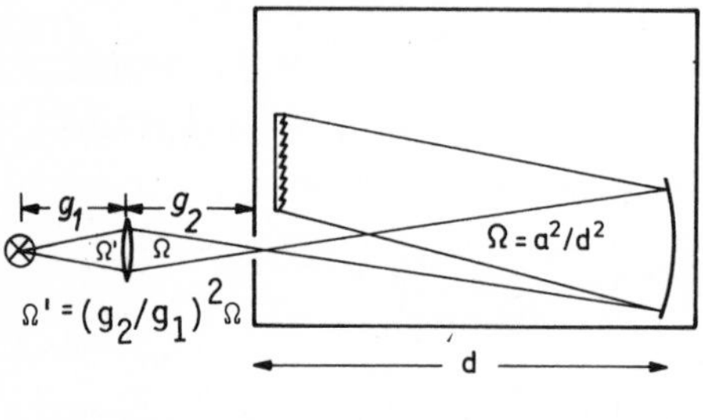

Fig.4.4. Optimized imaging of a light source onto the entrance slit of a spectrometer is achieved when the solid angle Ω of the incoming light matches the acceptance angle $(a/d)^2$ of the spectrometer

2) The Spectral Transmission

For prism spectrometers, the spectral transmission depends on the material of the prism and lenses. Using fused quartz the accessible spectral range spans from about 180 to 3000 nm. Below 180 nm (vacuum-ultraviolet region) the whole spectrograph has to be evacuated and lithium fluoride or calcium

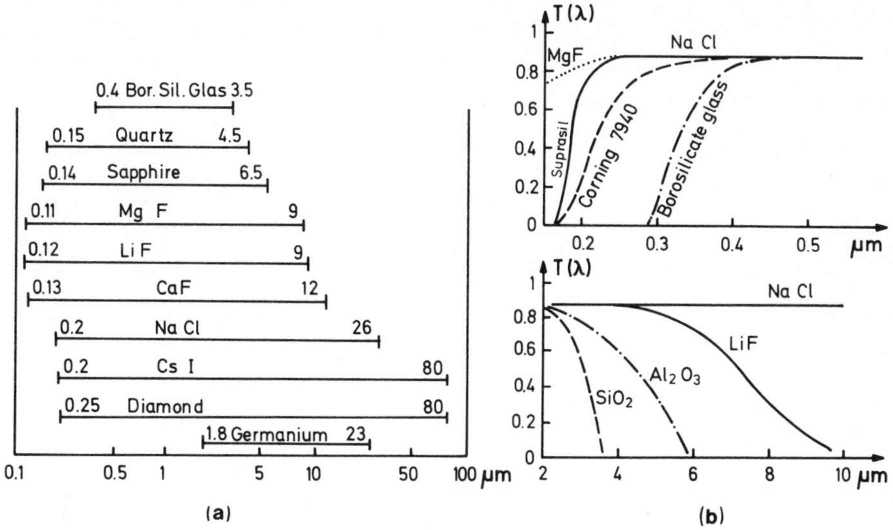

Fig.4.5. Useful spectral ranges of different optical materials (a) and transmittance of different materials with thicknesses 1 cm (b). [4.5b]

fluoride has to be used for the prism and the lenses, although most vuv spectrometers are equipped with reflection gratings and mirrors.

In the infrared region, several materials (for example, CaF_2 NaCl, and KBr crystals) are transparent up to 30 μm (see Fig.4.5). However, because of the high reflectivity of metallic coated mirrors and gratings in the infrared region, grating spectrometers with mirrors are preferentially used rather than prism spectrographs.

Many vibrational-rotational transitions of molecules such as H_2O or CO_2 fall within the range between 3-10 μm causing selective absorption of the transmitted radiation. Infrared spectrometers therefore have to be either evacuated or filled with dry nitrogen. Because dispersion and absorption are closely related (see Sect.2.6), prism materials with low absorption losses also show low dispersion, resulting in a limited resolving power (see below).

Since the ruling or holographic production of high-quality gratings has nowadays reached a high technological standard, most spectrometers used today are equipped with diffraction gratings rather than prisms. The spectral transmission of grating spectrometers reaches from the vuv region into the far infrared. The design and the coatings of the optical components and the geometry of the optical arrangement are optimized according to the specified wavelength region.

3) The Spectral Resolving Power

The spectral resolving power of any dispersing instrument is defined by the expression

$$R = |\lambda/\Delta\lambda| = |\nu/\Delta\nu| \quad , \tag{4.3}$$

where $\Delta\lambda = \lambda_1 - \lambda_2$ stands for the minimum separation of the central wavelengths λ_1 and λ_2 of two closely spaced lines which are considered to be just resolved. It is possible to recognize that an intensity distribution is composed of two lines with intensity profiles $I_1(\lambda - \lambda_1)$ and $I_2(\lambda - \lambda_2)$ if the total intensity $I(\lambda) = I_1(\lambda - \lambda_1) + I_2(\lambda - \lambda_2)$ shows a pronounced dip between two maximum (Fig.4.6). The intensity distribution $I(\lambda)$ depends of course on the ratio I_1/I_2 and on the profiles of both components and therefore the minimum resolvable interval $\Delta\lambda$ will differ for different profiles.

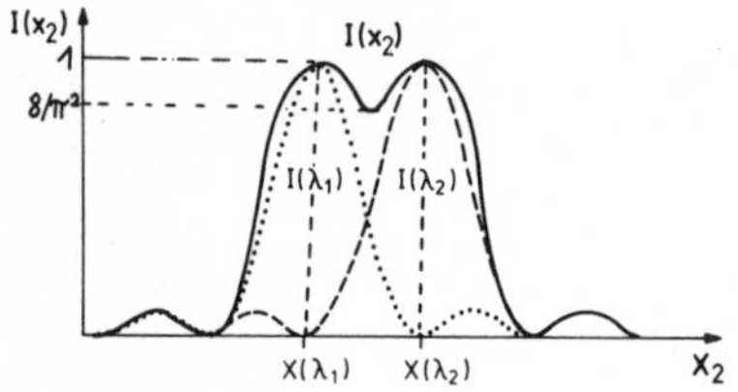

Fig.4.6. Rayleigh's criterion for the resolution of two nearly overlapping lines

Lord Rayleigh has somewhat arbitrary introduced a criterion of resolution in connection with prism and grating spectrometers, where the line profiles at the maximum attainable resolution are determined by diffraction and are of the form $I(\lambda - \lambda_0) = I(\lambda_0)[\sin((\lambda - \lambda_0)/2)(\lambda - \lambda_0)/2]^2$ (see below). In this case two lines are considered to be just resolved if the central diffraction maximum of $I_1(\lambda - \lambda_1)$ coincides with the first minimum of $I_2(\lambda - \lambda_2)$ [4.5a]. As shown below, the total intensity distribution $I(\lambda)$ exhibits in this case for $I_1 = I_2$ a dip between the two maxima which drops to $(8/\pi^2)$ of the maximum intensity I_{max} (Fig.4.6). Generalizing this *Rayleigh criterion* to arbitrary line profiles, we may define the resolving power for any dispersing instrument by defining two lines with equal intensities to be just resolved if the dip between the two maxima drops to $(8/\pi^2) \approx 0.8$ of I_{max}. With this general definition the minimum resolvable interval $\Delta\lambda$ depends of course on the line profiles.

The attainable spectral resolving power of a spectrometer is determined by its angular dispersion. When passing the dispersing element (prism or grating), a parallel beam composed of two monochromatic waves with wave-

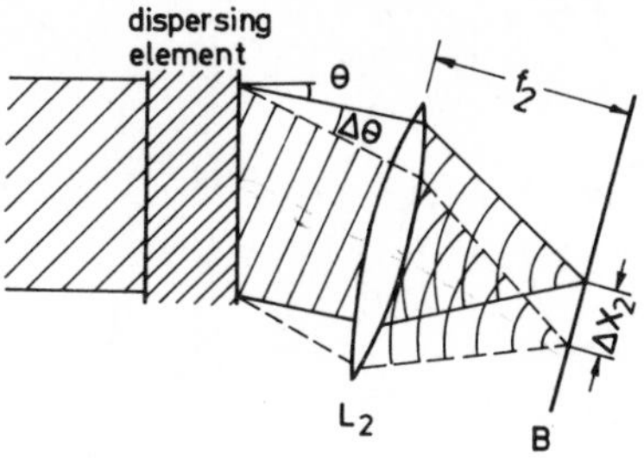

Fig.4.7. Angular dispersion of a parallel beam

lengths λ and λ + Δλ is split into two partial beams with angular deviations θ and θ + Δθ from their initial direction (see Fig.4.7).

$$\Delta\theta = (d\theta/d\lambda)\Delta\lambda \quad , \tag{4.4}$$

where $d\theta/d\lambda$ is called the *angular dispersion* [rad/nm]. Since the camera lens with focal length f_2 images the entrance slit S_1 into the plane B (see Fig.4.1) the distance Δx_2 between the two images $S_2(\lambda)$ and $S_2(\lambda + \Delta\lambda)$ is, according to Fig.4.7,

$$\Delta x_2 = f_2 \cdot \Delta\theta = f_2 \frac{d\theta}{d\lambda} \Delta\lambda = \frac{dx}{d\lambda} \Delta\lambda \ . \tag{4.5}$$

The factor $dx/d\lambda$ *is called the linear dispersion* of the instrument. It is generally measured in mm/A. In order to resolve two lines at λ and λ + Δλ, the separation Δx_2 in (4.5) has to be at least the sum $\delta x_2(\lambda) + \delta x_2(\lambda + \Delta\lambda)$ of the widths of the two slit images. Since the width x_2 is related to the width of the entrance slit by

$$\delta x_2 = (f_2/f_1)\delta x_1 \quad , \tag{4.6}$$

the resolving power $\lambda/\Delta\lambda$ can be increased by decreasing δx_1. Unfortunately there is a theoretical limitation set by diffraction. Because of the fundamental importance of this resolution limit we discuss this point in more detail.

When a parallel light beam passes a limiting aperture with diameter a, a Fraunhofer diffraction pattern is produced in the plane of the focussing lens L_2 (see Fig.4.8). The intensity distribution $I(\varphi)$ as a function of the angle φ with the optical axis of the system is given by the well-known formula [2.3]

$$I(\varphi) = I_0 \left[\frac{\sin(a\pi\sin\varphi/\lambda)}{(a\pi\sin\varphi)/\lambda}\right]^2 \approx I_0 \left[\frac{\sin(a\pi\varphi/\lambda)}{a\pi\varphi/\lambda}\right]^2 \quad . \tag{4.7}$$

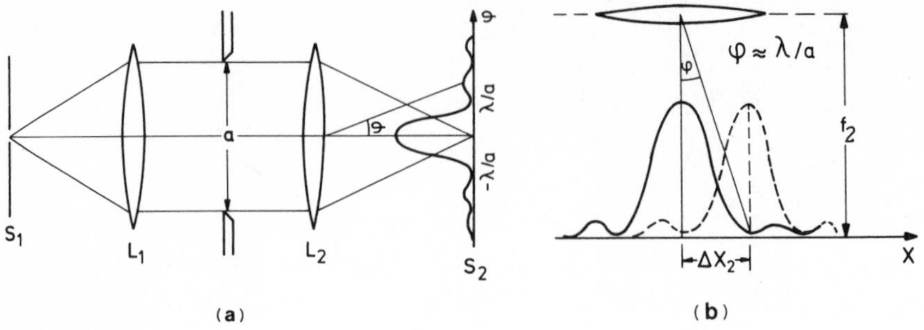

Fig.4.8. (a) Diffraction in a spectrometer by the limiting aperture with diameter a. (b) Limitation of spectral resolution by diffraction

The first two diffraction minima at $\varphi = \pm\lambda/a$ are symmetrical to the central maximum (zeroth diffraction order) at $\varphi = 0$. The central maximum contains about 90% of the total intensity.

When the entrance slit S_1 of a spectrometer with width δx_1 is illuminated by a light source and is imaged into the plane B, the image width is $\delta x_2 = (f_2/f_1)\delta x_1$ if there were no diffraction. Because of diffraction by the aperture a (which may be the lens holder or prism edges), the slit image S_2 is broadened to the intensity distribution (4.7). Even an infinitesimally small entrance slit therefore produces a slit image of width

$$\delta x_s^{\text{diffr.}} = f_2(\lambda/a) \qquad (4.8)$$

defined as the distance between the central diffraction maximum and the first minimum.

According to the Rayleigh criterion two equally intense spectral lines with wavelengths λ and $\lambda + \Delta\lambda$ are just resolved if the central diffraction maximum of $S_2(\lambda)$ coincides with the first minimum of $S_2(\lambda + \Delta\lambda)$ (see above). From (4.7) one can compute that in this case both lines partly overlap with a dip of $(8/\pi^2)I_{max}$ between the two maxima. The distance between the centers of the two slit images is then from (4.8) (see Fig.4.8b)

$$\Delta x_2 = f_2(\lambda/a) \quad . \qquad (4.9)$$

With (4.5) one therefore obtains the fundamental limit on the resolving power,

$$|\lambda/\Delta\lambda| \leqq a(d\theta/d\lambda) \quad , \qquad (4.10)$$

which clearly depends only on the size a of the limiting aperture and on the angular dispersion of the instrument.

Note, that the spectral resolution is limited, *not* by the diffraction by the entrance slit, but by the diffraction caused by the much larger aperture a, determined by the size of prism or grating.

Although it does not influence the spectral resolution, the much larger diffraction by the entrance slit imposes a limitation on the transmitted intensity at small slit widths. This can be seen as follows. When illuminated with parallel light, the entrance slit with width b produces a Fraunhofer diffraction pattern analogous to (4.7) with a replaced by b. The central diffraction maximum extends between the angles $\delta\varphi = \pm\lambda/b$ (see Fig.4.9) and can completely pass the limiting aperture a only if $2\lambda/b$ is smaller than the acceptance angle a/f_1 of the spectrometer. This imposes a lower limit to the useful width b_{min} of the entrance slit,

$$b_{min} \geqq 2\lambda f_1/a \quad . \tag{4.11}$$

In all practical cases, the incident light is divergent, which demands that the sum of divergence angle and diffraction angle has to be smaller than a/f and the minimum slit width b correspondingly larger.

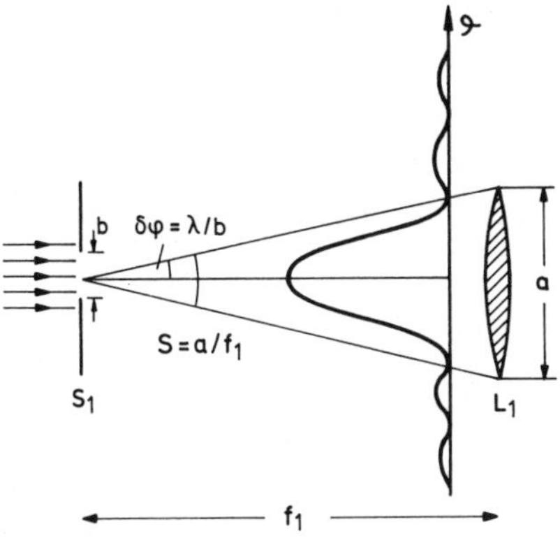

Fig.4.9. Diffraction by the entrance slit

Figure 4.10a illustrates the intensity distribution I(x) in the plane B for different slit widths b. The peak intensity $I(b)_{x=0}$ is plotted in Fig.4.10b as a function of the slit width. According to (4.2) the transmitted radiation flux $\phi(\lambda)$ depends on the product $U = A\Omega$ of entrance slit area A and acceptance angle $\Omega = (a/f_1)^2$. The flux in B would therefore depend linearly on the slit width b, if diffraction were not present (dotted line in Fig.4.10b). For monochromatic radiation the peak intensity (W/m^2) in the plane B should then be constant, for a spectral continuum it should

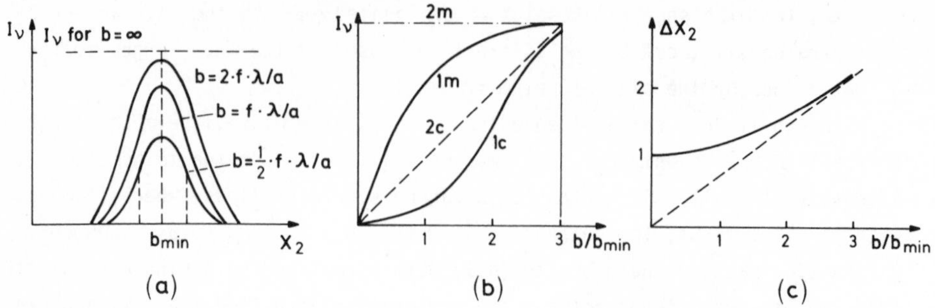

Fig.4.10. (a) Diffraction-limited intensity distribution $I(X_2)$ in the plane B for different widths b of the entrance slit. (b) Peak intensity $I(X_2 = 0)$ as a function of b for monochromatic incident light (m) and for a continuous spectrum (c) without diffraction (dashed curves 2m and 2c) and including diffraction (solid curves 1m and 1c). (c) Width $\Delta x_2(b)$ of the slit image S_2 including diffraction by the aperture a

decrease linearly with the slit width. Because of the diffraction by S_1 the intensity loss decreases more than linearly with the slit width b. Note the steep decrease for $b < b_{min}$. Figure 4.10c shows the dependence of the width $\Delta x_2(b)$ of the slit image S_2 taking into account the diffraction caused by the aperture a. This demonstrates that the resolution cannot be increased much by decreasing b below b_{min}.

Considering the finite width b of the entrance slit S_1, the minimum wave length interval $\Delta\lambda$ which can be resolved by a spectrograph with slit width b and $f_1 = f_2 = f$ can be easily derived from the discussion above to be

$$\Delta\lambda = (f\lambda/a + b)d\lambda/dx \quad . \tag{4.12}$$

Substituting $b = b_{min} = 2f\lambda/a$, the practical limit for $\Delta\lambda$ imposed by diffraction by S_1 and by the limiting aperture with width a is

$$\Delta\lambda = 3f(\lambda/a)d\lambda/dx \quad . \tag{4.13}$$

Instead of the theoretical limit (4.10) given by the diffraction through the aperture a, a practically attainable resolving power limited by the intensity loss for a slit width b below b_{min} is obtained from (4.13),

$$\boxed{R = \lambda/\Delta\lambda = (a/3)d\theta/d\lambda} \quad . \tag{4.14}$$

Example

For a = 10 cm, $\lambda = 5\times10^{-5}$ cm, f = 100 cm, $d\lambda/dx = 10$ Å/mm, with b = 10 μm, $\Rightarrow \Delta\lambda = 0.15$ Å, with b = 5 μm, $\Rightarrow \Delta\lambda = 0.10$ Å. However, from Fig.4.10, one can see that the transmitted intensity with b = 5 μm is only 25% of that with b = 10 μm.

Note

For photographic detection of line spectra, it is actually better to use the lower limit b_{min} for the width of the entrance slit, because the density of the photographic plate depends only on the spectral irradiance [W/m^2] rather than on the radiation power [W]. Increasing the slit width beyond the diffraction limit b_{min}, in fact does not significantly increase the density contrast on the plate, but does decrease the spectral resolution.

Using photoelectric recording, the detected signal depends on the radiation power $\phi_\lambda d\lambda$ transmitted through the spectrometer and therefore increases with increasing slit width. In the case of completely resolved line spectra, this increase is proportional to the slit width b since $\phi_\lambda \propto b$. For continuous spectra it is even proportional to b^2 because the transmitted spectral interval $d\lambda$ also increases proportional to b and therefore $\phi_\lambda d\lambda \propto b^2$.

The obvious idea of increasing the product of ΩA without loss of spectral resolution by keeping the width b constant but increasing the height h of the entrance slit is of limited value because imaging defects of the spectrometer cause a curvature of the slit image which again decreases the resolution (see below).

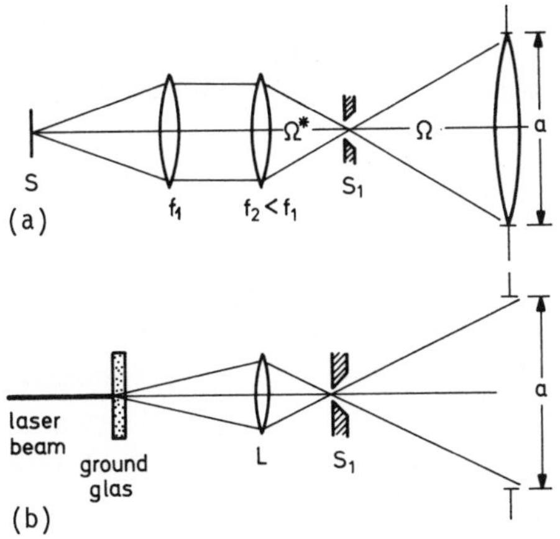

Fig.4.11. (a) Imaging of an extended light source onto the entrance slit of a spectrometer. (b) Correct imaging optics for laser wavelength measurements with a spectrometer. The laser light, scattered from the ground glass forms the source which is imaged onto the entrance slit

As has been discussed above, the optimum imaging of an extended incoherent source onto the entrance slit of the spectrometer is achieved when the solid angle of the incident light matches the acceptance angle of the spectrometer (Fig.4.11a).

Often the wavelength of lasers is measured with a spectrometer. In this case it is not recommendable to shine the laser beam directly onto the entrance slit, because the prism or grating would be not uniformely illuminated. This decreases the spectral resolution. Furthermore, the symmetry of the optical path with respect to the spectrometer axis is not guaranteed with such an arrangement. It is better to illuminate a groundglass plate with the laser and to use the incoherently scattered laser light as a secondary source which is imaged in the usual way (Fig.4.11b).

4) The Free Spectral Range

The free spectral range of a spectrometer is that wavelength interval $\delta\lambda$ of the incident radiation for which a one-valued relation exists between λ and the position $x(\lambda)$ of the entrance slit image. While for prism spectrometers the free spectral range covers the whole region of normal dispersion of the prism material, for grating spectrometers $\delta\lambda$ is determined by the diffraction order m. For light, vertically incident on a grating with a groove distance d, constructive interference occurs in the direction β for all wavelengths λ_m, which satisfy the condition

$$m\lambda_m = d\sin\beta \quad .$$

This shows that two wavelengths $\lambda_1 = d\sin\beta/m$ and $\lambda_2 = d\sin\beta/(m+1)$ appear at the same angle β in the spectrometer output. The free spectral range

$$\delta\lambda = d\sin\beta\left(\frac{1}{m} - \frac{1}{m+1}\right) = \frac{d\sin\beta}{m(m+1)}$$

therefore decreases with increasing order m.

Interferometers, which are generally used in very high orders ($m = 10^4 - 10^8$), have a high spectral resolution but a small free spectral range. For unambiguous wavelength determination they need a preselector, which allows one to measure the wavelength within $\delta\lambda$ of the high-resolution instrument.

4.1.2 Prism Spectrometer

When passing a prism, a light ray is refracted by an angle θ which depends on the prism angle ε, the angle of incidence α_1, and the refractive index n of the prism material (see Fig.4.12). The minimum deviation θ is obtained when the ray passes the prism parallel to the base g (symmetrical arrangement with $\alpha_1 = \alpha_2 = \alpha$). In this case one can derive the equation [4.5]

$$\sin\frac{\theta+\varepsilon}{2} = n\sin(\varepsilon/2) \quad . \tag{4.15}$$

From (4.15) the derivation $d\theta/dn = (dn/d\theta)^{-1}$ is

$$\frac{d\theta}{dn} = \frac{2\sin(\varepsilon/2)}{\cos((\theta+\varepsilon)/2)} = \frac{2\sin(\varepsilon/2)}{\sqrt{1-n^2\sin^2(\varepsilon/2)}} \, . \tag{4.16}$$

The *angular dispersion* $d\theta/d\lambda = (d\theta/dn)(dn/d\lambda)$ is therefore

$$\frac{d\theta}{d\lambda} = \frac{2\sin\varepsilon/2}{\sqrt{1-n^2\sin^2(\varepsilon/2)}}\frac{dn}{d\lambda} \quad . \tag{4.17}$$

This shows that the angular dispersion increases with the prism angle ε *but does not depend on the size of the prism.*

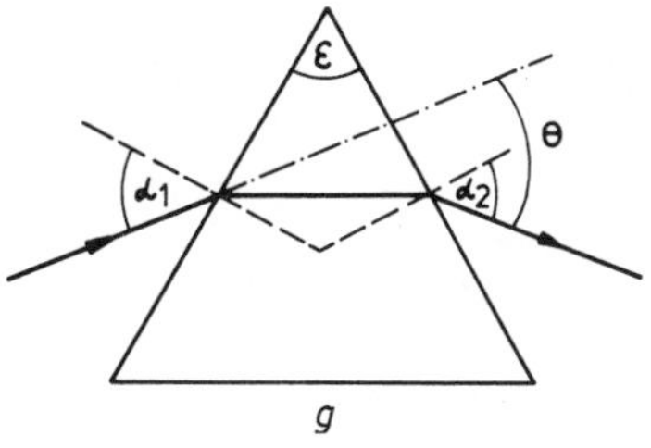

Fig.4.12. Refraction of light by a prism at minimum deviation where $\alpha_1 = \alpha_2 = \alpha$ and $\theta = 2\alpha - \varepsilon$

For the deviation of laser beams with small beam diameters small prisms can therefore be used without losing angular dispersion. In a prism spectrometer, however, the size of the prism determines the limiting aperture a and has to be large in order to achieve a large spectral resolving power (see previous section). For given angular dispersion, an equilateral prism with $\varepsilon = 60^\circ$ uses the smallest quantity of prism material (which might be quite expensive). Because $\sin 30^\circ = 1/2$, (4.17) then reduces to

$$\frac{d\theta}{d\lambda} = \frac{dn/d\lambda}{\sqrt{1-(n/2)^2}} \quad . \tag{4.18}$$

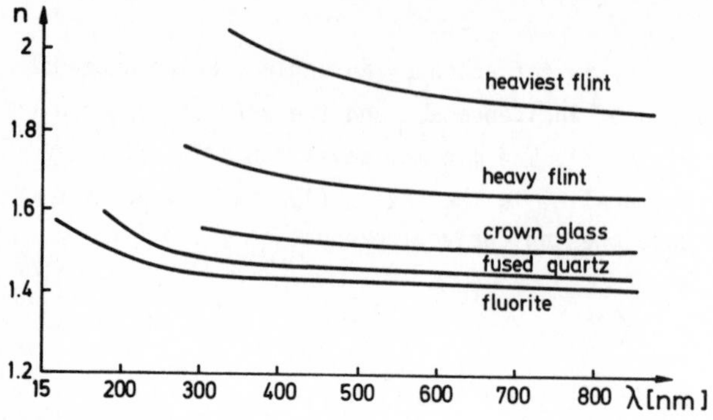

Fig.4.13. Refraction index $n(\lambda)$ for some prism materials

The spectral dispersion $dn/d\lambda$ is a function of prism material and wavelength λ. Figure 4.13 shows dispersion curves $n(\lambda)$ for some materials commonly used for prisms. Since the refractive index increases rapidly in the vicinity of absorption lines (see Fig.2.16), glass has a larger dispersion in the visible and near ultraviolet regions than quartz which, on the other hand, can be used advantageously in the uv down to 180 nm. In the vacuum-ultraviolet range CaF, MgF, or LiF prisms are sufficiently transparent. Table 4.1 gives a summary of the optical characteristics and useful spectral ranges of some prism materials.

Table 4.1.

Material	Useful spectral range [μm]	Refractive index n	Dispersion $dn/d\lambda$ [nm^{-1}]
Glass (BK7)	0.35 - 3.5	1.516	4.6×10^{-5} at 589 nm
		1.53	1.1×10^{-4} at 400 nm
Heavy flint	0.4 - 2	1.755	1.4×10^{-4} at 589 nm
		1.81	4.4×10^{-4} at 400 nm
Fused quartz	0.15 - 4.5	1.458	3.4×10^{-5} at 589 nm
		1.470	1.1×10^{-4} at 400 nm
NaCl	0.2 - 26	1.79	6.3×10^{-3} at 200 nm
		1.38	1.7×10^{-5} at 20 μm
LiF	0.12 - 9	1.44	6.6×10^{-4} at 200 nm
		1.09	8.6×10^{-5} at 10 μm

Examples

a) Suprasil (fused quartz) has a refractive index $n = 1.47$ at $\lambda = 400$ nm and $dn/d\lambda = 1100\ cm^{-1}$. This gives $d\theta/d\lambda = 1.6\times10^{-4}$ rad/nm.

b) For heavy flint glass at 400 nm $n = 1.81$ and $dn/d\lambda = 4400\ cm^{-1}$, giving $d\theta/d\lambda = 1.2\times10^{-3}$ rad/nm. This is about 8 times larger than for quartz. With a focal length $f = 100$ cm for the camera lens one achieves a linear dispersion $dx/d\lambda = 0.12$ mm/Å with the flint prism, but only 0.015 mm/Å with the quartz prism.

The resolving power $\lambda/\Delta\lambda$ is according to (4.10)

$$\lambda/\Delta\lambda \leqq a(d\theta/d\lambda) \quad .$$

The diameter a of the limiting aperture in a prism spectrograph is (see Fig.4.14)

$$a = d\ \cos\alpha_1 = g\ \cos\alpha/(2\ \sin\varepsilon/2) \quad . \qquad (4.19)$$

Substituting $d\theta/d\lambda$ from (4.17) gives

$$\lambda/\Delta\lambda = \frac{g\ \cos\alpha_1}{\sqrt{1 - n^2\ \sin^2\varepsilon/2}}\ \frac{dn}{d\lambda} \quad . \qquad (4.20)$$

At minimum deviation, (4.15) gives $n\ \sin\varepsilon/2 = \sin(\theta + \varepsilon)/2 = \sin\alpha_1$ and therefore (4.20) reduces to

$$\boxed{\lambda/\Delta\lambda = g(dn/d\lambda)} \quad . \qquad (4.21a)$$

According to (4.21a), the theoretical maximum resolving power depends solely on the base length g and on the spectral dispersion of the prism material. Because of the finite slit width $b \geqq b_{min}$ the resolution, reached in practice, is somewhat lower and the corresponding resolving power can be derived from (4.12) to be at most

$$R = \lambda/\Delta\lambda \leqq \frac{1}{3}\ g(dn/d\lambda) \quad . \qquad (4.21b)$$

Image defects may further decrease the resolution. One of these defects is the curvature of the image of a straight entrance slit (Fig.4.15). Rays from the ends of the entrance slit pass the prism at a small inclination to the principal axis. This causes a larger angle of incidence α_2 which exceeds that of minimum deviation. These rays are therefore refracted by a larger angle θ and the image of a straight slit becomes curved towards shorter wavelengths.

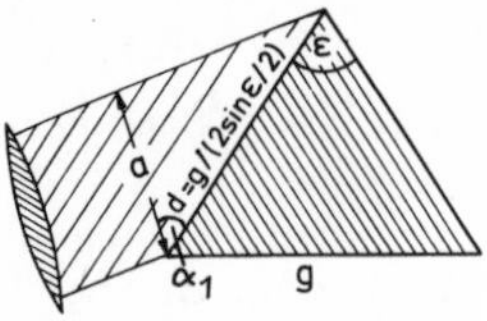

Fig.4.14. Limiting aperture in a prism spectrometer

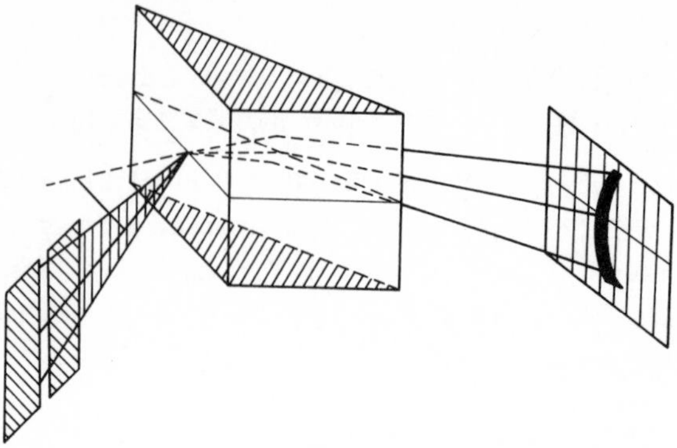

Fig.4.15. Curvature of the slit image

Since the deviation in the plane B is equal to $f_2\theta$, the radius of curvature is of the same order of magnitude as the focal length of the camera lens and increases with increasing wavelength because of the decreasing spectral dispersion.

If achromatic lenses (which are expensive in the infrared and ultraviolet region) are not used, the focal length of the two lenses decreases with the wavelength. This can be partly compensated by inclining the plane B against the principal axis in order to bring it at least approximately into the focal plane of L_2 for a large wavelength range (see Fig.4.1).

4.1.3 Grating Spectrometer

In a grating spectrometer (Fig.4.2) the collimating lens L_1 is replaced by a spherical mirror M_1 with the entrance slit S_1 in the focal plane of M_1. The collimated, parallel light is reflected by M_1 onto a reflection grating consisting of many straight grooves (about 10^5) parallel to the entrance slit. The grooves have been ruled onto an optically smooth glass substrate or have been produced by holographic techniques [4.6]. The whole grating surface is coated with a highly reflecting layer (metal or dielectric film). The light reflected from the grating is focussed by a spherical mirror M_2 onto the exit slit S_2 or onto a photographic plate in the focal plane of M_2.

The many grooves, which are illuminated coherently, can be regarded as small radiation sources, each of them diffracting the light incident onto this small groove with a width of about the wavelength λ, into a large range of angles β around the direction of geometrical reflection. The total reflected light consists of a coherent superposition of these many partial contributions. Only in those directions where all partial waves, emitted from

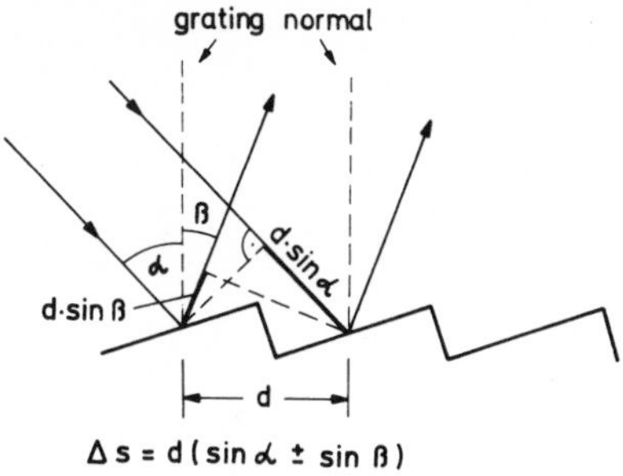

Fig.4.16. Illustration of the grating equation (4.22)

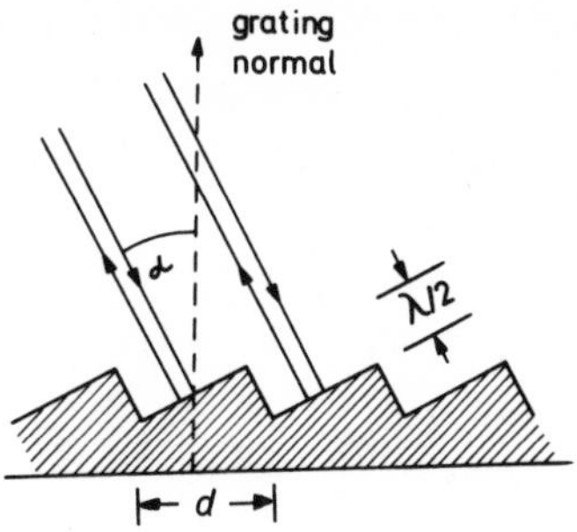

Fig.4.17. Littrow mount of a grating

the different grooves, are in phase will constructive interference result in a large total intensity, while in all other directions the different contributions cancel by destructive interference.

Figure 4.16 shows a parallel light beam incident onto two adjacent grooves. At an angle of incidence α to the grating normal (which is normal to the grating surface, but not necessary to the grooves) one obtains constructive interference for those directions β of the reflected light for which

$$\boxed{d(\sin\alpha \pm \sin\beta) = m\lambda} \quad , \qquad (4.22)$$

the plus sign has to be taken if β and α are on the same side of the grating normal; otherwise the minus sign, which is the case shown in Fig.4.16.

In laser spectroscopic applications the case $\alpha = \beta$ often occurs, which means that the light is reflected back into the direction of the incident light. For such an arrangement, called a *Littrow-grating mount* (shown in Fig.4.17), the condition (4.22) for constructive interference reduces to

$$2d \sin\alpha = m\lambda \quad . \qquad (4.23)$$

The Littrow grating acts as a wavelength-selective reflector because light is only reflected if the incident wavelength satisfies the condition (4.23).

We now examine the intensity distribution $I(\beta)$ of the reflected light, when a monochromatic plane wave is incident onto the grating. At normal incidence ($\alpha = 0$) of a plane wave

$$E = A\, e^{i(\omega t - kz)} \quad ,$$

the path difference between partial waves reflected by adjacent grooves is $\Delta s = d \sin\beta$, and the corresponding phase difference is

$$\delta = (2\pi d \sin\beta)/\lambda \quad . \tag{4.24}$$

The superposition of the amplitudes reflected from all N grooves in the direction β gives the total reflected amplitude

$$A_R = \sqrt{R} \sum_{m=0}^{N} A_g\, e^{-im\delta} = \sqrt{R}\, A_g \frac{1 - e^{-iN\delta}}{1 - e^{-i\delta}} \quad , \tag{4.25}$$

where $R(\beta)$ is the reflectivity of the grating, which depends on the reflection angle β, and A_g is the amplitude of the partial wave incident onto each groove. Because the intensity of the reflected wave is related to its amplitude by $I_R = \varepsilon_0 c A_R A_R^*$ [see (2.30c)], we find from (4.25)

$$I_R = RI_0 \frac{\sin^2(N\delta/2)}{\sin^2(\delta/2)} \quad \text{with} \quad I_0 = c\varepsilon_0 A_g A_g^* \quad . \tag{4.26}$$

This intensity distribution is plotted in Fig.4.18 for two different values of the total groove number N. The principal maxima occur for $\delta = 2m\pi$, which is, according to (4.24) equivalent to

$$d \sin\beta = m\lambda \quad . \tag{4.22a}$$

This is the grating equation (4.22) for the special case $\alpha = 0$ and means that the path difference between partial beams from adjacent grooves is an integer multiple of the wavelength. The integer m is called the order of the interference. The function (4.26) has (N-1) minima with $I_R = 0$ between to successive principal maxima. These occur at values of δ for which $N\delta/2 = l\pi$, $l = 1,2 \ldots N-1$, and mean that for each groove of the grating another one can be found which emits light into the direction β with a phase shift π, such that all pairs of partial waves just cancel.

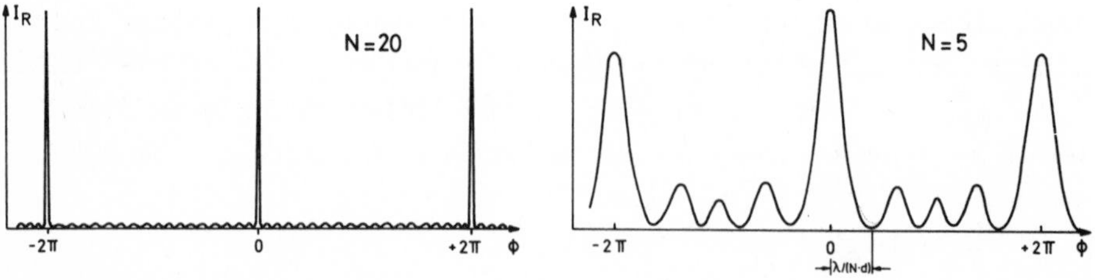

Fig.4.18. Intensity distribution $I(\beta)$ for two different numbers of grooves

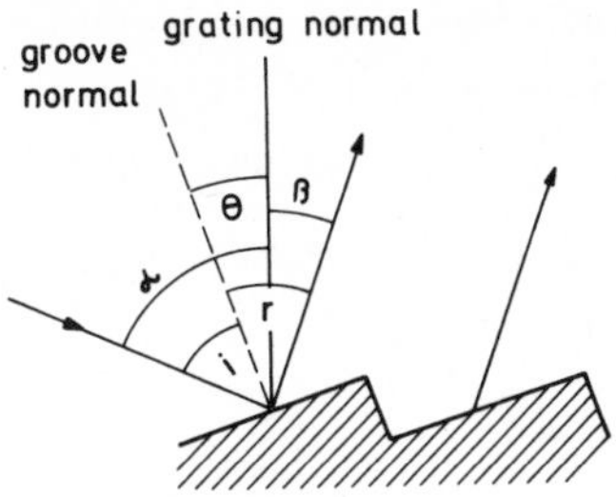

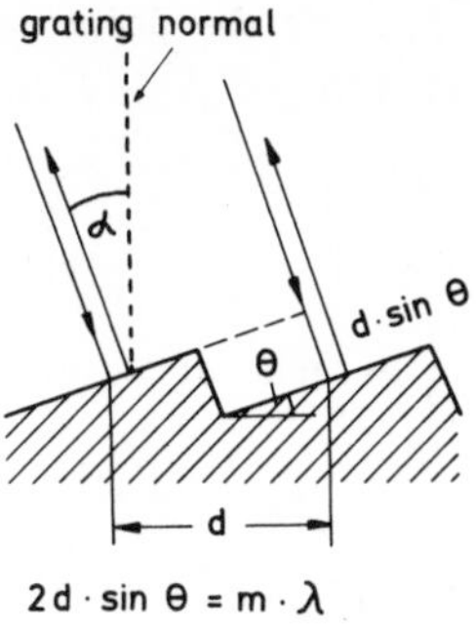

Fig.4.19a,b. Illustration of blaze angle. (a) Declined incidence for commonly used grating mounts, (b) normal incidence for Littrow mount

The intensity of the N-2 small maxima, which are caused by incomplete destructive interference, decreases proportional to 1/N with increasing groove number N. Figure 4.18 illustrates that for gratings used in practical spectroscopy, with groove numbers of about 10^5, the reflected intensity $I_R(\lambda)$ at a given wavelength λ has very sharply defined maxima only in those directions β as defined by (4.22). The small side maxima are completely negligible at such large values of N, provided the distance d between the grooves is exactly constant over the whole grating area.

The reflectivity $R(\beta,\theta)$ of a ruled grating depends on the slope θ of the grooves. If the diffraction angle β coincides with the angle r of specular reflection from the groove surfaces (see Fig.4.19), $R(\beta,\theta)$ reaches its optimum value, which depends on the reflectivity of the groove coating. From Fig.4.19 one infers in the case of specular reflection, i = r, with $i = \alpha - \theta$ and $r = \theta + \beta$ the condition for the blaze angle θ,

$$\theta = (\alpha - \beta)/2 \quad . \tag{4.27}$$

Because of the diffraction of each partial wave into a large angular range the reflectivity $R(\beta)$ will not have a sharp maximum at $\beta = \alpha - 2\theta$ but will rather show a broad distribution around this optimum angle. The angle of incidence α is determined by the particular construction of the spectrometer and the angle β for which constructive interference occurs depends on the wavelength λ. Therefore the blaze angle θ has to be specified for the desired spectral range and the spectrometer type.

Often it is advantageous to use the spectrometer in second order (m = 2) which increases the spectral resolution by a factor 2 without losing much intensity, if the blaze angle θ is correctly choosen to satisfy (4.27) and (4.22) with m = 2.

The line profile $I(\beta)$ of the principal maximum of order m can be derived from (4.26) by substituting $\beta = \beta_0 + \varepsilon$. Because for large N, $I(\beta)$ is very sharply centered around β_0, we can assume $\varepsilon \ll \beta_0$. With the relation

$$\sin(\beta_0 + \varepsilon) = \sin\beta_0 \cos\varepsilon + \cos\beta_0 \sin\varepsilon \approx \sin\beta_0 + \varepsilon\cos\beta_0 \quad ,$$

we obtain from (4.24) because of $(2\pi d/\lambda)\sin\beta_0 = 2m\pi$

$$\delta = 2m\pi + 2\pi(d/\lambda)\varepsilon\cos\beta_0 \quad ,$$

and (4.26) can be written, using again a trigonometric approximation valid for $\varepsilon \ll 1$,

$$I_R = RI_0N^2 \frac{\sin^2(N\delta/2)}{(N\delta/2)^2} \quad , \tag{4.28}$$

with $N \cdot \delta/2 = \pi N(d/\lambda)\varepsilon\cos\beta_0$. The first two minima on both sides of the central maximum at β_0 are at $N\delta = \pm 2\pi \Rightarrow \varepsilon_{1,2} = \pm\lambda/(Nd\cos\beta_0)$. The central maximum of mth order therefore has a line profile (4.28) with a base halfwidth $\Delta\beta = \lambda/(Nd\cos\beta_0)$. This corresponds to a diffraction pattern produced by an aperture with width $b = Nd\cos\beta_0$, which is just the size of the whole grating projected onto a direction normal to β_0.

Differentiating the grating equation (4.22) with respect to λ we obtain at a given angle α the angular dispersion

$$d\beta/d\lambda = m/(d\cos\beta) \quad . \tag{4.29}$$

Substituting from (4.22) $(m/d) = (\sin\alpha \pm \sin\beta)/\lambda$, we find

$$d\beta/d\lambda = \frac{\sin\alpha \pm \sin\beta}{\lambda\cos\beta} \quad . \tag{4.30a}$$

This illustrates that the angular dispersion is determined solely by the angles α and β and not by the number of grooves. For the Littrow mount with $\alpha = \beta$, we obtain

$$d\beta/d\lambda = 2\tan\alpha/\lambda \quad . \tag{4.30b}$$

The resolving power can be immediately derived from (4.30a) and the base width $\Delta\beta = \lambda/(Nd\cos\beta)$ of the principal diffraction maximum (4.28) if we apply the Rayleigh criterion (see above) that two lines λ and $\lambda + \Delta\lambda$ are just resolved when the maximum $I(\lambda)$ falls into the adjacent minimum for $I(\lambda + \Delta\lambda)$. One obtains from the condition $(d\beta/d\lambda)\Delta\lambda = \lambda/(Nd\cos\beta)$,

$$\frac{\lambda}{\Delta\lambda} = \frac{Nd(\sin\alpha \pm \sin\beta)}{\lambda} \quad , \tag{4.31}$$

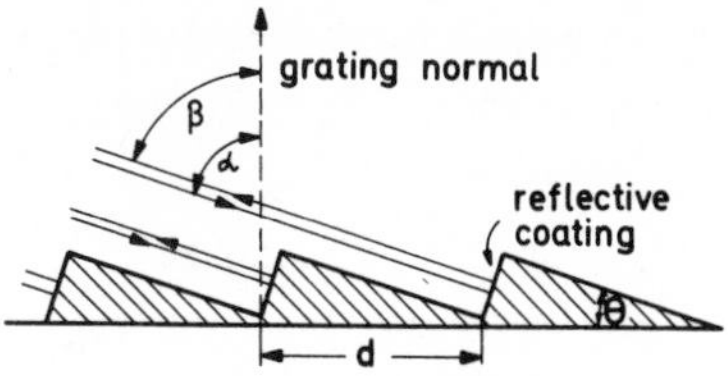

Fig.4.20. Echelle grating

which reduces with (4.22) to

$$\boxed{R = \frac{\lambda}{\Delta\lambda} = mN} \quad . \tag{4.32}$$

The spectral resolving power is the product of the diffraction order m with the total number N of grooves.

A special design is the so-called echelle grating, which has very widely spaced grooves forming right-angled steps (see Fig.4.20). The light is incident normal to the small side of the grooves. The path difference between two reflected partial beams incident on two adjacent grooves with an angle of incidence $\alpha = 90^\circ - \theta$ is $\Delta s = 2d \cos\theta$ and the grating equation (4.22) gives for the angle β of the mth diffraction order

$$d(\cos\theta + \sin\beta) = m\lambda \quad , \tag{4.33}$$

where β is close to $\alpha = 90^\circ - \theta$.

With $d \gg \lambda$ the grating is used in a very high order ($m \approx 10 - 100$) and the resolving power is according to (4.32) very high. Because of the larger distance d between the grooves, the relative ruling accuracy is higher and large gratings (up to 30 cm) can be ruled. The disadvantage of the echelle is the small spectral range $\delta\lambda = \lambda/m$ between successive diffraction orders.

Example

$N = 3 \times 10^4$, $d = 10$ µm, $\theta = 30^\circ$, $\lambda = 500$ nm, $m = 34$. The spectral resolving power is $R = 10^6$, but the free spectral range is only $\delta\lambda = 15$ nm.

Minute deviations of the distance d between adjacent grooves, caused by inaccuracies during the ruling process, may result in constructive interference from parts of the grating for "wrong" wavelengths. Such unwanted maxima, which occur for a given angle of incidence α into "wrong" directions β, are called *grating ghosts*. Although the intensity of these ghosts is generally very small, intense incident radiation at a wavelength λ_i may cause ghosts with intensities comparable to those of other weak lines in the spectrum. This problem is particularly serious in laser spectroscopy when the intense light at the laser wavelength which is scattered by cell walls or windows, reaches the entrance slit of the monochromator.

In order to illustrate the problematics of achieving that ruling accuracy which is required to avoid these ghosts, let us assume that the carriage of the ruling engine expands by only 0.1 μm during the ruling of a 10×10 cm grating, e.g., due to temperature drifts. The groove distance d in the second half of the grating differs then from that of the first half by 10^{-5} d. With $N = 10^5$ grooves the waves from the second half are then completely out of phase with those from the first half, and the condition (4.22) is fulfiled for different wavelengths in both parts of the grating, giving rise to unwanted wavelengths at wrong positions β. Such ghosts are particularly troublesome in laser Raman spectroscopy (see Chap.10) or low-level fluorescence spectroscopy, where very weak lines have to be detected in the presence of extremely strong excitation lines. The ghosts from these excitation lines may overlap with the fluorescence or Raman lines and complicate the assignment of the spectrum.

Although modern ruling techniques with interferometric length control have greatly improved the quality of ruled gratings [4.7] the most satisfactory way of producing completely ghost-free gratings is with holography. The production of holographic gratings proceeds as follows. A photosensitive layer on the grating blank surface is illuminated by two coherent plane waves with wavevectors $\underline{k}_1$ and $\underline{k}_2$ ($|\underline{k}_1| = |\underline{k}_2|$) which form angles α and β against the surface normal (Fig.4.21). The intensity distribution of the superposition in the plane $z = 0$ of the photolayer consists of parallel dark and bright fringes imprinting an ideal grating into the layer which becomes visible after developing the photo-emulsion. The grating constant depends on the wavelength $\lambda = 2\pi/|k|$ and on the angles α and β. Such holographic gratings are essentially free of ghosts. Their reflectivity R, however, is lower than that of ruled gratings and is furthermore strongly dependent on the polarization of the incident wave.

Summarizing the considerations above we find that the grating acts as a wavelength-selective mirror, reflecting light of a given wavelength only into definite directions β_m, called the mth diffraction orders, which are defined by (4.28). The intensity profile of a diffraction order corresponds to the diffraction profile of a slit with width $b = Nd \cos\beta_m$ representing the size of the whole grating projection as seen in the direction β_m. *The spectral resolution is therefore limited by the effective size of the grating measured in units of the wavelength* [see (4.31)].

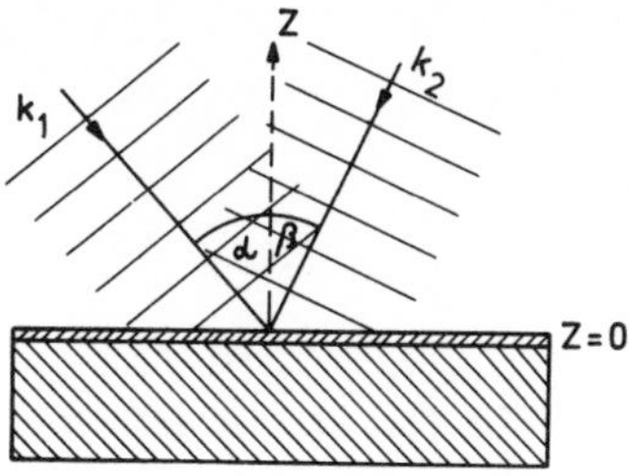

Fig.4.21. Photographic production of a holographic grating

For a more detailed discussion of special designs of grating monochromators such as the concave gratings used in vuv spectroscopy, the reader is referred to the special literature on this subject [4.9-11]. An excellent account of the production and design of ruled gratings can be found in [4.6].

4.2 Interferometers

For the investigation of the various line profiles discussed in Chap.3, interferometers are preferentially used because, in respect of spectral resolving power, they are superior even to large spectrometers. In laser spectroscopy the different types of interferometers not only serve to measure emission—or absorption—line profiles, but they are also essential devices for narrowing the spectral width of lasers, monitoring the laser linewidth, and controlling and stabilizing the wavelength of single-mode lasers (see Chap.6).

In this section we discuss some basic properties of interferometers illustrated by some examples [4.12]. The characteristics of the different types of interferometer that are essential for spectroscopic applications are discussed in more detail. Since laser technology is inconceivable without dielectric coatings for mirrors, interferometers, and filters, an extra section deals with such dielectric multilayers.

4.2.1 Basic Concepts

The basic principle of all interferometers may be summarized as follows (see Fig.4.22). The incident lightwave with intensity I_0 is divided into two or more partial beams with amplitudes A_k which pass different optical path lengths $S_k = nx_k$ (n = refractive index) before they are again superimposed at the exit of the interferometer. Since all partial beams come from the

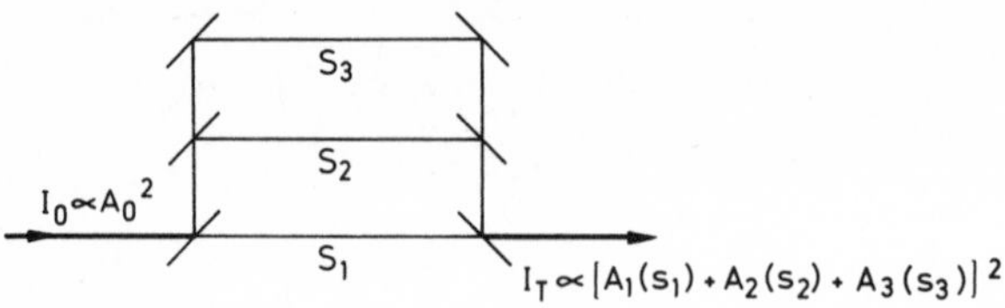

Fig.4.22. Schematic illustration of the basic principle for all interferometers.

same source, they are coherent as long as the maximum path difference does not exceed the coherence length (see Sect.2.10). The total amplitude of the transmitted wave, which is the superposition of all partial waves, depends on the amplitudes A_k and on the phases $\varphi_k = \varphi_0 + 2\pi S_k/\lambda$ of the partial waves. *It is therefore sensitively dependent on the wavelength* λ.

The maximum transmitted intensity is obtained when all partial waves interfere constructively. This gives the condition for the optical path difference $\Delta s_{ik} = s_i - s_k$,

$$\Delta s_{ik} = m\lambda \quad , \quad m = 1, 2, 3 \ldots . \tag{4.34}$$

According to (2.30), the transmitted intensity I_T is proportional to the square of the total amplitude,

$$I_T \propto \left|\sum_k A_k\right|^2 \quad . \tag{4.35}$$

Examples of devices in which only *two* partial beams interfere, are the Michelson interferometer and the Mach-Zehnder interferometer. *Multiple*-beam interference is used for instance in the Fabry-Perot interferometer and in multilayer dielectric coatings of highly reflecting mirrors.

Some interferometers use the optical birefringence of some crystals to produce two partial waves with mutually orthogonal polarization. The phase difference between the two waves is generated by the different refractive index for the two polarizations. An example of such a "polarization interferometer" is the *Lyot filter* [4.13] used in dye lasers to narrow the spectral linewidth (see Sect.7.3).

The condition (4.34) for maximum transmission of the interferometer applies not only to a single wavelength λ but to all λ_m for which

$$\lambda_m = \Delta s/m \quad (m = 1,2,3, \ldots) \quad .$$

The wavelength interval

$$\delta\lambda = \Delta s/m - \Delta s/(m + 1) = \Delta s/(m^2 + m) \tag{4.36a}$$

is called the *free spectral range* of the interferometer. It is more conveniently expressed in terms of frequency. With $\nu = c/\lambda$, (4.34) yields $\Delta s = mc/\nu$ and the free spectral frequency range

$$\boxed{\delta\nu = c/\Delta s} \tag{4.36b}$$

becomes independent of the order m.

Note

From an interferometric measurement alone one can only determine λ modulo $m\delta\lambda$ because all wavelengths $\lambda = \lambda_0 + m\delta\lambda$ are equivalent with respect to the transmission of the interferometer. One therefore has to determine λ within one free spectral range using other measurements.

The mathematical treatment of interferometers [4.12] is generally performed assuming a *plane* incident wave. For laser spectroscopic applications, a nearly parallel laser beam may often be approximated by a plane wave as long as the beam diameter is large compared with the wavelength. However, the following essential aspect should not be disregarded when discussing characteristics of interferometers. Because of the limited beam diameter, the different partial beams may not completely overlap in the interferometer, particularly if they have been shifted by the beam splitter (see Fig.4.39). This incomplete overlap results in imperfect interference of the partial beams which may alter the transmitted intensity considerably. The limited beam aperture furthermore causes diffraction effects which may influence the angular divergence and with it the interference pattern.

A correct treatment has to take into account the spatial intensity profiles of the incident light beam and of the split partial beams to obtain the real interference pattern. The electric field of such a wave travelling in the z directions is written in our complex notation,

$$E(x,y,z) = A(x,y)e^{i[\omega t-\varphi(x,y)-kz]} \quad , \tag{4.37}$$

where the phase fronts are not necessarily planes. The profile of a laser beam for instance, emitted from a laser which oscillates in the fundamental mode, shows a Gaussian intensity distribution $|(A(x,y)|^2$ (see Sect.5.3).

Bearing this restriction in mind, we shall use in the following the plane-wave approximation for the discussion of interferometers, calling attention to cases where it cannot be used.

4.2.2 Michelson Interferometer

The basic principle of the Michelson interferometer (M.I.) is illustrated in Fig.4.23. The incident plane wave

$$E = A_0\, e^{i(\omega t - kx)}$$

is split by the beam splitter S (of reflectivity R and transmittance T) into two waves

$$E_1 = A_1 \exp[i(\omega t - kx)] \quad \text{and} \quad E_2 = A_2 \exp[i(\omega t - ky + \varphi)] \quad .$$

If the beam splitter has negligible absorption (R + T = 1), the amplitudes A_1 and A_2 are determined by $A_1 = \sqrt{T}A_0$ and $A_2 = \sqrt{R}A_0$ with $A_0^2 = A_1^2 + A_2^2$.

After being reflected at the plane mirrors M_1 and M_2 the two waves are superposed in the plane of observation B. In order to compensate for the dispersion which beam 1 suffers by passing twice through the glass plate of beam splitter S, often a corresponding compensation plate P is placed in one side arm of the interferometer. The amplitudes of the two waves in the plane B are $(T\,R)^{\frac{1}{2}}A_0$, because each wave has been transmitted and reflected once at the beam splitter S. The phase difference δ between the two waves is

$$\delta = (2\pi/\lambda)2(SM_1 - SM_2) + \Delta\varphi \quad , \tag{4.38}$$

where $\Delta\varphi$ accounts for additional phase shifts which may be caused by reflection. The total complex field amplitude in the plane B is then

$$E = \sqrt{RT}A_0\, e^{[i(\omega t + \varphi_0)]}(1 + e^{i\delta}) \quad . \tag{4.39}$$

The detector in B cannot follow the rapid oscillations with frequency ω but measures the time-averaged intensity $\bar{I}$, which is according to (2.30d)

$$\begin{aligned} \bar{I} &= \frac{1}{2}\, c\varepsilon_0 A_0^2 RT(1 + e^{i\delta})(1 + e^{-i\delta}) \\ &= c\varepsilon_0 RTA_0^2(1 + \cos\delta) \\ &= \frac{1}{2}\, \bar{I}_0(1 + \cos\delta) \quad \text{for} \quad R = T = \frac{1}{2} \quad \text{and} \quad \bar{I}_0 = \frac{1}{2}\, c\varepsilon_0 A_0^2 \quad . \end{aligned} \tag{4.40}$$

If mirror M_2 (which is mounted on a carriage) moves along a distance Δy, the optical path difference changes by $\Delta s = 2n\Delta y$ (n is the refractive index between S and M_2) and the phase difference δ changes by $2\pi\Delta s/\lambda$. Figure 4.24 shows the intensity $I_T(\delta)$ in the plane B as a function of δ for a monochromatic incident plane wave. For the maxima at $\delta = 2m\pi (m = 0, 1, 2, \ldots)$ the transmitted intensity I_T becomes equal to the incident intensity I_0, which

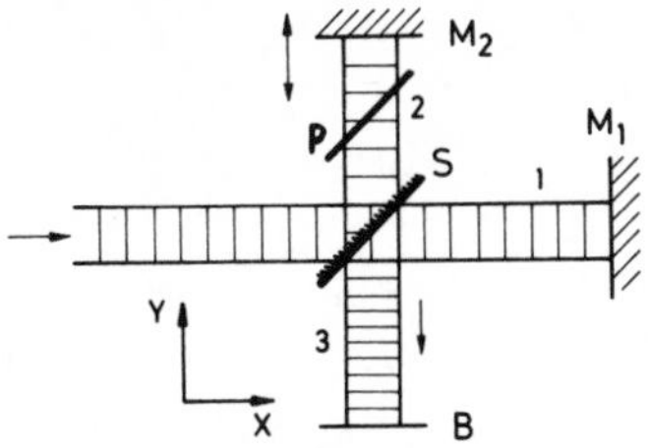

Fig.4.23. Two-beam interference in a Michelson interferometer

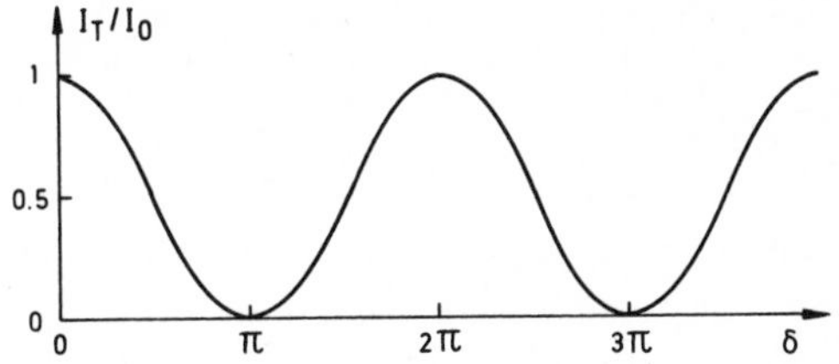

Fig.4.24. Intensity transmitted through the Michelson interferometer in dependence on the phase difference between the two partial beams for T = R = 0.5

means that the transmission of the interferometer is $T_I = 1$ for $\delta = 2m\pi$. In the minima for $\delta = (2m + 1)\pi$ the transmitted intensity I_T is zero! The incident plane wave is being reflected back into the source.

This illustrates that the M.I. can be regarded either as a wavelength-dependent filter for the transmitted light, or as a wavelength-selective reflector. In the latter function it is often used for mode selection in lasers (Fox-Smith selector, see Sect.6.4).

For divergent incident light the path difference between the two waves depends on the inclination angle (see Fig.4.25). In the plane B an interference pattern of circular fringes, concentric to the symmetry axis of the system, is produced. Moving the mirror M_2 causes the ring diameter to change. The intensity $I(\theta,\Delta s)$ behind a small aperture still follows approximately the function $I(\Delta s)$ in Fig.4.24. With parallel incident light, but slightly tilted mirrors M_1 or M_2, the interference pattern consists of parallel fringes which move into a direction perpendicular to the fringes when Δs is changed.

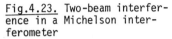

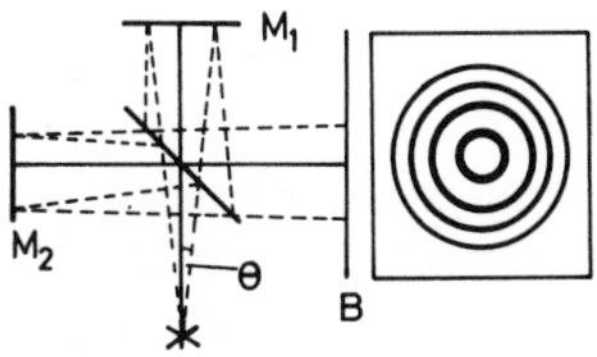

Fig.4.25. Circular fringe pattern produced by the Michelson interferometer with divergent incident light

The M.I. can be used for absolute wavelength measurements by counting the number N of maxima in B when the mirror M_2 is moved along a known distance Δy. The wavelength λ is then obtained from

$$\lambda = 2n\Delta y/N \quad . \tag{4.41}$$

This technique has been applied to very precise determinations of laser wavelengths (see Sect.4.4).

The M.I. may be described in another, equivalent way, which is quite instructive. Assume that the mirror M_2 in Fig.4.23 moves with a constant velocity $v = \Delta y/\Delta t$. A wave with frequency ω and wave vector $\underline{k}$ incident on the moving mirror suffers a Doppler shift

$$\Delta\omega = \omega - \omega' = 2\underline{k}\cdot\underline{v} = (4\pi/\lambda)v \tag{4.42}$$

on reflection.

Inserting the path difference $\Delta s = \Delta s_0 + 2vt$ and the corresponding phase difference $\delta = (2\pi/\lambda)\Delta s$ into (4.40) gives, with (4.42) and $\Delta s_0 = 0$,

$$\bar{I} = \frac{\bar{I}_0}{2}(1 + \cos\Delta\omega t) \quad . \tag{4.43}$$

We recognize (4.43) as the time-averaged beat signal, obtained from the superposition of two waves with frequencies ω and $\omega' = \omega - \Delta\omega$, giving an intensity of

$$\bar{I} = \bar{I}_0(1 + \cos\Delta\omega t)\cos^2[(\omega + \omega')t/2] = (\bar{I}_0/2)(1 + \cos\Delta\omega t) \quad .$$

Note that the frequency $\omega = (c/v)\Delta\omega/2$ of the incoming wave can be measured from the beat frequency $\Delta\omega$, provided the velocity v of the moving mirror is known. The M.I. with uniformly moving mirror can be therefore regarded as a device which transforms the high frequency ω $(10^{14} - 10^{15}\ s^{-1})$ into an easily accessible audio range $(v/c)\omega$. This property is used in *Fourier spectroscopy* (see next section) to measure spectral lines in the infrared region.

Example

$v = 3$ cm/s $\Rightarrow (v/c) = 10^{-10}$. A frequency $\omega = 2\times 10^{14}\ s^{-1}$ ($\lambda = 9\ \mu$m) is transformed to $\Delta\omega = 40$ KHz.

The maximum path difference Δs which still gives interference fringes in B is limited by the coherence length of the incident radiation (see Sect. 2.10). Using spectral lamps the coherence length is limited by the Doppler width of the spectral lines and is typically a few cm. With stabilized single-mode lasers, however, coherence lengths of several kilometers can be achieved. In this case the maximum path difference in the M.I. is in general not restricted by the source but by technical limits imposed by laboratory facilities.

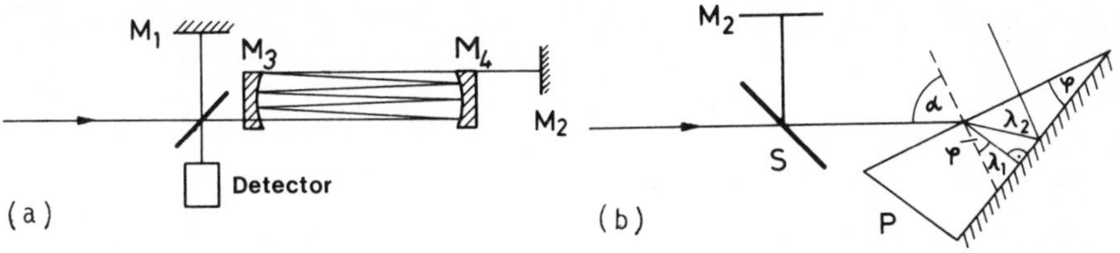

Fig.4.26. (a) Michelson interferometer with optical delay line, allowing a large path difference between the two interfering beams. (b) Michelson interferometer with Littrow prism as preselector

The attinable path difference Δs can be considerably increased by an *optical delay line*, placed in one arm of the interferometer (Fig.4.26a). It consists of a pair of mirrors M_3, M_4 which reflect the light back and forth many times. In order to keep diffraction losses small, spherical mirrors are preferable which compensate by collimation the divergence of the beam caused by diffraction. With a stable mounting of the whole interferometer, optical path differences up to 350 m could be realized [4.14], allowing a spectral resolution of $\nu/\Delta\nu \approx 10^{11}$. This was demonstrated by measuring the linewidth of a He-Ne laser oscillating at $\nu = 5\times 10^{14}\ s^{-1}$ as a function of discharge current. The accuracy obtained was better than 5 KHz.

Often the incident wave includes several components with different wavelengths λ_k, with a separation that is large compared with the free spectral range of the interferometer. If the superposition of these components in the transmitted or reflected light is unwanted, they can be separated by an additional dispersing element. This may be realized for instance by replacing one of the interferometer mirrors by a Littrow prism with reflecting back surface (see Fig.4.26b). Such an arrangement has been used to select one from many lines of an argon ion laser and to achieve single-mode operation of the laser on this line [4.15]. If the angle of incidence α is choosen such that $\sin\alpha = n(\lambda_1)\sin\varphi$, a light wave with wavelength λ_1 is being refracted parallel to the normal of the coated back surface and is therefore reflected back into itself, while all other wavelengths are reflected into other directions. In order to minimize additional reflection losses at the front surface, the angle of incidence α should be close to the Brewster angle. This demands that $\alpha + \varphi = 90^\circ$ or $\tan\varphi = n(\lambda_1)$ where n is the refractive index of the prism material.

4.2.3 Fourier-Transform Spectroscopy

The M.I. described in the previous section has been used very successfully for high-resolution spectroscopy in the infrared region. The basic idea, already discussed above, is the transfer of the high frequencies ω of the incident radiation to audio frequencies $\Delta\omega = 2(v/c)\omega$ of the interference pattern by the uniform motion of mirror M_2 with constant velocity v.

Assume that the incoming radiation is composed of several components with frequencies ω_k. The total amplitude in the plane B of the detector is the sum of all interference amplitudes (4.39),

$$\underline{E} = \sum_k A_k \, e^{i(\omega_k t+\varphi_k)}(1 + e^{i\delta_k}) \quad . \tag{4.44}$$

A detector with a time constant large compared with the maximum period $1/\omega_k$ does not follow the rapid oscillations of the amplitude at frequencies ω_k or at the difference frequencies $(\omega_i - \omega_k)$ but gives a signal proportional to the sum of the intensities I_k in (4.40) or (4.43)

$$\bar{I} = \sum_k (\bar{I}_{k0}/2)(1 + \cos\delta_k) = \sum_k (\bar{I}_{k0}/2)(1 + \cos\Delta\omega_k t) \quad , \tag{4.45}$$

where the audio frequencies $\Delta\omega_k = 2\omega_k v/c$ are determined by the frequencies ω_k of the components and by the velocity v of the moving mirror. Measurements of these frequencies $\Delta\omega_k$ allows one to reconstruct the spectral components of the incoming wave with frequencies ω_k as shown below.

Example

When the incoming wave consists of two components with frequencies ω_1 and ω_2, the interference pattern will vary with time according to

$$\bar{I} = (\bar{I}_{10}/2)[1 + \cos 2\omega_1(v/c)t] + (I_{20}/2)[1 + \cos(2\omega_2(v/c)t]$$

$$= I_0\{1 + \cos[(\omega_1 - \omega_2)vt/c]\cos[(\omega_1 + \omega_2)(v/c)t]\} \quad ,$$

where we have assumed $I_{10} = I_{20} = I_0$. This is a beat signal, where the amplitude of the interference signal at $(\omega_1 + \omega_2)(v/c)$ is modulated at the difference frequency $(\omega_1 - \omega_2)v/c$ (see Fig.4.27).

In order to determine the two frequencies ω_1 and ω_2, one has to measure at least over one modulation period $T = 2\pi c/[v(\omega_1 - \omega_2)] = [(\nu_1 - \nu_2)v/c]^{-1}$. *Numerical example*: $\lambda_1 = 10$ µm, $\lambda_2 = 9.8$ µm $\Rightarrow (\nu_2 - \nu_1) = 6 \times 10^{11}$ s^{-1}; with $v = 1$ cm/s, $\Rightarrow T = 50$ ms.

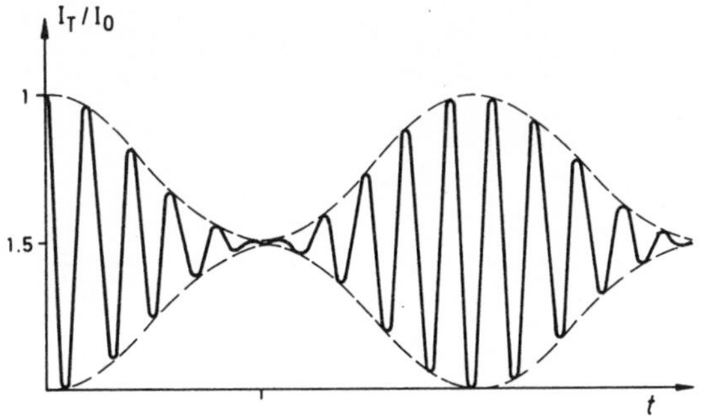

Fig.4.27. Interference signal from an incoming wave with two components ω_1 and ω_2

The spectral resolution can be roughly estimated as follows. If Δy is the path difference travelled by the moving mirror, the number of interference maxima which are counted by the detector is $N_1 = 2\Delta y/\lambda_1$ for an incident wave with wavelength λ_1, and $N_2 = 2\Delta y/\lambda_2$ for $\lambda_2 < \lambda_1$. The two wavelengths can be clearly distinguished, when $N_2 = N_1 + 1$. This yields for the spectral resolving power

$$\frac{\lambda}{\Delta\lambda} = \frac{2\Delta y}{\lambda} = N \quad . \tag{4.46}$$

Example

a) $\Delta y = 5$ cm, $\lambda = 10\ \mu m \Rightarrow N = 10^4$,
b) $\Delta y = 100$ cm, $\lambda = 0.5\ \mu m \Rightarrow N = 4 \times 10^6$
where the latter example can be realized only with lasers, which have sufficiently large coherence length (see Sect.4.4).

We now consider the general case where the spectral distribution of the incoming radiation may be a superposition of an arbitrary number of spectral components [4.16]. The amplitude of such a component at frequency ω is represented by

$$E(\omega,x,t) = E_0(\omega)e^{i(\omega t-kx)} \quad . \tag{4.47}$$

According to (4.40) the interference of the two split waves of this component gives the mean intensity

$$\bar{I}(\omega,\delta) = c\varepsilon_0 RT|E_0(\omega)|^2[1 + \cos\delta(\omega)] \quad , \tag{4.48}$$

where the phase difference $\delta(\omega)$ depends of course on the frequency ω (see 4.38). The total flux received by the detector at all frequencies ω is therefore

$$\bar{I}(\delta) = \int_0^\infty \bar{I}(\omega,\delta)d\omega = c\varepsilon_0 RT\left[\int |E_0(\omega)|^2 d\omega + \int |E_0(\omega)|^2 \cos\delta d\omega\right] \quad . \qquad (4.49)$$

For $\delta = 0$, (4.49) yields $I(0) = 2c\varepsilon_0 RT \int_0^\infty |E_0(\omega)|^2 d\omega$, and (4.49) can be written as

$$|\bar{I}(\delta) - \tfrac{1}{2}\bar{I}(0)| = c\varepsilon_0 RT \int_0^\infty |E_0(\omega)|^2 \cos\delta d\omega \quad . \qquad (4.50)$$

The Fourier cosine transform of (4.50) gives [4.16]

$$I(\omega) = c\varepsilon_0 |E_0(\omega)|^2 = \frac{1}{\pi RT} \int_0^\infty \left[\bar{I}(\delta) - \tfrac{1}{2}\bar{I}(0)\right]\cos\delta d\delta \quad . \qquad (4.51)$$

Equation (4.51) shows that from the measurement of $\bar{I}(\delta)$ the spectrum $I(\omega)$ of the incoming radiation can be obtained, provided the interference pattern $\bar{I}(\delta)$ is measured from $\delta = 0$ up to large values of δ. In practice a displacement Δy of a few cm is sufficiently large and produces phase shifts of $\delta \approx 10^4\pi$ at $\lambda = 10$ µm.

A schematic experimental arrangement [4.17] is shown in Fig.4.28. The moving mirror M_2 forms the common part of two interferometers. The right M.I. receives the unknown radiation, the left M.I. serves as control unit. The movement of mirror M_2 is controlled by the interference pattern of a single-mode He-Ne laser, which gives an accurate time base. White light incident on the left M.I. gives an interference maximum only if the path difference in the left M.I. is zero. This maximum provides a trigger signal which marks a definite position of the moving mirror M_2.

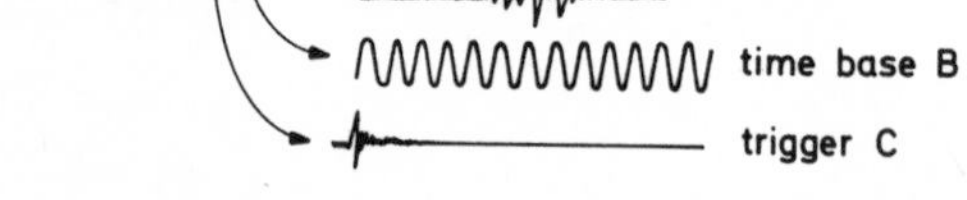

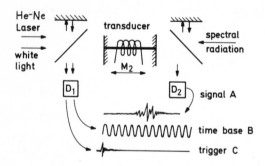

Fig.4.28. Schematic arrangement of a Fourier spectrometer (after [4.17])

In practice it is very difficult to construct an interferometer which is perfectly balanced over an extended wavelength range. Inevitable phase shifts in actual Fourier spectrometers produce besides the pure cosine components also sine components in the interferogram. Instead of (4.51), a complex representation has to be used and the spectral distribution $I(\omega)$ is obtained from the interferogram $\bar{I}(\delta)$ by

$$I(\omega) = C \int_{-\infty}^{+\infty} \left[\bar{I}(\delta) - \frac{1}{2} I(0)\right] e^{-i\omega\delta} d\delta \quad . \qquad (4.52)$$

Compared with conventional spectroscopy using dispersing spectrometers, Fourier spectroscopy has some definite advantages [4.17,18]:

a) The whole spectral range $\Delta\omega$ of interest is recorded simultaneously, whereas in a conventional spectrometer each of the $M = \Delta\omega/\delta\omega$ resolved intervals $\delta\omega$ is scanned across the detector in sequence. In a conventional spectrometer the irradiance of each interval is only measured for a fraction (T/M) of the total recording time T. Since the signal is proportional to the time, but the noise is proportional to the square root of the sampling time, the signal-to-noise ratio S/N is proportional to $(T/M)^{\frac{1}{2}}$ for a dispersing spectrometer, but is $M^{\frac{1}{2}}$ times larger in Fourier spectroscopy. For M = 1000, for example, Fourier spectroscopy gives a thirty-three-times better signal-to-noise ratio for the same total observation time, or else gives the same S/N ratio as a conventional spectrometer in a 1000 times shorter sampling time! (Fellgett's advantage).

b) The throughput or étendue (see Sect.4.1) of the M.I. can be much larger than that of a monochromator with comparable resolution. Diameters of 50 mm of the limiting aperture in the M.I. should be compared with slit widths of a few μm in monochromators. Although the acceptance angle of the M.I. is smaller than for monochromators, this does not limit the étendue if the radiation source is sufficiently small.

c) The spectrum, calculated by a minicomputer from the recorded Fourier transform, is obtained in digital form and is therefore easy to process. Ratio spectra, scale expanded sections of the spectrum, logarithmic presentation, and other modes of processing can be readily performed by the same computer.

Because of these advantages Fourier spectroscopy has rapidly developed and has become a major technique in the infrared and recently also in the visible region. The most serious disadvantage is the high price of the instrument. For a more detailed treatment the reader is referred to the literature [4.16-21].

4.2.4 Mach-Zehnder Interferometer

Analogous to the Michelson interferometer, the Mach-Zehnder interferometer is based on two-beam interference by amplitude splitting of the incoming wave. The two waves travel along different paths (Fig.4.29a). Inserting a transparent object into one arm of the interferometer alters the optical path difference between the two beams. This results in a change of the interference pattern, which allows a very accurate determination of the refractive index of the sample and its local variation. The Mach-Zehnder interferometer may be regarded therefore as a sensitive refractometer.

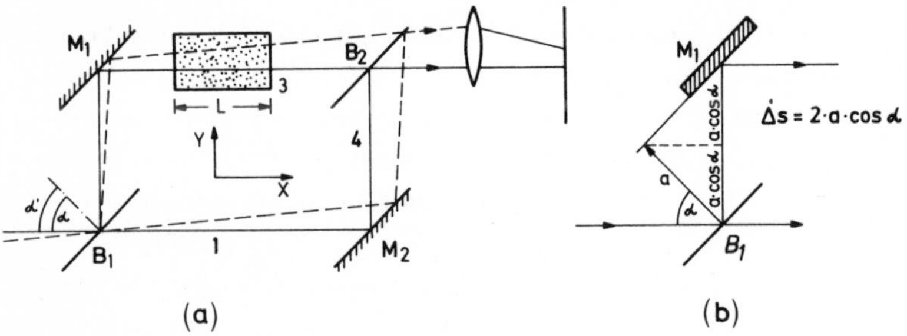

Fig.4.29a,b. Mach-Zehnder interferometer. (a) Schematic arrangment. (b) Path difference between the two parallel beams

If the beam splitters B_1, B_2 and the mirrors M_1, M_2 are all strictly parallel, the path difference between the two split beams does not depend on the angle of incidence α (see Fig.4.29b). This means that the interfering waves in the symmetric interferometer (without sample) experience the same path difference on the solid path in Fig.4.29a as on the dashed path, because the path difference $\Delta_1 = B_1M_1 = 2a\cos\alpha$ between beam 1 and 3 is exactly compensated by the same path length between M_2 and B_2. Without the sample the total path difference is therefore zero, and it is $\Delta s = (n-1)L$ *with* the sample with refractive index n in one arm of the interferometer.

Expanding the beam on path 3 gives an extended interference fringe pattern, which reflects the local variation of the refractive index. Using a laser as a light source with large coherence length, the path lengths in the two interferometer arms can be made different without losing the contrast of the interference pattern (Fig.4.30). With a beam expander (lenses L_1 and L_2), the laser beam can be expanded up to 10-20 cm and large objects can be tested. The interference pattern can either be photographed or may be viewed directly

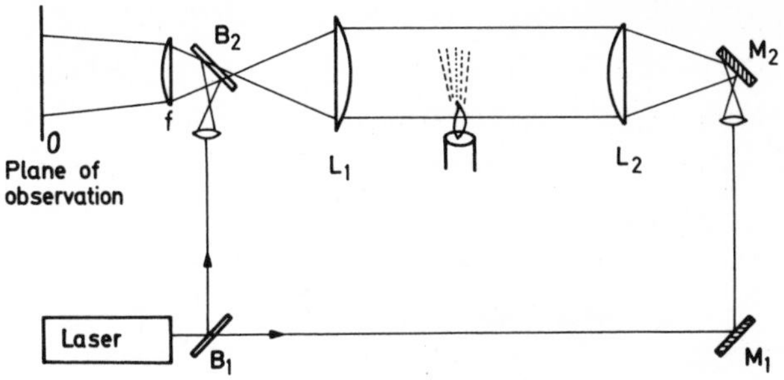

Fig.4.30. Laser interferometer for sensitive measurements of local variations of the index of refraction in extended samples

with the naked eye or even with a television camera [4.22]. Such a laser interferometer has the advantage that the laser beam diameter can be kept small everywhere in the interferometer, except between the two expanding lenses. The illuminated part of the mirror surfaces should not deviate from an ideal plane by more than $\lambda/2$ in order to obtain good interferograms. Therefore, the smaller the beam diameter the more easily and less expensively this demand can be met.

The Mach-Zehnder interferometer has found a wide range of applications. Density variations in laminar or turbulent gas flows can be seen with this technique and the optical quality of mirror substrates or interferometer plates can be tested with high sensitivity.

In order to get quantitative information of the local variation of the optical path through the sample, it is useful to generate a fringe pattern for calibration purposes by slightly tilting the plates B_1, M_1 and B_2, M_2 in Fig.4.29, which makes the interferometer slightly asymmetric. Assume B_1 and M_1 are tilted clockwise around the z direction by a small angle β and the pair B_2M_2 is tilted counterclockwise by the same angle β. The optical path between B_1 and M_1 is then $\Delta_1 = 2a\cos(\alpha+\beta)$ whereas $\overline{B_2M_2} = \Delta_2 = 2a\cos(\alpha-\beta)$. After being recombined, the two beams therefore have a path difference

$$\Delta = \Delta_2 - \Delta_1 = 2a[\cos(\alpha-\beta) - \cos(\alpha+\beta)] = 4a\sin\alpha\sin\beta \quad , \qquad (4.53)$$

which depends on the angle of incidence α. In the plane of observation, an interference pattern of parallel fringes is observed with an angular separation between fringes m and m + 1 given by $(\sin\alpha_m - \sin\alpha_{m+1}) = \lambda/(4a\beta)$.

A sample in path 3 introduces an additional path difference $\Delta s(\beta) = (n-1)L/\cos\beta$ depending on the local refractive index n and the path length

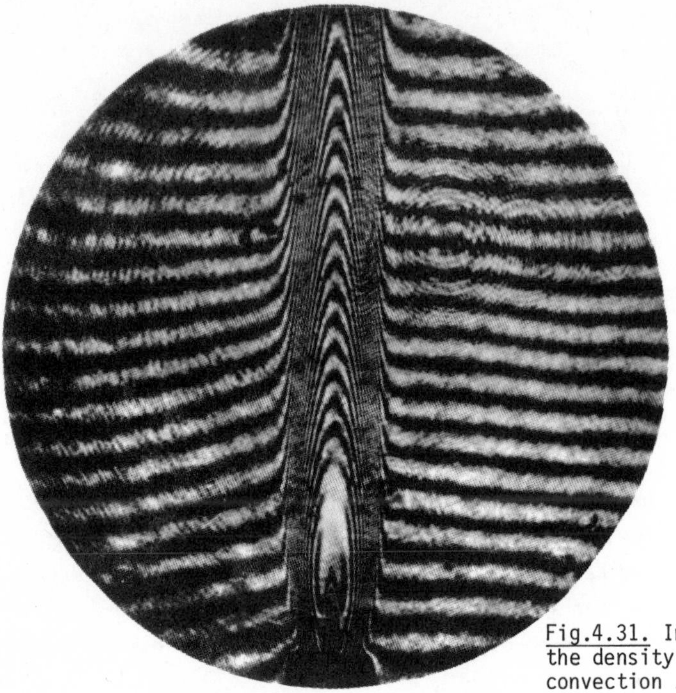

Fig.4.31. Interferogram of the density profile in the convection zone above a candle flame [4.22]

through the sample. The resulting phase difference shifts the interference pattern by an angle $\gamma = \Delta s/(4a\beta)$. Using a lens with a focal length f, which images the interference pattern onto a plane 0, the linear shift is $\Delta y = f\Delta s/(4a\beta)$. Figure 4.31 shows for illustration the interferogram of the convection zone of hot air above a candle flame, placed below one arm of the laser interferometer in Fig.4.30. It can be seen that the optical path through this zone changes by many wavelengths.

The Mach-Zehnder interferometer has been used to measure the refractive index of atomic vapors in the vicinity of spectral lines (see Sect.2.7.2). The experimental arrangement (Fig.4.32) consists of a combination of the spectrograph and the interferometer, where the plates B_1,M_1 and B_2,M_2 are tilted in such a direction that without the sample the parallel interference fringes with a separation $\Delta y(\lambda) = \lambda f/(4a\beta)$ are perpendicular to the entrance slit. The spectrograph disperses $\Delta y(\lambda)$ in the x direction. Because of the wavelength-dependent refractive index $n(\lambda)$ of the atomic vapor [see (2.59)], the fringe shift follows a dispersion curve in the vicinity of the spectral line (Fig.4.33). The dispersed fringes look like hooks around an absorption line, which gave this technique the name *hook method*. To compen-

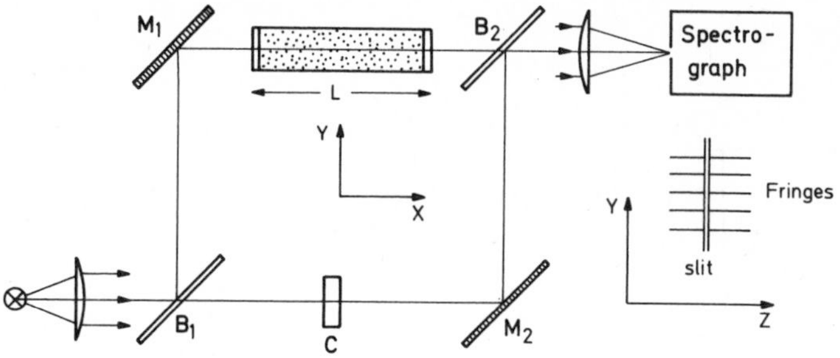

Fig.4.32. Combination of Mach-Zehnder interferometer and spectrograph used in the hook method

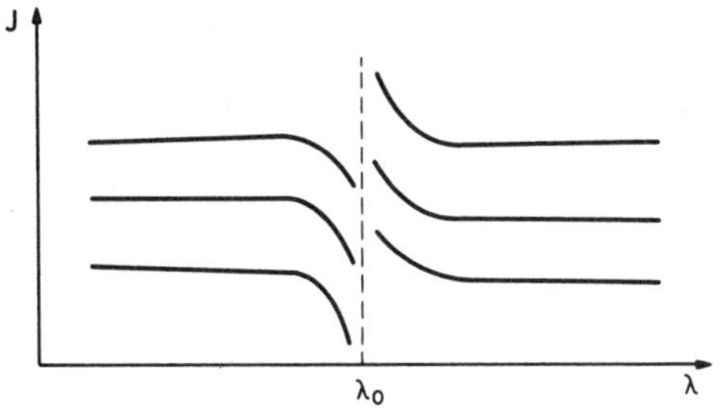

Fig.4.33. Position of fringes as a function of wavelength as observed behind the spectrograph

sate for background shifts caused by the windows of the absorption cell, a compensating plate is placed into the second arm. For more details see [2.17,23].

4.2.5 Multiple Beam Interference

Assume that a plane wave $\underline{E} = \underline{A}_0 \exp i(\omega t - kx)$ is incident at the angle α on a plane transparent plate with two parallel partially reflecting surfaces (Fig.4.34). At each surface the amplitude A_i is split into a reflected component $A_R = A_i\sqrt{R}$ and a refracted component $A_T = A_i\sqrt{1-R}$, neglecting absorption. The reflectivity $R = I_R/I_i$ depends on the angle of incidence α and on the polarization of the incident wave. Provided the refractive index n is known, R can be calculated from Fresnel's formulas [2.3]. From Fig. 4.34, the following relations are obtained for the amplitudes A_i of waves reflected at the upper surface, B_i of refracted waves, C_i of waves reflected at the lower surface, and D_i of transmitted waves

$$|A_1| = \sqrt{R}\,|A_0| \quad ; \quad |B_1| = \sqrt{1-R}\,|A_0| \quad ; \quad |C_1| = \sqrt{R(1-R)}\,|A_0| \quad ;$$

$$D_1 = (1-R)\,|A_0| \quad ;$$

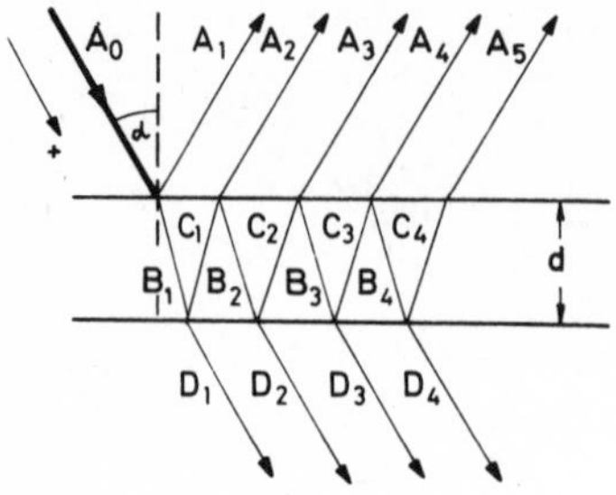

Fig.4.34. Multiple beam interference at two plane parallel partially reflecting surfaces

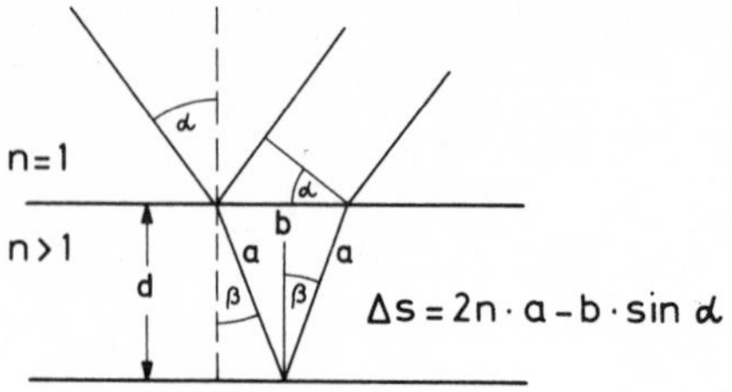

Fig.4.35. Optical path difference between two beams being reflected from the two surfaces of a plane parallel plate

$$|A_2| = \sqrt{1-R}\,|C_1| = (1-R)\sqrt{R}\,|A_0| \quad ; \qquad |C_2| = R\sqrt{R(1-R)}\,|A_0| \quad ;$$

$$|A_3| = \sqrt{(1-R)}\,|C_2| = R^{3/2}(1-R)\,|A_0| \;\dots \quad . \tag{4.54}$$

This scheme can be generalized to the equations

$$|A_{i+1}| = R\,|A_i| \qquad i \geqq 2 \tag{4.55a}$$

$$|D_{i+1}| = R\,|D_i| \qquad i \geqq 1 \quad . \tag{4.55b}$$

Two successive partial waves E_i and E_{i+1} have the optical path difference (see Fig.4.35)

$$\Delta s = (2nd/\cos\beta) - 2d\,\tan\beta\,\sin\alpha \quad . \tag{4.56a}$$

Because $\sin\alpha = n\sin\beta$, this can be reduced to

$$\Delta s = 2nd\,\cos\beta = 2d\sqrt{n^2 - \sin^2\alpha} = 2dn\sqrt{1 - \sin^2\beta} \tag{4.56b}$$

if the refractive index within the plane parallel plate is $n > 1$ and outside the plate $n = 1$. This path difference causes a corresponding phase difference

$$\delta = 2\pi\Delta s/\lambda + \Delta\varphi \quad , \tag{4.56c}$$

where $\Delta\varphi$ takes into account possible phase changes caused by the reflections. For instance, the incident wave with amplitude A_1 suffers a phase jump $\Delta\varphi = \pi$ while being reflected at the medium with $n > 1$. Including this phase jump, we can write

$$A_1 = \sqrt{R}\,A_0\,\exp(i\pi) = -\sqrt{R}\,A_0 \quad .$$

The total amplitude A of the reflected wave is obtained by summation over all partial amplitudes A_i taking into account the different phase shifts,

$$A = \sum_{m=1}^{p} A_m e^{i(m-1)\delta} = -\sqrt{R}\, A_0[1 - (1 - R)]e^{i\delta} \sum_{m=0}^{p-2} R^m e^{im\delta} \quad . \tag{4.57}$$

For vertical incidence ($\alpha = 0$), or for an infinitely extended plate, we have an infinite number of reflections. The geometrical series (4.57) has for $p \to \infty$ the limit

$$A = -\sqrt{R}\, A_0 \frac{1 - e^{i\delta}}{1 - R\, e^{i\delta}} \quad . \tag{4.58}$$

The intensity of the reflected wave is therefore

$$I_R = 2c\varepsilon_0 AA^* = I_0 R \frac{2 - 2\cos\delta}{1 + R^2 - 2R\cos\delta} \quad , \quad \text{with} \quad I_0 = 2c\varepsilon_0 A_0 A_0^* \quad . \tag{4.59}$$

Using the relation $(1 - \cos\delta) = 2\sin^2(\delta/2)$, this can be reduced to

$$\boxed{I_R = I_0 R \frac{4\sin^2(\delta/2)}{(1 - R)^2 + 4R\sin^2(\delta/2)}} \quad . \tag{4.60}$$

In an analogous way, we find for the total transmitted amplitude

$$D = \sum_{m=1}^{\infty} D_m e^{i(m-1)\delta} = (1 - R)A_0 \sum_{0}^{\infty} R^m e^{im\delta} \quad , \tag{4.61}$$

which gives the total transmitted intensity

$$\boxed{I_T = I_0 \frac{(1 - R)^2}{(1 - R)^2 + 4R\sin^2(\delta/2)}} \quad . \tag{4.62}$$

Equations (4.60,62) are called the *Airy formulas*. Since we have neglected absorption, we should have $I_R + I_T = 1$, as can easily be verified from (4.60) and (4.62).

The abbreviation $F = 4R/(1 - R)^2$ is often used, which allows the Airy equations to be written in the form

$$\boxed{I_R = I_0 \frac{F\sin^2(\delta/2)}{1 + F\sin^2(\delta/2)}} \tag{4.63}$$

$$\boxed{I_T = I_0 \frac{1}{1 + F\sin^2(\delta/2)}} \quad . \tag{4.64}$$

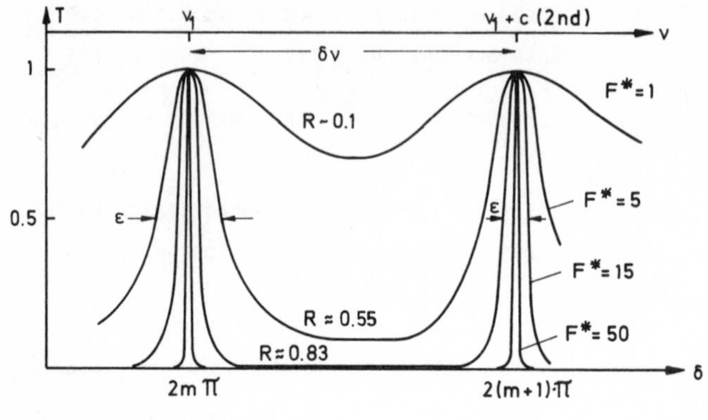

Fig.4.36. Transmittance of an absorption-free multiple beam interferometer as a function of the phase difference δ for different values of the finesse F^*

Figure 4.36 illustrates (4.64) for different values of the reflectivity R. The maximum transmittance is T = 1 for $\delta = 2m\pi$. At these maxima $I_T = I_0$, and the reflected intensity I_R is zero. The frequency range $\delta\nu$ between two maxima is the *free spectral range of the interferometer*. With $\delta = 2\pi\Delta s/\lambda$ and $\lambda = c/\nu$, we obtain

$$\delta\nu = \frac{c}{\Delta s} = \frac{c}{2d\sqrt{n^2 - \sin^2\alpha}} \quad . \tag{4.65a}$$

For vertical incidence ($\alpha = 0$) the free spectral range becomes

$$\boxed{\delta\nu_{\alpha=0} = \frac{c}{2nd}} \quad . \tag{4.65b}$$

The full halfwidth $\varepsilon = |\delta_1 - \delta_2|$ of the transmission maxima in Fig.4.36 with $I(\delta_1) = I(\delta_2) = I_0/2$ can be calculated from (4.62) to be

$$\varepsilon = 4 \arcsin[(1 - R)/2\sqrt{R})] \quad , \tag{4.66}$$

which reduces for $(1 - R) \ll R$ to

$$\varepsilon = 2(1 - R)/\sqrt{R} = 4/\sqrt{F} \quad .$$

The ratio of free spectral range $\delta\nu$ to the halfwidth $\Delta\nu = (\varepsilon/2\pi)\delta\nu$ is called the *finesse* F^* of the interferometer,

$$F^* = \frac{\text{FSR}}{\text{halfwidth}} = \frac{\delta\nu}{\Delta\nu} \quad . \tag{4.67a}$$

Since a free spectral range corresponds to a phase difference of 2π, the finessee can be written

$$\boxed{F^* = \frac{2\pi}{\varepsilon} = \frac{\pi\sqrt{R}}{1 - R}} \quad . \tag{4.67b}$$

Since we have assumed an ideal plane parallel plate with a perfect surface quality, the finesse (4.67b) is determined only by the reflectivity R of the surfaces. In practice, however, deviations of the surfaces from an ideal plane, and slight inclinations of the two surfaces, cause imperfect superposition of the interfering waves. This results in a decrease and a broadening of the transmission maxima which decreases the total finesse. If, for instance, a reflecting surface deviates by an amount λ/q from an ideal plane, the finesse cannot be larger than q. One can define a total finesse F_t^* of the interferometer by

$$1/F_t^* = \left(\sum 1/F_i^{*2}\right)^{1/2} \quad , \tag{4.68}$$

where the different terms F_i^* give the contributions to the decrease of the finesse caused by the different imperfections of the interferometer.

Example

A plane, nearly parallel plate may have a diameter D = 5 cm, a thickness d = 1 cm and a wedge angle of 0.2". The two reflecting surfaces have a reflectivity R = 95%. The surfaces are flat to within $\lambda/50$, which means that no point of the surface deviates from an ideal plane by more than $\lambda/50$. The different contributions to the finesse are:

Reflectivity finesse: $F_R^* = \pi\sqrt{R}/(1 - R) \approx 60$;

Surface finesse: $F_s \approx 50$;

Wedge finesse: With a wedge angle of 0.2" the optical path between the two reflecting surfaces changes by about 0.1λ ($\lambda = 0.5$ µm) across the diameter of the plate. This causes for a monochromatic incident wave imperfect interference and broadens the maxima corresponding to a finesse of about 20. The total finesse is then $F^* = 1/(1/60 + 1/50 + 1/20) \approx 12$. This illustrates that high-quality optical surfaces are necessary to obtain a high total finesse [4.24a]. It makes no sense to increase the reflectivity without a corresponding increase of the surface finesse. In our example the imperfect parallelism was the main cause for the low finesse. Decreasing the wedge angle to 0.1" increases the wedge finesse to 40 and the total finesse to 16.

A much larger finesse can be achieved, using spherical mirrors because the demand for parallelism is dropped. With sufficiently accurate alignment and high reflectivities, values of F^* up to 500 are possible (see Sect.4.2.8).

The spectral resolution $\nu/\Delta\nu$ or $\lambda/\Delta\lambda$ of an interferometer is determined by the free spectral range and by the finesse. Two incident waves with frequencies ν_1 and $\nu_1 - \Delta\nu_1$ can still be resolved if their interference maxima are separated at least by ε. For equal intensities $I_{01}(\nu_1) = I_{01}(\nu_2) = I_0$ the total transmitted intensity $I(\nu) = I_1(\nu) + I_2(\nu)$ can be obtained from (4.64)

$$I(\nu) = I_0\left[\frac{1}{1 + F\sin^2(\pi\nu/\delta\nu)} + \frac{1}{1 + F\sin^2[\pi(\nu + \Delta\nu)/\delta\nu]}\right] , \qquad (4.69)$$

where the phase shift δ has been replaced by the free spectral range $\delta\nu$ using the relation $\delta = 2\pi\Delta s\nu/c = 2\pi\nu/\delta\nu$. $I(\nu)$ is plotted in Fig.4.37 around the transmission maximum $\delta(\nu) = 2m\pi$ for the minimum still-resolvable frequency separation,

$$\Delta\nu = \varepsilon\delta\nu/2\pi = \delta\nu/F^* \quad . \qquad (4.70)$$

Inserting (4.70) into (4.69) yields for $I(\nu = m\delta\nu) \approx 1.2\ I_0$; $I(\nu + \varepsilon\delta\nu/4\pi) \approx I_0$ and $I(\nu + \varepsilon\delta\nu/2\pi) \approx 1.2\ I_0$. This just corresponds to the Rayleigh criterion for the resolution of two spectral lines. The spectral resolving power of the interferometer is therefore

$$\boxed{\nu/\Delta\nu = (\nu/\delta\nu)F^*} \Rightarrow \Delta\nu = \delta\nu/F^* \quad . \qquad (4.71)$$

This can be also expressed by the optical path differenence

$$\boxed{\frac{\nu}{\Delta\nu} = \frac{\lambda}{\Delta\lambda} = F^*\,\frac{\Delta s}{\lambda}} \quad . \qquad (4.72)$$

The resolving power of an interferometer is the product of finesse F^* *and optical path difference* $\Delta s/\lambda$ *in units of the wavelength* λ.

Example

$d = 1$ cm, $n = 1.5$, $R = 0.98$, $\lambda = 500$ nm. An interferometer with negligible wedge and high-quality surfaces, where the finesse is mainly determined by the reflectivity, achieves with $F^* = \pi\sqrt{R}/(1 - R) = 150$ a resolving power of $\lambda/\Delta\lambda = 10^7$. This means that the instrument linewidth is about $\Delta\lambda \approx 5\times10^{-5}$ nm or, in frequency units $\Delta\nu = 60$ MHz.

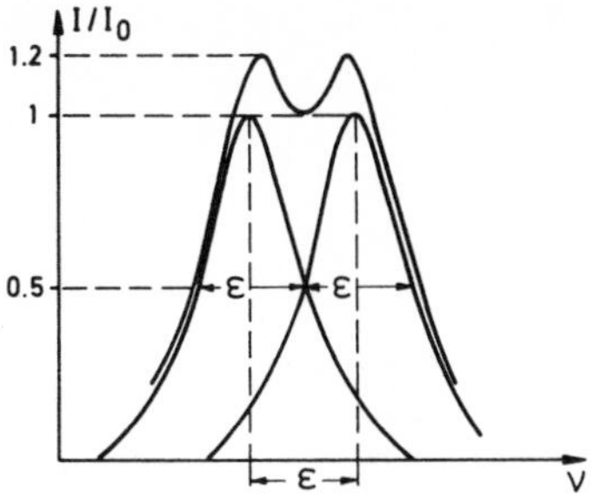

Fig.4.37. Interference intensity as a function of phase difference δ or frequency ν for two closely spaced spectral lines at the limit of resolution

Taking into account the absorption $A = (1 - R - T)$ of each reflective surface, (4.64) has to be modified to

$$I_T = I_0 \frac{T^2}{(A + T)^2} \frac{1}{[1 + F \sin^2(\delta/2)]} \quad , \tag{4.73}$$

where $T^2 = T_1 T_2$, T_1, T_2 being the transmittance of the two reflecting surfaces. The absorption causes two effects. (a) The transmittance is decreased by a factor

$$T^2/(1 - R)^2 = T^2/(A + T)^2 < 1 \quad \text{for } A > 0 \quad .$$

Note that a small absorption of each reflecting surface results in a drastic reduction of the total transmittance. For $A = 0.05$, $R = 0.9 \Rightarrow T = 0.05$ and $T^2/(1 - R)^2 = 0.25$. (b) The absorption causes a phase shift $\Delta\varphi$ at each reflection, which depends on wavelength λ, polarization, and angle of incidence α [2.3]. The first effect causes a *decrease* and a *broadening* of the interference maximum, the second effect a wavelength-dependent *shift* of the maxima.

The practical realization of the multiple beam interference discussed in this section may use either a solid plane parallel glass or fused quartz plate with two coated reflecting surfaces (etalon) (Fig.4.38a) or two separate wedged-quartz plates, where one surface of each plate is coated with a reflection layer, the other with an antireflection layer. The two reflecting surfaces oppose each other and are aligned to be as parallel as achievable (Fig.4.38b). The latter device is called *Fabry-Perot interferometer* (F.P.I.). The transmission maxima of the etalon can be tuned within certain limits by tilting the etalon which alters the angle of incidence, while the F.P.I. can be tuned by altering the air pressure between the two reflection surfaces. Both devices can be used for parallel as well as for divergent incident light. We now discuss them in some more detail.

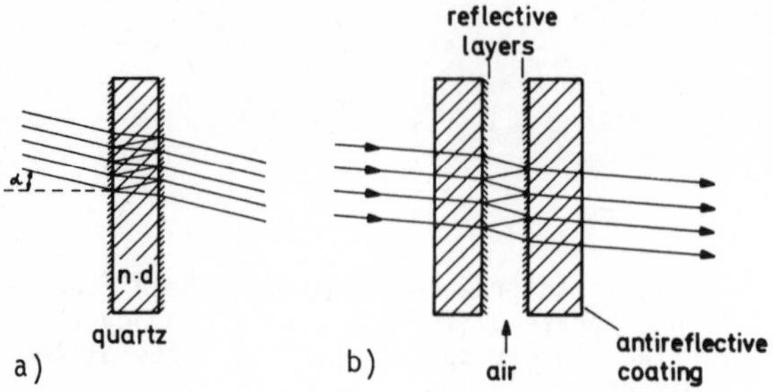

Fig.4.38. Etalon (a) and Fabry-Perot interferometer (b)

4.2.6 Etalons

Plane parallel plates of glass or fused quartz with reflecting surfaces are called etalons. In laser spectroscopy they are mainly used as wavelength-selective transmission filters within the laser resonator to narrow the laser bandwidth (see Sect.6.4). The wavelength λ_m or frequency ν_m for the transmission maximum of mth order, where the optical path between successive beams is $\Delta s = m\lambda$, can be deduced from (4.56b) and Fig.4.35 to be

$$\lambda_m = \frac{2d}{m}\sqrt{n^2 - \sin^2\alpha} = \frac{2nd}{m}\cos\beta \tag{4.74a}$$

$$\nu_m = \frac{mc}{2nd\ \cos\beta} \quad . \tag{4.74b}$$

For all wavelengths $\lambda = \lambda_m$ (m = 0, 1, 2 ...) in the incident light, the phase difference between the transmitted partial waves becomes $\delta = 2m\pi$ and the transmitted intensity (4.73) is

$$I_T = \frac{T^2}{(1-R)^2} I_0 = \frac{T^2}{(A+T)^2} I_0 \quad ,$$

where A is the absorption of the etalon (substrate absorption plus absorption of one reflecting surface). The reflected waves interfere destructively for $\lambda = \lambda_m$ and the reflected intensity becomes zero.

Note, however, that this is only true for infinitely extended plane waves where the different reflected partial waves completely overlap. If the incident wave is a laser beam with the finite diameter D, the different reflected partial beams do *not* completely overlap because they are laterally shifted by $b = 2d\tan\beta$ (see Fig.4.39). The fraction b/D of the reflected partial waves does not overlap and cannot interfere destructively. This means that even for maximum transmission the reflected intensity is not zero but a

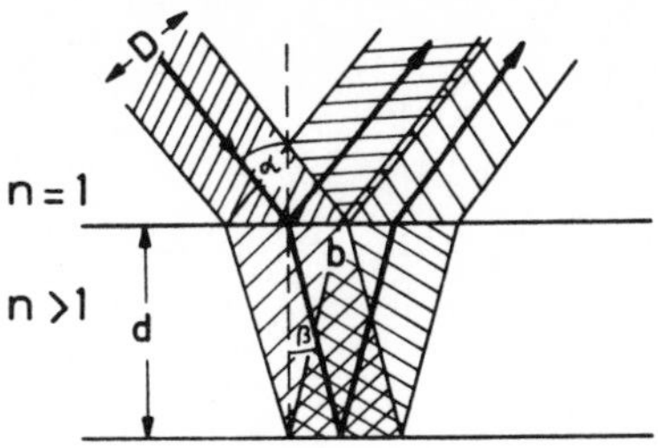

Fig.4.39. Incomplete overlap of interfering beams with diameter D

background reflection

$$I_R \approx (b/D)RI_0 = 2(d/D)RI_0 \tan\beta \tag{4.75}$$

remains, which is missing in the transmitted light. A parallel light beam with diameter D passing a plane parallel plate with an angle of incidence α therefore suffers, besides eventual absorption losses, reflection losses which increase with α and which are proportional to the ratio d/D of etalon thickness and beam diameter (walk-off losses).

Example

$d = 1$ cm, $D = 0.1$ cm, $R = 0.3$, $\tan\beta = 10^{-2}$ $I_R/I_0 = 0.06$, which means 6% walk-off losses.

The walk-off losses limit the tuning range of tilted etalons within a laser resonator. With increasing angle β the losses may become intolerably large [4.24b].

Illuminating the etalon with divergent monochromatic light (e.g., from an extended source or from a laser beam, behind a diverging lens), a continuous range of incident angles α is offered to the etalon, which transmits those directions α_m that obey (4.74a). In the transmitted light we then observe an interference system of bright rings (Fig.4.40). Since the reflected intensity $I_R = I_0 - I_T$ is complementary to the transmitted one, a corresponding system of dark rings appears in the reflected light at the same angles of incidence α_m.

4.2.7 Multilayer Dielectric Coatings

The constructive interference found for the reflection of light from plane parallel interfaces between two regions with different refractive indices can be utilized to produce highly reflecting, essentially absorption-free mirrors. The improved technology of such dielectric mirrors has greatly supported the development of visible and ultraviolet laser systems.

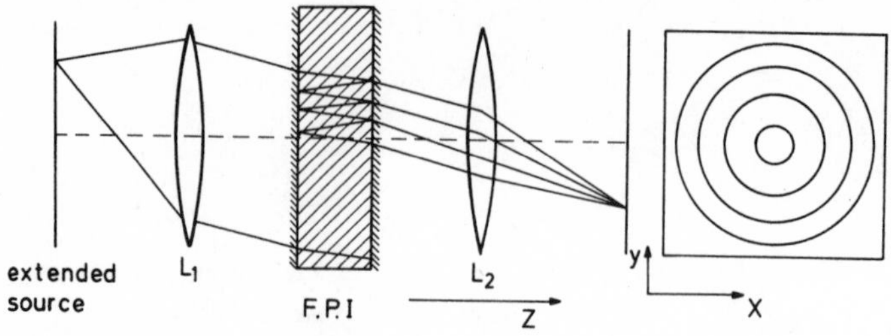

Fig.4.40. Ring system of the transmitted intensity caused by interference of divergent incident light

The reflectivity R of a plane interface between two regions with complex refractive indices $n_1 = n_1' - i\kappa_1$ and $n_2 = n_2' - i\kappa_2$ can be calculated from Fresnel's formulas (see, e.g., [2.3]). It depends on the angle of incidence and on the direction of polarization. For illustration $R(\alpha)$ is shown in Fig.4.41 for three different materials for incident light polarized parallel and perpendicular to the plane of incidence.

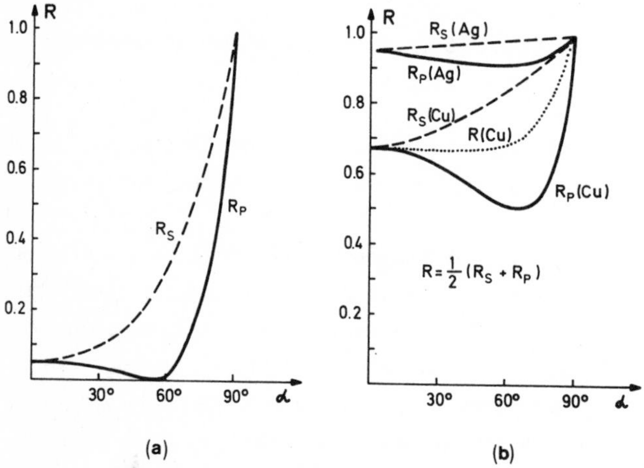

Fig.4.41a,b. Reflectivities R_p and R_s for the two polarization components parallel and perpendicular to the plane of incidence as a function of the angle of incidence. (a) Boundary air-glass ($n = 1.5$). (b) Air-metal boundary, for Cu($n' = 0.76$, $\kappa = 2.46$) and Ag($n' = 0.055$, $\kappa = 3.32$)

For vertical incidence ($\alpha = 0$) one obtains from Fresnel's formulas for both polarizations

$$R_{\alpha=0} = \left(\frac{n_1 - n_2}{n_1 + n_2}\right)^2 \quad . \tag{4.76}$$

Since this case represents the most common situation for laser mirrors, we shall restrict the following discussion to vertical incidence.

To achieve maximum reflectivities, the numerator $(n_1 - n_2)^2$ should be maximized and the denominator minimized. Since n_1 is always larger than one, this implies that n_2 should be as large as possible. Unfortunately the dispersion relations (2.55,56) imply that a large value of n also causes large absorption. For instance, highly polished metal surfaces have a maximum reflectivity R = 0.95 in the visible spectral range. The residual 5% of the incident intensity is absorbed and is therefore lost.

The situation can be improved by selecting reflecting materials with low absorption (which then necessarily have also low reflectivity) but using many layers with alternating high and low refractive index n. Choosing the proper optical thickness nd of each layer, allows constructive interference between the different reflected amplitudes to be achieved. Reflectivities of up to R = 0.999 have been reached [4.25].

Figure 4.42 illustrates such constructive interference for the example of a two-layer coating. The layers with refractive indices n_1, n_2 and thicknesses d_1, d_2 are evaporated onto an optically smooth substrate with refractive index n_3. The phase differences between all reflected components have to be $\delta_m = 2m\pi$ (m = 1, 2, 3, ...) for constructive interference. Taking into account the phase shift $\delta = \pi$ at a reflection from an interface with larger refractive index, we obtain the conditions

$$n_1 d_1 = \lambda/4 \quad \text{and} \quad n_2 d_2 = \lambda/2 \tag{4.77a}$$

for $n_1 > n_2 > n_3$, and

$$n_1 d_1 = n_2 d_2 = \lambda/4 \tag{4.77b}$$

for $n_1 > n_2$; $n_3 > n_2$. The reflected amplitudes can be calculated from Fresnel's formulas. The total reflected intensity is obtained by summation over all reflected amplitudes taking into account the correct phase. The refractive indices are now selected such that $\sum_i A_{iR}$ becomes a maximum. The calculation is still feasible for our example of a two-layer coating, and yields for the three reflected amplitudes (double reflections are neglected)

$$A_1 = \sqrt{R_1}\, A_0 \quad ; \quad A_2 = \sqrt{R_2}\,(1 - \sqrt{R_1})\, A_0 \quad ; \quad A_3 = \sqrt{R_3}\,(1 - \sqrt{R_2})(1 - \sqrt{R_1})\, A_0 \quad .$$

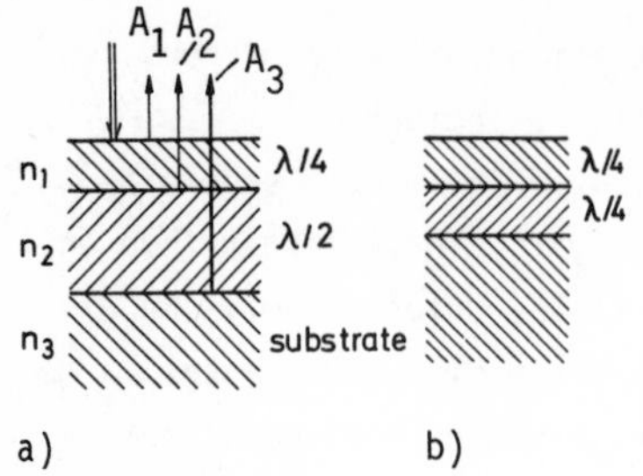

Fig.4.42a,b. Reflection of light with wavelength λ at a two-layer coating. (a) $n_1 > n_2 > n_3$. (b) $n_1 > n_2$; $n_2 < n_3$

Example

$|n_1| = 1.5$; $|n_2| = 1.2$; $|n_3| = 1.45$. $A_1 = 0.2\ A_0$; $A_2 = 0.14\ A_0$; $A_3 = 0.065\ A_0$. $A_R = \sum A_{iR} = 0.4\ A_0 \Rightarrow I_R = 0.16\ I_0 \Rightarrow R = 0.16$, provided the path differences have been choosen correctly.

This examples illustrates that for materials with low absorption, many layers are necessary to achieve a high reflectivity. Figure 4.43a shows schematically the composition of a dielectric multilayer mirror. The calculation and optimization of multilayer coatings with up to 20 layers becomes very tedious and time consuming, and is therefore performed using computer programs [4.26,27]. Figure 4.43b illustrates the reflectivitiy R(λ) of a high-reflectance mirror with 17 layers.

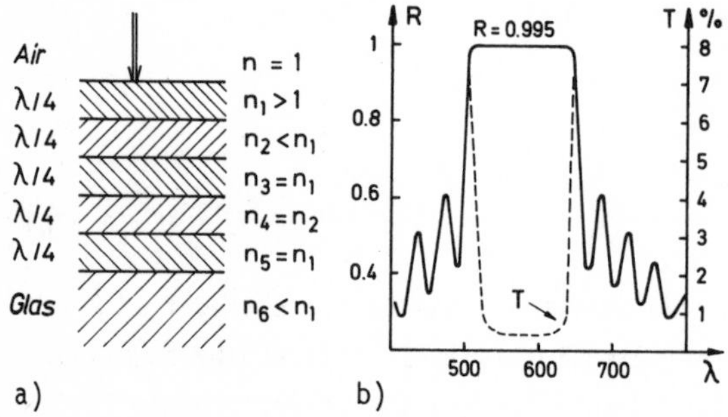

Fig.4.43a,b. Dielectric multilayer mirror. (a) Composition of multilayer coating. (b) Reflectivity of a high-reflectance multilayer mirror as a function of the incident wavelength

By proper selection of different layers with slightly different optical path lengths, one can achieve a high reflectivity over an extended spectral range. Nowadays such "broad-band reflectors" are available which have a reflectivity of $R \geq 0.99$ within a spectral range of $(\lambda_0 \pm 0.15\ \lambda_0)$ while the absorption losses are less than 0.3% [4.25]. At such low absorption losses the scattering of light from imperfect mirror surfaces may become the major loss contribution. When total losses of less than 0.5% are demanded, the mirror substrate has to be of high optical quality (better than $\lambda/20$), the dielectric layers have to be evaporated very uniformly, and the mirror surface must be clean and free of dust or dirty films.

Instead of maximizing the reflectivity of a dielectric multilayer coating through *constructive* interference, it is of course also possible to minimize it by *destructive* interference. Such *antireflection coatings* are commonly used to minimize unwanted reflections from the many surfaces of multiple-lens camera objectives, which would otherwise produce an annoying background illumination of the photomaterial. In laser spectroscopy such coatings are important for minimizing reflection losses of optical components inside the laser resonator and for avoiding reflections from the back-surface output mirrors which would introduce undesirable couplings, causing frequency instabilities of single-mode lasers.

Using a single layer (Fig.4.44a) the reflectivity reaches a minimum only for a selected wavelength λ (Fig.4.45b). We obtain $I_R = 0$ for $\delta = (2m + 1)\pi$, if the two amplitudes A_1 and A_2 reflected by the interfaces (n_1, n_2) and (n_2, n_3) are equal. For vertical incidence this gives the condition

$$R_1 = \left(\frac{n_1 - n_2}{n_1 + n_2}\right)^2 = R_2 = \left(\frac{n_2 - n_3}{n_2 + n_3}\right)^2 \quad , \tag{4.78}$$

which can be reduced to

$$n_2 = \sqrt{n_1 n_3} \quad . \tag{4.79}$$

For a single layer on a glass substrate the values are $n_1 = 1$ and $n_3 = 1.5$. According to (4.79) n_2 should be $n_2 = \sqrt{1.5} = 1.23$. Durable coatings with such low refractive indices are not available. One often uses MgF_2 with $n_2 = 1.38$, giving a reduction of reflection from 4% to 1.2% (see Fig.4.45).

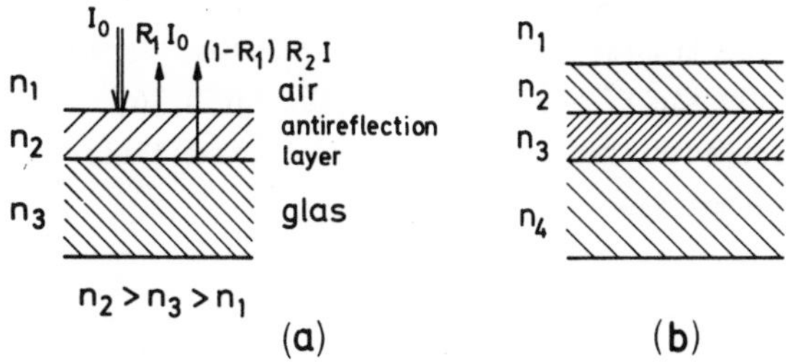

Fig.4.44a,b. Antireflection coatings. (a) Single layer. (b) Multilayer coatings

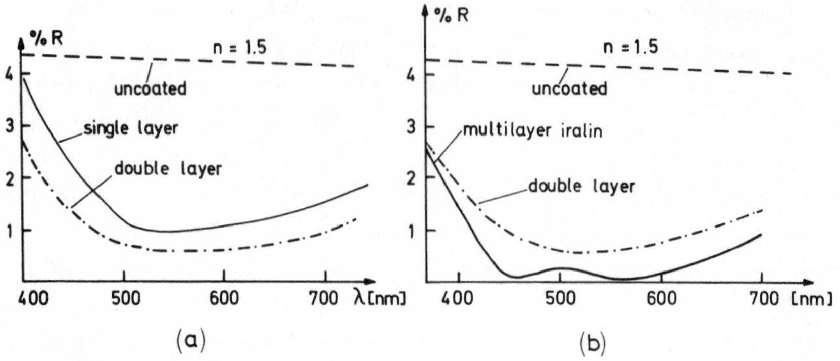

Fig.4.45a,b. Wavelength dependence of the reflectivity $R(\lambda)$ at normal incidence on a substrate with $n = 1.5$ at 550 nm, for uncoated substrate, single, double, and multilayer antireflection coatings

With multilayer antireflection coatings the reflectivity can be decreased below 0.2% for an extended spectral range [4.29]. For instance, with three $\lambda/4$ layers (MgF_2, SiO, and CeF_3) the reflection drops to below 1% for the whole range between 4200 and 8400 Å [4.30].

4.2.8 Interference Filters

Interference filters are used for the selective transmission of a narrow spectral range. Incident radiation of wavelength outside this transmission range is either reflected or absorbed. One distinguishes between line filters and band-pass filters.

A line filter is essentially a Fabry-Perot etalon with a very small optical path nd between the two reflecting surfaces. The technical realization uses the evaporation of two highly reflecting coatings (either silver coatings or dielectric multilayer coatings) which are separated by a nonabsorbing layer with low refractive index (Fig.4.46). For $nd = 0.5$ μm for instance, the transmission maxima at vertical incidence are obtained from (4.74a) at $\lambda_1 = 1$ μm, $\lambda_2 = 0.5$ μm, $\lambda_3 = 0.33$ μm, etc. In the visible range this filter has therefore only one transmission peak at $\lambda = 500$ nm, with a halfwidth which depends on the finesse $F^* = \pi\sqrt{R}/(1 - R)$ (see Fig.4.36).

The interference filter is characterized by the following quantities:

1) The wavelength λ_m at peak transmission.
2) The maximum transmission.
3) The contrast factor, which gives the ratio of maximum to minimum transmission.
4) The bandwidth at half transmitted peak intensity.

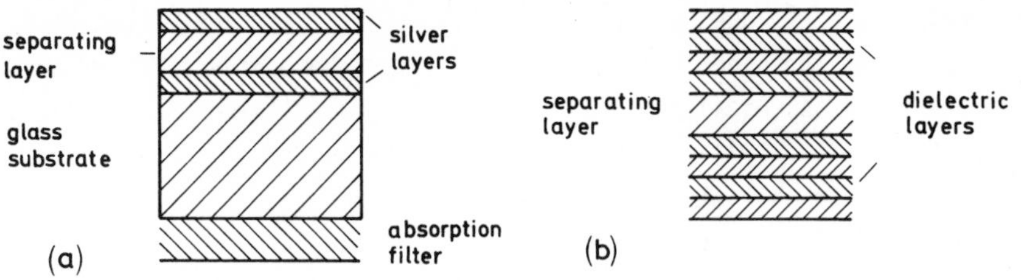

Fig.4.46a,b. Interference filter of the Fabry-Perot type. (a) With two single silver layers. (b) With dielectric multilayer coatings

The maximum transmission is according to (4.36) $T^*_{max} = T^2/(1 - R)^2$. Using thin silver or aluminum coatings with R = 0.8, T = 0.1, and A = 0.1 the transmission of the filter is only $T^* = 0.25$ and the finesse $F^* = 15$. For our example this means a halfwidth of 660 cm^{-1} at a free spectral range of 10^4 cm^{-1}. At λ = 500 nm this corresponds to a halfwidth of about 16 nm. For many applications in laser spectroscopy the low peak transmission of interference filters with absorbing metal coatings is not tolerable. One has to use absorption-free dielectric multilayer coatings (Fig.4.46b) with high reflectivity, which allows a large finesse and therefore a smaller bandwidth and a larger peak transmission (see Fig.4.47).

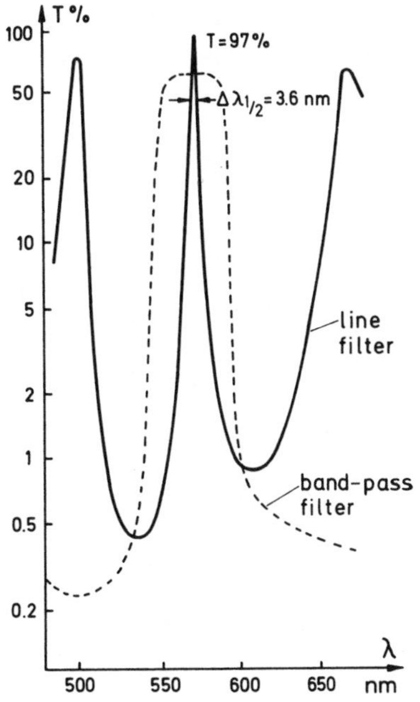

Fig.4.47. Spectral transmission of some interference filters. Solid line - line filter; dashed line - band-pass filter

A higher finesse F^*, due to larger reflectivities of the reflecting films, not only decreases the bandwidth but also increases the contrast factor which is according to (4.64) equal to $(1 + F) = 1 + 4F^{*2}/\pi^2$. With $R = 0.95$, ~ $F = 4R/(1 - R)^2 = 1.7 \times 10^3$, which means that the intensity at the transmission minimum is less than 1‰ of the peak transmission.

The bandwidth can be further decreased by using two interference filters in series. However, it is preferable to construct a double filter which consists of three highly reflecting systems, separated by two nonabsorbing layers of the same optical thickness. If the thickness of these two layers is made slightly different, a band-pass filter results which has a flat transmission peak but step slopes to both sides. Nowadays commercial interference filters are available with a peak transmission of at least 90% and a bandwidth of less than 2 nm [4.31]. Special narrow-band filters even reach 0.3 nm, however, with reduced peak transmission.

The wavelength λ_m of the transmission peak can be shifted to lower values by tilting the interference filter, which increases the angle of incidence α [see (4.74a)]. The tuning range is, however, restricted, because the reflectivity of the multilayer coatings also depends on the angle α and is in general optimized for $\alpha = 0$. The transmission bandwidth increases for divergent incident light with the divergence angle.

In the ultraviolet region where the absorption of most materials, used for interference filters, becomes large, the selective *reflectance* of interference filters can be utilized to achieve narrow-band filters with low losses. For more detailed treatment see [4.26-31].

4.2.9 Plane Fabry-Perot Interferometer

The widely used plane Fabry-Perot interferometer (F.P.I.) is applied to absolute wavelength measurements and high-resolution studies of spectral line profiles. Since its spectral resolving power may exceed 10^7, spectral profiles of Doppler-broadened or pressure-broadened lines can be studied with this instrument.

Different from the solid etalon, which is a plane parallel plate coated on both sides with reflecting layers, the plane F.P.I. consists of two wedged plates, each having one high-reflection and one antireflection coating (Figs.4.38 and 48). The wedge prevents undesirable interference from the back surfaces. The finesse of the F.P.I. depends, apart from the reflectivity R and the optical surface quality, critically on the parallel alignment of the two reflecting surfaces. The advantage of the F.P.I., that any desired

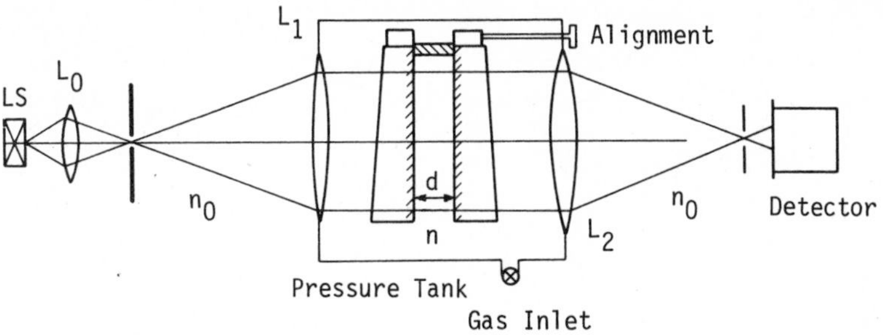

Fig.4.48. Use of a plane F.P.I. with parallel incident light from a point source and photoelectric recording of the transmitted intensity I(nd)

free spectral range can be realized by choosing the corresponding plate separation d, must be paid for by the inconvenience of careful alignment.

The F.P.I. can be used with divergent incident light from extended sources and also with parallel light from a point source. In the former case the transmitted intensity shows an interference pattern consisting of concentric rings (see Fig.4.40) which can be photographed for wavelength measurements. When θ is the angle of inclination to the interferometer axis, the transmitted intensity is maximum for

$$m\lambda = 2nd\cos\theta \quad , \tag{4.80}$$

where n is the refractive index between the plates. Let us number the rings by the integer p, beginning with p = 0 for the central ring. With $m = m_0 - p$, we can rewrite (4.80) for small angles θ_p as (see Fig.4.49)

$$(m_0 - p)\lambda = 2nd\cos\theta_p \approx 2nd(1 - \theta_p^2/2) = 2nd[1 - (n_0/n)^2\theta_p'^2/2] \quad . \tag{4.81}$$

When the interference pattern is imaged by a lens with focal length f into the plane of the photoplate, we obtain for the ring diameters the relations

$$\begin{aligned}(m_0 - p)\lambda &= 2nd[1 - (n_0/n)^2 D_p^2/(8f^2)]\\ (m_0 - p - 1)\lambda &= 2nd[1 - (n_0/n)^2 D_{p+1}^2/(8f^2)] \quad .\end{aligned} \tag{4.82}$$

Subtracting the second equation from the first yields

$$\boxed{D_{p+1}^2 - D_p^2 = \frac{4nf^2}{n_0^2 d}\lambda} \quad . \tag{4.83}$$

Provided the distance d is accurately known (n = 1 if the F.P.I. is evacuated), the wavelength λ can be obtained from the ring diameters. However, the wavelength is determined by (4.83) only modulo a free spectral range 2nd. This

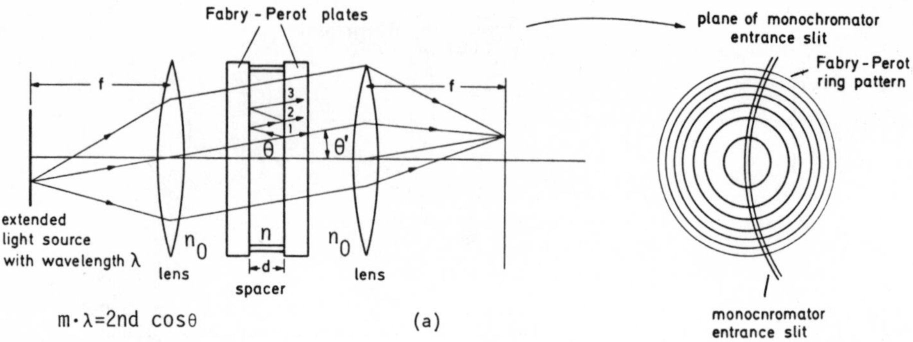

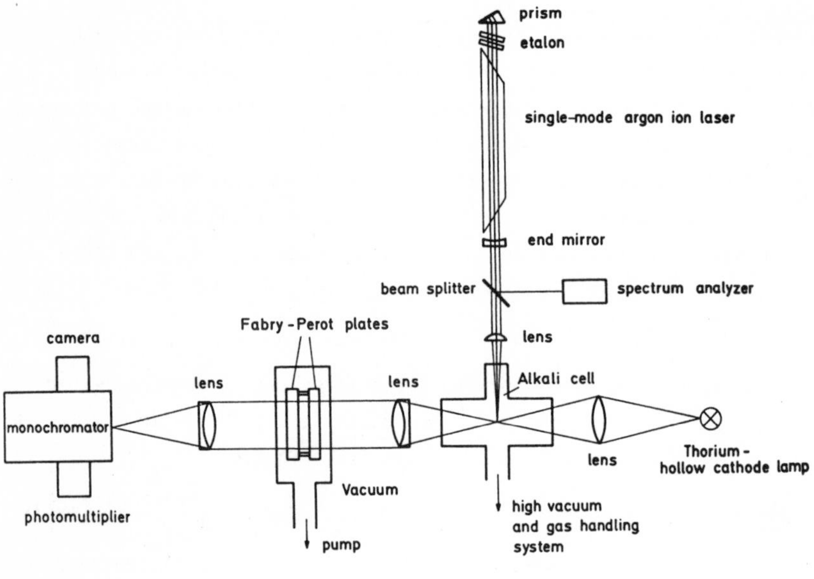

Fig.4.49a,b. Combination of F.P.I. and spectrograph for high-resolution spectroscopy of laser-induced fluorescence. (a) Imaging of the ring system onto the entrance slit of the spectrograph. (b) Experimental arrangement

means that all wavelengths λ_m differing by m free spectral ranges produce the same ring systems.

The experimental solution of this ambiguity in the determination of λ utilizes a combination of F.P.I. and spectrograph in a so-called crossed arrangement (Fig.4.49) where the ring system of the F.P.I. is imaged onto the entrance slit of a spectrograph. The spectrograph disperses the slit images s(λ) with a medium dispersion in the x direction (see Sect.4.1) and the F.P.I. provides high dispersion in the y direction. The resolution of the spectrograph must be only sufficiently high to separate the images of two wavelengths differing by one free spectral range of the F.P.I. Figure 4.50 shows for illustration a section of the Na_2 fluorescence spectrum excited by an argon laser line. The ordinate corresponds to the F.P.I. dispersion and the abscissa to the spectrograph dispersion [4.32].

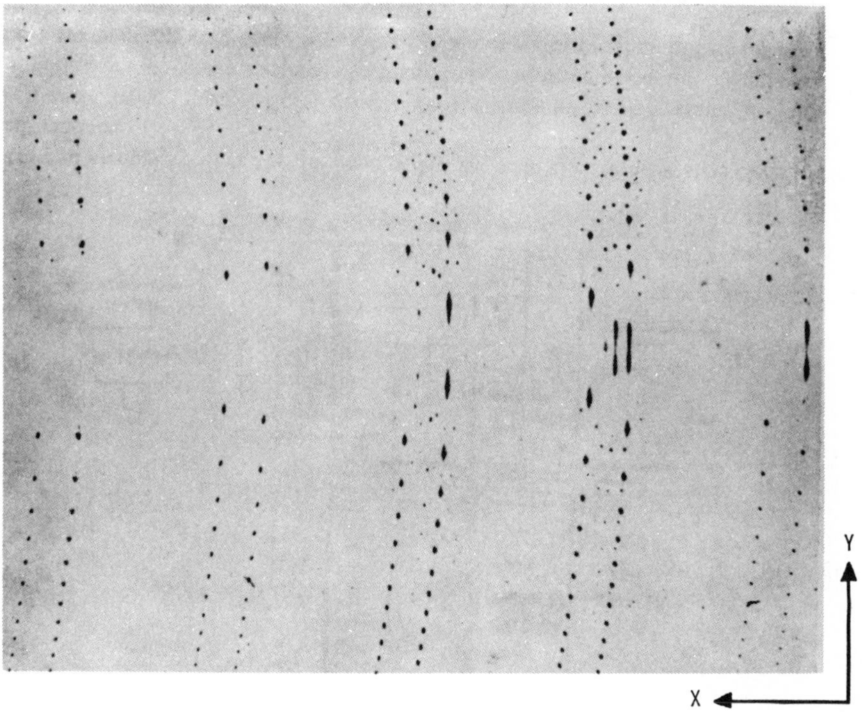

Fig.4.50. Section of the laser-excited Na_2 fluorescence spectrum obtained with the crossed arrangement of F.P.I. and spectrograph, shown in Fig.4.49

The angular dispersion of the F.P.I. can be deduced from (4.80)

$$\frac{d\theta}{d\lambda} = \left(\frac{d\lambda}{d\theta}\right)^{-1} = \frac{m}{2nd\ \sin\theta} = \frac{1}{\lambda_m \sin\theta} \quad \text{with} \quad \lambda_m = 2nd/m \quad . \tag{4.84}$$

Equation (4.84) shows that the angular dispersion becomes infinite for $\theta \rightarrow 0$. The linear dispersion of the ring system on the photoplate is

$$\frac{dD}{d\lambda} = f\,\frac{d\theta}{d\lambda} = \frac{f}{\lambda_m \sin\theta} \quad . \tag{4.85}$$

Example

f = 50 cm, λ = 0.5 µm. At a distance of 1 mm from the ring center is $\theta = 0.1/50$ and we obtain a linear dispersion of $dD/d\lambda$ = 50 mm/Å. This is at least one order of magnitude larger than the dispersion of a large spectrograph.

With photoelectric recording, the large dispersion at the ring center can be demonstrated. The light source is now imaged onto a small pinhole which serves as the point source in the focal plane of L_2 (Fig.4.48). The parallel light beam passes the F.P.I. and the transmitted intensity is imaged onto another pinhole in front of the detector. All light rays around $\theta = 0$ contribute to the central fringe. If the optical path length nd is tuned, the different transmission orders with $m = m_0, m_0 + 1, m_0 + 2 \ldots$ are successively transmitted for a wavelength λ according to $m\lambda = 2nd$.

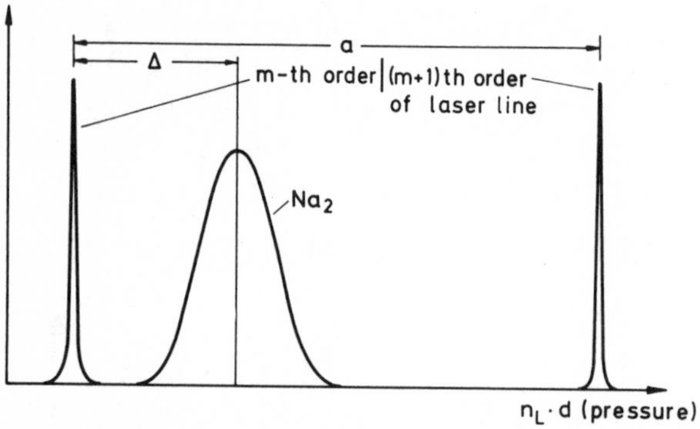

Fig.4.51. Photoelectric recording of a laser-excited Na_2 fluorescence line and its profile. The F.P.I. is pressure scanned through one free spectral range for the reference line λ = 488.0 nm of a single-mode argon laser

Tuning of nd can be achieved either by changing d with piezoelectric tuning of the plate separation or by altering the refractive index by a pressure change in the container enclosing the F.P.I. The wavelength measurement of an unknown wavelength λ relies on known calibration lines. Figure 4.51 shows a measurement of a Doppler-broadened Na_2 fluorescence line with a reference line from a single-mode argon laser. When the pressure within the F.P.I. is scanned, the Na_2 line appears between two successive orders of the reference line λ_R. From Fig.4.51 we obtain with $nd = d^*$ the relations

$$\left.\begin{aligned} 2d^* &= m_1\lambda_R \\ 2(d^* + a) &= (m_1 + 1)\lambda_R \\ 2(d^* + \Delta) &= m_2\lambda_x \end{aligned}\right\} \Rightarrow \lambda_x = \frac{m_1 + \Delta/a}{m_2}\lambda_R \quad ,$$

which shows that λ_x can be directly obtained from measurements of Δ and a, provided that the interference orders m_1 and m_2 are known (see above). The width of the measured profile of the single-mode laser in Fig.4.51 is determined by the finesse $F^* = 75$ of the F.P.I., whereas the profile of the Na_2 line reflects the Doppler broadening.

4.2.10 Confocal Fabry-Perot interferometer

A confocal F.P.I., often called a spherical interferometer [4.33], consists of two spherical mirrors M_1, M_2 with equal curvatures (radius r) which oppose each other at a distance d = r (Fig.4.52a). These interferometers have gained great importance in laser-physics: firstly, as high-resolution spectrum analyzers for detecting the mode structure and linewidth of lasers [4.34]; and secondly, in the nearly confocal form, as laser resonators. We discuss the former application in this section, while laser resonators will be treated in a more general way in Sect.6.3.

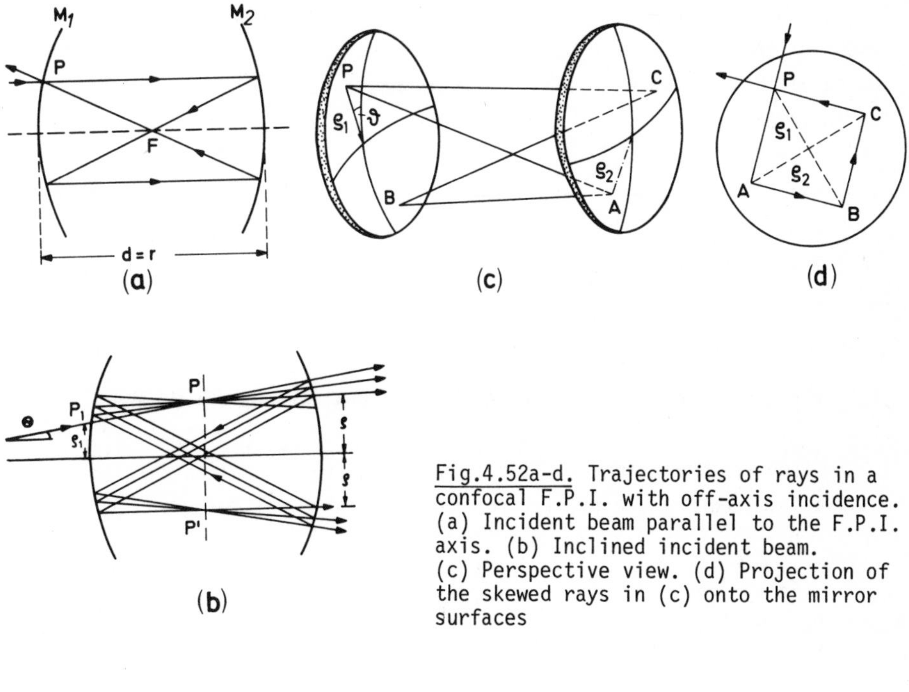

Fig.4.52a-d. Trajectories of rays in a confocal F.P.I. with off-axis incidence. (a) Incident beam parallel to the F.P.I. axis. (b) Inclined incident beam. (c) Perspective view. (d) Projection of the skewed rays in (c) onto the mirror surfaces

Neglecting spherical aberration, all light rays entering the F.P.I. parallel to the interferometer axis would pass through the focal point F and would reach the entrance point $\bar{P}$ again after having passed the F.P.I. four times. Figure 4.52b,c illustrate the general case of a ray which enters the F.P.I. at a small inclination ϑ and passes the successive points P, A, B, C, P shown in Fig.4.52d in a projection. ϑ is the skew angle of the entering ray.

Because of spherical aberration rays with different distances ρ_1 from the axis will not all go through F but will intersect the axis at different positions F' depending on ρ_1 and ϑ. Also each ray will not exactly reach the entrance point P_1 after four passages through the F.P.I. since it is slightly shifted at successive passages. It can be shown, however [4.33], that for sufficiently small angles ϑ, all rays intersect in the vicinity of two points P and P', located in the central plane of the F.P.I. at a distance $\rho(\rho_1,\vartheta)$ from the axis (Fig.4.52b).

The optical path difference Δs between two successive rays passing through P can be calculated from geometrical optics. For $\rho_1 \ll r$ and $\vartheta \ll 1$ one obtains [4.33a]

$$\Delta s = 4r + \rho_1^2\rho_2^2 \cos 2\theta/r^3 + \text{higher-order terms} \quad . \tag{4.86}$$

An incident light beam with diameter $D = 2\rho_1$ therefore produces in the central plane of the F.P.I an interference pattern of concentric rings. Analogous of the treatment in Sect.4.2.5, the intensity $I(\rho,\lambda)$ is obtained by adding all amplitudes with their correct phases $\delta = \delta_0 + (2\pi/\lambda)\Delta s$. According to (4.63) we get

$$I(\rho,\lambda) = \frac{I_0 T}{(1-R)^2 + 4R\sin^2[(\pi/\lambda)\Delta s]} \quad , \tag{4.87}$$

where $T = 1 - R - A$ is the transmission of M_1. The intensity has maxima for $\delta = 2m\pi$, which is equivalent to

$$4r + \rho^4/r^3 = m\lambda \tag{4.88}$$

when we neglect the higher order terms in (4.86) and set $\theta = 0$, and $\rho^2 = \rho_1\rho_2$.

The free spectral range $\delta\nu$, i.e., the frequency separation between successive interference maxima, is, for the confocal F.P.I. with $\rho \ll r$,

$$\delta\nu = \frac{c}{4r + \rho^4/r^3} \quad , \tag{4.89}$$

which is *different* from the expression $\delta\nu = c/2d$ for the plane F.P.I!

The radius ρ_m of the mth order interference ring is obtained from (4.88),

$$\rho_m = [(m\lambda - 4r)r^3]^{1/4} \quad . \tag{4.90}$$

If $4r = m\lambda$ is an integer multiple of λ, the radius of the central ring becomes zero. We can number the outer rings by the integer p and obtain with $m = m_0 + p$ for the radius of the pth ring the expression

$$\rho_p = (p\lambda r^3)^{1/4} . \tag{4.91}$$

The radial dispersion deduced from (4.90),

$$d\rho/d\lambda = (mr^3/4)[(m\lambda - 4r)r^3]^{-3/4} \quad , \tag{4.92}$$

becomes infinite for $m\lambda = 4r$, which means $\rho = 0$.

The total finesse of the confocal F.P.I. is in general higher than that of a plane F.P.I. for the following reasons:

1) The alignment of spherical mirrors is by far less critical than that of plane mirrors, because tilting of the spherical mirrors does not change (to a first approximation) the optical path length 4r through the confocal F.P.I. which remains approximately the same for all incident rays (see Fig.4.53). For the plane F.P.I., however, the pathlength increases for rays below the interferometer axis, but decreases for rays above the axis.

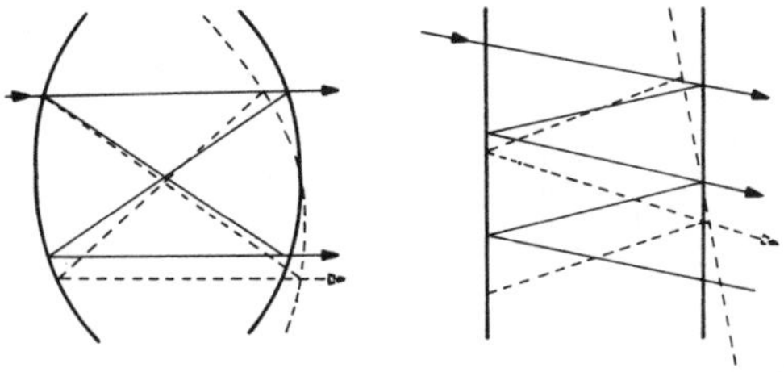

Fig.4.53. Illustration of the different sensitivities of the plane F.P.I. and the confocal F.P.I. to misalignment

2) Spherical mirrors can be polished to a higher precision than plane mirrors. This means that the deviations from an ideal sphere are less for spherical mirrors than those from an ideal plane for plane mirrors.

Furthermore such deviations do not wash out the interference structure but cause only a distortion of the ring system because a change of r allows according to (4.86) the same path difference Δs for another value of ρ.

The total finesse of a confocal F.P.I. is therefore mainly determined by the reflectivity R of the mirrors. For R = o.99 a finesse

$F^* = \pi\sqrt{R}/(1 - R) \approx 300$ can be achieved, which is much higher than that obtainable with a plane F.P.I. With a mirror separation $r = d = 3$ cm, the free spectral range is $\delta\nu = 2.5$ GHz and the spectral resolution would be $\Delta\nu = 10$ MHz at a finesse $F^* = 250$. This is already sufficient to measure the natural linewidth of many optical transitions.

If the central plane of the F.P.I. is imaged by a lens onto a circular aperture with sufficiently small radius b, only the central interference order is transmitted to the detector while all other orders are stopped (Fig.4.54). Because of the large radial dispersion for small ρ one obtains a high spectral resolving power. With this arrangement the instrumental line profile can be measured, when an incident monochromatic wave (from a stabilized single-mode laser) is used. The mirror separation $d = 4r + \varepsilon$ is varied by a small amount ε and the power $P(\lambda, b, \varepsilon)$, transmitted through the aperture, is measured as a function of ε, at fixed values of λ and b,

$$P(\lambda,b,\varepsilon) = 2\pi \int_{\rho=0}^{b} \rho \cdot I(\rho,\lambda,\delta) d\rho \quad . \tag{4.93}$$

The integrand $I(\rho,\lambda,\delta)$ can be obtained from (4.87), and the phase difference $\delta = 2\pi\Delta s/\lambda$ is related to the path difference

$$\Delta s = 4(r + \varepsilon) + \rho^4/(r + \varepsilon)^3 \approx 4(r + \varepsilon) + \rho^4/r^3 \quad . \tag{4.94}$$

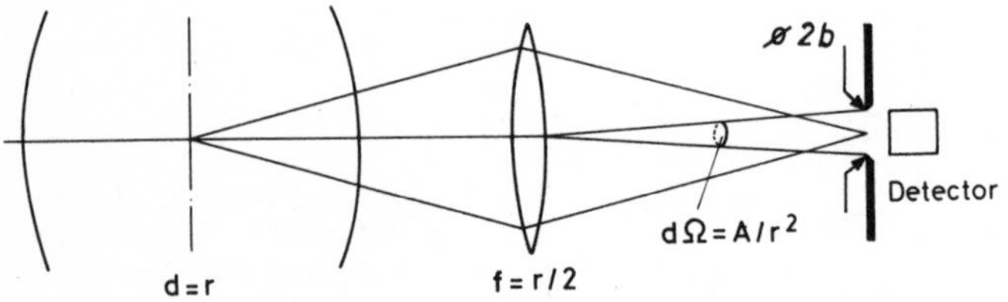

Fig.4.54. Photoelectric recording of the transmitted power of a scanning confocal F.P.I.

The optimum choice for the radius of the aperture is based on a compromise between spectral resolution and transmitted intensity. When the interferometer has a finesse F^*, the spectral halfwidth of the transmission peak is $\delta\nu/F^*$ [see (4.70)] and the maximum spectral resolving power becomes $F^*\Delta s/\lambda$ (4.72). For a radius $b = (r^3\lambda/F^*)^{1/4}$ of the aperture, which is just $1/F^*$ of the radius ρ_1 of a fringe with $p = 1$ in (4.91), the spectral resolving power is reduced to about 70% of its maximum value. This can be verified by inserting this value of b into (4.93) and calculating the halfwidth of the transmission peak $I(\lambda_1, F^*, \varepsilon)$.

We now briefly compare the confocal F.P.I. with the plane F.P.I. In Sect.4.1, we discussed several basic characteristics of spectrometers which determine the selection of a special device for particular applications. Of these we mainly consider here the *light gathering power* (êtendue) U and the spectral resolution $\nu/\Delta\nu = F^*\nu/\delta\nu$. From Fig.4.54 we see that the solid angle accepted by the detector behind the aperture with radius $b = (r^3\lambda/F^*)^{1/4}$ is $\Omega = \pi b^2/r^2$. The êtendue U is therefore (see Sect.4.1.1)

$$U = A\Omega = \pi b^2 \pi b^2/r^2 = \pi^2 r\lambda/F^* \quad . \qquad (4.95)$$

For a given finesse F^*, the êtendue increases with the mirror separation $d = r$. Since the free spectral range $\delta\nu = c/4r$ decreases with increasing r, the spectral resolving power $(\nu/\Delta\nu) = F^* 4r/\lambda$ also increases with r. With a given êtendue $U = \pi^2 r\lambda/F^*$ the spectral resolving power is

$$\boxed{(\nu/\Delta\nu) = [2F^*/(\pi\lambda)]^2 U} \qquad \text{(confocal F.P.I.)} \quad . \qquad (4.96)$$

Let us compare this with the situation in case of a plane F.P.I. with plate diameter D and separation d, which is illuminated with nearly parallel light (Fig.4.50). According to (4.80) the path difference between a ray parallel to the axis and a ray with an inclination θ is for small θ given by $\Delta s = nd\theta^2$. To achieve a finesse F^*, this variation of path length should not exceed λ/F^*, which restricts the solid angle $\Omega = \theta^2$ acceptable by the detector to $\Omega \leqq \lambda/(dF^*)$. The êtendue is therefore

$$U = A\Omega = \pi(D^2/4)\lambda/(dF^*) \quad . \qquad (4.97)$$ (4.97)

Inserting the value of d given by this equation in the spectral resolving power, $\nu/\Delta\nu = 2dF^*/\lambda$, we obtain

$$\boxed{\nu/\Delta\nu = \pi D^2/(2U)} \qquad \text{(plane F.P.I.)} \quad . \qquad (4.98)$$

While the spectral resolving power is proportional to U for the confocal F.P.I. it is inversely proportional to U for the plane F.P.I. The reason for this is that the êtendue increases with the mirror separation d for the confocal F.P.I. but decreases proportional to 1/d for the plane F.P.I. For a mirror radius $r > D^2/4d$ the êtendue of the confocal F.P.I. is larger than that of a plane F.P.I. with equal spectral resolution.

4.2.11 Lyot Filter

The basic principle of the Lyot filter [4.35] is founded on the interference of polarized light after having passed a birefringent crystal. Assume that a linearly polarized plane wave $\underline{E} = \underline{A}_0 \cos(\omega t - kx)$ with

$$\underline{A} = \{0, A_y, A_z\} \quad ; \quad A_y = |A|\sin\alpha \quad ; \quad A_z = |A|\cos\alpha$$

is incident on the birefringent crystal. The electric vector $\underline{E}$ makes an angle α with the optical axis, which points in the z direction. Within the crystal, the wave is split into an ordinary beam with wave number $k_0 = n_0 k$ and phase velocity $v_0 = c/n_0$, and an extraordinary beam with $k_e = n_e k$ and $v_e = c/n_e$. The partial waves have mutually orthogonal polarizations in directions parallel to the y and z axis, respectively.

The elementary Lyot filter consists of a birefringent crystal placed between two linear polarizers (Fig.4.55). Assume that the two polarizers are both parallel to the electric vector $\underline{E}(0)$ of the incoming wave. Let the crystal with length L be placed between $x = 0$ and $x = L$. Because of the different refractive indices n_0 and n_e for the ordinary and the extraordinary beams, the two partial waves at $x = L$,

$$E_y(L) = A_y \cos(\omega t - k_e L)$$

$$E_z(L) = A_z \cos(\omega t - k_0 L) \quad ,$$

show a phase difference of

$$\delta = k(n_0 - n_e)L = (2\pi/\lambda)\Delta nL \quad . \tag{4.99}$$

The superposition of these two waves results in general in elliptically polarized light, except for phase differences $\delta = 2m\pi$, where linearly polarized light with $\underline{E}(L)||\underline{E}(0)$ is obtained. For $\delta = (2m + 1)\pi$ and $\alpha = 45^\circ$, the transmitted wave is also linearly polarized but now $\underline{E}(L) \perp \underline{E}(0)$.

The second polarizer parallel to $\underline{E}(0)$ transmits only the projections

$$A_1 = A_y \sin\alpha = |A|\sin^2\alpha$$

$$A_2 = A_z \cos\alpha = |A|\cos^2\alpha$$

of the two components onto the direction of $\underline{E}(0)$, which interfere with each other. Because A_1 and A_2 have a phase difference δ, the total intensity behind the polarizer is, with $\bar{I} = I\cos^2\omega t = (I/2)$

$$\bar{I} = \frac{1}{2} c\epsilon_0 (A_1 + A_2)^2 = \bar{I}_0(\sin^4\alpha + \cos^4\alpha + 2\sin^2\alpha\cos^2\alpha\cos\delta) \quad . \tag{4.100}$$

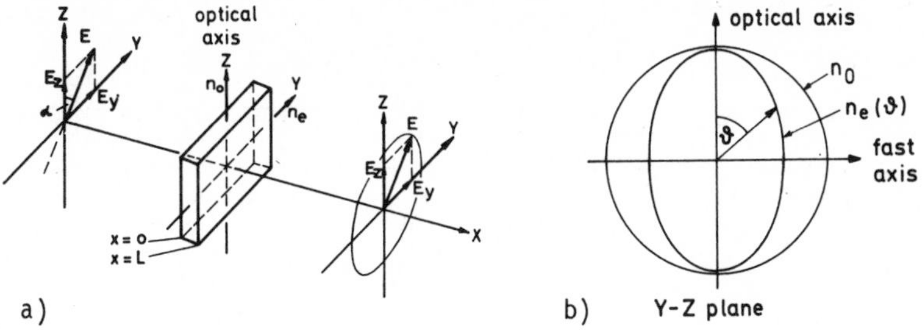

Fig.4.55a,b. Lyot filter. (a) Schematic arrangement. (b) Linearly polarized light passing through a birefringent crystal

Using the relation $\cos\delta = 1 - 2\sin^2\delta/2$, this reduces to

$$\bar{I} = \bar{I}_0(1 - \sin^2(\delta/2)\,\sin^2 2\alpha) \quad ; \tag{4.101a}$$

which gives for $\alpha = 45^\circ$

$$\bar{I} = \bar{I}_0(1 - \sin^2\delta/2) = \bar{I}_0\cos^2(\delta/2) \quad . \tag{4.101b}$$

The transmission of the Lyot filter is therefore a function of the phase retardation δ. Taking into account absorption and reflection losses, the maximum transmission $I_T/I_0 = T_0$ is smaller than unity,

$$T(\lambda) = I_T/I_0 = T_0\cos^2\frac{\pi\,\Delta n\,L}{\lambda} \quad . \tag{4.101c}$$

Within a small wavelength interval, the difference $\Delta n = n_0 - n_e$ can be regarded as constant and (4.101) gives the wavelength-dependent transmission function $\cos^2\delta$ typical of a two-beam interferometer (Fig.4.23). For extended spectral ranges the different dispersion of $n_0(\lambda)$ and $n_e(\lambda)$ has to be considered, which causes a wavelength dependence, $\Delta n(\lambda)$.

The free spectral range $\delta\nu$ is obtained from (4.101) with $\nu = c/\lambda$

$$\delta\nu = \frac{c}{(n_0 - n_e)L} \quad . \tag{4.102}$$

Example

For a crystal of potassium dihydrogen phosphate (KDP), $n_e = 1.31$, $n_0 = 1.47 \Rightarrow \Delta n = 0.04$ at $\lambda = 600$ nm. A crystal with L= 2 cm has then a free spectral range $\delta\nu = 3.75\times10^{11}\ \mathrm{s}^{-1} \leftrightarrow 37.5\ \mathrm{cm}^{-1} \leftrightarrow 1.35$ nm.

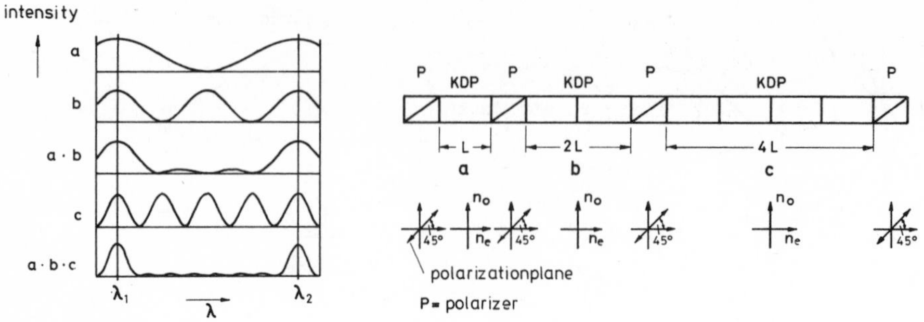

Fig.4.56. (a) Lyot filter, composed of three crystals with lengths L, 2L, and 3L. (b) Transmitted intensity

If N elementary Lyot filters with different lengths L_m are placed in series, the total transmission T is the product of the different transmissions T_m,

$$T(\lambda) = \prod_{m=1}^{N} \left(T_{0m} \cos^2 \frac{\pi \Delta n L_m}{\lambda} \right) \quad . \tag{4.101d}$$

Figure 4.56 shows a possible experimental arrangement and the corresponding transmission for a Lyot filter composed of three components with lengths $L_1 = L$, $L_2 = 2L$, and $L_3 = 4L$ [4.36]. The free spectral range $\delta\nu$ of this filter equals that of the shortest component; the halfwidth $\Delta\nu$ of the transmission peaks is, however, mainly determined by the longest component. When we define, analogous to the Fabry-Perot interferometer, the finesse F of the Lyot filter as the ratio of the free spectral range $\delta\nu$ to the halfwidth $\Delta\nu$, we obtain for a composite Lyot filter with N elements of lengths $L_m = 2^{m-1} L_1$ a finesse which is approximately $F^* = 2^N$.

The wavelength of the transmission peak can be tuned electro-optically by using the different dependence of the refractive indices n_0 and n_e on an external electric field [4.37]. This "induced birefringence" depends on the orientation of the crystal axis in the electric field. A common arrangement uses a KDP crystal with an orientation where the electric field is parallel to the optical axis (z axis) and the wave vector $\underline{k}$ of the incident wave is perpendicular to the z direction (transverse electro-optic effect) (Fig.4.57). Two opposite sides of the rectangular crystal with side length d are coated with gold electrodes and the electric field $E = U/d$ is controlled by the applied voltage.

In the external electric field the uniaxial crystal becomes biaxial. Additional to the natural birefringence of the uniaxial crystal, a field-in-

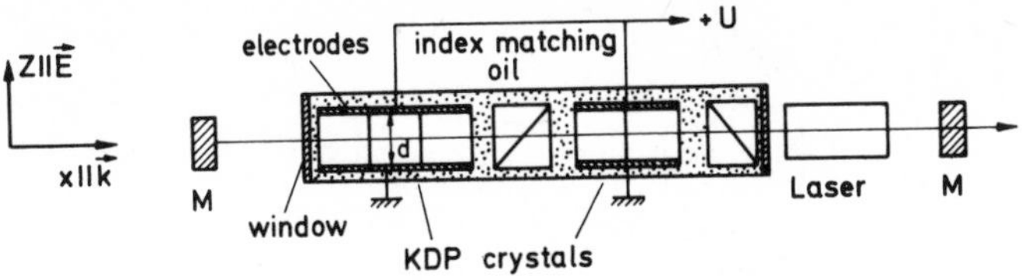

Fig.4.57. Electro-optic tuning of Lyot filter

duced birefringence is generated which is approximately proportional to the field strength [4.38]. The refraction index $n = \sqrt{\varepsilon}$ is determined by the dielectric tensor

$$\varepsilon_{ij} = \varepsilon_0 \left[1 + \chi_{ij}(E = 0) + \sum_k \chi_{ijk} E_k\right] , \qquad (4.102a)$$

where the second term gives the birefringence of the crystal without external field.

The field-induced susceptibility tensor χ_{ijk} has 27 components which, however, reduce to 18 because symmetry requires that $\chi_{ijk} = \chi_{ikj}$. These 18 components which are arranged in a reduced 3×6 matrix depend on the symmetry class of the crystal [4.38a]. For KDP only three components are nonzero and the field-induced contribution to the refractive index for the electric field component E_z is

$$\Delta n(E_z) = \frac{1}{2} n_1^3 d_{36} E_z , \qquad (4.102b)$$

where d_{36} is the component $\chi_{z,(x,y)}$ in the reduced Voigt notation [4.38a].

The electro-optical coefficient for KDP is $d_{36} = -10.7 \times 10^{-12}$ m/V.

Maximum transmittance is obtained for

$$(\Delta n + \Delta n_{el})L/\lambda = m \quad (m = 0, 1, 2 \ldots)$$

which gives the wavelength λ at the maximum transmittance

$$\lambda = (\Delta n + 0.5\, n_1^3 d_{36} E_z)L/m \qquad (4.103)$$

as a function of the applied field.

While this electro-optic tuning of the Lyot filter allows rapid switching of the peak transmission, for many applications, where a high tuning speed is not demanded, mechanical tuning is more convenient. This can be achieved with tilted birefringent plates which are used as tuning elements within the resonator of cw dye lasers to narrow the spectral width of the laser output [4.39]. If the tilting angle is chosen to be equal to the Brewster angle, the plates act both as retarding and polarizing elements and no additional polarizers are necessary (Fig.4.58). When the incoming beam enters

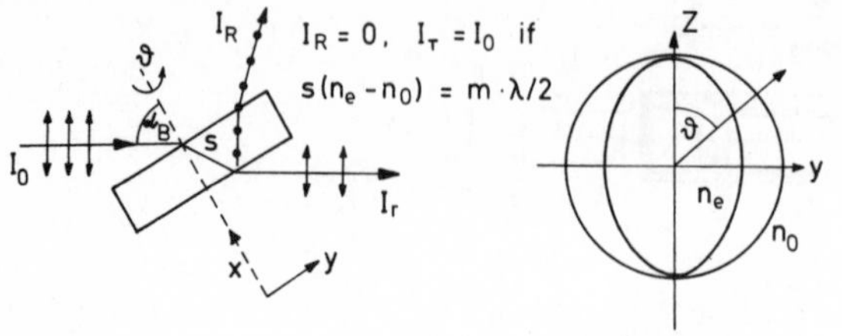

Fig.4.58a,b. Mechanical tuning of a Lyot filter. (a) Tilted birefringent plate used as Lyot filter within a laser resonator. (b) Corresponding index ellipsoid

the birefringent plate whose optic axis is out of the plane defined by the p polarization of the Brewster surfaces, it will be split into an ordinary and an extraordinary wave. While the ordinary refractive index n_0 in an uniaxial crystal is the same for all directions ϑ against the optical axis, the extraordinary beam has a refractive index n_e depending on ϑ. This is illustrated with the index ellipsoid in Fig.4.55b, which gives both refractive indices for a given wavelength as a function of ϑ. The difference $\Delta n = n_0 - n_e$ therefore depends on ϑ. The two axes of the ellipsoid with minimum n_e and maximum n_0 are often called the *fast* and the *slow* axes. Turning the crystal around the dashed x axis, which is perpendicular to the yz plane of Fig.4.55b, results in a continuous change of Δn and a corresponding tuning of the peak transmission wavelength λ.

4.2.12 Tunable Interferometers

For many applications in laser spectroscopy it is advantageous to have a high-resolution interferometer which is able to scan in a given time interval Δt through a limited spectral range $\Delta\nu$. The scanning speed depends on the method used for tuning and the spectral range $\Delta\nu$ is limited by the free spectral range $\delta\nu$ of the instrument. All techniques for tuning the wavelength $\lambda_m = 2nd/m$ at the transmission peak of an interferometer are based on a continuous change of the optical path nd. This can be achieved in different ways.

a) Change of the refractive index n by altering the pressure between the reflecting plates of a F.P.I. (pressure-scanned F.P.I.).

b) Change of the distance d with piezoelectric or magnetostrictive elements.

c) Tilting of solid etalons with a given thickness d against the direction of the incoming plane wave.

d) Change of the optical path difference $\Delta s = \Delta nL$ in birefringent crystals by electro-optic tuning or by turning the optical axis of the crystal (Lyot filter).

The electro-optic tuning methods b) and d) have the advantage of fast tuning rates. With a commercial "spectrum analyzer", (which is a confocal F.P.I. with a piezoceramic distance holder) the transmitted wavelength λ can be repetetively scanned over more than one free spectral range with a saw-tooth voltage applied to the piezoelectric crystal [4.40,41]. Scanning rates up to several kilohertz are possible. Although the finesse of such devices may exceed 10^3, the hysteresis of piezoelectric crystals limits the accuracy of absolute wavelength calibration. Here a pressure-tuned F.P.I. may be advantageous. The pressure change has to be sufficiently slow to avoid turbulence and temperature drifts. With a digitally pressure-scanned F.P.I., where the pressure of the gas in the interferometer chamber is changed by small discrete steps, repetitive scans are reproduced within about 10^{-3} of the free spectral range [4.42].

For fast wavelength tuning of dye lasers, Lyot filters with electro-optic tuning are used within the laser resonator. A tuning range of a few nm can be repetitively scanned with repetition rates up to 10^5 per second [4.43].

4.3 Comparison Between Spectrometers and Interferometers

When comparing the advantages and disadvantages of different dispersing devices for spectroscopic analysis, the characteristic instrument properties discussed in the foregoing sections, such as *spectral resolving power, étendue, spectral transmission,* and *free spectral* range, are important for the selection of the optimum instrument. Of equal significance is the question of how accurately the wavelengths of spectral lines can be measured. To answer this question, further specifications are necessary, such as the backlash of monochromator drives, imaging errors in spectrographs, hysteresis in piezo-tuned interferometers, etc. In this section we shall include these points in the comparison of different devices in order to give the reader an impression of the capabilities and limitations of these instrument.

4.3.1 Spectral Resolving Power and Light-Gathering Power

The spectral resolving power discussed for the different instruments in the previous sections can be expressed in a more general way which applies to all devices with spectral dispersion, based on interference effects. Let

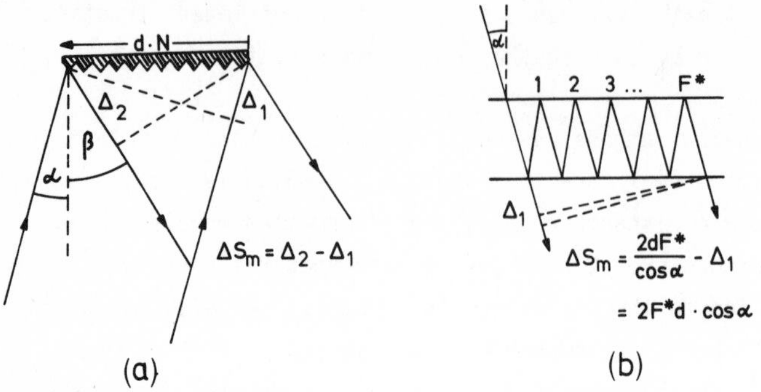

Fig.4.59a,b. Optical path difference and spectral resolving power. (a) In a grating spectrometer. (b) In a Fabry-Perot interferometer

Δs_m be the maximum path difference between interfering waves in the instrument, e.g., between the rays from the first and the last groove of a grating (Fig.4.59a) or between the direct beam and a beam reflected m times in a Fabry-Perot interferometer. Two wavelengths λ_1 and $\lambda_2 = \lambda_1 + \Delta\lambda$ can still be resolved if the number of wavelengths on this maximum path difference

$$m\lambda_2 = (m + 1)\lambda_1 = \Delta s_m \tag{4.104a}$$

differs for the two wavelengths at least by one. In this case an interference maximum for λ_1 coincides with the first minimum for λ_2. From (4.104a) we obtain the theoretical upper limit for the resolving power,

$$\boxed{\frac{\lambda}{\Delta\lambda} = \frac{\Delta s_m}{\lambda}} \quad , \tag{4.104b}$$

which is equal to the maximum path difference measured in units of the wavelength λ.

With the maximum time difference $\Delta T_m = \Delta s_m/c$ for the traversing times on two paths with the path difference Δs_m, we obtain with $\nu = c/\lambda$ from (4.104) for the minimum resolvable frequency interval $\Delta\nu = -|c/\lambda^2|\Delta\lambda$,

$$\Delta\nu = 1/\Delta T_m \quad . \tag{4.105}$$

The product of the minimum resolvable frequency interval $\Delta\nu$ and the maximum difference of traversing times through the spectral apparatus is equal to one.

Examples

a) Grating Spectrometer

The maximum path difference is according to (4.31) and Fig.4.59a
$\Delta s = Nd(\sin\alpha - \sin\beta) = mN\lambda$.
The upper limit for the resolving power is therefore

$$R = \lambda/\Delta\lambda = mN \quad . \quad (m = \text{diffraction order}, N = \text{number of grooves})$$

For $m = 2$ and $N = 10^5$ this gives $R = 2\times10^5$, or $\Delta\lambda = 5\times10^{-6}\lambda$. Because of diffraction (see Sect.4.1.2), the realizable resolving power is 3-4 times lower. This means that at $\lambda = 500$ nm two lines with $\Delta\lambda \geqq 0.1$ Å can be still resolved.

b) Michelson Interferometer

The path difference Δs between the two interfering beams is changed from $\Delta s = 0$ to $\Delta s = \Delta s_m$, and the numbers of interference maxima are counted for the two components λ_1 and λ_2 (see Sect.4.2.3). A distinction between λ_1 and λ_2 is possible if the number $m_1 = \Delta s/\lambda_1$ differs at least by one from $m_2 = \Delta s/\lambda_2$. This immediately gives (4.104). With modern design, maximum path differences Δs up to several meters have been realized for wavelength measurements of stabilized lasers (see Sect.4.5.3). For $\lambda = 500$ nm and $\Delta s = 1$ m, we obtain $\lambda/\Delta\lambda = 2\times10^6$, which is one order of magnitude better than for the grating spectrometer.

c) Fabry-Perot Interferometer

Here the path difference is determined by the optical path difference 2nd between successive partial beams times the effective number of reflections which can be expressed by the reflectivity finesse $F^* = \pi\sqrt{R}/(1 - R)$. With ideal reflecting planes and perfect alignment, the maximum path difference would be $\Delta s_m = 2ndF^*$ and the spectral resolving power

$$\lambda/\Delta\lambda = F^* 2nd/\lambda \quad . \tag{4.72}$$

Because of imperfections of alignment and deviations from ideal planes, the effective finesse is lower than the reflectivity finesse. With a value of $F^*_{eff} = 50$, which can be achieved, we obtain for nd = 1 cm

$$\lambda/\Delta\lambda = 2\times10^6 \quad ,$$

which is comparable with the Michelson interferometer with $\Delta s_m = 100$ cm. However, with a confocal F.P.I., a finesse of $F^*_{eff} = 200$ can be achieved. With r = d = 4 cm we then obtain

$$\lambda/\Delta\lambda = F^* 4d/\lambda \approx 10^8 \quad ,$$

which means that for $\lambda = 500$ nm, two lines with $\Delta\lambda = 5\times10^{-5}$ Å ($\Delta\nu = 5$ MHz at $\nu = 5\times10^{14}$ s^{-1}) are still resolvable, provided that their linewidth is sufficiently small.

The *light-gathering power*, or *étendue*, has been defined in Sect.4.1.1 as the product $U = A\Omega$ of entrance area A and solid angle of acceptance, Ω, of the spectral apparatus. For most spectroscopic applications it is desirable to have an étendue as large as possible to gain intensity, and also to reach a maximum resolving power R. However, the two quantities U and R are not indepedent of each other but are related as can be seen from the following examples.

Examples

a) Spectrometer

The area of the entrance slit with width b and heights h is $A = bh$. The acceptance angle $\Omega = (a/f)^2$ is determined by the focal length of the collimating lens or mirror and the diameter a of the limiting aperture in the spectrometer (see Fig.4.60a). Using typical figures for a medium-sized spectrometer ($b = 10$ µm, $h = 0.5$ cm, $a = 10$ cm, $f = 100$ cm) we obtain $\Omega = 0.01$, $A = 5\times10^{-4}$ $cm^2 \Rightarrow U = 5\times10^{-6}$ cm^2 ster. According to (4.12) and (4.29) the spectral resolving power for a slitwidth b large compared to the minimum width $b_{min} = 2f\lambda/a$ is given by

$$R = \lambda/\Delta\lambda \approx (f\lambda/b)d\beta/d\lambda = f\lambda m/(bd) \quad . \text{(d = distance between adjacent grooves)}$$

The product

$$RU = \lambda mhaN/f \tag{4.106}$$

increases with the diffraction order m, the size a of the grating, the number of grooves N, and the slit height h (as long as imaging errors can be neglected). For $m = 1$, $N = 10^5$, $\lambda = 500$ nm and the above figures for h, a, f we obtain

$$RU = 0.25\ cm^2 \quad .$$

We can write the étendue,

$$U = bha^2/f^2 = (bh/f^2)a^2 = \Omega^* a^2 \quad ,$$

as the product of the area $S = a^2$ of the limiting aperture and the solid angle $\Omega^* = bh/f^2$ under which the entrance slit is seen from the collimating

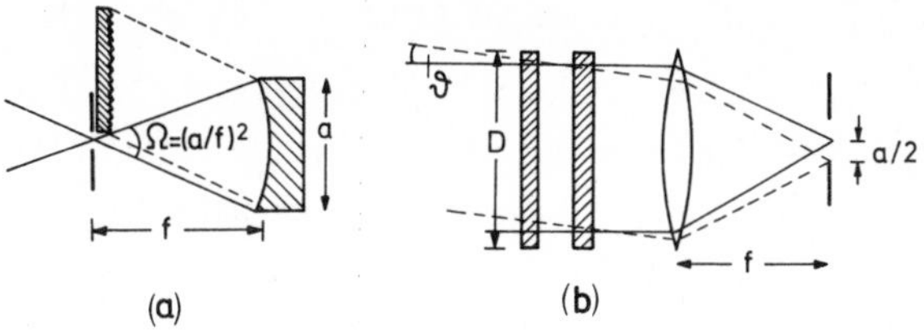

Fig.4.60. Acceptance angle of a spectrometer (a) and a Fabry-Perot interferometer (b)

mirror. This illustrates that the entrance slit forms the effective source area which is imaged into the plane of observation.

b) Interferometer

For the Michelson and Fabry-Perot interferometers, the allowable acceptance angle for photoelectric recording is limited by the aperture in front of the detector, which selects the central circular fringe. From Figs.4.54 and Fig.4.60b we see that the fringe images at the center and at the edge of the limiting aperture with diameter a are produced by incoming beams which are inclined by an angle ϑ against each other. With $a/2 = f\vartheta$ the solid angle accepted by the F.P.I. is $\Omega = \pi a^2/(4f^2)$. For a plate diameter D the étendue is $U = \pi(D^2/4)\vartheta^2$. According to (4.98) the spectral resolving power $R = \nu/\Delta\nu$ is correlated with the étendue U by $R = \pi D^2/(2U)$. The product RU is therefore

$$RU = \pi D^2/2 \quad . \tag{4.107}$$

For D = 5 cm, RU is about 40 cm^2 and therefore two orders of magnitude larger than for a grating spectrometer.

In section 4.2.10 we have seen that the spherical F.P.I. has for a given resolving power a larger étendue for mirror separations $r > D^2/4d$. For the example given above with D = 5 cm, d = 1 cm, the confocal F.P.I. therefore gives the largest product RU of all interferometers for r > 6 cm. Because of the higher total finesse, however, the confocal F.P.I. may be superior to all other instruments for these, as well as for smaller, mirror separations.

In summary, we can say that interferometers have a larger light-gathering power than spectrometers at comparable resolving power. However, the need for high-quality optical surfaces for mirrors and beam splitters and the inconvenience of careful alignment represent a certain disadvantage.

4.3.2 Precision and Accuracy of Wavelength Measurements

Resolving power and light-gathering power are not the only criteria by which a wavelength-dispersing instrument should be judged. A very important question is the attainable *precision* and *accuracy* of absolute wavelength measurements.

To measure a physical quantity always means to *compare* it with a reference standard. This comparison involves statistical and systematic errors. Measuring the same quantity n times will yield values X_i which scatter around a mean value

$$\bar{X} = \frac{1}{n}\sum_{i=1}^{n} X_i \quad .$$

The attainable *precision* for such a set of measurements is determined by statistical errors and is mainly limited by the signal-to-noise ratio for a single measurement and by the number n of measurements (i.e., by the total measuring time). The precision can be characterized by the *standard deviation* (see for instance [4.44])

$$\sigma = \sqrt{\sum_{i=1}^{n} \frac{(\bar{X} - X_i)^2}{n}} \quad . \tag{4.108}$$

The adopted mean value $\bar{X}$, averaged over many measured values X_i, is claimed to have a certain *accuracy*, which is a measure of the reliability of this value, expressed by its probable deviation $\Delta\bar{X}$ from the unknown "true" value X. A stated accuracy of $\bar{X}/\overline{\Delta X}$ means a certain confidence that the true value X is within $\bar{X} \pm \overline{\Delta X}$. Since the accuracy is determined not only by statistical errors but particularly by systematic errors of apparatus and measuring procedure, it is always lower than the precision. It is also influenced by the precision with which the reference standard can be measured and by the accuracy of its comparison with the value $\bar{X}$. Although the attainable accuracy depends on the experimental efforts and expenditures, the skillfulness, imagination, and critical judgement of the experimentalist always have a major influence on the finally achieved and stated accuracy.

We shall characterize precision and accuracy by the ratios $\sigma/\bar{X}$, $\overline{\Delta X}/\bar{X}$, respectively. Note that often both quantities are inversely defined as being inversely proportional to σ or $\overline{\Delta X}$. Although the latter definition has the advantage that a high precision or a high accuracy means a *small* uncertainty in accordance with the common meaning of both expressions, we shall follow the more common tradition where, e.g., a standard deviation $\sigma = 10^{-8} \cdot \bar{X}$ is characterized by a precision of 10^{-8}.

Let us now briefly examine the attainable precision and accuracy of wavelength measurements with the different instruments discussed above. Although both quantities are correlated with the resolving power and the attainable signal-to-noise ratio, they are furthermore influenced by many other instrumental conditions, such as backlash of the monochromator drive, or asymmetric line profiles caused by imaging errors, or shrinking of the photographic film during the developing process. Without such additional error sources the precision could be much higher than the resolving power, because the center of a symmetric line profile can be measured to a small fraction ε of the halfwidth. The value of ε depends on the attainable signal-to-noise ratio, which is determined apart from other factors by the êtendue of the spectrometer. We see that for the precision of wavelength measurements the product RU, discussed in the previous section, plays an important role.

For scanning monochromators with photoelectric recording the main limitation for the attainable accuracy is the backlash of the grating drive and nonuniformities of the gears, which limit the reliability of linear extrapolation between two calibration lines. Carefully designed monochromators have errors due to the drive which are less than 0.1 cm^{-1}, allowing an accuracy of about 10^{-5} in the visible range.

A serious source of error with scanning spectrometers is the distortion of the line profile and the shift of the line center caused by the time constant of the recording device. If the time constant τ is comparable with the time $\Delta t = \Delta\lambda / v_{sc}$ needed to scan through the halfwidth $\Delta\lambda$ of the line profile (which depends on the spectral resolution), the line becomes broadened, the maximum decreases, and the center wavelength is shifted. The line shift S depends on the scanning speed v_{sc}[Å/min] and is approximately $S = v_{sc}\tau$ [4.1].

Example

With a scanning speed of 100 Å/min and a time constant of the recorder of τ = 1s the line shift is already S = 1.5 Å!

Because of the additional line broadening the resolving power is reduced. If this reduction is to be less than 10%, the scanning speed must be below $v_{sc} < 0.24\Delta\lambda/\tau$.

Example

$\Delta\lambda = 0.2$ Å, $\tau = 1\text{s} \Rightarrow v_{sc} < 3$ Å/min.

Photographic recording avoids these problems and allows therefore a more accurate wavelength determination at the expense of inconvenient developing process of the photoplate and the following measuring procedure to determine the line positions. A typical figure for the standard deviation for a 3 m spectrograph is 0.01 cm^{-1}. Imaging errors causing curved lines, asymmetric line profiles due to misalignment, and backlash of the microdensitometer used for measuring the line positions on the photoplate are the main sources of errors.

HÄNSCH has proposed and verified an elegant improvement of wavelength measurements using a self-calibrating grating [4.45]. Assume that a grating is ruled in such a way that most of the grooves have a length L_1 (say $L_1 = 0.5$ a) but every tenth groove has a length $L_2 > L_1$ (say $L_2 = 0.6$ a); every hundredth, $L_3 > L_2$ ($L_3 = 0.7$ a); every thousandth, $L_4 = 0.8$ a, etc. (see Fig.4.61a). Illuminating such a grating with light from a reference source (e.g., a stabilized He-Ne laser) will produce a line pattern in the plane of observation, which is caused by interference of partial waves diffracted from the different parts of the grating with groove separations d, 10 d, 100 d, 1000 d. According to (4.22), the angles β_m for the different interference orders m are given by

$$m\lambda_R = D(\sin\alpha - \sin\beta_m) \quad \text{with} \quad D = 10^k d \quad , \quad k = 0, 1, 2 \ldots \quad . \tag{4.109}$$

The linear separation Δx of these "ghosts" produced by the special grating for a monochromatic incident wave are, for $\alpha = 0$,

$$\Delta x = f_2 \sin\beta_m = f_2 m\lambda_R/D \quad . \tag{4.110}$$

These spacings represent a calibration scale in the form of a decimal ruler. The unknown wavelength λ_x of a laser or an incoherent source illuminating the grating simultaneously with the calibration laser (Fig.4.61) can be read directly on this ruler with an accuracy which is not affected by the error sources discussed above, because the calibration scale is directly produced by the same grating which diffracts the unknown line. Compared to conventional photographic recording this method offers the advantage of uniformly spaced calibration lines with nearly equal intensities, which increases the accuracy. Since the ruler and the unknown line are produced simultaneously, thermal expansion of the instrument does not influence the accuracy.

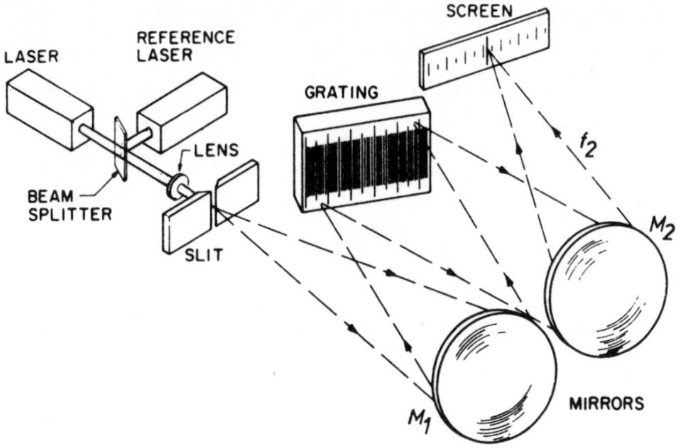

Fig.4.61. Schematic arrangement of wavelength determination with a self-calibrating grating (after [4.45])

4.4 New Techniques of Wavelength Measurements

With the ultrahigh resolution which can in principle be achieved with single-mode tunable lasers (see Chaps.6, 10), the accuracy of absolute wavelength measurements attainable with conventional techniques may not be satisfactory. New methods have been developed which are mainly based on interferometric measurements of laser wavelengths. For applications in molecular spectroscopy, the laser can be stabilized on the center of a molecular transition. Measuring the wavelength of such a stabilized laser yields simultaneously the wavelength of the molecular transition with a comparable accuracy. We shall briefly discuss some of these devices, often called "wavemeters" which measure the unknown laser wavelength by comparison with a reference wavelength λ_R of a stabilized reference laser. Most proposals use for reference a He-Ne laser, stabilized on a hyperfine component of a molecular iodine line, which has been measured by direct comparison with the primary wavelength standard to an accuracy of better than 10^{-8} [4.46].

Another method measures the absolute frequency ν_L of a stabilized laser and deduces the wavelength λ_L from the relation $\lambda_L = c/\nu_L$ using the most recent experimental value of the speed of light [4.47]. *Defining* this value of c as the speed of light would allow the much higher accuracy of frequency measurements to apply directly to wavelength determinations [4.48]. This method will be discussed in Sect.6.10.

The different types of wavemeters are based on modifications of the Michelson interferometer, the Fizeau interferometer [4.4], or on combinations of several Fabry-Perot interferometers with different free spectral ranges [4.50]. The wavelength is measured either by monitoring the spatial distribution of the interference pattern by photographic recording or with photodiode arrays, or by using travelling devices with electronic counting of the interference fringes.

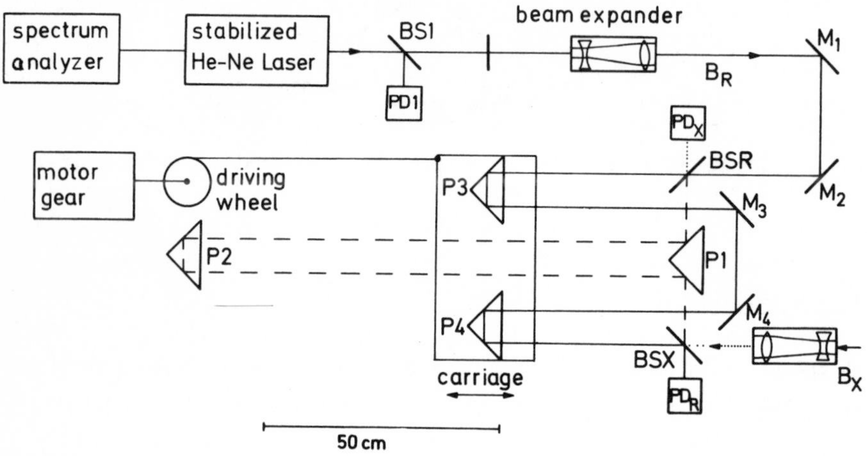

Fig.4.62. Travelling Michelson interferometer for accurate measurement of wavelengths of cw lasers

Figure 4.62 illustrates the principle of a travelling interferometer as used in our laboratory. Such a wavemeter was first demonstrated in a slightly different version by HALL and LEE [4.51] and by KOWALSKI et al. [4.52]. The Stanford version uses identical light paths for reference laser beam B_R and laser beam B_X with unknown wavelength λ_X; these are traversed by the two beams in opposite directions. The moving end mirror in one arm of the Michelson interferometer has been replaced by a corner-cube reflector which guarantees that the incoming light beam is always reflected exactly parallel to its incident direction, irrespective of slight misalignments or movements of the travelling reflector. The two partial beams (BSR-P_3-M_3-M_4-P_4-BSX) and BSR-P1-P2-P1-BSX for the reference laser interfere at the detector PD_R, and the two beams BSX-P4-M4-M3-P3-BSR and BSX-P1-P2-P1-BSR from the unknown laser interfere at the detector PD_X. When the carriage is moving at a speed v, the phase difference between the two interfering beams changes as

$\delta(t) = 2\pi\Delta s/\lambda = 2\pi \cdot 4\Delta x/\lambda = 8\pi vt/\lambda$, Δx = travelled distance

where the optical path difference Δs has been doubled by introducing two corner-cube reflectors. The rates of interference maxima which occur for $\delta = m2\pi$ are counted by PD_X and PD_R for the unknown wavelength λ_X and the reference wavelength λ_R. From the ratio of both counting rates the unknown wavelength λ_X can be obtained if proper corrections are made for the dispersion $n(\lambda_R) - n(\lambda_X)$ of air.

For a maximum optical path difference $\Delta s = 4$ m, the number of counts for $\lambda = 500$ nm is 8×10^6, which allows a precision of about 10^{-7}, if the counting error is not larger than one. Provided the signal-to-noise ratio is sufficiently good, the attainable precision can be, however, enhanced by interpolation between two successive counts using a *phase-locked loop* [4.53]. This is an electronic device which multiplies the frequency of the incoming signal by a factor M while always being locked to the phase of the incoming signal. Assume that both counters behind PD_R and PD_X are started simultaneously. The counting frequency $f_R = 4v/\lambda_R$ of PD_R may be multiplied by M. Both counters are simultaneously stopped as soon as PD_X has reached a preset value N_0. At this time t_0 the counter behind PD_R has counted $N_R = Mf_Rt_0$ counts, with $t_0 = N_0 4v/\lambda_x$. The unknown wavelength is then, including dispersion corrections, obtained from

$$N_0\lambda_x = (N_R/M)\lambda_R \frac{n(\lambda_x,p,T)}{n(\lambda_R,p,T)} \quad . \qquad \begin{aligned} p &= \text{atmospheric pressure} \\ T &= \text{temperature} \end{aligned} \tag{4.111}$$

For M = 100 this allows measurement of the ratio λ_x/λ_R with a precision $N_R = MN_0\lambda_x/\lambda_R$, which is about $1{:}(8 \times 10^8)$ for our example above, provided the signal-to-noise ratio is larger than M.

The attainable accuracy, however, is in general lower, because it is influenced by several sources of systematic errors. One of them is a misalignment of the interferometer, which causes both beams to travel slightly different path lengths. Another point which has to be considered is the curvature of the wavefronts in the diffraction-limited Gaussian beams (see Sect.5.3). This curvature can be reduced by expanding the beams through telescopes (see Fig.4.62). With careful alignment and with a good optical quality of all mirrors and beam splitters an accuracy of a few MHz can be achieved. This means for $\nu_L = 6 \times 10^{14}$ a relative accuracy of about 10^{-8}.

While the travelling Michelson wavemeter is restricted to cw lasers, a motionless Michelson interferometer has been designed by JACQUINOT et al. [4.54], which includes no moving parts and can be used for cw as well as for

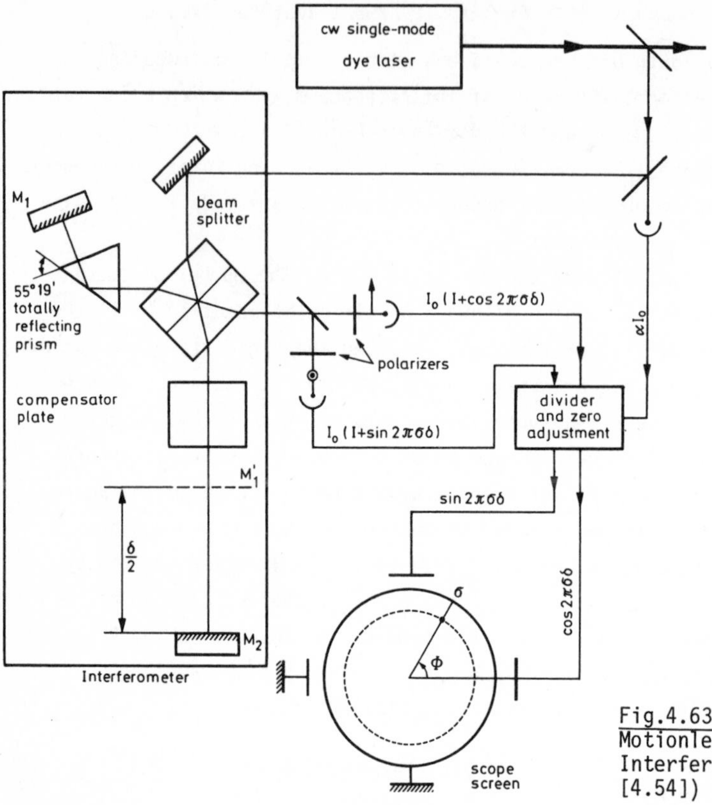

Fig.4.63. Motionless Michelson-Interferometer (from [4.54])

pulsed lasers. Figure 4.63 illustrates its operation. The basic element is a Michelson with a *fixed* path difference Δs. The laser beam enters the interferometer polarized at 45° with respect to the plane of Fig.4.63. When inserting a prism into one arm of the interferometer, where the beam is totally reflected at the prism base, a phase difference $\Delta\varphi$ is introduced between the two components polarized parallel and perpendicular to the totally reflecting surface. The value of $\Delta\varphi$ depends, according to Fresnel's formulas [2.3], on the incidence angle α and can be made $\pi/2$ for $\alpha = 55^{\circ}19'$ and $n = 1.52$. The interference signal at the exit of the interferometer is recorded separately for the two polarizations and one obtains, because of the phase shift $\pi/2$, $I_{\parallel} = I_0(1 + \cos 2\pi\Delta s/\lambda)$ and $I_{\perp} = I_0(1 + \sin 2\pi\Delta s/\lambda)$. From these signals it is possible to deduce the wave number $\sigma = 1/\lambda$ modulo $1/\Delta s$, since all wave numbers $\sigma_m = \sigma_0 + m/\Delta s$ $(m = 1, 2, 3, \ldots)$ give the same interference signals. Using several interferometers of the same type, with a common mirror M_1, which have path differences in geometric ratios, such as 50 cm,

5 cm, 0.5 cm, and 0.05 cm, the wave number $\bar{\nu}$ can be deduced unambiguously with an accuracy determined by the interferometer with the highest path difference. The actual path differences Δs_i are calibrated with a reference line and are servo-locked to this line. The precision obtained with this instrument is about 6 MHz, which is comparable with that of the travelling Michelson. The measuring time, however, is much smaller, since the different Δs_i can be measured simultaneously.

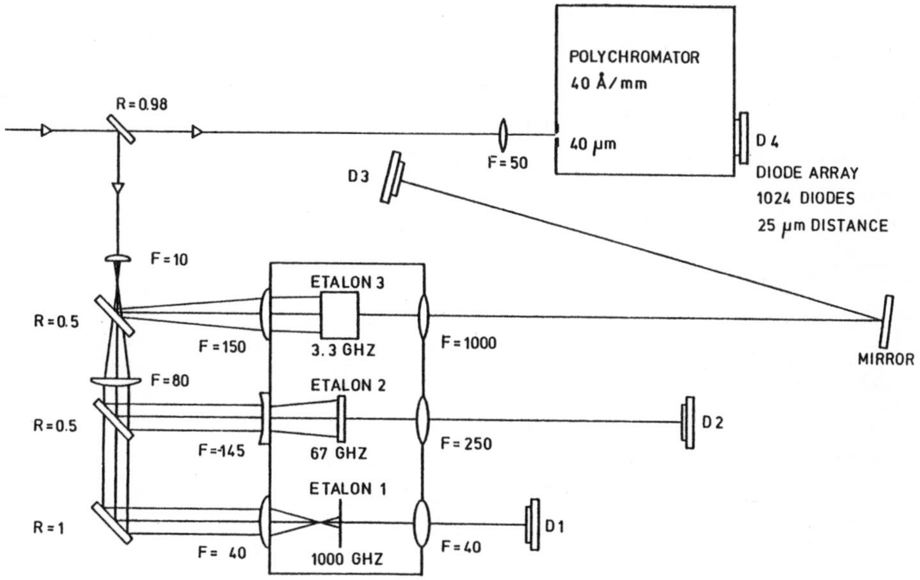

Fig.4.64. Wavemeter for pulsed and cw-lasers, based on a combination of a small polychromator and three Fabry-Perots with widely different free spectral ranges [50a]

Another approach to accurate wavelength measurements of pulsed and cw lasers, which can be also applied to incoherent sources, relies on a combination of grating monochromator and three Fabry-Perot etalons [4.50]. The incoming laser beam is sent simultaneously through a small monochromator and three temperature-stabilized Fabry-Perot etalons with free spectral ranges of

1000 GHz, 67 GHz and 3.3 GHz. The monochromator output and the fringe patterns behind the three FPI are projected onto photodiode-arrays (see Sect. 4.5.9). The first diode array behind the monochromator monitors the approximate line position to ±1 Å, which is about 15% of the free spectral range of the first thin etalon. Each of the other three diode arrays measures the diameters of 3 FPI fringes and a microprocessor calculates the fractional fringe order ε to ±1%.

The device is calibrated with different lines from a cw dye laser which are simultaneously measured with the travelling Michelson wavemeter (see above). The unknown wavenumber $\bar{\nu} = 1/\lambda = 2(m + \varepsilon)/d$ can then be deduced from the measured value of ε, and the distance d known from the calibration. The interference order m_1 for the thin etalon is known from the monochromator reading. The more accurate wavelength determination from this etalon provides the order m_2 for the second FPI, etc. The fringe center of the thick etalon can be measured to ±30 MHz. This wavemeter allows single shot observation and has the further advantage that line positions and lineprofiles can be measured simultaneously.

4.5 Detection of Light

For many applications in spectroscopy the sensitive detection of light and the accurate measurement of its intensity are of crucial importance for the successful performance of an experiment [4.55a]. The selection of the proper detector for optimum *sensitivity* and *accuracy* for the detection of radiation must take into account the following characteristic properties which may differ for the various detector types:

1) The spectral relative response $R(\lambda)$ of the detector, which determines the wavelength range where the detector can be used. The knowledge of $R(\lambda)$ is essential for the comparison of relative intensities $I(\lambda_1)$ and $I(\lambda_2)$ at different wavelengths.

2) The absolute sensitivity $S(\lambda) = V_S/P$, which is defined as the ratio of output signal V_S to incident radiation power P. If the output is a voltage, as in photovoltaic devices or in thermocouples, the sensitivity is expressed in units of volts per watt. In case of photocurrent devices, such as photomultipliers, $S(\lambda)$ is given in amperes per watt. With the detector area A the sensitivity S can be expressed in terms of the irradiance I,

$$S(\lambda) = V_S/(AI) \quad . \tag{4.112}$$

3) The attainable signal-to-noise ratio V_s/V_n which is in principle limited by the noise of the incident radiation, but may in practice be further reduced by inherent noise of the detector. The detector noise is often expressed by the *noise equivalent input power* (NEP), which means an incident radiation power which generates the same noise level as the detector itself yielding a signal to noise ratio S/N = 1. In infrared physics a figure of merit for the infrared detector is the detectivity,

$$D^* = \frac{\sqrt{A\Delta f}}{P}\,\frac{V_s}{V_n} \quad . \tag{4.113}$$

The detectivity D^* gives the obtainable signal to noise ratio V_s/V_n, multiplied by the square root of detector area A and, detector bandwidth Δf and divided by the incident radiation power P.

4) The maximum intensity range where the detector response is linear, which means that the output signal V_s is proportional to the incident radiation power P. This point is particularly important for applications where a wide range of intensities is covered. Examples are output power measurements of pulsed lasers, Raman spectroscopy, and spectroscopic investigations of line broadening.

5) The time response of the detector, characterized by its time constant τ. When the incident radiation intensity is modulated at a frequency f, the output signal $V_s(f)$ will in general fall with increasing f. Many detectors show a frequency response which can be described by the model of a capacitor, which is charged through a resistor R_1 and discharged through R_2 (Fig.4.65). The output signal of such a device is characterized by

$$V_s(f) = \frac{V_s(0)}{\sqrt{1+(2\pi f\tau)^2}} \quad , \tag{4.114}$$

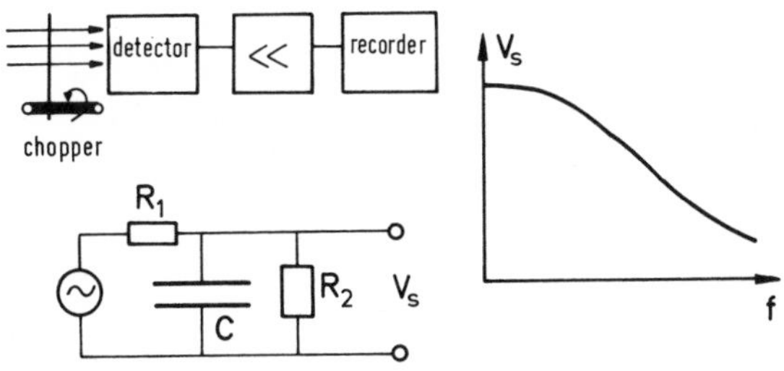

Fig.4.65. Schematic arrangement for detection of radiation, model of equivalent electronic circuit and frequency response of detectivity

where $\tau = C \cdot R_1R_2/(R_1 + R_2)$. At a modulation frequency $f = 1/(2\pi\tau)$, the output signal has decreased to $1/\sqrt{2}$ of its dc value. The knowledge of the detector time constant τ is essential for all applications where fast transient phenomena are to be monitored, such as atomic lifetimes or the time dependence of fast laser pulses (see Chap.12).

6) The price of a detector is another factor that cannot be ignored, since unfortunately it often restricts the optimum choice.

In this section we briefly discuss some detectors which are commonly used in laser spectroscopy. The different types can be divided into two categories, *thermal detectors* and *direct photodetectors*. In thermal detectors, the energy absorbed from the incident radiation raises the temperature and causes changes in temperature-dependent properties of the detector which can be monitored. Direct photodetectors are based either on emission of photoelectrons from photocathodes, or on changes of the conductivity of semiconductors due to incident radiation, or on photovoltaic devices where a voltage is generated by internal photoeffect. While thermal detectors have a *wavelength-independent* sensitivity, photodetectors show a spectral response which depends on the work function of the emitting surface or on the band gap in semiconductors.

During recent years the development of image intensifiers, image converters, and vidicon detectors has made impressive progress. At first pushed by military demands, these devices are now coming into use for light detection at low levels, e.g., in Raman spectroscopy, or for monitoring the faint fluorescence of spurious molecular constituents. Because of their increasing importance we give a short survey of the principles of these devices and their application in laser spectroscopy. In time-resolved spectroscopy, subnanosecond detection can be now performed with fast phototubes and scan converter tubes in connection with transient digitizers, which resolve time intervals of less than 100 ps. Since such time-resolved experiments in laser spectroscopy are discussed in Chap.12, we confine ourselves here to discussing these modern devices from the point of view of spectroscopic instrumentation [4.56]. For a review on photodetection techniques relevant in laser physics, see also [4.66].

4.5.1 Thermal Detectors

Because of their wavelength-independent sensitivity, thermal detectors are useful for calibration purposes, e.g., for absolute measurements of the radiation power of cw lasers or of the output energy of pulsed lasers. In the rugged form of medium-sensitivity calibrated calorimeters, they are convenient devices for any laser laboratory. With more sophisticated and delicate design, they have been developed as sensitive detectors for the whole spectral range and particularly for the infrared region, where other sensitive detectors are less abundant than in the visible range.

For a simple estimate of the sensitivity and its dependence on the detector parameters, such as the heat capacitance and thermal losses, we shall consider the following model [4.57]. Assume that a fraction η of the incident radiation power P is absorbed by a thermal detector with heat capacity H, which is connected to a heat sink at constant temperature T_s (Fig.4.66a. When G is the thermal conductivity of the link between the detector and the heat sink, the temperature T of the detector under illumination can be obtained from the equation

$$\eta P = H\frac{dT}{dt} + G(T - T_s) \quad . \tag{4.115}$$

In case of constant radiation power P, the temperature T rises from its initial value T_s to its stationary value (dT/dt = 0),

$$T = \eta P/G + T_s \quad , \tag{4.116}$$

which shows that the final temperature T is determined solely by the heat losses G and *not* by the heat capacity H.

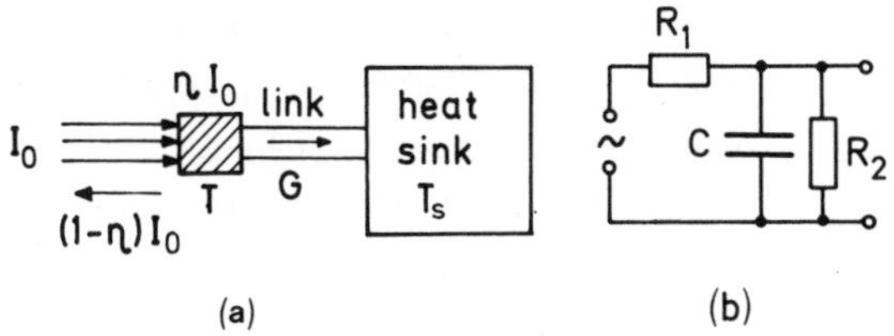

Fig.4.66. Model of a thermal detector

In general, P will be time dependent. When we assume a periodic dependence

$$P = P_0(1 + a\cos\omega t) \quad , \tag{4.117}$$

we obtain, inserting (4.117) into (4.115), a detector temperature of

$$T(\omega) = T_s + \frac{\eta P_0 a\cos(\omega t + \varphi)}{\sqrt{G^2 + \omega^2 H^2}} = T_s + \Delta T\cos(\omega t + \varphi) \quad , \tag{4.118}$$

which depends on the modulation frequency ω, and which shows a phase lag φ determined by

$$\tan\varphi = \omega H/G \quad . \tag{4.119}$$

At the frequency $\omega = G/H$, the amplitude ΔT decreases by a factor $\sqrt{2}$ compared to its dc value.

Note

The problem is equivalent to the analogous case of charging a capacitor ($C \leftrightarrow H$) through a resistor $R_1(R_1 \leftrightarrow P)$ which discharges through $R_2(R_2 \leftrightarrow 1/G)$. The ratio $\tau = H/G$ $(H/G \leftrightarrow R_2C)$ determines the time constant of the device (Fig.4.66b).

We learn from (4.118) that the sensitivity $S = \Delta T/P_0$ becomes large if G and H are made as small as possible. For modulation frequencies $\omega > G/H$ the amplitude ΔT will decrease approximately inversely to ω. Since the time constant $\tau = H/G$ limits the frequency response of the detector, *a fast and sensitive detector should have a minimum heat capacity H.*

For calibration of the output power of cw lasers, the demand for high sensitivity is not as relevant, since in general sufficiently large radiation power is available. Figure 4.67a shows a calorimeter and its circuit diagram. The radiation falls into a metal cone with a black inner surface. Because of the many reflections, the light has only a small chance of leaving the cone, ensuring that all light is absorbed. The absorbed power heats a thermocouple or a temperature-dependent resistor (thermistor), embedded in the cone. For calibration purposes, the cone can be heated by an electric heating wire. If the detector represents one part of a bridge (see Fig. 4.67c) which is balanced for an electric imput $W = UI$, but without incident radiation, the heating power has to be reduced by $\Delta W = P$ to maintain the balance with an incident radiation power P. A system with higher accuracy uses two identical cones, where only one is irradiated (Fig.4.67b).

For the measurement of output energies from pulsed lasers, the calorimeter should integrate the absorbed power at least over the pulse duration. From (4.115) we obtain

$$\int_0^{t_0} \eta P dt = H\Delta T + \int_0^{t_0} G(T - T_s)dt \quad . \tag{4.120}$$

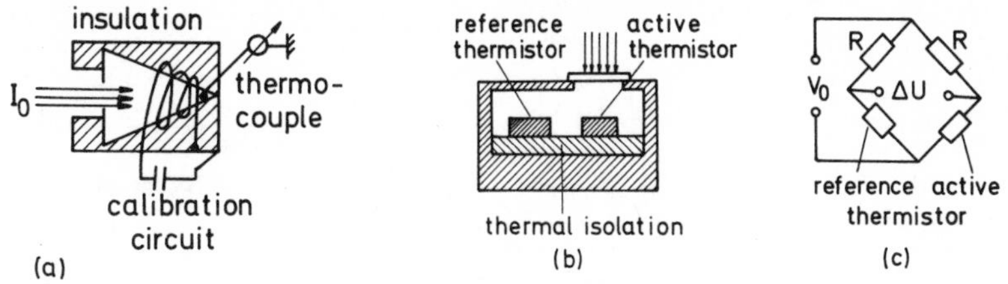

Fig.4.67. Calorimeter to measure the output energy of pulsed lasers or the output power of cw lasers. (a) Experimental design. (b) Calorimeter with active irradiated thermistor and non irradiated reference thermistor. (c) Balanced bridge circuit

When the detector is thermally isolated, the heat conduction G is small, and the second term may be completely neglected for sufficiently short pulse durations t_0. The temperature rise,

$$\Delta T = \frac{1}{H} \int \eta P dt \quad ,$$

is then directly proportional to the input energy. For calibrations, a charged capacitor C is discharged through the heating coil (Fig.4.66b). If the discharge time is matched to the laser pulse time, the heat conduction is the same for both cases, and does not enter into the calibration. If the temperature rise caused by the discharge of the capacitor equals that caused by the laser pulse, the pulse energy is $CU^2/2$.

For more sensitive detection of low incident powers, *bolometers* and *Golay cells* are used. A bolometer is a detector constructed from a material with a large temperature coefficient $\alpha = (dR/dT)/R$ of resistance R. If a constant current I is fed through R (see Fig.4.68), the incident power P which causes a temperature increase ΔT produces an output signal

$$U = IR\alpha\Delta T = [V_0 \, R/(R + R_1)]\alpha\Delta T \qquad (4.121)$$

where ΔT is determined from (4.118) as $\Delta T = \eta P(G^2 + \omega^2 H^2)^{-\frac{1}{2}}$. The response $\Delta U/P$ of the detector is therefore proportional to I, R, α and decreases with increasing H and G. Since the input impedance of the following amplifier has to be larger than R, this puts an upper limit to R. Because any fluctuation of I causes a noise signal, the current I through the bolometer has to be extremely constant. This, and the fact that the temperature rise by Joule heating should be small, limits the maximum current through the bolometer.

Equation (4.121) shows again that small values of G and H are desirable. Even with a perfect thermal isolation heat radiation is still present and

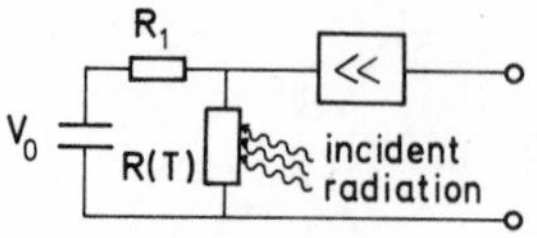

Fig.4.68. Bolometer. Schematic circuit diagram

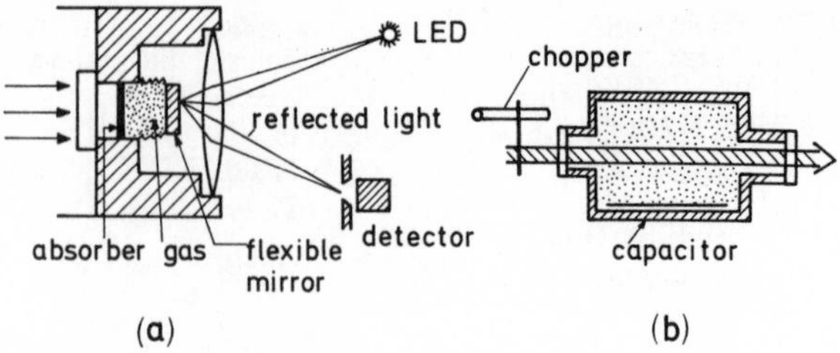

Fig.4.69. Golay cell. (a) Using deflection of light by a flexible mirror. (b) Using the capacitance change of a capacitor with flexible membrane (spectraphone)

limits the lower value of G. At a temperature difference ΔT from the surroundings, the Stefan-Boltzmann law gives for the net radiation flux $\Delta\phi$ to the surroundings from the detector with receiving area A and emissivity ε,

$$\Delta\phi = 4A\varepsilon\sigma T^3 \Delta T \quad . \tag{4.122}$$

The minimum thermal conductivity is therefore

$$G_m = 4A\sigma\varepsilon T^3 \quad . \tag{4.123}$$

This limits the detection sensitivity to a minimum input radiation of about 10^{-10} W for detectors operating at room temperatures and with a bandwidth of 1 Hz. It is therefore advantageous to cool the bolometer which furthermore decreases the heat capacity.

Another method of thermal detection of radiation is used in the Golay cell [4.58], which is based on the heating of an inert gas, such as xenon, by absorption of radiation in a closed capsule. According to the ideal gas law, the temperature rise ΔT causes a pressure rise $\Delta p = n(R/V)\Delta T$ (n = number of moles) which expands a flexible membrane on which a mirror is mounted (Fig.4.69a). The movement of the mirror is monitored by observing the deflection of a light beam from a light emitting diode. In modern devices the flexible membrane is part of a capacitor with the other plate fixed. The pressure rise causes a corresponding change of the capacitance which can be converted to an ac voltage (Fig.4.69b). This sensitive detector, which is essentially a *capacitor microphone* is now widely used in photoacoustic spectroscopy (see Sect.8.3) to detect the absorption spectrum of molecular gases by the pressure rise proportional to the absorption coefficient.

A recently developed thermal detector for infrared.radiation is based on the pyroelectric effect [4.59]. Pyroelectric materials are good electrical insultors which possess an internal macroscopic electric dipole moment, depending on the temperature. The electric field of this dielectric polarization is neutralized by an extrinsic surface-charge distribution. A change of the internal polarization caused by a temperature rise will produce a measurable change in surface charge, which can be monitored by a pair of electrodes applied to the sample. While the sensitivity of good pyroelectric detectors is comparable to that of Golay cells or high-sensitivity bolometers, they are more robust and therefore less delicate to handle [4.60,61].

4.5.2 Photoemissive Detectors

Photoemissive detectors, such as the photocell or the photomultiplier, are based on the external photoeffect. The photocathode of such a detector is covered with one or several layers of materials with a low work function ϕ (e.g., alkali metal compounds or semiconductor compounds). Under illumination with monochromatic light of wavelength $\lambda = c/\nu$ the emitted photoelectrons leave the photocathode with a kinetic energy given by the Einstein relation

$$E_{kin} = h\nu - \phi \quad . \tag{4.124}$$

They are further accelerated by the voltage V_0 between anode and cathode and are collected at the anode. The resultant photocurrent is measured either directly or by the voltage drop across a resistor (Fig.4.70a).

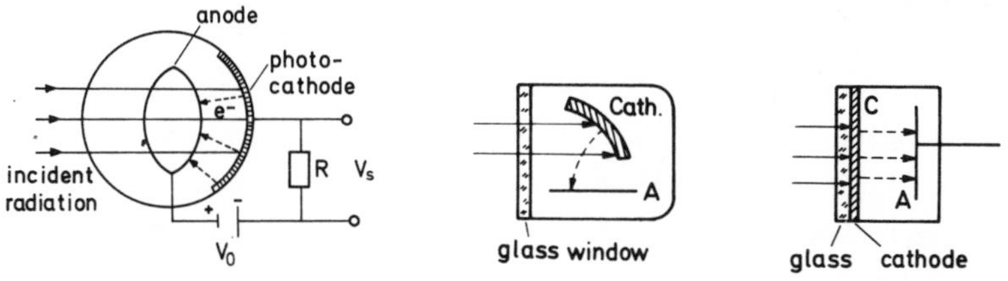

(a) (b) (c)

Fig.4.70a. Vacuum photocell. b) Opaque photocathode. c) Semitransparent photocathode

The quantum efficiency $\eta = n_e/n_{ph}$ is defined as the ratio of the rate of production of photoelectrons, n_e, to the rate of incidence of photons, n_{ph}. It depends on the cathode material, on the form and thickness of the photoemissive layer, and on the wavelength λ of the incident radiation. The quantum efficiency $\eta = n_a n_b n_c$ can be represented by the product of three factors. The first factor n_a gives the probability that an incident photon is actually absorbed. For materials with a large absorption coefficient, such as pure metals, the reflectivity R is high (e.g., for metallic surfaces $R \geq 0.8$-0.9 in the visible range), and the factor n_a cannot be larger than (1-R). For semitransparent photocathodes of thickness d on the other hand, the absorption must be large enough to ensure that $\alpha d \gg 1$. The second factor gives the probability that the absorbed photon really produces a photoelectron instead of heating the cathode material. Finally, the third factor stands for the probability that this photoelectron reaches the surface and will be emitted instead of being backscattered into the interior of the cathode.

Two types of photoelectron emitters are manufactured: opaque layers, where light is incident on the same side of the photocathode from which the photoelectrons are emitted (Fig.4.70b); and semitransparent layers (Fig. 4.70c), where light enters at the opposite side to the photoelectron emission and is absorbed throughout the thickness d of the layer. Because of the two factors n_a and n_c, the quantum efficiency of semitransparent cathodes and its spectral change are critically dependent on the thickness d and only reach that of the reflection-mode cathode if the value of d is optimized.

Figure 4.71 shows the spectral sensitivity $S(\lambda)$ of some typical photocathodes, scaled in milliamperes of photocurrent per watt incident radiation. For comparison, the quantum efficiency curves for $\eta = 0.01$ and 0.1 are also drawn (dashed curves). Both quantities are related by

$$S = I/P = (n_e e)/(n_{ph} h\nu) \Rightarrow S = \eta e/(h\nu) = \eta e\lambda/(hc) \quad . \tag{4.125}$$

For most emitters, the threshold wavelength for photoemission is below 0.85μ m, corresponding to a work function $\phi \geq 1.4$ eV. An example for such a material with $\phi \approx 1.4$ eV is NaKSb with a surface skin of NaKSb [4.62]. Only some complex cathodes consisting of two or more separate layers have an extended sensitivity up to about $\lambda \leq 1.2$ μm. The spectral response of the most commonly fabricated photocathodes is designated by a standard nomenclature, using the symbols S1 to S20. Some newly developed types are labelled by special numbers which differ for the different manufacturers.

Recently a new type of photocathode has been developed which is based on photoconductive semiconductors whose surface has been treated to obtain a state of Negative Electron Affinity (NEA). In this state an electron at the bottom of the conduction band inside the semiconductor has a higher energy than the zero energy of a free electron in vacuum [4.63]. When an electron is excited by absorption of a photon into such an energy level

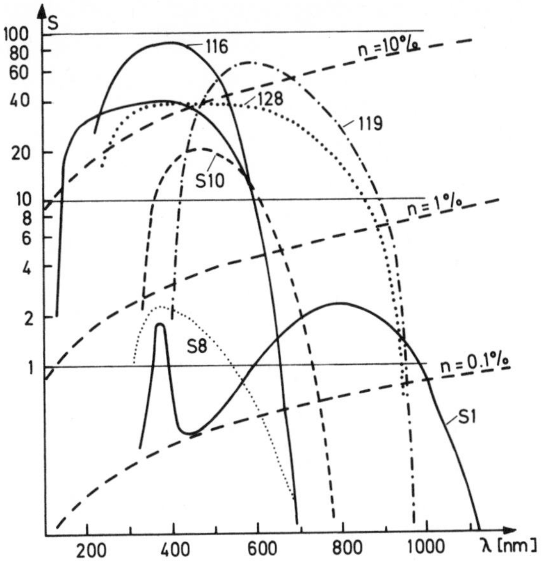

Fig.4.71. Spectral response of some commonly used photo cathodes

within the bulk, it may travel to the surface and leave the photocathode. These NEA cathodes have the advantage of high sensitivity which is fairly constant over an extended spectral range and even reaches into the infrared up to about 1.2 μm. Since these cathodes represent cold electron emission devices the dark current is very low. Until now their main disadvantage has been the complicated fabrication procedure and the resulting high price.

Three different devices of photoemissive detectors are of major importance in modern spectroscopy. These are the *vacuum phototube*, the *photomultiplier*, and the *image intensifier*.

4.5.3 Photocells and Photomultipliers

In the vacuum photocell (Fig.4.70) the photocathode is generally of the reflection mode. The photoelectrons are collected on a ring anode with a voltage between a few volts and several kilovolts. With a quantum efficiency of $\eta = 0.2$, we obtain from an incident radiation power of 10^{-9} W a photocurrent of 10^{-10} A, which can be conveniently measured with a picoampere meter or an operational amplifier.

The rise time of the anode voltage under pulsed illumination can be made short because of the small stray capacitance of the anode. In a photocell of biplanar geometry (Fig.4.72) where the semitransparent cathode has a distance from the anode of only a few millimeters, an electric field of about

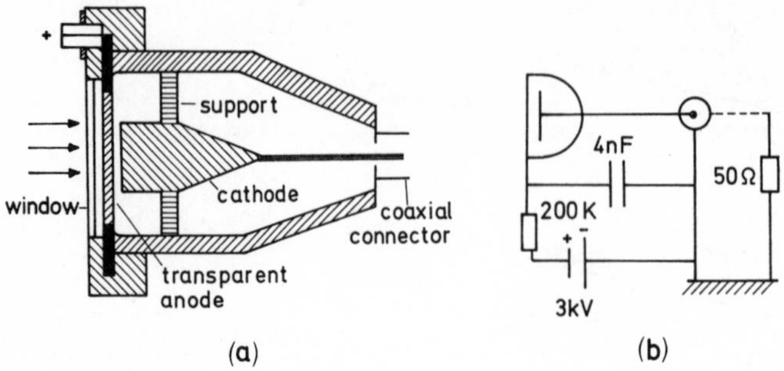

Fig.4.72. Fast photocell with biplanar geometry and coaxial 50 Ω output. (ITL) (a) Schematic construction diagram. (b) Electric circuit

1 kV/mm allows large photocurrent pulses up to several amperes without distortion by space charges. With a coaxial design, matched to a 50 Ω cable, rise times of 100 ps can be achieved for pulses with peak heights of some volts, which can be directly viewed on a fast sampling scope [4.64]. Such fast photocells are used to monitor subnanosecond laser pulses.

The minimum radiation power which can still be conveniently detected with photocells is mainly limited by two sources of noise [4.65,66].

a) The shot noise of the photoelectron current i. At a bandwidth Δf of the detection system, the rms value of the noise current is

$$\langle i_n \rangle_s = (2ei\Delta f)^{\frac{1}{2}} \quad . \tag{4.126}$$

b) The Johnson noise of the load resistor R at a temperature T which gives, according to the Nyquist formula, an rms noise current

$$\langle i_n \rangle_J = (4kT\Delta f/R)^{\frac{1}{2}} \quad . \tag{4.127}$$

The total rms noise voltage of a photodiode is therefore

$$\langle V_n \rangle = R(\langle i_n \rangle_s^2 + \langle i_n \rangle_J^2)^{\frac{1}{2}} = [eR\Delta f(2Ri + 4kT/e)]^{\frac{1}{2}} \quad . \tag{4.128}$$

At room temperature (300 K) the second term 4 kT/e is about 0.1 V. Since the Johnson noise should not be predominant, the first term 2Ri has to be larger than 0.1 V. For a maximum tolerable load resistor of $10^9 \Omega$, this puts a lower limit of $i > 5 \times 10^{-11}$ A to the photocurrent. With a quantum efficiency $\eta = 0.1$, the minimum radiation power conveniently detectable with photocells is then about 3×10^8 photons/s which corresponds to 10^{-10} W at $\lambda = 500$ nm.

For the detection of lower radiation powers, photomultipliers are necessary; these overcome this noise limitation by internal amplification of the

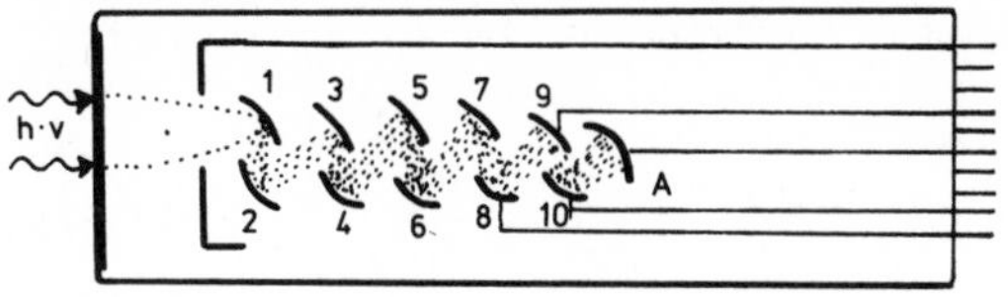

Fig.4.73. Photomultiplier

photocurrent using secondary electron emission from internal dynodes to multiply the number of photoelectrons (Fig.4.73). The photoelectrons emitted from the cathode are accelerated by a voltage of a few hundred volts and are focussed onto the metal surface (e.g., Cu-Be) of the first "dynode" where each impinging electron releases on the average q secondary electrons. These electrons are further accelerated to a second dynode where each secondary electron again produces about q tertiary electrons, and so on. The amplification factor q depends on the accelerating voltage U, on the incidence angle α, and on the dynode material. Typical figures for U = 200 V are q = 3-5. A photomultiplier with 10 dynodes therefore has a total current amplification of $G = q^{10} \approx 10^5 - 10^7$. Each photoelectron in a photomultiplier with N dynodes produces a charge avalanche at the anode of $Q = q^N e$ and a corresponding voltage pulse of

$$V = Q/C = q^N e/C = Ge/C \quad , \tag{4.129}$$

where C is the capacitance of the anode (including connections).

Example

$G = 2 \times 10^6$, $C = 30$ pf $\Rightarrow V = 10$ mV.

For experiments demanding high time resolution, the rise time of this anode pulse should be as small as possible. Let us consider which effects may contribute to the anode pulse rise time, caused by the spread of transit times for the different electrons [4.67]. Assume that a single photoelectron is emitted from the photocathode, and is accelerated to the first dynode. The initial velocities of the secondary electrons vary because these electrons are released at different depths of the dynode material and their initial energies, when leaving the dynode surface, are between 0 and 5 eV. The transit time between two parallel electrodes with distance d and potential difference V is

$$t = d\sqrt{2m/eV} \tag{4.130}$$

for electrons with mass m starting with zero initial energy. Electrons with initial energy E_{kin} reach the next electrode earlier by a time difference

$$\Delta t_1 = (E_{kin}/eV)^{\frac{1}{2}} t = (d/V)\sqrt{\frac{2mE_{kin}}{e}} \quad . \tag{4.131}$$

For typical values of $E_{kin} = 0.5$ eV, d = 1 cm, V = 250 V we obtain $\Delta t = 0.3$ ns. The electrons travel slightly different path lengths through the tube, which causes an additional time spread

$$\Delta t_2 = \Delta d\sqrt{2m/eV} \quad , \tag{4.132}$$

which is of the same magnitude. The rise time of an anode pulse, started by a single photoelectron, therefore decreases with increasing voltage proportional to $V^{-\frac{1}{2}}$, and depends on the geometry and form of the dynode structures.

When a short intense light pulse produces many photoelectrons simultaneously, the time spread is further increased by two effects:

a) The initial velocities of the emitted photoelectrons differ, e.g., for a cesium antimonide S5 cathode between 0 and 2 eV. This spread depends on the wavelength of the incoming light [4.68].

b) The time of flight between cathode and first dynode strongly depends on the location of the spot on the cathode where the photoelectron is emitted [4.69]. The resulting time spread may be larger than that from the other effects, but may be reduced by a focussing electrode between cathode and first dynode with careful optimization of its potential. Typical anode rise times of photomultipliers range from 1-20 ns. For specially designed tubes or with optimized commercial side-on types [4.70], rise times of 0.4 ns have been achieved.

For low-level light detection, the question of noise mechanisms in photomultipliers is of fundamental importance. There are three main sources of noise:

a) Photomultiplier dark current.
b) Noise of the incoming radiation.
c) Shot noise and Johnson noise caused by fluctuations of the amplification and by noise of the load resistor.

We shall discuss these separately.

a) When a photomultiplier is operated in complete darkness, electrons are still emitted from the cathode. This dark current is mainly due to thermionic emission, and only partly caused by cosmic rays or by radioactive decay of spurious radioactive isotopes in the multiplier material. According to Richardson's law the thermionic emission current,

$$i = C_1 T^2 e^{(-C_2\phi/T)} \quad , \tag{4.133}$$

strongly depends on the cathode temperature T and on its work function ϕ.

If the spectral sensitivity is to extend to long wavelengths, the work function ϕ should be as small as possible. In order to decrease the dark current, the temperature T of the cathode must be reduced. For instance, cooling a cesium-antimony cathode from 20° C to 0° C already reduces the dark current by a factor of about 10. The optimum operation temperature depends on the cathode type (because of ϕ). For S1 cathodes, e.g., which have a high infrared sensitivity and therefore a low workfunction ϕ, it is advantageous to cool the cathode down to liquid nitrogen temperatures. For other types with maximum sensitivity in the green, cooling below -40° C gives no significant improvement because the thermionic part of the dark current has then already dropped below other contributions, e.g., caused by high-energy β particles from disintegration of ^{40}K nuclei in the window material. Excessive cooling can even cause undesirable effects such as reduction of signal photocurrent or voltage drops across the cathode because the resistance of the cathode film increases with decreasing temperature [4.71].

For many spectroscopic applications only a small fraction of the cathode area is illuminated, e.g., for photomultipliers behind the exit slit of a monochromator. In such cases the dark current can be further reduced either by using photomultipliers with small effective cathode area or by placing small magnets around an extended cathode. The magnetic field defocusses electrons from the outer parts of the cathode area. These electrons cannot reach the first dynode and do not contribute to the dark current.

b) The shot noise of the photocurrent (4.127) is amplified in a photomultiplier by the gain factor G. The rms noise voltage across the anode load resistor R is therefore

$$\begin{aligned} \langle V\rangle_s &= GR(2ei_c\Delta f)^{1/2} \qquad & i_c &= \text{cathode current} \\ &= R(2eGi_a\Delta f)^{1/2} \quad . & i_a &= \text{anode current} \end{aligned} \tag{4.134}$$

The gain factor G itself is not constant but shows fluctuations due to random variations of the secondary emission coefficient q which is a small integer. This contributes to the total noise and multiplies the rms shot noise voltage by a factor $a > 1$ which depends on the mean value of q [4.72].

c) From (4.128) we obtain with (4.134) for the sum of shot noise and Johnson noise across the anode load resistor R at room temperature where $4\ kT/e \approx 0.1$ V

$$\langle V\rangle_{J+S} = eR\Delta f(2Ga^2Ri_a + 0.1)^{\frac{1}{2}} \quad . \tag{4.135}$$

For $GRi_a a^2 \gg 0.05$ V, the Johnson noise can be neglected. With a gain factor $G = 10^6$ and a load resistor of $R = 10^5\ \Omega$ this implies that the anode current i_a should be larger than 5×10^{-13} A. Since the anode dark current is already much larger than this limit, we see that *the Johnson noise does not contribute to the total noise of photomultipliers.*

A significant improvement of the signal-to-noise ratio in detection of low levels of radiation can be achieved with single photon counting techniques which enable spectroscopic investigations to be performed at incident radiation fluxes down to 10^{-16} W. This technique will be discussed in the next section. More details about photomultipliers and optimum conditions of performance can be found in the excellent introductions issued by EMI [4.66] or RCA [4.73]. An extensive review of photoemissive detectors has been given by ZWICKER [4.62].

4.5.4 Photon Counting

At very low incident radiation powers it is advantageous to use the photomultiplier for counting single photoelectrons emitted at a rate n per second rather than to measure the photocurrent $\bar{i} = neG\Delta t/\Delta t$ averaged over a period Δt. The charge avalanches with $q = Ge$, generated by single photoelectron, produce voltage pulses $U = eG/C$ at the anode with capacitance C. With $C = 1.5 \times 10^{-11}$ F, $G = 10^6 \Rightarrow U = 10$ mV. These pulses with rise times of about 1 ns trigger a fast discriminator which delivers a TTL-norm pulse of 5 volts to a counter or to a digital-analogue converter driving a rate meter with variable time constant [4.74] (see Fig.4.74).

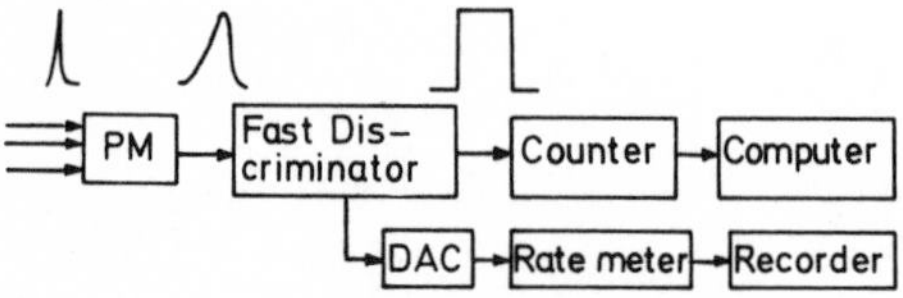

Fig.4.74. Schematic block-diagram of photon-counting electronics

Compared with the conventional analogue measurement of the anode current, the photon-counting technique has the following advantages:

1) Fluctuations of the photomultiplier gain G, which contribute to the noise in analogue measurements [see (4.135)], are not significant here, since each photoelectron induces the same normalized pulse from the discriminator as long as the anode pulse exceeds the discriminator threshold.

2) Dark current generated by thermal electrons from the various dynodes can be suppressed by setting the discriminator threshold correctly. This discrimination is particularly effective in photomultipliers with a large conversion efficiency q at the first dynode, covered with a GaAsP layer.

3) Leakage currents between the leads in the photomultiplier socket contribute to the noise in current measurements but are not counted by the discriminator if it is correctly biased.

4) High-energy β particles from the disintegration of radioactive isotopes in the window material and cosmic ray particles cause a small, but nonnegligible, rate of electron bursts from the cathode with a charge ne of each burst ($n \gg 1$). The resulting large anode pulses cause additional noise of the anode current. They can, however, be completely suppressed by a window discriminator used in photon counting.

5) The digital form of the signal facilitates its further processing. The discriminator pulses can be directly fed into a computer which analyzes the data and may control the experiment.

The upper limit of the counting rate depends on the time resolution of the discriminator, which may be below 10 ns. This allows counting of randomly distributed pulse rates up to about 10 MHz without essential counting errors.

The lower limit is set by the dark pulse rate [4.75]. With selected low-noise photomultipliers and cooled cathodes, the dark pulse rate may be below 1 per second. Assuming a quantum efficiency of $\eta = 0.2$ it should be therefore possible to achieve, within a measuring time of 1 s, a signal-to-noise ratio of unity already at a photon flux of 5 photons/s. At these low photon fluxes, the probability p(N) of N photo electrons being detected within the time interval Δt follows a Poisson distribution,

$$p(N) = \frac{\bar{N}^N \exp(-\bar{N})}{N!} \qquad (4.136)$$

when $\bar{N}$ is the average number of photo electrons detected within Δt [4.65]. If the probability that at least one photo electron will be detected within Δt shall be 99%, then $1 - p(0) = 0.99$ and

$$p(0) = \exp(-\bar{N}) = 0.01 \quad , \tag{4.137}$$

which yields $\bar{N} \geq 4.6$. This means that we can expect a pulse with 99% certainty only if at least 20 photons fall during the observation time onto the photocathode with a quantum efficiency of $\eta = 0.2$. For longer detection times, however, the detectable photoelectron rate may be even lower than the dark current rate if, for instance, lock-in detection is used. It is not the dark pulse rate N itself which limits the signal-to-noise ratio, but rather its fluctuations which are proportional to $N^{\frac{1}{2}}$.

4.5.5 Photoelectric Image Intensifiers

Image intensifiers consist of a photocathode, an electro-optical imaging device, and a fluorescent screen where an intensified image of the irradiation pattern at the photocathode is reproduced by the accelerated photoelectrons. Either magnetic or electric fields can be used for imaging the cathode pattern onto the fluorescent screen. Instead of the intensified image being viewed on a phosphor screen, the electron image can be used in a television camera tube to generate picture signals which can be reproduced on the television screen and can be stored either photographically or on recording tape [4.76].

For applications in spectroscopy the following characteristic properties of image intensifiers are important:

a) The intensity magnification factor M, which gives the ratio of output intensity to input intensity.

b) The dark current of the system, which limits the minimum detectable input power.

c) The spatial resolution of the device which is generally given as the maximum number of parallel lines per mm of a pattern at the cathode which can still be resolved in the intensified output pattern.

d) The time resolution of the system which is essential for recording of fast transient input signals.

Figure 4.75 shows a simple single-stage image intensifier with a magnetic field parallel to the accelerating electric field. All photoelectrons starting from a point P at the cathode follow helical paths around the magnetic field lines and are focussed into P' at the phosphor screen after a few revolutions. The location of P' is to a first approximation independent of the directions β of the initial photoelectron velocities. To get a rough

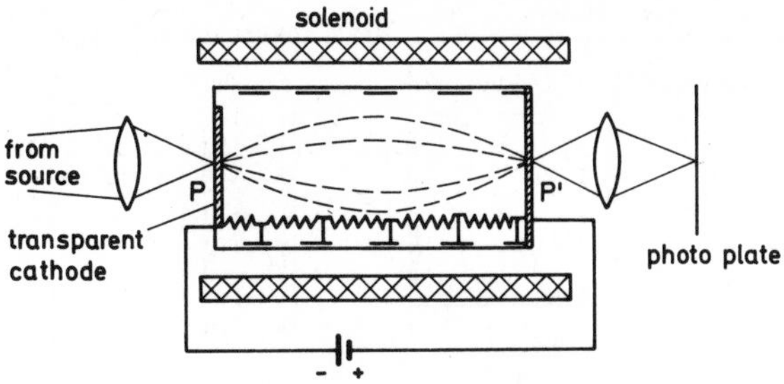

Fig.4.75. Single-stage image intensifier with magnetic focussing

idea about the possible magnification factor M, let us assume a quantum efficiency of 20% for the photocathode and an accelerating potential of 10 kV. With an efficiency of 20% for the conversion of electron energy to light energy in the phosphor screen, each electron will produce about 1000 photons with $h\nu = 2$ eV. The amplification factor M giving the number of output photons per incoming photon is then M = 200. However, the light from the phosphor is emitted into all directions and only a small fraction of it can be collected by an optical system. This reduces the total gain factor.

The collection efficiency can be enhanced when a thin mica window is used to support the phosphor screen and photographic contact prints of the image are made. Another way is the use of fiber optic windows.

Larger gain factors can be achieved with cascade intensifier tubes (Fig. 4.76), where two or more stages of simple image intensifiers are coupled in series [4.77]. The critical components of this design are the phosphor photocathode sandwich screens which influence sensitivity and spatial resolution. Since the light emitted from a spot around P on the phosphor should release photoelectrons from the opposite spot around P' of the photocathode, the distance between P and P' should be as small as possible in order not to spoil the spatial resolution. Therefore a thin layer of phosphor (a few μm) of very fine grain is deposited by electrophoresis on a mica sheet with a few μm thickness. An aluminum foil reflects the light from the phosphor back onto the photocathode (Fig.4.76b) and prevents optical feedback to the preceding cathode.

The spatial resolution depends on the imaging quality which is influenced by the thickness of the phosphor-screen photocathode sandwiches, by the homogenity of the magnetic field, and by the lateral velocity spread of the photoelectrons. Red-sensitive photocathodes generally have a lower spatial resolution since the initial velocities of the photoelectrons are larger. The resolution is highest at the center of the screen and decreases towards the edges. Table 4.2 compiles some typical data of a commercial three-stage image intensifier [4.78].

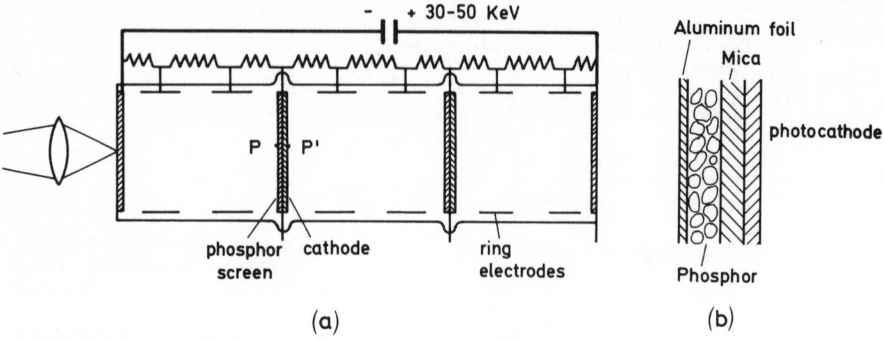

Fig.4.76a,b. Cascade image intensifier. (a) Schematic diagram with cathodes, fluorescence screens, and accelerating ring electrodes. (b) Detail of phosphor cathode sandwich

Table 4.2.

Type	Useful diameter [mm]	Resolution [line pairs/mm]	Gain
RCA 4550	18	32	3×10^4
C33085DP RCA	38	40	6×10^5
EMI 9794	48	50	2×10^5

Image intensifiers can be advantageously employed behind a spectrograph for the sensitive detection of extended spectral ranges [4.79]. Let us assume a linear dispersion of 10 Å/mm of a medium-sized spectrograph. An image intensifier with a useful cathode size of 30 mm and a spatial resolution of 30 lines/mm allows simultaneous detection of a spectral range of 300 Å with a spectral resolution of 0.3 Å with a sensitivity exceeding that of a photographic plate by many orders of magnitude. With cooled photocathodes, the thermal noise can be reduced to a level comparable with that of a photomultiplier and incident radiation powers of a few photons can be detected. A combination of image intensifiers and vidicons or special diode arrays has been developed (optical multichannel analyzer) which has proved to be very useful for fast and sensitive measurements of extended spectral ranges (see Sect.4.5.9).

4.5.6 Photoconductors

Photoconductive detectors are based on the increase of electric conductivity in pure or doped semiconductors caused by absorption of light. This increase of the conductivity

$$\sigma = e(n_+\mu_+ + n_-\mu_-) \qquad (4.138)$$

can be achieved either by increasing the densities n^- and n^+ of electrons in the conduction band (respectively, holes in the valence band) or by enhancing the mobilities μ_- and μ_+. Figure 4.77 illustrates different possible excitation mechanisms. In semiconductors with a band gap E_G direct optical excitation causing a band-band transition is only possible for $h\nu > E_G$ or $\lambda < hc/E_G$. The spectral dependence of the absorption coefficient $\alpha_{intr}(\nu)$ for this intrinsic absorption is determined by the density of states in the conduction band. One obtains [4.80]

$$\begin{aligned} \alpha_{intr}(\nu) &= C(h\nu - E_G)^{1/2} \quad \text{for} \quad h\nu > E_G \ , \\ &= 0 \qquad\qquad\qquad \text{for} \quad h\nu \leqq E_G \ . \end{aligned} \qquad (4.139)$$

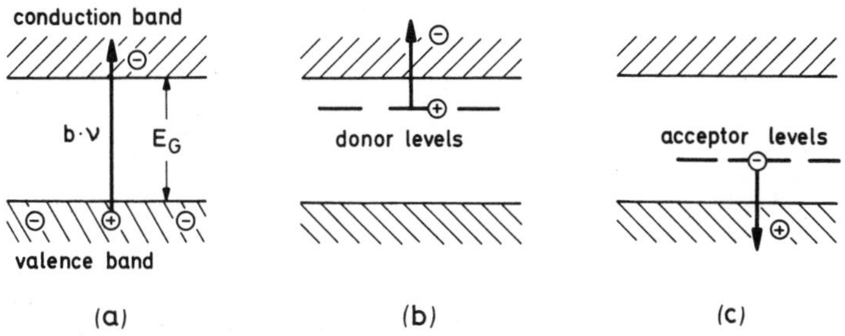

Fig.4.77. (a) Direct band-band absorption in a semiconductor and photoabsorption due to transitions between donor levels and conduction band (b) or between acceptor levels and the valence band (c)

In semiconductors doped with donor or acceptor atoms, optical transitions between discrete levels of these atoms and conduction or valence band are possible. Since these levels are within the band gap, close to the conduction band or valence band, respectively, this extrinsic photoexcitation is already possilbe at low photon energies (Fig.4.77b). Because of these low excitation energies the donor or acceptor levels can easily be ionized by thermal excitation which would decrease the population density and therefore the absorption coefficient. Such extrinsic photoconductors which are sensitive in the infrared up to wavelengths of about 30 μm, have to be operated at low temperatures to prevent thermal ionization. The absorption coef-

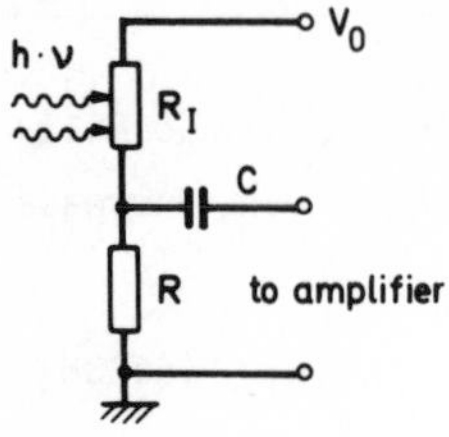

▲ Fig.4.78. Circuit diagram of a photoconductive detector for chopped incident light

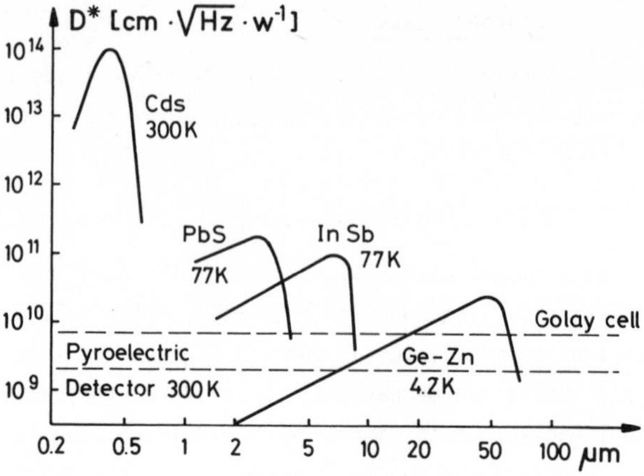

Fig.4.79 ►

Fig.4.79. Spectral dependence of detectivity D^* for several photoconductor materials (after [4.55a])

ficient for extrinsic excitation is determined by the wavelength-dependent absorption cross sections σ_D or σ_A of donors or acceptors and by the excess density $(N_D - N_-)$ [respectively, $(N_A - N_+)$] of donor atoms over electrons in the conduction band (respectively, acceptor atoms over holes in the valence band)

$$\alpha_{extr}(\nu) = \sigma_D(\nu)(N_D - N_-) \quad \text{or} \quad \sigma_A(\nu)(N_A - N_+) \quad . \tag{4.140}$$

Photoconductive detectors are used as resistors which change their resistance by $\Delta R = R_D - R_I$ from a "dark value" R_D without illumination to a lower value R_I under illumination. From Fig.4.78, which shows a typical circuit diagram, we obtain with a supply voltage V_0 the signal voltage V_s,

$$V_s = \frac{V_0 R_D}{R_D + R} - \frac{V_0 R_I}{R_I + R} = \frac{R \Delta R}{(R + R_D)(R + R_I)} V_0 \quad . \tag{4.141}$$

The optimum signal at a given incident radiation power is achieved for a load resistance of

$$R = R_D \sqrt{1 - \Delta R / R_D} = \sqrt{R_D R_I} \tag{4.142}$$

which gives approximately $R = R_D$ for $\Delta R \ll R_D$ (weak illumination).

Most intrinsic photoconductors are made of indium antimonide (InSb), cadmium sulfide (CdS), or lead sulfide (PbS). Figure 4.79 shows the spectral sensitivity of these materials. While PbS detectors can be used also at room temperature with detectivities of 5×10^{10} cm Hz W^{-1}, InSb detectors require cooling to 77 K where they reach a sensitivity of about 10^4 V/W and detectivities of about 10^{10} - 10^{11} cm$\sqrt{\text{Hz}}$/W.

The time response of a photoconductive detector is limited by the carrier lifetimes. PbS detectors for instance have time constants between 0.1 - 1 ms while InSb detectors reach a few μs. For measurements of fast transient signals photodiodes are more suitable (see Sect.4.5.8).

4.5.7 Photovoltaic Detectors

While photoconductive detectors are passive elements which need an external power supply, photovoltaic detectors are active devices which generate their own voltage when illuminated. They convert part of the incident radiation energy into electric energy. Their basic principle can be explained from Fig.4.80. Around the p-n junction of a p-type and an n-type semiconductor, a volume charge builds up due to diffusion of electrons and holes into the minority region. This results in a potential difference V_D between the p region and n region [4.81]. When incident photons generate electron-hole pairs in the vicinity of the p-n junction, the electrons in the p region are driven into the n region by the potential difference V_D, while the opposite is true for the holes. This separation of the photo-generated carriers decreases the voltage V_D by an amount ΔV, which is measured as the photovoltage at the ends of the semiconductor. If these ends are short circuited, the photogenerated carriers drift across the junction to the ends and a photocurrent I_{ph} is generated which has the same direction as the backward current. The photovoltage ΔV and the photocurrent I_{ph} have different signs. The voltage-current characteristics of an illuminated photovoltaic detector, connected to an external voltage U is given by [4.81] (Fig.4.81)

$$I = CT^2\, e^{(-eV_D/kT)} \left[e^{(eU/kT)} - 1\right] - I_{ph} \quad . \tag{4.143}$$

For I = 0, the open-circuit voltage becomes

$$U_{ph}(I = 0) = \frac{kT}{e} \ln(I_{ph}/I_s + 1) \quad , \tag{4.144}$$

where $I_s = CT^2 \exp(-e\, V_D/kT)$ is the saturated reverse current without illumination. The photocurrent I_{ph} is proportional to the difference of photoproduction rate minus recombination rate of electron-hole pairs. Note that $e\Delta V$ is always smaller than the energy gap E_G. Semiconductors with a large band gap can therefore deliver a higher photo voltage ΔV.

Materials used for photovoltaic detectors are, e.g., silicon, cadmium sulfide (CdS), and gallium arsenide (GaAs). Si detectors deliver photovoltages

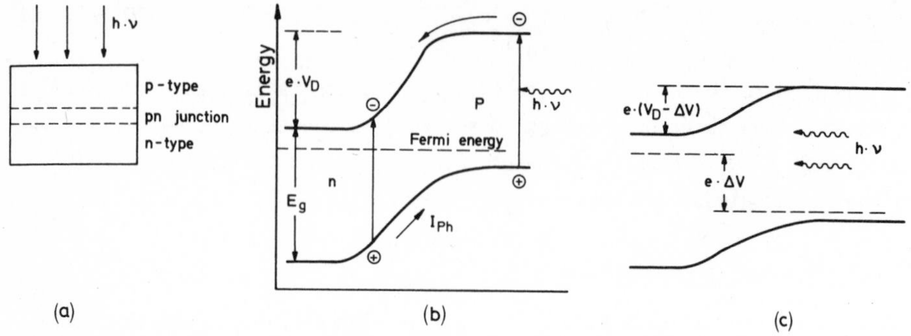

Fig.4.80a-c. Photovoltaic element. (a) Schematic structure of photodiode. (b) Photoproduction of electron-hole pairs at the p-n junction. (c) Reduction of V_D for an open external circuit

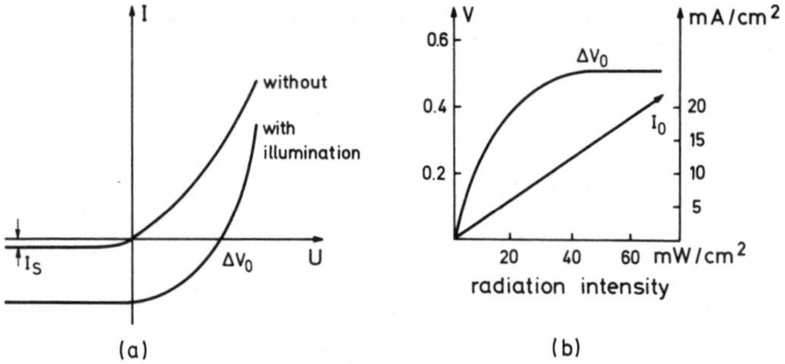

Fig.4.81a. Voltage-current characteristic of a p-n junction connected to an external voltage U, with and without illumination. b) Open circuit voltage ΔV_0 and short circuited current I_0 as a function of incident intensity for a diode without external voltage

up to 550 mV and photocurrents up to 40 mA/cm^2 [4.82]. The efficiency $\eta = P_{el}/P_{ph}$ of energy conversion reaches 10-14%. GaAs detectors show larger photovoltages up to 1 V but slightly lower photocurrents of about 20 mA/cm^2.

Note

When using photovoltaic detectors for measuring radiation powers, the load resistor R has to be sufficiently low to keep the output voltage always below its saturation value. Otherwise the output signal is no longer proportional to the input power.

4.5.8 Photodiodes

Photodiodes are photovoltaic or photoconductive semiconductors which are operated at a large negative reverse-bias voltage U. When the p-n junction is irradiated, a photocurrent proportional to the absorbed radiation power flows throughout the diode even though it is reverse biased. From (4.143) we obtain for the total current (saturated reverse current I_s and photocurrent I_{ph}) at large negative values of U where $\exp(eU/kT) \ll 1$

$$I = - I_s - I_{ph} \quad , \tag{4.145}$$

which shows that I becomes independent of the bias voltage U. The photocurrent generates a signal voltage $V_s = R_L I_{ph}$ across the load resistor R_L (see Fig.4.82) which is proportional to the absorbed radiation power over a large intensity range of several decades. From the circuit diagram in Fig.4.82 with the capacitance C_s of the semiconductor and its series and parallel resistances R_s and R_p one obtains for the upper frequency limit [4.83]

$$f_{max} = \frac{1}{2\pi C_s(R_s + R_L)(1 + R_s/R_p)} \quad , \tag{4.146}$$

which reduces for diodes with large R_p and small R_s to

$$f_{max} = \frac{1}{2\pi C_s R_L} \quad . \tag{4.147}$$

With a small resistor R_L, a high-frequency response can be achieved which is limited only by the drift time of the carriers through the boundary layer of the p-n junction. Using diodes with large bias voltages and a 50 Ω load resistor matched to the connecting cable, rise times in the subnanosecond range can be obtained.

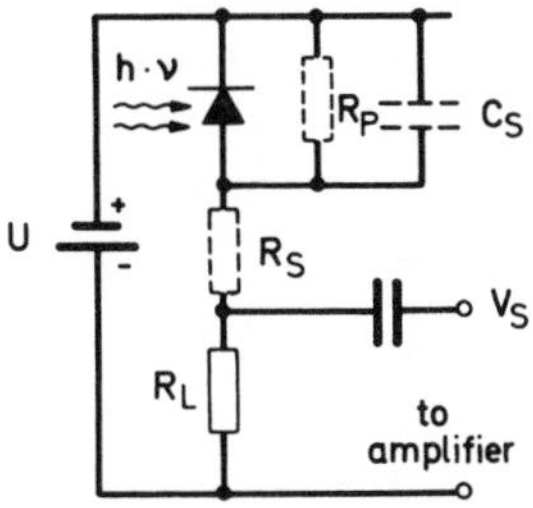

Fig.4.82. Equivalent circuit of a photodiode with internal capacity C_S, series resistor R_S, parallel resistor R_P, and external load resistor R_L

For photon energies hν close to the band gap, the absorption coefficient decreases [see (4.139)]. The penetration depth of the radiation, and with it the volume from which carriers have to be collected, becomes large. De-

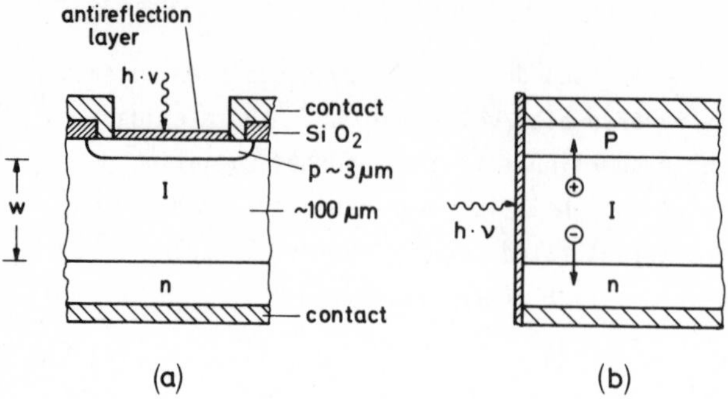

Fig.4.83. PIN photodiode with head on (a) and side on (b) illumination

finite collection volumes can be achieved in the *PIN diodes* where an undoped zone I of an intrinsic semiconductor separates the p and n regions (Fig. 4.83). Since no space charges exist in the intrinsic zone, the bias voltage applied to the diode will cause a constant electric field which accelerates the carriers. The intrinsic region may be made quite wide which results in a low capacitance of the p-n junction and provides the basis for a very fast and sensitive detector. The limit for the response time is, however, also set by the transit time $\tau = w/v_{th}$ of the carriers in the intrinsic region which is determined by the width w and the thermal velocity v_{th} of the carriers. Silicon PIN diodes with a 700 μm I zone have response times of about 10 ns and a sensitivity maximum at $\lambda = 1.06$ μm, while diodes with a 10 μm I zone reach 100 ps with a sensitivity maximum at a shorter wavelength-around $\lambda = 0.6$ μm [4.84]. Fast response combined with high sensitivity can be achieved when the incident radiation is focussed from the side into the I zone (Fig.4.83b). The only experimental disadvantage is the critical alignment necessary to hit the small active area.

Internal amplification of the photocurrent can be achieved with *avalanche diodes*, which are reverse-biased semiconductor diodes, where the free carriers acquire sufficient energy in the accelerating field to produce additional carriers on collisions with the lattice (Fig.4.84). The multiplication factor M, defined as the average number of electron-hole pairs after avalanche multiplication initiated by a single photoproduced electron-hole pair, increases with the reverse-bias voltage. Values of M up to 10^6 have been reported in silicon which allows sensitivities comparable with those of a photomultiplier. The advantage of these avalanche diodes is their fast response time which decreases with increasing bias voltage. In this device

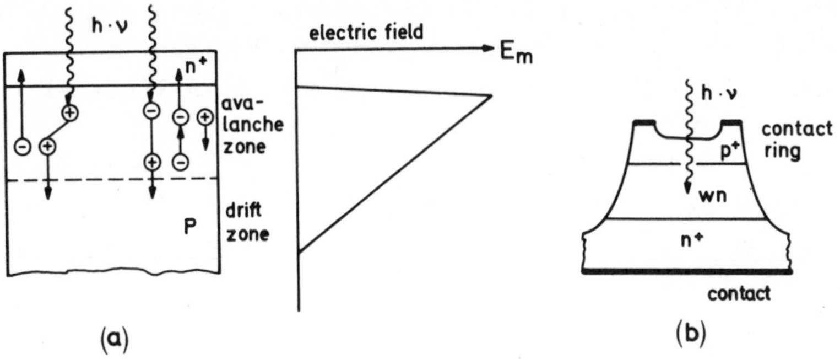

Fig.4.84. Schematic illustration of an avalanche photodiode (a) and mesa structure of an avalanche diode. (b) (w_n = weakly doped, n^+ = strongly doped n region

the product of gain times bandwidth may exceed 10^{11} Hz if the breakdown voltage is sufficiently high [4.85].

Very fast response times can be reached by using the photoeffect at the metal-semiconductor boundary known as a Schottky barrier [4.86]. Because of the different work functions ϕ_m and ϕ_s of the metal and the semiconductor, electrons can tunnel from the material with low ϕ to that with high ϕ (Fig.4.85) causing a space-charge layer and a potential barrier

$$V_B = \phi_B/e \quad \text{with} \quad \phi_B = \phi_m - \chi \tag{4.148}$$

between metal and semiconductor. χ is the electron affinity and equal to $\chi = \phi_s - (E_c - E_F)$. If the metal absorbs photons with $h\nu > \phi_B$ the metal electrons gain sufficient energy to overcome the barrier and "fall" into the semiconductor, which thus acquires a negative photovoltage. The *majority* carriers are responsible for the photocurrent, which ensures fast response time.

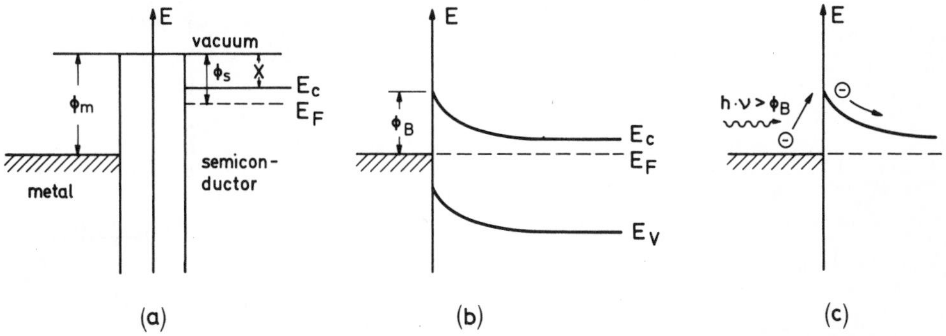

Fig.4.85a-c. (a) Work functions ϕ_m of a metal and ϕ_s of a semiconductor and electron affinity χ. E_C is the energy at the bottom of the conduction band and E_F is the Fermi-energy. (b) Schottky barrier at the contact layer between metal and n-type semiconductor. (c) Generation of a photocurrent

For measurements of optical frequencies, ultrafast metal-metal diodes have been developed [4.87], which have been operated up to 88 THz (λ = 3.39 μm). In these diodes, a 25 μm diameter tungsten wire, electrochemically etched to a point less than 200 nm in radius, serves as the point element and antenna, while the optically polished end of a pure nickel wire of diameter 2 mm forms the base element of the diode. A response time of 10^{-14} s or better has been demonstrated by the measurement of the 88 THz emission from the third harmonic of the CO_2 laser.

4.5.9 Optical Multichannel Analyzer

Optical multichannel analyzers (OMA), often called optical spectrum analyzers (OSA) [4.88] use a silicon target with a microscopic array of up to 10^7 photodiodes as detectors with spatial resolution. The target is a thin crystal wafer with a microscopic array of p-n junctions grown on it, which have a common cathode and isolated anodes. The array is irradiated on one side by the image to be detected (see Fig.4.86) and the absorbed photons produce electron-hole pairs. The diodes are reverse biased and the holes which diffuse into the p zone partially discharge the p-n junction. If the other side of the array is scanned with a focussed electron beam (20 μm diameter), the electrons recharge the anodes until they have nearly cathode potential. This recharging, which is equal to the discharge by the incident photons, can be measured as a current pulse in the external circuit and is measured as the video signal. The whole system is called a vidicon [4.89a].

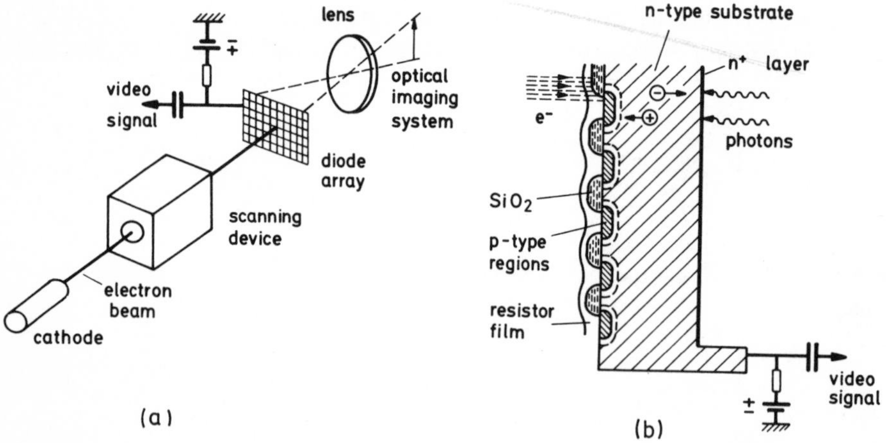

Fig.4.86a,b. Vidicon. (a) Schematic arrangement. (b) Section of the diode array

The maximum charge which can be stored in one diode,

$$Q = C\Delta V = eA\sqrt{\frac{2\varepsilon}{e} N(V_D - V_s)} \quad , \qquad (4.149)$$

is limited by the breakdown voltage ($V_D - V_s$). The integration time for the photo-induced decharging must be smaller than the time $t_s = Q_s/I_D$ which the thermal dark current needs to discharge the diode. This time for which an image can be stored is about 100 ms at room temperature but already some hours at dry-ice temperatures.

The sensitivity is limited by the smallest change ΔQ of the photodiode charge which can be still measured, and is influenced by charge leakage from the surface of the target due to thermal dark current, and also by the amplifier noise. The lower limit for detectable signals is about 2000 photo-electrons per video count (which means per photodiode) which is equivalent to about 2500 photons per diode at a quantum efficiency of about 80%.

The sensitivity can be greatly enhanced using an image intensifier in combination with the vidicon. With an S20 photocathode a maximum quantum efficiency of 16% can be obtained. The photoelectrons are accelerated in the image intensifier and create typically 1500 electron-hole pairs in the vidicon target which replaces the phosphor screen in the usual image intensifier (see Sect.4.5.5). Compared with the direct photoproduction in the vidicon diodes, this represents a net gain of $0.16 \times 1500 \approx 250$. This reduces the minimum detectable signal to about 15 photons producing 2 photoelectrons. With an additional image intensifier in front of the combined system (Fig. 4.87) the sensitivity can be further enhanced up to the limit which is given by thermal dark current of the cooled photocathode of the first intensifier.

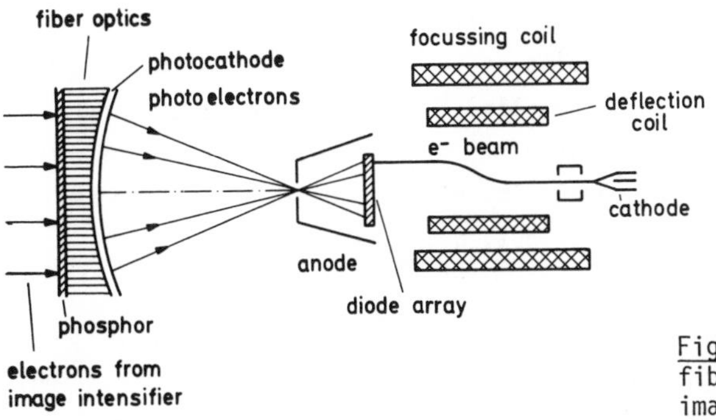

Fig.4.87. OSA system with fiber optic cathode and image intensifier

In this version the OMA competes in its sensitivity with low-noise photo-multipliers but has several definite advantages:

a) The vidicon targets store optical signals and allow integration over an extended period, whereas photomultipliers respond only while the radiation falls on the cathode.

b) All channels of the vidicon acquire optical signals simultaneously. Mounted behind a spectrometer, the OMA can measure an extended spectral range simultaneously while the photomultiplier accepts only the radiation passing through the exit slit, which defines the resolution. With a spatial resolution of 30 lines per mm and a linear dispersion of 5 Å/mm of the spectrometer, the spectral resolution is 0.17 Å. A vidicon target with a length of 16 mm can detect a spectral range of 80 Å simultaneously.

c) The signal readout is performed electronically in digital form. This allows computers to be used for signal processing and data analyzing. The dark current of the OMA, for instance, can be automatically subtracted, or the program can correct for background radiation superimposed on the signal radiation.

d) Photomultipliers have an extended photocathode where the dark current from all cathode elements is summed up and adds to the signal. In the image intensifier in front of the vidicon, only a small spot of the photocathode is imaged onto a single diode. Thus the whole dark current from the cathode is distributed over the spectral range, covered by the OMA.

The image intensifier can be gated and allows detection of signals with high time resolution [4.89b]. If the time dependence of a spectral distribution is to be measured, the gate pulse can be applied with variable delay and the whole system acts like a boxcar integrator with additional spectral display. The two-dimensional diode arrays also allow the time dependence of single pulses and their spectral distribution to be displayed, if the light entering the entrance slit of the spectrometer is swept (e.g., by a rotating mirror) parallel to the slit. The OMA or OSA systems therefore combine the advantages of high sensitivity, simultaneous detection of extended spectral ranges, and the capability of time resolution. These merits have led to their being increasingly applied in spectroscopy [4.90].

4.5.10 Measurements of Fast Transient Events

Many spectroscopic investigations require the observation of fast transient events. Examples are lifetime measurements of excited atomic or molecular states, investigations of collisional relaxation and studies of fast laser pulses or the transient response of molecules when the incident light frequency is switched into resonance with molecular eigenfrequencies (see Chap.11). Several techniques are used to observe and to analyze such events and recently developed instruments help to optimize the measuring procedure. We briefly present three examples of such equipment: the *boxcar integrator* with signal averaging, the *transient recorder*, and the *fast transient digitizer* with subnanosecond resolution.

The boxcar integrator measures the signal amplitudes integrated over a specific sampling interval Δt on a repetitive waveform several times and computes the average value of those measurements. With a synchronized trigger signal it can be assured that one looks each time at the identical time interval of each sampled waveform. A delay circuit permits the sampled time interval Δt (called aperture) to be shifted to any portion of the waveform under investigation. Figure 4.88 illustrates a possible way to perform this sampling and averaging. The aperture delay is controlled by a ramp generator which is synchronized to the signal repetition rate and which provides a sawtooth voltage at the signal repetition frequency. A slow aperture scan ramp shifts the gating time interval Δt, where the signal is sampled between two successive signals by an amount τ, which depends on the slope of the ramp. This slope has to be sufficiently slow in order to permit a sufficient number of samples to be taken in each segment of the waveform. A following signal averager [4.91] allows the output signal to be averaged over several scans of the time ramp. This increases the signal-to-noise ratio and smooths the dc output, following the shape of the waveform under study.

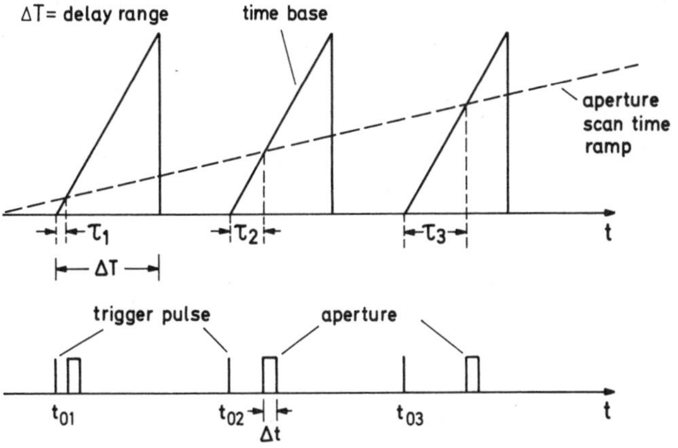

Fig.4.88. Synchronization of aperture delay range ΔT and delay time τ in a boxcar integrator

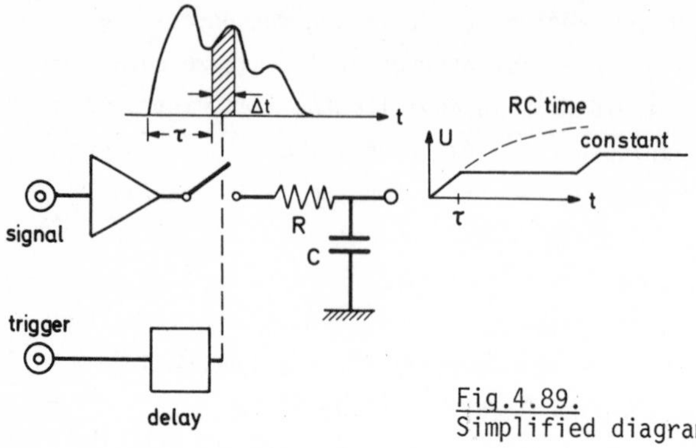

Fig.4.89.
Simplified diagram of boxcar operation

The integration of the input signal $U_s(t)$ over the sampling time interval Δt can be performed by charging a capacitance C through a resistor R which permits a current $I(t) = U_s(t)/R$. The output is then

$$U(\tau) = \frac{1}{C}\int_{\tau}^{\tau+\Delta t} I(t)dt = \frac{1}{RC}\int_{\tau}^{\tau+\Delta t} U_s(t)dt \quad . \tag{4.150}$$

For repetitive scans, the voltages $U(\tau)$ can be summed (Fig.4.89). Because of inevitable leakage currents, however, unwanted discharge of the capacitance occurs if the signal under study has a low duty factor and the time between successive samplings becomes large. This difficulty may be overcome by a digital output, consisting of a two-channel analog-to-digital-to-analog converter. After a sampling switch opens, the acquired charge is digitized and loaded into a digital storage register. The digital register is then read by a digital-to-analog converter producing a dc voltage equal to the voltage $U(\tau) = Q(\tau)/C$ on the capacitor. This dc voltage is fed back to the integrator to maintain its output potential until the next sample is taken.

The boxcar integrator needs repetitive waveforms because it samples each time only a small time interval Δt of the input pulse and composes the whole period of the repetitive waveform by adding many sampling points with different delays. For many spectroscopic applications, however, only single-shot signals are available. Examples are shock tube experiments or spectroscopic studies in laser-induced fusion. In such cases, the boxcar integrator is not useful and a *transient recorder* is a better choice. This instrument uses digital techniques to record a preselected section of an analog signal as it varies with time. The wave shape during the selected period of time is recorded and held in the instrument's memory until the operator instructs the instrument to make a new recording. The operation of a transient recorder is illustrated in Fig.4.90 [4.92]. A trigger, derived from the input signal

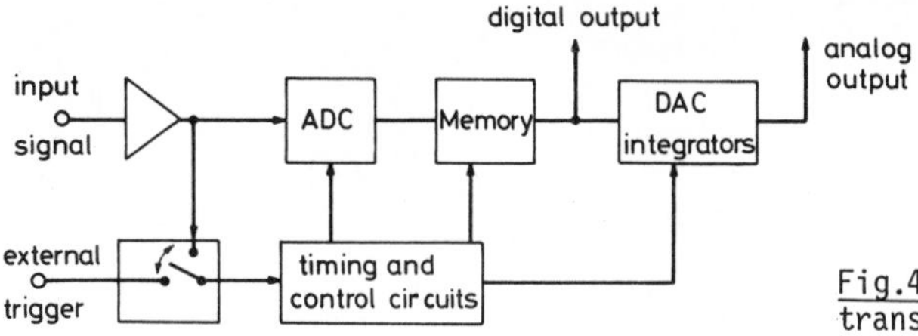

Fig.4.90. Block diagram of a transient recorder

or provided externally, initiates the sweep. The amplified input signal is converted at equidistant time intervals to its digital equivalent by an ADC and stored in a semiconductor memory in different channels. With 100 channels for instance, a single shot signal is recorded by 100 equidistant sampling intervals. The time resolution depends on the sweep time and is limited by the frequency response of the transient recorder. Sample intervals between 10 ns up to 20 s can be selected, e.g., in the Biomation model 8100 which has an input bandwidth of 25 MHz and utilizes a high-speed 8-bit analog-to-digital converter. This instrument allows sweep times of 20 μs to 5 hours for 2000 sampling points.

Acquisition and analysis beyond 500 MHz has become possible by combining the features of a transient recorder with the fast response time of an electron beam which writes and stores information on a diode matrix target in a scan converter tube. Figure 4.91 illustrates the basic principle of the transient digitizer [4.93]. The diode array of about 640 000 diodes is scanned by the reading electron beam, which charges all reverse-biased p-n junctions until the diodes reach a saturation voltage. The writing electron beam impinges on the other side of the 10 μm thick target and creates electron-hole pairs which diffuse to the anode and partially discharge it (see Sect.4.5.8). When the reading beam hits a discharged diode, it becomes recharged, and a current signal is generated at the target lead which can be digitally processed.

The instrument can be used in a nonstoring mode where the operation is similar to that of a conventional television camera with a video signal, which can be monitored on a conventional TV monitor. In the digital mode the target is scanned by the reading beam in discrete steps. The addresses of points on the target are transferred and stored in memory only when a trace has been written at those points on the target. This transient digitizer allows one to monitor fast transient signals with a time resolution of 100 ps and to process the data in digital form in a computer. It is, for instance, possible to obtain the frequency distribution of the studied sig-

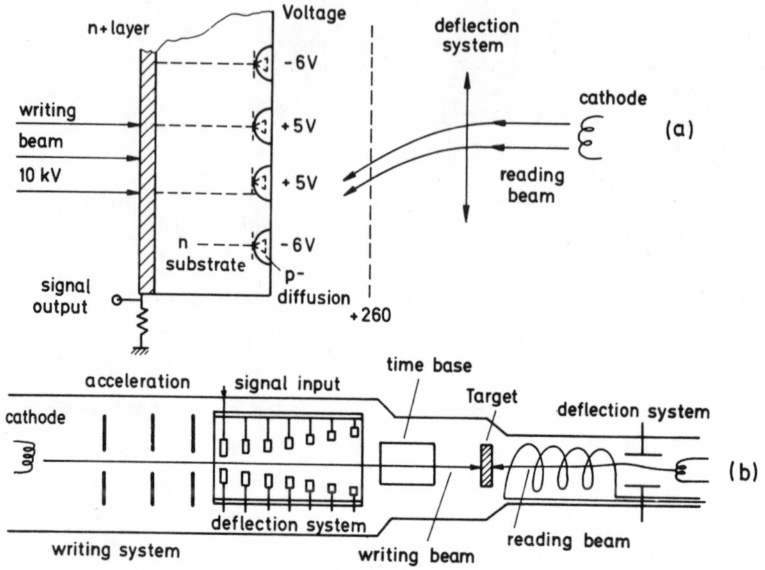

Fig.4.91a,b. Fast transient digitizer. (a) Silicon diode array target. (b) Writing and reading gun. [4.81]

nal from its time distribution by a Fourier transformation performed by the computer.

4.6 Conclusions

The direct of the foregoing sections was to provide a general background in spectroscopy and its instrumentation, to summarize some basic ideas of spectroscopy, and to present some important relations between spectroscopic quantities. This background should be helpful in understanding the following chapters which deal with the main subject of this book; the applications of lasers to the solution of spectroscopic problems. Although up to now, we have only dealed with general spectroscopy, the examples given above have been selected with special emphasis on laser spectroscopy. This is especially true in this chapter, which is of course not a complete account of spectroscopic equipment but is intended to give a survey on modern instrumentation used in laser spectroscopy.

There are several excellent and more detailed presentations of special instruments and spectroscopic techniques, such as spectrometers, interferometry, and Fourier spectroscopy. Besides the references given in the various sections, several series on optics [4.94], optical engineering [4.95], and advanced optical techniques [4.96] may help to give more extensive information about special problems.

Problems to Chapter 4

4.1. Calculate the spectral resolution of a grating spectrometer with an entrance slit width of 10 μm, focal lengths $f_1 = f_2 = 2$ m of mirrors M_1 and M_2, and a grating with 1800 grooves/mm. What is the useful minimum slit width if the size of the grating is 100×100 mm?

4.2. The spectrometer in problem 4.1 shall be used in first order for a wavelength range around 500 nm. What is the optimum blaze angle, if the geometry of the spectrometer allows an angle of incidence α around 20°?

4.3. Calculate the number of grooves/mm for a Littrow grating for 25° incidence at $\lambda = 488$ nm (i.e., the first diffraction order is being reflected back into the incident beam at an angle $\alpha = 25^\circ$ to the grating normal).

4.4. A prism can be used for expansion of a laser beam, if the incident beam is nearly parallel to the prism surface. Calculate the angle of incidence α for which an He-Ne laser beam ($\lambda = 632.8$ nm) transmitted through a 60° equilateral flint glass prism is tenfold expanded.

4.5. Assume that a signal-to-noise ratio of 50 has been achieved in measuring the fringe pattern of a Michelson interferometer with one continuously moving mirror. Estimate the minimum path length ΔL which the mirror has to travel in order to reach an accuracy of 10^{-4} nm in the measurement of a laser wavelength at $\lambda = 600$ nm.

4.6. The dielectric coatings of a Fabry-Perot interferometer have the following specifications: $R = 0.98$; $A = 0.3\%$. The flatness of the surfaces is $\lambda/100$ at $\lambda = 500$ nm. Estimate the finesse, the maximum transmission, and the spectral resolution of the F.P.I. for a plate separation of 5 mm.

4.7. A fluorescence spectrum shall be measured with a spectral resolution of 0.1 Å. The experimenter decides to use a crossed arrangement of grating spectrometer and an F.P.I. with coatings of $R = 0.98$ and $A = 0.3\%$. Estimate the optimum combination of spectrometer slit width and F.P.I. plate separation to obtain unambiguous line positions with an accuracy of 0.003 Å, if the dispersion of the spectrometer is 10 Å/mm and the spectrometer accuracy 0.1 A.

4.8. An interference filter shall be designed with peak transmission at $\lambda = 550$ nm and a bandwidth of 5 nm. Estimate the reflectivity R of the dielectric coatings and the thickness of the etalon, if no further transmission maximum is allowed between 350 and 750 nm.

4.9. A confocal F.P.I. shall be used as optical spectrum analyzer, with a free spectral range of 3 GHz. Calculate the mirror separation d and the finesse which is necessary to resolve spectral features in the laser output within 10 MHz. What is the minimum reflectivity R of the mirrors, if the surface finesse is 500?

4.10. Calculate the transmission of a Lyot filter with two plates ($d_1 = 1$ nm, $d_2 = 4$ nm) with $n = 1.40$ in the fast axis and $n = 1.45$ in the slow axis

a) as a function of λ for $\vartheta = 45^{\circ}$

b) as a function of ϑ for a fixed wavelength λ.

4.11. A thermal detector has a heat capacity $H = 10^{-8}$ J/K and a thermal conductivity to a heat sink of $G = 10^{-9}$ W/K. What is the temperature increase ΔT for 10^{-7} W incident cw radiation if the efficiency $\eta = 0.8$? If the radiation is switched on at a time $t = 0$, how long does it take before the detector reaches a temperature increase $\Delta T(t) = 0.9\ \Delta T_{\infty}$? What is the time constant of the detector and at which modulation frequency ω of the incident radiation has the response decreased to 0.5 of its dc value?

4.12. The anode of a photomultiplier tube is connected by a resistor of $R = 1\ \mathrm{K\Omega}$ to ground. The stray capacitance is 10 pf, the current amplification 10^6, and the anode rise time 1.5 ns. What is the peak amplitude and the half-width of the anode output pulse produced by a single photoelectron? What is the dc output current produced by 10^{-12} W cw radiation at $\lambda = 500$ nm, if the quantum efficiency of the cathode is $\eta = 0.2$? Estimate the necessary voltage amplification of a preamplifier

a) to produce 1 V pulses for single-photon counting?

b) to read 1 V on a dc meter for the cw radiation?

4.13. A manufacturer of a two-stage optical image intensifier states that incident intensities of 10^{-17} W at λ = 500 nm can still be "seen" on the phosphor screen of the output stage. Estimate the minimum intensity amplification, if the quantum efficiency of the cathodes and the conversion efficiency of the phosphor screens are both 0.2 and the collection efficiency of light emitted by the phosphor screens is 0.1. The human eye needs at least 20 photons/s to observe a signal.

4.14. Estimate the maximum output voltage of a photovoltaic detector at room temperature under 10 μW irradiation when the photocurrent of the shortened output is 50 μA and the dark current 50 nA.

5. Fundamental Principles of Lasers

In this chapter we summarize basic concepts of lasers with regard to their applications in spectroscopy. A well-founded knowledge of some subjects in laser physics, such as passive and active optical cavities and their mode spectra, amplification of light and saturation phenomena, mode competition and the frequency spectrum of laser emission, will help the reader to gain a deeper understanding of many problems in laser spectroscopy and to achieve optimum performance of an experimental setup. A more detailed treatment of laser physics and an extensive discussion of various types of lasers can be found in textbooks on lasers (see, for instance, [1.1-3, 5.1-4]). For more advanced presentations based on a quantum mechanical description of lasers, the reader is referred to [1.4,5, 5.5-7].

5.1 Basic Elements of a Laser

A laser consists essentially of three components (Fig.5.1):

1) The active medium which amplifies an incident E.M. wave.
2) The energy pump which selectively pumps energy into the active medium to populate selected levels and to achieve population inversion.
3) The optical resonator, composed of two opposite mirrors, which stores part of the induced emission, concentrated within a few resonator modes.

The energy pump (e.g., flash lamps, gas discharges, or other lasers) generates a population distribution N(E) in the laser medium which strongly deviates from the Boltzmann distribution (2.18) that exists for thermal equilibrium. At sufficiently large pump powers the population density $N(E_k)$ of a specific level E_k may exceed that of a lower level E_i (Fig.5.2).

For such a population inversion, the induced emission rate $N_k B_{ki} \rho(\nu)$ on a transition $E_k \rightarrow E_i$ exceeds the absorption rate $N_i B_{ik} \rho(\nu)$. An E.M. wave,

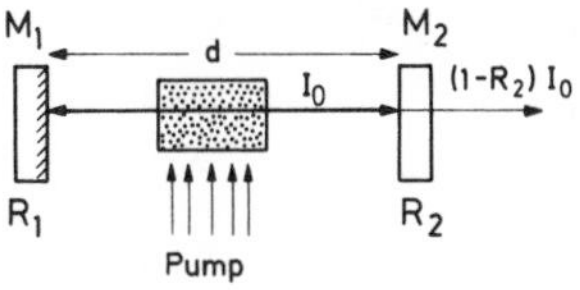

Fig.5.1. Schematic composition of a laser

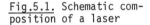

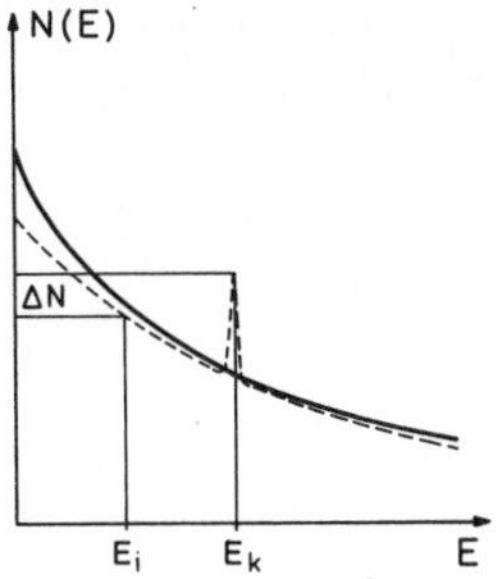

Fig.5.2. Population inversion, compared with a Boltzmann distribution at thermal equilibrium

passing through this active medium, is amplified instead of being attenuated (2.67).

The function of the optical resonator is the selective feedback of radiation emitted from the excited molecules of the active medium. Above a certain pump threshold this feedback converts the laser *amplifier* into a laser *oscillator*. When the resonator is able to store the E.M. energy of induced emission within a few resonator modes, the spectral energy density $\rho(\nu)$ in these modes may become very large. This will enhance the induced emission into these modes since, according to (2.23), the induced emission rate already exceeds the spontaneous rate for $\rho(\nu) > h\nu$. In Sect.5.3 we shall see that this concentration of induced emission into a small number of modes can be achieved with open resonators, which act as spatial-selective and frequency-selective optical filters.

5.2 Threshold Condition

When a monochromatic electromagnetic wave with frequency ν travels in the z direction through a medium of molecules with energy levels E_i and E_k and $(E_k - E_i)/h = \nu$, the intensity $I(\nu,z)$ is, according to (2.53), given by

$$I(\nu,z) = I(\nu,0)\, e^{-\alpha(\nu)z} \quad , \tag{5.1}$$

where the frequency-dependent absorption coefficient,

$$\alpha(\nu) = \left[N_i - (g_i/g_k)N_k\right]\sigma(\nu) \quad , \tag{5.2}$$

is determined by the cross section $\sigma(\nu)$ for the transition $(E_i \to E_k)$ and by the population densities N_i, N_k in the energy levels E_i, E_k with statistical weights g_i, g_k [see (2.68)]. We see from (5.2) that for $N_k > (g_k/g_i)N_i$, the absorption coefficient $\alpha(\nu)$ becomes negative and the incident wave is amplified instead of attenuated.

If the active medium is placed between two mirrors (Fig.5.1) the wave is reflected back and forth and traverses the amplifying medium many times, which increases the total amplification. With a length L of the active medium the total gain per single round trip is without losses

$$G(\nu) = I(\nu,2L)/I(\nu,0) = e^{-2\alpha(\nu)L} \quad . \tag{5.3}$$

A mirror with reflectivity R reflects only the fraction R of the incident intensity. The wave therefore suffers at each reflection a fractional reflection loss of (1-R). Furthermore, absorption in the windows of the cell containing the active medium, diffraction by apertures, and scattering due to dust particles in the beam path or due to imperfect surfaces, introduce additional losses. When we summarize all these losses by a loss factor β which gives the fractional energy loss per second, $(dW/dt = -\beta W)$, the intensity decreases per round trip (if we assume the losses to be equally distributed along the resonator length d) as

$$I = I_0\, e^{(-\beta 2d/c)} = I_0\, e^{-\gamma} \quad \text{with} \quad \gamma = 2\beta d/c \quad . \tag{5.4}$$

Including the amplification by the active medium with length L we obtain for the intensity after a single round trip through the resonator with length d

$$I(\nu,2d) = I(\nu,0)e^{(-2\alpha(\nu)L-\gamma)} \quad . \tag{5.5}$$

The wave is amplified if the gain overcomes the losses per round trip. This implies that

$$-2L\alpha(\nu) > \gamma \quad . \tag{5.6}$$

With the absorption cross section $\sigma(\nu)$ from (5.2) this can be written as the *threshold condition* for the population difference,

$$\boxed{\Delta N = N_k(g_i/g_k) - N_i > \Delta N_{thr} = \gamma/[2\sigma(\nu)L]} \quad . \tag{5.7}$$

If the inverted population difference ΔN of the active medium is larger than ΔN_{thr}, a wave, reflected back and forth between the mirrors, will be amplified in spite of the losses, and its intensity will increase. The wave is initiated by spontaneous emission from the excited atoms in the active medium.

Those spontaneously emitted photons which travel into the right direction (namely parallel to the resonator axis) have the longest path through the active medium and therefore the greater chance of creating new photons by induced emission. Above threshold they induce a photon avalanche which grows until the depletion of the population inversion just compensates the repopulation by the pump. Under steady state conditions the inversion has decreased to the threshold value ΔN_{thr}, the saturated net gain is zero and the laser intensity limits itself to a finite intensity I_L which is determined by the pump power, the losses γ and the gain coefficient $\alpha(\nu)$ (see Sects.5.7 and 5.9).

The frequency dependence of the gain coefficient $\alpha(\nu)$ is related to the line profile $g(\nu-\nu_0)$ of the amplifying transition. Without saturation effects (i.e. for small intensities), $\alpha(\nu)$ directly reflects this line profile, for homogeneous as well as for inhomogeneous profiles. According to (2.73) we obtain with the Einstein-coefficient B_{ik}

$$\alpha(\nu) = \Delta N(h\nu/c)B_{ik}g(\nu-\nu_0) \quad , \tag{5.8}$$

which shows that the amplification is largest at the line center. For large intensities, saturation of the inversion occurs, which is different for homogeneous than for inhomogeneous line profiles (see Sect.2.9).

The loss factor γ will also depend on the frequency ν because the resonator losses are strongly dependent on ν. The frequency spectrum of the laser therefore depends on a number of parameters which we discuss in more detail in the next section.

5.3 Optical Resonators

In Sect.2.1 it was shown that in a closed cavity a radiation field exists with a spectral energy density $\rho(\nu)$ which is determined by the temperature T of the cavity walls and by the eigenfrequencies of the cavity modes. In the optical region, where the wavelength λ is small compared with the dimension L of the cavity, we obtained the *Planck distribution (2.13)* at thermal equilibrium for $\rho(\nu)$. The number of modes per unit volume,

$$n(\nu)d\nu = 8\pi(\nu^2/c^3)d\nu \quad ,$$

within the spectral interval $d\nu$ of a molecular transition turned out to be very large [see example a) in Sect.2.1]. When a radiation source is placed inside the cavity, its radiation energy will be distributed among all modes, and the system will, after a short time, again reach thermal equilibrium at a correspondingly higher temperature. Because of the large number of modes in such a closed cavity, the mean number of photons per mode (which gives the ratio of induced to spontaneous emission rate in a mode) will be very small in the optical region (see Fig.2.5). *Closed cavities with* $L \gg \lambda$ *are therefore not suitable as laser resonators.*

In order to achieve a concentration of the radiation energy into a small number of modes, the resonator should exhibit a strong feedback for these modes but large losses for all other modes. This would allow an intense radiation field to be built up in the modes with low losses but would prevent to reach the oscillation threshold in the modes with high losses.

Assume the kth resonator mode with a loss factor β_k contains the radiation energy W_k. The energy loss per second in this mode is then

$$dW_k/dt = -\beta_k W_k \quad . \tag{5.9}$$

Under stationary conditions the energy in this mode will build up to a stationary value where the losses equal the energy input. If the energy input is switched off at t = 0 the energy W_k will decrease exponentially, since integration of (5.9) yields

$$W_k(t) = W_k(0)e^{-\beta_k t} \quad . \tag{5.10}$$

When we define the quality factor Q_k of the kth cavity mode as 2π times the ratio of energy stored in the mode to energy loss per oscillation period $T = 1/\nu$

$$Q_k = -2\pi\nu W_k/(dW_k/dt) \quad , \tag{5.11}$$

we can relate the loss factor β_k and the quality factor Q_k by

$$Q_k = -2\pi\nu/\beta_k \quad . \tag{5.12}$$

After a time $T = 1/\beta_k$, the energy stored in the mode has decreased to 1/e of its value at t = 0. This time can be regarded as the mean lifetime of a photon in this mode. If the cavity has large loss factors for most modes but a small β_k for a selected mode, the number of photons in this mode will be larger than in the other modes, even if at t = 0, the radiation energy in all modes had been the same. If the unsaturated gain $\alpha(\nu)L$ of the active medium is larger than the loss factor β_k but smaller than the losses of all other modes, the laser will oscillate only in this selected mode.

Such a resonator, which concentrates the radiation energy of the active medium into a few modes, can be realized with *open* cavities, consisting of two plane or curved mirrors which are aligned in such a way that a light ray travelling parallel to the resonator axis will be reflected back and forth between the mirrors. Such a ray traverses the active medium many times, resulting in a larger total gain, while rays inclined to the resonator axis leave the resonator after a few reflections before the intensity has reached a noticeable level.

Besides these *walk-off losses* also *reflection losses* cause a decrease of the energy stored in the resonator modes. With a reflectivity R_1 and R_2 of the resonator mirrors M_1 and M_2, the intensity of a wave in the passive resonator has decreased after a single round trip to

$$I = R_1 R_2 I_0 = I_0 \exp(-\gamma_R) \quad , \tag{5.13}$$

with $\gamma_R = -\ln(R_1 R_2)$. Since the round trip time is $T = 2d/c$, the decay constant β in (5.10) is $\beta = \gamma_R c/2d$, and the mean lifetime of a photon in the resonator becomes without any additional losses

$$\tau = 1/\beta = 2d/(\gamma_R c) = (2d/(c \ln R_1 R_2) \ . \tag{5.14}$$

These open resonators are in principle the same as the Fabry-Perot interferometers discussed in the previous chapter, and we shall see that several relations derived in Sect.4.2 can be used. However, there is an essential difference with regard to the geometrical dimensions. While in a common F.P.I. the distance between both mirrors is small compared with their diameter, the relation is generally reversed in laser resonators. The mirror diameter 2a is small compared with the mirror separation d. This implies that diffraction losses of the wave, being reflected back and forth between the mirrors, may play a major role in laser resonators while they can be completely neglected in the conventional F.P.I.

In order to estimate the magnitude of diffraction losses let us make use of a simple example. A plane wave incident onto a mirror with diameter 2a exhibits, after being reflected, a spatial intensity distribution which is determined by diffraction and which is completely equivalent to the intensity distribution of a plane wave passing through an aperture with diameter 2a. (Fig.5.3). The central diffraction maximum at $\theta = 0$ lies between the two first minima at $\theta = \pm\lambda/2a$. (For circular apertures a factor 1.2 has to be included, see, e.g., [2.3]). About 15% of the total intensity transmitted through the aperture is diffracted into higher orders with $|\theta| > \lambda/2a$. Because of the diffraction the outer part of the reflected wave misses the second mirror M_2 and is therefore lost. This example shows that the diffrac-

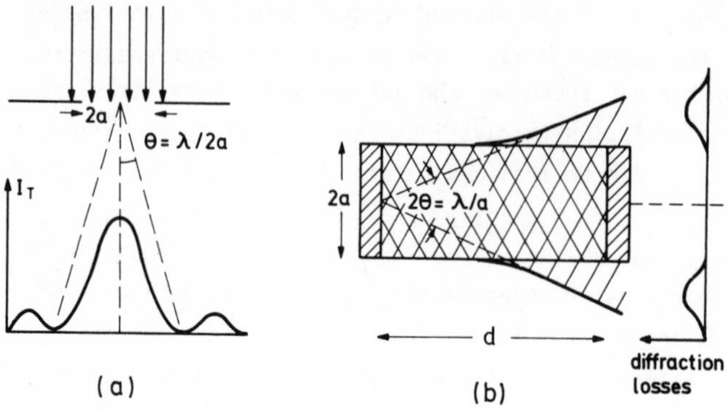

Fig.5.3. Equivalence of diffraction at an aperture (a) and at the mirrors of an open resonator (b). The case $\theta d = a \rightarrow N = 0.5$ is shown.

tion losses depend on the values of a, d, λ, and on the amplitude distribution A(x,y) of the incident wave across the mirror surface. The influence of diffraction losses can be characterized by the Fresnel number

$$N = a^2/(\lambda d) \quad . \tag{5.15}$$

If the number of transits which a photon makes through the resonator, is m, the maximum diffraction angle 2θ should be smaller than a/(md). With $2\theta = \lambda/a$ we obtain the condition

$$N > m \quad , \tag{5.16}$$

which states *that the Fresnel number of a resonator with negligible diffraction losses should be larger than the number of transits through the resonator.*

Examples

1) Plane Fabry-Perot interferometer:
d = 1 cm, a = 3 cm, λ = 500 nm ⇒ $N = 1.8 \times 10^5$.

2) Resonator of a gas laser with plane mirrors:
d = 50 cm, a = 0.1 cm, λ = 500 nm ⇒ N = 4.

The fractional energy loss per transit for a plane wave reflected back and forth between the two plane mirrors is approximately given by

$$\gamma_{Diffr} \approx \frac{1}{N} \quad .$$

For our first example the diffraction losses of the plane F.P.I. are about 5×10^{-6} and therefore completely negligible while, for the second example, they reach 25% and may already exceed the gain for many laser transitions. A plane wave would not reach threshold in such a resonator. However, these high diffraction losses cause nonnegligible distortions of a plane wave and the amplitude A(x,y) is no longer constant across the mirror surface (see Sect.5.4), but decreases towards the mirror edges. This decreases the diffraction losses, which become, for example, $\gamma_{Diffr} \leq 0.01$ for $N \geq 20$.

It can be shown [5.8] *that all resonators with plane mirrors which have the same Fresnel number also have the same diffraction losses, independent of the special choice of* a, d, *or* λ.

Resonators with curved mirrors exhibit much lower diffraction losses than the plane mirror resonator because they can refocus the divergent diffracted waves (see Sect.5.4).

5.4 Spatial Field Distributions in Open Resonators

In Sect.2.1 we have seen that any stationary field configuration in a closed cavity (called a mode) can be composed of plane waves. Because of diffraction effects, plane waves cannot give stationary fields in open resonators, since the diffraction losses depend on the coordinates (x,y) and increase from the axis of the resonator towards its edges. This implies that the distribution A(x,y), which is independent of x and y for a plane wave, will be altered with each round trip for a wave travelling back and forth between the resonator mirrors until it approaches a stationary distribution. Such a stationary field configuration, called a *mode of the open resonator*, is reached when A(x,y) no longer changes its form, although of course the losses result in a decrease of the total amplitude, if they are not compensated by the gain of the active medium.

The mode configurations of open resonators can be obtained by an iterative procedure using the Kirchhoff-Fresnel diffraction theory [5.9]. The resonator with two plane square mirrors is replaced by the equivalent arrangement of apertures with size $(2a)^2$ and a distance d between successive apertures (Fig.5.4). When an incident plane wave is travelling into the z direction its amplitude distribution is successively altered by diffraction, from a constant amplitude to the final stationary distribution $A_n(x,y)$. The spatial distribution $A_n(x,y)$ in the plane of the nth aperture is determined by the distribution $A_{n-1}(x,y)$ across the previous aperture.

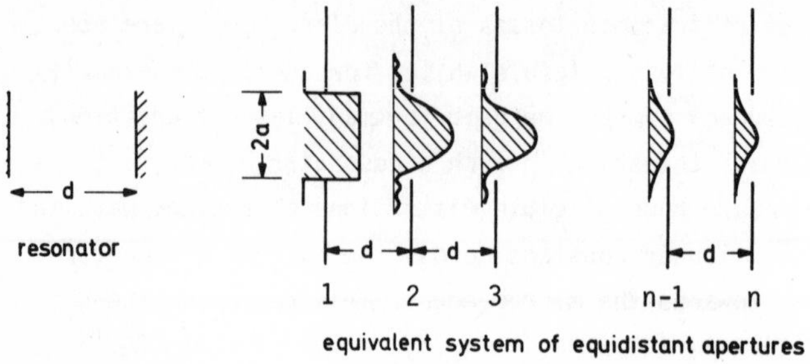

Fig.5.4. The diffraction of a plane incident wave at successive apertures is equivalent to the diffraction by successive reflections in a plane mirror resonator

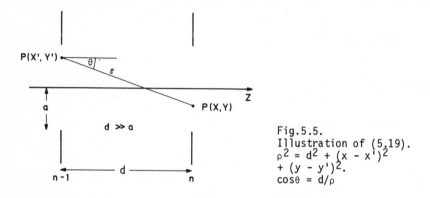

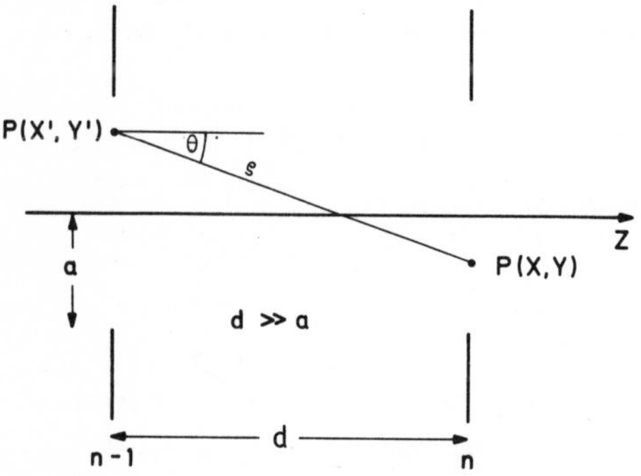

Fig.5.5. Illustration of (5.19). $\rho^2 = d^2 + (x - x')^2 + (y - y')^2$. $\cos\theta = d/\rho$

From Kirchhoff's diffraction theory we obtain (see Fig.5.5)

$$A_n(x,y) = -\frac{i}{2\lambda}\int\int A_{n-1}(x',y')\,\frac{1}{\rho}\,e^{-ik\rho}(1+\cos\theta)dx'dy' \quad . \tag{5.19}$$

A stationary field distribution is reached if

$$A_n(x,y) = CA_{n-1}(x,y) \quad . \tag{5.20}$$

The attenuation factor C, which does not depend on x and y, represents the diffraction losses and can be expressed by

$$C = (1-\gamma_D)^{\frac{1}{2}}\,e^{i\varphi} \quad , \tag{5.21}$$

where γ_D gives the fractional intensity loss due to diffraction and φ the corresponding phase shift.

Inserting (5.20) into (5.19) gives the following integral equation for the stationary field configurations

$$A(x,y) = \frac{i}{2\lambda}(1 - \gamma_D)^{-\frac{1}{2}} e^{-i\varphi} \iint A(x',y') \frac{1}{\rho} e^{-ik\rho} (1 + \cos\theta) dx'dy' . \quad (5.22)$$

Because the arrangement of successive apertures is equivalent to the plane mirror resonator, the solutions of this integral equation also represent the stationary modes of the open resonator. The diffraction-dependent phase shifts φ for the modes are determined by the condition of resonance, which requires that the mirror separation d equals an integer multiple of $\lambda/2$.

The general integral equation (5.22) cannot be solved exactly and one has to look for approximation methods. For two identical plane mirrors with a square size $(2a)^2$, (5.22) can be split into two one-dimensional equations, one for each coordinate x and y, if the Fresnel number $N = a^2/(d\lambda)$ is small compared with $(d/a)^2$, which means if $a \ll (d^3\lambda)^{\frac{1}{4}}$. Such numerical iterations of the "infinite strip" resonator case have been performed by FOX and LI [5.10], who showed that stationary field configurations do exist and who calculated the field distributions of these modes, their phase shifts and their diffraction losses.

The computations have been generalized by BOYD, GORDON, and KOGELNIK to confocal resonators with curved mirrors [5.11,12] and by FOX and LI to general laser resonators [5.13]. For the confocal resonator case (mirror separation d is equal to the radius of curvature b), the integral equation (5.22) can be solved with the acceptable approximation $a \ll d$, which implies $\rho \approx d$ in the denominator and $\cos\theta \approx 1$. In the phase term $\exp(-ik\rho)$, the distance ρ cannot be replaced by d, since the phase is already sensitive to small changes in the exponent. One can, however, expand ρ into a power series of xx'/d^2 and yy'/d^2. For the confocal case (d = b) one obtains [5.5]

$$\rho \approx b[1 - (xx' + yy')/b^2] \quad . \quad (5.23)$$

Inserting (5.23) into (5.22) allows the two-dimensional equation to be separated into two one-dimensional homogeneous Fredholm equations which can be solved analytically [5.11].

From the solutions, the stationary amplitude distribution in a plane $z = z_0$ vertical to the resonator axis is obtained. For the confocal resonator it can be represented by the product of Hermitian polynomials, a Gaussian function, and a phase factor,

$$A_{mn}(x,y,z_0) = C^* H_m(x^*) H_n(y^*) e^{-(x^{*2}+y^{*2})} e^{-i\phi(w,z_0)} \quad . \quad (5.24)$$

The function H_m is the Hermitian polynomial of mth order. The last factor gives the phase $\phi(w,z_0)$ in the plane $z = z_0$ at a distance $w = (x^2 + y^2)^{\frac{1}{2}}$

from the resonator axis. The arguments x^* and y^* depend on the mirror separation $d = b$ and are related to the coordinates x, y, and z_0 by

$$x^* = x\left(\frac{2\pi}{b\lambda(1+\xi^2)}\right)^{\frac{1}{2}} , \quad y^* = y\left(\frac{2\pi}{b\lambda(1+\xi^2)}\right)^{\frac{1}{2}} , \quad \text{with} \quad \xi = \frac{2z_0}{b} \quad . \tag{5.25}$$

The factor $C^* = [(1+\xi^2)/2]^{-\frac{1}{2}}C'(m,n)$ depends on z_0, b, m, and n. For the phase ϕ one obtains [5.11]

$$\phi(w,z_0) = k\left[\frac{b}{2}(1+\xi^2) + \frac{w^2}{b}\frac{\xi}{1+\xi^2}\right] + (1+m+n)\frac{\pi}{2} - \arctan\frac{1-\xi}{1+\xi} \quad . \tag{5.26}$$

From the definition of the Hermitian polynomial [5.14], one can see that the indices m and n give the number of nodes for the amplitude A(x,y) in the x (or the y) direction. Figure 5.6 and 5.7 illustrate some of these Transverse Electro-Magnetic standing waves which are called $TEM_{m,n}$-modes. The diffraction effects do not essentially influence the transverse character of the waves. While Fig.5.6 shows the one-dimensional amplitude distribution A(x) for some modes, Fig.5.7 illustrates the two-dimensional field amplitude A(x,y) in Cartesian coordinates and $A(r,\vartheta)$ in polar coordinates.

Modes with $m = n = 0$ are called *fundamental modes* or *axial modes* (often also zero-order transverse modes), while configurations with $m > 0$ or $n > 0$ are transverse modes of higher order. The intensity distribution of the fundamental mode can be derived from (5.24). With $H_0(x^*) = H_0(y^*) = 1$ and $x^{*2} + y^{*2} = w^2\{2\pi/[b\lambda(1+\xi^2)]\}$, we obtain

$$A_{00}(x,y,z_0) = C\exp\left[-w^2\left(\frac{2\pi}{b\lambda(1+\xi^2)}\right)\right] \quad . \tag{5.27}$$

The fundamental modes have a *Gaussian profile*. For $w = w_s$ with

$$w_s = \left(\frac{\lambda b}{2\pi}\right)^{\frac{1}{2}}\left(1 + 4z_0^2/b^2\right)^{\frac{1}{2}} \quad , \tag{5.28}$$

the amplitude decreases to 1/e of its maximum value C on the axis ($w = 0$). This value w_s is called the *beam radius* or mode radius. The smallest beam radius w_0 within the confocal resonator is the *beam waist*, which is located at the center $z_0 = 0$. From (5.28) we obtain

$$w_0 = (\lambda b/2\pi)^{\frac{1}{2}} \quad . \tag{5.29}$$

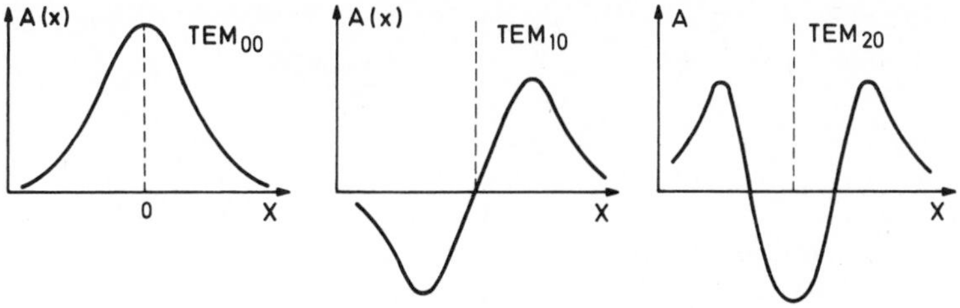

Fig.5.6. Stationary one-dimensional amplitude distributions $A_m(x)$ in a confocal resonator for some values of the index m

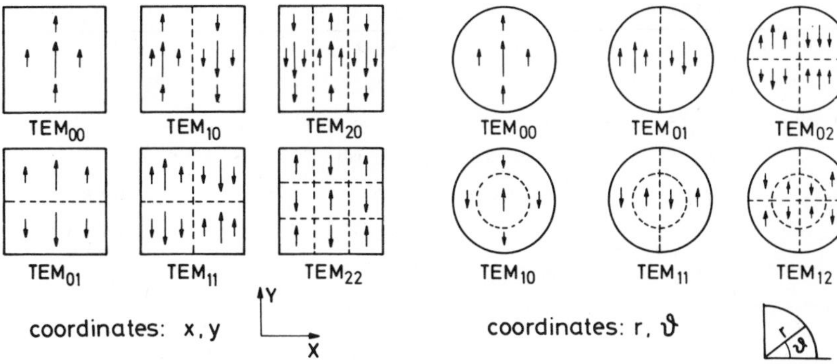

Fig.5.7. Two-dimensional presentation of linearly polarized resonator modes for square and circular apertures

Note that w_0 and w_s do not depend on the mirror size. Increasing the mirror width 2a reduces, however, the diffraction losses as long as no other limiting aperture with a diameter $D < 2a$ exists inside the resonator.

In terms of the beam radius w_s the field distributions (5.24) may be written in the reduced form

$$A_{m,n}(x,y,z_0) = \frac{\text{const}}{w_s} e^{[-(w/w_s)^2]} H_m\left(\sqrt{2}\ \frac{x}{w_s}\right) H_n\left(\sqrt{2}\ \frac{y}{w_s}\right) e^{-i\phi} \quad . \tag{5.30}$$

The equations (5.24-26) show that the field distributions $A_{mn}(x,y)$ and the form of the phase fronts depend on the location z_0 within the resonator. For the fundamental mode the constant phase surface can be directly obtained from (5.26). Since any point (x, y, z) on this surface must have the same phase as the intersection point $(0, 0, z_0)$ on the axis, the condition

$$\frac{kb}{2}\ (1 + \xi^2) + \frac{kw^2}{b}\ \frac{\xi}{1 + \xi^2} = \frac{kb}{2}\ (1 + \xi_0^2) \tag{5.31}$$

must be satisfied, when we neglect the small variation of ϕ with z in the argument of the arctan term. This gives, close to the axis,

$$z_0 - z = \frac{x^2 + y^2}{b} \; \frac{\xi}{1 + \xi^2} \;, \tag{5.32}$$

which is a spherical surface with a radius of curvature

$$b' = \left| \frac{1 + \xi^2}{2\xi} \right| b \;. \tag{5.33}$$

At the mirror surfaces is $\xi = \pm 1 \curvearrowright b' = b$, which means that the phase front is identical with the mirror surface. [Due to diffraction this is not quite true at the mirror edges at larger distances from the axis, where the approximation (5.32) is not correct]. At the center of the resonator is $z_0 = 0$ $\curvearrowright b'$ becomes infinite. *At the beam waist the constant phase surface becomes a plane* z = 0. This is illustrated in Fig.5.8 which shows the phase fronts and intensity profiles of the fundamental mode at different locations inside a confocal resonator. It can be shown [5.15] that also in nonconfocal resonators with large Fresnel numbers N the field distribution of the fundamental mode can be described by the Gaussian profile (5.27).

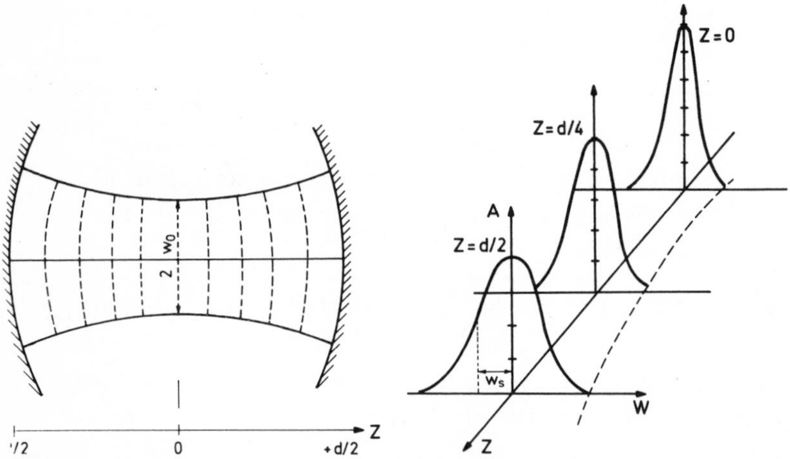

Fig.5.8. Phase fronts and intensity profiles of the fundamental mode at several locations z_0 in a confocal resonator

The confocal resonator can be replaced by other mirror configurations without changing the field configurations if the radius b^* of each mirror at a position z equals the radius b' of the wavefront in (5.33) at this position. This means that any two surfaces of constant phase can be replaced by reflectors, which have the same radius of curvature as the wavefront. From (5.33) we find with $z_0 = d/2 \rightarrow \xi = d/b$,

$$2db' = b^2 + d^2 \quad . \tag{5.34}$$

This allows two possible mirror separations,

$$d_{1,2} = b' \pm \sqrt{b'^2 - b^2} \quad . \tag{5.35}$$

According to (5.28) the beam radius w_s' at the mirrors (often called the spot size) is with $z_0 = \pm d/2$,

$$w_s' = \left(\frac{d\lambda}{\pi}\right)^{\frac{1}{2}}\left[\frac{2d}{b'} - \left(\frac{d}{b'}\right)^2\right]^{-\frac{1}{4}} \quad . \tag{5.36}$$

The second factor achieves a maximum of unity as a function of b' for b' = d which shows *that of all symmetrical resonators with a given mirror separation d the confocal resonator has the smallest spot sizes at the mirrors.*

The diffraction losses of a resonator depend on its Fresnel number $N = a^2/d\lambda$ (see Sect.5.3) and also on the field distribution $A(x,y,z = \pm d/2)$ at the mirror. The fundamental mode, where the field energy is concentrated near the resonator axis, has the lowest diffraction losses while the higher transverse modes, where the field amplitude has larger values towards the mirror edges, exhibit larger diffraction losses. Figure 5.9 shows the diffraction losses of a confocal resonator as a function of the Fresnel number N for the fundamental mode and some higher transverse modes. For comparison, the much higher diffraction losses of a plane mirror resonator are also shown in order to illustrate the advantages of curved mirrors which refocus the waves diverging by diffraction. From Fig.5.9 one may see that higher transverse modes can be suppressed by choosing a resonator with a suitable Fresnel number, which may be realized for instance by limiting apertures with variable diameter 2a inside the laser resonator. If the losses exceed the gain for these modes, they do not reach threshold and the laser oscillates only in the fundamental mode.

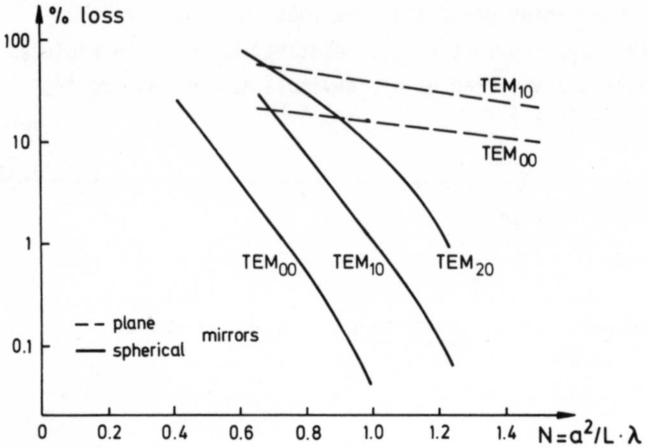

Fig.5.9. Diffraction losses of some modes in a confocal and in a plane mirror resonator

5.5 Frequency Spectrum of Passive Resonators

The stationary field configurations of open resonators, discussed in the previous section, have an eigenfrequency spectrum which can be directly derived from the condition that the phase fronts at the reflectors have to be identical with the mirror surfaces. Because these stationary fields represent standing waves in the resonators, the mirror separation d must be an integer multiple of $\lambda/2$ and the phase factor in (5.24) becomes unity at the mirror surfaces, which implies that the phase ϕ has to be an integer multiple of π. Inserting the condition $\phi = q\pi$ into (5.26) gives the eigenfrequencies ν_r of the confocal resonator,

$$\nu_r = \frac{c}{2d}\left[q + \frac{1}{2}(m + n + 1)\right] \quad . \tag{5.37}$$

The fundamental modes (m = n = 0) have the frequencies $\nu = (q + \frac{1}{2})c/2d$ and the frequency separation of the adjacent axial modes is $\delta\nu = c/2d$. Equation (5.37) shows that the frequency spectrum of the confocal resonator is degenerated because the transverse modes with $q = q_1$ and $m + n = 2p$ have the same frequency as the axial mode with $m = n = 0$ and $q = q_1 + p$. The free spectral range of a confocal resonator is therefore

$$\delta\nu_{confocal} = c/4d \quad . \tag{5.38}$$

If the mirror separation d deviates slightly from the radius of mirror curvature b, the degeneracy is removed. We obtain from (5.26) with $\phi = q\pi$ and $\xi = 2z_0/b = d/b \neq 1$ for a nonconfocal resonator with two equal mirror radii $b_1 = b_2 = b$

$$\nu_r = \frac{c}{2d}\left[q + \frac{1}{2}(m + n + 1)\left(1 + \frac{4}{\pi}\arctan\frac{d - b}{d + b}\right)\right] . \tag{5.39}$$

Now the higher transverse modes are no longer degenerate with axial modes. The frequency separation depends on the ratio $(d - b)/(d + b)$. Figure 5.10 illustrates the frequency spectrum of the confocal resonator (b = d) and of a nonconfocal resonator where d is slightly larger than b.

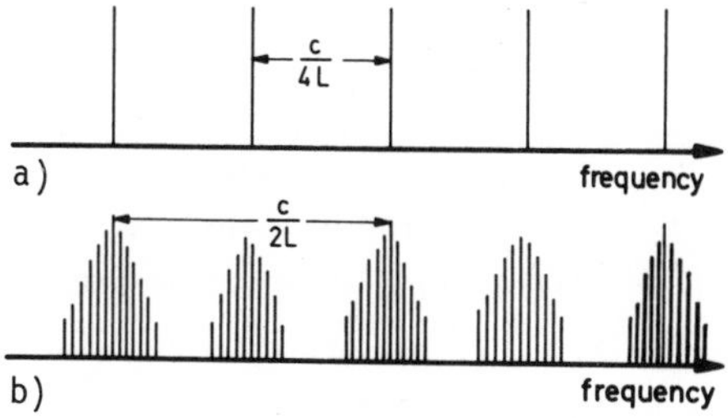

Fig.5.10. Eigenfrequency spectrum of a confocal resonator (a) and a nearly confocal resonator with d > b (b)

As has been shown in [5.12] the frequency spectrum of a general resonator with mirror curvatures b_1 and b_2 can be represented by

$$\nu = \frac{c}{2d}\left[q + \frac{1}{\pi}(m + n + 1)\cos^{-1}\sqrt{\left(1 - \frac{d}{b_1}\right)\left(1 - \frac{d}{b_2}\right)}\right] . \tag{5.40}$$

The eigenfrequencies of the axial modes (m = n = 0) are no longer at $(c/2d)(q + \frac{1}{2})$ but are slightly shifted. The free spectral range, however, is again $\delta\nu = c/2d$.

Examples

a) Nonconfocal resonator: $b_1 = b_2 = 75$ cm, d = 100 cm. The free spectral range $\delta\nu$, which is the frequency separation of the adjacent axial modes q and q + 1, is $\delta\nu = (c/2d) = 150$ MHz. The frequency separation $\Delta\nu$ between the (q, 0, 0) mode and the (q, 1, 0) mode is from (5.39) $\Delta\nu \approx 85$ MHz.

b) Confocal resonator: b = d = 100 cm. The frequency spectrum consists of equidistant frequencies with $\delta\nu = 75$ MHz. If, however, the transverse modes are suppressed, only axial modes oscillate with a frequency separation $\delta\nu = 150$ MHz.

We now briefly discuss the spectral width $\Delta\nu$ of the resonator resonances. The problem will be approached in two different ways.

Since the laser resonator is a Fabry-Perot interferometer, the spectral distribution of the transmitted intensity follows the Airy formula (4.64). With an incident intensity I_0, a transmission factor T, and a reflectivity R of each resonator mirror, the intensity stored within the resonator is

$$I = I_0 \frac{T}{(1-R)^2} \frac{1}{1+F\sin^2\delta/2} \quad . \tag{5.41}$$

For the eigenfrequencies ν_r the phase shift is $\delta = 2q\pi$. According to (4.70), the halfwidth $\Delta\nu$ of the resonances, expressed in terms of the free spectral range $\delta\nu$, is $\Delta\nu = \delta\nu/F^*$. If diffraction losses can be neglected, the finesse F^* is mainly determined by the reflectivity R of the mirrors, and the halfwidth of the resonance becomes

$$\Delta\nu = \delta\nu/F^* = \frac{c}{2d}\frac{1-R}{\pi\sqrt{R}} \quad . \tag{5.42}$$

With R = 0.98 this gives $F^* = 150$. With $\delta\nu = 150$ MHz, the halfwidth becomes $\Delta\nu = 1$ MHz. Generally other losses, such as diffraction, absorption, and scattering, decrease the total finesse. Realistic values are $F^* = 50$-100, giving for the example above a resonance halfwidth of the passive resonator of about 2 MHz.

The second approach to the estimate of the resonance width starts from the quality factor Q of the resonator. With total losses β per second, the energy W stored in a mode of a passive resonator decays exponentially according to (5.10). The Fourier transform of (5.10) yields the frequency spectrum of this mode, which gives a Lorentzian (see Sect.3.1) with a halfwidth $\Delta\nu = \beta/2\pi$. With the mean lifetime $T = 1/\beta$ of a photon in the resonator mode, the frequency width can be written as

$$\Delta\nu = 1/(2\pi T) \quad . \tag{5.43}$$

If reflection losses give the main contribution to the loss factor, the photon lifetime is, with $R = \sqrt{R_1R_2}$ [see (5.14)], $T = -d/(c\ln R)$, and the width $\Delta\nu$ of the resonator mode becomes

$$\Delta\nu = \frac{c|\ln R|}{2\pi d} = \delta\nu\frac{|\ln R|}{\pi} \ , \tag{5.44}$$

which gives, with $|\ln R| \approx 1-R$, the same result as (5.42) apart from a factor $\sqrt{R} \approx 1$. The slight difference of the two results stems from the fact, that in the second estimation we distributed the reflection losses uniformly over the resonator length.

The examples above illustrate that the resonance halfwidth of typical resonators for gas lasers is very small compared with the linewidth of laser transitions, which is generally determined by the Doppler width. The active medium inside a resonator compensates the losses of the passive resonator resulting in an exceedingly high-quality factor Q. The linewidth of the laser oscillator should be therefore much smaller than the passive resonance width. This question is discussed in Chapt.6.

For frequencies between the resonator resonances, the losses are high. In the case of a Lorentzian resonance profile, for instance, at frequencies which are $3\Delta\nu$ away from the resonance center ν_0, the loss factor has increased to about 10 times $\beta(\nu_0)$.

5.6 Active Resonators and Laser Modes

Introducing the amplifying medium into the resonator changes the refractive index between the mirrors and with it the eigenfrequencies of the resonator. We obtain the frequencies of the active resonator by replacing the mirror separation d in (5.39) by

$$d^* = (d - L) + n(\nu)L = d + (n - 1)L \quad , \tag{5.45}$$

where $n(\nu)$ is the refractive index in the active medium with length L. The refractive index $n(\nu)$ depends on the frequency ν of the oscillating modes, which are within the gain profile of a laser transition where anomalous dispersion is found (see Sect.2.6). Let us at first consider how laser oscillation builds up in an active resonator.

If the pump power is increased continuously, threshold is reached first at those frequencies which have a maximum net gain. According to (5.5) the net gain per round trip,

$$G(\nu,2d) = e^{[-2\alpha(\nu)L - \gamma(\nu)]} \quad , \tag{5.46}$$

is determined by the amplification factor $\exp[-2\alpha(\nu)L]$ which has the frequency dependence of the gain profile (5.8) and by the loss factor $\exp(-2\beta d/c) = \exp[-\gamma(\nu)]$ per round trip. While absorption or diffraction losses do not strongly depend on the frequency within the gain profile of a laser transition, the transmission losses exhibit a strong frequency dependence, which can be obtained from (5.41). The frequency spectrum of the laser is therefore closely connected to the eigenfrequency spectrum of the resonator. This can be illustrated as follows.

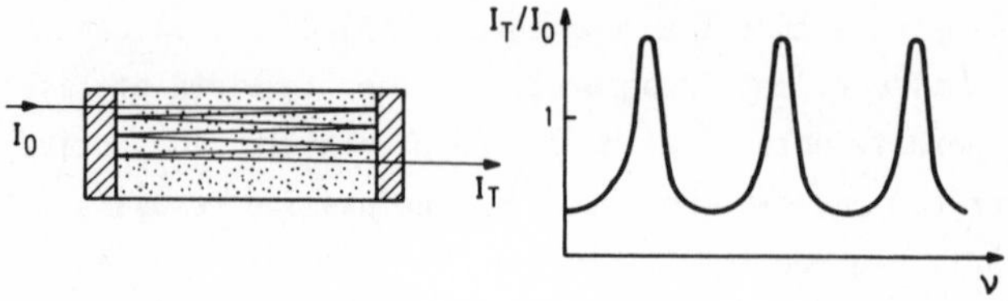

Fig.5.11. Transmission of an incident wave through an active resonator

Assume that a wave with the spectral intensity distribution $I_0(\nu)$ traverses an interferometer with two mirrors, each having the reflectivity R and transmission factor T (Fig.5.11). For the passive interferometer we obtain a frequency spectrum of the transmitted intensity according to (4.73). With an amplifying medium inside the resonator the incident wave experiences an amplification factor per round trip

$$G_0(\nu) = \exp[-2\alpha(\nu)L] \quad .$$

If the total losses per round trip are $\exp(-\gamma)$, the net gain is $G(\nu) = G_0 \exp(-\gamma)$ and we obtain, analogous to (4.62) by summation over all interfering amplitudes, the total transmitted intensity

$$I_T = I_0 \frac{T^2 G(\nu)}{[1 - G(\nu)]^2 + 4G(\nu)\sin^2\delta/2} \quad . \tag{5.47}$$

The total amplification I_T/I_0 has maxima for $\delta = 2q\pi$, which corresponds to the condition for the eigenfrequencies of the resonator [see (5.26)] with the modification (5.45). For $G(\nu) \rightarrow 1$, the total amplification I_T/I_0 becomes infinite for $\delta = 2q\pi$. This means that already an infinitesimally small input signal results in a finite output signal. Such an input is always provided, for instance, by the spontaneous emission of the excited atoms in the active medium. *For* $G(\nu) = 1$ *the laser amplifier converts to a laser oscillator*. This condition is equivalent to the threshold condition (5.7). Because of gain saturation (see Sect.5.7), the amplification remains finite and the total output power is determined by the pump power rather than by the gain (see Sect.5.10).

According to (5.8) the gain profile $G_0(\nu) = \exp[-2\alpha(\nu)L]$ depends on the line profile $g(\nu - \nu_0)$ of the molecular transition $E_i \rightarrow E_k$. The threshold condition can be illustrated graphically by subtracting the frequency dependent losses from the gain profile. Laser oscillation is possible at all frequencies ν_L, where this subtraction gives a positive net gain (Fig.5.12).

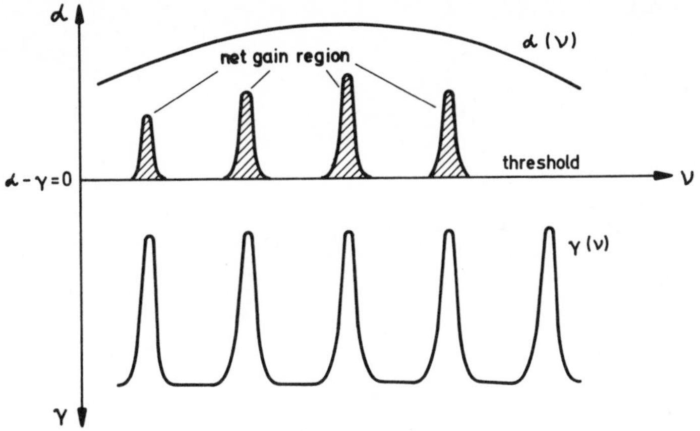

Fig.5.12. Gain profile of a laser transition with the eigenfrequencies of axial resonator modes and those frequencies where the gain exceeds the losses

In gas lasers, the gain profile is the Doppler-broadened profile of a molecular transition (see Sect.3.2) and shows therefore a Gaussian distribution with the Doppler width $\Delta\nu_D$,

$$\alpha(\nu) = \Delta N B_{ik}(h\nu/c)g(\nu - \nu_0) = \alpha(\nu_0)e^{[-2,77(\nu-\nu_0)^2/\Delta\nu_D^2]} \quad . \tag{5.48}$$

Solid-state or liquid lasers generally exhibit broader gain profiles because of additional broadening mechanisms (see Sect.3.7).

From (5.46) we obtain for the halfwidth $\Delta\nu$ of the resonances of the active resonator with a free spectral range $\delta\nu$ the expression

$$\Delta\nu = \delta\nu \frac{1 - G(\nu)}{2\pi\sqrt{G(\nu)}} = \delta\nu/F^* \quad .$$

The finesse F^* of the active resonator approaches infinity for $G(\nu) \to 1$. Although the laser linewidth $\Delta\nu_L$ may become smaller than the halfwidth of the passive resonator, it does not approach zero. This will be discussed in Chap.6.

We now briefly discuss the frequency shift or mode pulling of the passive resonator frequencies by the presence of the active medium [5.16]. The phase shift for a wave with frequency ν_a travelling through the active medium between the mirrors with distance d is

$$\phi_a = \omega t = 2\pi\nu_a n(\nu)d/c \quad . \tag{5.49}$$

In the passive resonator the phase shift is

$$\phi_p = 2\pi\nu_p d/c \quad . \tag{5.50}$$

The resultant change of the phase shift between active and passive resonator

$$\Delta\phi = \frac{2\pi d}{c}(n\nu_a - \nu_p) \quad , \tag{5.51}$$

is caused by the refractive index n of the active medium. From the Kramer-Kronig relations (2.55b,56b) we obtain the relation between $n(\nu)$ and the absorption coefficient $\alpha(\nu) = (4\pi/\lambda)\kappa(\nu)$

$$n(\nu) = \frac{\nu - \nu_0}{\Delta\nu_m} \frac{c}{2\pi\nu} \alpha(\nu) \quad , \tag{5.52}$$

where $\Delta\nu_m$ is the linewidth of the amplifying transition in the active medium. Under stationary conditions, the total gain per pass, $\alpha(\nu)2L$, saturates to the threshold value which equals the total losses γ. These losses determine the resonance width $\Delta\nu_p = \gamma/2\pi$ of the passive cavity [see (5.43)]. With the abbreviations $\nu_r = \nu_p/n$ and $\Delta\nu_r = \Delta\nu_p/n$, we obtain from (5.49-52) the final result for laser transitions with homogeneous line broadening

$$\nu_a = \frac{\nu_r\Delta\nu_m + \nu_0\Delta\nu_r}{\Delta\nu_m + \Delta\nu_r} \quad . \tag{5.53}$$

The resonance width $\Delta\nu_r$ of gas laser resonators is of the order of 1 MHz, while the Doppler width of the amplifying medium is about 1000 MHz. Therefore, when $\Delta\nu_r \ll \Delta\nu_m$, (5.53) reduces to

$$\boxed{\nu_a = \nu_r + \frac{\Delta\nu_r}{\Delta\nu_m}(\nu_0 - \nu_r)} \quad . \tag{5.54}$$

This shows that the mode pulling effect increases proportional to the difference of cavity resonance frequency ν_r and central frequency ν_0 of the amplifying medium. At the slopes of the gain profile, the laser frequency is pulled towards the center.

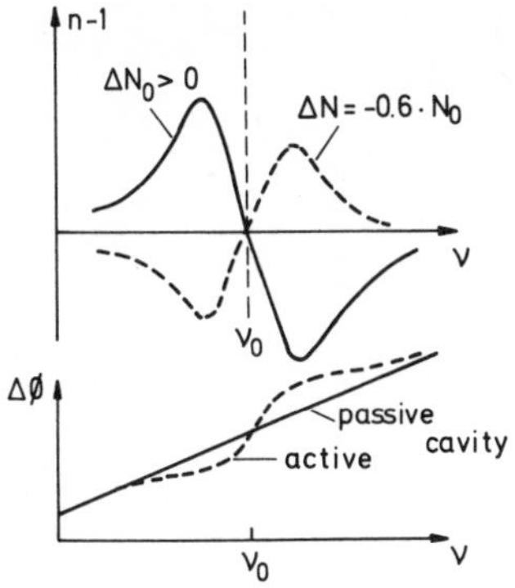

Fig.5.13. Dispersion curves for absorbing transitions ($\Delta N < 0$) and amplifying transitions ($\Delta N > 0$)

5.7 Gain Saturation and Mode Competition

When the pump power of a laser is increased beyond its threshold value, laser oscillation will start at first at a frequency ν where the difference between total gain minus total losses has a maximum. During the buildup time of laser oscillation the gain is larger than the losses and the stimulated wave inside the resonator is amplified at each round trip until the radiation energy is sufficiently large to deplete the population inversion ΔN down to the threshold value ΔN_{thr}. Under stationary conditions the increase of ΔN due to pumping is just compensated by its decrease due to stimulated emission. The gain of the active medium saturates from a value $G_0(I = 0)$ at small intensities to a threshold value

$$G_{thr} = e^{[-2L\alpha_{sat}(\nu)]} = e^{(+\gamma)} \quad , \tag{5.55}$$

where the gain equals the total losses per round trip. This gain saturation is different for homogeneous and for inhomogeneous transitions (see Sect.3.6).

In the case of a homogeneous laser transition all molecules in the upper level can contribute to stimulated emission at the laser frequency ν_a with a probability $B_{ik}\rho g(\nu_a - \nu_0)$ where $g(\nu - \nu_0)$ is the normalized line profile. Although the laser may oscillate only on a single frequency ν, the whole homogeneous gain profile

$$\alpha(\nu) = \Delta N\sigma(\nu) \quad ,$$

saturates until the inverted population difference ΔN has decreased to the threshold value ΔN_{thr} (see Fig.5.14). The saturated amplification coefficient $\alpha_{sat}(\nu)$ at an intracavity laser intensity I is, according to Sect.2.9,

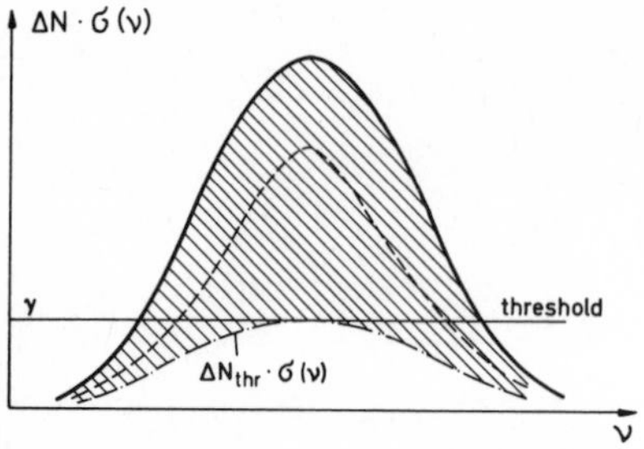

Fig.5.14. Saturation of a homogeneous gain profile. The dashed curve is the line profile during the buildup time; the dotted-dashed curve gives the stationary saturated profile

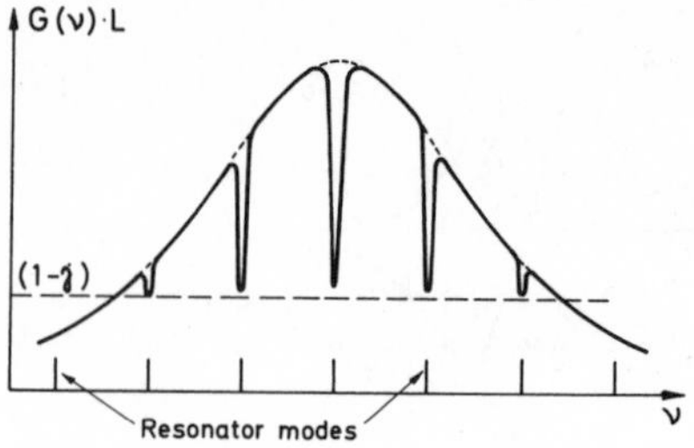

Fig.5.15. Gain saturation of an inhomogeneous laser transition

$$\alpha_s^{hom}(\nu) = \frac{\alpha_0(\nu)}{1 + S} = \frac{\alpha_0(\nu)}{1 + I/I_s} \quad , \tag{5.56}$$

where $I = I_s$ is that intensity for which the saturation parameter $S = 1$, which means that the induced transition rate equals the relaxation rate.

In the case of inhomogeneous laser transitions, the whole line profile can be divided into homogeneously broadened subsections with a spectral width $\Delta\nu^{hom}$ where $\Delta\nu^{hom}$ is, for example, the natural linewidth or the pressure- or power-broadened linewidth. Only those molecules in the upper laser level which belong to the subgroup in the spectral interval $\nu_L \pm \Delta\nu^{hom}/2$, centered at the laser frequency ν_L, can contribute to the amplification of the laser wave. A monochromatic laser wave therefore causes selective saturation of this subgroup and burns a hole into the inhomogeneous distribution $\Delta N(\nu)$ (see Fig.5.15). At the bottom of the hole the inversion $\Delta N(\nu_L)$ has decreased to the threshold value ΔN_{thr} but several homogeneous widths $\Delta\nu^{hom}$ away from ν_L, ΔN remains unsaturated. According to (3.72), the homogeneous width $\Delta\nu^{hom}$ of this hole increases with increasing saturating intensity as

$$\Delta\nu_s = \Delta\nu_0\sqrt{1 + S} = \Delta\nu_0\sqrt{1 + I/I_s} \quad . \tag{5.57}$$

This implies that with increasing saturation *more* molecules from a larger spectral interval $\Delta\nu_s$ can contribute to the amplification. The gain factor decreases by a factor $1/(1 + S)$ due to the decrease of ΔN caused by saturation. It increases by a factor $(1 + S)^{\frac{1}{2}}$ due to the increased homogeneous width. The combination of both effects gives

$$\alpha_s^{inh}(\nu) = \alpha_0(\nu)\,\frac{\sqrt{1+S}}{1+S} = \frac{\alpha_0(\nu)}{\sqrt{1+I/I_s}} \quad . \tag{5.58}$$

This different gain saturation of homogeneous and inhomogeneous transitions strongly affects the frequency spectrum of multimode lasers, as can be understood from the following arguments. Let us first consider a laser transition with a purely *homogeneous* line profile. The resonator mode which is next to the center of the gain profile will start oscillating when the pump power exceeds threshold. Since this mode experiences the largest net gain, its intensity will grow faster than that of the other laser modes. This causes partial saturation of the whole gain profile (Fig.5.14) mainly by this strongest mode. This saturation, however, decreases the gain for the other, weaker modes and their amplification will be slowed down, which further increases the differences in amplification and favors the strongest mode even more. This mode competition of different laser modes within a homogeneous gain profile will finally lead to a complete suppression of all but the strongest mode (Fig.5.16). Provided that no other mechanism disturbs the predominance of the strongest mode, this saturation coupling results in single-frequency oscillation of the laser, even if the homogeneous gain profile is broad enough to allow in principle simultaneous oscillation of several resonator modes [5.17].

In fact, such single-mode operation without further frequency-selecting elements in the laser resonator can be observed only in a few exceptional cases because there are several effects, such as spatial hole burning, frequency jitter, or time-dependent gain fluctuations, which interfere with the pure case of mode competition discussed above. These effects, which will be discussed below, prevent the unperturbed growth of one definite mode, introduce time-dependent coupling phenomena between the different modes, and cause in many cases a frequency spectrum of the laser which consists of a random superposition of many modes that fluctuate in time.

In the case of a purely *inhomogeneous* gain profile, the different laser modes do not share the same molecules for their amplification, and no mode competition occurs. Therefore all laser modes within that part of the gain profile which is above threshold, can oscillate simultaneously. The laser output consists of all axial and transverse modes for which the total losses are less than the gain (Fig.5.17).

Real lasers do not represent these pure cases but exhibit a gain profile which is a convolution of inhomogeneous and homogeneous broadening and it is the ratio of mode spacing $\delta\nu$ to the homogeneous width $\Delta\nu^{hom}$ which

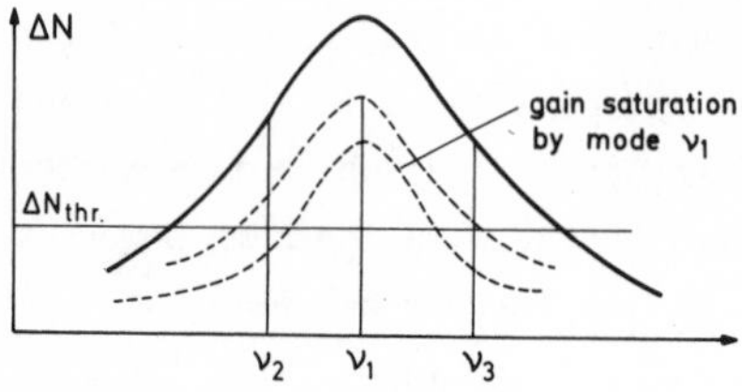

Fig.5.16. Mode competition due to gain saturation of a homogeneous gain profile

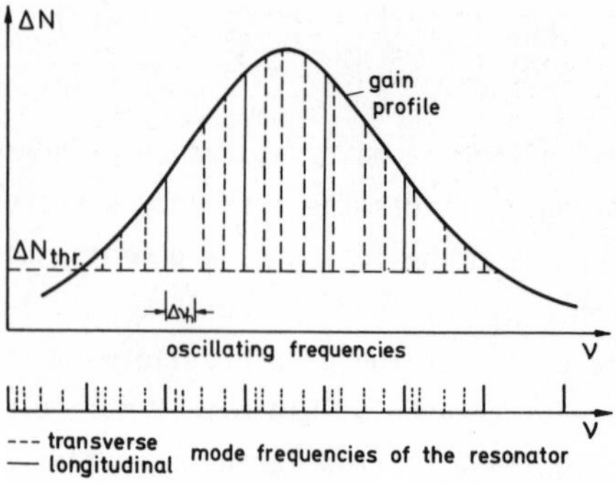

Fig.5.17. Multimode oscillation in case of a predominantly inhomogeneous gain profile. The lengths of the lines give the unsaturated gain

governs the strength of mode competition. We illustrate this by some examples.

1) *He-Ne Laser* at $\lambda = 632.8$ nm.

The Doppler width of the Ne transition is about 1500 MHz, and the width of the gain profile above threshold, which depends on the pump power, may be 1000 MHz. With a resonator length of d = 100 cm, the spacing of the longitudinal modes is $\delta\nu = c/2d = 150$ MHz. If the higher transverse modes are suppressed by an aperture inside the resonator, 7 longitudinal modes reach threshold. The homogeneous width $\Delta\nu^{hom}$ is determined by several factors: the natural linewidth $\Delta\nu_n = 20$ MHz; a pressure broadening of about the same magnitude; and power broadening which depends on the laser intensity in the different modes. With $I/I_s = 10$, for example, we obtain with $\Delta\nu_0 = 30$ MHz a power-broadened linewidth of about 100 MHz which is still smaller than the longitudinal mode spacing. The modes will therefore not compete strongly and simultaneous oscillation of all longitudinal modes above threshold is possible. This is illustrated by Fig.5.18, which shows the spectrum of a He-Ne laser with d = 1 m, monitored with a spectrum analyzer and integrated over a time interval of 1 second.

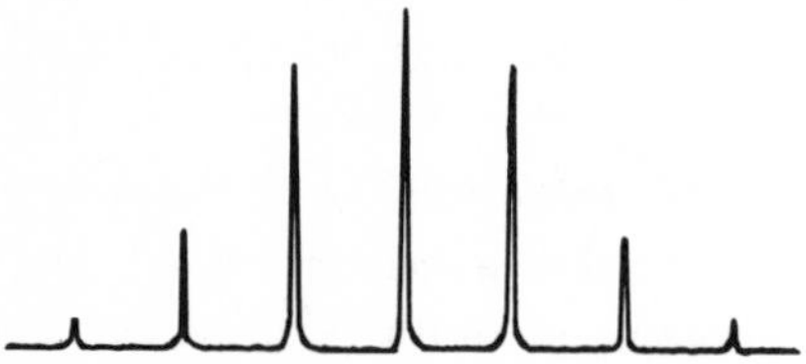

Fig.5.18. Multimode spectrum of a He-Ne laser, oscillating simultaneously on 7 longitudinal modes. The spectrum was monitored by a spectrum analyzer. Exposure time 1 s

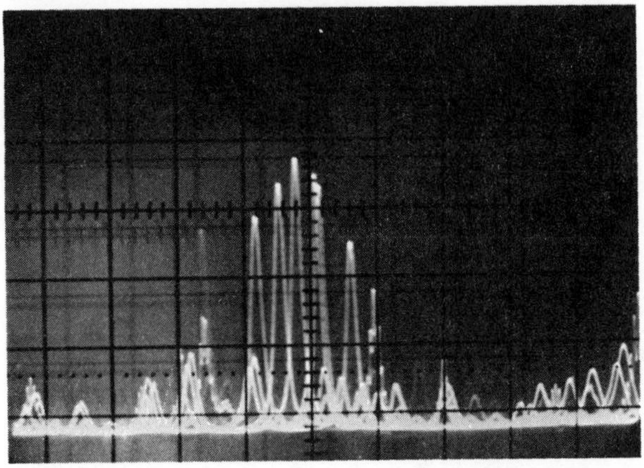

Fig.5.19. Two short-time exposures of the mode spectrum of a multimode argon laser, superimposed on the same film to demonstrate the randomly fluctuating mode distribution

2) Argon Laser

Because of the high temperature in the high-current discharge (about 10^3 A/cm^2), the Doppler width of the Ar^+ transitions is very large (about 8-10 GHz). The homogeneous width $\Delta\nu^{hom}$ is also much larger than for the He-Ne laser for two reasons. The long-range Coulomb interaction causes a large pressure broadening from electron-ion collisions and the high laser intensity (100 mW - 10 W) in a mode results in appreciable power broadening. Both effects generate a homogeneous linewidth which is large compared to the mode spacing $\delta\nu$ = 125 MHz at a commonly used resonator length of d = 120 cm. The resultant mode competition in combination with the perturbations mentioned above, cause the observed randomly fluctuating mode spectrum of the multimode argon laser. Figure 5.19 illustrates this by two short-time exposures of the oscilloscope display of a spectrum analyzer taken at two different times.

5.8 Spatial Hole Burning

A laser mode represents a standing wave in the laser resonator with a z-dependent field amplitude E(z) as illustrated in Fig.5.20a. Since the saturation of the inversion ΔN, discussed in the previous section, depends on the intensity $I \propto |E|^2$, the inversion saturated by a single laser mode will exhibit a spatial modulation $\Delta N(z)$ as drawn in Fig.5.20c. Even for a completely homogeneous gain profile, there are always regions of unsaturated inversion

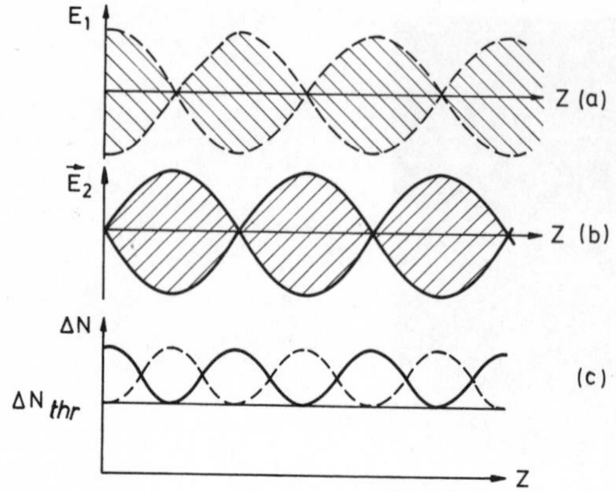

Fig.5.20a-c. Spatial hole burning: Field distributions of two standing waves with slightly different wave lengths along the resonator axis and the resultant spatial modulation of the inversion due to gain saturation

at the nodes of the standing wave $E_1(z)$, which may give sufficient gain for another laser mode (Fig.5.20b) or for a third mode with a shift of $\lambda/3$ of its amplitude maximum.

If the mirror separation d changes by only one wavelength (e.g., caused by acoustical vibrations of the mirrors), the maxima and nodes of the standing waves are shifted and the gain competition, governed by spatial hole burning, is altered. Therefore every fluctuation of the laser wavelength caused by changes of the refractive index or the cavity length d results in a corresponding fluctuation of the coupling strength between the modes and changes the gain relations and the intensities of the simultaneously oscillating modes.

If the length L of the active medium is small compared to the resonator length (e.g., in cw dye lasers), it is possible to minimize the spatial hole-burning effect by placing the active medium close to one cavity mirror (Fig.5.21). Consider two standing waves with wavelengths λ_1 and λ_2, and with maxima in the active medium which are shifted by λ/p (p = 2, 3 ...). Since all standing waves must have nodes at the mirror surface, we obtain for two waves with the minimum wavelength differences $\Delta\lambda = \lambda_1 - \lambda_2$ the relation

$$m\lambda_1 = a = (m + 1/p)\lambda_2 \quad , \tag{5.59}$$

or, for their frequencies,

$$\nu_1 = \frac{cm}{a} \quad ; \quad \nu_2 = \frac{c}{a}(m + 1/p) \Rightarrow \delta\nu_{sp} = \frac{c}{ap} \quad . \tag{5.60}$$

In terms of the spacing $\delta\nu = c/2d$ of the longitudinal resonator modes, the spacing of spatial hole-burning modes is

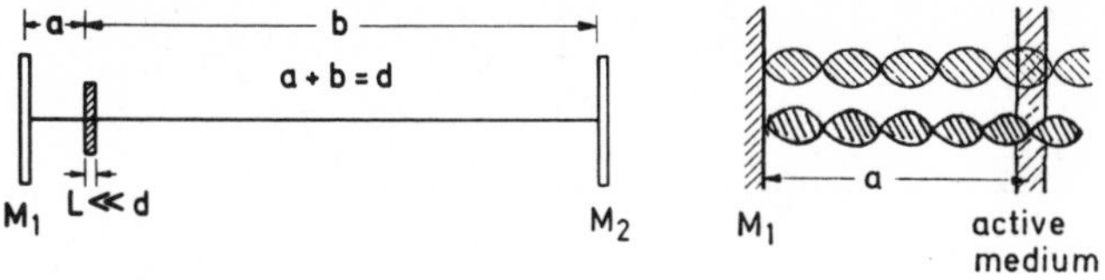

Fig.5.21. Spatial hole-burning modes when the active medium with $L \ll d$ is placed close to one resonator mirror

$$\delta\nu_{sp} = \frac{2d}{ap}\,\delta\nu \quad . \tag{5.61}$$

Even when the net gain is sufficiently large to allow competitions of e.g. up to three spatially separated standing waves (p = 1, 2, 3) only one mode can oscillate if the spectral width of the homogeneous gain profile is smaller than $(2/3)(d/a)\delta\nu$ [5.18].

Example

$d = 100$ cm, $L = 0.1$ cm, $a = 5$ cm, $p = 3$, $\delta\nu = 150$ MHz, $\delta\nu_{sp} = 1000$ MHz. Single-mode operation could be achieved if the homogeneous part of the gain profile is smaller than 1000 MHz.

In gas lasers the effect of spatial hole burning is partly averaged out by the diffusion of excited molecules from the nodes to the maxima of a standing wave. It is, however, important in solid-state and in liquid lasers such as the ruby laser or the dye laser. Spatial hole burning can be completely avoided in unidirectional ring lasers (see Sect.5.10) where no standing waves exist but waves propagating only into one direction can saturate the whole spatially distributed inversion.

5.9 Output Power and Optimum Output Coupling

Maximum output power of a laser is one of the major demands for most applications. At a first sight one might assume that a large gain factor of the active medium would also ensure a large output power of the laser. However, this is not necessarily true and we shall briefly discuss both quantities and their dependence on the pump power and the total losses.

We start with the rate equations which are illustrated in Fig.5.22. The pumping rate P gives the number of atoms per unit volume which are pumped per second into the upper laser level 2. The relaxation rates R_1 and R_2

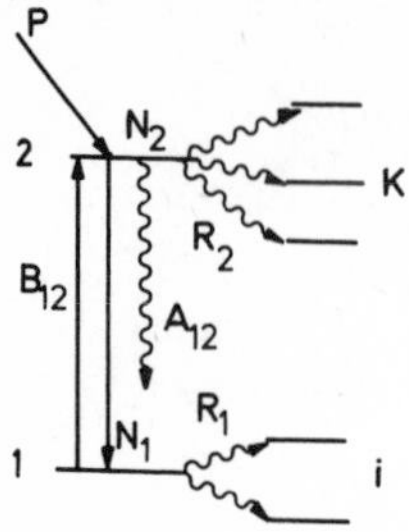

Fig.5.22.
Illustration of rate equations

describe the sum of all possible relaxation mechanisms which may depopulate levels 1 and 2 (e.g., collisions, spontaneous emission, etc.) and which result in transitions to other levels, labelled by i and k in Fig.5.22. With the spontaneous emission rate A_{12}, the photon density n in the oscillating mode, and the relative photon loss rate κ (κn gives the number of photons which are lost per unit volume and per second), we obtain the rate equations which couple the population densities N_1 and N_2 with the photon density n,

$$dN_1/dt = (N_2 - N_1)B_{12}nh\nu + N_2A_{12} - N_1R_1 \quad , \tag{5.62}$$

$$dN_2/dt = P - (N_2 - N_1)B_{12}nh\nu - N_2A_{12} - N_2R_2 \quad , \tag{5.63}$$

$$dn/dt = (N_2 - N_1)B_{12}n - \kappa n \quad . \tag{5.64}$$

Under stationary conditions, $dN_1/dt = dN_2/dt = dn/dt = 0$. Adding (5.62) and (5.63) gives

$$P = N_1R_1 + N_2R_2 \quad , \tag{5.65}$$

which illustrates that the pump rate just equals the sum of the loss rates of upper and lower level populations. Multiplying (5.62) with R_2 and (5.63) with R_1 and subtracting (5.63) from (5.62) yields with (5.64) for the population difference (inversion density) $\Delta N = N_2 - N_1$ with $\rho = nh\nu$

$$\Delta N = \frac{(R_1 - A_{12})P}{B_{12}\rho(R_1 + R_2) + A_{12}R_1 + R_1R_2} \quad . \tag{5.66}$$

We can conclude from (5.66) that a steady-state inversion $\Delta N > 0$ can be maintained only for $R_1 > A_{12}$. In the previous section we have seen that the photon density n increases up to a stationary value where ΔN has decreased to its saturation value ΔN_{thr}. From (5.64) we obtain with $dn/dt = 0$ the relation

$$\kappa = B_{12}\Delta N_{thr} \quad . \tag{5.67}$$

Since $\kappa h\nu = \gamma c/2d$ this is equivalent to the threshold condition (5.7). From (5.62,63) we deduce with (5.67) the stationary photon density

$$n_{stat} = \left(\frac{(R_1 - A_{12})PB_{12}}{\kappa(A_{12}R_1 + R_1R_2)} - 1\right)\frac{A_{12}R_1 + R_1R_2}{B_{12}(R_1 + R_2)} \quad . \tag{5.68}$$

With the unsaturated inversion ΔN_0, obtained from (5.66) with $n = 0$, we can write (5.68) as

$$n_{stat} = \frac{A_{12}R_1 + R_1R_2}{B_{12}(R_1 + R_2)}\left(\frac{\Delta N_0}{\Delta N_{thr}} - 1\right) \quad . \tag{5.69}$$

The output power P_L of a laser with active volume V and resonator length d through a mirror with transmittance T is

$$P_L = nh\nu cVT/d \quad . \tag{5.70}$$

This illustrates that the output power is proportional to the active volume V and to the relative inversion excess $(\Delta N_0 - \Delta N_{thr})/\Delta N_{thr}$ over the threshold inversion.

There is essential difference between homogeneous and inhomogeneous laser transitions. For homogeneously broadened transitions, the whole area under the gain profile of Fig.5.14 above the threshold line can be converted into induced emission at a single frequency ω. The induced emission power is according to (2.67)

$$\Delta NB_{12}n\hbar\omega = (nc/\hbar\omega)\int \alpha(\omega)d\omega \quad ,$$

where the integral extends over the whole homogeneous line profile. In case of inhomogeneous laser transitions only that part of the total population inversion ΔN which corresponds to the area of a homogeneous hole in Fig.5.15 can contribute to the output power of the laser at a single frequency. The total inversion of the whole gain profile can only be converted into induced emission when the laser oscillates simultaneously on many modes producing homogeneously broadened holes which overlap.

The active volume V which contributes to the total output power is

$$V \leq \sum_k V_k \quad , \tag{5.71}$$

where V_k is the mode volume in the active medium covered by the kth mode. The mode volume of a fundamental mode,

$$V_k \approx \pi\bar{w}_k^2 L \quad , \tag{5.72}$$

depends on the length L of the active medium and on an averaged spot size $\bar{W}_k$, which is determined by the resonator parameters. In general, multimode lasers utilize a larger total active volume V unless a special resonator design assures that the fundamental mode already fills the total active volume.

The laser output at a given intracavity power is proportional to the mirror transmission T, but the intracavity power decreases at a given pump power with increasing transmission losses. The problem therefore occurs how to find the optimum value of T. The maximum output power as a function of T can be found from (5.70) by differentiation of P_L with respect to T,

$$dP_L/dT = (h\nu cV/d)(n + T\,dn/dT) \quad ; \tag{5.73}$$

putting $dP_L/dT = 0$ gives the optimum transmission,

$$T_{opt} = -n/(dn/dT) \quad . \tag{5.74}$$

When we summarize all other losses apart from transmission losses (e.g., absorption, scattering, diffraction, etc.) by γ_a the total loss factor becomes $\gamma = \gamma_a + T$. Under stationary conditions the total losses are related to the threshold inversion ΔN_{thr} by [see (5.7) and (5.8)]

$$\Delta N_{thr} = (\gamma_a + T)c/(2dh\nu B_{12}) \quad . \tag{5.75}$$

From (5.69) we can therefore obtain dn/dT which finally gives the optimum output transmission

$$T_{opt} = -\gamma_a + \sqrt{\Delta N_0 \gamma_a B_{12} 2dh\nu/c} \quad . \tag{5.76}$$

For a homogeneous gain profile, the unsaturated gain per round is $G_0 = \Delta N_0 B_{12} 2dh\nu/c$, and we can write (5.76) as

$$\boxed{T_{opt} = -\gamma_a + \sqrt{\gamma_a G_0}} \quad . \tag{5.77}$$

Figure 5.23 shows the optiumum transmittance for several values of γ_a as a function of G_0/G_{thr}.

Finally, we discuss how much the output power of a laser at a given pump power P_p decreases when the losses γ are increased by a small amount $\Delta\gamma$. The dependence $I(\gamma, P_p)$ of laser intensity on losses is different for homogeneous and for inhomogeneous gain profiles. The saturated gain G_s is, according to (5.56,58),

$$G_s = G_0/(1 + I/I_s) \qquad \text{(homogeneous case)} \quad , \tag{5.56}$$

$$G_s = G_0/\sqrt{1 + I/I_s} \qquad \text{(inhomogeneous case)} \quad . \tag{5.58}$$

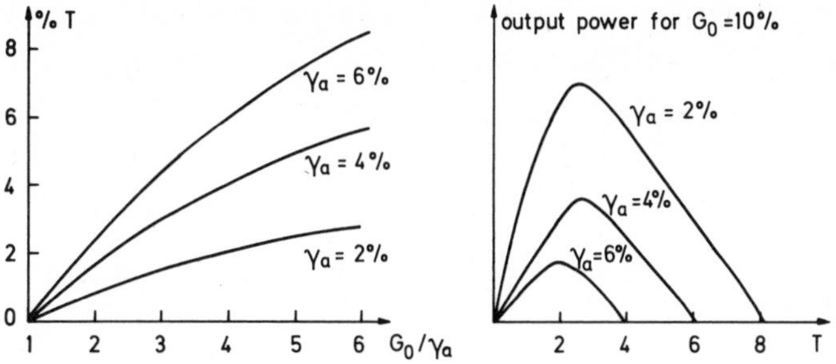

Fig.5.23. Optimum output coupling as a function of G_0/G_{thr} for several values of the losses γ_a

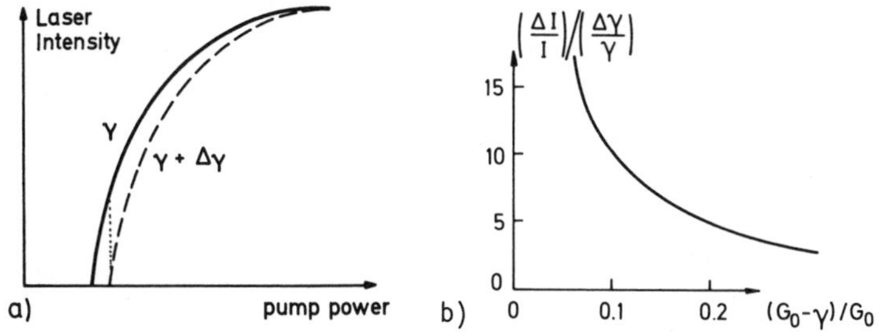

Fig.5.24. (a) Laser intensity as a function of pump power for two different total losses γ and $\gamma + \Delta\gamma$. (b) Relative intensity changes of laser output for a small increase $\Delta\gamma$ of the losses as a function of $(G_0 - \gamma)/G_0$

The unsaturated gain $G_0 = \exp[-2\Delta N\sigma(\omega)L]$ is determined by the pump power P_p which produces the inversion ΔN and by the absorption cross section $\sigma(\nu)$. The saturated gain G_s equals the total losses γ. The intracavity laser intensity I is therefore

$$I = I_s\left(\frac{G_0}{\gamma} - 1\right) \quad \text{homogeneous case} \quad , \tag{5.78a}$$

$$I = I_s\left(\frac{G_0^2}{\gamma^2} - 1\right) \quad \text{inhomogeneous case} \quad . \tag{5.78b}$$

If the losses γ are increased by a small amount $\Delta\gamma$ (Fig.5.24a), the intensity I changes by

$$\Delta I^{hom} = I_s G_0 \frac{\Delta\gamma}{\gamma(\gamma + \Delta\gamma)} \quad , \tag{5.79a}$$

$$\Delta I^{inh} = I_s G_0^2 \frac{2\Delta\gamma}{\gamma(\gamma + \Delta\gamma)^2} \quad . \tag{5.79b}$$

For the relative intensity changes, which are equal to the relative change of the laser output power, we obtain from (5.78,79)

$$\frac{\Delta I}{I} = \frac{G_0}{G_0 - \gamma}\,\frac{\Delta\gamma}{\gamma + \Delta\gamma} \qquad \text{(homogeneous gain profile)} \quad , \tag{5.80a}$$

$$\frac{\Delta I}{I} = \frac{G_0^2}{G_0^2 - \gamma^2}\,\frac{2\gamma\Delta\gamma}{(\gamma + \Delta\gamma)^2} \qquad \text{(inhomogeneous gain profile)} \quad . \tag{5.80b}$$

This shows that already small changes of the losses may result in large intensity changes, particularly if the laser operates close above threshold ($G_0 - \gamma \ll G_0$) (see Fig.5.24b). This can be used for sensitive detection of small concentrations of absorbing molecules inside the cavity (see Sect.8.2).

If we split the total losses $\gamma = \gamma_a + T$ into transmission losses T and all other losses γ_a, the laser output power is $P_L \approx TIA$ where I is the intracavity intensity and A the mean area $\pi\bar{W}_s^2$ of the mode. For homogeneous gain profiles we obtain from (5.78)

$$P_L = AI_s\,\frac{G_0 - \gamma_a - T}{1 + \gamma_a/T} \quad , \tag{5.81}$$

which gives, for the optimum transmission $T_{opt} = -\gamma_a + \sqrt{\gamma_a G_0}$, an optimum output power of

$$\boxed{P_L^{opt} = AI_s(\sqrt{G_0} - \sqrt{\gamma_a})^2} \quad . \tag{5.82}$$

5.10 Ring Lasers

A ring laser [5.19] utilizes an optical resonator consisting of at least three reflecting surfaces which can be provided by mirrors or prisms. Figure 5.25 shows two possible arrangements. Instead of the standing waves in a Fabry-Perot type resonator, the ring resonator allows travelling waves which may run clockwise or counterclockwise through the resonator. The unidirectional ring laser has the advantage that spatial hole burning which impedes single-mode oscillation of lasers (see Sect.5.8) can be avoided. In the case of homogeneous gain profiles the ring laser can utilize the total population inversion within the active mode volume. One therefore expects larger output powers in single-mode operation than from standing wave cavities at comparable pump powers.

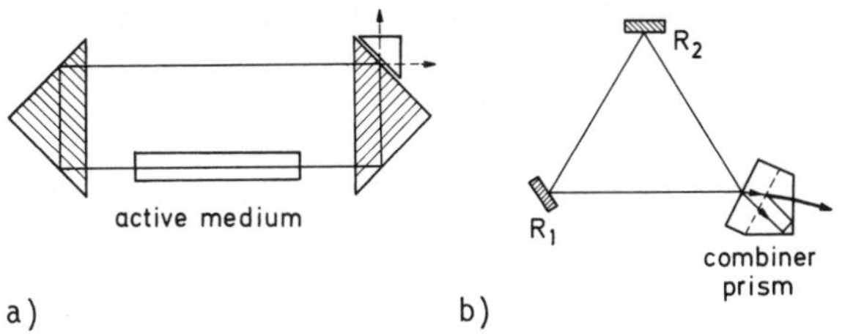

Fig.5.25a,b. Two examples of possible ring laser configurations. (a) Using total reflection in corner-cube reflectors and frustrated total reflection for output coupling. (b) Three mirror arrangements with beam combining prism

A unidirectional ring laser also avoids the unwanted hole-burning effects in inhomogeneous gain profiles of active media in Fabry-Perot type resonators, which are caused by the fact that the wave travelling into the z direction is absorbed by molecules with a velocity component $v_z = +(\nu - \nu_0)c/\nu_0$, but the reflected wave by other molecules with $v_z = -(\nu - \nu_0)c/\nu_0$ (see Sects. 5.7, 10.2). The experimental difficulty in the practical performance of unidirectional ring lasers is the suppression of one of the running waves since small fractions of backscattered light, e.g., from windows or imperfections in the ring, already couple the two counterpropagating waves, which form at simultaneous oscillation a closed standing-wave mode rather than the desired travelling-wave mode. This coupling can be avoided by increasing the losses or the gain for the clockwise-running wave but not for the counterclockwise wave. This may be achieved in various ways. One method uses a Faraday rotator with polarizers which acts as "optical diode" (see Sect.7.3).

If the whole ring laser is placed on a revolving platform the frequency of the two waves running in opposite directions is split due to the Sagnac effect [5.20]. The frequency difference $\Delta\nu = \nu_+ - \nu_-$ is proportional to the angular velocity of rotation. Such devices, called laser gyros, are sensitive detectors for slow rotations and are therefore developed and optimized in many laboratories [5.21]. Ring dye lasers are discussed in Sect.7.3.5.

5.11 Gaussian Beams

In Sect.5.4 we saw that the radial intensity distribution of a laser oscillating in the fundamental mode has a Gaussian profile. The laser beam emitted through the output mirror therefore also exhibits this Gaussian intensity profile. Although such a nearly parallel laser beam is in many respects similar to a plane wave, it shows several features which are different but which are important when the Gaussian beam is imaged by optical elements, such as lenses or mirrors. Often the problem arises of how to match the laser output to the fundamental mode of a passive resonator, such as a confocal spectrum analyzer (see Sect.4.3). We therefore briefly discuss some properties of Gaussian beams; our presentation follows that of the recommendable review of KOGELNIK and LI [5.22].

A laser beam travelling into the z direction can be represented by the field amplitude

$$E = A(x,y,z)e^{-i(\omega t - kz)} \quad . \tag{5.83}$$

While A(x,y,z) is constant for a plane wave, it is a slowly varying complex function for a Gaussian beam. Since every wave obeys the general wave equation

$$\Delta E + k^2 E = 0 \quad , \tag{5.84}$$

we can obtain the amplitude A(x,y,z) of our particular wave (5.83) by inserting (5.83) into (5.84), we make the trial solution

$$A = e^{\{-i[\varphi(z) + (k/2q)r^2]\}} \quad , \tag{5.85}$$

where $r^2 = x^2 + y^2$, and $\varphi(z)$ represents a complex phase shift. In order to understand the physical meaning of the complex parameter q(z) we express it in terms of two real parameters w(z) and R(z),

$$\frac{1}{q} = \frac{1}{R} - i\frac{\lambda}{\pi w^2} \quad . \tag{5.86}$$

With (5.86) we obtain from (5.85) the amplitude A(x,y,z) in terms of R, w, and φ,

$$A = e^{(-r^2/w^2)}\, e^{\{-i[kr^2/R(z)] - i\varphi(z)\}} \quad . \tag{5.87}$$

This illustrates that R(z) represents the radius of curvature of the wavefronts intersecting the axis at z, and w(z) gives the distance $r = (x^2 + y^2)^{1/2}$ from the axis where the amplitude has decreased to 1/e of its value on the axis (see Sect.5.4 and Fig.5.26). Inserting (5.87) into (5.84) and comparing terms of equal power in r yields the relations

$$dq/dz = 1 \quad \text{and} \quad d\varphi/dz = -i/q \quad , \tag{5.88}$$

which can be integrated and gives, with $R(z{=}0) = \infty$,

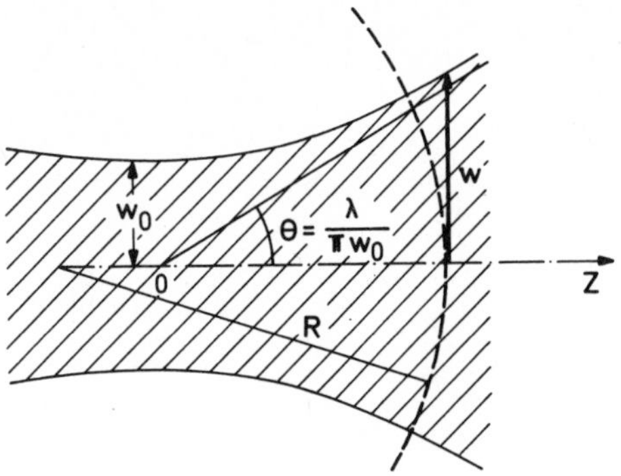

Fig.5.26. Gaussian beam with beam waist w_0, and phase front curvature R(z) (after [5.22])

$$q = q_0 + z = \frac{i\pi w_0^2}{\lambda} + z \quad , \tag{5.89}$$

where $w_0 = w(z=0)$. When we measure z from the beam waist at z = 0, we obtain

$$w^2(z) = w_0^2\left[1 + \left(\frac{\lambda z}{\pi w_0^2}\right)^2\right] \quad , \tag{5.90}$$

$$R(z) = z\left[1 + \left(\frac{\pi w_0^2}{\lambda z}\right)^2\right] \quad . \tag{5.91}$$

Integration of the phase relation (5.88),

$$d\varphi/dz = -i/q = -i/(z + i\pi w_0^2/\lambda) \quad ,$$

yields the z-dependent phase factor

$$i\varphi(z) = \ln\sqrt{1 + (\lambda z/\pi w_0^2)} - i\arctan(\lambda z/\pi w_0^2) \quad . \tag{5.92}$$

Having found the relations between φ, R, and w, we can finally express the Gaussian beam (5.83) by the real beam parameters R and w. From (5.92) and (5.87), we get

$$E = C\,\frac{w_0}{w}\,e^{(-r^2/w^2)}\,e^{[-ik(z + r^2/2R) - i\phi]}e^{-i\omega t} \quad . \tag{5.93}$$

The first exponential factor gives the radial Gaussian distribution, the second the phase which depends on z and r. We have used the abbreviation

$$\phi = \arctan(\lambda z/\pi w_0^2) \quad .$$

The factor C is a normalization factor. When we compare (5.93) with the field distribution (5.24) of the fundamental mode in a laser resonator, we see that both formulas are identical for m = n = 0.

The radial intensity distribution is

$$I(r) = EE^* = C^2 \frac{w_0^2}{w^2} e^{(-2r^2/w^2)} \quad . \tag{5.94}$$

The normalization factor C allows

$$\int_{r=0}^{\infty} 2\pi r I(r) dr = P_0 \tag{5.95}$$

to be normalized, which yields $C^2 = (2/\pi w_0^2)P_0$ where P_0 gives the total power in the beam.

When the Gaussian beam is sent through an aperture with diameter 2a, the fraction

$$P_t/P_i = \frac{2}{\pi w^2} \int_{r=0}^{a} 2r\pi e^{(-2r^2/w^2)} dr = 1 - e^{(-2a^2/w^2)} \tag{5.96}$$

of the incident power is transmitted through the aperture. Figure 5.27 illustrates this fraction as a function of a/w. For $a = (3/2)w$, already 99%, and for $a = 2w$ more than 99.9% of the incident power is transmitted. In this case diffraction losses are therefore negligible.

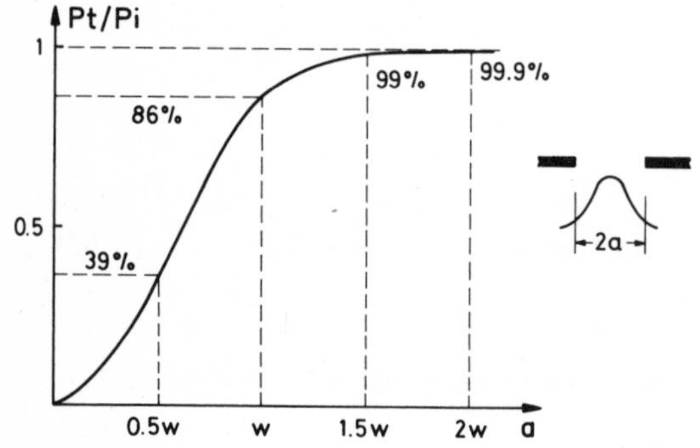

Fig.5.27. Intensity of a Gaussian beam with radius w through an aperture with radius a (after [1.1])

A Gaussian beam can be imaged by lenses or mirrors and the imaging equations are similar to those of spherical waves. When a Gaussian beam passes through a focussing thin lens with focal length f, the spot size w_s is the same on both sides of the lens (Fig.5.28). The radius of curvature R of the phase fronts changes from R_1 to R_2 in the same way as for a spherical wave, so that

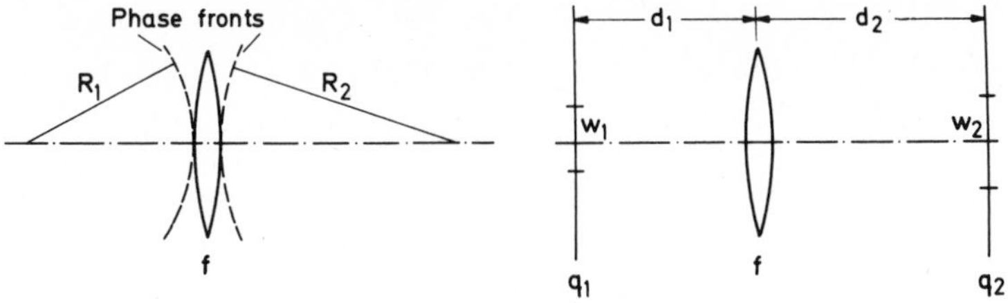

Fig.5.28. Imaging of a Gaussian beam by a thin lens

$$\frac{1}{R_2} = \frac{1}{R_1} - \frac{1}{f} \ . \tag{5.97}$$

The beam parameter q therefore satisfies the imaging equation

$$\frac{1}{q_2} = \frac{1}{q_1} - \frac{1}{f} \ . \tag{5.98}$$

If q_1 and q_2 are measured at distances d_1 and d_2 from the lens, we obtain from (5.98) and (5.89) the relation

$$q_2 = \frac{(1 - d_2/f)q_1 + (d_1 + d_2 - d_1d_2/f)}{(1 - d_1/f) - q_1/f} \ , \tag{5.99}$$

which allows the spot size w and radius of curvature R at any distance d_2 behind the lens to be calculated.

If, for instance, the laser beam is focussed into the interaction region with absorbing molecules, the beam waist of the laser resonator has to be transformed into a beam waist located in this region. The beam parameters in the waists are purely imaginary,

$$q_1 = i\pi w_1^2/\lambda \quad ; \quad q_2 = i\pi w_2^2/\lambda \ . \tag{5.100}$$

The beam diameters in the waists are $2w_1$ and $2w_2$, and the radius of curvature is infinite. Inserting (5.100) into (5.99) and equating the imaginary and the real parts yields the two equations

$$\frac{d_1 - f}{d_2 - f} = \frac{w_1^2}{w_2^2} \ , \tag{5.101}$$

$$(d_1 - f)(d_2 - f) = f^2 - f_0^2 \quad \text{with} \quad f_0 = \pi w_1 w_2/\lambda \ . \tag{5.102}$$

Since $d_1 > f$ and $d_2 > f$ this shows that any lens with $f > f_0$ can be used. For a given f, the position of the lens is determined by solving the two equations for d_1 and d_2,

$$d_1 = f \pm \frac{w_1}{w_2}\sqrt{f^2 - f_0^2} \quad , \tag{5.103}$$

$$d_2 = f \pm \frac{w_2}{w_1}\sqrt{f^2 - f_0^2} \quad . \tag{5.104}$$

From (5.101) we obtain the beam waist radius w_2 in the collimated region,

$$w_2 = w_1\left(\frac{d_2 - f}{d_1 - f}\right)^{\frac{1}{2}} \quad . \tag{5.105}$$

When the Gaussian beam is mode matched to another resonator, the beam parameter q_2 at the mirrors of this resonantor must match the curvature R and the spot size w (5.36). From (5.99), the correct values of f, d_1, and d_2 can be calculated.

We define the collimated or *waist region* as the region $|z| \leqq z_R$ around the beam waist at z = 0 where at $z = \pm z_R$ the spot size w(z) has increased by $\sqrt{2}$ compared with the value w_0 at the waist. Using (5.90), we obtain

$$w(z) = w_0\left[1 + \left(\frac{\lambda z_R}{\pi w_0^2}\right)^2\right]^{\frac{1}{2}} = \sqrt{2}w_0 \quad , \tag{5.106}$$

which yields for the *waist length* or *Rayleigh range*

$$z_R = \pi w_0^2/\lambda \quad . \tag{5.107}$$

The waist region extends about one Rayleigh range on either side of the waist (Fig.5.29). The length of the Rayleigh range depends on the spot size and therefore on the focal length of the focussing lens. Figure 5.30 shows the dependence on w_0 of the full length $2z_R$ of the collimated beam region.

At large distances $z \gg z_R$ from the waist, the Gaussian beam wavefront is essentially a spherical wave emitted from a point source at the waist. This region is called the *far field*. The divergence angle θ (far field half angle) of the beam can be obtained from (5.107) and Fig.5.26 with $z \gg z_R$ as

$$\theta = \frac{w(z)}{z} = \frac{\lambda}{\pi w_0} \quad . \tag{5.108}$$

Note, however, that in the near field region the center of curvature *does not* coincide with the center of the beam waist (Fig.5.26). When a Gaussian

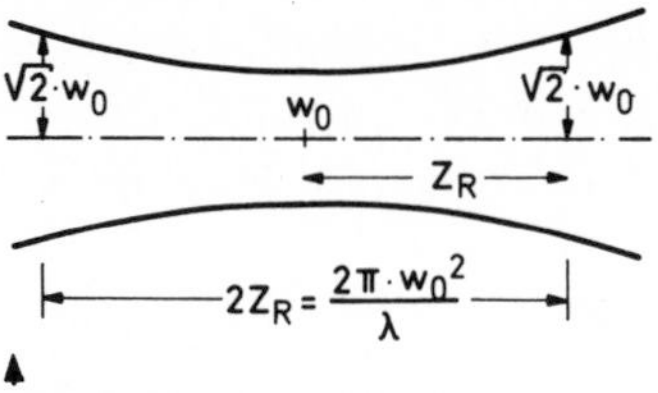

Fig.5.29. Waist region and Rayleigh range

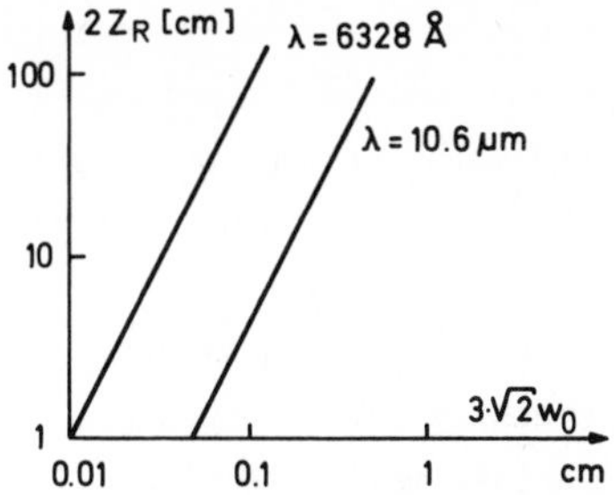

Fig.5.30. Rayleigh range as a function of the beam waist radius w_0, for the He-Ne laser at λ = 6328 Å and the CO_2 laser at λ = 10.6 μm

beam is focussed by a lens or a mirror with focal length f, the spot size in the beam waist is for $f \gg w_s$

$$w_0 = \frac{f\lambda}{\pi w_s} , \qquad (5.109)$$

where w_s is the spot size at the lens (Fig.5.31).

To avoid diffraction losses the diameter of the lens should be $d \geqq 3w_s$.

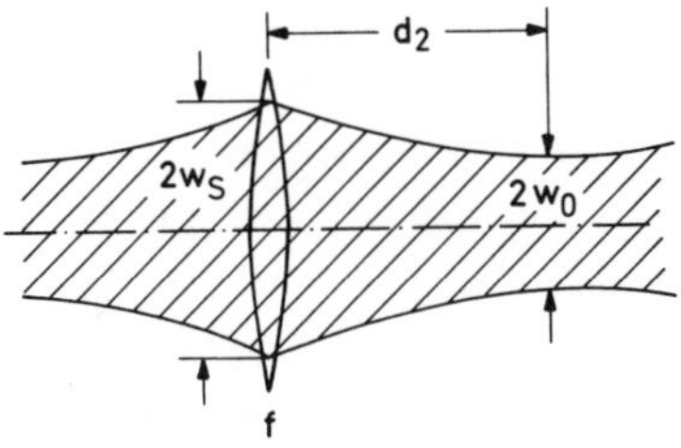

Fig.5.31. Focussing of a Gaussian beam by a lens

Problems to Chapter 5

5.1. Calculate the necessary threshold inversion of a gas laser transition at λ = 500 nm with a transition probability $A_{ik} = 5\times10^7\ s^{-1}$ and a homogeneous linewidth $\Delta\nu_{hom}$ = 20 MHz. The active length is L = 20 cm and the resonator losses per round trip are 5%.

5.2. A laser medium has a Doppler-broadened gain profile of halfwidth 2 GHz. The homogeneous width is 50 MHz, and the transition probability $A_{ik} = 1\times10^8\ s^{-1}$. Assume that one of the resonator modes (L = 40 cm) coincides with the center frequency ν_0 of the gain profile. What is the threshold inversion of the central mode, and at which inversion does oscillation start on the two adjacent longitudinal modes if the resonator losses are 10%?

5.3. The frequency of a passive resonator mode ($L = 15$ cm) lies $0.5\ \Delta\nu_D$ away from the center of the Gaussian gain profile of a gas laser at $\lambda = 632.8$ nm. Estimate the mode pulling if the cavity resonance width is 2 MHz and $\Delta\nu_D = 1$ GHz.

5.4. Assume a laser transition with a homogeneous width of 100 MHz while the inhomogeneous width of the gain profile is 1 GHz. The resonator length is 200 cm and the active medium is placed 20 cm from one end mirror. Estimate the spacing of the spatial hole burning modes. How many modes can oscillate simultaneously, if the unsaturated gain at the line center exceeds the losses by a factor of 10?

5.5. Estimate the optimum transmission of the laser output mirror if the unsaturated gain is 2 and the internal resonator losses are 10%.

5.6. The output beam from an He-Ne laser with a confocal resonator ($R = L = 30$ cm) is focussed by a lens of $f = 30$ cm, 50 cm away from the output mirror. Calculate the location of the focus, the Rayleigh length and the beam waist in the focal plane.

5.7. A nearly parallel Gaussian beam is expanded by a telescope with two lenses of focal lengths $f_1 = 1$ cm and $f_2 = 10$ cm. The spot size at the entrance lens is $w = 1$ mm. An aperture in the common focal plane of the two lenses acts as a spatial filter to improve the quality of the wavefront in the expanded beam (why?) What is the diameter of this aperture, if 95% of the intensity shall be transmitted?

6. Lasers as Spectroscopic Light Sources

Having summarized in the previous chapter some basic characteristics of lasers and optical resonators, we shall now discuss those properties which make the laser such an interesting and useful light source in spectroscopy. We shall describe the experimental techniques that are necessary for achieving optimal results in spectroscopic applications. These techniques comprise *mode selection* in lasers, *wavelength* and *intensity stabilization* of single-mode lasers, and experimental realizations of *controlled wavelength tuning*. Furthermore we briefly discuss the interesting question of why a lower limit exists for the laser linewidth. At the end of this chapter some methods of *relative and absolute frequency measurements* in the optical region will be presented.

6.1 Advantages of Lasers in Spectroscopy

The fact that lasers are replacing conventional spectral lamps to an increasing extent in numerous applications demonstrates their superiority over incoherent light sources for many experiments. To illustrate the advantages and limitations of lasers in the present state of the art, we pick out five characteristic properties.

1) The large spectral power density $\rho_\nu(\nu)$ attainable from many laser types may exceed that of incoherent light sources by many orders of magnitude. This may significantly reduce noise problems caused by detector noise or background radiation. Furthermore, the large intensity allows new nonlinear spectroscopic techniques such as saturation spectroscopy (Sect.10.2), or multiphoton processes (Sect.10.5), which open new possibilities of studying molecular transitions not accessible to linear spectroscopy.

2) The small divergence of collimated laser beams brings about a number of experimental advantages for the spectroscopist. For measurements of small absorption coefficients, for instance, long path lengths through the absorbing sample can be realized. Perturbing background noise due to light scattered from cell walls or windows can be more readily reduced than with divergent beams from incoherent sources. The interaction zone of the sample molecules within the Rayleigh length around the beam waist of a focussed laser beam (see Sect.5.11) can be efficiently imaged onto the entrance slit of a spectrograph. Since this zone is the source of Raman-scattered light or of laser-excited fluorescence, the high collection efficiency attainable is particularly important in Raman spectroscopy (Chap.9) or in low-level fluorescence spectroscopy. In the latter case, most of the laser-excited molecules radiate while having travelled a mean distance $\bar{d} = \bar{v}/\tau$ determined by their mean velocity $\bar{v}$ and their mean lifetime τ. With typical values of $\bar{v} = 5 \times 10^4$ cm/s and $\tau = 10^{-8}$ s, we obtain $\bar{d} = 5 \times 10^{-4}$ cm = 0.5 μm, which shows that the fluorescence from excited molecules with lifetimes $\tau < 10^{-7}$ s is essentially emitted from the interaction zone where the molecules are excited. Diffusion of the excited molecules out of this zone becomes noticeable only for $\tau > 10^{-6}$ s. In these cases either the diffusion has to be reduced by increasing the pressure or the field of observation has to be enlarged.

3) Of particular advantage for high-resolution spectroscopy is the extremely small spectral linewidth of lasers which can be achieved with special techniques. This allows a spectral resolution which may exceed that of the largest spectrographs by several orders of magnitude. In laser spectroscopy, the resolution is often no longer limited by the instrumental bandwidth but rather by the spectral linewidth of the absorbing or emitting molecules.

4) The possibility of continuously tuning the wavelength of such narrow-band lasers has certainly opened a new area of spectroscopy. A single-mode tunable laser represents a device which is a combination of an intense light source and an ultrahigh resolution spectrometer. Tunable lasers and tuning techniques are therefore covered in a separate chapter.

5) The capability of pulsed or mode-locked lasers to deliver intense and short light pulses with pulse widths down to the subpicosecond range allows the study of ultrafast transient phenomena such as short spontaneous lifetimes or fast relaxation processes in gases, liquids, or solids (see Chap.11).

The examples given in Chaps.8-14 will illustrate in more detail these and other advantages of lasers in spectroscopy. Although coherent light sources can already cover the whole spectral range from the vacuum ultraviolet to the far infrared region (see Chap.7), there are some wavelength ranges where no direct laser oscillation has yet been achieved and which can be only reached by frequency-mixing techniques. The main disadvantages of using lasers in such unfavorable spectral regions are the experimental expenditure and the costs necessary to achieve sufficient intensity at a given spectral bandwidth. In many cases, however, the use of lasers may be less expensive than that of incoherent light sources because tunable lasers may replace radiation sources *and* spectrometers.

6.2 Fixed-Frequency Lasers and Tunable Lasers

In Sect.5.6 we saw that the different wavelengths λ_i on which laser oscillation is possible, are in principle determined by two factors: the gain profile of the amplifying medium, and the eigenresonances of the laser resonator. Without additional wavelength-selecting elements inside the resonator we would therefore expect laser oscillation on all resonator modes with wavelengths within the spectral gain profile above threshold, because on these modes the gain exceeds the total losses. However, as has already been discussed in Sect.5.7, this is only true for completely inhomogeneous gain profiles where no gain competition occurs between the different modes. The actual number of simultaneously oscillating modes depends on the spectral width $\Delta\nu_g$ of the gain profile, on the homogeneous width $\Delta\nu_h$, and on the free spectral range $\delta\nu = c/2d^*$ of the cavity, where d^* is the optical pathlength between the resonator mirrors. For $\Delta\nu_g \gg \delta\nu \gg \Delta\nu_h$, the laser will simultaneously oscillate on all resonator modes within the gain profile.

In many cases, the spectral width of the gain profile is comparable with that of spectral lines from incoherent spectral lamps, as, for example, is the case for gas lasers with gain profiles determined by the Doppler width of the amplifying transition between two discrete states of excited atoms or molecules in the active medium. Also, in many solid-state lasers, where the active medium is composed of excited impurity atoms or ions, diluted at low concentrations in a host crystal lattice, the linewidth of the amplifying transition is often small compared with the commonly found broad absorption bands of solids. In such cases the laser wavelength is restricted

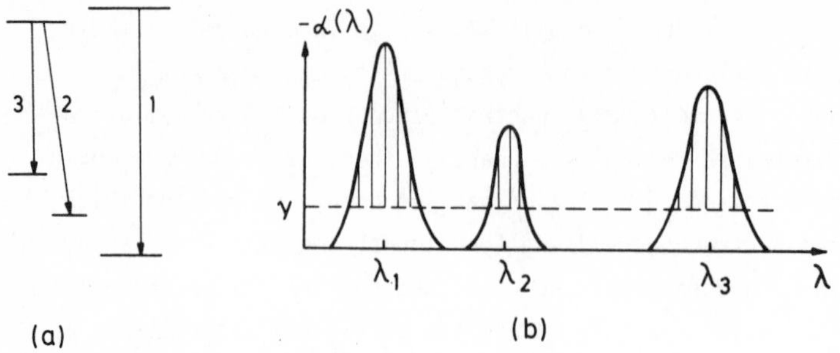

Fig.6.1a,b. Level scheme and oscillating wavelengths of a multiline laser with inhomogeneous gain profiles at each transition

to this small gain region and these lasers are therefore called *fixed-wavelength* or *fixed-frequency lasers*.

Often the active medium exhibits gain on several transitions and the laser can oscillate simultaneously on many lines. As long as the different gain regions are narrow and do not overlap, the wavelength of each line is still restricted to its narrow gain range, and the laser is a *multiline* fixed wavelength laser (Fig.6.1). Examples are the argon-ion laser or the CO_2 laser.

On the other hand, there are several cases where the gain profile extends over a broad spectral range. The most important representative of this type is the dye laser, where stimulated emission from the excited state to many vibronic levels of the electronic ground state is possible (Fig.6.2). Because of the strong interaction of the dye molecules with the solvent, these levels are broadened so that the linewidth exceeds the level separation. The absorption and fluorescence of dye molecules in liquid solvents therefore exhibits a broad continuous spectral distribution (Fig.6.2b), and the gain profile extends over a large spectral range of several hundred nanometers. Another example is given by molecular transitions from discrete bound states to repulsive dissociative states in excimer molecules (see Figs.2.13,14). The fluorescence and the gain profile show a continuous distribution. In such cases the laser wavelength can be tuned continuously over a larger spectral range (see Chap.7) and the lasers are therefore called *tunable lasers*.

Strictly speaking there is only a gradual difference between fixed wavelength lasers and tunable lasers because the wavelength of any single-mode laser can be tuned over the spectral range of the gain profile, as will be shown in Sect.6.8. The difference between the two types lies only in the extention of the tuning range which is narrow for the "fixed wavelength" lasers and broader for the commonly called "tunable lasers."

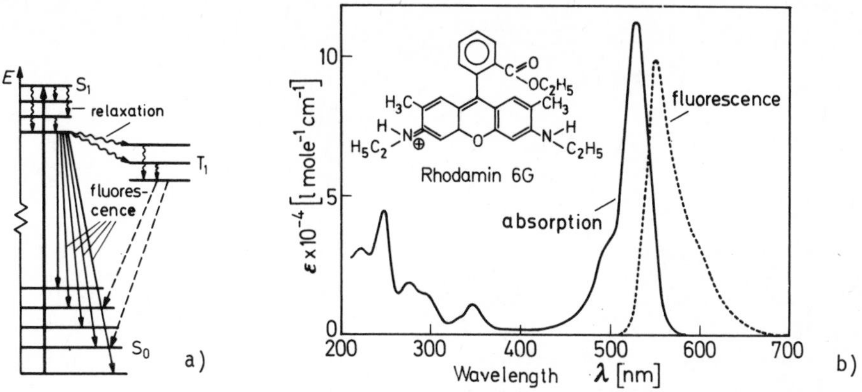

Fig.6.2. Schematic level diagram of the dye laser (a) and absorption and fluorescence spectrum of the dye rhodamine 6G, in a glycol solution (b)

There are many spectroscopic applications where tuning of the wavelength is not required. Then fixed-frequency lasers, which often deliver higher output powers, are the better choice. The numerous existing laser lines compiled in [6.1] have been used, for instance, to study spectroscopic properties of laser media, and excitation or relaxation processes.

Often, optical pumping of atoms or molecules by strong laser lines has been used to achieve population changes in selectively excited or depleted levels. Because diatomic or polyatomic molecules have so many rotational and vibrational levels, generally one or even more molecular transitions are superposed by fortuitous coincidences on a laser line in a spectral region of an electronic band spectrum. Fixed-frequency lasers have therefore been used for a long time for optical pumping of molecules (see Chap.8) in the whole spectral range from the ultraviolet to the far infrared.

Compared to widely tunable lasers, fixed frequency lasers with a narrow gain profile have the advantage that their wavelength is well defined. In case of low-pressure gas lasers, such as the He-Ne or the CO_2 laser, one can stabilize the wavelength onto the center of the Doppler profile by using the narrow saturation dip burnt into the center of the inhomogeneous gain profile (Sect.3.6). This Lamb-dip stabilization is discussed in Sect.10.2.3. These stabilized lasers can serve as wavelength and frequency standards; the whole spectral range may be covered by using frequency mixing techniques and the higher harmonics of the stabilized wavelength (see Sect.6.10).

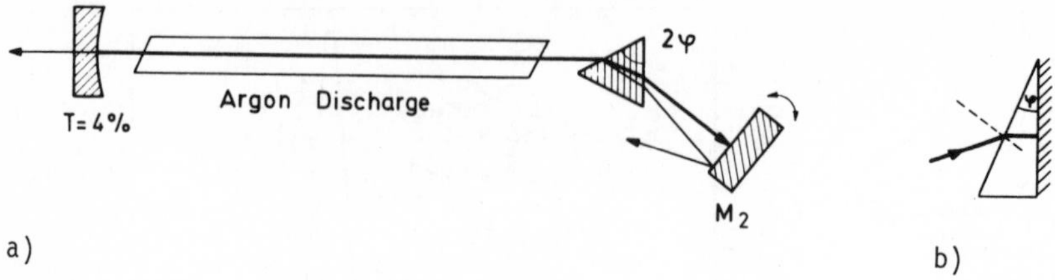

Fig.6.3. (a) Line selection in a gas laser by use of a prism with refractive index n and prism angle 2φ where $\tan 2\varphi = 1/n$. (b) Littrow prism reflector which combines line selector and end reflector

In order to achieve single-line oscillation in laser media which exhibit gain for several transitions, wavelength-selecting elements outside or inside the laser resonator can be used. If the different lines are widely separated in the spectrum, the selective reflectivity of the dielectric mirrors may already be sufficient to select a single transition. In the case of broad-band reflectors or closely spaced lines, prisms, gratings, or Lyot filters are commonly used for wavelength selection. Figure 6.3 shows line selection by a prism in an argon laser. The different lines are refracted by the prism, and only that line which is vertically incident upon the end mirror, is reflected back into itself and can reach oscillation threshold, while all other lines are reflected out of the resonator. Turning the end reflector M_2 allows the desired line to be selected. To avoid reflection losses at the prism surfaces, a Brewster prism with $\tan 2\varphi = 1/n$ is used, with the angle of incidence for both prism surfaces being Brewster's angle. The prism and the end mirror can be combined by coating the endface of a Brewster prism reflector (Fig.6.3b). Such a device is called a Littrow prism.

Because most prism materials such as glass or quartz absorb in the infrared region, it is more convenient to use a Littrow grating (see Sect.4.1) as wavelength selector in this wavelength range. Figure 6.4 illustrates the line selection in a CO_2 laser which can oscillate on many rotational lines of a vibrational transition. Often the laser beam is expanded by a mirror configuration in order to cover a larger number of grating grooves, thus

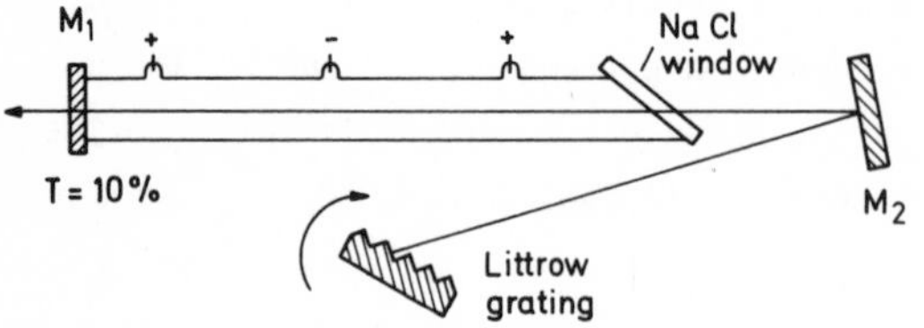

Fig.6.4. Selection of different rotational lines in a CO_2 laser by a Littrow grating

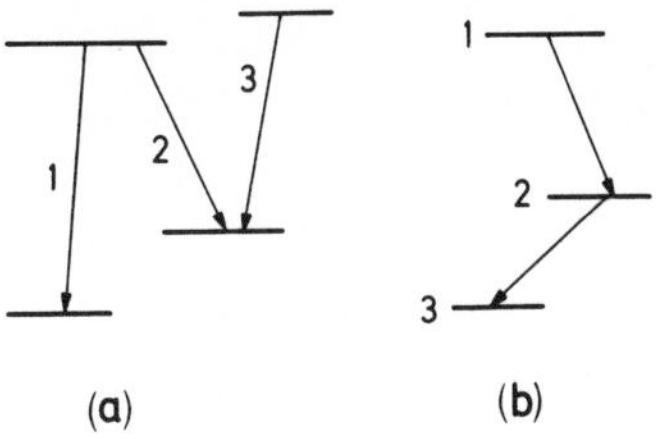

Fig.6.5a,b. Schematic level diagram for a laser with simultaneously oscillating lines. (a) On transitions sharing a common upper or lower level. (b) On cascade transitions

increasing the spectral resolution (see Sect.4.1). This has the further advantage that the power density is lower and damage of the grating is less likely.

If some of the simultaneously oscillating laser transitions share a common upper or lower level, such as the lines 1, 2, and 3 in Fig.6.5a, gain competition will diminish the output of each line. In this case it is advantageous to use *internal* line selection in order to suppress all but one of the competing transitions. Sometimes, however, the laser may oscillate on cascade transitions (Fig.6.5b). In such a case, the laser transition 1 → 2 increases the population of level 2 and therefore enhances the gain for the transition 2 → 3. Apparently it is then more favorable to allow multiline oscillation and to select a single line by an external prism or grating. Using a special mounting design, it can be arranged for no deflection of the output beam to occur when the multiline output is tuned from one line to the other [6.2].

6.3 Frequency Spectrum of Multimode Lasers

Even when single-line operation of a multiline laser has been achieved by spectral preselection with a prism or grating, there are generally still several resonator modes within the gain profile of the laser transition for which the gain exceeds the total losses. In Sect.5.7,8 we discussed how gain competition and spectral hole burning affects the coupling between these modes and determines the frequency spectrum of multimode lasers. We may summarize the results of these considerations as follows:

a) For laser transitions with inhomogeneous gain profiles, stable multimode oscillation of all laser modes above threshold can be achieved without special precautions, provided the frequency jitter of the different modes (e.g., due to acoustic vibrations of the laser mirror) is less than their spacing $\delta\nu = c/(2d^*)$ where d^* is the optical pathlength between the mirrors.

b) For laser transitions with purely homogeneous gain profiles, single-mode operation is possible if spatial hole burning (see Sect.5.8) can be avoided, and the eigenfrequency of the active resonator is sufficiently stable. In general, however, fluctuations of the optical path length d^* between the resonator mirrors and gain competition between oscillating modes result in random fluctuations of number, amplitudes and phases of the oscillating modes. The spectral intensity distribution of the laser output is the superposition

$$I(\omega,t) = |\sum A_k(t)\cos[\omega_k t + \varphi_k(t)]|^2 \tag{6.1}$$

of these competing modes, where the phases $\varphi_k(t)$ are randomly fluctuating in time.

c) The time average of the spectral distribution of the output intensity,

$$I(\omega) = \frac{1}{T}\int_0^T |\sum A_k(t)\cos[\omega_k t + \varphi_k(t)]|^2 dt \quad , \tag{6.2}$$

reflects the gain profile of the laser transition. The necessary averaging time T depends on the buildup time of the laser modes which is determined by the unsaturated gain and the strength of the mode competition. In case of gas lasers, the average spectral width $\langle\Delta\nu\rangle$ corresponds to the Doppler width of the laser transition and the coherence length of such a multimode laser is comparable to that of a conventional spectral lamp where a single line has been filtered out. For lasers with a broad spectral gain profile, the laser linewidth depends on the preselecting elements inside the laser resonator. Some examples, besides that given in Sect.5.7, illustrate the situation:

i) He-Ne laser

The He-Ne laser is probably the most thoroughly investigated gas laser [6.3]. From the level scheme in Fig.6.6 which uses the Paschen notation [6.3a], we see that the transitions at λ = 3.39 μm and λ = 0.6328 μm share a common *upper* level. Suppression of the 3.390 μm line therefore enhances the output power at 0.6328 μm. The 1.15 μm and the 0.6328 μm lines, on the other hand, share a common *lower* level, and they also compete for gain, since both laser transitions increase the lower-level population and therefore decrease the inversion. If the 3.3903 μm transition is suppressed, e.g., by placing an absorbing CH_4 cell inside the resonator, the population of the upper $3s_2$ level increases, and a new line at λ = 3.3913 μm reaches threshold.

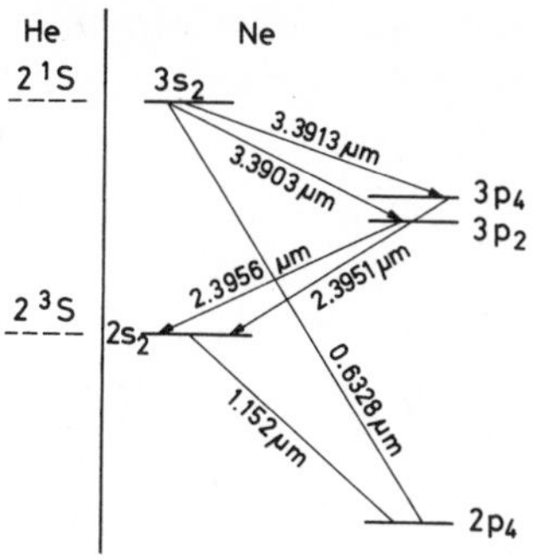

Fig.6.6. Level diagram of the He-Ne system in Paschen notation with the main laser transitions

This laser transition populates the $3P_4$ level, and produces gain for another line at $\lambda = 2.3951$ μm. This last line only oscillates together with the 3.3913 μm, which acts as pumping source. This is an example of cascade transitions in laser media [6.4] as depicted in Fig.6.5b.

The homogeneous width of the laser transitions is determined by pressure broadening and power broadening. At total pressures of about 3 torr, the homogeneous linewidth for the transition $\lambda = 632.8$ nm is about 300 MHz, which is still small compared with the Doppler width $\Delta\nu_D = 1500$ MHz. In single-mode operation, one can obtain about 20% of the multimode power [6.5] which corresponds to the ratio $\Delta\nu_h/\Delta\nu_D$ of homogeneous to inhomogeneous linewidth. The mode spacing $\delta\nu = c/(2d^*)$ equals the homogeneous linewidth for $d^* = c/2\Delta\nu_h$. For $d^* < c/2\Delta\nu_h$, stable multimode oscillation is possible; for $d^* > c/2\Delta\nu_h$, mode competition occurs.

ii) Argon Laser

The discharge of a cw argon laser exhibits gain for more than 15 different transitions. Figure 6.7 shows part of the energy-level diagram, illustrating the coupling of different laser transitions. Since the lines 514.5 nm,

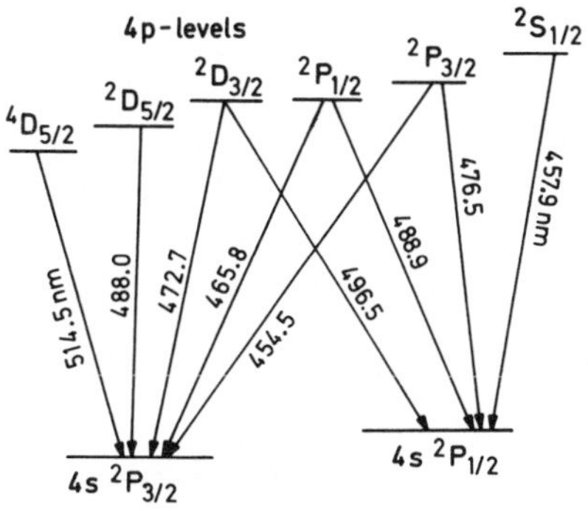

Fig.6.7. Energy level diagram and coupling of laser transitions in the argon-ion laser

488.0 nm, and the 465.8 nm share the same lower level, suppression of the competing lines will enhance the inversion and the output power of the selected line. The mutual interaction of the various laser transitions has been therefore extensively studied [6.6,7] in order to optimize the output power. Line selection is generally achieved with an internal Brewster prism (Fig.6.3). The homogeneous width $\Delta\nu_h$ is mainly caused by collision broadening due to electron-ion collisions. Additional broadening and shifts of the ion lines result from ion drifts in the field of the discharge. At intracavity intensities of 350 W/cm^2, which corresponds to about 1 W output power, appreciable saturation broadening increases the homogeneous width, which may exceed 1000 MHz. This explains why the output at single-mode operation may reach 30% of the multimode output on a single line [6.8].

iii) CO_2 Laser

A section of the level diagram is illustrated by Fig.6.8. The vibrational levels (v_1, v_2^l, v_3) are characterized by the number of quanta in the three normal vibrational modes. The upper index of the degenerate vibration ν_2 gives the quantum number of the corresponding angular momentum l [6.9]. Laser oscillation is achieved on many rotational lines within two vibrational transitions $(v_1, v_2^l, v_3) = 00^01 \rightarrow 10^00$ and $00^01 \rightarrow 02^00$ [6.9a-c]. Without line selection, generally only the band around 961 cm^{-1} (10.6 μm) appears because these transitions exhibit larger gain. The laser oscillation depletes the population of the 00^01 vibrational level and suppresses laser oscillation on the second transition, due to gain competition. With internal line selection (see Fig.6.4), many more lines can be successively optimized by turning the wavelength-selection grating. The output power of each line is higher than

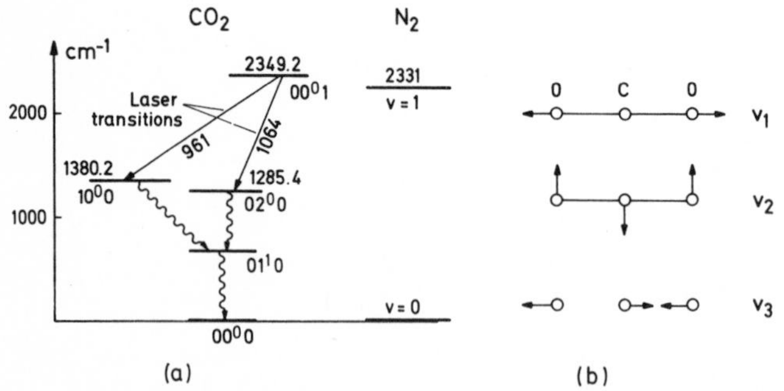

Fig.6.8a,b. Level diagram and laser transitions in the CO_2 molecule (a) and normal vibrations (b)

that of the same line in multiline operation. Because of the small Doppler width (66 MHz), the free spectral range $\delta\nu = c/2d^*$ is already larger than the width of the gain profile for $d < 200$ cm. For such resonators, the mirror separation d has to be adjusted to tune the resonator eigenfrequency $\nu_R = qc/2d^*$ (q = integer) to the center of the gain profile. If the resonator parameters are properly chosen to suppress higher transverse modes, the CO_2 laser will then oscillate on a single longitudinal mode.

6.4 Mode Selection in Lasers

In Chap.5 we saw that in an active laser resonator, all those TEM_{mnq} modes for which the gain exceeds the total losses can participate in laser oscillation. The selection of one of these many modes implies that the losses of all other modes must be increased beyond the gain so that these modes do not reach threshold, while the losses for the wanted mode should be as low as possible. The suppression of the higher *transverse* modes can be achieved with other means than the selection of one of the *longitudinal* modes because the transverse modes differ in their radial field distribution, while the different longitudinal modes have the same radial field profile but are separated in frequency by the free spectral range.

Let us first consider the selection of *transverse* modes. In Sect.5.4 it was shown that the higher transverse TEM_{mnq} modes have radial field distributions which are less and less concentrated along the resonator axis with increasing transverse order n or m. This means that their diffraction losses are much higher than that of the fundamental modes TEM_{mnq} when an aperture is inserted inside the resonator. The field distribution of the modes and therefore also their diffraction losses depend on the resonator parameters such as the radii R_i of curvature of the mirrors, the mirror separation d, and of course the Fresnel number N (see Sect.5.3). Only those resonators which fulfil the stability condition [6.10]

$$0 < (1 - d/R_1)(1 - d/R_2) < 1 \qquad (6.3a)$$

have finite spot sizes of the field distributions inside the resonator. With the abbreviations $g_1 = (1 - d/R_1)$ and $g_2 = (1 - d/R_2)$, the stability condition can be written as

$$0 < g_1g_2 < 1 \quad . \qquad (6.3b)$$

A stability diagram in the g_1g_2 plane, as shown in Fig.6.9, allows the stable and unstable regions to be identified. In Fig.6.10, the ratio γ_{10}/γ_{00} of the diffraction losses for the TEM_{10} and the TEM_{00} modes is plotted for different values of g as a function of the Fresnel number N. From this diagram one can obtain for any given resonator the diameter 2a of an aperture which suppresses the TEM_{10} mode but still has sufficiently small losses for the fundamental TEM_{00} mode. In gas lasers, the diameter 2a of the discharge tube generally forms the limiting aperture. One has to choose the resonator parameters in such a way that $a \approx 3w/2$ (see Sect.5.11) because this assures that the fundamental mode nearly fills the whole active medium but still suffers less than 1% diffraction losses.

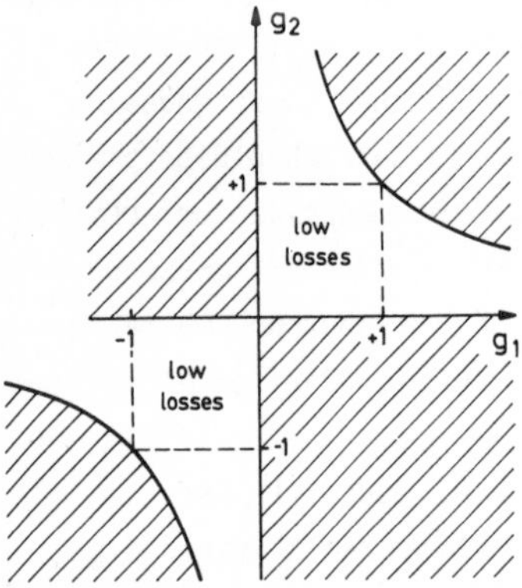

Fig.6.9. Stability diagram of laser resonators with stable (dashed areas) and unstable regions

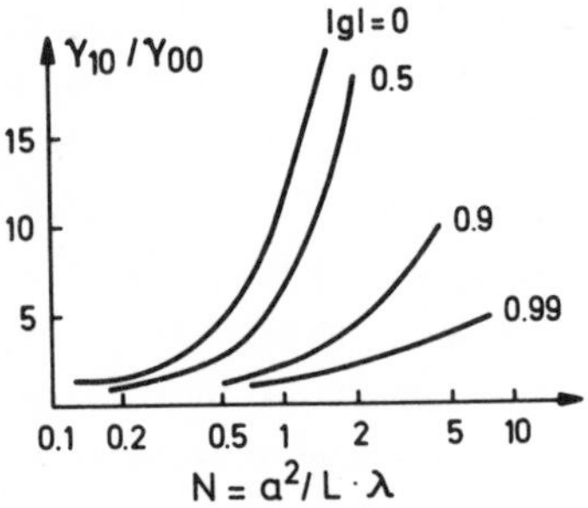

Fig.6.10. Ratio γ_{10}/γ_{00} of the diffraction losses for the TEM_{10} to that of the TEM_{00} mode, as a function of the Fresnel number N for a resonator with two mirrors of equal curvatures where $g_1 = g_2 = g$

Because the frequency separation of the transverse modes is small and the TEM_{10q}-mode frequency is separated from the TEM_{00q} frequency by less than the homogeneous width of the gain profile, the fundamental mode can partly saturate the inversion at distances r from the axis, where the TEM_{10q}-mode has its field maximum. The resultant transverse mode competition (Fig.6.11) reduces the gain for the higher transverse modes and may suppress their oscillation even if the unsaturated gain exceeds the losses. The restriction for the maximum allowed aperture diameter is therefore less stringent. The resonator geometry of many commercial lasers has already been designed in

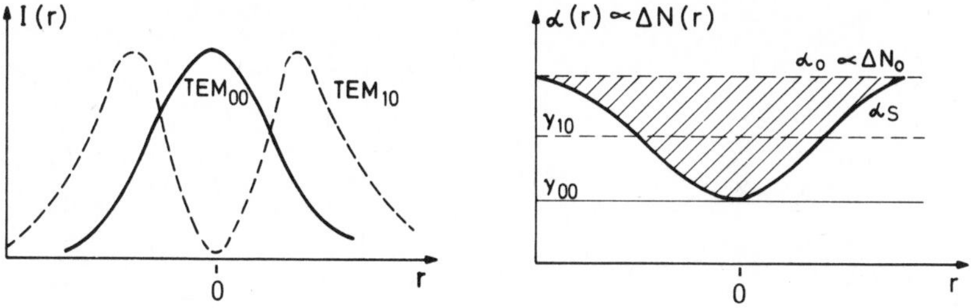

Fig.6.11. Transverse gain competition between the TEM_{00} mode and the TEM_{10} mode

such a way that "single-transverse-mode" operation is obtained. The laser can, however, still oscillate on several longitudinal modes, and for true single-mode operation, the next step is to suppress all but one of the longitudinal modes.

From the discussion in Sect.5.7 it should have become clear that simultaneous oscillation on several longitudinal resonator modes is possible when the inhomogeneous width $\Delta\nu_g$ of the gain profile exceeds the mode spacing $c/2d^*$ (Fig.6.12). A simple way to achieve single-mode operation is therefore the reduction of the resonator length d below a value d_{max} such that the width $\Delta\nu_g$ of the gain profile above threshold becomes smaller than the free spectral range $\delta\nu = c/2d^*$ [6.11].

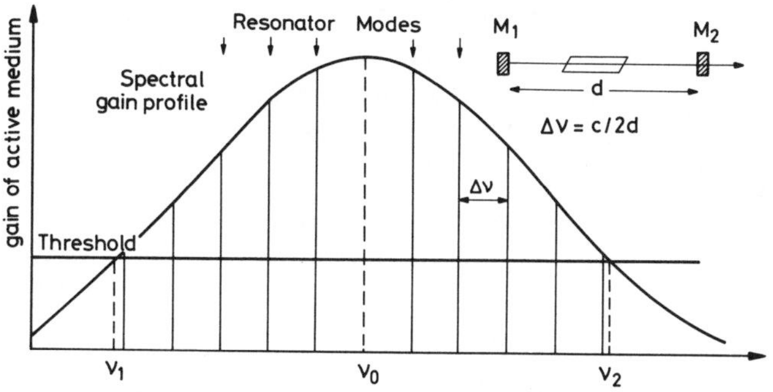

Fig.6.12. Longitudinal resonator modes within the spectral gain profile of a laser transition

Example

He-Ne laser: $\delta\nu = 1200$ MHz $\Rightarrow d^* = 13$ cm.

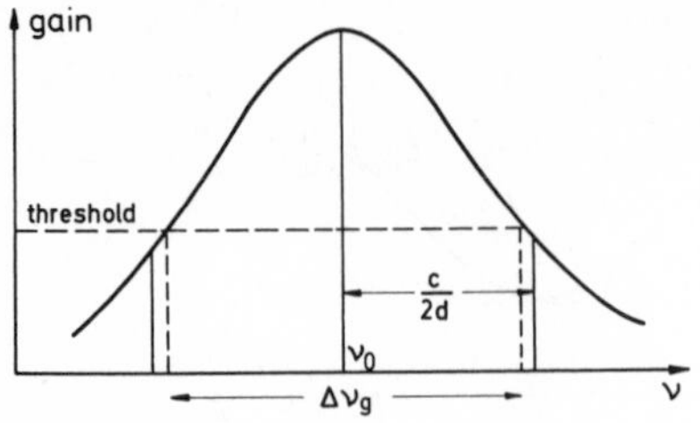

Fig.6.13. Single-mode operation by reducing the cavity length d to a value where the cavity mode spacing exceeds half of the gain profile width above threshold

If the resonator frequency can be tuned to the center of the gain profile, single-mode operation can be achieved even with the double length $2d^*$, because then the two neighboring modes just do not reach threshold (Fig.6.13). This solution for the achievement of single-mode operation has, however, several drawbacks. Since the length L of the active medium cannot be larger than d ($L \leq d$), threshold can only be reached for transitions with a high gain. The output power, which is proportional to the active mode volume, is also small in most cases. For single-mode lasers with higher output powers, other methods are therefore preferable. We distinguish between *external* and *internal* mode selection.

When the output of a multimode laser passes through an external spectral filter, such as an interferometer or spectrometer, a single mode can be selected. For perfect selection, however, high suppression by the filter of the unwanted modes and high transmission of the wanted mode are required. This technique of external selection has the further disadvantage that only part of the total laser output power can be used. *Internal* mode selection with spectral filters completely suppresses the unwanted modes already when the losses exceed the gain. Furthermore, the output power of a single-mode laser is generally higher than the power in this mode at multimode oscillation because the total inversion $V \cdot \Delta N$ in the active volume V is no longer shared by many modes as is the case in multimode operation with gain competition.

In single-mode operation we can expect roughly the fraction $\Delta\nu_{hom}/\Delta\nu_g$ of the multimode power, where $\Delta\nu_{hom}$ is the homogeneous width within the inhomogeneous gain profile. This width $\Delta\nu_{hom}$ becomes even larger for single-mode operation because of power broadening by the more intense mode. In an argon laser, for example, one can obtain up to 30% of the multimode power in a single mode with internal mode selection.

This is the reason why virtually all single-mode lasers use *internal* mode selection. We discuss in the next section some experimental possibilities which allow stable single-mode operation of lasers.

6.5 Experimental Realization of Single-Mode Lasers

As pointed out in the previous section, all methods of achieving single-mode operation are based on mode suppression by increasing the losses beyond the gain for all but the wanted mode. A possible realization of this idea is illustrated in Fig.6.14, which shows longitudinal mode selection by a

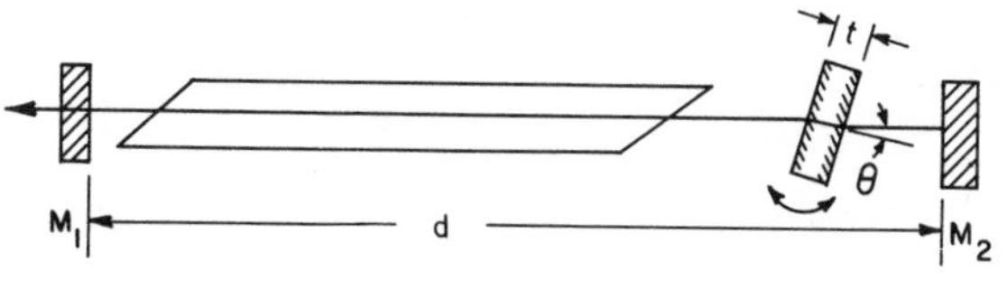

Fig.6.14. Single-mode operation by inserting a tilted etalon inside the resonator

tilted plane-parallel etalon (thickness t and refractive index n) inside the laser resonator [6.12]. In Sect.4.2.6, it was shown that such an etalon has transmission maxima at those wavelengths λ_m for which

$$m\lambda_m = 2nt\cos\theta \quad . \tag{6.4}$$

If the free spectral range of the etalon,

$$\delta\lambda = 2nt\cos\theta\left(\frac{1}{m} - \frac{1}{m+1}\right) = \frac{\lambda}{m+1} \quad , \tag{6.5}$$

is larger than the spectral width $|\lambda_1 - \lambda_2|$ of the gain profile above threshold, only a single mode can oscillate (Fig.6.15). Since the wavelength λ is also determined by the resonator length d, the tilting angle θ has to be adjusted so that

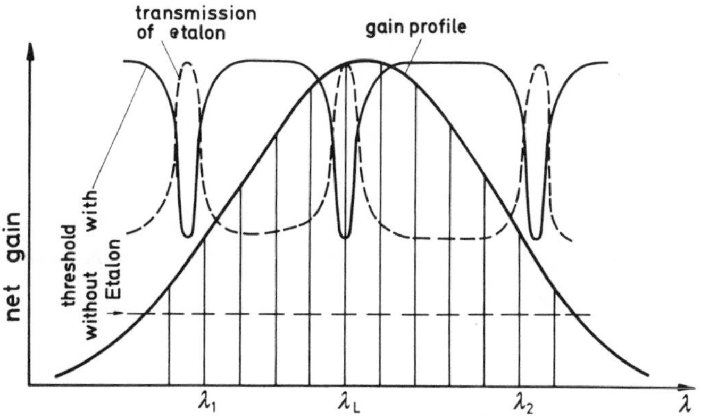

Fig.6.15. Gain profile, resonator modes, and spectral transmission of the etalon tuned for single-mode operation

$$2nt\cos\theta/m = 2d/q \quad , \quad q = \text{integer} \tag{6.6}$$

which means that the transmission peak of the etalon has to coincide with an eigenresonance of the laser resonator.

Example

In the argon-ion laser, the width of the gain profile is about 8 GHz. With a free spectral width $\Delta\nu = c/(2nt) = 10$ GHz of the intracavity etalon, single-mode operation can be achieved.

The finesse F^* of the etalon has to be sufficiently high to introduce losses for the modes adjacent to the selected mode which overcome their gain (Fig.6.14). Fortunately in many cases this gain is already reduced by the oscillating mode due to gain competition. This allows the less stringent demand that the losses of the etalon must exceed the unsaturated gain at a distance $\Delta\nu_{hom}$ away from the transmission peak.

Often a Michelson interferometer is used for mode selection, coupled by a beam splitter St to the laser resonator (Fig.6.16). The free spectral range of this "Fox-Smith cavity" [6.13], which is $\Delta\lambda = c/[2(L_2+L_3)]$, has

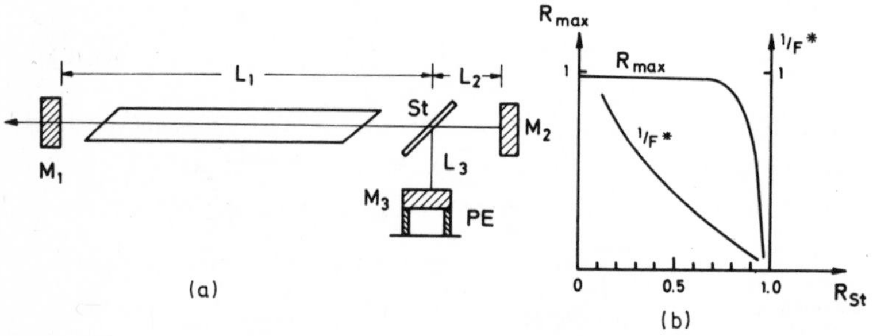

Fig.6.16a,b. Mode selection with a Fox-Smith selector. (a) Experimental arrangement. (b) Maximum reflectivity R and finesse F^* of the Fox-Smith cavity as a function of the reflectivity R_{St} of the beam splitter for $R_2 = R_3 = 0.99$ and $A_s = 0.5\%$

again to be larger than the width of the gain profile. With a piezo element PE, the mirror M_3 can be translated by a few μm to obtain resonance between the two coupled resonators. For resonance

$$(L_1 + L_2)/q = (L_2 + L_3)/m \cdot , \quad m, q = \text{integers} \quad , \tag{6.7}$$

the partial wave $M_1 \rightarrow St$ reflected by St and the partial wave $M_3 \rightarrow St$, transmitted through St, interfere destructively. This means that for the resonance condition (6.7) the reflection losses by St have a minimum (in the ideal case they are zero). For all other wavelengths, however, these losses are larger and single-mode oscillation is achieved if they exceed the gain [6.14].

In a more detailed discussion the absorption losses A_{St}^2 of the beam splitter St cannot be neglected, since they cause the maximum reflectance R of the Fox-Smith cavity to be less than 1. Similar to the derivation of (4.73), the reflectance of the Fox-Smith selector, which acts as wavelength-selecting laser reflector, can be calculated to be [6.15]

$$R = \frac{T_{St}^2 R_2 (1 - A_{St})^2}{1 - R_{St}\sqrt{R_2 R_3} + 4R_{St}\sqrt{R_2 R_3}\,\sin^2\delta/2} \quad . \tag{6.8}$$

Figure 6.16b shows the reflectance R_{max} for $\delta = 2m\pi$ and the additional losses of the laser resonator introduced by the Fox-Smith cavity as a function of the beam-splitter reflectance R_{St}. The finesse F^* of the selecting device is also plotted for $R_2 = R_3 = 0.99$ and $A_{St} = 0.5\%$. The spectral width $\Delta\nu$ of the reflectivity maxima is determined by

$$\Delta\nu = \delta\nu/F^* = c/[2F^*(L_2 + L_3)] \quad . \tag{6.9}$$

There are several other resonator coupling schemes which can be used for mode selection. Figure 6.17 compiles some of them together with their frequency-selective losses [6.16].

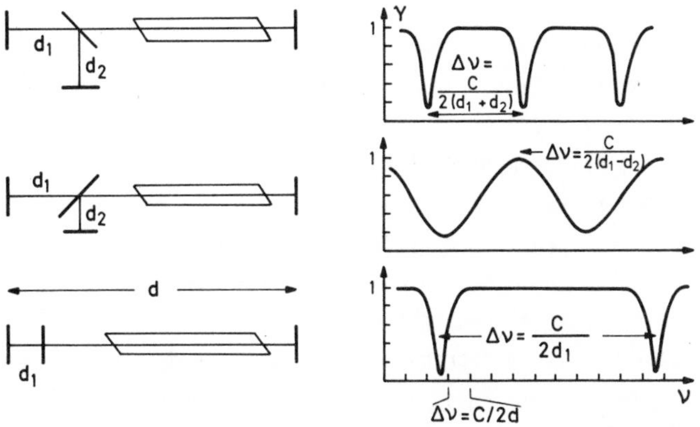

Fig.6.17. Some possible coupled resonator schemes for longitudinal mode selection, with the corresponding frequency-dependent losses. For comparison the eigenresonances of the long laser cavity with a mode spacing $\Delta\nu = c/2d$ are indicated

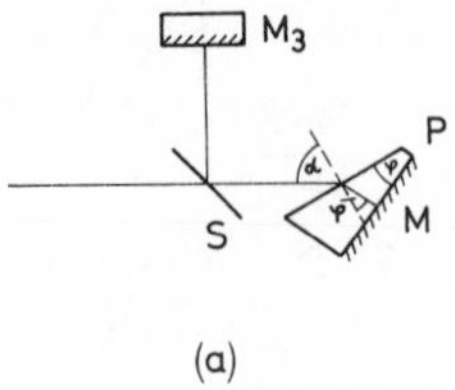

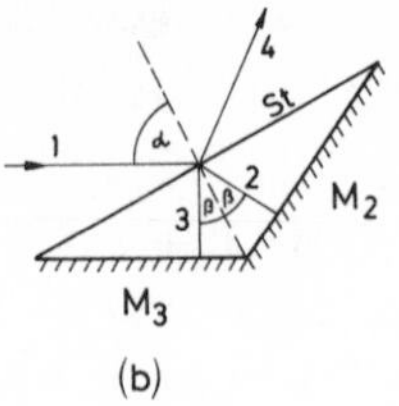

Fig.6.18a,b. Simultaneous line selection and mode selection by a comination of prism and Michelson type interferometers

In case of multiline lasers (e.g., argon or krypton lasers), line selection and mode selection can be simultaneously achieved by a combination of prism and Michelson interferometers. Figure 6.18 illustrates two possible realizations. The first replaces mirror M_2 in Fig.6.16 by a Littrow prism reflector (Fig.6.18a). In Fig.6.18b, the front surface of the prism acts as beam splitter, and the two coated back surfaces replace the mirrors M_2 and M_3. The incident wave is split into the partial beams 4 and 2. After being reflected by M_2, beam 2 is again split into 3 and 1. Destructive interference between beams 4 and 3 after reflection from M_3 occurs if the optical path difference $\Delta s = 2n(S_2 + S_3) = m\lambda$. If both beams have equal amplitudes, no light is emitted in the direction of 4. This means that all the light is reflected back into the incident direction and the device acts as a wavelength-selective reflector analogous to the Fox-Smith-cavity [6.17]. Since the wavelength λ depends on the optical path length $n(L_2 + L_3)$, the prism has to be temperature stabilized to achieve a stable wavelength in single-mode operation. The whole prism is therefore embedded in a temperature-stabilized oven.

For lasers with a broad gain profile, one wavelength-selecting element alone may not be sufficient to achieve single-mode operation, and one has to use a proper combination of different dispersing elements. With preselectors, such as prisms, gratings, or Lyot filters, the spectral range of the effective gain profile is narrowed down to a width which is comparable to that of the Doppler width of fixed-frequency lasers. Figures 6.19,20 represent two possible ways which have been realized in practice. The first uses two prisms as preselector to narrow the spectral width of a cw dye laser [6.18], and two etalons with different thicknesses t_1 and t_2 to achieve stable single-mode operation. Figure 6.19b illustrates mode selection, depicting schematically the gain profile narrowed by the prisms and the spectral transmission curves of the two etalons. In case of the dye laser with its homogeneous gain profile, not every resonator mode can oscillate but only those which draw gain from the spatial hole-burning effect (see Sect. 5.8). The "suppressed modes" at the bottom of Fig.6.19 represent these spa-

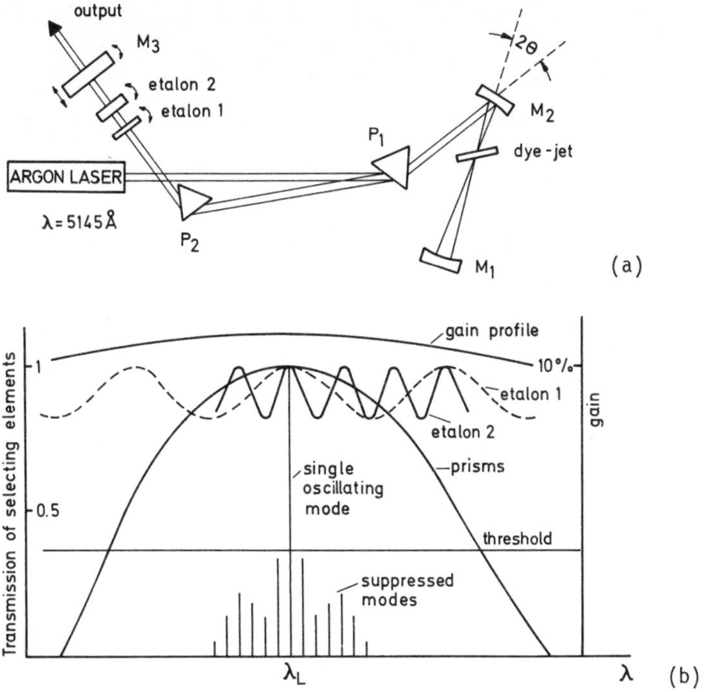

Fig.6.19. Mode selection in case of broad gain profiles with preselection by heavy flint prisms. (a) Experimental arrangement for a cw dye laser. (b) Gain profile with spectral narrowing by the prisms and transmission curves of the two etalons

tial hole-burning modes which would simultaneously oscillate without the etalons. The transmission maxima of the two etalons have of course to be at the same wavelength λ_L. This can be achieved by choosing the correct tilting angles θ_1 and θ_2 such that

$$nt_1 \cos\theta_1 = m_1\lambda_L \quad ,$$

$$nt_2 \cos\theta_2 = m_2\lambda_L \quad . \qquad (6.10)$$

Figure 6.20 shows the experimental arrangement for single-mode operation of an N_2-laser-pumped dye laser. The laser beam is expanded to fill the whole grating. Because of the higher spectral resolution of the grating (compared with a prism) and the wider mode spacing due to the short cavity, a single etalon may be sufficient to achieve single-mode operation [6.19].

There are many more experimental possibilities of achieving single-mode operation. For details, the reader is referred to the extensive special

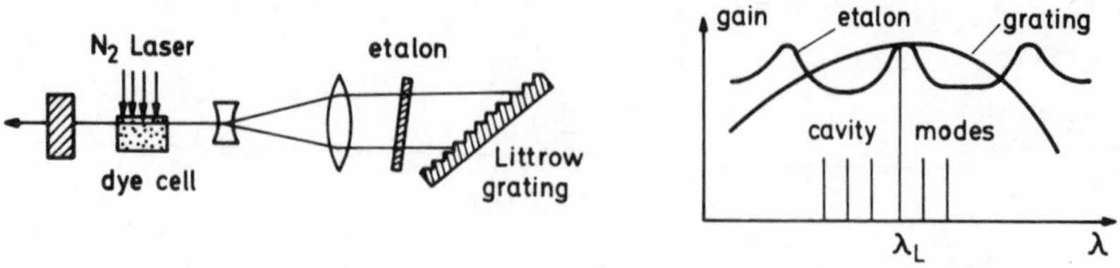

Fig.6.20. Preselection with a grating and mode selection with one etalon used to achieve single-mode operation of a dye laser pumped by a nitrogen laser

literature on this subject which can be found, for instance, in the excellent reviews on mode selection and single-mode lasers by SMITH [6.16] or GOLDSBOROUGH [6.20].

6.6 Wavelength Stabilization

For many applications in high-resolution laser spectroscopy, it is essential that the laser wavelength stay as stable as possible at a preselected value λ_0. This means that the fluctuations $\Delta\lambda$ around λ_0 should be smaller than the molecular linewidths which are to be resolved. For such experiments only *single-mode* lasers can in general be used, because in most multimode lasers the momentary wavelengths fluctuate and only the time-averaged envelope of the spectral output profile is defined, as has been discussed in the previous sections. This stability of the wavelength is important both for fixed-wavelength lasers where the laser wavelength has to be kept on a time-independent value λ_0 as well as for tunable lasers where the fluctuations $\Delta\lambda = |\lambda_L - \lambda_R(t)|$ around a controlled tunable wavelength $\lambda_R(t)$ have to be smaller than the resolvable spectral interval.

In this section we discuss some methods of wavelength stabilization with their advantages and drawbacks. Since the laser frequency $\nu = c/\lambda$ is directly related to the wavelength, one often speaks about *frequency* stabilization, although for most methods in the visible spectral region, it is not the frequency but the wavelength which is directly measured and compared with a reference standard. There are, however, some stabilization methods in the infrared which do rely directly on absolute frequency measurements (see Sect. 6.10).

In Sect.5.6, we saw that wavelength λ or frequency ν of a longitudinal mode in the active resonator is determined by the mirror separation d and the refractive indices n_2 of the active medium with length L, and n_1, outside the amplifying region

$$q\lambda = 2n_1(d - L) + 2n_2L \quad . \tag{6.11}$$

For simplicity, we shall assume that the active medium fills the whole region between the mirrors. Thus (6.11) reduces, with $L = d$ and $n_2 = n_1 = n$, to

$$q\lambda = 2nd \quad , \quad \text{or} \quad \nu = qc/(2nd) \quad . \tag{6.12}$$

Any fluctuation of n or d will cause a corresponding change of λ and ν. We obtain from (6.12)

$$\frac{\Delta\lambda}{\lambda} = \frac{\Delta d}{d} + \frac{\Delta n}{n} \quad \text{or} \quad -\frac{\Delta\nu}{\nu} = \frac{\Delta d}{d} + \frac{\Delta n}{n} \quad . \tag{6.13}$$

To illustrate the demands of frequency stabilization, let us assume that we want to keep the frequency $\nu = 6 \times 10^{14}\ s^{-1}$ of an argon laser constant within 1 MHz. This means a relative stability of $\Delta\nu/\nu = 1.6 \times 10^{-9}$, and implies that the mirror separation of d = 1 m has to be kept constant within 1.6 nm.

From this example it is evident that the requirements for such stabilisation are by no means trivial. Before we discuss possible experimental solutions let us consider the causes of fluctuations or drifts in the resonator length d or the refractive index n. If we were able to reduce or even to eliminate these causes, we would already be well on the way to achieving a stable laser frequency. We shall distinguish between *long-term drifts* of d and n which are mainly caused by temperature drifts or slow pressure changes; and *short-term fluctuations*, caused, for example, by acoustic vibrations of mirrors, by acoustic pressure waves which modulate the refractive index, or by fluctuations of the gas discharge in gas lasers or of the jet flow in dye lasers.

To illustrate the influence of long-term drifts, let us make the following estimate. If α is the thermal expansion coefficient of the material (e.g., quartz or invar rods) which defines the mirror separation d, the relative change $\Delta d/d$ for a temperature change ΔT is, under the assumption of linear thermal expansion,

$$\Delta d/d = \alpha\Delta T \quad . \tag{6.14}$$

Table 6.1 compiles the thermal expansion coefficients for some commonly used materials. For invar, with $\alpha = 1 \times 10^{-6}\ K^{-1}$, we obtain from (6.14), even for $\Delta T = 0.1$ K a relative distance change $\Delta d/d = 10^{-7}$ which, in our example above, gives a frequency drift of 60 MHz.

Table 6.1. Linear thermal expansion coefficients of some materials at room temperature T = 20°C

Material	$\alpha\,[10^{-6}\,K^{-1}]$
Aluminium	23
Brass	19
Steel	11-15
Titanium	8.6
Tungsten	4.5
Al_2O_3	5
BeO	6
Invar	1.2
Soda-Glass	5-8
Pyrex Glass	3
Fused quartz	0.4-05
Cerodur	< 0.1

If the laser wave inside the laser resonator travels a path length d-L through air at atmospheric pressure, any change Δp of air pressure results in a change

$$\Delta s = (d-L)(n-1)\Delta p/p \quad \text{with} \quad \Delta p/p = \Delta n/(n-1) \tag{6.15}$$

of the optical path length between the resonator mirrors. With n = 1.00027 and d-L = 0.2 d, which is typical for gas lasers, we obtain from (6.15) and (6.11) for pressure changes $\Delta p = 3$ mbar (which can readily occur during one hour, particularly in air-conditioned rooms),

$$\Delta\lambda/\lambda = -\Delta\nu/\nu \approx (d-L)\Delta n/(nd) \geqq 1.5\times 10^{-7} \quad .$$

For our example above, this means a frequency change of $\Delta\nu \geqq 90$ MHz. In cw dye lasers, the length L of the active medium is negligible compared with the resonator length d, and we can take d-L = d. This implies for the same pressure change, a frequency drift which is even five times larger than estimated above.

To keep these long-term drifts as small as possible, one has to choose distance holders for the resonator mirrors with a minimum thermal expansion coefficient α. A good choice is, for example, the recently developed cerodur-quartz composition with a temperature-dependent $\alpha(T)$ which can be made zero at room temperature [6.21]. Often massive granite blocks are used as support for the optical components; these have a large heat capacity with a time constant of several hours to smoothen temperature fluctuations. To minimize pressure changes, the whole resonator must be enclosed by a pressure-tight container, or the ratio (d-L)/d must be chosen as small as possible. However, we shall see that such long-term drifts can be mostly compensated by electronic servocontrol if the laser wavelength can be bound to a constant reference wavelength standard.

A more serious problem arises from the short-term fluctuations, since these may have a broad frequency spectrum, depending on their causes, and the frequency response of the electronic stabilization control must be adapted to this spectrum. The main contribution comes from acoustical vibrations of the resonator mirrors. The whole setup of a wavelength-stabilized laser should be therefore as far as possible vibrationally isolated. Figure 6.21 shows a possible table mount for the laser system as used in our laboratory. The optical components are mounted on a heavy granite plate which rests in a flat container filled with sand to damp the eigenresonances of the granite block. Styropor blocks and acoustic damping elements prevent room vibrations being transferred to the system. The optical system is protected against direct sound waves through the air, air turbulence and dust by a dust-free solid cover resting on the granite plate.

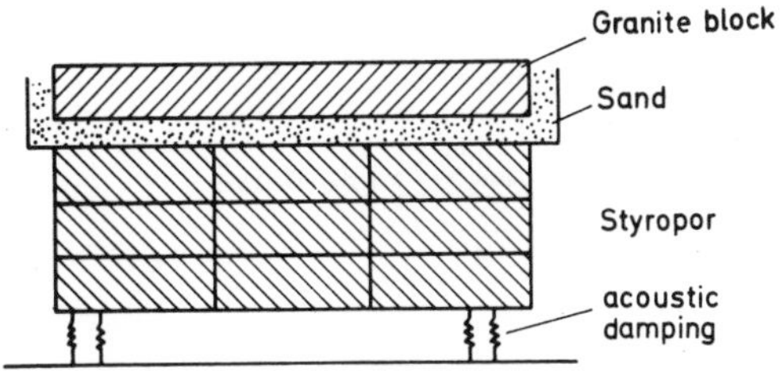

Fig.6.21. Experimental realization of an acoustically isolated table for a wavelength-stabilized laser system

The high-frequency part of the noise spectrum is mainly caused by fast fluctuations of the refractive index in the discharge region of gas lasers or in the liquid jet of cw dye lasers. These perturbations can be only partly reduced by choosing the optimum discharge conditions in gas lasers. In jet-stream dye lasers, density fluctuations in the free jet, caused by small air bubbles or by pressure fluctuations of the jet pump and by surface waves along the jet surfaces, are the main causes of fast laser-frequency fluctuations. Careful fabrication of the jet nozzle and filtering of the dye solution are essential to minimize these fluctuations.

All the perturbations discussed above cause fluctuations of the optical path length inside the resonator which are typically in the nanometer range. In order to keep the laser wavelength stable, these fluctuations must be compensated by corresponding changes of the resonator length d. For such controlled and fast length changes in the nm range piezoceramic elements are mainly used [6.22]. They consist of a piezoelectric material whose length in an external electric field changes proportionally to the field strength. Either cylindrical plates are used, where the end faces are covered by silver coatings which provide the electrodes; or hollow cylinders, where the coatings cover the inner and outer wall surface. Typical parameters of such piezo elements are a few nm length change per volt. When a resonator mirror is mounted on such a piezo element (Fig.6.22), the resonator length can be controlled within a few μm by the voltage applied the electrodes of the piezo element.

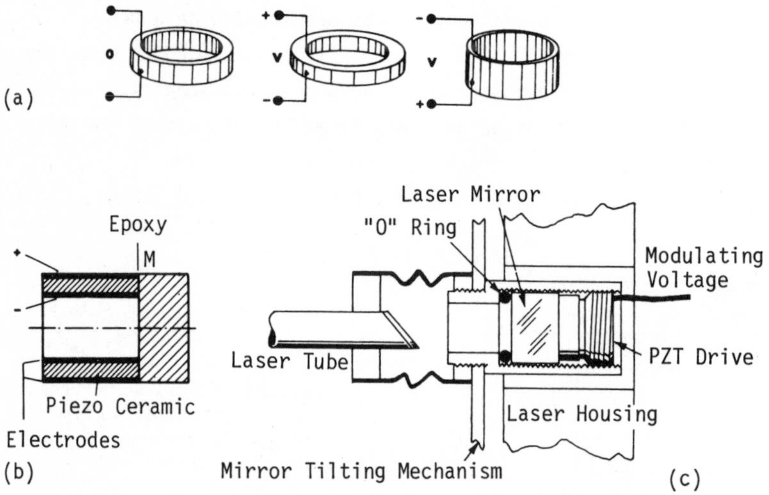

Fig.6.22a-c. Resonator end mirror, mounted on a piezo element. (a) Length change of a piezo element with applied voltage. (b) Mirror epoxyed on the PZT. (c) The PZT presses the mirror onto a rubber ring (Jodon, Inc.)

The frequency response of this length control is limited by the inertial mass of the moving system consisting of the mirror and the piezo element and by the eigenresonances of this system. Using small mirror sizes and carefully selected piezos, one may reach the 100 kHz range. For the compensation of faster fluctuations, an optical anisotropic crystal, such as KDP, can be used inside the laser resonator. The optical axis of this crystal must be oriented in such a way that a voltage applied to the crystal electrodes changes its refractive index along the resonator axis. This allows nd, and therefore the laser wavelength to be controlled with a frequency response up into the megahertz range.

The wavelength stabilization system consists essentially of three elements:

i) The wavelength reference standard with which the laser wavelength is compared. One may, for example, use the wavelength λ_R at the maximum or at the slope of the transmission peak of a Fabry-Perot interferometer which is maintained in a controlled environment (temperature and pressure stabilization); or the wavelength of an atomic or molecular transition may serve as reference. Sometimes another stabilized laser is used as standard and the laser wavelength is locked to this standard wavelength.

ii) The controlled system, which is in our case the resonator length nd defining the laser wavelength λ_L.

iii) The electronic control system with the servo loop, which measures the deviation $\Delta\lambda = \lambda_L - \lambda_R$ of the laser wavelength λ_L from the reference value λ_R and which tries to bring $\Delta\lambda$ to zero as fast as possible.

The stability of the laser wavelength can, of course, never exceed that of the reference wavelength. Generally it is worse because the control system is not ideal. Deviations $\Delta\lambda(t) = \lambda_L(t) - \lambda_R$ cannot be compensated immediately because the system has a finite frequency response and the inherent time constants always cause a phase lag between deviation and response.

Figure 6.23 shows how the electronic system can be designed to optimize the response over the whole frequency spectrum of the input signals. In principle three operational amplifiers with a different frequency response are put parallel. The first is a common *proportional* amplifier, with an upper frequency determined by the resonance frequency of the mirror piezo system. The second is an integral amplifier with an output

$$U_{out} = \frac{1}{RC}\int_0^T U_{in}(t)dt \quad .$$

This amplifier is necessary to bring the signal ($U_{in} \sim \Delta\lambda$), which is proportional to the wavelength deviation, really back to zero, which cannot be performed with a proportional amplifier. The third amplifier is a differentiating device which takes care of fast peaks in the perturbations. All three functions can be combined in a system called PID control [6.23].

A schematic diagram of a commonly used stabilization system is shown in Fig.6.24. A few percent of the laser output are sent from the two beam splitters St_1 and St_2 into two interferometers. The first F.P.I. 1 is a

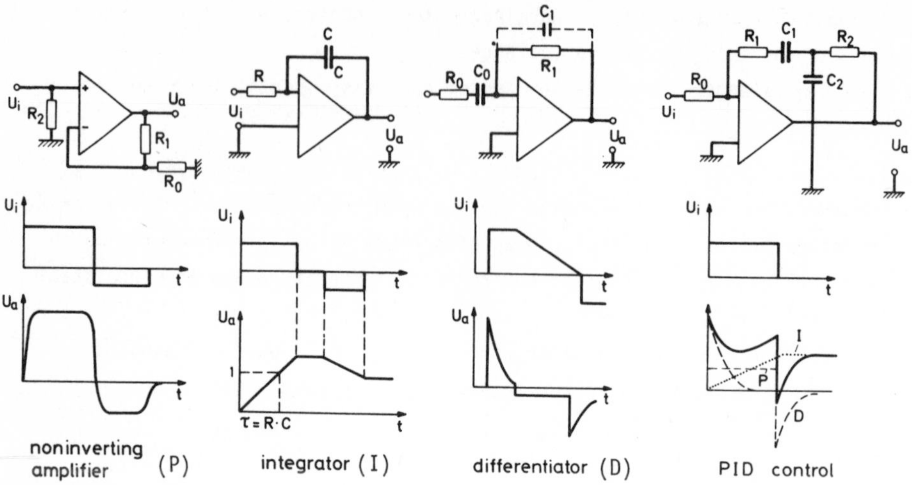

Fig.6.23. Schematic diagram of noninverting proportional amplifier, integrator, differentiator, and PID control with corresponding time response to a step function at the input

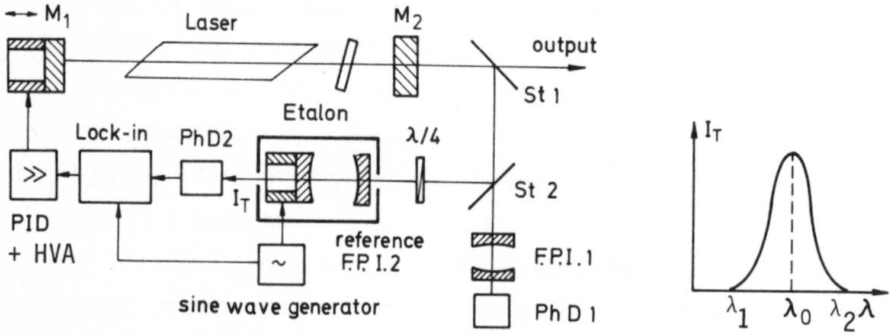

Fig.6.24. Schematic diagram for wavelength stabilization with a stable Fabry-Perot interferometer as reference

scanning confocal resonator, and serves as spectrum analyzer for monitoring the mode spectrum of the laser. The second interferometer F.P.I. 2 is the wavelength reference, and is therefore placed in a pressure-tight and temperature-controlled box to keep the optical path nd between the interferometer mirrors and with it the wavelength $\lambda_R = 2nd/m$ at the transmission peak as stable as possible (see Sect.4.2). One of the mirrors is mounted on a piezo element. If a small ac voltage with frequency f is fed to the piezo, the transmission peak of F.P.I. 2 is periodically shifted around the center wavelength λ_0 which we take as the required reference wavelength λ_R. If the laser wavelength λ_L is within the transmission range λ_1 to λ_2 in Fig.6.24, the photodiode PhD 2 behind F.P.I. 2 delivers a dc signal which is modulated at the frequency f. The modulation amplitude depends on the slope of the transmission curve $dI_T/d\lambda$ of F.P.I. 2 and the phase depends on the sign of $\lambda_L - \lambda_0$. Whenever the laser wavelength λ_L deviates from the reference wavelength λ_R, the photodiode delivers an ac amplitude which increases as the difference $\lambda_L - \lambda_R$ increases, as long as λ_L stays within the turning points of $I_T(\lambda)$. This signal is fed to a lock-in amplifier, where it is rectified, passes a PID control, and a high-voltage amplifier HVA. The output of the HVA is connected with the piezo element, which moves the resonator mirror until the laser wavelength λ_L is brought back to the reference value λ_R.

Instead of using the maximum λ_0 of the transmission peak of $I_T(\lambda)$ as reference wavelength, one may also choose the wavelength λ_t at the turning point of $I_T(\lambda)$ where the slope $dI_T(\lambda)/d$ has its maximum. This has the advantage that a modulation of the F.P.I. transmission curve is not necessary and the lock-in amplifier can be dispensed with. The cw laser intensity $I_T(\lambda)$ transmitted through F.P.I. 2 is compared with a reference intensity I_R split by BS_2 from the same partial beam (Fig.6.25). The output signals S_1 and S_2 from the two photodiodes D 1 and D 2 are fed into a difference amplifier, which is adjusted so that its output voltage becomes zero for $\lambda_L = \lambda_t$. If the laser wavelength λ_L deviates from $\lambda_R = \lambda_t$, S_1 becomes smaller or, larger, depending on the sign of $\lambda_L - \lambda_R$, and the output of the difference amplifier is, for small differences $\lambda - \lambda_R$, proportional to the deviation. The output signal again passes a PID control and a high-voltage amplifier, and is fed to the piezo of the resonator mirror. The advantages of this difference method are the larger bandwidth of the difference amplifier (compared with a lock-in amplifier), and the simpler and less expensive composition of the whole electronic control system. Its drawback lies in the fact that different dc-voltage drifts in the branches of the difference amplifier result in a dc output, which shifts the zero adjustment, and, with it, the reference wavelength λ_R. Such dc drifts are much more critical in dc amplifiers than in the ac-coupled devices used in the first method.

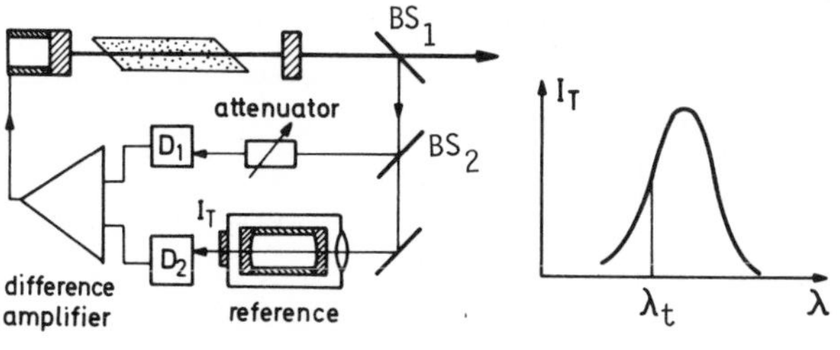

Fig.6.25. Wavelength stabilization on the slope of the reference F.P.I. using a difference amplifier

Both methods use a stable F.P.I. as reference standard. This has the advantage that the reference wavelength λ_0 or λ_t can be tuned by applying a voltage to the piezo element in the reference F.P.I. 2. This means that the laser can be stabilized onto any desired wavelength within the gain profile. Because the signal from the photodiodes PhD 1 and PhD 2 has sufficiently large amplitude, the signal-to-noise ratio is good, and the method is suitable for correcting short-term fluctuations of the laser wavelength.

For long-term stabilization, however, stabilization onto an external F.P.I. has its drawbacks. In spite of the temperature stabilization of F.P.I. 2, small drifts of the transmission peak cannot be eliminated completely. With a thermal expansion coefficient $\alpha = 10^{-6}$ of the distance holder for the F.P.I. mirrors, even a temperature drift of 0.01°C causes, according to (6.13), a relative frequency drift of 10^{-8}, which gives 6 MHz for a laser frequency $\nu_L = 6 \times 10^{14}\ s^{-1}$. For this reason, an atomic or molecular laser transition is more suitable as a long-term frequency standard. The accuracy with which the laser wavelength can be stabilized onto the center of such a transition depends on the linewidth of the transition and on the attainable signal-to-noise ratio of the stabilization signal. Doppler-free line profiles are therefore preferable. Figure 6.26 illustrates a possible arrangement. The laser beam is crossed perpendicularly with a collimated molecular beam. The Doppler width of the absorption line is reduced by a factor depending on the collimation ratio (see Sect.10.1). The intensity $I_F(\lambda_L)$ of the laser-excited fluorescence serves a monitor for the deviation $\lambda_L - \lambda_c$ from the line center λ_c. The output signal of the fluorescence detector after amplification could be fed directly to the piezo of the laser resonator. However, in case of small fluorescence intensities, the signal-to-noise ratio may be not good enough to achieve satisfactory stabilization. It is therefore advantageous to continue to lock the laser to the reference F.P.I. 2, but to use the fluorescence signal to lock the F.P.I. 2 to the molecular line. In this double servo-control system, the short-term fluctuations of λ_L are compensated by the fast servo loop with the F.P.I. 2 as reference, while the slow drifts of the F.P.I. are stabilized by being locked to the molecular line. To decide whether λ_t drifts to lower or to higher wavelengths, one must either modulate the laser frequency or use a digital servo control which shifts the laser frequency in small steps, and monitors at each step whether the fluorescence intensity has increased or decreased. A simple program can ensure that the laser wavelength remains within one step around the maximum of the molecular line. This can be performed, for instance, as follows: A pulse-generator activates a for-

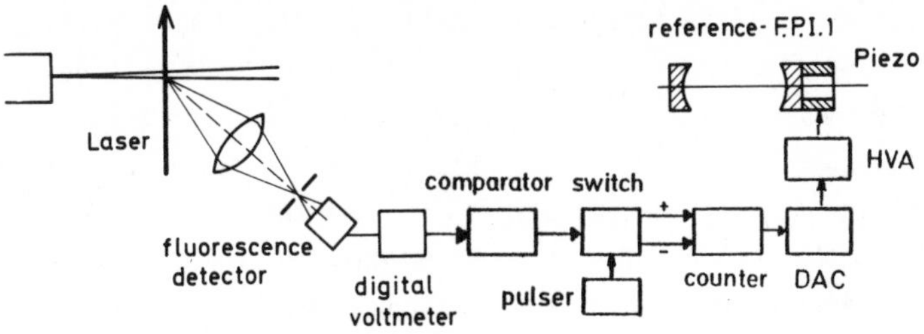

Fig.6.26. Long-term stabilization of the laser wavelength locked to a molecular transition

ward-backward counter, which generates by a DAC a voltage, fed to the piezo of the F.P.I. This voltage shifts the reference wavelength of the F.P.I. and with it the laser wavelength and therefore also the fluorescence intensity is altered. A comparator compares whether the intensity has increased or decreased by the last step and activates accordingly the switch. Since the drift of the reference F.P.I. is slow, the second servo control can also be slow, and the fluorescence intensity can be integrated. This allows the laser to be stabilized for a whole day, even onto faint molecular lines where the detected fluorescence intensity is less than 100 photons/s [6.24].

Since the accuracy of wavelength stabilization increases with decreasing molecular linewidth, spectroscopists have looked for particularly narrow lines which could be used for extremely well-stabilized lasers. The Doppler width can be eliminated either by using collimated molecular beams, or by employing the small saturation dip (Lamb dip) at the center of a Doppler-broadened line. An interesting proposal is based on Doppler-free two-photon transitions, which has the additional advantage that the lifetime of the upper state can be very long, and the natural linewidth may become extremely small. An example of a promising candidate is the 1s → 2s transition of atomic hydrogen, which should have a natural linewidth of less than 1 Hz.

A good reference wavelength should be reproducible and essentially independent of external perturbations, such as electric or magnetic fields and temperature or pressure changes. Therefore transitions in atoms or molecules without permanent dipole moments, such as CH_4 or noble gas atoms, are best suited to serve as reference wavelength standards (see Chap.10).

So far we have only considered the stability of the laser resonator itself. In the previous section we saw that wavelength-selecting elements inside the resonator are necessary for single-mode operation to be achieved, and that their stability and the influence of their thermal drifts on the laser wavelength must be considered. We illustrate this with the example of single-mode selection by a tilted intracavity etalon. If the transmission peak of the etalon is shifted by more than 1/2 of the cavity mode spacing the total gain becomes more favorable for the next cavity mode and the laser wavelength will jump to the next mode. This implies that the optical pathlength of the etalon nt must be kept stable, so that the peak transmission drifts by less than c/4d, which is about 50 MHz for an argon laser. One can use either an air-spaced etalon with distance holders with very small thermal expansion or a solid etalon in a temperature-stabilized oven. The air-spaced etalon is simpler but has the drawback that changes of the air pressure influence the transmission peak wavelength.

The actual stability obtained for a single-mode laser depends on the laser system, on the quality of the electronic servo loop, and on the design of the resonator and mirror mounts. With moderate efforts, a frequency stability of about 1 MHz can be achieved, while extreme precautions and sophisticated equipment allow a stability of 1 Hz to be achieved for some laser types [6.27]. A statement about the stability of the laser frequency depends on the averaging time and on the kind of perturbations. The stability against short-term fluctuations, of course, becomes better if the averaging time is increased, while long-term drifts increase with the sampling time. Figure 6.27 illustrates the stability of a single-mode argon laser, stabilized with the arrangement of Fig.6.26. With more expenditure, a stability of better than 3 kHz has been achieved for this laser [6.28].

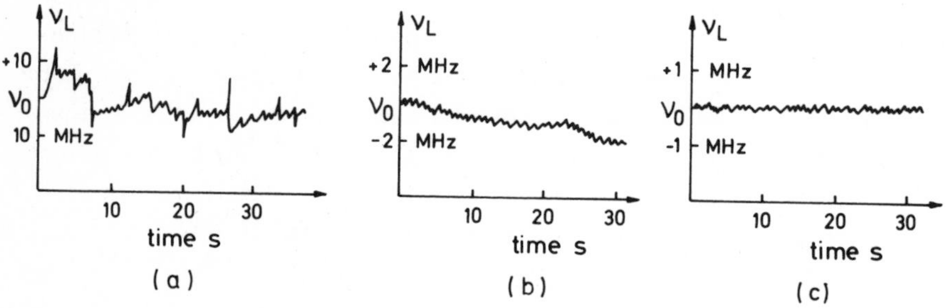

Fig.6.27a-c. Frequency fluctuations of a single-mode argon laser. (a) Unstabilized. (b) Stabilized with the arrangement of Fig.6.26 on an external F.P.I. (c) Additional long-term stabilization onto a molecular transition

The residual frequency fluctuations of a stabilized laser can be represented in an Allen plot. The Allen variance [6.29]

$$\sigma = \frac{1}{\nu}\sqrt{\sum_{i=1}^{N} \frac{(\Delta\nu_i - \Delta\nu_{i-1})^2}{2(N - 1)}} \tag{6.16}$$

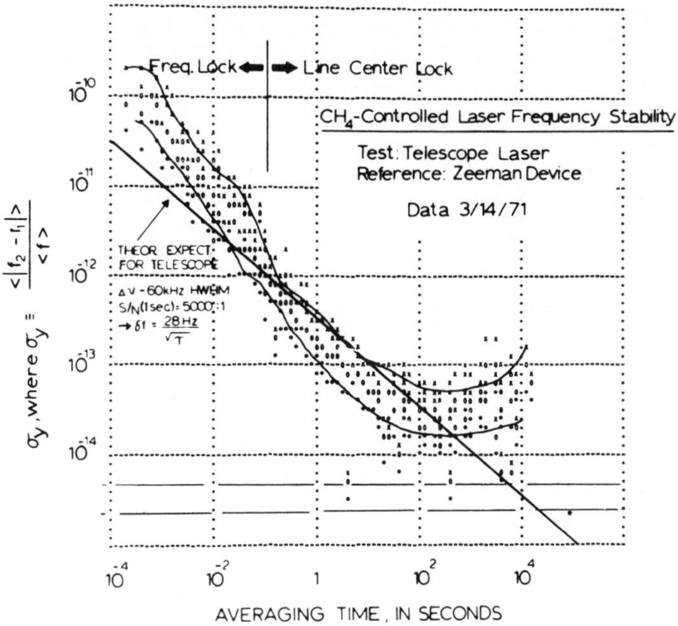

Fig.6.28. Plot of the Allen variance to illustrate the short-term stability of a He-Ne laser stabilized onto a CH_4 transition [6.30]

is comparable to the relative standard deviation. It is determined by measuring N times at equal time intervals $t_i = t_0 + i\Delta t$ the frequency difference $\Delta\nu_i$ between two lasers stabilized onto the same reference frequency ν_R. Figure 6.28 illustrates the Allen variance for a He-Ne laser, stabilized at $\lambda = 3.3\ \mu m$ onto a vibrational-rotational transition of the CH_4 molecule [6.30].

Such extremely stable lasers are of great importance in metrology since they can provide high-quality wavelength or frequency standards with an accuracy approaching, or even surpassing, that of present-day standards [6.29a]. For most applications in high-resolution laser spectroscopy, a frequency stability of 100 kHz to 1 MHz is sufficiently good because most spectral linewidths exceed that value by several orders of magnitude.

For a more complete survey of wavelength stabilization, the reader is referred to the reviews by BAIRD and HANES [6.25] and TOMLINSON and FORK [6.26].

6.7 Intensity Stabilization

The intensity I(t) of a cw laser is not completely constant but shows periodic and random fluctuations and also, in general, long-term drifts. The reasons for these fluctuations are manifold and may be, for example, an insufficiently filtered power supply which results in a ripple on the discharge current of the gas laser and a corresponding intensity modulation. Other noise sources are instabilities of the gas discharge, dust particles diffusing through the laser beam inside the resonator, vibrations of the resonator mirrors, and, in multimode lasers, internal effects as well, such as mode competition. In cw dye lasers, density fluctuations in the dye jet stream and air bubbles are the main cause of intensity fluctuations.

Long-term drifts of the laser intensity may be caused by slow temperature or pressure changes in the gas discharge, by thermal detuning of the resonator, or by increasing degradation of the optical quality of mirrors, windows and other optical components in the resonator. All these effects give rise to a noise level which is well above the theoretical lower limit set by the photon noise. Since these intensity fluctuations lower the signal-to-noise ratio, they may become very troublesome in many spectroscopic applications, and one should consider steps which reduce these fluctuations by stabilizing the laser intensity.

Of the various possible methods, we shall discuss two which are often used for intensity stabilization. They are schematically depicted in Fig. 6.29. In the first method, a small fraction of the output power is split by the beam splitter St to a detector (Fig.6.29a). The detector output V_D is compared with a reference voltage V_R and the difference $\Delta V = V_D - V_R$ is amplified and fed to the power supply of the laser, where it controls the discharge current. The servo loop is effective in a range where the laser intensity increases with increasing current.

The upper frequency limit of this stabilization loop is determined by the capacitances and inductances in the power supply and by the time lag between current increases and the resultant increase of the laser intensity. The lower limit for this time delay is given by the time the gas discharge needs to reach a new equilibrium after the current has been changed. It is therefore not possible with this method to stabilize the system against fluctuations of the gas discharge. For most applications, however, this stabilization technique is sufficient; it provides an intensity stability where the fluctuations are less than 0.5%.

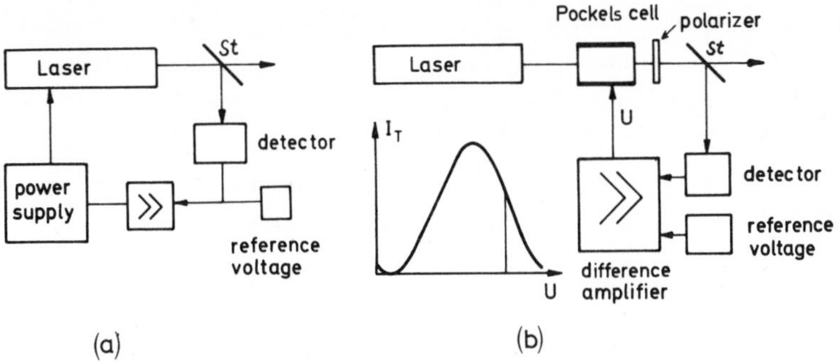

Fig.6.29a,b. Intensity stabilization of lasers. (a) With a servo loop to control the pumping power of the laser. (b) By controlling the transmission of a Pockels cell

To compensate fast intensity fluctuations, another technique, illustrated in Fig.6.29b, is more suitable. The output from the laser is sent through a Pockels cell, which consists of an optically anisotropic crystal, placed between two linear polarizers. An external voltage applied to the electrodes of the crystal causes optical birefringence which rotates the polarization plane of the transmitted light and therefore changes the transmittance through the second polarizer. If part of the transmitted light is detected, the amplified detector signal can be used to control the voltage at the Pockels cell. Together with a PID amplifier, any change of the transmitted intensity can be compensated by a transmission change of the Pockels cell. This stabilization control works up to frequencies in the megahertz range. Its disadvantage is an intensity loss of 20% - 50% because one has to bias the Pockels cell to work on the slope of the transmission curve (Fig.6.29b).

For spectroscopic applications of dye lasers, where the dye laser has to be tuned through a large spectral range, the intensity change, caused by the

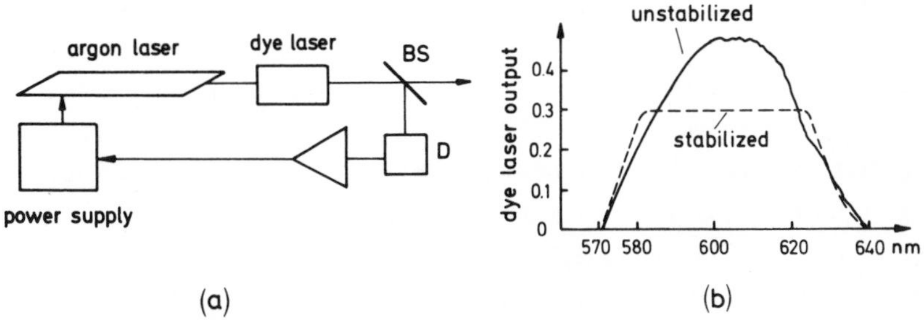

Fig.6.30a,b. Intensity stabilization of a cw dye laser by controlling the argon laser pump power. (a) Experimental arrangement. (b) Stabilized and unstabilized dye laser intensity $I(\lambda)$ when tuning the dye laser across its gain profile

decreasing gain at both ends of the gain profile, may be inconvenient. The elegant way to avoid this change of $I_L(\lambda)$ with λ stabilizes the dye laser output to control the argon laser power (Fig.6.30). Since the servo control must not be too fast, the stabilization scheme of Fig.6.29 can be employed. Figure 6.30b demonstrates how effectively this method works if one compares the stabilized with the unstabilized intensity profile $I(\lambda)$ of the dye laser.

6.8 Controlled Wavelength Tuning

Although fixed-wavelength lasers have proved their importance for many spectroscopic applications (see Sect.6.3 and Chap.5,8 and 9), it was the development of continuously tunable lasers which really revolutionized the whole field of spectroscopy. This is demonstrated by the avalanche of publications on tunable lasers and their applications (see, e.g., [1.10]). We shall therefore treat in this section some basic techniques of controlled tuning of single-mode lasers, while the next chapter gives a survey on tunable coherent sources, developed in various spectral regions.

Since the laser wavelength λ_L of a single-mode laser is determined by the optical path length nd between the resonator mirrors,

$$q\lambda = 2nd \quad ,$$

either the mirror separation d or the refractive index n can be continuously varied to obtain a corresponding tuning of λ_L. This can be achieved, for example, by a linear voltage ramp $U = U_0 + at$ in the piezo elements on which the resonator mirror is mounted, or by a continuous pressure variation in a tank containing the resonator or parts of it. However, as has been discussed in Sect.6.6, most lasers need additional wavelength-selecting elements inside the laser resonator to ensure single-mode operation. When the resonator length is varied, the frequency ν of the oscillating mode is tuned away from the transmission maximum of these elements (see Fig.6.15). During this tuning the neighboring resonator mode (which is not yet oscillating) approaches this transmission maximum and its losses may now become smaller than those of the oscillating mode. As soon as this mode reaches threshold, it will start to oscillate and will suppress the former mode because of mode competition (see Sect.5.7). This means that the single-mode laser will jump from one resonator mode to the next. In this way, the continuous tuning

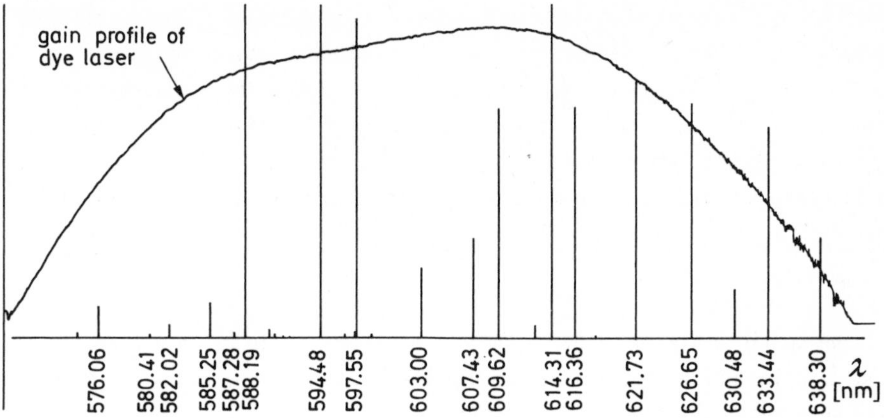

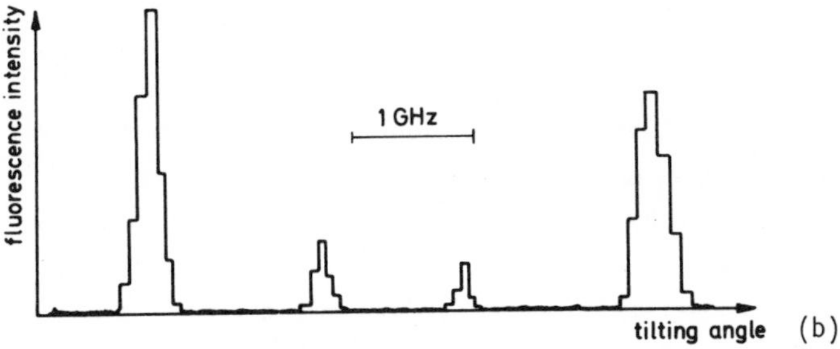

Fig.6.31a,b. Discontinuous tuning of lasers. (a) Part of the Ne spectrum excited by a single-mode dye laser with Doppler-limited resolution, which hides the cavity mode hops of the laser. (b) Excitation of Na_2 lines by a single-mode argon laser in a weakly collimated molecular beam. The intra-cavity etalon was continuously tilted but the resonator length was kept constant

range is restricted to about half the free spectral range $\delta\nu = c/2t$ of the selecting interferometer with thickness t.

Similar mode hops occur when the wavelength-selecting elements are continuously tuned but the resonator length is not controlled but kept constant. Such a discontinuous tuning of the laser wavelength will be sufficient if the mode hops $\delta\nu = c/2d$ are small compared with the spectral linewidths under investigation. As illustrated by Fig.6.31a, which shows part of the neon spectrum excited in a He-Ne gas discharge with a discontinuously tuned single-mode dye laser, the mode hops are barely seen and the spectral resolution is limited by the Doppler width of the neon lines. In sub-Doppler spectroscopy, however, the mode jumps appear as steps in the line profiles, as is shown in Fig.6.31b, where a single-mode argon laser is tuned with mode hops through some absorption lines of Na_2 molecules in a slightly collimated molecular beam where the Doppler width is reduced to about 200 MHz.

In order to enlarge the tuning range, the transmission maxima of the wavelength selectors have to be tuned synchronously with the tuning of the resonator length. When a tilted etalon with thickness t and refractive index n is used, the transmission maximum λ_m, which, according to (6.4) is given by

$$m\lambda_m = 2nt\cos\theta \quad ,$$

can be continuously tuned by changing the tilting angle θ. In all practical cases, θ is very small, and we can use the approximation $\cos\theta \approx 1 - \theta^2/2$. The wavelength shift $\Delta\lambda = \lambda_0 - \lambda$ is

$$\Delta\lambda = \frac{2nt}{m}(1 - \cos\theta) \approx \lambda_0\theta^2/2 \quad , \qquad \lambda_0 = \lambda(\theta = 0) \quad . \tag{6.17}$$

Equation (6.17) shows that the wavelength shift $\Delta\lambda$ is proportional to θ^2 but is independent of the thickness t. Two etalons with different thicknesses t_1 and t_2 can be mounted on the same tilting device, which may be simply a lever which is tilted by a micrometer screw driven by a small motor gearbox. The motor simultaneously drives a potentiometer which provides a voltage proportional to the tilting angle θ. This voltage is electronically squared, amplified, and fed to the piezo element of the resonator mirror. With properly adjusted amplification, one can achieve an exact synchronization of the resonator wavelength shift $\Delta\lambda_L = \lambda_L\Delta d/d$ with the shift $\Delta\lambda$ of the etalon transmission maximum.

For many applications in high-resolution spectroscopy, it is desirable that the fluctuations of the laser wavelength λ_L around the programmed tunable value $\lambda(t)$ are kept as small as possible. This can be achieved by stabilizing λ_L to the reference wavelength λ_R of a stable external F.P.I. (see Sect.6.7), and to tune this reference wavelength λ_R synchronously with the transmission peaks of the selecting etalons. The amplified potentiometer voltage is then fed not to the resonator end mirror, but to a piezo which controls the mirror separation of the F.P.I. The laser end mirror is stabilized by an extra servo loop to the F.P.I. Figure 6.32 illustrates schematically the complete design of a controlled tunable stabilized single mode argon laser.

Tuning of the etalon transmission by tilting the etalons has the disadvantage that the reflection losses increase drastically with increasing θ. This results from the finite diameter of the laser beam, which causes an

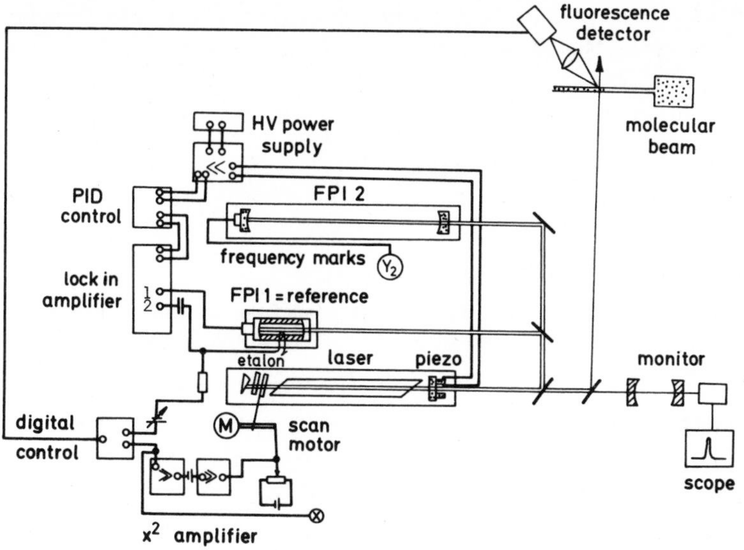

Fig.6.32. Continuous tuning of a single-mode argon laser by tilting the mode selection etalon. The laser wavelength is always stabilized onto the transmission peak of F.P.I. 1 while F.P.I. 2 provides frequency marks. Lock-in input 1 receives the signal from the reference beam, output 2 gives the reference voltage for modulation of the reference F.P.I. 1.

incomplete overlap of the partial beams reflected from front and back surfaces of the etalon (see Fig.4.39), and which can therefore not completely interfere. This implies that even at the transmission maximum and apart from absorption losses, the transmission $T < 1$ and part of the incident intensity is reflected out of the resonator. The relative reflection losses $\Delta I_R/I$ per transit for a laser beam diameter D and a reflectance R of the etalon surfaces can be calculated to be [6.12]

$$\Delta I_R/I \approx 4tR\theta/(nD) \quad . \tag{6.18}$$

Examples

For an uncoated etalon, with R = 0.04, t = 1 cm, and D = 0.1 cm, we obtain, for θ = 10 mrad, a relative loss of 1% per transit. For R = 0.5, we already obtain 13%. The losses increase with $t\theta$, and they limit the tuning range of this tilting method.

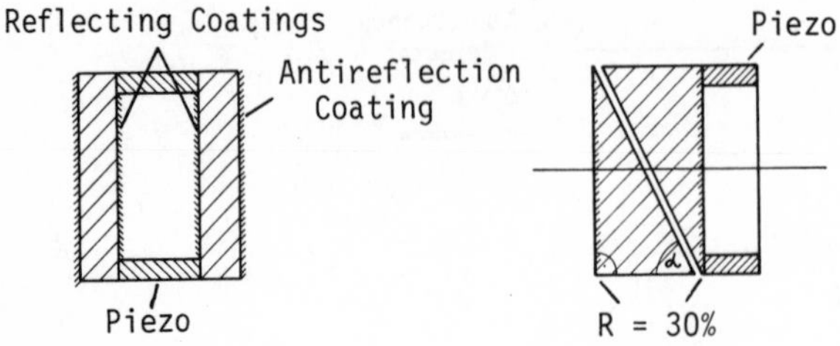

(a) (b)

Fig.6.33a,b. Piezo-tuned mode selective etalons. (a) Plane parallel plates with antireflection coatings. (b) Two Brewster prisms which allow a very narrow air spacing

For the realization of larger tuning ranges one can use, for example, air-spaced etalons which are tuned at a fixed tilting angle θ by changing the distance between the two plates. The reflection losses are now constant when tuning the wavelength of the transmission peak. However, because of the four surfaces the reflection losses are higher and the back surface should be antireflection coated (Fig.6.33a) to avoid unnecessary reflection losses. A disadvantage compared with the tilting of solid etalons is the inconvenience of careful alignment which is necessary to make both plates sufficiently parallel. For solid etalons this alignment has already been performed by the manufacturer.

To minimize fluctuations and drifts of the transmission maximum by air-pressure changes, the air space between the etalon plates should be as small as possible. An elegant solution is shown in Fig.6.33b where the etalon is formed by two Brewster prisms which are coated on a cathete surface and which also avoid reflection losses on the other surfaces due to the Brewster angle α.

A method which automatically yields the correct synchronization of the tuning speed for all elements uses pressure scanning (Fig.6.34). The wavelength-determining elements are enclosed in a pressure-tight tank. When the pressure in the tank is continuously increased, the refractive index n increases proportionally, and the change of the transmission peak of all air-spaced etalons is proportional to the change of the resonator wavelength. Also the preselector (e.g., the grating) responds to the incident wavelength $n\lambda$ [6.31].

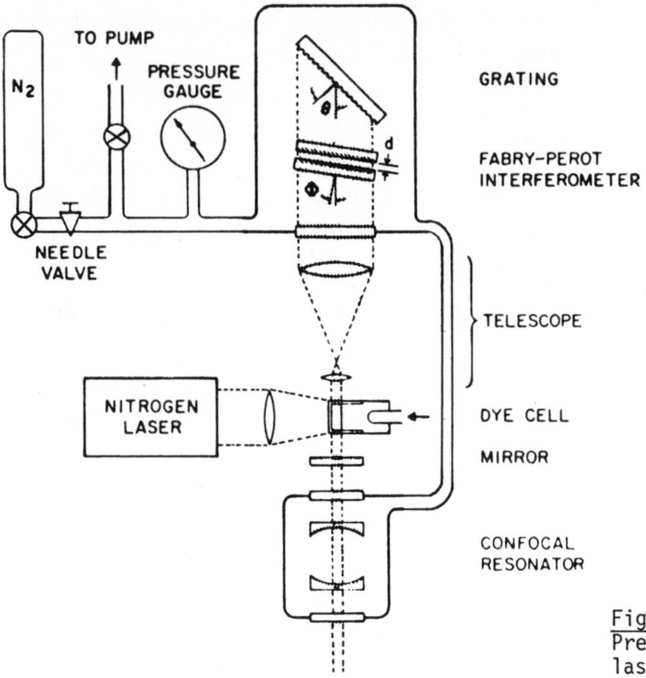

Fig.6.34. Pressure-tuned dye laser [6.31]

Example

For nitrogen, which has n = 1.000278 at 760 torr, a tuning rate of 188 MHz/torr is expected at λ = 600 nm. This means that a pressure change over 760 torr will tune the laser frequency over 142 GHz which corresponds to 4.7 cm^{-1} or nearly 0.2 nm.

6.9 Wavelength Calibration

One of the major goals of high-resolution spectroscopy is the accurate determination of atomic and molecular energy levels and the splittings and shifts of these levels under the influence of external fields or other perturbations. This implies that the wavelength distances between the centers of different absorption lines have to be accurately measured when the laser is tuned through the spectrum. If the laser is kept on the center of the spectral line, either by hand or by an electronic servo loop, the measurement of the laser wavelength simultaneously yields the wavelength of the spectral line. There are several experimental solutions of the problem of accurately measuring wavelengths (see Sect.4.4).

Often a long pressure-tight and temperature-stabilized Fabry-Perot interferometer with a fixed mirror separation is used, and a small fraction of the laser output is split into the F.P.I. (see Fig.6.32). Each time the wavelength λ_L of the tuned laser reaches a transmission maximum of the F.P.I., the photodiode PhD 3 gives a signal which serves as wavelength marker. According to (4.80), the wavelength distance between successive markers is $\Delta\lambda = \lambda^2/2nd$, which corresponds to a free spectral range $\delta\nu = c/(2nd)$. These wavelength or frequency markers can be recorded on a two-pen recorder simultaneously with the spectral lines (see Fig.10.39). For a confocal F.P.I., the free spectral range is $\delta\nu = c/(4nd)$ (see Sect.4.2.10), and we obtain with d = 125 cm, for example, $\delta\nu$ = 60 MHz. Between these markers the frequencies are generally deduced by linear interpolation. If the length change Δd of the piezo element which tunes the laser wavelength is not strictly proportional to the applied voltage, this linear interpolation may lead to inaccuracies which can be avoided by the following elegant method.

That part of the laser intensity which is split into the stabilizing F.P.I. is modulated by a Pockels cell, at a modulation frequency f. This generates, besides the carrier wave at the laser frequency ν_L, two sidebands at $\nu_L \pm f$ (Fig.6.35). The reference F.P.I. is kept constant on one of the two sidebands, and it stabilizes the laser to a frequency ν_L which is shifted by the amount f against the reference frequency $\nu_R = \nu_L \pm f$. By controlling the modulation frequency f, the laser frequency ν_L can be continuously tuned at a fixed reference frequency ν_R. The accuracy of the determination of the frequency difference $\nu_1 - \nu_2$ between two spectral lines is determined by the very high accuracy of measuring $f_1 - f_2$. The continuous tuning range of this method is determined by the maximum possible modulation frequency which is limited to some hundred megahertz [6.32].

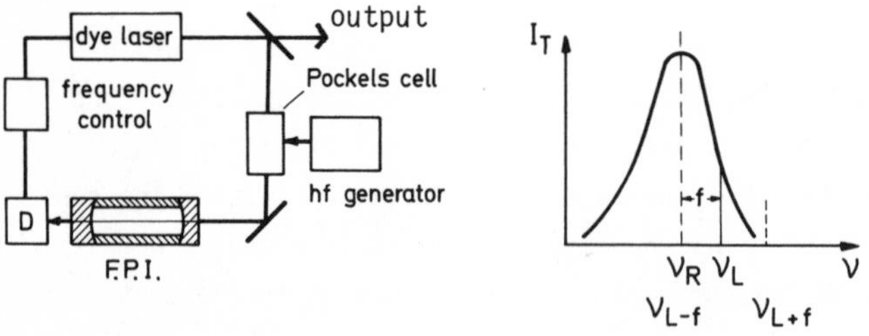

Fig.6.35. Stabilized tuning of a single-mode laser by sideband generation

This controlled frequency shift of the laser frequency against a reference frequency can be also realized by electronic elements in the servo loop. This dispenses with the modulation by the Pockels cell of the previous method. Such a frequency-offset locking technique has been demonstrated by HALL [6.33], who locked a tunable single-mode laser with a variable frequency offset to an ultrastable He-Ne laser, stabilized onto the center of a CH_4 line.

A very precise method of measuring the frequency separation of two closely spaced spectral lines is based on a heterodyne technique. In this method, two single-mode lasers are used. Each laser is stabilized onto the centers ν_1, ν_2 of two spectral lines. The output of both lasers is superimposed onto a detector with nonlinear response generating the difference frequency $\nu_1 - \nu_2$, which is electronically filtered from the frequencies ν_1, ν_2, and $\nu_1 + \nu_2$, and which directly yields the desired line separation. This heterodyne method is illustrated by several examples in Sect.10.6.

6.10 Absolute Laser-Frequency Measurements

The methods discussed above are well suited to measuring separations of closely spaced spectral lines, involving the determination of small *differences* of two or more wavelengths. *Absolute* wavelength measurements are all based on interferometric comparisons with primary or secondary wavelength standards. Such measurements may be performed in a conventional way with Fabry-Perot interferometers. The laser wavelength is stabilized onto the center of the spectral line under investigation, and part of the laser beam is sent through a F.P.I. If the beam is slightly divergent (which can be achieved by a defocusing lens), an interference ring system is produced (see Sect.4.2.9) which can be photographed and compared with the ring system produced on the same photoplate by a stabilized reference laser with known wavelength. The advantage of such measurements over those with incoherent spectral lines is the smaller linewidth and the larger intensity which allows a better signal-to-noise ratio and a more precise determination of the fringe centers. The accuracy $\lambda/\Delta\lambda$ of wavelength measurements may exceed 10^8 which means that a wavelength at $\lambda = 500.0$ nm can be measured to within 5×10^{-6} nm.

Several other recently developed techniques for absolute determination of the wavelengths of single-mode lasers have already been discussed, in Sect.4.4. A very precise method competing with these wavelength measurements is based on the determination of absolute optical *frequencies*, which will now be discussed.

This method [6.34] relies on the fact that of all physical quantities, it is the *frequency* which can be measured most accurately. With modern fast counters, frequencies up to 500 MHz can be counted directly and calibrated against frequency standards. At higher frequencies, a heterodyne technique may be used whereby the difference between the unknown frequency ν_x and a known, nearly equal frequency is generated by a nonlinear detector and can be counted directly. The known frequency is synthesized from two or more known lower frequencies by a nonlinear device which can generate harmonics or which can sum different frequencies [6.35].

Several of these nonlinear devices have been developed to generate sum frequencies or higher harmonics. Examples are fast silicon diodes which respond to frequencies up to 10^{12} Hz or metal-insulator-metal diodes which consist of an oxidized nickel base and a tungsten cat's whisker fabricated from a 10-25 μm diameter tungsten wire electrochemical sharpened to a tip of radius of about 50 nm (see Sect.4.5.8).

Assume that two lower frequency radiations, with known frequencies ν_1 and ν_2 (e.g., the radiation from two far infrared lasers), impinge on the diode, together with the laser of unknown frequency ν_x. The nonlinear response of the diode generates harmonics $m\nu_1$ and $n\nu_2$ of the lower frequency radiations, and by difference-frequency generation, the beat frequency

$$\nu_b = \pm \nu_x \pm m\nu_1 \pm n\nu_2 \qquad (6.19)$$

is produced. If ν_b can be measured directly, the unknown laser frequency ν_x can be obtained from (6.19).

Starting from microwave frequencies which can be readily compared with the frequency standard, a chain of frequencies can be built up which uses, for example, the HCN laser, the H_2O laser, the CO_2 laser, and the He-Ne laser, and their harmonics and which has been extended up to 197 THz (1.52 μm). Recently even optical frequencies in the visible range have been directly measured by this synthesis technique. Figure 6.36 shows such a laser frequency synthesis chain as used by EVENSON et al. [6.34].

An interesting question of fundamental metrology is raised by the experimental progress in measuring wavelengths and absolute optical frequencies with high accuracy. Since the wavelength λ and frequency ν are related by $c = \nu\lambda$, the accuracy of the determination of the speed of light could be improved 100-fold by measuring the wavelength and frequency of a stabilized laser. Because frequency is the more accurately measurable quantity, there

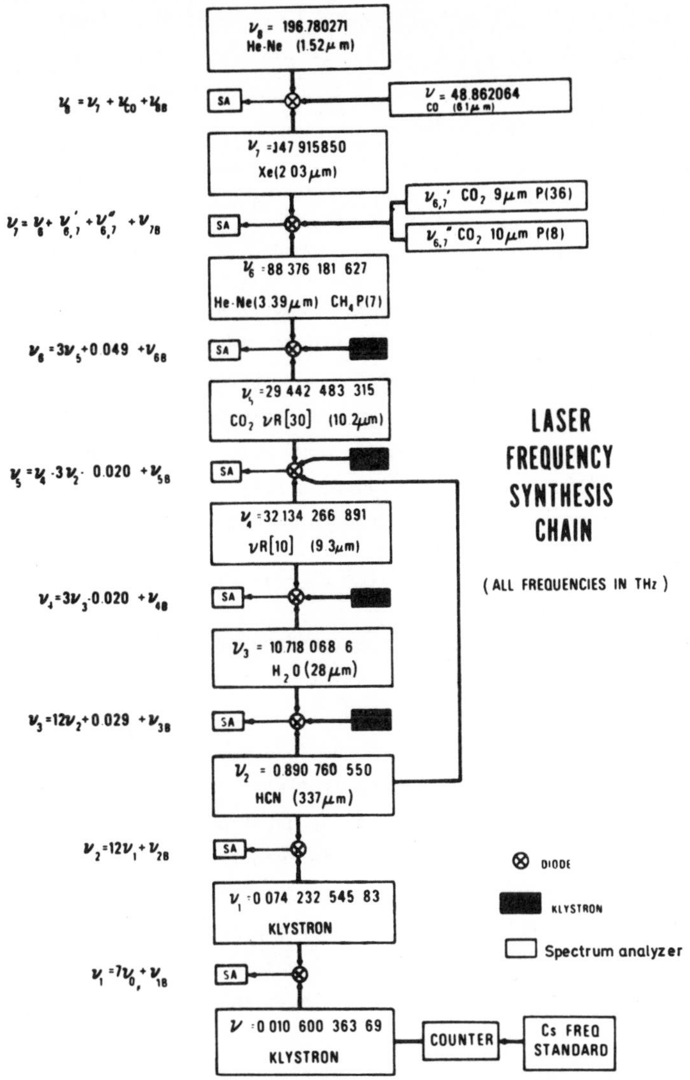

Fig.6.36. Laser frequency synthesis chain from the Cs frequency standard to a He-Ne laser at ν = 197 THz (from [6.34])

is now the possibility of fixing the value of c, *defining* the speed of light by, for example, the value

$$c = 299\ 792\ 458\ \mathrm{m/s}\quad ,$$

which is the value recommended by the Commission of Standards [6.35]. The meter would then no longer be an independent unit but would have to be redefined, using this *fixed* value of c and a *measured* frequency. This redefinition has the advantage that the accuracy of absolute wavelength determinations is no longer limited by the inadequate accuracy with which the wavelength standards can be measured [6.36].

6.11 Linewidths of Single-Mode Lasers

In the previous sections we have seen that the frequency fluctuations of single-mode lasers caused by fluctuations of the product nd of the refractive index n and the resonator length d can be greatly reduced by appropriate stabilization techniques. The output beam of such a single-mode laser can be regarded for most applications as a *monochromatic wave* with a radial Gaussian amplitude profile [see (5.87)]. The field amplitude $E(r,z)$ at an axial distance z from the beam waist and at a radial distance r from the beam axis is

$$E(r,z) = E_0 \exp(-r^2/w^2)\exp[-ikr^2/R(z)]\exp\{i[\omega t - \varphi(z)]\} \quad . \qquad (6.20)$$

For some problems in ultrahigh-resolution spectroscopy, the residual finite linewidth $\Delta\nu_L$, which may be small but nonzero, still plays an important role and has therefore to be known. Furthermore the question *why* there is an ultimate lower limit for the linewidth of a laser is of fundamental interest, since this leads to basic problems of the nature of electromagnetic waves. Any fluctuation of amplitude, phase, or frequency of our "monochromatic" wave (6.20) results in a finite linewidth, as can be seen from a Fourier analysis of such a wave (see the analogous discussion in Sects.3.1,3). Besides the "technical noise" caused by fluctuations of the product nd there are essentially three noise sources of a *fundamental* nature, which cannot be eliminated, even by an ideal stabilization system. These noise sources are responsible for the residual linewidth of a single-mode laser.

The first contribution to the noise results from the spontaneous emission of excited atoms in the upper laser level E_i. The total intensity I_{sp} of the fluorescence spontaneously emitted on the transition $E_i \rightarrow E_k$ is, according to Sect.2.3, proportional to the population density N_i, the active mode volume V_m and the transition probability A_{ik}

$$I_{sp} = N_i V_m A_{ik} \quad . \qquad (6.21)$$

This fluorescence is emitted into all modes of the EM field within the spectral width of the fluorescence line. According to the example in Sect.2.2, there are about 3×10^8 modes/cm^3 within the Doppler-broadened linewidth $\Delta\nu_D = 10^9\ s^{-1}$ at $\lambda = 500$ nm. The mean number of fluorescence photons per mode is therefore small compared to unity. Furthermore only a small fraction of the total fluorescence is emitted into the small solid angle of the Gaussian beam. With $d\Omega = 1$ mrad, this fraction becomes, for example, $\eta = 10^{-7}$.

When the laser reaches threshold, the number of photons in the laser mode increases rapidly by stimulated emission and the narrow laser line grows from the weak but Doppler-broadened background radiation (see Fig. 6.37). Far above threshold the laser intensity is larger than this background by many orders of magnitude and we may therefore neglect this noise contribution.

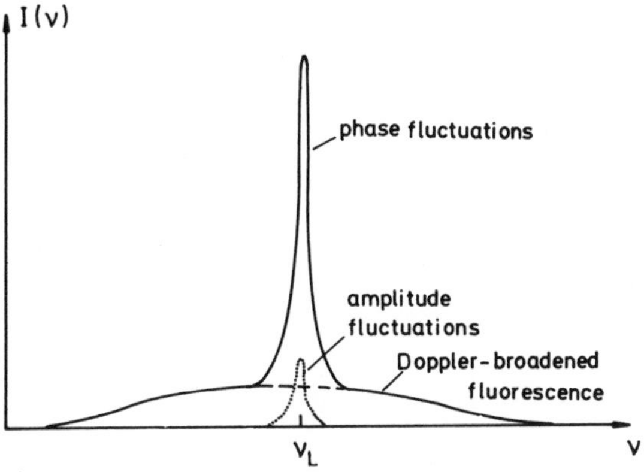

Fig.6.37. Linewidth of a single-mode laser with Doppler-broadened background from spontaneous emission

The second contribution to the noise resulting in line broadening is due to amplitude fluctuations caused by the statistical distribution of the number of photons in the oscillating mode. At a laser output power P, the average number of photons which are transmitted per second through the output mirror is $\bar{n} = P/h\nu$. With $P = 1$ mW and $h\nu = 2$ eV ($\hat{=}\ \lambda = 600$ nm), we obtain $\bar{n} = 8 \times 10^{15}$. If the laser operates far above threshold, the probability p(n) that n photons are emitted per second is given by the Poisson distribution [6.37]

$$p(n) = \frac{e^{-\bar{n}}(\bar{n})^n}{n!} \quad . \tag{6.22}$$

As was discussed in Sect.5.9, the average number of photons n is mainly determined by the pump power P_p, and, at a given value of P_p, the field amplitude of the single-mode laser is constrained to fluctuate about its steady-state value $\bar{E} \sim (\bar{n})^{\frac{1}{2}}$.

The main contribution to the residual laser linewidth comes from *phase fluctuations*. Each photon which is spontaneously emitted into the laser mode can be amplified by induced emission and this amplified contribution is superposed on the oscillating wave. This does not essentially change the total amplitude of the wave because, due to gain saturation, these additional photons decrease the gain for the other photons such that the average photon number $\bar{n}$ remains constant. However, the phases of these spontaneously initiated photon avalanches show a random distribution, and so does the phase of the total wave. There is no such stabilizing mechanism for the total phase as there is for the amplitude. In the course of time a "phase diffusion" occurs which can be described in a thermodynamic model by a diffusion coefficient D [6.37,38].

For the spectral distribution of the laser emission in the ideal case in which all technical fluctuations of nd are totally eliminated this model yields a Lorentzian line profile

$$|E(\nu)|^2 = A_0^2 \frac{1}{(\nu - \nu_0)^2 + (D/2)^2} \quad \text{with} \quad A_0 = E_0(\nu = 0)D/2 \quad , \tag{6.23}$$

which may be compared with the Lorentzian line profile of a classical oscillator broadened by phase-perturbing collisions.

The full halfwidth $\Delta\nu = D$ of this line $|E(\nu)|^2$ decreases with increasing output power because the contributions of the spontaneously initiated photon avalanches to the total amplitude and phase becomes less and less significant with increasing total amplitude. Furthermore, the halfwidth $\Delta\nu_c$ of the resonator resonance must influence the laser linewidth, because it determines the spectral interval where the gain exceeds the losses. The smaller $\Delta\nu_c$, the smaller is the number of spontaneously emitted photons (which are emitted within the full Doppler width) which find enough gain to build up a photon avalanche. When all these factors are taken into account, one obtains for the theoretical lower limit $\Delta\nu_L = D$ for the laser linewidth the relation

$$\Delta\nu_L = \frac{\pi h\nu_L(\Delta\nu_c)^2(N_{sp} + N_{th} + 1)}{2P_L} \quad , \tag{6.24}$$

where N_{sp} is the number of photons spontaneously emitted per second into the oscillating laser mode, N_{th} is the number of photons in this mode due to the thermal radiation field, and P_L is the laser output power. At room temperatures in the visible region, $N_{th} \ll 1$ (see Fig.2.5). With $N_{sp} = 1$ (at threshold, the induced emission is comparable to the spontaneous emission and the saturated gain equals the threshold gain) we obtain from (6.24) the famous Schwalow-Townes relation [6.39],

$$\boxed{\Delta\nu_L = \frac{\pi h \nu_L \Delta\nu_c^2}{P_L}} \quad . \tag{6.25}$$

Example

a) For a He-Ne laser with $\nu_L = 5 \times 10^{14}\ s^{-1}$, $\Delta\nu_c = 1$ MHz, P = 1 mW, we obtain $\Delta\nu_L = 5 \times 10^{-4}\ s^{-1}$.
b) For an argon laser with $\nu_L = 6 \times 10^{14}\ s^{-1}$, $\Delta\nu_c = 3$ MHz, P = 1 W, the theoretical lower limit at the linewidth is $\Delta\nu_L = 3 \times 10^{-5}\ s^{-1}$.

However, even for lasers with a very sophisticated stabilization system, the residual uncompensated fluctuations of nd cause frequency fluctuations which are large compared with this theoretical lower limit. With moderate efforts, laser linewidths of $\Delta\nu_L = 10^4 - 10^6\ s^{-1}$ have been realized for gas lasers and dye lasers. With very high expenditure, laser linewidths of a few Hertz can be achieved [6.40,41]. There are, however, several proposals as to how the theoretical lower limit may be approached more closely.

Problems to Chapter 6

6.1. A He-Ne laser with an unsaturated gain of $G_0(\nu_0) = 1.3$ at the center of the Gaussian gain profile has a resonator length of d = 50 cm and total losses of 4%. Single-mode operation at ν_0 shall be achieved with a coated tilted etalon inside the resonator. Design the optimum combination of etalon thickness and finesse.

6.2. An argon laser with resonator length d = 100 cm and two mirrors with radius $r_1 = \infty$ and $r_2 = 400$ cm has an intracavity circular aperture close to the spherical mirror to prevent oscillation on transversal modes. Estimate the maximum diameter of the aperture which introduces losses $\gamma_{diffr} < 1\%$ for the TEM_{00} mode, but prevents oscillation of higher transverse modes which have, without the aperture, a net gain of 10%.

6.3. A single-mode He-Ne laser with resonator length L = 15 cm shall be tuned by moving a resonator mirror mounted on a piezo. Estimate the maximum tuning range before a mode hop will occur, assuming an unsaturated gain of 10% at line center and resonator losses of 3%. Which voltage has to be applied to the piezo (expansion 1 nm/V) for this tuning range?

6.4. Estimate the frequency drift of a laser oscillating at λ = 500 nm due to thermal expansion of the resonator at a temperature drift of 1°C/h, when the resonator mirrors are mounted on distance holder rods a) made of invar b) made of fused quartz.

6.5. Mode selection in an argon laser is often accomplished with an intra-cavity etalon. What is the frequency drift of the transmission maximum

a) for a solid fused quartz etalon with thickness d = 1 cm due to a temperature change of 2°C?

b) for an air-space etalon due to an air-pressure change of 2 torr?

c) Estimate the average time between two mode hops (cavity length L = 100 cm) for a temperature drift of 1°C/h or a pressure drift of 1 torr/h.

6.6. Assume the output power of a laser shows random fluctuations of about 5%. Intensity stabilization is accomplished by a Pockels cell with a half-wave voltage of 600 V. Estimate the ac output voltage of the amplifier driving the Pockels cell that is necessary to stabilize the transmitted intensity if the Pockels cell is operated around the maximum slope of the transmission curve.

6.7. A single-mode laser is frequency stabilized onto the slope of the transmission maximum of an external reference Fabry-Perot interferometer, made of invar, with a free spectral range of 8 GHz. Estimate the frequency stability of the laser

a) against temperature drifts, if the F.P.I. is temperature stabilized within 0.01°C,

b) against acoustic vibrations of the mirror distance d in the F.P.I. with amplitudes of 1 nm.

c) Assume the *intensity* fluctuations are compensated to 1% by a difference amplifier. Which *frequency* fluctuations are still caused by the residual intensity fluctuations, if a FPI with a free spectral range of 10 GHz and a finesse of 50 is used for frequency stabilization at the slope of the FPI-transmission peak?

7. Tunable Coherent Light Sources

In this chapter experimental realizations of some tunable coherent sources are discussed which are of particular relevance for spectroscopic applications. In the different spectral regions different tuning methods have been developed which will be illustrated by several examples. While semiconductor lasers, spin-flip Raman lasers, and optical parametric oscillators are to date the most widely used tunable *infrared* sources, the dye laser in its various modifications is by far the most important tunable laser in the *visible* region. The development of color center lasers seems to be very promising for achieving a tunable device in the near infrared, which competes in its outstanding properties with the dye laser. In the ultraviolet region the last years have brought great progress in the development of new types of lasers as well as in the generation of coherent uv radiation by frequency doubling or mixing techniques. Meanwhile the whole spectral range from the far infrared to the vacuum ultraviolet can be covered by a variety of tunable coherent sources.

This chapter can give only a brief survey on those tunable devices which have proved to be of particular importance for spectroscopic applications. For a more detailed discussion of the different techniques the reader is referred to the literature cited in the corresponding sections. A review on tunable lasers which covers the development up to 1974 has been given by COLLES and PIDGEON [7.1], a more recent survey on infrared spectroscopy with tunable lasers by MCDOWELL [7.1a].

7.1 Basic Concepts

Tunable coherent light sources can be realized in different ways. One possibility which has already been discussed in Sect.6.8 relies on lasers with a *broad gain profile*. Wavelength-selecting elements inside the laser resonator restrict laser oscillation to a narrow spectral interval and the laser wavelength may be continuously tuned across the gain profile by varying the transmission maxima of these elements. Dye lasers and excimer lasers are examples for this type of tunable devices.

Another possibility of wavelength tuning is based on the shift of energy levels in the active medium by external perturbations, which cause a corresponding spectral shift of the gain profile and therefore of the laser wavelength. This level shift may be affected by an external magnetic field (spin-flip Raman lasers and Zeeman-tuned gas lasers) or by temperature or pressure changes (semiconductor lasers).

The third possibility of generating coherent radiation with tunable wavelength uses the principle of optical frequency mixing. When the radiations from two lasers with frequencies ω_1 and ω_2 are superimposed in a medium with a sufficiently large nonlinear part of its susceptibility, each atom is induced to forced oscillations and generates radiation with the sum frequency $\omega_1 + \omega_2$ and the difference frequency $\omega_1 - \omega_2$. All these secondary waves generated at the location of each atom will have the correct phase to interfere constructively with each other, if the phase velocities of primary wave and secondary waves are identical (phase-matching condition). The resultant macroscopic wave with frequency $\omega_1 + \omega_2$ or $\omega_1 - \omega_2$ is emitted into a direction where the phase-matching condition is fulfilled. In favorable cases the intensity of the secondary wave may reach up to 50% of the incident intensity. If the frequency ω_1 or ω_2 of the incident radiation can be tuned, the difference or sum frequency exhibits the same absolute tuning range, provided the phase-matching condition can be always fulfilled over this tuning range.

Instead of two incident waves also three or more waves with equal or different frequencies may be mixed, either in nonlinear anisotropic crystals or in homogeneous gas mixtures. Furthermore frequency doubling or tripling has been used to transform the tunable radiation of visible or infrared lasers into other spectral regions. Such techniques are of particular importance for the generation of tunable ultraviolet radiation. Some examples in Sect. 7.5 will illustrate these methods.

An interesting development, based on frequency mixing in crystals with nonlinear susceptibility, is represented by the optical parametric oscillator. A pump wave with frequency ω_p and wave vector $\underline{k}_p$ splits into two waves with frequencies ω_i and ω_s and wave vectors $\underline{k}_i$ and $\underline{k}_s$ such that energy and momentum are conserved, which means that $\omega_p = \omega_i + \omega_s$ and $\underline{k}_p = \underline{k}_i + \underline{k}_s$. By choosing the proper angle of incidence against the optical axis of the birefringent crystal or by controlling the refractive index n through the crystal temperature, the idler and signal frequencies ω_i and ω_s can be varied within wide ranges.

Instead of the nonlinear crystal a molecular gas can be used to produce through the Raman effect the parametric frequency splitting of $\omega_p = \omega_R \pm m\omega_M$ where ω_M corresponds to a molecular eigenfrequency and ω_R to the Raman line. Tuning of ω_p allows a tunable ω_R which may cover a wide range in the infrared, if ω_p is in the visible region (Raman laser).

The experimental realization of these tunable coherent light sources is of course determined by the spectral range for which they are to be used. For the particular spectroscopic problem, one has to decide which of the possibilities, summarized above, represents the optimum choice. The experimental expenditure depends substantially on the desired tuning range, on the achieveable output power, and last but not least on the realized spectral bandwidth $\Delta\nu$. Coherent light sources with bandwidths $\Delta\nu \approx 100$ MHz - 30 GHz (0.001 - 1 cm^{-1}), which can be quasicontinuously tuned over a larger range, are already commercially available. In the visible region single-mode dye lasers are offered with a bandwidth down to about 1 MHz, which are continuously tunable over a restricted tuning range of about 30 GHz (1 cm^{-1}), while larger tuning ranges are generally realized with quasicontinuous tuning, where mode hops occur. However, the rapid development of different tunable devices will certainly lead to a great variety of coherent sources with larger tuning ranges and with true continuous tuning.

We now briefly discuss the most important tunable coherent sources, arranged according to their spectral region. Figure 7.1 illustrates the spectral ranges covered by the different devices.

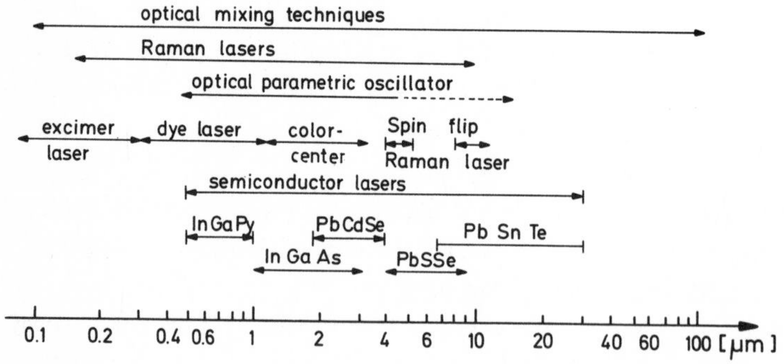

Fig.7.1. Spectral ranges of different tunable coherent sources

7.2 Tunable Infrared Lasers

Many of the most widely used tunable coherent infrared sources use various semiconductor materials, either directly as the active laser medium (semiconductor lasers) or as the nonlinear mixing device (spin-flip Raman laser, frequency-difference generation). Alkali halides with different types of color centers provide the material for the color center lasers while birefringent crystals, such as lithium niobate are used in parametric oscillators and near infrared optical frequency mixing devices. Gases at high pressure offer another possibility of gain media for continuously tunable infrared gas lasers.

We start with the semiconductor diode laser, which has to date provided the major contribution to applications in high-resolution infrared spectroscopy [7.2].

7.2.1 Semiconductor Diode Lasers

The basic principle of semiconductor lasers [7.3] may be summarized as follows. When an electric current is sent in the forward direction through a p-n semiconductor diode, the electrons and holes can recombine within the p-n junction and may emit the recombination energy in the form of E.M. radiation (Fig.7.2). The linewidth of this spontaneous emission amounts to several cm^{-1} and the wavelength is determined by the energy difference between the energy levels of electrons and holes, which is essentially determined by the band gap. The spectral range of spontaneous emission can

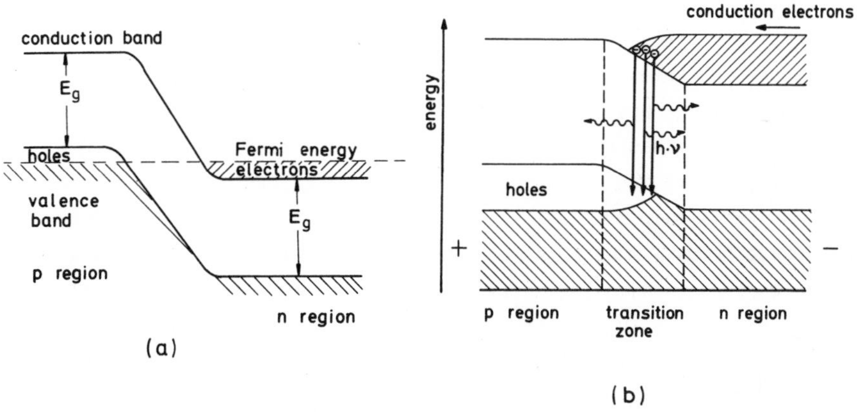

Fig.7.2a,b. Schematic level diagram of a semiconductor diode laser. (a) Unbiased, (b) with a forward voltage applied

be therefore varied within wide limits (about 0.8 - 40 μm) by the proper selection of the semiconductor and its impurity concentration (see Fig.7.3).

Above a certain threshold current, determined by the particular semiconductor diode, the radiation field in the junction becomes sufficiently intense to make the induced emission rate exceed the spontaneous or radiationless recombination processes. The radiation can be amplified by multiple reflections from the plane end faces of the crystal and may become strong enough that induced emission occurs in the p-n junction before other relaxation processes deactivate the population inversion.

The wavelengths of the laser radiation are determined by the spectral gain profile and by the eigenresonances of the laser resonator (see Chap.5). If the polished end faces of the crystal are used as resonator mirrors, the free spectral range

$$\delta\nu = \frac{c}{2nd\left(1 + \frac{\nu}{n}\frac{dn}{d\nu}\right)} \quad \text{or} \quad \delta\lambda = \frac{\lambda^2}{2nd\left(1 - \frac{\lambda}{n}\frac{dn}{d\lambda}\right)}$$

is very large, because of the short resonator length d. Note that $\delta\nu$ depends not only on d but also on the dispersion $dn/d\nu$ of the active medium.

Example

With d = 0.5 mm, n = 2.5 and $(\nu/n)dn/d\nu = 1{,}5$ the free spectral range becomes $\delta\nu = 48$ GHz $\hat{=}$ 1,6 cm^{-1} or $\delta\lambda = 1.6$ Å at $\lambda = 1$ μm.

This illustrates that only a few resonator modes fit within the gain profile, which has a spectral width of several cm^{-1} (see Fig.7.4).

For wavelength tuning, all those parameters which determine the energy gap between the upper and lower laser level may be varied. A temperature change produced by an external cooling system or by a current change is most frequently used to generate a wavelength shift. Often also an external magnetic field or a mechanical pressure applied to the semiconductor is employed for wavelength tuning. In general, however, no truly continuous tuning over the whole gain profile is possible. After a continuous tuning over about one wave number, mode hops occur because the resonator length is not altered synchronously with the maximum of the gain profile (see Fig.7.4). In case of temperature tuning this can be seen as follows.

A temperature change ΔT changes the energy difference $\Delta E = E_1 - E_2$ and also the index of refraction by $\Delta n = (dn/dT)\Delta T$. The resultant shift $\Delta\nu = \nu\Delta n/n$ of the resonator eigenfrequency ν amounts, however, only to about 10-20% of the shift $\Delta\nu_g = \Delta E/h$. As soon as the maximum of the gain profile

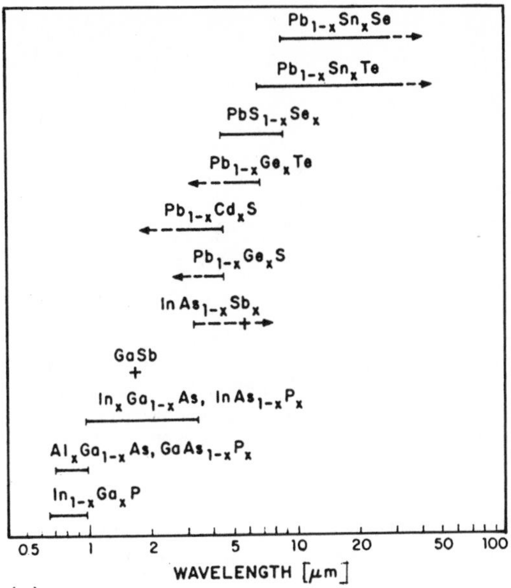

(a)

Fig.7.3. (a) Spectral ranges of laser emission for several semiconductor materials [7.4]. (b) Dependence of emission frequency on composition for Pb-salt lasers of various alloy systems, temperature = 10 K. (c) Temperature tuning of a PbSnSe laser (Courtesy Spectra Physics)

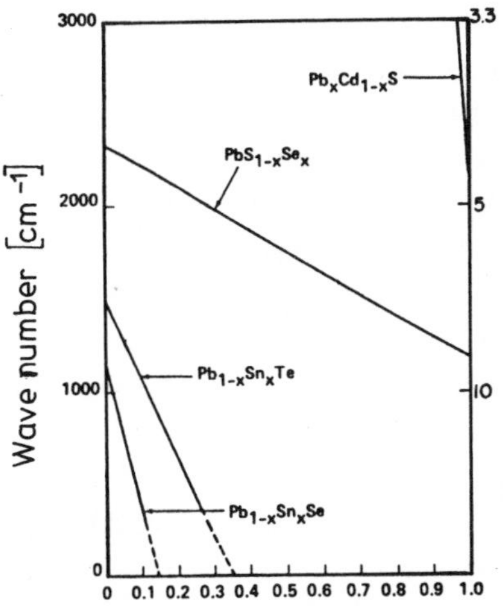

(b)

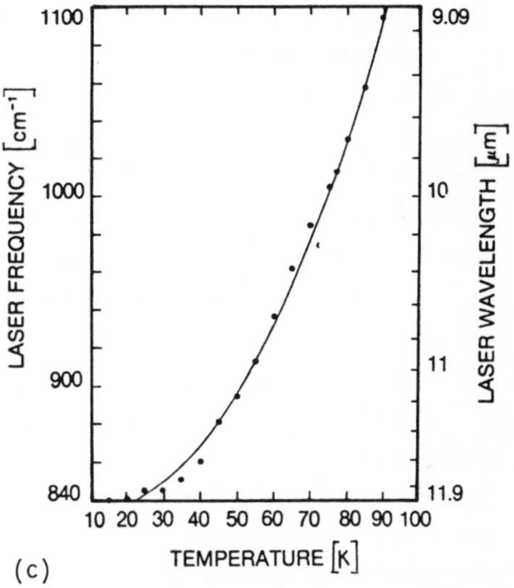

(c)

reaches the next resonator mode, the gain for this mode becomes larger than that of the oscillating one and the laser frequency jumps to this mode (see Fig.7.4c).

For the realization of continuous tuning it is therefore necessary to use external resonator mirrors with a distance d which can be independently controlled. This implies, however, a larger distance d and therefore a

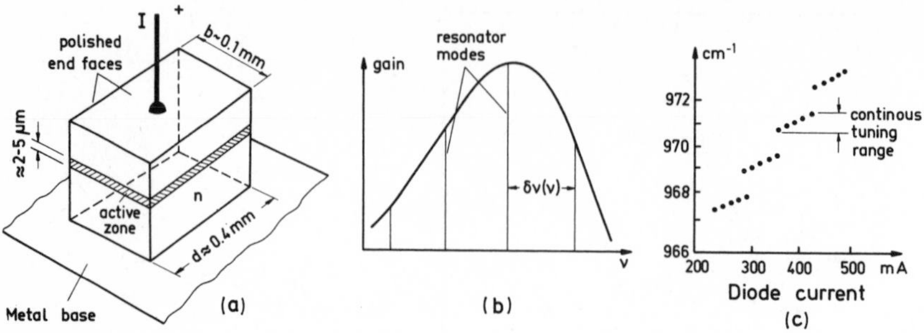

Fig.7.4a-c. Schematic diagram of a diode laser. (a) Injection laser structure. (b) Mode spectrum within the gain profile. (c) Mode hops of a quasi-continuously tunable cw Pb Sn Te diode laser in an He cryostat. The laser frequency is tuned by changing the diode current. The points correspond to the transmission maxima of an external Ge etalon with a free spectral range of 1.955 GHz [7.2]

smaller free spectral range. To achieve single-mode oscillation, an additional etalon has to be inserted into the resonator. Furthermore the end faces of the semiconductor must be antireflection coated because the large reflection coefficient of the uncoated surfaces (with n = 2.5 the reflectivity already becomes 0.2) causes large reflection losses. Such single-mode semiconductor lasers have been built already [7.5a].

By a combined variation of temperature and external magnetic field the continuous tuning range can be greatly increased. The drawback of temperature changes induced by altering the electric current is the variation of the laser output power with current. It is therefore advantageous to use lasers with a closed refrigeration system which allows alteration of the temperature over a wide range without changing the current through the semiconductor. With specially designed heterostructure semiconductor diode lasers [7.6] cw operation at room temperature has been achieved [7.7].

In its advanced form the semiconductor laser represents a compact ir laser spectrometer which combines an intense light source and a high-resolution monochromator. The cw output powers of good lasers range from microwatts up to several milliwatts and the spectral resolution reaches from several MHz into the kilohertz range, depending on the expenditure and efforts with the frequency stabilization. Despite the expensive closed-cycle refrigerator which can cool down to liquid helium temperatures, such a compact laser spectrometer (see Fig.7.5) can easily compete with a Fourier spectrometer regarding price, signal-to-noise ratio, and spectral resolving power. There are numerous examples where semiconductor lasers have been successfully used in high-resolution infrared spectroscopy (see Sect.8.4 and [7.2,8, 8.39a]).

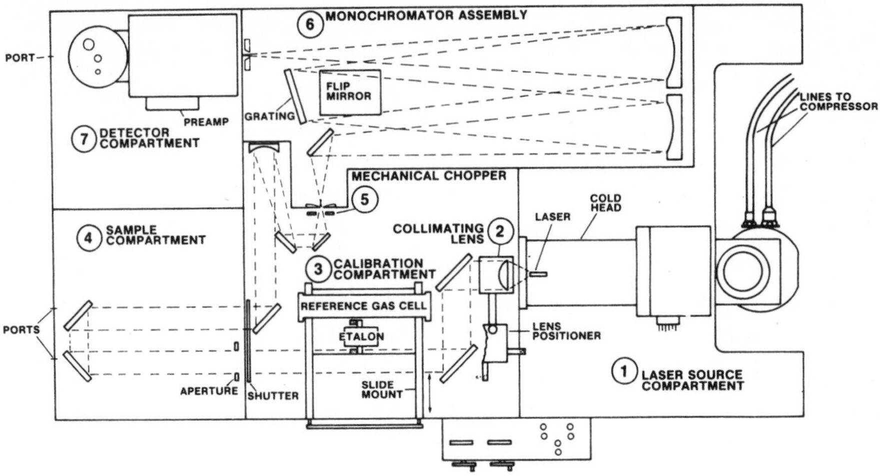

Fig.7.5. Schematic diagram of a tunable diode laser instrument for high-resolution spectrometry (Courtesy Spectra Physics)

7.2.2 Spin-Flip Raman Lasers (SFRL)

The basic principle of the SFRL [7.9,10] is based on stimulated Raman scattering (see Sect.9.3) of pump radiation from a fixed-frequency laser by conduction electrons in low-temperature semiconductors subject to a homogeneous magnetic field. The twofold degenerate energy levels E_n^0 of electrons in the conduction band are split in an external magnetic field into two sublevels with electron spins oriented parallel and antiparallel to the field. The splitting is analogous to the Zeeman splitting of free atoms in ^2S states

$$E^{\pm} = E^0 \pm g^* \mu_B B \quad , \tag{7.1}$$

where μ_B is the Bohr magneton. However, the Landé factor g^* which gives the effective gyromagnetic ratio of the conduction electrons in the semiconductor is quite different from g = 2.002 for the free electron. In InSb, for instance, $g^* = -50$. When the crystal is illuminated by radiation from the pump laser with frequency ω_p, the interaction with the electrons will cause "spin-flips" and the Raman-scattered light is frequency shifted to

$$\omega = \omega_p \pm g^* \mu_B B/\hbar \quad . \tag{7.2}$$

The cross section for this light-induced spin-flip process may become very large because the Thomson cross section of an electron in a semiconductor

$\sigma_t = (e^2/m^*c^2)^2$ exceeds that of a free electron in the ratio $(m/m^*)^2$ where m^* is the effective mass. With infrared pump lasers, where the frequency of the exciting photons approaches the energy of the direct band gap of several semiconductors (e.g., InSb, PbTe), the light scattering cross section can approach one million times that of the free electron. The SFR cross section furthermore depends on the magnetic field strength.

The polished end faces of the crystal can serve as resonator mirrors and at sufficiently high pump powers the stimulated emission will exceed the losses and oscillation starts. Because the SFR scattering cross section is influenced by the magnetic field, there is generally a lower and an upper limit for the magnetic field strength at which lasing will occur. These limits depend on the carrier concentration in the semiconductor. For InSb, which is the most widely used medium for SFR lasers, the minimum field strength is about 20 KG at electron concentrations of 10^{16} cm^{-3}. At lower concentrations the lower field limit can be decreased at the expense of lower output powers.

The frequency ω of the SFR laser can be tuned according to (7.2) by the external field strength B. The tuning rate depends on the effective Landé factor g^*. Typical values are 2 cm^{-1}/KG for InSb and 3.5 cm^{-1}/KG for HgCdTe. This implies that tuning ranges of up to 200 cm^{-1} can be achieved for either the Stokes or the anti-Stokes lines with magnetic fields up to 100 KG. For such high fields, generally superconducting magnets are used. Because g^* depends on the field strength B the frequency tuning is not strictly linear with B.

The most widely used pumping sources for InSb-SFR lasers are the CO laser at λ = 5 μm or the CO_2 laser at λ = 9-10 μm. Since the 5 μm pump radiation is close to the band gap, the resonance enhancement of the SFR cross section allows cw operation of InSb lasers with output powers of about 1 W. For $Hg_{0.77}Cd_{0.23}Te$ the band gap is at 9.4 μm which favors the CO_2 laser as pumping source, while InAs can be pumped with an HF laser. In favorable cases threshold pump powers as low as 5 mW have been observed in CO laser-pumped InSb lasers. The tuning range of InSb-SFR lasers extends from 5.0 to 6.5 μm when pumped by a CO laser and from 9.0 to about 14 μm when a CO_2 laser is used as pumping source.

The linewidth of the SFR laser depends on the free carrier concentration and on local temperature variations in the crystal, on the properties of the cavity, and on the stabilities of pump frequency and magnetic field strength. Note that at tuning rates of 2 cm^{-1}/KG a relative change $\Delta I/I = 10^{-6}$ of the magnet current already causes a shift $\Delta\nu = 10^6$Hz of the SRL frequency ν. With sufficient efforts for frequency stabilization a spectral linewidth of 30 kHz has been achieved for periods of several minutes [7.11]. In order to realize continuous tuning of a single-mode cw laser, external resonator mirrors and an additional etalon inside the resonator are necessary, quite analogous to the diode laser [7.12].

A drawback of the SFR laser is the relatively expensive experimental equipment since a pump laser, a refrigeration system for liquid helium, and a strong magnet are needed. If only a small tuning range is required, a strong permanent magnet may be used where the field is tuned by an additional pair of Helmholtz coils [7.13]. Figure 7.6 is a schematic diagram of a SFR laser used for molecular absorption spectroscopy (see Sect.8.2).

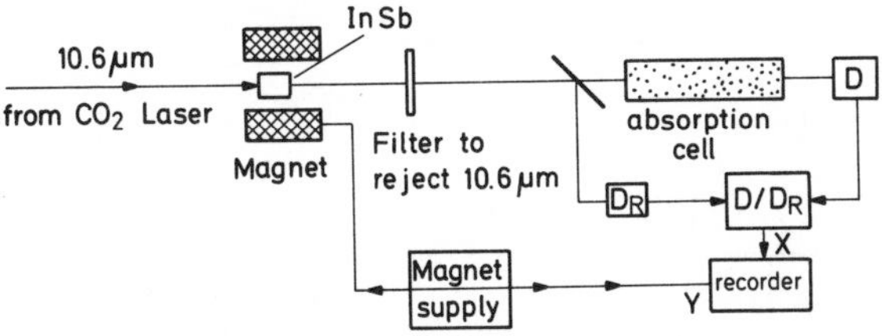

Fig.7.6. Molecular absorption spectroscopy with an SFR laser

Numerous applications of SFR lasers to high-resolution infrared spectroscopy have demonstrated the capabilities of this tunable laser type regarding spectral resolution and achievable signal-to-noise ratio (see Sect.8.5 and [7.12,14]).

7.2.3 Tunable Infrared Gas Lasers

The gain profiles of infrared transitions in low pressure gas lasers are essentially determined by Doppler broadening. The continuous tuning range of single-mode lasers is therefore limited to the spectral interval $\delta\nu_D$ within the Doppler width which amounts to about $\pm$ 0.002 cm^{-1} at $\lambda = 10\ \mu m$. The tuning range of gas lasers may be greatly increased by two methods. The first uses the Zeeman shift of laser transitions in external magnetic fields and the second pressure broadening of the gain profiles to obtain a continuous overlap between adjacent transitions.

Historically the Zeeman tuned gas lasers have been the first continuously tunable lasers used in high-resolution molecular spectroscopy [7.15a,b]. The discharge tube of a gas laser is placed in an axial magnetic field which causes a splitting of the gain profiles for the laser transitions into two groups of Zeeman components with opposite circular polarization σ^+ and σ^- and with frequencies

$$\omega = \omega_0 + [m_1(g_1 - g_2) \pm g_1]\mu_B B/\hbar \quad , \tag{7.3}$$

where g_1, g_2 are the Landé factors of upper and lower level and m_1 the magnetic quantum number. Generally g_1 and g_2 are not very different and the components in each group overlap within their Doppler width. We then have two gain components separated by $\Delta\omega = 2g\mu_B B/\hbar$.

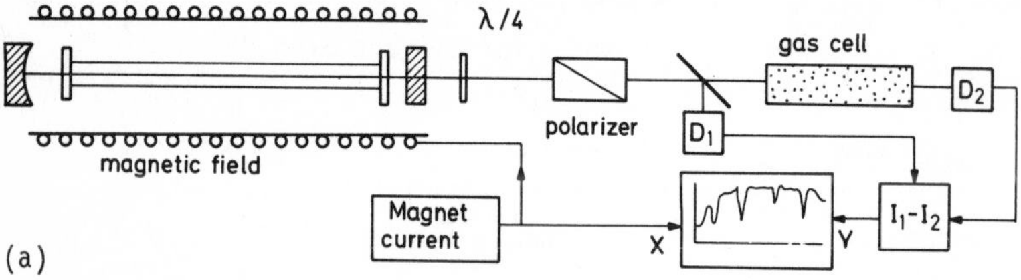

(a)

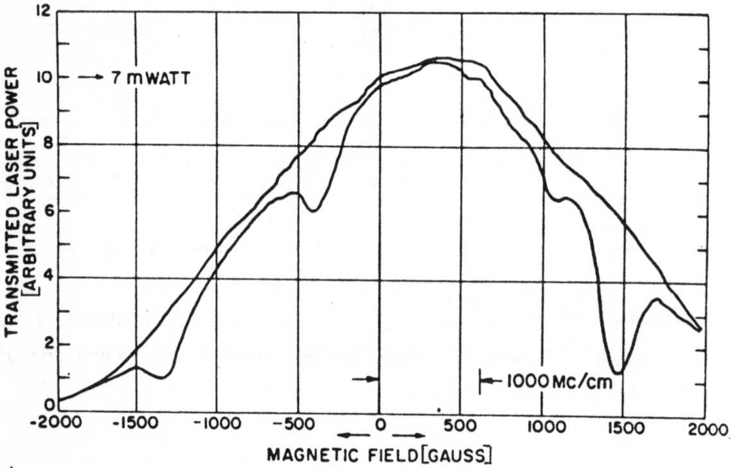

(b)

Fig.7.7. (a) Experimental arrangement of a Zeeman tuned gas laser. (b) Tuned laser spectrogram of CH_3F. Upper curve: empty cell, lower curve: cell filled with CH_3F [7.15b]

Either one of the components can be selected with a $\lambda/4$ plate and a polarizer. Unfortunately the magnetic field affects the gas discharge and the gain of the active laser medium. The laser intensity therefore depends on the magnetic field and drops to zero at some maximum field strength B_{max}. This restricts the tuning range to

$$\Delta\omega_{max} = 2g_1\mu_B B_{max}/\hbar \quad . \tag{7.4}$$

Although tuning ranges of as much as ± 3.5 cm^{-1} have been demonstrated for transitions in He-Ne and He-Xe lasers [7.16], for most gas laser lines the tuning range is restricted to $\Delta\nu/c \leq 1\ cm^{-1}$. For tunable single-mode operation an intracavity etalon can be used which is tilted synchronously with the magnetic gain tuning. To avoid mode hops the cavity length has of course to be also adjusted synchronously (see Sect.6.8).

A number of infrared gas lasers can oscillate on many closely spaced rotational lines of a vibrational transition. Examples are CO_2, N_2O, HF, DF, or H_2O lasers. If the pressure in the gas discharge can be made sufficiently large to increase the pressure-broadened linewidth beyond the spacings between adjacent lines, a quasicontinuous gain profile is formed which allows continuous tuning over a larger spectral interval [7.17].

The self-broadening coefficient for CO_2, for example, is about 0.2 cm^{-1}/atm. Since the different rotational transitions are separated by roughly 1 - 3 cm^{-1}, pressures of about 10 atm are necessary to provide adequate overlap for continuous tuning. Although mixing of different isotopes increases the line density and thus reduces the minimum required pressure, transverse excitation with high-energy electrons (up to 1 MeV) is generally used to achieve a uniform discharge and sufficient gain [7.18]. Tuning ranges up to 70 cm^{-1} have been achieved for a 15 atm CO_2 laser [7.19]. With an intracavity etalon a linewidth of 0.03 cm^{-1} could be reached. With different isotopic gas mixtures of $^{12}CO_2$ and $^{13}CO_2$ or $^{12}CS_2$ and $^{13}CS_2$ or N_2O the spectral region between 9.1 and 12.5 μm was nearly completely covered [7.20].

Continuous operation at medium pressures with smaller tuning ranges of up to 0.1 - 0.2 cm^{-1} is possible with CO_2 *wave guide lasers* [7.21]. Here the dis-

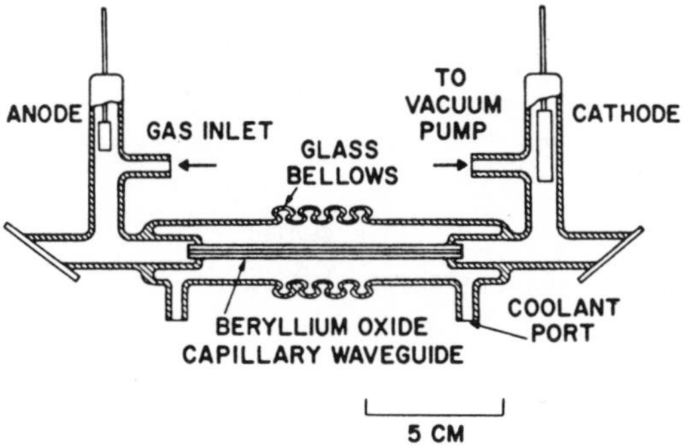

Fig.7.8. A beryllium oxide waveguide CO_2 laser tube. The waveguide bore is 1 mm [7.21]

charge is contained in a small bore of about 1 mm diameter which acts as a waveguide for the laser radiation. At pressures of up to 1 atm, cw output powers of several watts at linewidths down to 100 kHz have been observed [7.22]. Although these waveguide lasers have a smaller tuning range than the high-pressure TEA lasers, they have the advantage of small size, low price, and very good line stability, which makes them useful for high-resolution spectroscopy in restricted spectral intervals.

7.2.4 Color Center Lasers

Color centers in alkali halide crystals are based on a halide ion vacancy in the crystal lattice of rock salt structure (Fig.7.9a). If a single electron is trapped at such a vacancy, its energy levels result in new absorption lines in the visible spectrum, broadened to bands by the interaction with phonons. Since these visible absorption bands which are caused by the trapped electrons and which are absent in the spectrum of the ideal crystal lattice, make the crystal appear colored, these imperfections in the lattice are called F centers (from the German word "Farbe" for color) [7.23]. If *one* of the six positive metal ions that immediately surround the vacancy is foreign (e.g., a Na^+ ion in a KCl crystal), the F center is specified as F_A center [7.24], while F_B centers are surrounded by *two* foreign ions (Fig.7.9b,c).

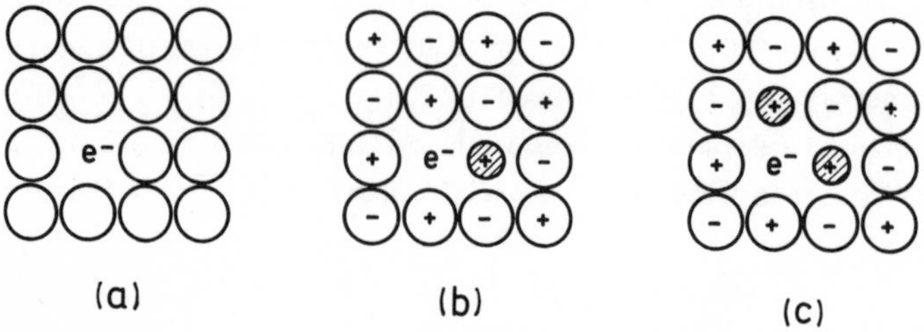

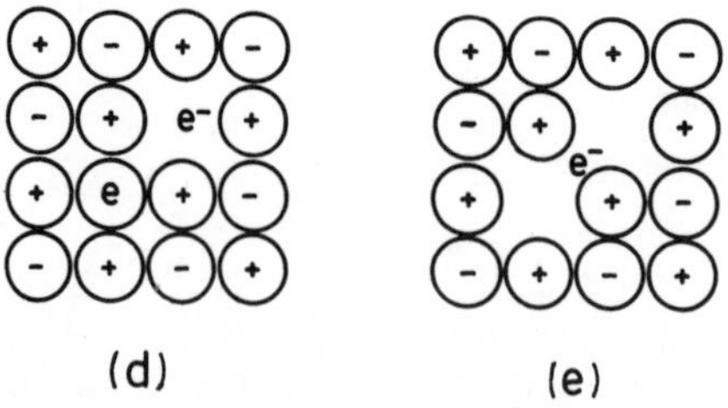

Fig.7.9. (a) F center, (b) F_A center, (c) F_B center, (d) F_2 center, (e) F_2^+ center

These F_A and F_B centers can be further classified into two categories according to their relaxation behavior following optical excitation. While centers of type I retain the single vacancy and behave in this respect like ordinary F centers, the type II centers relax to a double-well configuration (Fig.7.10) with energy levels completely different from the unrelaxed counterpart. The oscillator strength for an electric dipole transition between upper and lower level in the relaxed double-well configuration is quite large. The relaxation times T_{R1} and T_{R2}, for the transitions to the upper level E_k and from the lower level E_i back to the initial configuration are below 10^{-12} s. All these facts make the F_AII and F_BII type color centers very suited for tunable laser action [7.25].

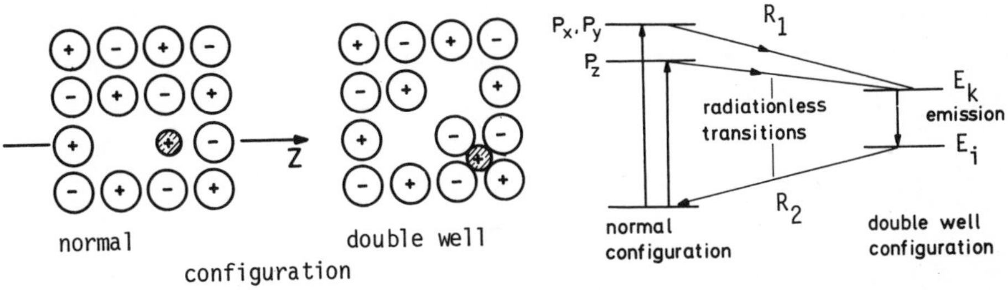

Fig.7.10. Level diagram for illustration of laser action in F_A(II) centers [7.25]

The quantum efficiency η of F_A(II) center luminescence decreases with increasing temperature. For KCl:Li crystal, for example, η amounts to 40% at liquid nitrogen temperatures (77 K) and approaches zero at room temperature (300 K). This implies that color center lasers should be operated at low temperatures.

The tuning range of color center lasers extends over a broad spectral range. Figure 7.11 shows the homogeneously broadened luminescence bands of

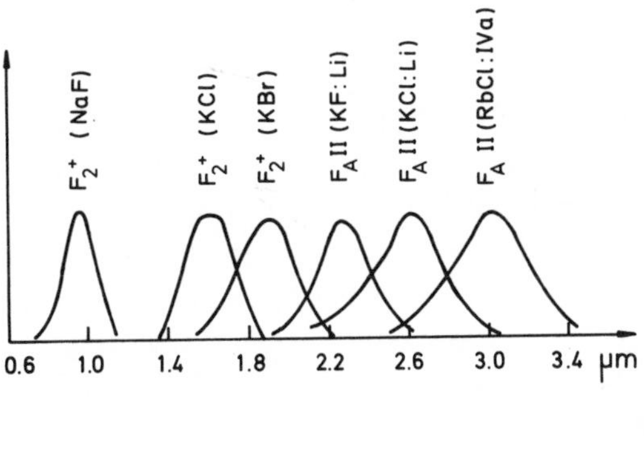

Fig.7.11. Luminescence bands of different color center types in alkali halides [7.25]

some types of F centers in different alkali halides. When all hosts are included, an overall spectral range 0.85 μm $\leq \lambda \leq$ 3.6 μm can be covered. Color center lasers may be therefore regarded as the "dye lasers for the near infrared". The pump powers required are generally much lower than those of cw dye lasers and furthermore no bleaching or aging effects have been observed under normal operation [7.30c].

The experimental arrangement of a tunable color center laser is shown in Fig.7.12 [7.26]. The folded astigmatically compensated three-mirror cavity design is identical to that of cw dye lasers of the Kogelnik type [7.27] (see Sect.7.3). A collinear pump geometry allows optimum overlap between pump beam and the waist of the fundamental resonator mode in the crystal. The mode-matching parameter (i.e., the ratio of pump beam waist to resonator mode waist) can be chosen by appropriate mirror curvatures. The optical density of the active medium, which depends on the preparation of the F_A centers [7.25], has to be carefully adjusted to achieve optimum absorption of the pump wavelength. The crystal is mounted on a cold finger cooled with liquid nitrogen to achieve a high quantum efficiency η.

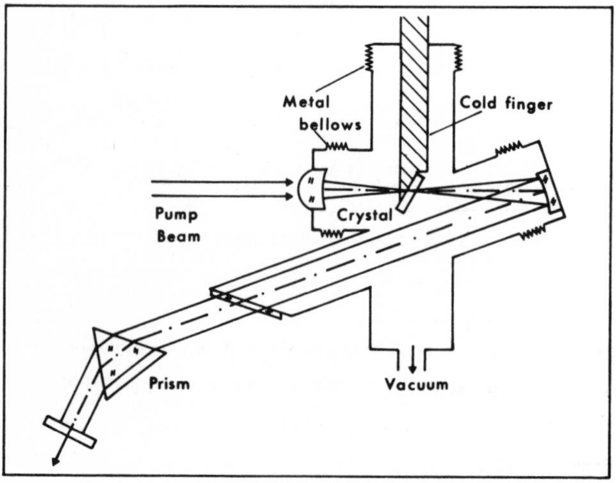

Fig.7.12. Design of a tunable cw color center laser [7.26]

Coarse wavelength tuning can be accomplished by a sapphire Brewster prism. Because of the homogeneous broadening of the gain profile single-mode operation would be expected without any further selecting element (see Sect.5.7). This is in fact observed except that neighboring hole-burning modes appear, which are separated from the main mode by

$$\Delta\nu = c/4z$$

where z is the distance between end mirror and crystal (see Sect.5.8). With one Fabry-Perot etalon of 5 mm thickness and a reflectivity of 60 - 80%, stable single-mode operation without other spatial hole-burning modes can be achieved [7.28]. With careful design of low-loss optical components inside the cavity (made, e.g., of sapphire or CaF_2), single-mode powers up to 75% of the multimode output can be achieved, due to the homogeneous gain profile.

Table 7.1 compiles some of the main characteristics of cw color center lasers [7.29,30a,c].

Table 7.1. Characteristics of different cw color center lasers

Crystal Type of color center	KCl:Li $F_A(II)$	RbCl:Li $F_A(II)$	KCl:Na $F_B(II)$	RbCl:Na $F_B(II)$	KF F_2^+	LiF F_2^+	NaF F_2^+
Pump laser	Kr^+,Ar^+	Kr^+	Ar^+,Kr^+	Kr^+	Nd:YAG	Kr^+	Kr^+
Pump wavelength [nm]	647,514	647,752	514,568,647	647,676,752	1064		752
Threshold pump power [mW]	13	60	20	26	50		40
Output power [mW]	230	55	35	6	30	1000	
Slope efficiency [%]	9.1	2.5	2.3	2.1	40	60	10
Tuning range [μm]	2.5-2.9	2.6-3.33	2.25-2.65	2.5-2.9	1.26-1.48	0.82-1.07	0.88-1.0

Two F centers along a (110) crystal axis constitute an F_2 center and its single ionized counterpart is a F_2^+ center. In contrast to the $F_A(II)$ center laser the F_s^+ center laser apparently shows inhomogeneous contributions to the broadening of the gain profile which results in many simultaneously oscillating modes in the free-running laser. The selection of a single mode demands more selecting elements causing a significant reduction of output power. However, using a ring resonator with an unidirectional travelling wave allows one to obtain nearly 0.5 W single-mode output power [7.30,30b].

The linewidth $\Delta\nu$ of a single-mode laser is mainly determined by fluctuations of the optical path length in the cavity (see Sect.6.6). Besides the contribution $\Delta\nu_m$ caused by mechanical instabilities of the resonator, temperature fluctuations in the crystal, caused by pump power variations or by temperature variations of the cooling system, further increase the linewidth by adding contributions $\Delta\nu_p$ and $\Delta\nu_t$. Since all three contributions are independent, we obtain for the total frequency fluctuations

$$\Delta\nu = \sqrt{\Delta\nu_m^2 + \Delta\nu_p^2 + \Delta\nu_t^2} \quad . \tag{7.5}$$

The linewidth of the unstabilized single-mode laser was measured to be smaller than 260 kHz, which was the resolution limit of the measuring system [7.28]. An estimated value for the overall linewidth $\Delta\nu$ is 25 kHz [7.29]. This extremely small linewidth is ideally suited to perform high-resolution Doppler-free spectroscopy (see Chap.10).

7.3 Dye Lasers

In the visible range dye lasers [7.30d] in their various modifications are by far the most widely used types of tunable lasers. Their active media are organic dye molecules solved in liquids, which display strong broad-band fluorescence spectra under excitation by visible or uv light. With different dyes the overall spectral range, where cw or pulsed laser operation has been achieved, extends from 300 nm to 1.2 μm. In this section we briefly summarize the basic physical background and the most important experimental realizations of dye lasers, used in high-resolution spectroscopy. For a more extensive treatment the reader is referred to the laser literature (e.g., [7.31,32b]).

7.3.1 Physical Background

When dye molecules in a liquid solvent are irradiated with visible or ultraviolet light, higher vibrational levels of the first excited singlet state S_1 are populated by optical pumping from thermally populated rovibronic levels in the S_0 ground state (Fig.7.13a). Induced by collisions with solvent molecules, the excited dye molecules undergo very fast radiationless transitions into the lowest vibrational level v_0 of S_1 with relaxation times of 10^{-11} - 10^{-12} s. This level is depopulated either by spontaneous emission into the different rovibronic levels of S_0 or by radiationless transitions into a lower triplet state T_1 (intersystem crossing). Since the levels populated by optical pumping are generally above v_0 and since many fluorescence transitions terminate at higher rovibronic levels of S_0, the fluorescence spectrum of a dye molecule is red shifted against its absorption spectrum. This is shown in Fig.7.13b for rhodamine 6G, the most widely used laser dye.

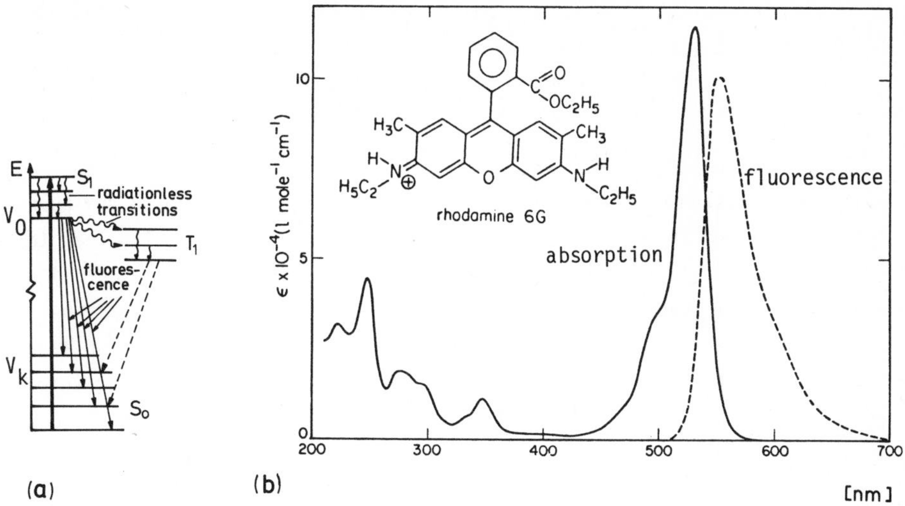

Fig.7.13. (a) Schematic energy level diagram and pumping cycle in dye molecules. (b) Absorption and fluorescence spectrum of rhodamine 6G, solved in ethanol

Because of the strong interaction of dye molecules with the solvent, the closely spaced rovibronic levels are collision broadened to such an extent that the different fluorescence lines completely overlap. The absorption and fluorescence spectra therefore consist of a broad continuum, which is homogeneously broadened (see Sect.3.3).

At sufficiently high pump intensity, population inversion may be achieved between the level v_0 in S_1 and higher rovibronic levels v_k in S_0 which have a negligible population at room temperature, due to the small Boltzmann factor $\exp[-E(v_k)/kT]$. As soon as the gain on the transition $v_0(S_1) \rightarrow v_k(S_0)$ exceeds the total losses, laser oscillation starts. The lower level $v_k(S_0)$, which now becomes populated by stimulated emission, is depleted very rapidly by collisions with the solvent molecules. The whole pumping cycle can be therefore described by a four-level system.

According to Sect.5.2 the spectral gain profile $G(\nu)$ is determined by the population difference $N(v_0) - N(v_k)$, the absorption cross section $\sigma_{0k}(\nu)$ at the frequency $\nu = E(v_0) - E(v_k)/h$, and the length L of the active medium. The negative absorption coefficient at the frequency ν is therefore

$$\alpha(\nu)L = -[N(v_0) - N(v_k)] \int \sigma_{0k}(\nu - \nu')d\nu' \quad . \tag{7.6}$$

The spectral profile of $\sigma(\nu)$ is essentially determined by the Franck-Condon factors for the different transitions ($v_0 \rightarrow v_k$).

The total losses are determined by resonator losses (mirror transmission and absorption in optical components) and by absorption losses in the active dye medium. The latter are mainly caused by two effects.

a) The intersystem crossing transitions $S_1 \rightarrow T_1$ not only diminish the population $N(v_0)$ and therefore the attainable inversion, but they also lead to an increased population $N(T_1)$ of the triplet state. The triplet absorption spectrum $T_1 \rightarrow T_m$ partly overlaps with the singlet fluorescence spectrum which results in additional absorption losses $N(T_1)\alpha_T(\nu)L$ for the dye laser radiation. Because of the long lifetimes of molecules in this lowest triplet state, which can only relax into the S_0 ground state by slow phosphorescence or by collisional deactivation, the population density $N(T_1)$ may become undesirably large. One therefore has to take care that these triplet molecules are removed from the active zone as quickly as possible. This may be accomplished by mixing "triplet-quenching additives" to the dye solution. These are molecules which quench the triplet population effectively by spin-exchange collisions enhancing the intersystem crossing rate $T_1 \rightarrow S_0$. Examples are O_2 or cyclo-octotetraene (COT). Another solution of the triplet problem is "mechanical quenching", used in cw dye lasers. This means that the triplet molecules are transported very rapidly through the active zone. The transit time should be much smaller than the triplet lifetime. This is achieved for example by fast flowing free jets, where the molecules pass the active zone in the focus of the pump laser in about 10^{-6} s.

b) For many dye molecules the absorption spectra $S_1 \rightarrow S_m$, corresponding to transitions from the optically pumped singlet state S_1 to still higher states S_m, partly overlap with the gain profile of the laser transition $S_1 \rightarrow S_0$. These inevitable losses often restrict the spectral range where the net gain is larger than the losses [7.31a].

The essential characteristics of dye lasers is their broad homogeneous gain profile. Under ideal experimental conditions the homogeneous broadening allows *all* excited dye molecules to contribute to the gain at a single frequency. This implies that under single-mode operation the output power should be not much lower than the multimode power (see Sects.5.7-9), provided that the selecting intracavity elements do not introduce large additional losses.

The experimental realizations of dye lasers use either flashlamps, pulsed lasers, or cw lasers as pumping sources. Recently several experiments on pumping of dye molecules in the gas phase by high-energy electrons have been reported [7.32a]. Up to now, however, no electron-pumped gas phase dye laser could be realized. With optical pumping by a N_2 laser, however, stimulated emission from various dyes in the vapor phase has been observed [7.32c]. We now present the most important types of dye lasers used in practice for high-resolution spectroscopy.

7.3.2 Laser-Pumped Pulsed Dye Lasers

Since the absorption bands of many laser dyes reach from the blue-green region down to the near ultraviolet, the N_2 laser at $\lambda = 337$ nm is well suited as a pumping source. Sometimes frequency-doubled neodymium YAG or ruby lasers are employed. In order to avoid triplet losses, the pumping time T_p should be shorter than $T_{IC} = 1/R(S_1 \rightarrow T_1)$, where R is the intersystem crossing rate. This demand is readily met by N_2 laser pulses ($T_p = 10^{-9} - 10^{-8}$ s) or by giant pulses from ruby and Nd-YAG lasers. Sometimes ultrashort pulses from mode-locked lasers (see Chap.11) are used.

The short wavelength $\lambda = 337$ nm of the nitrogen laser permits pumping of dyes with fluorescence spectra from the near uv up to the near infrared. The high pump power available from this laser sources allows us to achieve sufficient inversion even in dyes with lower quantum efficiency. The shortest wavelength of a dye laser reported so far is at $\lambda = 350$ nm [7.33]. Wavelength tuning can be accomplished with prisms, gratings, or Lyot-filters.

Various pumping geometries and resonator designs have been proposed or demonstrated [7.31]. In transverse pumping (Fig.7.14) the N_2 laser beam is focussed by a cylindrical lens into the dye cell. Since the absorption coefficient for the pump radiation is large, the pump beam is strongly attenuated and the maximum inversion in the dye cell is reached in a thin layer directly behind the entrance window along the focal line of the cylindrical lens. This geometrical restriction to a small gain zone gives rise to large diffraction losses and beam divergence.

In the longitudinal pumping schemes (Fig.7.15) the pump beam enters the dye laser resonator through one of the mirrors which are transparent for the pump wavelength. Although this arrangement avoids the drawback of nonuniform pumping, present in the transverse pumping scheme, most resonator designs today are based on transverse pumping because of the more convenient geometrical arrangement.

If wavelength selection is performed with a grating, it is preferable to expand the dye laser beam for two reasons.

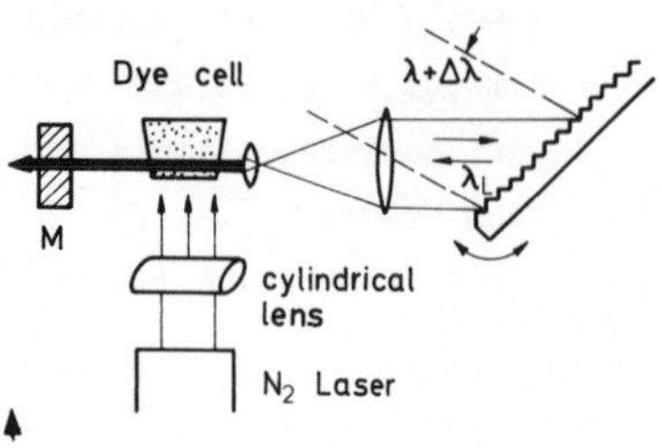

Fig.7.14. Transverse pumping of dye lasers, and beam expansion by a telescope

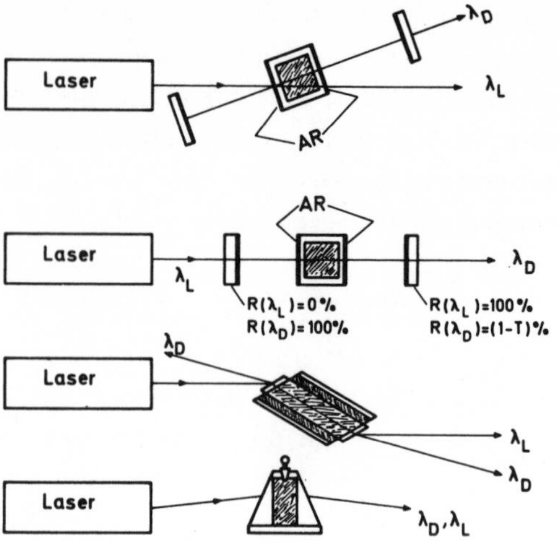

Fig.7.15. Longitudinal pumping configurations [7.31]

1) The resolving power of a grating is proportional to the product Nm of the number of grooves times the diffraction order m (see Sect.4.1). The more grooves are hit by the laser beam, the better is the resolution and the smaller is the resulting laser linewidth.

2) The power density without beam expansion might be high enough to damage the grating surface.

The enlargement of the beam can be accomplished either with a beam-expanding telescope (Hänsch type laser [7.34]) (Fig.7.14) or by using grazing incidence under an angle $\alpha \approx 90^{\circ}$ against the grating normal (Fig.7.16).

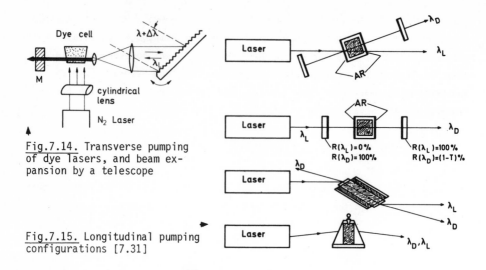

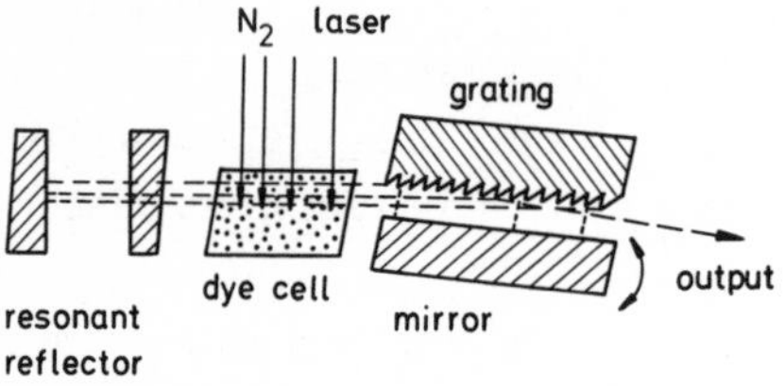

Fig.7.16. Short dye laser cavity with grazing incidence grating. Wavelength tuning is accomplished by turning the end mirror. The resonant reflector, composed of two wedges coated on their inner surfaces, acts as Fabry-Perot interferometer [7.36]

The latter arrangement [7.35] allows very short resonator lengths (below 10 cm). This has the advantage that even for short pump pulses the induced dye laser photons can make several transits through the resonator during the pumping time. A further, very important advantage is the large spacing $\delta\nu = c/2d$ of the resonator modes, which allows single-mode operation with only one etalon [7.36] or even without any etalon but with two gratings (Fig.7.17) [7.37]. At a wavelength λ the first diffraction order is reflected from the grazing incidence grating into the direction β determined by the grating equation (4.22)

$$\lambda = d(\sin\alpha + \sin\beta) \approx d(1 + \sin\beta) \quad , \tag{7.7}$$

for $d = 4 \times 10^{-5}$ cm (2500 lines/mm) and $\lambda = 400$ nm $\Rightarrow \beta = 0^{\circ}$, which means that the first diffraction order is reflected normal to the grating surface. The second grating is a *Littrow* grating (see Sect.4.1) which reflects the first diffraction order back into the incident direction. With the arrangement in Fig.7.17 a single-shot linewidth of less than 300 MHz and a time-averaged linewidth of 750 MHz have been achieved. Wavelength tuning is accomplished by tilting the end mirror M_2 or the Littrow grating. In order to avoid mode hops, the cavity length must be synchronously tuned. The single-mode conversion efficiency of the laser is 2% using rhodamine 6G dye [7.37].

Typical output powers of nitrogen laser pumped dye lasers range from one KW up to a few hundred KW. In the single-mode version of Fig.7.26, 1.5 KW output was achieved at 50 KW input with pump pulse durations of 12 ns. With oscillator-amplifier combinations the output power can be significantly increased.

The short duration ΔT of the laser output pulse puts a principal lower limit to the laser linewidth. Even with single-mode lasers the Fourier-limited bandwidth $\Delta\nu \geqq 1/(2\pi\Delta T)$ cannot be surpassed. For multimode operation the linewidth is of course still larger. It may be narrowed, however, by an external Fabry-Perot interferometer [7.40], which filters a small fraction of the spectral laser output profile. Figure 7.18 shows an arrangement with F.P.I. and grating inside the cavity and a further confocal F.P.I. filter outside the resonator. All wavelength-determining elements are synchronously tuned by controlling the pressure in a tank which contains these elements.

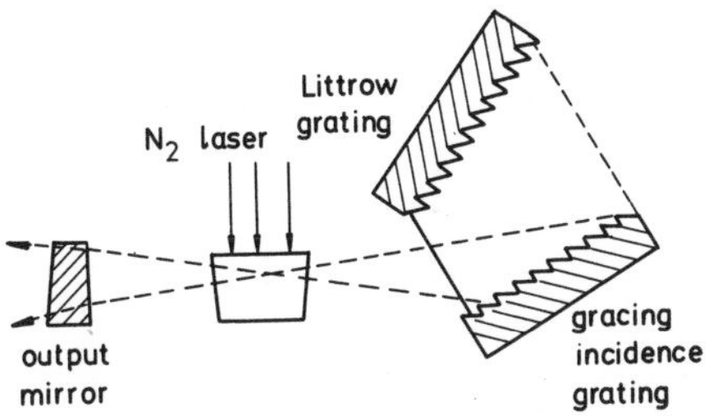

Fig.7.17. Single-mode dye laser with grazing incidence and Littrow grating. No etalon is needed [7.37]

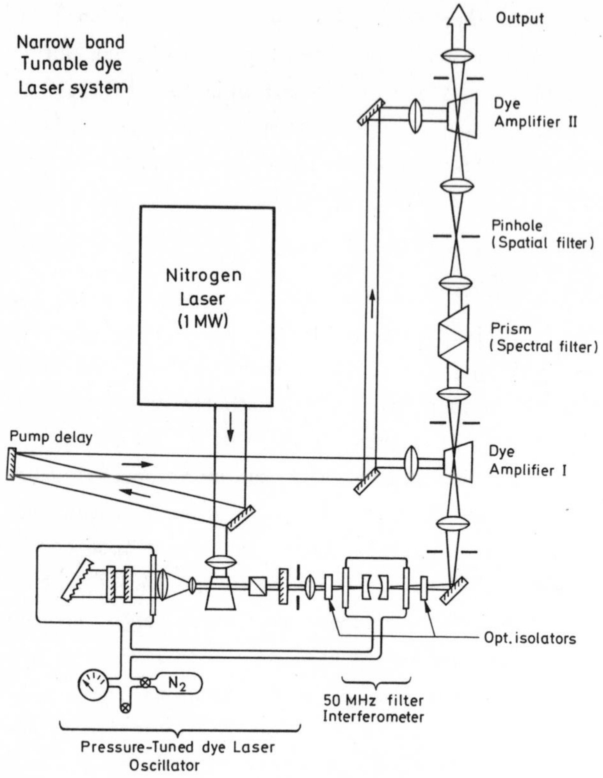

Fig.7.18. Pressure-tuned dye laser oscillator-amplifier system. The external confocal F.P.I. serves as narrow band spectral filter. The two amplifiers are pumped by the same N_2 laser [7.40]

The advantage of the long cavity in Fig.7.18 is the lack of discrete axial cavity mode structure since the transit time through the cavity is longer than the pump pulse of 10 ns. This allows smooth continuous tuning of the laser wavelength without tracking of the laser cavity length. For nitrogen as a scan gas, a tuning rate of 188 MHz/torr (142.5 GHz/atm) is achieved at λ = 600 nm [7.38].

The external confocal F.P.I. (free spectral range $\delta\nu$ = 2 GHz, finesse 200) serves as an ultranarrow passband filter with a linewidth of about $\Delta\nu_F$ = 10 MHz. The pulse duration of the transmitted laser intensity is of course lengthened to $T_E = 1/(2\pi\Delta\nu_F)$. The F.P.I. acts like a resonator with the transient time T_E. If the linewidth $\Delta\nu_L$ of the laser output incident onto the F.P.I. is much larger than $\Delta\nu_F$, only the small fraction $\Delta\nu_F/\Delta\nu_L$ is being transmitted through the filter. The intensity loss can be compensated by one or several amplifiers which consist of dye cells pumped by the same N_2 laser as the dye oscillator (Fig.7.18). Saturation of the amplification factor may be profitably used for intensity stabilization of the dye laser radiation [7.39]. Peak powers of 50 KW at linewidths down to 6×10^{-4} Å ($\hat{=}$ 25 MHz at λ = 500 nm) have been generated with such a system [7.40] pumped by a 1 MW N_2 laser.

7.3.3 Flashlamp-Pumped Dye Lasers

Flashlamp-pumped dye lasers [7.41] have the advantage that they do not need any expensive pump laser. Figure 7.19 shows a commonly used pumping arrangement. The linear flashlamp, filled with xenon, is placed in one of the focal lines of an elliptical cylindric reflector and the flowing liquid dye solution is pumped through a glass tube in the second focal line. The useful maximum

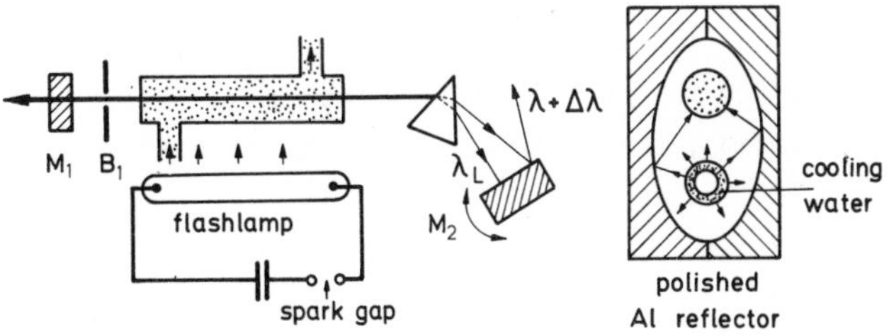

Fig.7.19. Elliptical reflector geometry for pumping of a flowing dye solution by a linear Xe flashlamp

pumping time is again limited by the triplet conversion rate. Low-inductance capacitor circuits have been designed to achieve short flashlamp pulses below 1 μs. By using additives as triplet quenchers, the triplet absorption is greatly reduced and long pulse emission has been obtained. A pulse-forming network of several capacitors is superior to the single energy storage capacitor because it matches the circuit impedance to that of the lamps, and a constant flashlight intensity over a period of 60-70 μs can be achieved. With two linear flashlamps in a double elliptical reflector a reliable Rhodamine 6G laser with 60 μs pulse duration, a repetition rate up to 100 Hz, and an *average* power of 4 W has been demonstrated [7.42]. With the pumping geometry of Fig.7.20 a very high collection efficiency for the pump light is achieved. The light rays parallel to the plane of the figure are collec-

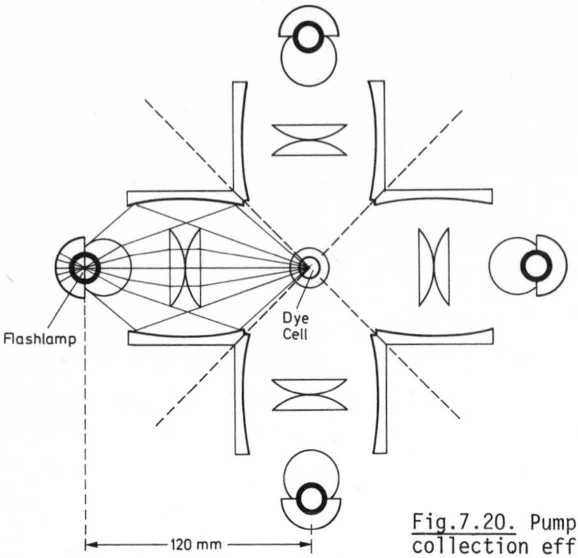

Fig.7.20. Pumping geometry with high light collection efficiency [7.43]

ted into an angle of about 85° by the rear reflector, the aplanatic lens directly in front of the flashlamp, the condensor lens, and the cylindrical mirrors. An average laser output capability of 100 W is possible with this design [7.43].

Similar to the laser-pumped dye lasers, reduction of the linewidth and wavelength tuning can be accomplished by prisms, gratings, interference filters [7.44], Lyot filters [7.45], and interferometers [7.46]. A combination of Lyot filter and F.P.I., composed of a birefringent crystal with mirror coatings on both end faces, inserted between two polarizers parallel to each other, allows tuning with a higher finesse and therefore smaller linewidth [7.47]. With an electro-optically tunable Lyot filter (see Sect. 4.2.11) in combination with a grating, a spectral bandwidth of below 0.01 Å was achieved [7.45] where the laser wavelength could be conveniently tuned by the voltage applied to the birefringent KDP crystals.

Electro-optic tuning [7.48] has the advantage that the laser wavelength can be tuned in a short time over a large spectral range [7.49]. This is of particular importance for the spectroscopy of fast transient species, such as radicals formed in intermediate stages of chemical reactions. The single-element electro-optical birefringent filter can be used to tune a flashlamp-pumped dye laser across the entire dye emission band. The relatively large linewidth can be drastically reduced without losing the tunability by injecting the output of an electro-optically tuned cw dye laser into the flashlamp-pumped laser [7.50].

One drawback of flashlamp-pumped dye lasers is the bad optical quality of the dye solution during the pumping process. Local variations of the refractive index due to schlieren in the flowing liquid, and temperature gradients due to absorption of the pump light deteriorate the optical homogenity. The frequency jitter of narrow-band flashlamp-pumped dye lasers is therefore generally larger than the linewidth obtained in a single shot. With three F.P.I. inside the laser cavity, single-mode operation of a flashlamp-pumped dye laser has been reported [7.50a]. The linewidth achieved was 4 MHz, stable to within 12 MHz.

7.3.4 Continuous Wave Dye Lasers

Besides color center lasers the tunable single-mode cw dye laser is probably the most promising laser type for sub-Doppler laser spectroscopy. Great efforts have therefore been expended in many laboratories to increase output power, tuning range, and frequency stability. Various resonator configurations, pump geometries, and designs of the dye flow system have been successfully tried to realize optimum dye laser performance. In this section we can only present some examples of dye laser arrangements used in high-resolution spectroscopy.

Figure 7.21 illustrates two possible resonator configurations. The pump beam from an argon or krypton laser enters the resonator either collinearly through the semitransparent mirror M_1 and is focussed by L_1 into the dye, or pump beam and dye laser beam are separated by a prism. In another commonly used arrangement the pump beam crosses the dye medium under a small angle against the resonator axis (Fig.7.23b). The active zone consists of the dye solution streaming in a laminar flow of about 1 mm thickness between two flat windows, or in a free jet which is formed through a carefully designed polished nozzle. At flow velocities of 10 m/s the time of flight for the dye molecules through the focus of the pump laser (about 10 μm) is about 10^{-6} s. During this short time the intersystem crossing rate cannot build up a large triplet concentration and the triplet losses are therefore small. The threshold pump power depends on the size of the pump focus and on the resonator losses, and varies between 1 mW and 1 W. Pump efficiencies (dye laser output/pump power input) up to $\eta = 25\%$ have been achieved yielding dye output powers of 1 W at 4 W pump power.

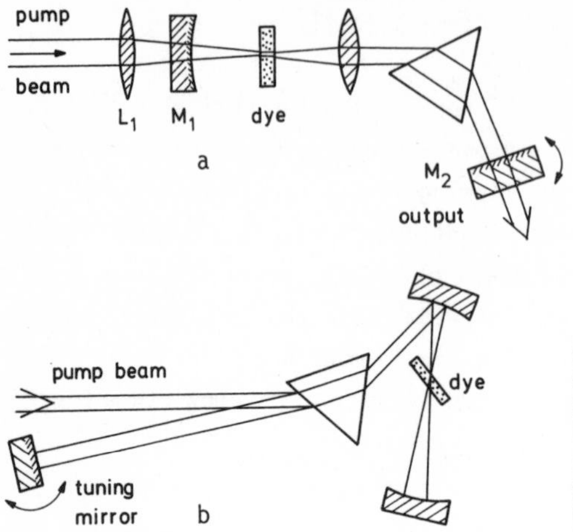

Fig.7.21. Two possible resonator configurations used for cw dye lasers: a: collinear pumping geometry b: folded astigmatically compensated resonator of the Kogelnik type [7.53], where a prism is used to separate pump beam and dye laser beam

For free-running dye jets the viscosity of the liquid solvent must be sufficiently large to ensure laminar flow necessary for a high optical quality of the gain zone. Most jet-stream dye lasers use high alcohols such as ethylene glycol or propylene glycol as solvents. Since these alcohols decrease the quantum efficiency of several dyes and also do not have optimum thermal properties, the use of water-based dye solutions with viscosity-raising additives can improve power efficiency and frequency stability of jet stream cw dye lasers [7.51]. Output powers of more than 30 W have been reported for cw dye lasers [7.52].

In order to achieve a symmetric beam waist profile of the dye laser mode in the active medium, the astigmatism produced in the folded cavity design by the spherical folding mirror has to be compensated by the plane parallel liquid slab of the dye jet, which is tilted under the Brewster angle against the resonator axis [7.53]. The folding angle for optimum compensation depends on the optical thickness of the jet and on the curvature of the folding mirror.

Coarse tuning of the dye laser wavelength can be accomplished with a Lyot filter (see Sect.4.8) or a prism inside the resonator. With a three-element Lyot filter a bandwidth $\Delta\lambda$ below 0.1 nm is obtained. If the Lyot filter is continuously turned by a slow gear drive, the wavelength of the dye laser can be continuously tuned across the whole gain profile, which amounts for Rhodamine 6G, for example, to about 90 nm. Such a dye laser represents a device with medium resolution which combines intense light source *and* monochromator. Figure 7.22 shows the tuning ranges for different

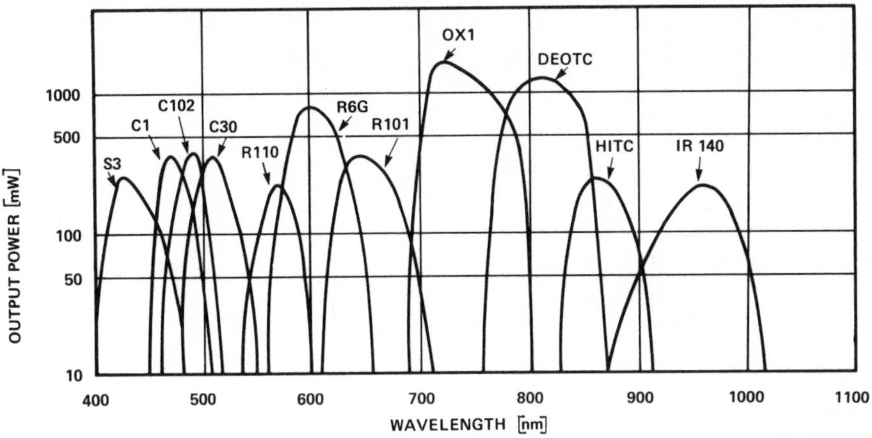

Fig.7.22. Spectral gain profiles of different laser dyes illustrated by the output power of cw dye lasers (from Coherent Radiation information sheet)

laser dyes, which illustrates that the whole spectral region between 400 and 900 nm is already accessible to cw dye lasers.

For single-mode operation additional selecting elements have to be inserted in the resonator (see Sect.6.5). In most designs two F.P.I. etalons with different free spectral ranges are used. Continuous tuning of the single-mode laser demands synchronous control of cavity length and transmission maxima of the selecting elements (see Sect.6.8). The optical path length of the cavity can be conveniently tuned by turning a tilted plane parallel glass plate inside the resonator (galvo plate). If the tilting range is restricted within a small interval around the Brewster angle, the reflection losses remain negligible. The piezo-tuned prism F.P.I. etalon in Fig.7.23 (see also Sect.6.5) with a free spectral range of about 10 GHz can be locked to the oscillating cavity eigenfrequency by a servo loop. If the transmission maximum ν_T of the F.P.I. is slightly modulated by an ac voltage fed to the piezo element, the laser intensity will show this modulation with a phase depending on the difference $\nu_c - \nu_T$ between the cavity resonance ν_c and the transmission peak ν_T. This phase-sensitive error signal can be used to keep the difference $\nu_c - \nu_T$ always zero. If only one F.P.I. is tuned synchronously with the cavity length, tuning ranges of about 30 GHz ($\hat{=}$ 1 cm^{-1}) can be covered without mode hops. For larger tuning ranges the second etalon and the Lyot filter must be also tuned synchronously. This demands a more sophisticated servo system.

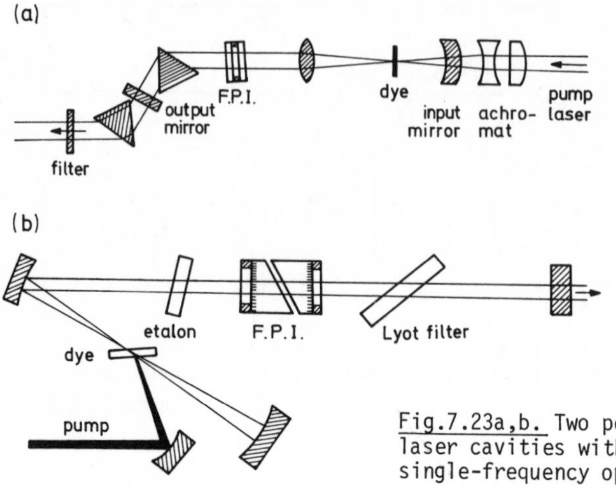

Fig.7.23a,b. Two possible designs of cw dye laser cavities with selecting elements for single-frequency operation

The frequency fluctuations of a single-frequency "free-running" laser without active stabilization system are mainly determined by density fluctuations of the dye jet and by mechanical jitter of the optical components, and the average "linewidth" is limited to about 10-100 MHz, depending on the special design. In order to achieve better stability, the laser frequency has to be locked to a stable but tunable reference (e.g., an external F.P.I.) (see Sect.6.6).

Figure 7.24 illustrates such a stabilized tunable "dye laser spectrometer" [7.55] with a frequency stability of better than 100 kHz. The intensity is stabilized to about 10^{-3} by a second servo loop acting upon the optical modulator. The reference cavity can be tuned in a very precise way by locking it with a controlled offset to the frequency of a He-Ne laser, stabilized onto an hfs component of an I_2 transition (see Sect.6.8).

Frequency tuning over at least 10 GHz has been achieved with a magnetic tuning system where the resonator output mirror is mounted on a loudspeaker membrane in a magnetic coil and is displaced linearly with the magnet current. The mechanical construction also allows synchronous tilting of the F.P.I. etalon [7.56].

7.3.5 Ring Dye Lasers

A severe drawback of standing wave resonator configurations for dye lasers is the spatial hole-burning effect (see Sect.5.8) which impedes single-mode operation and prevents all excited molecules within the active mode volume from contributing to the gain of a single mode. *Ring resonators* with uni-

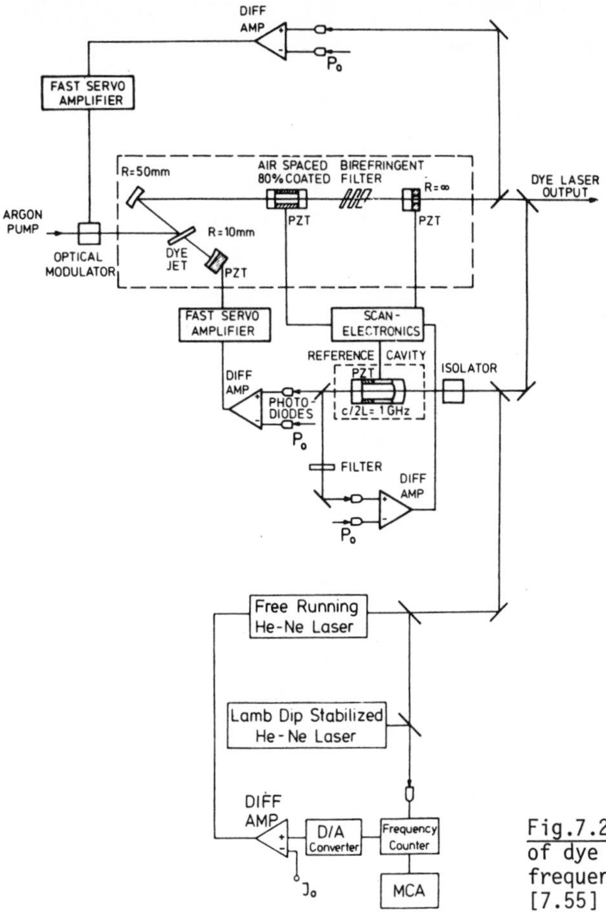

Fig.7.24. Schematic diagram of dye laser spectrometer and frequency calibration system [7.55]

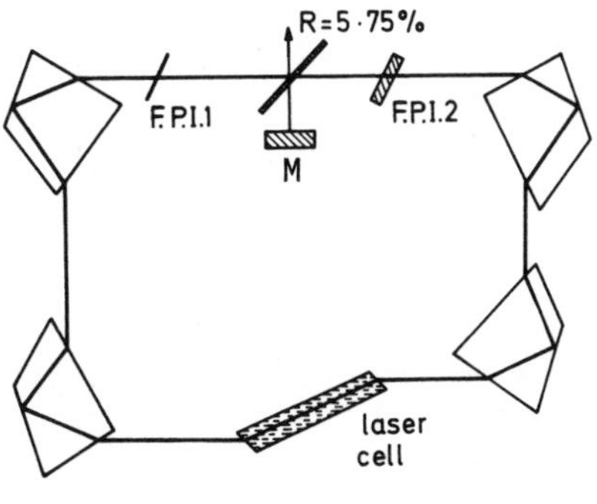

Fig.7.25. Ring laser resonator with Abbé prisms [7.58]

directional travelling waves overcome this drawback (see Sect.5.10). These ring lasers therefore allow single-mode operation with fewer selecting elements and higher output powers. Alignment and resonator design are, however, more sophisticated than in standard resonators of the Fabry-Perot type.

Figure 7.25 shows one of the early resonator designs [7.57], where the ring is formed by total internal reflection inside four Abbé prisms. The laser beam enters and leaves the prisms under the Brewster angle. Unidirectional operation is accomplished by reflection of the counterclockwise wave by the mirror M and tuning is performed by turning the prisms. Output coupling is achieved either through the beam splitter or by frustrated internal reflection devices [7.58a].

A better suppression of the unwanted direction can be achieved with an optical diode inside the ring resonator [7.58b]. This diode essentially consists of a Faraday rotator and a birefringent crystal which turns the Faraday rotation back to the linear polarization for the wave incident in one direction but increases the rotation for the other direction. The specific characteristics of a cw ring dye laser regarding output power and linewidth have been studied in [7.59]. A theoretical treatment of mode selection in Fabry-Perot type and in ring resonators can be found in [7.60]. Figure 7.26 shows the resonator configuration of commercially available ring dye lasers, which deliver more than 1 W single-mode output at 6 W pump power [7.61].

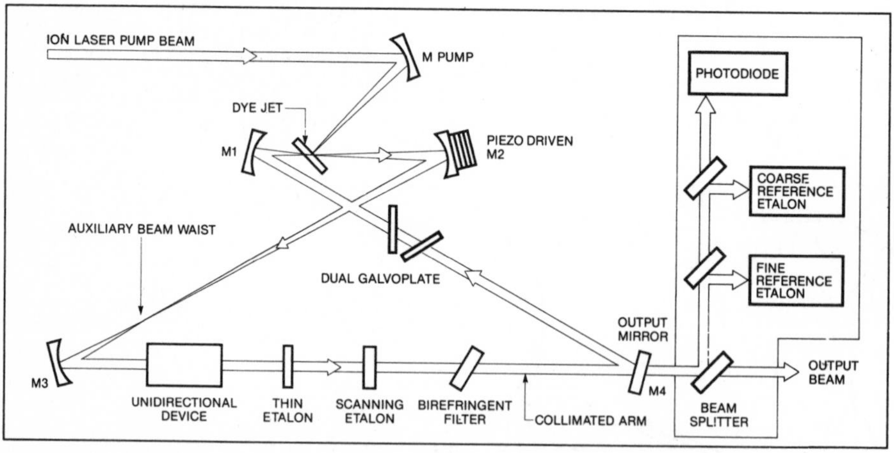

Fig.7.26. Ring dye laser (Courtesy Spectra Physics)

Table 7.2. Characteristic properties of dye lasers with different excitation sources

Pump	Tuning range [nm] with different dyes	Average output power [w]	Peak power [w]	Pulse duration [ns]	Linewidth [nm]	References
N_2 laser	350 - 1000	0.1 - 1	10^4-10^5	1 - 10	Fourier-limited	[7.37,38,40]
Excimer multiline laser	320 - 980	< 0.4	10^4-10^6	1 - 10	Fourier-limited	Lambda-Physics Information Sheet
Flash-lamp	400 - 800	0.1 - 100	10^5	10^2-10^5	multimode: 0.1-0.01 single-mode: 10^{-4}	[7.41,43,45,50]
cw argon laser	400 - 800	0.1 - 10	maximum cw power reported:40 W	cw	< 1 MHz if well stabilized	[7.54,55,59,61]
cw krypton laser	400 - 800	0.1 - 1	-	cw		
YAG laser λ/2=530 nm λ/3=355 nm	400 - 800	0.1 - 1	10^4-10^6	5 - 30	0.01	

Ring lasers have the additional advantage that by proper choice of mirror radii and distances a second beam waist may be generated at a convenient location within the ring resonator where a nonlinear crystal (such as ADA = $NH_4H_2AsO_4$ ammonium dihydrogen arsenide) can be placed for efficient intracavity frequency doubling. This gives a tunable ultraviolet source around λ = 250-300 nm (see Sect.7.5). Output powers of 50 mW multimode power and 4 mW single-mode uv power at argon laser pump powers of 7 W have already been achieved [7.62,62a].

7.4 Excimer Lasers

Excimers are molecules which are bound in excited states but are unstable in their electronic ground states. Examples are diatomic molecules composed of closed shell atoms with $^1S_0$ ground states such as the rare gases, which form stable *excited dimers* He_2^*, Ar_2^*, etc., but have a mainly repulsive potential in the ground state with a very shallow van der Waals minimum (Fig.7.27). The well depth ε of this minimum is small compared to the thermal energy kT at room temperature, which prevents a stable formation of ground state molecules. Mixed excimers such as KF or XeNa can be formed from combinations of closed shell-open shell atoms (e.g., $^1S + {}^2S$, $^1S + {}^2P$, $^1S + {}^3P$, etc.) which lead to repulsive ground state potentials [7.63].

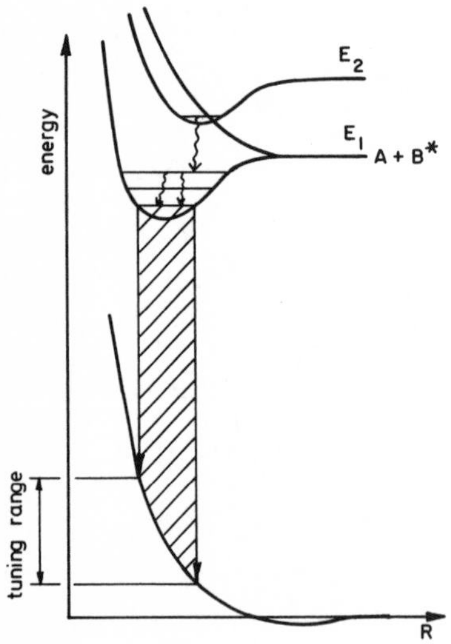

Fig.7.27. Schematic potential diagram of an excimer molecule

These excimers are ideal candidates for forming the active medium of tunable lasers since inversion between the upper bound state and the dissociating lower state is automatically maintained. The output power of these excimer lasers mainly depends on the excitation rate of the *upper* state. The lower state dissociates very rapidly ($\approx 10^{-12} - 10^{-13}$ s) and the frequently occurring bottleneck caused by a small depletion rate of the lower laser level is here prevented.

The tunability range depends on the slope of the repulsive potential and on the internuclear positions r_1 and r_2 of the classical turning points in the excited vibrational levels. The spectral gain profile is determined by the Franck-Condon factors for bound-free transitions. The corresponding intensity distribution $I(\omega)$ of the fluorescence shows a modulatory structure reflecting the r dependence $|\psi_{vib}(r)|^2$ of the vibrational wave function in the bound level [7.64] (see Figs.2.14 and 7.27).

The gain of the active medium at a frequency $\omega = (E_k - E_i)/\hbar$ is, according to (5.2), given by

$$\alpha(\omega) = [N_i - (g_i/g_k)N_k]\sigma(\omega) \quad , \qquad (7.8)$$

where the absorption cross section $\sigma(\omega)$ is related to the spontaneous transition probability $A_{ki} = 1/T_k$ (see [7.70]) by

$$\int_{\omega_1}^{\omega_2} \sigma(\omega)d\omega = (\lambda/2)^2 A_{ki} = (\lambda/2)^2/T_k \quad . \qquad (7.9)$$

Because of the broad spectral range $\Delta\omega = \omega_1 - \omega_2$ the cross section $\sigma(\omega)$ may be very small in spite of a large overall transition probability indicated by the short upper state lifetime T_k. Consequently a high population density N_k is necessary to achieve sufficient gain. Since the pumping rate R_p has to compete with the spontaneous transition rate which is proportional to the third power of the transition frequency ω, the pumping power $R_p\hbar\omega$ at laser threshold scales at least as the fourth power of the lasing frequency. Short wavelength lasers therefore require high pumping powers.

Pumping sources are provided by high-voltage-high-current electron beam sources, such as the FEBATRON [7.65] or by fast transverse discharges [7.66]. Since the excitation of the upper excimer states needs collisions between excited atoms (remember that there are no ground state excimer molecules), high densities are required to form a sufficient number N^* of excimers in the upper state by collisions between excited atoms or molecules. These high pressures impede a uniform discharge along the whole active zone in the

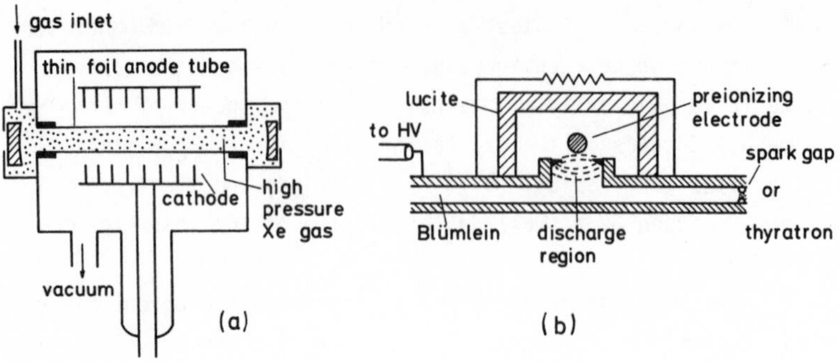

Fig.7.28. (a) Schematic diagram of an electron beam pumped laser. (b) Electric circuit and discharge zone of an excimer laser pumped by a fast transverse discharge

channel. Preionization by fast electrons or by ultraviolet radiation is required to achieve a sufficiently large and uniform density of excimers.

Up to now the rare gas halide excimers, such as KrF, ArF, or XeCl, form the active medium of the most advanced uv excimer lasers. Similar to the nitrogen laser, these rare gas halide lasers can be pumped by fast transverse discharges and lasers of this type are already on the commercial market. We illustrate their performance by taking the KrF laser as example.

The active medium of the KrF laser is a mixture of krypton, molecular fluorine F_2, and helium or argon as buffer gas. The whole gas system can be operated in a closed cycle where the gas is pumped from a reservoir through the laser channel. In the fast discharge, excited Kr and Ar atoms and ions are formed by electron impact. The excimers are formed by several reactions, for example,

$$
\begin{aligned}
&Ar^* + F_2 \rightarrow ArF^* + F \\
&Ar^+ + F^- + Ar \rightarrow ArF^* + F + Ar \\
&Kr^+ + F_2 \rightarrow KrF^* + F \\
&Kr^+ + F^- + Ar \rightarrow KrF^* + F + Ar \quad .
\end{aligned}
\tag{7.10}
$$

Since the density of the argon buffer gas is higher than that of krypton, more ArF excimers are formed. By a displacement reaction,

$$Kr + ArF^* \rightarrow KrF^* + Ar \quad , \tag{7.11}$$

KrF^* excimers can be efficiently produced. All these formation processes occur very rapidly in times typically of the order of 10^{-8} s and are very efficient in populating the upper laser level in KrF^* excimers [7.68]. There are also quenching processes which depopulate this level by forming new products.

$$KrF^* + Kr + Ar \rightarrow Kr_2F^* + Ar$$

$$KrF^* + F_2 \text{ , Ar, Kr} \rightarrow \text{products} \quad .$$

The formation of the KrF^* excimers therefore has to occur rapidly enough to allow a sufficiently rapid buildup of the laser field to compete with the quenching processes.

At a pressure of 2.25 atm, output pulses with more than 10 MW peak power and pulse widths of 125 ns have been achieved. The efficiency of 15% is remarkably high [7.69]. Figure 7.28 illustrates schematically the electrical circuit and the discharge zone of an excimer laser pumped by a fast transverse discharge.

The bound-free transitions from the upper laser level into the dissociation continuum of the lower state are homogeneously broadened. Without wavelength-selecting elements the induced emission starts at the wavelength with the largest transition probability. Table 7.3 gives the wavelengths of these maxima for several excimer lasers.

With selecting elements inside the cavity, the bandwidth of laser emission can be narrowed and the wavelength can be tuned over the gain profile. The optical components must be transparent to the laser radiation which demands in the vuv region sapphire, lithium fluoride, or barium fluoride prisms. Figure 7.29 shows the optical arrangement of a Xe_2 laser with a tuning prism and an etalon of BaF_2, which allows a tuning range from 170 to 176 nm [7.70]. Table 7.3 compiles some characteristic properties of different excimer lasers which represent typical figures achieved up to now. The development of new and more efficient excimer systems is in rapid progress and these lasers will certainly be of great importance for uv and vuv spectroscopy and photochemistry. For a more detailed treatment see several reviews [7.63,71,71a] and the recently published book on excimer lasers, edited by RHODES [7.72].

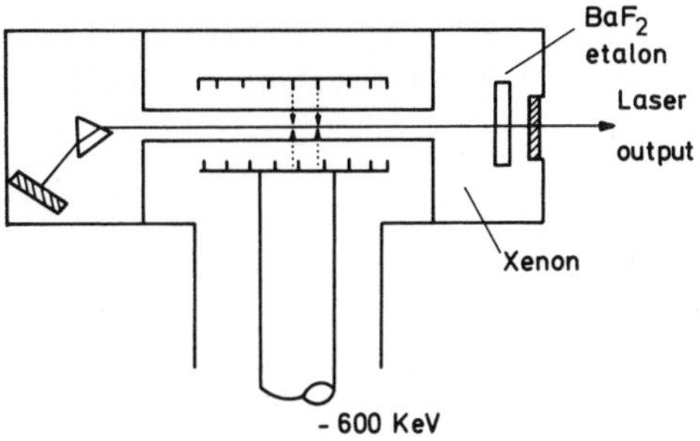

Fig.7.29. Optical arrangements of Xe_2 laser showing tuning prism and resonator mirrors [7.70]

Table 7.3. Survey on excimer laser characteristics

Excimer	Wavelength [nm]	Energy/pulse [mJ]	Peak power [MW]	Average power [W]	Excitation source
ArF	193	200	> 10	55	
KrCl	222	70	> 3,5	0.4	fast
KrF	248	350	> 15	6	
XeBr	282	17	> 2	0.1	transverse
XeCl	308	90	> 9	4	discharge
XeF	351	90	> 5	5	
Xe_2	170-175	30	5	-	high
Kr_2	146			-	energy
Ar_2	126			-	electron
XeF	351	1800/liter		-	beam

7.5 Nonlinear Optical Mixing Techniques

Besides the various types of tunable lasers, discussed in the foregoing sections, tunable coherent radiation sources have been developed which are based on the nonlinear interaction of intense radiation with atoms or molecules in crystals or in liquid and gaseous phases. Second harmonic generation, sum or difference frequency generation, parametric processes, or stimulated Raman scattering are examples of such nonlinear optical mixing techniques, which have succeeded in covering the whole spectral range from the vacuum ultraviolet (vuv) up to the far infrared with sufficiently intense tunable coherent sources. After a brief summary of the basic physics of these devices we exemplify their applications by presenting some experimentally realized systems (see also [7.72a]).

7.5.1 Physical Background

The dielectric polarization P of a medium with nonlinear susceptibility χ, subject to an electric field E, can be written as an expansion in powers of the applied field

$$P = \varepsilon_0[\chi^{(1)}E + \chi^{(2)}E^2 + \chi^{(3)}E^3 + \dots)] \quad , \tag{7.12}$$

where $\chi^{(k)}$ is the kth order susceptibility.

Consider an E.M. wave

$$E = E_1 \cos(\omega_1 t - k_1 z) + E_2 \cos(\omega_2 t - k_2 z) \qquad (7.13)$$

composed of two components incident on the nonlinear medium. The induced polarization at a fixed position (say z = 0) in the crystal is generated by the combined action of both components. The quadratic term $\chi^{(2)}E^2$ in (7.12) includes the contributions

$$E^2(z = 0) = E_1^2 \cos^2\omega_1 t + E_2^2 \cos^2\omega_2 t + 2E_1E_2 \cos\omega_1 t \cos\omega_2 t$$

$$= \frac{1}{2}(E_1^2 + E_2^2) + \frac{1}{2}E_1^2 \cos 2\omega_1 t + \frac{1}{2}E_2^2 \cos 2\omega_2 t$$

$$+ E_1E_2[\cos(\omega_1 + \omega_2)t + \cos(\omega_1 - \omega_2)t] \qquad (7.14)$$

which represent a dc polarization, ac components at the second harmonics $2\omega_1$, $2\omega_2$, and components at the sum or difference frequencies $\omega_1 \pm \omega_2$.

Taking into account that the field amplitudes E_1, E_2 are vectors and that the second-order susceptibility $\chi_{ijk}^{(2)}$ is a tensor with components depending on the symmetry properties of the nonlinear crystal [7.73], we can write the quadratic term in (7.12) in the explicit form for the components

$$P_i = \varepsilon_0 \sum_{j,k=1}^{3} \chi_{ijk}^{(2)} E_j E_k \quad , \quad i,j,k = 1,2,3 \quad , \qquad (7.15)$$

where P_i gives the ith component of the dielectric polarization $\underline{P} = \{P_x, P_y, P_z\}$.

Equation (7.15) shows that the components of the induced polarization P are determined by the tensor components χ_{ijk} and the components of the incident fields. For example, in KDP the only nonvanishing components of the susceptibility tensor are

$$\chi_{xyz}^{(2)} = d_{14} \;;\; \chi_{yxz}^{(2)} = d_{25} \text{ and } \chi_{zxy}^{(2)} = d_{36} \;;$$

which are often written in the reduced Voigt notation, using the symmetry conjecture that $\chi_{ijk} = \chi_{ikj}$ [7.74]. The components of the induced polarization are therefore with $d_{25} = d_{14}$

$$P_x = 2\varepsilon_0 d_{14}E_yE_z \;;\; P_y = 2\varepsilon_0 d_{14}E_xE_z \;;\; P_z = 2\varepsilon_0 d_{36}E_xE_y \;.$$

Suppose there is only one incident wave travelling in a direction $\underline{k}$ with a polarization vector $\underline{E}$ normal to the optical axis of an uniaxial birefringent crystal, which we choose to be the z axis (Fig.7.30). In this case

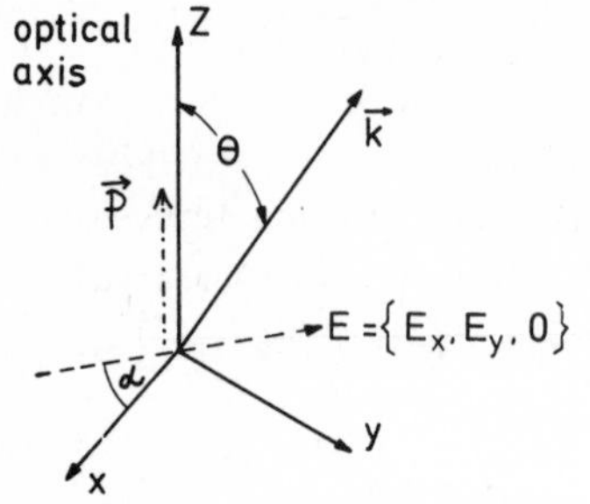

Fig.7.30. Coordinate system for description of nonlinear optics in a uniaxial birefringent crystal

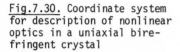

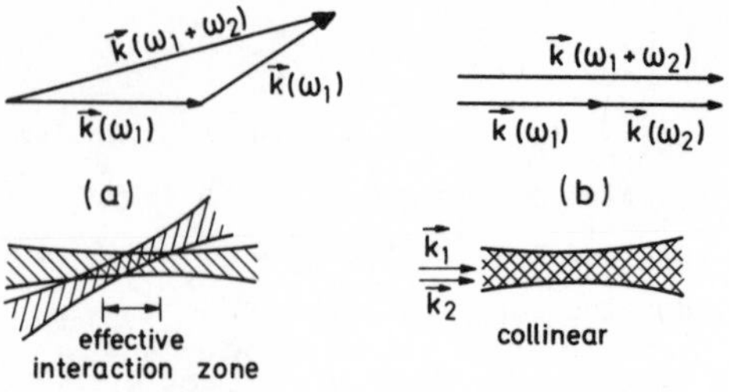

Fig.7.31. Phase matching condition for (a) noncollinear and (b) collinear propagation

$E_z = 0$ and the only nonvanishing component of $P(2\omega)$,

$$P_z(2\omega) = 2\varepsilon_0 d_{36} E_x(\omega) E_y(\omega) \quad ,$$

is perpendicular to the polarization plane of the incident wave.

The nonlinear polarization induced in an atom or molecule acts as a source of new waves at frequencies $\omega = \omega_1 \pm \omega_2$ which propagate through the nonlinear medium with a phase velocity $v_{ph} = \omega/k = c/n(\omega)$. However, the microscopic contributions generated by atoms at different positions (x,y,z) in the nonlinear medium can only add up to a macroscopic wave with appreciable intensity if the phase velocities of incident inducing waves and polarization waves are properly matched. This "phase-matching condition" can be written as

$$\underline{k}(\omega_1 \pm \omega_2) = \underline{k}(\omega_1) \pm \underline{k}(\omega_2) \quad , \tag{7.16}$$

which may be interpreted as momentum conservation for the three photons participating in the mixing process.

The phase-matching condition (7.16) is illustrated by Fig.7.31. If the angles between the three wave vectors are too large, the overlap region between focussed beams becomes too small and the efficiency of the sum or difference frequency generation decreases [7.75]. Maximum overlap is achieved for collinear propagation of all three waves. In this case we obtain with $c/n = \omega/k$, $\underline{k}_1 \,||\, \underline{k}_2 \,||\, \underline{k}_3$ and $\omega_3 = \omega_1 \pm \omega_2$ the condition

$$n_3\omega_3 = n_1\omega_1 \pm n_2\omega_2 \tag{7.17}$$

for the refractive indices n_1, n_2, and n_3.

This condition can be fulfilled in birefringent crystals which have two different refractive indices n_0 and n_e for the ordinary and the extraordinary waves. While the ordinary index n_0 does not depend on the propagation direction the extraordinary index n_e depends on both the directions of $\underline{E}$ and $\underline{k}$. The refractive indices can be illustrated by the index ellipsoid defined by the three principal axes of the dielectric susceptibility tensor. If these axis are aligned with the x, y, z axes we obtain

$$\frac{1}{\varepsilon_0}\left(\frac{x^2}{1+\chi_{xx}}+\frac{y^2}{1+\chi_{yy}}+\frac{z^2}{1+\chi_{zz}}\right)=1 \quad , \tag{7.18}$$

where $1+\chi_{ik}=\varepsilon_{ik}$. For uniaxial crystals, two of the principal axes are equal and the index ellipsoid becomes symmetric with respect to the optical axis. If we specify a propagation direction $\underline{k}$, we can describe the refractive indices n_0 and n_e experienced by the E.M. wave $E = E_0 \cos(\omega t - \underline{k}\cdot\underline{r})$ by a plane passing through the center of the index ellipsoid with its normal in the direction of $\underline{k}$ (Fig.7.32a). The trace of the refractive index on this

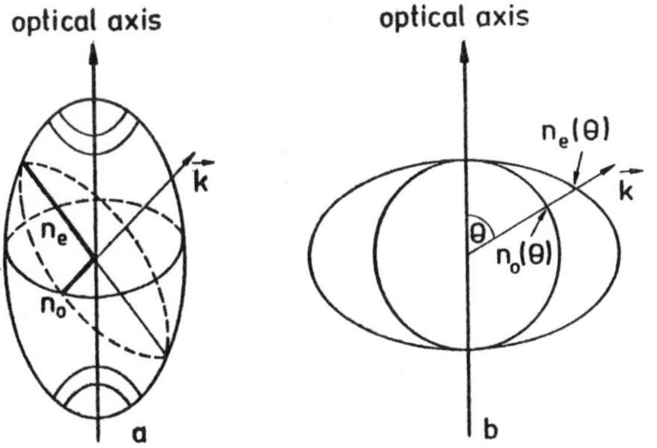

Fig.7.32. (a) Index ellipsoid and refraction indices n_0 and n_e for two modes of polarization in a plane perpendicular to the wave propagation. (b) Dependence of n_0 and n_e on the angle θ between optical axis and wavevector $\vec{k}$, for a positive birefringent uniaxial crystal

plane is an ellipse. Its two principal axes lie in the directions of the two normal modes of the electric induction $\underline{D} = \varepsilon{:}\underline{E}$ and their lenghts give the values n_0 and n_e of ordinary and extraordinary refractive indices. If the angle θ between $\underline{k}$ and the optical axis (which is assumed to coincide with the z axis) is varied, n_0 remains constant, while the extraordinary index changes according to (Fig.7.32b)

$$\frac{1}{n^2}=\frac{\cos^2\theta}{n_0^2}+\frac{\sin^2\theta}{n_e^2} \quad . \tag{7.19}$$

It is possible to find nonlinear crystals where the phase-matching condition (7.17) for collinear phase matching can be fulfilled if one of the three waves propagates as extraordinary wave and the others as ordinary waves through the crystal in a direction θ specified by (7.19) and (7.17) [7.76]. Let us now illustrate these general considerations by some specific examples.

7.5.2 Second Harmonic Generation (SHG)

For the case $\omega_1 = \omega_2 = \omega$ the phase-matching condition (7.16) for SHG becomes

$$k(2\omega) = 2k(\omega) \Rightarrow v_{ph}(2\omega) = v_{ph}(\omega) \quad , \tag{7.20}$$

which implies that *the phase velocities of incident and SH wave must be equal.* This can be achieved in a certain direction θ against the optical axis if in this direction the extraordinary refractive index $n_e(2\omega)$ for the SH wave equals the ordinary index $n_0(\omega)$ for the fundamental wave (Fig.7.33). When the incident wave propagates in this direction θ_p through the crystal, the local contributions of $P(2\omega,\underline{r})$ can all add up in phase and a macroscopic SH wave at the frequency 2ω can develop. The polarization direction of this SH wave is orthogonal to that of the fundamental wave.

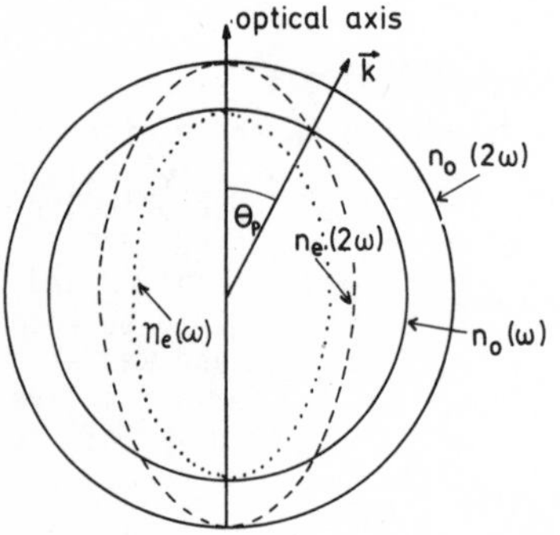

Fig.7.33. Index matching for SHG in a uniaxial negative birefringent crystal

In favorable cases phase matching is achieved for $\theta = 90^\circ$. This has the advantage that both the fundamental and the SH beam travel collinear through the crystal whereas for $\theta \neq 90^\circ$ the power flow direction of the extraordinary wave differs from the propagation direction $\underline{k}_e$, resulting in a decrease of the overlap region between both beams [7.73].

The nonlinear polarization $P(2\omega)$ generated at a position $\underline{r}$ by the driving field $E(\omega)\cos[\omega t - \underline{k}(\omega)\cdot r]$ can be deduced from (7.12) as

$$P(2\omega) = \varepsilon_0\chi^{(2)}_{eff}\, E^2(\omega)\cos\{[\underline{k}(\omega) - \underline{k}(2\omega)]\underline{r}\} \quad , \tag{7.21}$$

where the effective nonlinear coefficient $\chi_{eff}^{(2)}$ depends on the nonlinear crystal and the propagation direction. Assume the pump wave propagates in the z direction. If the field amplitude $E(2\omega)$ always remains small compared to $E(\omega)$ (low conversion efficiency), we may neglect the decrease of $E(\omega)$ with increasing z and obtain the total amplitude of the SH wave summed over the path length $z = 0$ to $z = L$ through the nonlinear crystal by integration over the microscopic contributions $dE(2\omega,z)$ generated by $P(2\omega,z)$. With (7.21), $\Delta k = |2\underline{k}(\omega) - \underline{k}(2\omega)|$ and $dE(2\omega) = [\omega/(\varepsilon_0 nc)]P(2\omega)$ (see [7.74]) one obtains

$$E(2\omega,L) = \int_{z=0}^{L} \chi_{eff}^{(2)}(\omega/nc)E^2(\omega)\cos(\Delta\underline{k}\cdot z)dz$$
$$= \chi_{eff}^{(2)}(\omega/nc)E^2(\omega)\,\frac{\sin\Delta kL}{\Delta k} \quad . \qquad (7.22)$$

The intensity $I = (nc\varepsilon_0/2)|E|^2$ of the SH wave is then

$$I(2\omega) = I^2(\omega)\,\frac{2\omega^2|\chi_{eff}^{(2)}|^2L^2}{n^3c^3\varepsilon_0}\;\frac{\sin^2(\Delta kL)}{(\Delta kL)^2} \quad . \qquad (7.23)$$

After the coherence length

$$L_{coh} = \pi/(2\Delta k) = \lambda/[4(n_{2\omega} - n_\omega)] \qquad (7.24)$$

the fundamental wave and the SH wave come out of phase and destructive interference begins which diminishes the amplitude of the SH wave. The difference $n_{2\omega} - n_\omega$ should therefore be sufficiently small to provide a coherence length larger than the crystal length L.

According to (7.23) the intensity $I(2\omega)$ of the SH wave is proportional to the square of the pump intensity $I(\omega)$. Most of the work on SHG has been therefore performed with pulsed lasers which offer high peak powers. The choice of the nonlinear medium depends on the wavelength of the pump laser and on its tuning range. For SHG of lasers around $\lambda = 1$ µm 90° phase matching can be achieved with $LiNbO_3$ crystals, while for SHG of dye lasers around $\lambda = 0.5 - 0.6$ µm KDP crystals or ADA can be used. Figure 7.34 illustrates the dispersion curves $n_0(\lambda)$ and $n_e(\lambda)$ of ordinary and extraordinary waves in KDP and $LiNbO_3$ which show that 90° phase matching can be achieved in $LiNbO_3$ for $\lambda_p = 1.06$ µm and in KDP for $\lambda_p \approx 515$ nm [7.77].

Focussing of the pump wave into the nonlinear medium increases the power density and therefore enhances the SHG efficiency. However, the resulting divergence of the focussed beam decreases the coherence length because the wave vectors $\underline{k}_p$ are spread out over an interval $\Delta\underline{k}_p$ which de-

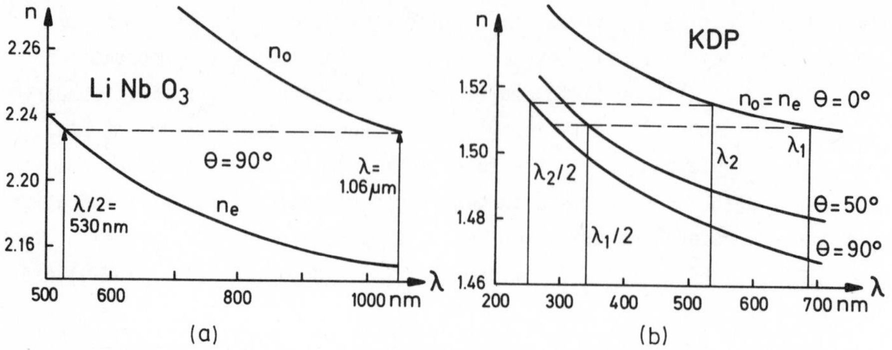

Fig.7.34. Refractive indices $n_0(\lambda)$ and $n_e(\lambda)$ in (a) $LiNbO_3$ (from [7.77]), and (b) KDP (from [7.76]). Collinear phase matching can be achieved in $LiNbO_3$ for $\theta = 90°$ and $\lambda = 1.06$ μm (Nd^+ laser), in KDP for $\theta = 50°$ at $\lambda = 694$ nm (ruby laser), or for $\theta = 90°$ at $\lambda = 515$ nm (argon laser)

pends on the divergence angle. The partial compensation of both effects leads to an optimum focal length of the focussing lens which depends on the angular dispersion $dn_e/d\theta$ of the refractive index n_e and on the spectral bandwidth $\Delta\omega_p$ of the pump radiation [7.75].

If the wavelength λ_p of the pump laser is tuned, phase matching can be maintained either by turning the crystal orientation θ against the pump beam propagation $\underline{k}_p$ (angle tuning) or by temperature control (temperature tuning) which relies on the temperature dependence $\Delta n(T) = n_e(\lambda,T) - n_0(\lambda,T)$. The tuning range $2\omega \pm \Delta_2\omega$ of the SH wave depends on that of the pump wave ($\omega \pm \Delta_1\omega$) and on the range where phase matching can be maintained. Generally $\Delta_2\omega < 2\Delta_1\omega$ because of the limited phase-matching range. With frequency-doubled pulsed dye lasers and different dyes the whole tuning range between $\lambda = 217$ to 450 nm can be completely covered. The strong optical absorption of most nonlinear crystals below 220 nm causes a low damage threshold, and the shortest wavelength achieved by SHG in currently available phase matching materials is $\lambda = 217$ nm [7.79].

Intracavity frequency doubling in KDP or ADA crystals placed at the auxiliary beam waist in cw dye laser ring resonators offers a convenient and efficient way to generate cw ultraviolet radiation with output powers in the 1-100 mW range, bandwidths below 10^{-3} Å, and wavelengths which are continuously tunable over 100 nm intervals. With different dyes the whole range between 220 and 440 nm can be covered [7.62].

7.5.3 Sum Frequency and Higher Harmonic Generation

In case of laser-pumped dye lasers it is often more advantageous to generate tunable uv radiation by optical mixing of the pump laser and the dye laser outputs rather than by frequency doubling of the dye laser. Since the intensity $I(\omega_1 + \omega_2)$ is proportional to the product $I(\omega_1)I(\omega_2)$, the larger intensity $I(\omega_1)$ of the pump laser allows enhanced uv intensity $I(\omega_1 + \omega_2)$. Furthermore it is often possible to choose the frequencies ω_1 and ω_2 in such a way that 90° phase matching can be achieved. The range $(\omega_1 + \omega_2)$ which can be covered by sum frequency generation is generally wider than that accessible to SHG. Radiation at wavelengths too short to be produced by frequency doubling can be generated by mixing of two different frequencies ω_1 and ω_2. This is illustrated by Fig.7.35 which shows possible wavelength combinations λ_1 and λ_2 which allow 90° phase-matched sum frequency mixing in KDP and ADP at room temperature or along the b axis of the biaxial KB5 crystal [7.80].

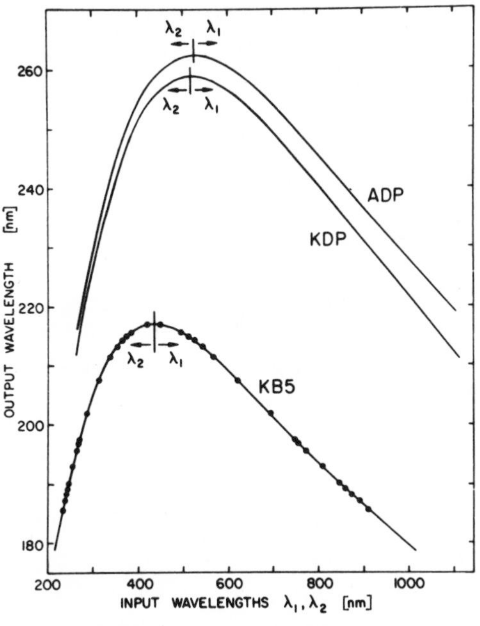

Fig.7.35. Possible combinations of wavelength pairs (λ_1,λ_2) which allow phase-matched sum-frequency generation in ADP, KDP, and KB5 for the phase matching conditions described in the text [7.80]

Some examples demonstrate experimental realizations of the sum frequency mixing technique.

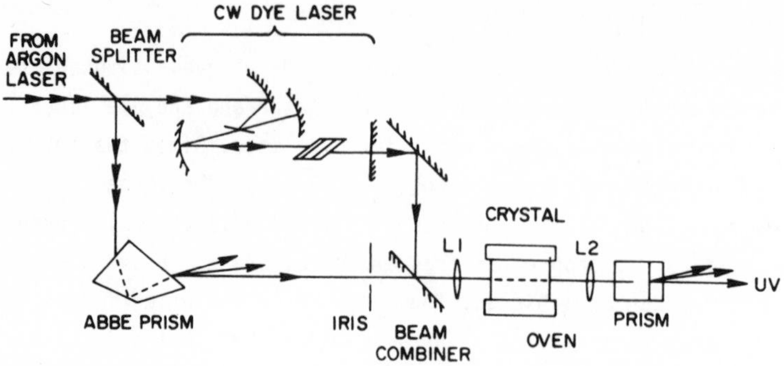

Fig.7.36. Experimental arrangement for sum frequency generation in KDP [7.81]

a) The output of a cw Rhodamine 6G dye laser is mixed with selected lines of the same argon laser which is used to pump the dye laser with 15 W on all lines (Fig.7.36). The superimposed beams are focussed into the temperature-stabilized KDP crystal. Tuning is accomplished by simultaneously tuning the dye laser wavelength and the orientation of the KDP crystal. The entire wavelength range 257-320 nm can be covered by using different argon lines but a single Rhodamine 6G dye laser without changing of dyes [7.81].

b) The generation of intense tunable radiation in the 240 - 250 nm range has been demonstrated by mixing in a temperature-tuned 90° phase-matched ADP crystal the second harmonic of a ruby laser with the output of an infrared dye laser pumped by the ruby laser's fundamental output [7.80].

c) uv radiation tunable between 208 and 234 nm has been generated efficiently by mixing the fundamental output of a neodymium YAG laser and the output of a frequency-doubled dye laser. Wavelengths down to 202 nm can be obtained with a refrigerated ADP crystal because ADP is particularly sensitive to temperature tuning [7.82].

Besides optical mixing in birefringent crystals sum frequency mixing or higher harmonic generation can be also achieved in homogeneous mixtures of rare gases and metal vapors. Because in centro-symmetric media the second-order susceptibility must vanish, SHG is not possible, but all third-order processes can be utilized for the generation of tunable ultraviolet radiation. Phase matching is achieved by a proper density ratio of rare gas atoms to metal atoms. Several examples illustrate the method.

a) Third harmonic generation of neodymium YAG laser lines around $\lambda = 1.05\ \mu m$ can be achieved in mixtures of xenon and rubidium vapor in a heat pipe.

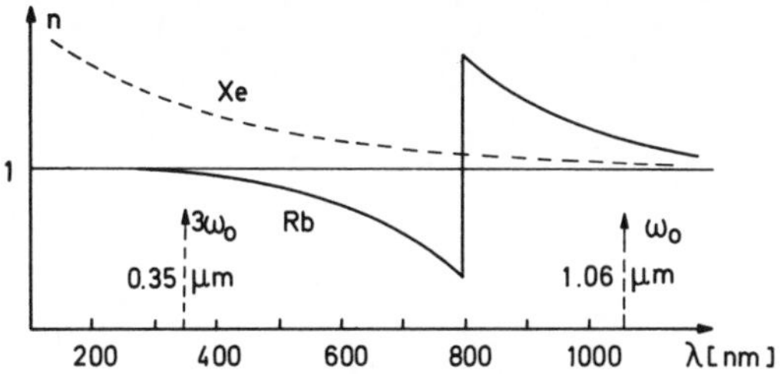

Fig.7.37. Schematic diagram of the refractive indices $n(\lambda)$ for rubidium vapor and xenon, illustrating phase matching for third harmonic generation

Figure 7.37 is a schematic diagram for the refractive indices $n(\omega)$ for Xe and rubidium vapor. Choosing the proper density ratio N(Xe)/N(Rb), phase matching is obtained for $n(\omega) = n(3\omega)$ where the refractive index $n = n(Xe) + n(Rb)$ is determined by the rubidium and Xe densities. Figure 7.37 illustrates that this method utilizes the compensation of the normal dispersion in Xe by the anomalous dispersion for rubidium [7.83].

Because of the lower densities of gases compared with solid crystals the efficiency $I(3\omega)/I(\omega)$ is much smaller than in crystals. However, there is no short wavelength limit as in crystals and the spectral range accessible by optical mixing can be extended far into the vuv range [7.83a,b]. The efficiency may be greatly increased by resonance enhancement, if for example a resonant two-photon transition $2\omega_1 = E_i \rightarrow E_k$ can be utilized as a first step of the sum frequency generation $\omega = 2\omega_1 + \omega_2$. This is demonstrated by an experiment shown in Fig.7.38. The orthogonally polarized outputs from two N_2 laser-pumped dye lasers are spatially overlapped in a Glan-Thompson prism. The collinear beams of frequencies ω_1 and ω_2 are then focussed into a heat pipe containing the atomic metal vapor. One laser is fixed at half the frequency of an appropriate two-photon transition and the other is tuned. For a tuning range of the dye lasers between 700 and 400 nm achievable with different dyes, tunable vuv radiation at frequencies $\omega = 2\omega_1 + \omega_2$ is generated which can be tuned over a large range [7.84]. Third harmonic generation can be eliminated in this experiment by using circularly polarized ω_1 and ω_2 radiation, since angular momentum will not be conserved for frequency tripling in an isotropic medium under these conditions. The sum frequency $\omega = 2\omega_1 + \omega_2$ corresponds to an energy level beyond the ionization limit.

The variety of possible frequency mixing and harmonic generation is illustrated by the work of Harris and his group [7.85]. Figure 7.39 shows the experimental arrangement of a "Photon Factory" with a mode-locked Nd:YAG laser and amplifier as the primary source of 1.06 µm radiation in the form

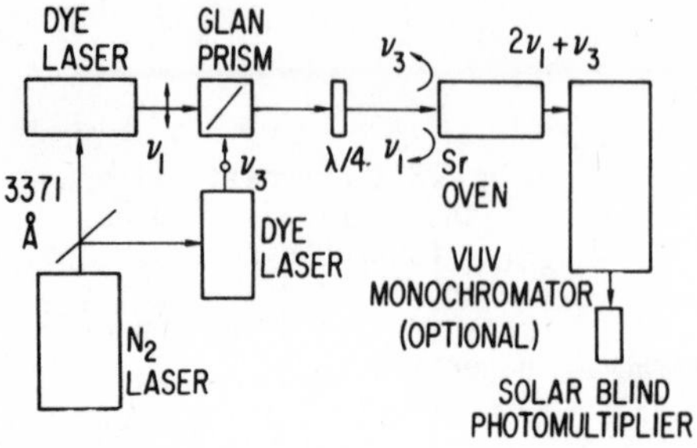

Fig.7.38. Generation of tunable vuv radiation by sum-frequency mixing $\omega = 2\omega_1 + \omega_2$ of two tunable dye lasers [7.84]

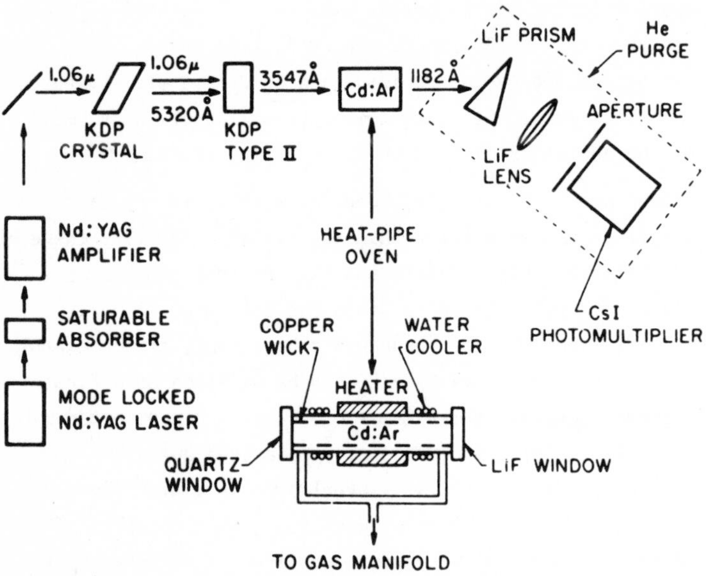

Fig.7.39. Experimental arrangement for optical mixing and harmonic generation to produce coherent tunable vuv radiation [7.86a]

of 50 ps pulses with peak powers of about 20 MW and pulse energies of 10 mJ. The primary radiation is frequency doubled by a KDP crystal to 532 nm with up to 80% efficiency. Sum frequency mixing of this SH wave with the fundamental gives radiation at $\lambda = 354.7$ nm while further frequency doubling yields 266 nm radiation. These high peak power radiations at four different wavelengths can now be used for further mixing experiments in rare gas-metal vapor cells. For example the tripling of 354.7 nm in a Cd-Ar mixture yields 118.2 Å radiation, tripling of 266 nm in a Xe-Ar mixture even allows generation of 88.7 nm radiation. If one of the involved radiation sources is tunable (e.g., a parametric oscillator or a dye laser pumped by one of the fixed frequency sources) tunable vuv radiation is generated [7.86a,b].

7.5.4 Difference Frequency Spectrometer

While generation of *sum frequencies* yields tunable ultraviolet radiation by mixing the output from two lasers in the visible range, the phase-matched generation of *difference* frequencies allows one to construct tunable coherent *infrared* sources. One example is the difference frequency spectrometer of PINE [7.87] which has proved to be very useful for high-resolution infrared spectroscopy.

Two collinear cw beams from a stable single-mode argon laser and a tunable single-mode dye laser are mixed in a $LiNbO_3$ crystal (Fig.7.40). For 90° phase matching of collinear beams the phase matching condition

$$\underline{k}(\omega_1 - \omega_2) = \underline{k}(\omega_1) - \underline{k}(\omega_2)$$

can be written as $|\underline{k}(\omega_1 - \omega_2)| = |\underline{k}(\omega_1)| - |\underline{k}(\omega_2)|$, which gives for the refractive index $n = c(k/\omega)$ the relation

$$n(\omega_1 - \omega_2) = \frac{\omega_1 n(\omega_1) - \omega_2 n(\omega_2)}{\omega_1 - \omega_2} \quad . \tag{7.25}$$

The whole spectral range of the difference spectrometer from 2.2 to 4.2 μm can be continuously covered by tuning the dye laser and the phase-matching temperature of the $LiNbO_3$ crystal ($-0.12^\circ C/cm^{-1}$). The infrared power is, according to (7.15) and (7.23), proportional to the product of the incident laser powers and to the square of the coherence length. For typical operating powers of 100 mW (argon laser) and 10 mW (dye laser) a few μW of infrared radiation is obtained. This is 10^4 to 10^5 times higher than the noise equivalent power of standard ir detectors.

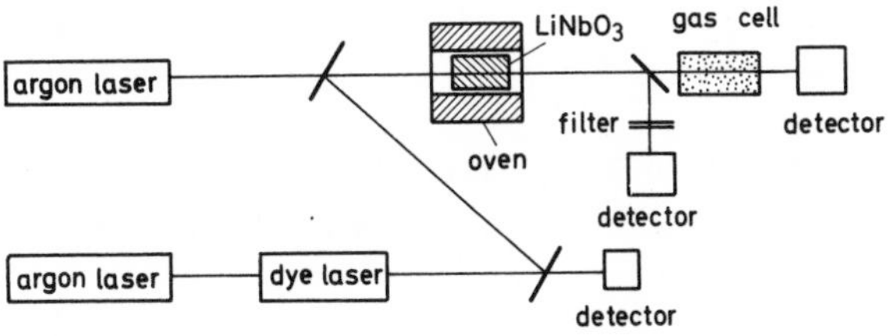

Fig.7.40. Difference-frequency spectrometer [7.87]

The spectral linewidth of the infrared radiation is determined by that of the two pump lasers. With frequency stabilization of the pump lasers, a linewidth of a few MHz has been reached for the difference frequency spectrometer. In combination with a multiplexing scheme, devised for calibration, monitoring, drift compensation, and absolute stabilization of the difference spectrometer, a continuous scan of 7.5 cm^{-1} has been achieved and a resetability of better than 10 MHz [7.88].

Pulsed difference frequency generation from the outputs of a ruby laser and a dye laser mixed in $LiNbO_3$ has achieved 6 KW of infrared power tunable between 3.1 and 4.5 μm [7.89]. Spectral narrowing of the dye laser output reduces the bandwidth to less than 1 cm^{-1}, and peak infrared powers of several hundred watts with repetition rates up to 30 s^{-1} have been generated with a long-term frequency stability of better than 1 GHz [7.90].

Of particular interest are tunable sources in the far infrared region where no microwave generators are available and incoherent sources are very weak. With selected crystals, such as proustite (Ag_3AsS_3) or HgS, phase matching for difference frequency generation can be achieved for the middle infrared. The search for new nonlinear materials will certainly enhance the spectroscopic capabilities in the whole infrared region.

7.6 Optical Parametric Oscillator (OPO)

The optical parametric oscillator [7.77,92] is based on the parametric interaction of a strong pump wave $E_p \cos(\omega_p t - \underline{k}_p \cdot \underline{r})$ with molecules in a crystal which has a sufficiently large nonlinear susceptibility. This interaction can be described as an inelastic scattering of a pump photon $\hbar\omega_p$ by a molecule where the pump photon is absorbed and two new photons $\hbar\omega_s$ and $\hbar\omega_i$ are generated. Because of energy conservation the frequencies ω_i and ω_s are related to the pump frequency ω_p by

$$\omega_p = \omega_i + \omega_s \quad . \tag{7.26}$$

Analogous to the sum frequency generation the parametrically generated photons ω_i and ω_i can add up to a macroscopic wave if the phase-matching condition

$$\underline{k}_p = \underline{k}_i + \underline{k}_s \tag{7.27}$$

is fulfilled, which may be regarded as the conservation of momentum for the three photons involved in the parametric process. Simply stated, parametric generation splits a pump photon into two photons which satisfies conservation of energy at every point in the nonlinear crystal. For a given wave vector $\underline{k}_p$ of the pump wave, the phase-matching condition (7.27) selects, out of the infinite number of possible combinations $\omega_1 + \omega_2$ allowed by (7.26), a single pair $(\omega_i, \underline{k}_i)$ and $(\omega_s, \underline{k}_s)$ which is determined by the orientation of the nonlinear crystal with respect to $\underline{k}_p$. The resultant two macroscopic waves $E_s \cos(\omega_s t - \underline{k}_s \cdot r)$ and $E_i \cos(\omega_i t - \underline{k}_i \cdot \underline{r})$ are called *signal* wave and *idler* wave. The most efficient generation is achieved for collinear phase matching where $\underline{k}_p \,||\, \underline{k}_i \,||\, \underline{k}_s$. For this case the relation (7.17) between the refractive indices gives

$$n_p\omega_p = n_s\omega_s + n_i\omega_i \quad . \tag{7.28}$$

If the pump is an extraordinary wave, collinear phase matching can be achieved for some angle θ against the optical axis, if $n_p(\theta)$, defined by (7.19), lies between $n_0(\omega_p)$ and $n_e(\omega_p)$.

The gain of signal and idler waves depends on the pump intensity and the effective nonlinear susceptibility. Analogous to the sum or difference frequency generation one can define a parametric gain coefficient

$$\Gamma^2 = \frac{\omega_i\omega_s|d^2||E_p|^2}{n_i n_s c^2} = \frac{2\omega_i\omega_s|d|^2 I_p}{n_i n_s n_p \varepsilon_0 c^3} \tag{7.29}$$

which is proportional to the pump intensity I_p and the square of the effective nonlinear susceptibility $|d| = \chi^{(2)}_{\text{eff}}$. For $\omega_i = \omega_s$ (7.29) becomes identical with the gain coefficient for SHG in (7.23).

If the nonlinear crystal which is pumped by the incident wave E_p is placed inside a resonator, oscillation on the idler or signal frequencies can start when the gain exceeds the total losses. The optical cavity may be resonant for both the idler and signal waves (doubly resonant oscillator) or only for one of the waves (singly resonant oscillator) [7.93].

Figure 7.41 shows schematically the experimental arrangement of a collinear optical parametric oscillator. Due to the much higher gains, pulsed operation is generally preferred where the pump is a Q-switched laser source. The threshold of a doubly resonant oscillator occurs when the gain equals the product of the signal and idler losses. If the resonator mirrors have high reflectivities for both the signal and idler waves, the losses are small and even cw parametric oscillators can reach threshold. For singly resonant cavities, however, the losses for the nonresonant waves are high and the threshold increases.

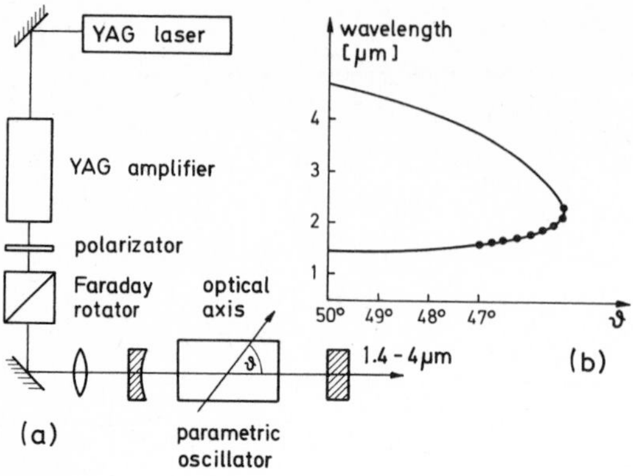

Fig.7.41a,b. Optical parametric oscillator. (a) Schematic diagram of experimental arrangement, (b) pairs of wavelengths (λ_1, λ_2) for collinear phase matching as a function of ϑ (after [7.91])

Example

For a 5 cm long 90° phase-matched $LiNbO_3$ crystal pumped at $\lambda_p = 0.532\ \mu m$, threshold is at 38 mW pump power for the doubly resonant cavity with 2% losses at ω_i and ω_s. For the singly resonant cavity threshold increases by a factor of 100 to 3.8 W [7.77].

Tuning of the OPO can be accomplished either by crystal rotation or by controlling the crystal temperature. The tuning range of a $LiNbO_3$ OPO, pumped by various frequency-doubled wavelengths of a Q-switched Nd:YAG laser, extends from 0.55 to about 4 μm. Turning the crystal orientation by 4° already covers a tuning range between 1.4 and 4.4 μm (see Fig.7.41b). Figure 7.42 shows temperature tuning curves for idler and signal generated in $LiNbO_3$ by different pump wavelengths. Angle tuning has the advantage of faster tuning rates than temperature tuning.

The frequency stability of idler and signal wave is influenced by that of the pump wave and by changes of the refractive index caused by temperature drifts. The bandwidth is determined by the dispersion $dn/d\omega$ and by the pump power. Typical figures are 0.1-5 cm^{-1}. Detailed spectral properties depend on the longitudinal mode structure of the pump and on the resonator mode spacing $\Delta\nu = (c/2L)$ for idler and signal standing waves. For the singly resonant oscillator the cavity has to be adjusted to only one frequency, while the nonresonant frequency can be adjusted so that $\omega_p = \omega_i + \omega_s$ is satisfied.

With a tilted etalon inside the resonator of a singly resonant cavity, single-mode operation can be achieved. Frequency stability of a few MHz has been demonstrated [7.94].

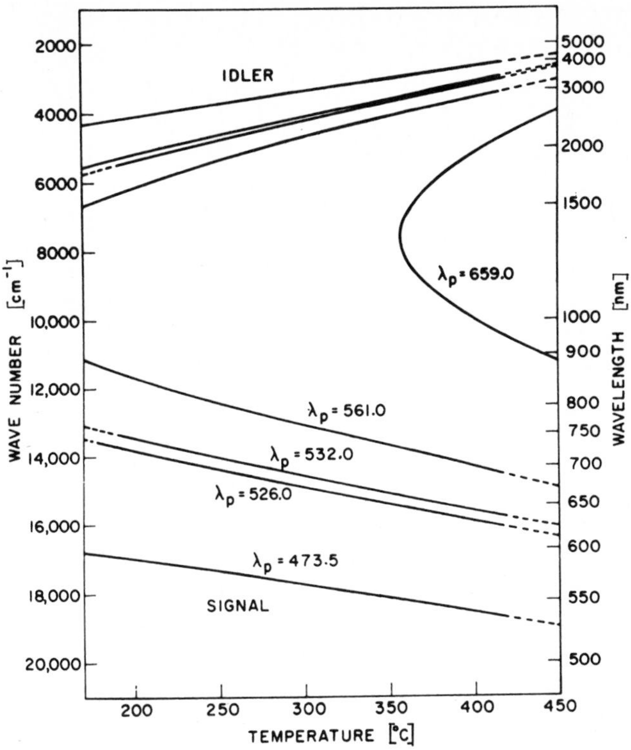

Fig.7.42. $LiNbO_3$ temperature tuning curves for doubled Nd:YAG laser pump wavelengths [7.77]

7.7 Tunable Raman Lasers

The tunable "Raman laser" may be regarded as a parametric oscillator based on stimulated Raman scattering. Since stimulated Raman scattering will be discussed in more detail in Sect.9.2, here we summarize only very briefly the basic concept of these devices.

The ordinary Raman effect can be described as an inelastic scattering of pump photons $\hbar\omega_p$ by molecules in the energy level E_i. The energy loss $\hbar(\omega_p - \omega_s)$ of the scattered "Stokes photons" $\hbar\omega_s$ is converted into excitation energy (vibrational, rotational, or electronic energy) of the molecules.

$$\hbar\omega_p + M(E_i) \rightarrow M^*(E_f) + \hbar\omega_s \quad ,$$

where $E_f - E_i = \hbar(\omega_p - \omega_s)$. For the vibrational Raman effect this process can be interpreted as parametric splitting of the pump photon $\hbar\omega_p$ into a Stokes photon $\hbar\omega_s$ and an optical phonon $\hbar\omega_v$ representing the molecular vibrations. The contributions $\hbar\omega_s$ from all molecules in the interaction region can add up to macroscopic waves, when the phase-matching condition

$$\underline{k}_p = \underline{k}_s + \underline{k}_v$$

for pump wave, Stokes wave, and phonon wave is fulfilled. In this case a strong Stokes wave $E_s \cos(\omega_s t - \underline{k}_s \cdot \underline{r})$ develops with a gain depending on the pump intensity and the Raman scattering cross section. If the active medium is placed in a resonator, oscillation arises on the Stokes component as soon as the gain exceeds the total losses. Such a device is called a Raman oscillator or "Raman laser", although, strictly speaking it is not a laser but a parametric oscillator.

Those molecules which are initially in *excited* vibrational levels can give rise to superelastic scattering of "anti-Stokes radiation", which has gained energy $(\hbar\omega_s - \hbar\omega_p) = (E_i - E_f)$ from the deactivation of vibrational energy.

The Stokes and the anti-Stokes radiation has a constant frequency shift against the pump radiation, which depends on the vibrational eigenfrequencies ω_v of the molecules in the active medium.

$$\omega_s = \omega_p - n\omega_v \quad , \quad n = 1,2,3 \ldots$$

$$\omega_{as} = \omega_p + n\omega_v \quad .$$

Tunable lasers as pumping sources therefore allow one to transfer the tunability range $(\omega_p \pm \Delta\omega)$ into other spectral regions $(\omega_p \pm \Delta\omega \pm n\omega_v)$.

Stimulated Raman scattering (SRS) of dye laser radiation in hydrogen gas can cover the whole spectrum between 185 and 880 nm without any gap, using three different laser dyes and frequency doubling the dye laser radiation [7.95]. A broadly tunable ir waveguide Raman laser pumped by a dye laser can cover the infrared region 0.7 to 7 μm without gaps, using SRS up to third Stokes order $(\omega_s = \omega_p - 3\omega_v)$ in compressed hydrogen gas. Energy conversion efficiencies of several percent are possible and output powers in excess of 80 KW for the third Stokes component $(\omega_p - 3\omega_v)$ have been achieved [7.96]. Figure 7.43 shows part of the experimental arrangement of an ir waveguide Raman laser.

For infrared spectroscopy Raman lasers pumped by the numerous intense lines of CO_2, CO, HF, or DF lasers may be advantageous. Besides the vibra-

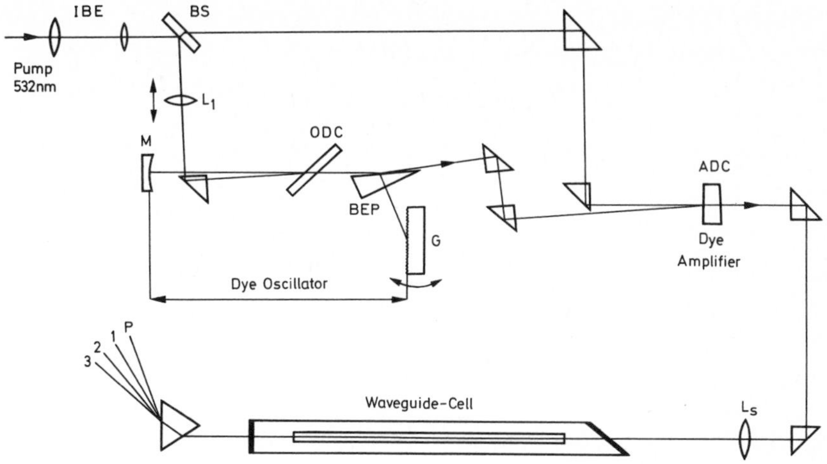

Fig.7.43. Infrared Raman waveguide laser in compressed H_2, pumped by a dye laser. The frequency doubled output beam from an Nd:YAG laser is reduced in diameter by an inverted beam expander and partioned between the oscillator and amplifier dye cell. The dye oscillator resonator is composed of mirror M, grating G, and beam-expanding prism BEP [7.96]

tional Raman scattering, the rotational Raman effect can be utilized, although the gain is much lower than for vibrational Raman scattering, due to the smaller scattering cross section. For instance H_2 and D_2 Raman lasers excited with a CO_2 laser could produce many Raman lines in the spectral range from 900 to 400 cm^{-1} while liquid N_2 and O_2 Raman lasers pumped with an HF laser would cover a quasicontinuous tuning range between 1000 and 2000 cm^{-1}. With high-pressure gas lasers as pumping sources the small gaps between the many rotational-vibrational lines could be closed by pressure broadening (see Sect.7.2.3) and a true continuous tuning range of ir Raman lasers in the far infrared region becomes possible.

A more detailed presentation of ir Raman lasers may be found in the recent review by GRASIUK et al. [7.97], an extensive treatment of nonlinear optics in [7.98-100].

Recently a cw tunable Raman oscillator has been realized which utilizes a 650 m long single mode silica fiber as active medium pumped by a 5 W cw Nd:YAG laser. The first Stokes radiation was tunable from 1.08 to 1.13 μm, the second Stokes from 1.15 to 1.175 μm [7.96a].

8. Doppler-Limited Absorption and Fluorescence Spectroscopy with Lasers

After having presented in the previous chapter the different realizations of tunable lasers, we now discuss their applications in absorption and fluorescence spectroscopy. At first those methods where the spectral resolution is limited by the Doppler width of the molecular absorption lines will be treated. This limit can in fact be reached if the laser linewidth is small compared with the Doppler width. In several examples, such as optical pumping or laser-induced fluorescence spectroscopy, multimode lasers may be employed, although in most cases single-mode lasers may be superior. In general, however, these lasers may not necessarily be frequency stabilized as long as the frequency jitter is small compared with the absorption linewidth. We compare several detection techniques of molecular absorption with regard to their sensitivity and their feasibility in the different spectral regions. Some examples illustrate the methods to give the reader a feeling of what has been achieved. After the discussion of the "Doppler-limited spectroscopy", Chap.10 gives an extensive treatment of various techniques which allow sub-Doppler spectroscopy.

8.1 Introduction

In order to illustrate the advantages of absorption spectroscopy with tunable lasers, we at first compare it with the conventional absorption spectroscopy which uses incoherent radiation sources. Figure 8.1 gives schematic diagrams of both methods.

In classical absorption spectroscopy, radiation sources with a *broad emission continuum* are preferred (e.g., high-pressure Hg arcs, Xe flash lamps, etc.). The radiation is collimated by a lens L_1 and passes through the absorption cell. Behind a dispersing instrument for wavelength selection (spectrometer or interferometer) the intensity $I_T(\lambda)$ of the transmitted light is measured as a function of the wavelength λ (Fig.8.1a). By comparison with a

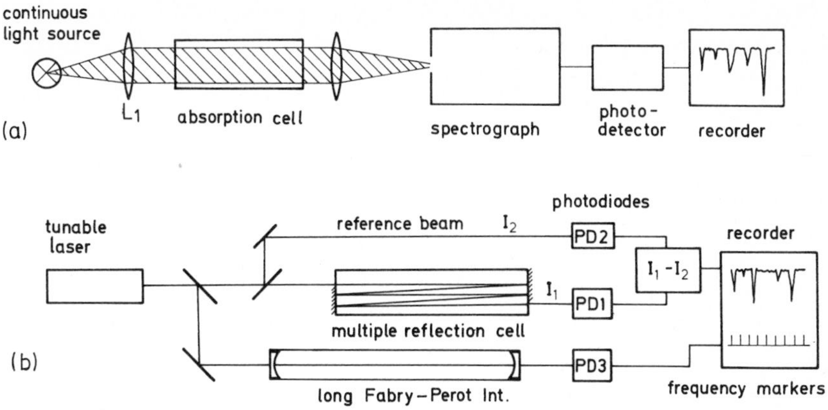

Fig. 8.1. Comparison between absorption spectroscopy with a broad-band incoherent source (a) and with a tunable single-mode laser (b)

reference beam $I_R(\lambda)$ (which can be realized, for instance, by shifting the absorption cell alternatively out of the light beam) the absorption spectrum

$$I_A(\lambda) = a[I_0(\lambda) - I_T(\lambda)] = a[bI_R(\lambda) - I_T(\lambda)]$$

can be obtained, where the constants a and b take into account wavelength-independent losses of I_R and I_T (e.g., reflections of the cell walls).

The *spectral resolution* is generally limited by the resolving power of the dispersing spectrometer. Only with large and expensive instruments (e.g., Fourier spectrometers) may the Doppler limit be reached [4.16].

The *detection sensitivity* of the experimental arrangement is defined by the minimum absorbed power which can still be detected. In most cases it is limited by the detector noise and by intensity fluctuations of the radiation source. Generally the limit of the detectable absorption is reached at relative absorptions $\Delta I/I \geqq 10^{-4} - 10^{-5}$. This limit can be pushed further down only in favorable cases by using special sources and lock-in detection or signal-averaging techniques.

Contrary to radiation sources with broad emission continua used in conventional spectroscopy, tunable lasers offer radiation sources in the spectral range from the uv to the ir with extremely narrow bandwidths and with spectral power densities which may exceed those of incoherent light sources by many orders of magnitude (see Chap.7).

In several regards laser absorption spectroscopy corresponds to microwave spectroscopy where clystrons or carcinotrons instead of lasers represent tunable coherent radiation sources. Laser spectroscopy transfers many of the techniques and advantages of microwave spectroscopy to the infrared, visible, and ultraviolet spectral ranges.

The advantages of absorption spectroscopy with tunable lasers may be summarized as follows.

1) No monochromator is needed since the absorption coefficient $\alpha(\omega)$ and its frequency dependence can be directly measured from the difference $\Delta I(\omega) = aI_R(\omega) - I_T(\omega)$ between the intensities of reference beam and transmitted beam (Fig.8.1b). The spectral resolution is higher than in conventional spectroscopy. With tunable single-mode lasers it is only limited by the linewidths of the absorbing molecular transitions. Using Doppler-free techniques (see Chap.10) even sub-Doppler resolution can be achieved.

2) Because of the high spectral power density of many lasers, the detector noise is generally negligible. Intensity fluctuations of the laser which limit the detection sensitivity, may be essentially suppressed by intensity stabilization (see Sect.6.7). This furthermore increases the signal-to-noise ratio and therefore enhances the sensitivity.

3) The detection sensitivity increases with increasing spectral resolution $\omega/\Delta\omega$ as long as $\Delta\omega$ is still larger than the linewidth $\delta\omega$ of the absorption line. This can be seen as follows.

The relative intensity attenuation per absorption path length x = 1 is

$$\Delta I/I = \int_{\omega_0-\delta\omega/2}^{\omega_0+\delta\omega/2} \alpha(\omega)I(\omega)d\omega \Big/ \int_{\omega_0-\delta\omega/2}^{\omega_0+\delta\omega/2} I(\omega)d\omega \quad . \tag{8.1}$$

If $I(\omega)$ does not change much within the internal $\delta\omega$ we can write

$$\int_{\omega_0-\Delta\omega/2}^{\omega_0+\Delta\omega/2} \cdot\ I(\omega)d\omega = \bar{I}\Delta\omega \quad \text{and} \quad \int \alpha(\omega)I(\omega)d\omega = \bar{I} \int \alpha(\omega)d\omega \ .$$

This yields

$$\Delta I/I = \frac{1}{\Delta\omega} \int_{\omega_0-\delta\omega/2}^{\omega_0+\delta\omega/2} \alpha(\omega)d\omega \simeq \bar{\alpha}\,\frac{\delta\omega}{\Delta\omega} \quad . \tag{8.2}$$

Decreasing the resolvable spectral interval $\Delta\omega$ from $10\delta\omega$ to $\delta\omega$ therefore increases the detection sensitivity approximately by a factor of 10.

4) Because of the good collimation of a laser beam, long absorption paths can be realized by multiple reflection back and forth through the absorption cell. Disturbing reflections from cell walls or windows which may influence the mea-

surements can be essentially avoided (for example by using Brewster end windows). Such long absorption paths enable the measurement of transitions even with small absorption coefficients. Furthermore pressure broadening can be reduced by using low gas pressure. This is especially important in the infrared region where the Doppler width is small and the pressure broadening may become the limiting factor for the spectral resolution (see Sect.3.3).

5) If a small fraction of the laser output is sent through a long Fabry-Perot interferometer with a separation d of the mirrors, the photodetector PD 3 receives intensity peaks each time the laser frequency ν_L is tuned to a transmission maximum at $\nu = mc/2d$ (see Sects.4.2 and 4.4). These peaks serve as accurate wavelength markers which allow one to calibrate the separation of adjacent absorption lines. With d = 1 m the frequency separation $\Delta\nu_p$ between successive transmission peaks $\Delta\nu_p = c/2d = 150$ MHz, corresponding to a wavelength separation of 10^{-4} nm at $\lambda = 550$ nm. With a semiconfocal F.P.I. the free spectral range is c/8d, which gives $\Delta\nu_p = 75$ MHz for d = 0.5 m.

6) The laser frequency may be stabilized onto the center of an absorption line. With the methods discussed in Sect.4.4 it is possible to measure the wavelength λ_L of the laser with an absolute accuracy of 10^{-8} or better. This allows determination of the molecular absorption lines with the same accuracy.

7) It is possible to tune the laser wavelength very rapidly over a spectral region where molecular absorption lines have to be detected. With electro-optical components, for instance, pulsed dye lasers can be tuned over several wavenumbers within a microsecond. This opens new perspectives for spectroscopic investigations of short-lived intermediate radicals in chemical reactions. The capabilities of classical flash photolysis may be considerably extended using such rapidly tunable laser sources.

8) An important advantage of absorption spectroscopy with tunable single-mode lasers stems from their capabilities to measure line profiles of absorbing molecular transitions with high accuracy. In case of pressure broadening the determination of line profiles allows one to derive information about the interaction potential of the collision partners (see Sect.3.3). In plasma physics this technique is widely used to determine electron and ion densities and temperatures.

9) In fluorescence spectroscopy and optical pumping experiments, the high intensity of lasers allows an appreciable population in selectively excited states to be achieved which may be comparable to that of the absorbing ground states. The small laser linewidth favors the selectivity of optical excitation and results in favorable cases in the exclusive population of single molecular levels. These advantageous conditions allow one to perform absorp-

tion and fluorescence spectroscopy of *excited* states and to transform spectroscopic methods, such as microwave or rf spectroscopy, up to now restricted to electronic ground states, also to excited states.

This short comprehension of some advantages of lasers in spectroscopy will be outlined in more detail in the following chapters and several examples will illustrate their relevance.

8.2 High-Sensitivity Detection Methods

The general method to measure absorption spectra is based on the determination of the absorption coefficient $\alpha(\omega)$ from the spectral intensity

$$I_T(\omega) = I_0 \exp[-\alpha(\omega)x] \tag{8.3}$$

which is transmitted through an absorbing path length x. For small absorption $\alpha x \ll 1$ we can use the approximation $\exp(-\alpha x) \approx 1 - \alpha x$ and (8.3) can be reduced to

$$I_T(\omega) \approx I_0[1-\alpha(\omega)x] \quad . \tag{8.4}$$

With a reference intensity $I_R = I_0$ as produced, for example, by a 50% beam splitter in Fig.8.1b with a reflectivity R = 0.5, one can measure the absorption coefficient

$$\alpha(\omega)x = [I_R-I_T(\omega)] / I_R \tag{8.5}$$

from the difference $I_R - I_T(\omega)$.

In the case of very small values of αx, this method cannot be very accurate since it measures a small difference of two large quantities and small fluctuations of the splitting ratio can already severely influence the measurement. Therefore several different techniques have been developed which allow increase of the sensitivity and accuracy of absorption measurements by several orders of magnitude. In the following we discuss some of these techniques.

The first method is based on a frequency modulation of the monochromatic incident wave. It has not been designed specifically for laser spectroscopy but was taken from microwave spectroscopy where it is a standard method. The laser frequency ω_L is modulated at a modulation frequency f, which sweeps ω_L periodically from ω_L to $\omega_L + \Delta\omega_L$. When the laser is tuned through the absorption spectrum, the difference $I_T(\omega_L) - I_T(\omega_L + \Delta\omega_L)$ is detected with a lock-in amplifier (phase-sensitive detector) tuned to the modulation frequency f. If the modulation sweep $\Delta\omega_L$ is sufficiently small, the first term of the Taylor expansion

$$I_T(\omega_L + \Delta\omega_L) - I_T(\omega_L) = (dI_T/d\omega)\Delta\omega_L + \frac{1}{2!}\frac{d^2I_T}{d\omega^2}\Delta\omega_L^2 + \dots \tag{8.6}$$

is dominant. This term is proportional to the first derivative of the absorption spectrum as can be seen from (8.5). When I_R is independent of ω we obtain

$$\frac{d\alpha(\omega)}{d\omega} = -\frac{1}{I_R x}\frac{dI_T}{d\omega} \quad . \tag{8.7}$$

The advantage of the frequency modulation is the possibility of phase-sensitive detection, which restricts the frequency response of the detection system to a narrow frequency interval centered at the modulation frequency f. Frequency-independent background absorption from cell windows and background noise due to fluctuations of the laser intensity or of the density of absorbing molecules are essentially reduced. Regarding signal-to-noise ratio and achievable sensitivity, the *frequency* modulation technique is superior to an *intensity* modulation of the incident radiation. The frequency of a single-mode laser can be readily modulated when an ac voltage is applied to the piezo on which a resonator mirror is mounted (see Sect.6.5).

Another very sensitive method directly monitors the absorbed energy rather than relying on a difference measurement $(I_R - I_T)$. The energy $I_0\alpha xA$ absorbed per second in a volume $V = Ax$ can either be converted into fluorescence energy and monitored with a fluorescence detection system (*excitation spectroscopy*) or it can be converted by collisions into thermal energy with a resultant temperature and pressure rise, which is monitored by a sensitive microphone (*photoacoustic spectroscopy*).

The third method is based on the sensitive dependence of the laser intensity on absorption losses *inside the laser resonator*. When the absorbing sample is placed inside the laser resonator, the laser intensity I_L decreases by $\Delta I_L = q^*\alpha(\omega)xI_L$ where the factor q^* which gives the enhancement of the intensity change compared with the same absorption outside the cavity, can become very large. This "intracavity absorption" technique will be discussed in Sect.8.2.3.

The most sensitive methods are capable of counting single absorbed photons. They use as a monitor the ionization of atoms and molecules from highly excited states which had been populated by absorption of laser photons. Compared to the conventional absorption measurements these sensitive detection methods represent remarkable progress. The sensitivity limit has been pushed down from detectable relative absorptions $\Delta\alpha/\alpha \approx 10^{-5}$ to about $\Delta\alpha/\alpha \leqq 10^{-17}$. We now discuss these different methods in more detail.

8.2.1 Excitation Spectroscopy

A very high sensitivity in the visible and ultraviolet regions can be achieved if the absorption of laser photons is monitored through the laser-excited fluorescence. When the laser wavelength λ_L is tuned to an absorbing molecular transition $E_i \rightarrow E_k$, the number of photons absorbed per s and pathlength Δx is

$$n_a = N_i \sigma_{ik} n_L \Delta x \quad , \tag{8.8}$$

where n_L is the number of incident laser photons, σ_{ik} the absorption cross section per molecule, and N_i the density of molecules in the absorbing state E_i (Fig.8.2).

The number of fluorescence photons emitted per second from the excited level E_k is

$$n_{fl} = N_k A_k = n_a \eta_k \quad , \tag{8.9}$$

where $A_k = \sum_m A_{km}$ stands for the total spontaneous transition probability (see Sect.2.7) to all levels with $E_m < E_k$. The quantum efficiency $\eta_k = A_k/(A_k + R_k)$ gives the ratio of the spontaneous transition rate to the total deactivation rate which may also include a radiationless transition rate R_k (e.g., collision-induced transitions). For $\eta_k = 1$ the number n_A of fluorescence photons

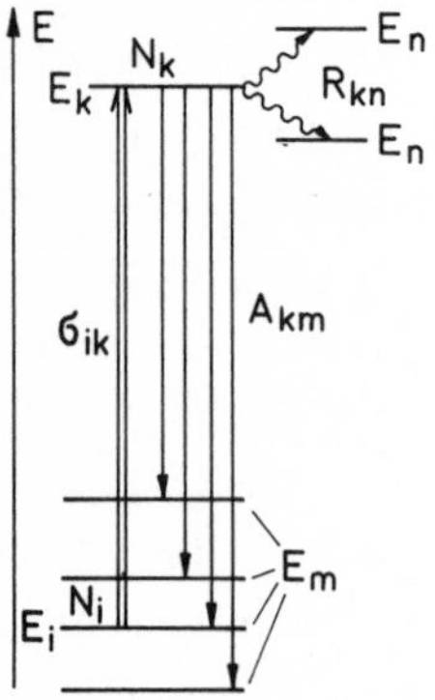

Fig. 8.2. Level scheme relevant for excitation spectroscopy

emitted per second equals the number n_a of photons absorbed per second under stationary conditions.

Unfortunately only a fraction δ of the fluorescence photons, emitted into all directions, can be collected on the photomultiplier cathode where again only the fraction $\eta_{Ph} = n_{Pe}/n_{Ph}$ of these photons produces on the average n_{Pe} photoelectrons. The quantity η_{Ph} is called the *quantum efficiency of the photocathode* (see Sect.4.5.2). The number n_{PE} of photoelectrons counted per second is then

$$n_{PE} = n_a \eta_k \eta_{Ph} \delta = N_i \sigma_{ik}\, n_L\, \Delta x \eta_k \eta_{Ph} \delta \quad . \tag{8.10}$$

Modern photomultipliers reach quantum efficiencies of $\eta_{Ph} = 0.2$. With carefully designed optics it is possible to achieve a collection factor $\delta = 0.1$. Using photon counting techniques and cooled multipliers (dark pulse rate ≤ 10 counts/s), counting rates of $n_{PE} = 100$ counts/s are already sufficient to obtain a signal-to-noise ratio $S/R \sim 8$ at integration times of 1s.

Inserting this figure for n_{PE} into (8.10) illustrates that with $\eta_k = 1$ absorption rates of $n_a = 5 \cdot 10^3$/s can already be measured quantitatively. Assuming a laser power of 1 W at a wavelength $\lambda = 500$ nm which corresponds to a photon flux of $n_L = 3 \cdot 10^{18}$/s, this implies *that it is possible to detect a relative absorption of* $\Delta I/I \leq 10^{-14}$. When placing the absorbing probe inside the cavity where the laser power is q times larger ($q \sim 10$ to 100, see Sect.8.2.3), this impressive sensitivity may be even further enhanced.

When the laser wavelength λ_L is tuned across the spectral range of absorption lines, the total fluorescence intensity $I_{Fl}(\lambda_L) \propto n_L \sigma_{ik} N_i$ monitored as a function of laser wavelength λ_L, represents an image of the absorption spectrum, called the *excitation spectrum*. According to (8.10) the photoelectron rate n_{PE} is directly proportional to the absorption coefficient $N_i \sigma_{ik}$ where the proportionality factor depends on the quantum efficiency η_{Ph} of the photomultiplier cathode and on the collection efficiency δ of the fluorescence photons.

Although the excitation spectrum directly reflects the absorption spectrum regarding the *position* of lines, the relative *intensities* of different lines $I(\lambda)$ are identical in both spectra only if the following conditions are guaranteed.

1) The quantum efficiency η_k must be the same for all excited states E_k. Under collision-free conditions, i.e., at sufficiently low pressures, the excited molecules radiate before they can collide and we obtain $\eta_k = 1$ for all levels E_k.

2) The quantum efficiency η_{Ph} of the detector should be constant over the whole spectral range of the emitted fluorescence. Otherwise the spectral distribution of the fluorescence, which may be different for different excited levels E_k, will influence the signal rate. Some modern photomultipliers can meet this requirement.

3) The geometrical collection efficiency δ of the detection system should be identical for the total fluorescence from different excited levels. This demand excludes, for example, excited levels with very long lifetimes, where the excited molecules may diffuse out of the observation region before they emit the fluorescence photon. Furthermore the fluorescence may not be isotropic, depending on the symmetry of the excited state.

However, even if these requirements are not strictly fulfiled, excitation spectroscopy is still very useful, to measure absorption lines with extremely high sensitivity, although their relative intensities may not be recorded accurately.

The technique of excitation spectroscopy has been widely used to measure very small absorptions. One example is the determination of absorption lines in molecular beams where both the pathlength Δx and the density N_i of absorbing molecules are small. Typical figures are $\Delta x = 0.1$ cm, $N_i = 10^7 - 10^{10}/\mathrm{cm}^3$. Figure 8.3 illustrates the technique by a small section of the excitation spectrum of the Na_2 molecule, excited by a tunable dye laser, around $\lambda = 604$ nm.

The extremely high sensitivity of this technique has been impressively demonstrated by FAIRBANKS et al. [8.1] who performed absolute density measurements of sodium vapor in the range $N = 10^2 - 10^{11}/\mathrm{cm}^3$ using the laser-excited fluorescence as the monitor. The lower detection limit of $N = 10^2/\mathrm{cm}^3$ was imposed by background stray light scattered out of the incident laser beam by windows and cell walls.

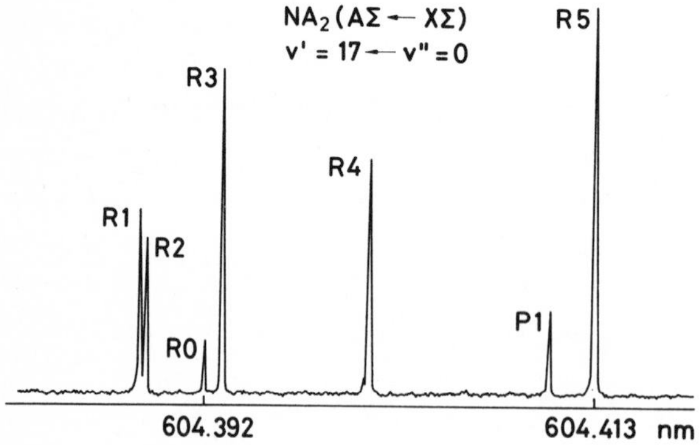

Fig. 8.3. Section of the excitation spectrum of Na_2 molecules obtained in a molecular beam with a tunable dye laser at $\lambda = 604$ nm

Because of its high sensitivity, excitation spectroscopy can be successfully used to monitor minute concentrations of radicals and short-lived intermediate products in chemical reactions [8.2]. Besides the measurements of small concentrations, detailed information on the internal state distribution $N_i(v_i'', J_i'')$ of reaction products can be extracted, since the fluorescence signal is, according to (8.15), proportional to the number N_i of absorbing molecules [8.3] (see Sect.8.8).

Excitation spectroscopy has its highest sensitivity in the visible, ultraviolet, and near infrared regions. With increasing wavelength λ the sensitivity decreases for the following reasons. Equation (8.10) shows that the detected photoelectron rate n_{PE} decreases with η_k, η_{Ph}, and δ. All these numbers generally decrease with increasing wavelength. The quantum efficiency η_{Ph} and the attainable signal-to-noise ratio are much lower for infrared than for visible photodetectors (see Sect.4.5). By absorption of infrared photons vibrational-rotational levels of the electronic ground state are excited with radiative lifetimes which are generally several orders of magnitude larger than those of excited *electronic* states. At sufficiently low pressures the molecules diffuse out of the observation region before they radiate. This diminishes the collection efficiency δ. At higher pressures the quantum efficiency η_k of the excited level E_k is decreased because collisional deactivation competes with radiative transitions. Under these conditions photoacoustic detection may become preferable.

8.2.2 Photoacoustic Spectroscopy

This sensitive technique of measuring small absorptions is applied mainly when minute concentrations of molecular species have to be detected in the presence of other components at higher pressure. An example is the detection of spurious pollutant gases in the atmosphere. Its basic principle may be summarized as follows:

The beam of a tunable laser is sent through the absorption cell (Fig.8.4). If the laser is tuned to an absorbing molecular transition $E_i \rightarrow E_k$, part of

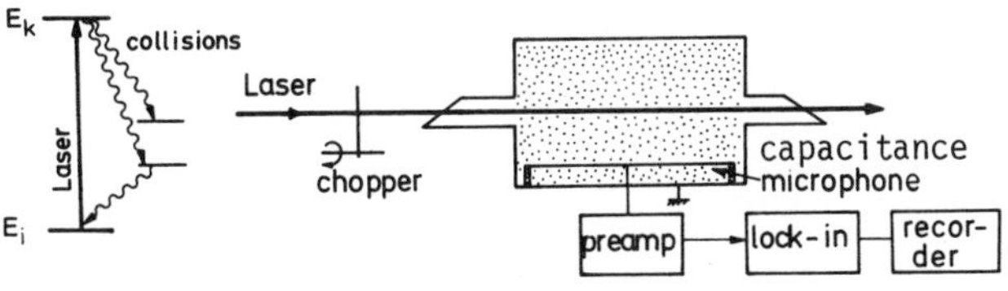

Fig. 8.4. Schematic experimental arrangement for opto-acoustic spectroscopy

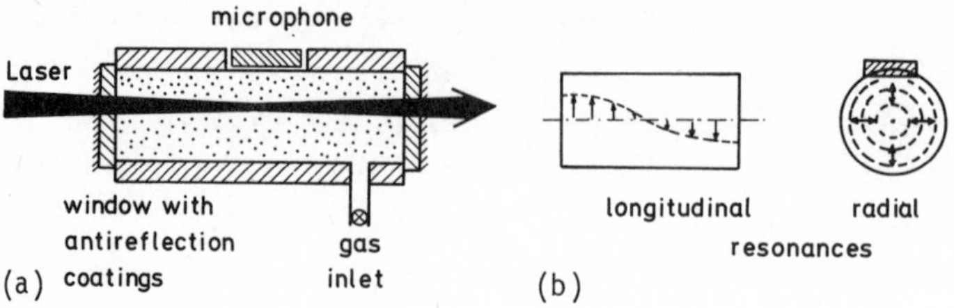

Fig. 8.5. (a) Spectraphone, with capacitance microphone, (b) longitudinal and radial acoustic resonance modes

the molecules in the lower level E_i will be excited into the upper level E_k. By collisions with other atoms or molecules in the cell these excited molecules may transfer their excitation energy $(E_k - E_i)$ completely or partly into translational, rotational, or vibrational energy of the collision partners. At thermal equilibrium this energy is randomly distributed onto all degrees of freedom, causing an increase of thermal energy and with it a rise of temperature and pressure at a constant density in the cell.

With infrared lasers the molecules are generally excited into higher vibrational levels of the electronic ground state. Assuming cross sections of $10^{-18} - 10^{-19}$ cm^2 for the collisional deactivation of vibrationally excited molecules, the equipartition of energy takes only about 10^{-5} s at pressures around 1 Torr. Since the spontaneous lifetimes of these excited vibrational levels are typically around $10^{-2} - 10^{-5}$ s, it follows that at pressures above 1 Torr the excitation energy, absorbed from the laser beam, will be almost completely transferred into thermal energy.

When the laser beam is chopped at frequencies below 10 KHz, periodical pressure variations appear in the absorption cell which can be detected with a sensitive microphone placed at the inner side of the cell. The output signal from the microphone is proportional to the absorbed laser energy and therefore allows determination of the absorption coefficient. Because this method uses the conversion of photon energy into periodical pressure variations it is called *photoacoustic spectroscopy* and the device is named *spectraphone*.

The idea of the spectraphone is very old and was already demonstrated by BELL and TYNDAL [8.4] in 1881. However, the impressive detection sensitivity obtained nowadays could only be achieved with the development of lasers, sensitive capacitance microphones, low-noise amplifiers, and lock-in techniques. Concentrations down to the ppb range (parts per billion = 10^{-9}) at total pressures of 1 Torr to several atmospheres are readily detectable with a modern spectraphone (Fig.8.5).

As far as saturation effects can be neglected, the acoustic signal S

$$S = CN_i\sigma_{ik}(\omega)\Delta x P_L(1-\eta_k)S_M \qquad (8.11)$$

is proportional to the density N_i of absorbing molecules in level E_i, the absorption cross section σ_{ik} and the path length Δx, the incident mean laser power P_L, and the sensitivity S_M of the microphone. The signal decreases with increasing quantum efficiency η_k (which gives the ratio of emitted fluorescence photons to absorbed laser photons). The factor C depends on the spectraphone parameters. Modern condensor microphones with low-noise FET preamplifier and phase-sensitive detection achieve signals of larger than 1 V/Torr with a background noise of 3×10^{-8} V at integration times of 1 s. This sensitivity allows detection of pressure amplitudes below 10^{-7} Torr, and is in general not limited by the electronic noise but by another disturbing effect. Laser light reflected from the cell windows or scattered by aerosols in the cell may be partly absorbed by the walls and contributes to a temperature increase. The resulting pressure rise is of course modulated at the chopping frequency and is therefore detected as background signal. There are several ways to reduce this effect. Antireflection coatings of the cell windows, or, in case of linearly polarized laser light, the use of Brewster windows minimize the reflections. An elegant solution chooses the chopping frequency to coincide with an acoustic resonance of the cell. This results in a resonant amplification of the pressure amplitude which may be as large as 100-fold. This experimental trick has the additional advantage that those acoustic resonances can be selected, which couple most efficiently to the laser beam profile but are less effectively excited by heat conduction from the walls. The background signal caused by wall absorption can thus be reduced and the true signal is enhanced. Figure 8.5b shows longitudinal, and radial acoustic resonances of a cylindrical cell.

The sensitivity can be further enhanced by frequency modulation of the laser (see Sect.8.1) and by intracavity absorption techniques. With the spectraphone inside the laser cavity the photoacoustic signal of nonsaturating transitions is increased by a factor q due to the q-fold laser intensity inside the resonator (see next section).

According to (8.11) the optoacoustic signal decreases with increasing quantum efficiency because the fluorescence carries energy away without heating the gas, as long as the fluorescence light is not absorbed within the cell. Since the quantum efficiency is determined by the ratio of spontaneous to collision-induced deactivation of the excited level, it decreases with increasing spontaneous lifetime and gas pressure. Therefore the optoacoustic method is particularly favorable to monitor vibrational spectra of molecules in the infrared region (because of the long lifetimes of excited vibrational levels) and to detect small concentrations of molecules in the presence of other gases at higher pressures (because of the large collisional deactivation rate). It is possible to use this technique for measuring even *rotational* spectra in the microwave region and also electronic molecular spectra in the visible or ultraviolet range where electronic states with short spontaneous lifetimes are excited. However, the sensitivity in these spectral regions is not quite as high and there are other methods which are superior.

Some examples illustrate this very useful spectroscopic technique. For a more detailed discussion of optoacoustic spectroscopy, its experimental tricks, and its various applications, the reader is referred to a recently published textbook [8.5] and to some reviews on this field [8.6-8a].

Examples:

a) The sensitivity of the spectraphone has been demonstrated by KREUTZER et al. [8.9]. At a total air pressure of 500 Torr in the absorption cell these authors could detect concentrations of ethylene down to 0.2 ppb, of NH_3 down to 0.4 ppb, and NO pollutants down to 10 ppb. The feasibility of determining certain important isotope abundances or ratios by simple and rapid infrared spectroscopy with the spectraphone and also the ready control of small leaks of polluting or poison gases has been demonstrated [8.10].

b) The optoacoustic method has been applied with great success to high-resolution spectroscopy of rotational-vibrational bands of numerous molecules [8.11]. Figure 8.6 illustrates as example a section of the absorption spectrum of the NO molecule showing the lambda doubling of the Q lines.

c) A general technique for the optoacoustic spectroscopy of *excited* molecular vibrational states has been demonstrated by PATEL [8.12]. This technique involves the use of vibrational energy transfer between two dissimilar molecules

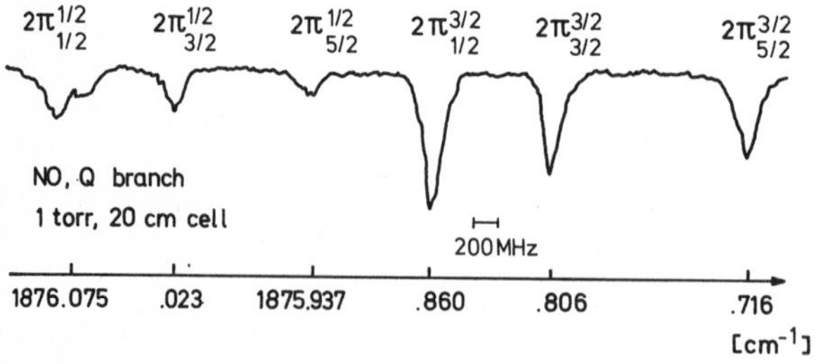

Fig. 8.6. Photoacoustic absorption spectrum of NO around $\bar{\nu}$ = 1875 cm^{-1} [8.11]

A and B. When A is excited to its first vibrational level by absorption of a laser photon $h\nu_1$ it can transfer its excitation energy by near resonant collision to molecule B. Because of the large cross section for such collisions, a high density of vibrationally excited molecules B can be achieved also for those molecules which cannot be pumped directly by existing powerful laser lines. The excited molecule B can absorb a photon $h\nu_2$ from a second, weak tunable laser, which allows spectroscopy of all accessible transitions (v=1→v=2). The technique has been proved for the NO molecule where the frequency of the four transitions in the $^2\Pi_{1/2}$ and $^2\Pi_{3/2}$ subbands of ^{15}NO and the Λ doubling for the $v=1\rightarrow 2$ transition have been accurately measured. The following scheme illustrates the method:

$$^{14}NO + h\nu_1(CO_2 \text{ laser}) \rightarrow {}^{14}NO^*(v=1)$$

$$^{14}NO^*(v=1) + {}^{15}NO(v=0) \rightarrow {}^{14}NO(v=0) + {}^{15}NO^*(v=1) + \Delta E(35\ cm^{-1})$$

$$^{15}NO^*(v=1) + h\nu_2(\text{spin-flip laser}) \rightarrow {}^{15}NO^*(v=2)$$

the last process is detected by optoacoustic spectroscopy.

d) The application of photoacoustic detection to the visible region has been reported by STELLA et al. [8.13]. These authors placed the spectraphone inside the cavity of a cw dye laser and scanned the laser across the absorption bands of CH_4 and NH_3 molecules. The high-quality spectra with resolving power of over $2\cdot10^5$ proved to be adequate to resolve single rotational features of the very weak vibrational overtone transitions of these molecules. The experimental results prove to be very useful for the investigation of the planetary atmospheres where such weak overtone transitions are induced by the sunlight.

e) An interesting application of photoacoustic detection lies in the measurement of dissociation energies of molecules [8.13a]. When the laser wavelength is tuned across the dissociation limit, the photoacoustic signal drops drastically because beyond this limit the absorbed laser energy is used for dissociation (which means that it is converted into potential energy and cannot be transferred into kinetic energy as in the case of excited state deactivation. Only the kinetic energy causes a pressure increase).

f) With a special design of the spectraphone, using a coated quartz membrane for the condensor microphone, even corrosive gases can be measured [8.14]. This extends the applications of optoacoustic spectroscopy to aggressive pollutant gases such as NO_2 or SO_2 which are important constituents of air pollution.

8.2.3 Intracavity Absorption

When the absorbing sample is placed inside the laser cavity (see Fig.8.7) the detection sensitivity can be considerably enhanced, in favorable cases by several orders of magnitude. Four different effects can be utilized to achieve this "amplified" sensitivity.

1) Assume the reflectivities of the two resonator mirrors to be $R_1 = 1$ and $R_2 = 1 - T_2$ (mirror absorption is neglected). At a laser output power P_{out} the power inside the cavity is $P_{int} = qP_{out}$ with $q = 1/T_2$. For $\alpha L \ll 1$ the laser power $\Delta P(\omega)$ absorbed at the frequency ω in the absorption cell (length L) is

$$\Delta P(\omega) = q\alpha(\omega)LP_{out} \quad . \tag{8.12}$$

If the absorbed power can be measured directly, for example through the resulting pressure increase in the absorption cell (see Sect.8.2.2) or through the laser-induced fluorescence (see Sect.8.2.1), the signal will be q times larger than for the case of single pass absorption outside the cavity. With $T_2 = 0.02$ (a figure which can be readily realized in practice) the enhancement factor becomes $q = 50$, as long as saturation effects can be neglected, and provided the absorption is sufficiently small that it does not noticeably change the laser intensity. This q-fold amplification of the sensitivity can be also understood from the simple fact that a laser photon travels on the average q times back and forth between the resonator mirrors before it leaves the resonator. It therefore has a q-fold chance to be absorbed in the sample.

This sensitivity enhancement in detecting small absorptions can be also realized in external passive resonators. If the laser output is mode matched (see Sect.5.11) by lenses or mirrors into the fundamental mode of the passive cavity containing the absorbing sample (Fig.8.8), the radiation power inside this cavity is q times larger. The enhancement factor q may become large, if the internal losses of the cavity can be kept low. The use of external passive resonators which represent an improvement over the generally used multi-

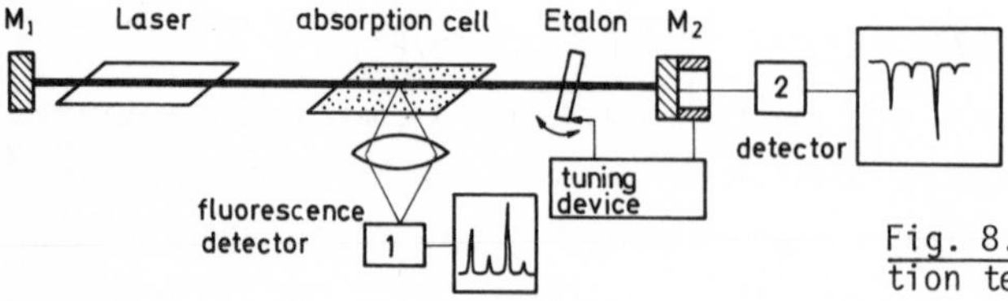

Fig. 8.7. Intracavity absorption technique

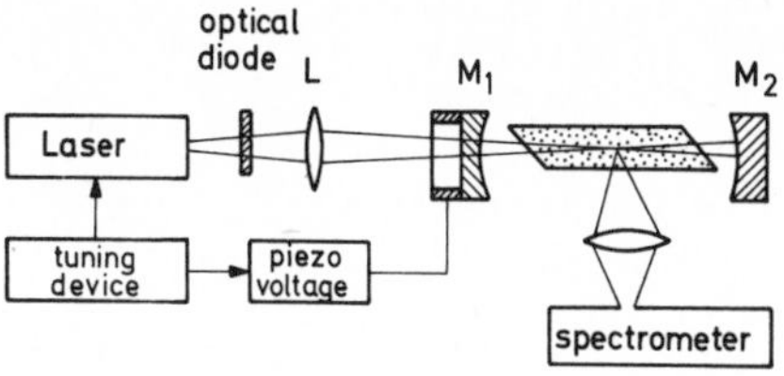

Fig. 8.8. Excitation spectroscopy inside an external resonator, mode matched to the laser output

pass absorption cells, may become advantageous when the absorption cell cannot be placed directly inside the active laser resonator. However, one has to take care to prevent optical feedback from the passive to the active cavity which causes a coupling of both cavities with resulting instabilities. The feedback can be prevented by an optical diode.

Since the radiation field inside the active resonator or inside the mode-matched passive cavity is concentrated within the region of the Gaussian beam (Sect.5.11), the laser-excited fluorescence can be effectively imaged onto the entrance slit of a spectrometer with a larger efficiency than in the commonly used multipass cells. The use of a spectrometer is necessary not only in fluorescence or Raman spectroscopy but also in absorption spectroscopy, if minute concentrations of an absorbing component have to be selectively detected in the presence of other constituents with overlapping absorption lines but different fluorescence spectra.

2) Another way of detecting intracavity absorption with very high sensitivity relies on the dependence of laser output power on absorption losses inside the laser resonator (detector 2 in Fig.8.7). At constant pump power, close above threshold, minor changes of the intracavity losses may already result in drastic changes of the laser output. In Sect.5.9 we saw that under steady-state conditions the laser intensity essentially depends on the pump power and reaches a value I_s where the saturated gain G equals the total losses γ. If additional losses $\Delta\gamma = \alpha(\omega)L$ are introduced by the absorbing sample, the relative change of the laser output power is, according to (5.80a),

$$\frac{\Delta P}{P} = \frac{\Delta I}{I} = \frac{G_0}{G_0-\gamma}\,\frac{\Delta\gamma}{\Delta\gamma+\gamma} \quad , \tag{8.13}$$

where G_0 is the unsaturated gain. Compared with the single pass absorption outside the resonator, where the relative intensity change is $\Delta I/I = -\alpha L = -\Delta\gamma$, this represents a sensitivity enhancement by a factor

$$Q = \frac{G_0}{G_0-\gamma}\,\frac{1}{\gamma+\Delta\gamma} \approx \frac{G_0/\gamma}{G_0-\gamma} \quad \text{for} \quad \Delta\gamma \ll \gamma \quad . \tag{8.14}$$

At pump powers far above threshold the unsaturated gain G_0 is large compared with the losses γ and (8.14) reduces to

$$Q \approx 1/\gamma \quad \text{for} \quad G_0 >> \gamma \quad . \tag{8.15}$$

If the resonator losses are mainly due to the transmission T_2 of the output mirror, the enhancement factor Q becomes $Q = 1/\gamma = 1/T_2 = q$, which is equal to the enhancement of the previous detection method.

Close above threshold, however, G_0 is only slightly larger than γ and the denominator in (8.14) becomes very small, which means that the enhancement factor Q may reach very large values. At first sight it might appear that the sensitivity could be made arbitrarily large for $G_0 \to \gamma$. However, there are experimental as well as fundamental limitations which restrict the maximum achievable value of Q. The increasing instability of the laser output, for instance, limits the detection sensitivity when the threshold is approached. Close at threshold, the spontaneous radiation which is emitted into the solid angle accepted by the detector cannot be neglected. It represents a constant background intensity, nearly independent of γ, which puts a principal upper limit for the relative change $\Delta I/I$ and thus for the sensitivity.

3) In the preceding discussion of the sensitivity enhancement by intracavity absorption we have implicitly assumed that the laser oscillates on a single mode. Even larger enhancement factors Q can be achieved, however, with lasers oscillating simultaneously on several competing modes. Pulsed or cw dye lasers without additional mode selection are examples of such lasers with mode competition. As discussed in Sect.7.3, the active dye medium exhibits a broad homogeneous spectral gain profile which allows the same dye molecules to simultaneously contribute to the gain of all modes with frequencies within the homogeneous linewidth (see the discussion in Sects.5.7 and 7.3). This means that the different oscillating laser modes may share the same molecules to achieve their gains. This leads to mode competition and exhibits the following mode coupling phenomena.

Assume that the laser oscillates simultaneously on N modes which may have equal gain and equal losses and therefore equal intensities. When the laser wavelength is tuned across the absorption spectrum of an absorbing sample inside the laser resonator, one of the oscillating modes may be tuned into resonance with an absorption line (frequency ω_k) of the sample molecules. This mode now suffers additional losses $\Delta\gamma = \alpha(\omega_k)L$ which cause a decrease ΔI of its intensity. Because of this decreased intensity the inversion of the active medium is less depleted by this mode and the gain at ω_k *increases*. Since the other (N-1) modes can participate in the gain at ω_k, their intensity will increase. This, however, again depletes the gain at ω_k and *decreases* the intensity of the mode oscillating at ω_k. With sufficiently strong coupling between the modes this mutual interaction may finally result in a total suppression of the absorbing mode. In dye lasers with standing wave resonators the spatial hole-burning effect (see Sect.5.8) limits the coupling between modes. Because of their slightly different wavelengths the maxima and nodes of the field distributions in the different modes are located at different positions in the active medium. This has the effect that the volumina of the active dye, from which the different modes extract their gain, overlap only partly. With sufficiently high pump power the absorbing mode has an adequate gain volume on its own and is not completely suppressed but suffers a large intensity loss.

A more detailed calculation (see for instance [8.15,16]) yields for the relative intensity change of the absorbing mode

$$\frac{\Delta I}{I} = \frac{G_0}{G_0 - \gamma} \frac{\Delta\gamma}{\gamma + \Delta\gamma} (1 + KN) \quad , \tag{8.16}$$

where $K (0 \leqq K \leqq 1)$ is a measure of the coupling strength.

Without mode coupling $(K = 0)$ (8.16) gives the same result as (8.13) for a single-mode laser. In the case of strong coupling $(K = 1)$ and a large number of modes $(N \gg 1)$, the relative intensity change of the absorbing mode increases proportional to the number of simultaneously oscillating modes.

If several modes are simultaneously absorbed, the factor N in (8.16) stands for the ratio of all modes to those absorbed. If all modes have equal frequency spacing, this number N gives the ratio of the spectral width of the homogeneous gain profile to the width of the absorption profile.

In order to detect the intensity change of one mode in the presence of many others, the laser output has to be dispersed by a monochromator or an interferometer. The absorbing molecules may have many absorption lines within the broad-band gain profile of a multimode dye laser. Those laser modes which overlap with absorption lines are attenuated or even completely quenched. This results in "spectral holes" in the output spectrum of the laser and allows the sensitive simultaneous recording of the whole absorption spectrum within the laser bandwidth, if the laser output is photographically recorded behind a spectrograph. ATKINSON et al. [8.17] demonstrated this technique by monitoring with a flash lamp-pumped dye laser the time-dependent concentration of the short-lived radicals NH_2 and HCO which had been formed by flash photolysis of NH_3.

For pulsed lasers it is the duration ΔT of the laser pulse which may put a limitation on the sensitivity if ΔT is shorter than the transient times for the energy exchange among the competing modes [8.18]. The enhancement factor Q is therefore in general smaller than that achieved with cw dye lasers.

4) In ring lasers (see Sect.5.6) the spatial hole-burning effect does not occur if the laser is oscillating on unidirectional travelling wave modes. If no optical diode is inserted into the ring resonator, the unsaturated gain is generally equal for clockwise and counterclockwise running waves. In such a bistable operational mode slight changes of the net gain which might be different for both waves because of their opposite Doppler shifts, may already switch the laser from clockwise to counterclockwise operation and vice versa. Such a bistable multimode ring laser with strong gain competition between the modes therefore represents an extremely sensitive detector for small absorptions inside the resonator [8.19].

The enhanced sensitivity of intracavity absorption may be used either to detect minute concentrations of absorbing components or to measure very weak forbidden transitions in atoms or molecules at sufficiently low pressures to study the unperturbed absorption line profiles. With intracavity absorption cells of less than 1 m, absorbing transitions have been measured which would demand a path length of several kilometers with conventional single-pass absorption at a comparable pressure [8.20,20a].

Some examples illustrate the various applications of the intracavity absorption technique.

a) With an iodine cell inside the resonator of a cw multimode dye laser an enhancement factor of $Q = 10^5$ could be achieved allowing the detection of I_2 molecules at concentrations down to $n \leq 10^8/cm^3$ [8.21]. This corresponds to a sensitivity limit of $\alpha L \leq 10^{-7}$. Instead of the laser output power the laser-induced fluorescence from a second iodine cell outside the laser resonator was monitored as a function of laser wavelength. This experimental arrangement (Fig.8.9) allows demonstration of the isotope specific absorption. When the laser beam passes through two external iodine cells filled with the isotopes $^{127}I_2$ and $^{129}I_2$ tiny traces of $^{127}I_2$ inside the laser cavity are already

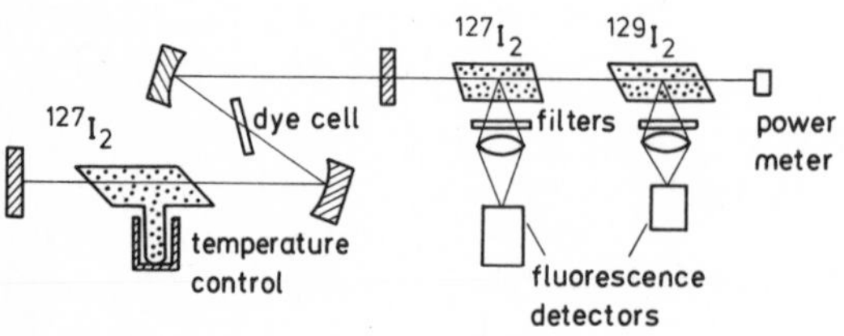

Fig. 8.9. Isotope selective intracavity absorption spectroscopy. The frequencies absorbed by the $^{127}I_2$ inside the cavity are missing in the laser output which therefore does not excite any fluorescence in the same isotope outside the resonator [8.21]

sufficient to completely quench the laser-induced fluorescence from the external $^{127}I_2$ cell while the $^{129}I_2$ fluorescence is not affected [8.22]. This demonstrates that those modes of the broad-band dye laser which are absorbed by the internal $^{127}I_2$ are completely suppressed.

b) Detection of absorbing transitions with very small oscillator strength (see Sect.2.6.2) has been demonstrated by BRAY et al. [8.23] who measured with a cw Rhodamine B dye laser (0.3 nm bandwidth) the extremely weak (2,0) absorption band of the red atmospheric system of molecular oxygen and also the (6,0) overtone band of HCl, using a 97 cm long intracavity absorption cell. Sensitivity tests indicated that even transitions with oscillator strengths down to $f \leqq 10^{-12}$ can be readily detected.

c) For the detection of short-lived radicals the transient response of the intracavity absorption has to be considered. When the frequency of a mode is tuned into resonance with an absorption line, this mode is not suppressed instantaneously but its intensity shows a transient behavior, with a buildup time dependent on the coupling parameter K in (8.16). The influence of pumping time on the buildup of the narrow spectral holes in the broad-band output of a pulsed dye laser has been quantitatively studied by ANTONOV et al. [8.24]. These authors chopped the argon laser and, varying the pumping time from 20 to 400 μs, they could determine the loss of sensitivity with decreasing pumping time.

8.2.4 Optogalvanic Spectroscopy

Optogalvanic spectroscopy is an excellent and simple technique to perform laser spectroscopy in gas discharges. Assume that the laser beam passes through part of the discharge volume. When the laser frequency is tuned to a transition $E_i \to E_k$ between two levels of atoms or ions in the discharge, the population densities $n_i(E_i)$ and $n_k(E_k)$ are changed by optical pumping. Because of the different ionization probabilities from the two levels, this population

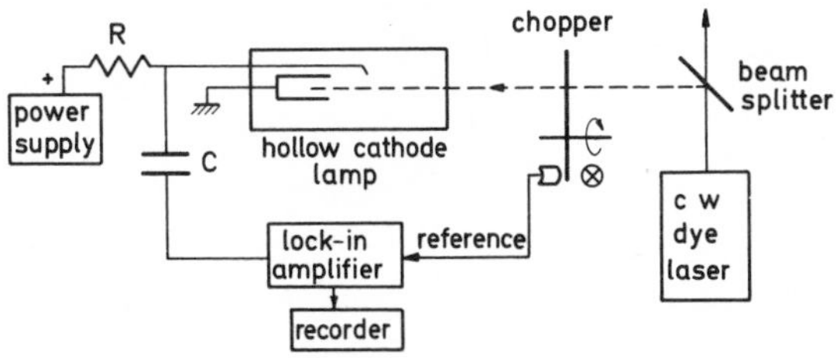

Fig. 8.10. Experimental arrangement of optogalvanic spectroscopy in a hollow cathode lamp

change will result in a change ΔI of the discharge current which is detected as a voltage change $\Delta U = R\Delta I$ across the ballast resistor R (Fig.8.10). When the laser intensity is chopped, an ac voltage is obtained which can be directly fed into a lock-in amplifier.

Already with moderate laser powers (a few milliwatts) large signals (μV to mV) can be achieved in gas discharges of several milliamperes [8.25]. Since the absorbed laser photons are detected by the optically induced current change, this very sensitive technique is called *optogalvanic spectroscopy* [8.26].

Both positive and negative signals are observed, depending on the levels E_i, E_k involved in the laser-induced transition $E_i \to E_k$. If $IP(E_i)$ is the total ionization probability of an atom in level E_i, the voltage change ΔU produced by the laser-induced population change $\Delta n_i = n_{i0} - n_{iL}$ is given by

$$\Delta U = R\Delta I = a[\Delta n_i IP(E_i) - \Delta n_k IP(E_k)] \quad . \tag{8.17}$$

There are several competing processes which may contribute to ionize atoms in level E_i, such as direct ionization by electron impact

$$A(E_i) + e \to A^+ + 2e \quad ,$$

collisional ionization by metastable atoms

$$A(E_i) + A^* \to A^+ + A + e \quad ,$$

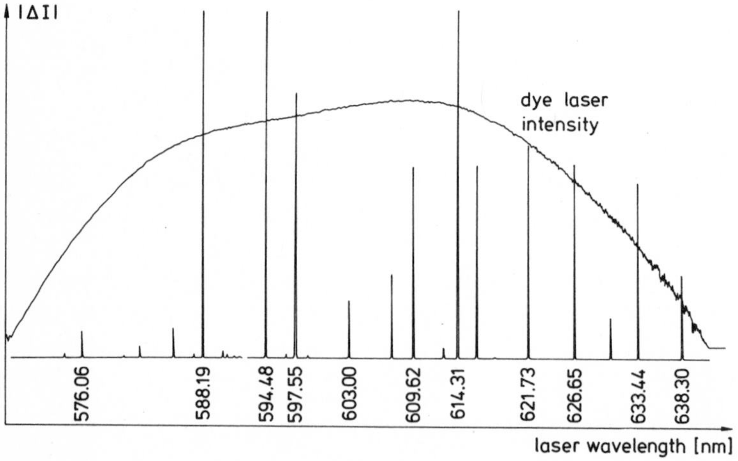

Fig.8.11. Optogalvanic spectrum of the neon discharge in a neon spectral lamp recorded with a lock-in, which detects $|\Delta U|$. The sign of the signals is therefore suppressed

or, in particular for highly excited levels, the direct photoionization by laser photons

$$A(E_i) + h\nu \rightarrow A^+ + e^- \quad .$$

The competition of these and other processes determines whether the population changes Δn_i and Δn_k cause an increase or a decrease of the discharge current. Figure 8.11 shows the optogalvanic spectrum of a Ne-discharge (5 mA) recorded in a fast scan with 0.1 s time constant. The good signal-to-noise ratio demonstrates the sensitivity of the method.

Besides its applications to studies of collision processes and ionization probabilities in gas discharges, this technique has the very useful aspect of simple wavelength calibration in laser spectroscopy [8.27]. A small fraction of the output from a tunable laser is split into a hollow cathode spectral lamp and the optogalvanic spectrum of the discharge is recorded simultaneously with the unknown spectrum under investigation (e.g., by a two-pen recorder). The numerous lines of thorium or uranium are nearly equally distributed throughout the visible and ultraviolet spectral regions and are recommended as secondary wavelength standards since they have been measured interferometrically to a high precision [8.27a,b]. They can therefore serve as convenient absolute wavelength markers, accurate to about 0.001 cm^{-1}. Figure 8.12 shows an optogalvanic spectrum of an uranium hollow cathode lamp which has been obtained by shining 10 mW from a cw dye laser into an uranium hollow cathode lamp filled with 1 Torr of argon. At discharge currents of less than 10 mA the ArI lines predominate (upper spectrum) while at higher currents the cathode material sputters more efficiently and the uranium lines become stronger (lower spectrum).

If the discharge cell has windows of optical quality, it can be placed inside the laser resonator taking advantage of the q-fold laser intensity (see Sect.8.2.3). With such an intracavity arrangement Doppler-free saturation spectroscopy can also be performed (see Sect.10.2 and [8.28]). Increased sensitivity can be achieved by optogalvanic spectroscopy in thermonic diodes under space-charge-limited conditions. Here the internal space charge amplification is utilized to generate signals in the mV to V range without further external amplification [8.29a,b].

For more details on optogalvanic spectroscopy see [8.25-29].

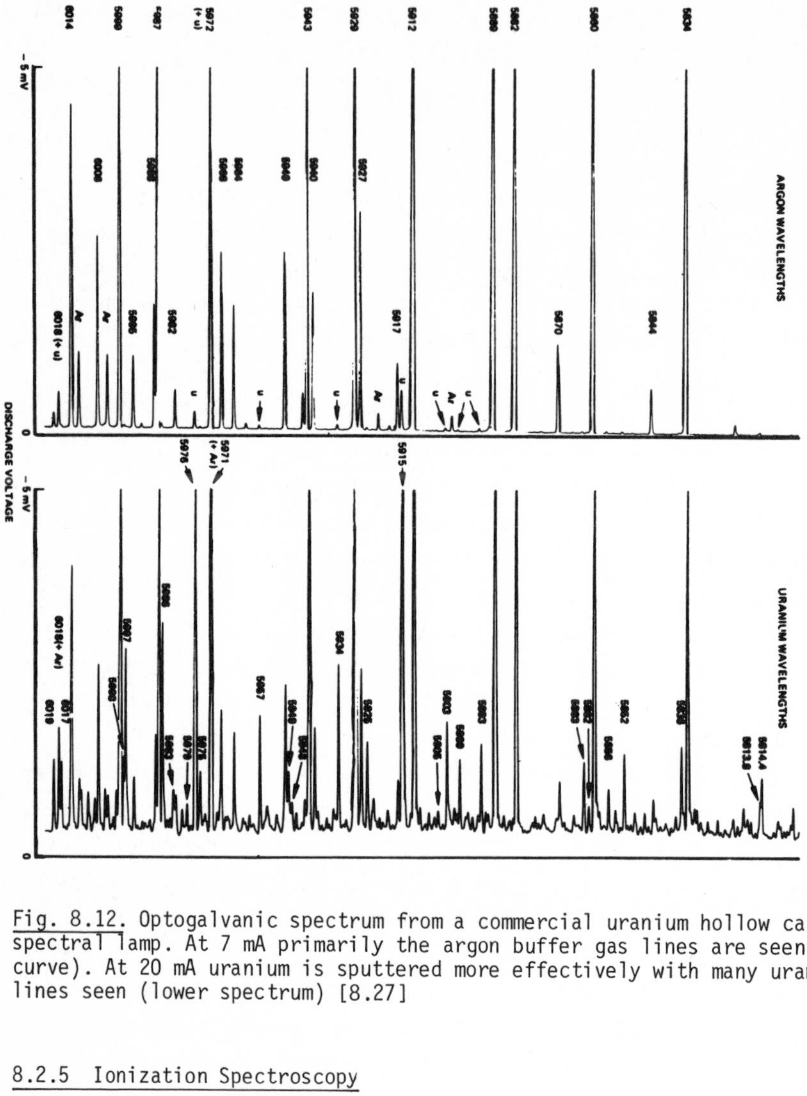

Fig. 8.12. Optogalvanic spectrum from a commercial uranium hollow cathode spectral lamp. At 7 mA primarily the argon buffer gas lines are seen (upper curve). At 20 mA uranium is sputtered more effectively with many uranium lines seen (lower spectrum) [8.27]

8.2.5 Ionization Spectroscopy

Ionization spectroscopy monitors the absorption of photons on a molecular transition $E_i \rightarrow E_k$ by detecting the ions or electrons, produced by some means, while the molecule is in its excited state E_k. The necessary ionization of the excited molecule may be performed in various ways.

a) *Photoionization*

$$M^*(E_k) + h\nu \rightarrow M^+ + e^- + E_{kin} \quad . \tag{8.18}$$

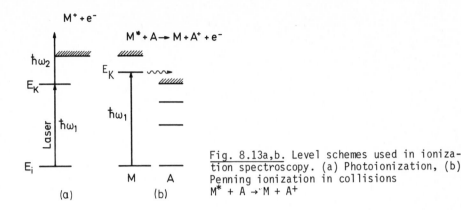

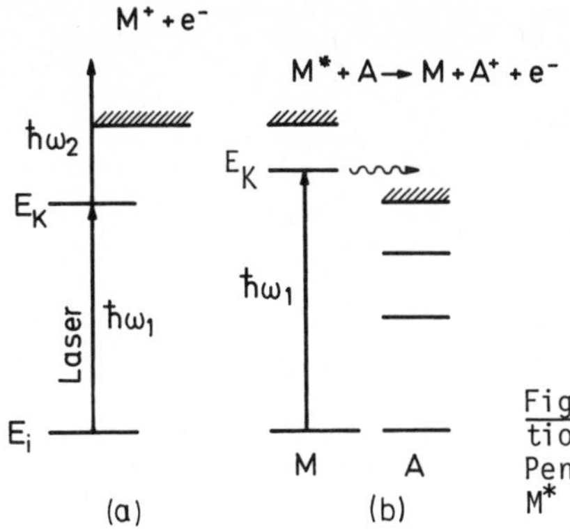

Fig. 8.13a,b. Level schemes used in ionization spectroscopy. (a) Photoionization, (b) Penning ionization in collisions $M^* + A \rightarrow M + A^+$

The ionizing photon may come either from the same laser which has excited the level E_k, or from a separate light source, which can be another laser or even an incoherent source (see Fig.8.13).

b) *Collision-induced ionization*

If the excited level E_k is not too far from the ionization limit, the molecule may be ionized by thermal collisions with other atoms. If E_k lies above the ionization limit of the collision partners A, Penning ionization becomes an efficient process [8.30] which proceeds as

$$M^*(E_k) + A \rightarrow M + A^+ + e^- \quad . \tag{8.19}$$

c) *Field ionization*

If the excited level E_k lies close below the ionization limit, the molecule $M^*(E_k)$ can be ionized by an external electric dc field. This method is particularly efficient if the excited level is a long-lived highly excited Rydberg state. The required minimum electric field can be readily estimated from Bohr's atomic model which gives a good approximation for atomic levels with large principal quantum number n. The ionization potential for the outer electron at a mean radius r from the nucleus is determined by the Coulomb field of the nucleus shielded by the inner electron core.

$$eV_{ion} = \int_r^\infty \frac{Z_{eff}e^2}{4\pi\varepsilon_0 r^2}\,dr = \int_r^\infty E_{eff}dr \quad , \tag{8.20}$$

where eZ_{eff} is the effective nuclear charge, i.e. the nuclear charge Ze partly screened by the electron cloud. If the external field $E_{ext}(r)$ becomes larger than $E_{eff}(r)$, field ionization will occur (Fig.8.14).

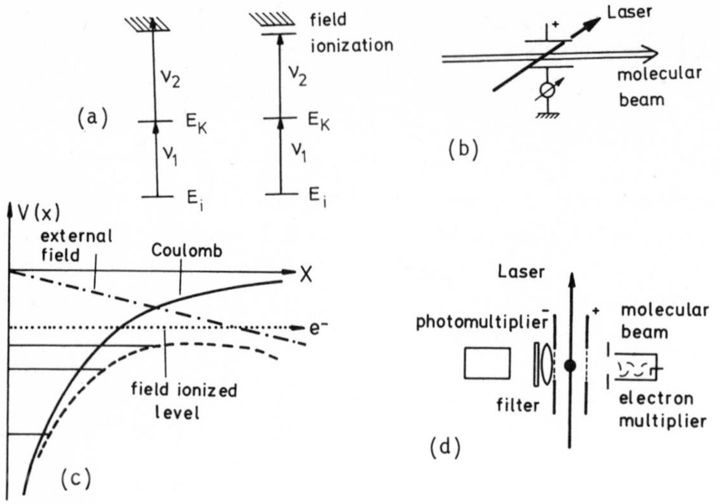

Fig.8.14a-d. Direct two-photon ionization (a) and two-photon excitation with subsequent field ionization (b). Schematic energy level scheme for field ionization of atoms (c), and experimental arrangement (d)

Example: For levels 10 meV below the ionization limit, (8.20) gives $E \geq 10^5$ V/m for the ionizing external field. However, because of the quantum mechanical tunnel effect the fields required for complete ionization are even lower.

With a proper design the *collection* efficiency δ for the ionized electrons or ions can reach 100%. If the electrons or ions are accelerated to several keV and detected by electron multipliers or channeltrons a *detection* efficiency of $\eta = 100\%$ can also be achieved.

The following estimation illustrates the possible sensitivity of ionization spectroscopy. Let N_k be the density of excited molecules in level E_k, P_{kI} the probability per second that a molecule in level E_k is ionized, and n_a the number of photons absorbed per second on the transition $E_i \to E_k$. If R_k is the total relaxation rate of level E_k, besides the ionization rate (spontaneous transitions plus collision-induced deactivation) the signal rate in counts per second for an absorption pathlength Δx and n_L incident laser photons is given by

$$S = N_k P_{kI} \delta\eta = n_a \frac{P_{kI}}{P_{kI}+R_k} \delta\eta = N_i n_L \sigma_{ik} \Delta x \frac{P_{kI}}{P_{kI}+R_k} \delta\eta \quad . \tag{8.21}$$

In the case of photoionization the ionization probability P_{kI} depends on the intensity of the ionizing radiation. Using intense lasers the ionization rate can be made large compared to the relaxation rate R_k. For the ideal case of

$\delta = \eta = 1$ and $P_{kI} \gg R_k$, (8.21) shows that a signal rate $S = n_a$ can be achieved which equals the absorption rate n_a of photons on the transition $E_i \rightarrow E_k$. This implies that *single absorbed photons* can be detected with an overall efficiency close to unity [8.31]. In experimental practice there are, of course, additional losses and sources of noise which limit the detection efficiency to a somewhat lower level. However, for all absorbing transitions $E_i \rightarrow E_k$, where the upper level E_k can be readily ionized, ionization spectroscopy is the most sensitive detection technique and superior to all other methods discussed so far [8.31a,b].

8.3 Laser Magnetic Resonance and Stark Spectroscopy

In all methods discussed in the previous sections of this chapter, the laser frequency ω_L was tuned across the constant frequencies ω_{ik} of molecular absorption lines. For molecules with permanent magnetic or electric dipole moments, it is often preferable to tune the absorption lines by means of external magnetic or electric fields across a fixed frequency laser line. This is particularly advantageous if intense lines of fixed-frequency lasers exist in the spectral region of interest but no tunable source with sufficient intensity is available. Such spectral regions of interest are, for example, the 3-5 μm and the 10 μm ranges, where numerous intense lines of HF, DF, CO, N_2O, and CO_2 lasers can be used. Since many of the vibrational bands fall into this spectral region, it is often called the fingerprint range of molecules.

Another spectral range of interest is the far infrared of the rotational lines of polar molecules. Here a large number of lines from H_2O or D_2O lasers (125 μm) and from HCN lasers (330 μm) provide intense sources. The successful development of numerous optically pumped molecular lasers [8.31c] has increased the number of FIR lines considerably.

8.3.1 Laser Magnetic Resonance

A molecular level E_0 with total angular momentum J splits in an external magnetic field B into (2J + 1) Zeeman components. The sublevel with magnetic quantum number M shifts from the energy E_0 at zero field to

$$E = E_0 - g\mu_0 BM \quad , \qquad (8.22)$$

where μ_0 is the Bohr magneton and g is the Landé factor which depends on the coupling scheme of the different angular momenta (electronic angular momentum,

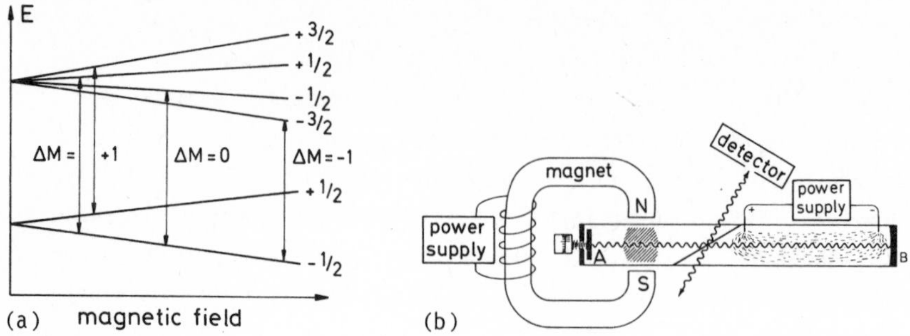

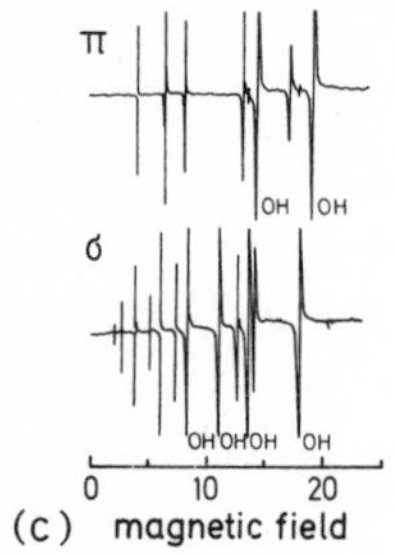

Fig. 8.15a-c. Laser magnetic resonance spectroscopy. (a) Energy level scheme, (b) experimental arrangement, (c) LMR spectrum of $CH(X^2\Pi)$ with several OH-lines in a low pressure oxygenacetylene flame, obtained with the H_2O laser [8.33]

electron spin, molecular rotation, and nuclear spin). The frequency ω of a transition $(v'',J'',M'') \rightarrow (v',J',M')$ is therefore tuned by the magnetic field from its unperturbed position ω_0 to

$$\omega = \omega_0 + \mu_0(g'M' - g''M'')B/\hbar \qquad (8.23)$$

and we obtain on the transition $(v'',J'',M'') \rightarrow (v',J',M')$ three groups of lines with $\Delta M = M'' - M' = 0, \pm 1$, which degenerate to three single lines if $g'' = g'$ (normal Zeeman effect). The tuning range depends on the magnitude of $g'' - g'$ and is larger for molecules with a large permanent dipole moment. In favorable cases a tuning range of up to 2 cm^{-1} can be reached in magnetic fields of 20 KG.

Figure 8.15a explains schematically the appearance of resonances between a fixed laser frequency ω_L and the different Zeeman components when the magnetic field B is tuned. The experimental arrangement is illustrated in Fig. 8.15b. The sample is placed inside the laser cavity and the laser output is monitored as a function of magnetic field. The sensitivity of this intracavity technique (see Sect.8.2.3) can be even enhanced by modulating the magnetic field, which yields the first derivative of the spectrum (see Sect.8.1). The cell is part of a flow system in which radicals are generated either directly in a microwave discharge or by adding reactants to the discharge close to the laser cavity. A polyethylene membrane beam splitter separates the laser medium from the sample. The beam splitter polarizes the radiation and transitions with either $\Delta M = 0$ or ± 1 can be selected by rotation about the laser axis. Concentrations of 2×10^8 molecules/cm^3 could be still detected with reasonable signal-to-noise ratio at a detector time constant of 1 s [8.32].

Because of its high sensitivity LMR spectroscopy is an excellent method to detect radicals in very low concentrations and to measure their spectra with high precision. If a sufficient number of resonances with laser lines can be found, the rotational constants, the fine structure parameters, and the magnetic moments can be determined very accurately. The identification of the spectra and the assignment of the lines are often possible even if the molecular constants have not been known before [8.33]. All radicals observed in interstellar space by radio astronomy have been found and measured with LMR spectroscopy.

Often a combination of LMR spectroscopy with a fixed laser frequency and absorption spectroscopy at zero magnetic field with a tunable laser is helpful for the identification of spectra. If the magnetic field is modulated around zero the phase of the zero field LMR resonances for $\Delta M = +1$ transitions is opposite to that for $\Delta M = -1$ transitions. The advantages of this zero field LMR spectroscopy have been proved for the NO molecule by URBAN et al. [8.34] using a spin-flip Raman laser.

8.3.2 Stark Spectroscopy

Analogous to the LMR technique, Stark spectroscopy utilizes the Stark shift of molecular levels in *electric* fields to tune molecular absorption lines into resonance with lines of fixed-frequency lasers. A number of small molecules with permanent electric dipole moments and sufficiently large Stark shifts have been investigated, in particular those molecules which have rotational spectra outside spectral regions accessible to conventional microwave spectroscopy [8.35].

To achieve large electric fields the separation of the Stark electrodes is made as small as possible (typically about 1 mm). This generally excludes an intracavity arrangement because the diffraction by this narrow aperture would introduce intolerably large losses. The Stark cell is therefore placed outside the resonator and for enhanced sensitivity the electric field is modulated while the dc field is tuned. This modulation technique is also common in microwave spectroscopy. Figure 8.16 is a block diagram of the experimental setup for laser Stark spectroscopy in the 10 μm region. The accuracy of 10^{-4} for the Stark field measurements allows a precise determination of

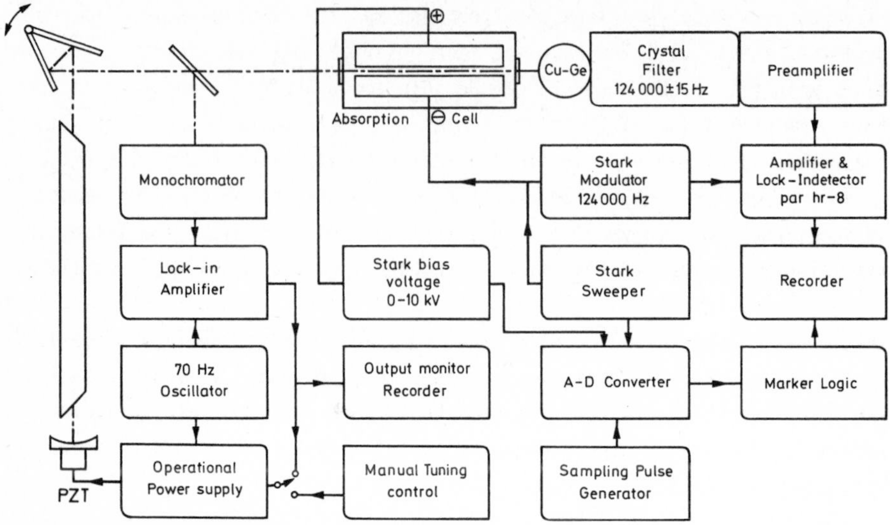

Fig. 8.16. Block diagram of the CO_2 and N_2O laser Stark spectrometer [8.35]

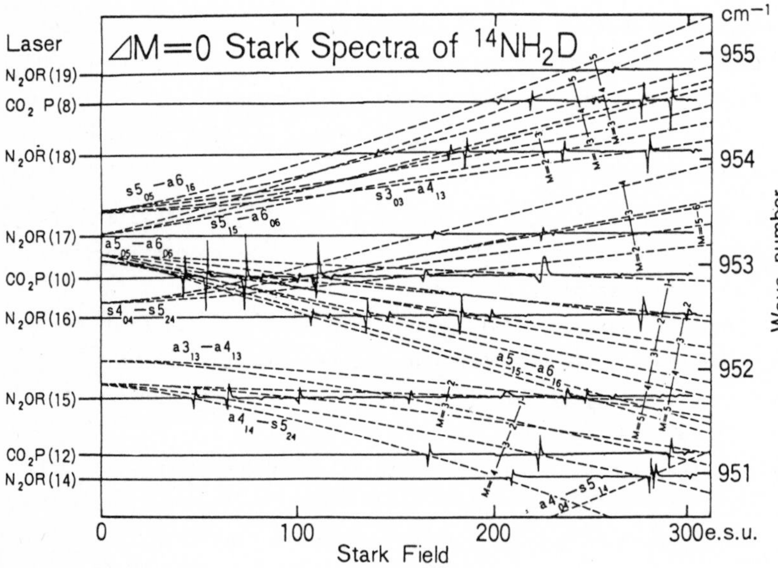

Fig. 8.17. $\Delta M = 0$ Stark spectra of $^{14}NH_2D$ displaced on an XY recorder. The Y coordinate was adjusted to scale the wave number of each laser line [8.35]

the absolute value for the electric dipole moment. The laser frequency is feedback stabilized at the maximum of the monochromator output.

Figure 8.17 illustrates the obtainable sensitivity by a $\Delta M = 0$ Stark spectrum of the ammonia isotope $^{14}NH_2D$ composed of measurements with several laser lines [8.35]. Since the absolute frequency of many laser lines has been accurately measured within 20-40 kHz (see Sect.6.10), the absolute frequency of the Stark components at resonance with the laser line can be measured with the same accuracy. The total accuracy in the determination of the molecular parameters is therefore mainly limited by the accuracy of 10^{-4} for the electric field measurements. To date numerous molecules have been measured with laser Stark spectroscopy [8.36]. The number of molecules accessible to this technique can be vastly enlarged if tunable lasers in the relevant spectral regions are developed which can be stabilized close to a molecular line with sufficient accuracy and long-term stability.

8.4 Comparison Between the Different Methods

The different sensitive techniques of Doppler-limited laser spectroscopy, discussed in the previous sections, supplement each other in an ideal way. In the *visible and ultraviolet* range, where *electronic* states of atoms or molecules are excited by absorption of laser photons *excitation spectroscopy* is generally the most suitable technique, particularly at low molecular densities. Because of the short spontaneous lifetimes of most excited electronic states E_k, the quantum efficiency η_k reaches 100% in many cases. For the detection of the laser-excited fluorescence, sensitive photomultipliers are available which allow, together with photon counting electronics (see Sect. 4.5), the detection of single fluorescence photons with an overall efficiency of 10^{-3}-10^{-1} (see Sect.8.2.1).

Excitation of very high-lying states close below the ionization limit, e.g., by ultraviolet lasers or by two-photon absorption, enables the detection of absorbed laser photons by monitoring the ions. Because of the high collection efficiency of these ions *ionization spectroscopy* represents the most sensitive detection method, superior to all other techniques in all cases where it can be applied.

In the *infrared* region excitation spectroscopy is less sensitive because of the lower sensitivity of infrared photodetectors and because of the longer lifetimes of excited vibrational levels. These long lifetimes bring about either at low pressures a diffusion of the excited molecules out of the observation region or at high pressures a collision-induced radiationless deactivation of the excited states. Here *photoacoustic spectroscopy* becomes superior, since this technique utilizes this collision-induced transfer of excitation energy into thermal energy. A specific application of this technique is the quantitative determination of small concentrations of molecular components in gases at higher pressures. Examples are measurements of air pollution, or of poisonous constituents in auto engine exhausts, where sensitivities in the ppb range have been successfully demonstrated.

Regarding detection sensitivity, LMR and Stark spectroscopy can compete with the other methods. However, their applications are restricted to molecules with sufficiently large permanent dipole moments to achieve the necessary tuning range. They are therefore mainly applied to the spectroscopy of free radicals with an unpaired electron. The magnetic moment of these radicals is predominantly determined by the electron spin and is therefore several orders of magnitude larger than that of stable molecules in $^1\Sigma$ ground states. The advantage of LMR or Stark spectroscopy is the direct determination of Zeeman or Stark splittings from which the Landé factors and therefore the coupling schemes of the different angular momenta can be deduced. A further merit is the higher accuracy of the *absolute* frequency of molecular absorption lines because the frequencies of fixed-frequency laser lines can be absolutely measured with a higher accuracy than is possible with tunable lasers.

For the spectroscopy of atoms or ions in gas discharges, *optogalvanic spectroscopy* is a very convenient and experimentally simple alternative to fluorescence detection. In favorable cases it may even reach the sensitivity of excitation spectroscopy.

All these methods can be performed within the laser resonator to enhance the sensitivity. Only Stark spectroscopy with high electric fields, which demand small separations of the Stark plates, is restricted to external sample cells.

8.5 Examples for Doppler-Limited Absorption Spectroscopy with Lasers

In Chap.7 we briefly presented the basic principles of different tunable coherent sources. This section illustrates their applications to Doppler-limited absorption spectroscopy performed either with conventional absorption measurements or based on one of the sensitive detection methods discussed above. Only a few examples could be selected from an abundant literature on this subject; the selection if of course more or less arbitrary.

The spectral range particularly important for molecular physics is the near and medium infrared region between 2 and 20 μm because here the vibrational bands of most molecules can be found. Doppler-limited laser spectroscopy already allows in many cases the selection of different single rotational-vibrational transitions and the resolution of their fine and hyperfine structures. Measurements of the Λ doubling give information about the coupling of the molecular rotation with the angular momentum of the electrons. The hyperfine structure in molecules is generally more complicated than in atoms because the nuclear spins of several nuclei, located at different positions with respect to the electron cloud, couple not only with the electronic angular momenta but also with those of the molecular rotation. The vibrations of the molecule further complicate the situation. For a thorough understanding of the molecular ground states many different experimental informations are needed. In particular the influence of vibrations on the line positions can be studied by infrared laser spectroscopy while the conventional microwave spectroscopy is mainly restricted to vibrational ground states.

In this spectral range it is up to now the semiconductor diode laser which has been mostly used for high-resolution laser spectroscopy. After HINKLEY [8.37] had first demonstrated in 1970 the superiority of laser spectroscopy over conventional techniques by a high-resolution absorption spectrum of SF_6 around 10 μm obtained with a Pb-Sn-Te laser, spectra of numerous molecules have been measured with different diode lasers. A review of the literature up to 1975 can be found in the article by HINKLEY et al. [8.38], more recent examples in [8.39] and [8.39a].

Because of the good signal-to-noise ratio the tunable diode lasers offer, besides the high spectral resolution, the further advantage of fast scans through extended spectral intervals. Figure 8.18 illustrates this by a section of the H_2CO absorption spectrum around 2800 cm^{-1}, where the laser was scanned in 10 ms over a spectral interval of 1 cm^{-1}. Other examples are completely resolved spectra of SO_2 between 1176 and 1265 cm^{-1} measured by ALLARIO et al. [8.40] with a PbSe laser. Frequency tuning was achieved by a combined control of diode current and magnetic field. With a Pb-Sn-Te laser 169 rotational lines in a vibrational band of ethylene around 942.2 cm^{-1} could be resolved. The line separation was measured within 0.008 cm^{-1} which allows accurate determination of the deviations of the C_2H_4 molecule from a symmetric top [8.41].

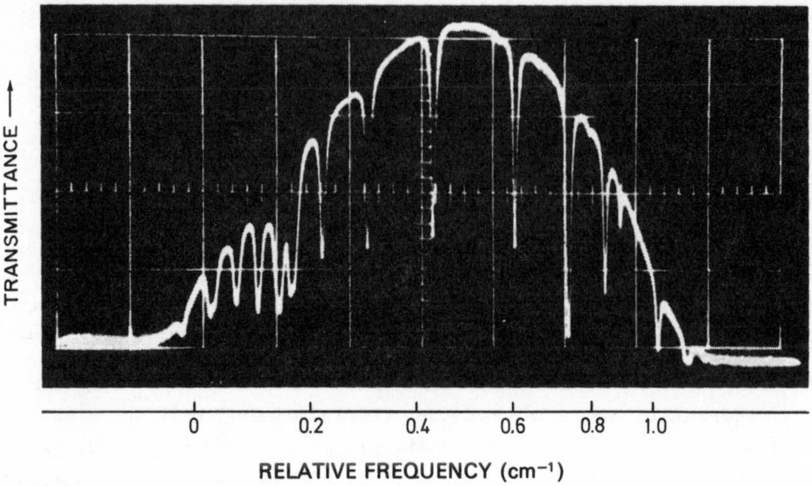

Fig. 8.18. Time-resolved absorption spectrum of H_2CO at 2800 cm^{-1}. The spectral interval of 1 cm^{-1} was scanned within 10 ms. Gas pressure of 5 Torr, length of the absorption cell 10 cm [8.38]

Coherent sources which are based on difference frequency generation (see Sect.7.2) have proved to be very successful in high-resolution infrared spectroscopy. One example is the *difference frequency spectrometer* developed by PINE [8.42,43], where the outputs from a single-mode argon laser and from a tunable single-mode cw dye laser are mixed in a $LiNbO_3$ crystal (see Sect. 7.5.4). The difference frequency radiation with output powers of a few μW at a linewidth of 15 MHz (5×10^{-4} cm^{-1}) could be continuously tuned from 2.2 to 4.2 μm simply by tuning the dye laser and the phase-matching temperature of the crystal. With this difference frequency spectrometer, absorption spectra of $^{12}CH_4$ and $^{13}CH_4$ have been measured with a resolution limited only by the Doppler width. The accuracy achieved was high enough to study in detail the Coriolis splitting of rotational levels and to decide which of two theoretical models of the Coriolis coupling best describes the molecule.

With a spin-flip Raman laser BUTCHER et al. [8.44] measured the optoacoustic spectrum of a NO band. The Doppler-limited linewidth of 127 MHz, achieved at low NO pressures, allowed the complete resolution of the Λ doubling of about 700 MHz and even the partial resolution of the hyperfine structure of each component. At a laser power of 10 mW the spectral interval of 2.5 cm^{-1} could be scanned within 10 min at a spectral resolution of 2×10^5 and a signal-to-noise ratio of 200:1. Measurements with comparable resolution have been performed on the ^{15}NO isotope [8.45] (see also Fig.8.6).

Immersion of the spectraphone in a magnetic field gives an instrument that can be used to study the Zeeman effect of molecules by optoacoustic spectroscopy. This magnetospectraphone has been used to measure the Zeeman splittings of NO [8.46].

Also in the visible and near ultraviolet region meanwhile numerous examples of Doppler-limited laser spectroscopy have been published. Here the tunable dye laser in its various modifications (pulsed or cw) is the dominant source, since up to now it has the widest tuning range, and laser linewidths can be achieved which are small compared to the Doppler width (see Sect.7.3). A few examples illustrate this. Using a cw dye laser (10 mW single mode with a linewidth of $\leqq$30 MHz) FIELD et al. [8.47] measured excitation spectra of CaF and deduced from the high-resolution spectra the molecular constants, including the Λ doubling, with high precision.

To demonstrate the superiority of laser spectroscopy GREEN et al. [8.48] compared the absorption spectrum of I_2 at λ = 589 nm as measured with a dye laser with the spectrum taken in a conventional way using a 7.3 m Ebert spectrograph with an echelle grating in 10th order (see Sect.4.2). Regarding spectral resolution as well as the time spent to take the spectrum, the laser spectrometer of course won the competition.

The visible spectrum of most polyatomic molecules is generally so complex that resolution of single lines is only seldom possible with conventional absorption spectroscopy. Even Doppler-limited laser spectroscopy may leave many partially overlapping lines. A good example where the transition region from a medium to a high line density has been studied in detail is the band head of the (1,00) → (0,00) band in the $X^2\Pi_{3/2} \rightarrow A^2\Pi_{3/2}$ transition of BO_2 [8.49], where (n_1, n_2, n_3) gives the number of vibrational quanta in the three normal vibrations of BO_2. Figure 8.19 shows a section of the excitation spectrum of this band head. While in conventional absorption spectroscopy several bands overlap in this region, excitation spectroscopy allows suppression of other than the desired bands by a simple trick. If the laser-excited fluorescence is observed through a monochromator which selects the R lines of the $(0,0,0)A^2\Pi_{3/2} \rightarrow (0,0,0)X^2\Pi_{3/2}$ transitions, only those excitation lines are monitored which excite the upper levels of these R fluorescence lines. The laser-induced fluorescence spectrum reveals, however, that in spite of this filtering technique each excitation line still consists of two rotational lines which overlap within their Doppler width. For a complete resolution therefore Doppler-free techniques are demanded (see Chap.10).

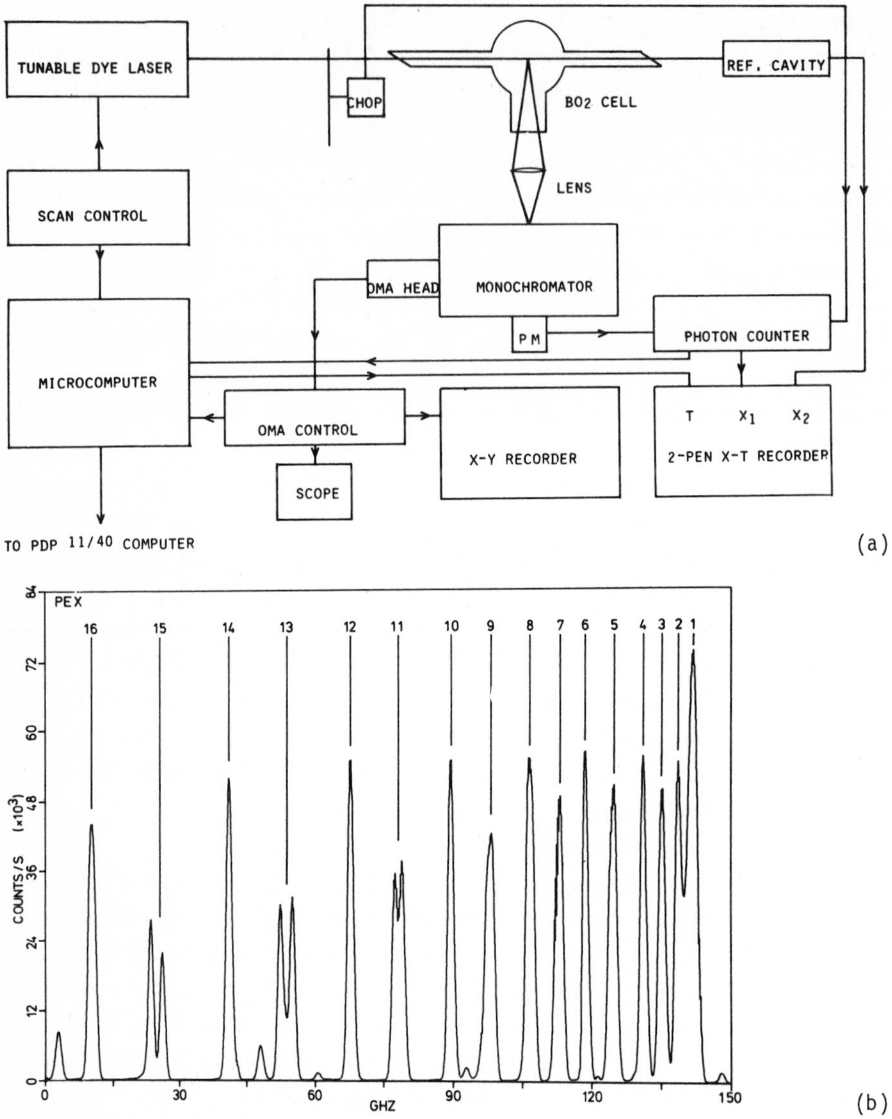

Fig. 8.19. (a) Experimental arrangement for combination of laser-induced fluorescence and photoexcitation spectroscopy. (b) Excitation spectrum of the $A^2\Pi_{3/2}(0,0,0) \rightarrow X^2\Pi_{3/2}(1,0,0)$ band of $^{11}BO_2$ at 579 nm. The numbering of the lines was only for internal use [8.49].

8.6 Optical Pumping with Lasers

Optical pumping means the selective population or depletion of atomic or molecular levels by absorption of radiation, resulting in a population change

ΔN in these levels which causes a noticeable deviation from the thermal equilibrium population. With intense atomic resonance lines, emitted from hollow cathode lamps or from microwave discharge lamps optical pumping had been successfully used for a long time in *atomic* spectroscopy, even before the invention of lasers [8.50,51]. However, the introduction of lasers as very powerful pumping sources with narrow linewidths has substantially increased the application range of optical pumping. In particular lasers have facilitated the transfer of this well-developed technique to *molecular* spectroscopy. While early experiments on optical pumping of molecules [8.52] were restricted to fortunate coincidences between molecular absorption lines and atomic resonance lines from incoherent sources, the possibility of tuning a laser to the desired molecular transition provides a much more selective and effective pumping process, and allows, because of the larger intensity, a much higher signal-to-noise ratio.

There are several different aspects of optical pumping which are related to a number of spectroscopic techniques based on optical pumping. The *first aspect* concerns the *increase or decrease of the population* in selected excited levels. With lasers as pumping sources large population densities in excited levels can be achieved which may become comparable to that of the absorbing ground states. This is particularly important in optical pumping of *molecules*, where these selectively populated levels emit fluorescence spectra which can be readily assigned and which allow determination of the molecular constants (see Sect.8.7). A sufficiently large population of the upper state furthermore allows the measurement of absorption spectra for transitions from this state to still higher lying levels (excited state spectroscopy, stepwise excitation; see Sect.8.8). The increased population of the upper state or the selectively depleted lower state facilitates all kinds of double-resonance experiments where a second E.M. field (optical, microwave, or radio frequency) induces transitions from E_1 or E_2 to other levels (see Sect.10.7). Figure 8.20 depicts schematically the different possible experiments which are based on population changes ΔN induced by optical pumping.

In this section we restrict the discussion mainly to optical pumping by cw or pulsed lasers with a linewidth larger than the Doppler width of the absorbing transition $E_1 \rightarrow E_2$. In this case all molecules in level E_1 within the thermal velocity distribution $N_1(v)$ can be pumped by the laser independent of their specific velocity v.

The degree to which the lower level can be depleted depends on the number of levels involved in the pumping-relaxation cycle and on the corresponding transition probabilities. Under stationary conditions the low level of a *two-*

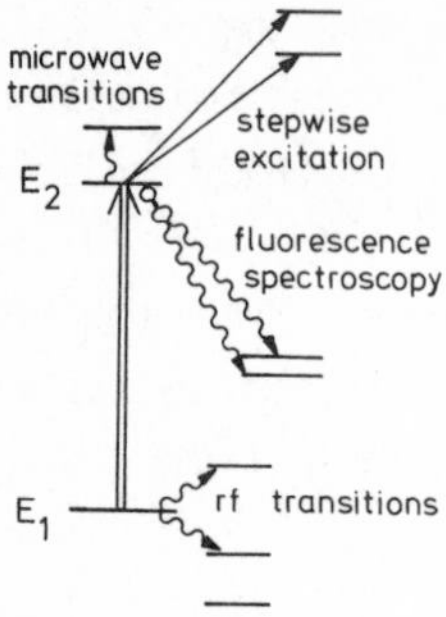

Fig.8.20. Schematic illustration of different spectroscopic techniques based on the population changes of selectively pumped molecular levels

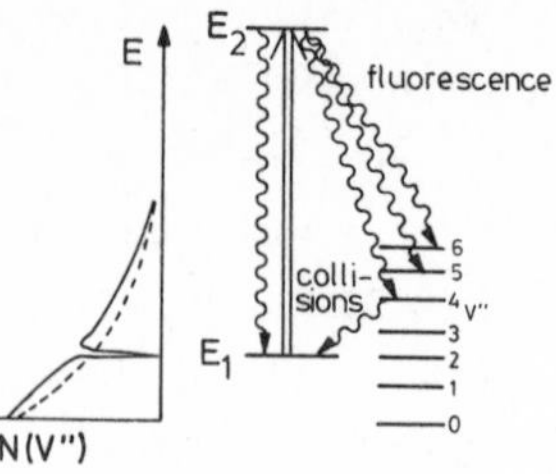

Fig.8.21. Term diagram for the depopulation of a molecular level by optical pumping, and population distribution of vibrational levels without (---) and with (——) optical pumping

level system cannot be depleted to more than 1/2 of its unsaturated population because stimulated emission starts to refill it (see Sect.2.8). In a multilevel system, however, complete depletion of the initial level can be achieved if the pump rate exceeds the rate of repopulation by relaxation processes. In optical pumping of molecules, for example, only a small fraction of the total fluorescence emitted from the excited level terminates on the initial state E_1 while most of it populates other rotational-vibrational levels (v",J") of the electronic ground state (Fig.8.21). This implies that a selected molecular level (v_i'', J_i'') can be completely depopulated.

The *second aspect* of optical pumping refers to the generation of *orientation* or of *alignment* in selectively pumped states. Atoms or molecules in a state (J,M) with total angular momentum J and magnetic quantum number M are called *oriented* if a nonuniform population of the different M sublevels has been produced. In a classical picture the angular momentum vector J in oriented states has a preferential direction and no longer a random orientation as under thermal equilibrium conditions.

Certain pumping schemes generate equal populations for each pair of +M and -M sublevels but different populations for levels with different $|M|$ values. They produce *alignment*. Note that orientation and alignment are generated both in the upper state by a nonuniform *increase* of the M level populations and in the lower state by a nonuniform *depletion*. Figure 8.22 illustrates these considerations by some *examples*.

On a transition $J_1 = 0 \rightarrow J_2 = 1$, optical pumping with left-hand circularly polarized light (σ^+ polarization) induces $\Delta M = +1$ transitions and only the sublevel M = +1 will be populated in the upper state. Linearly polarized light

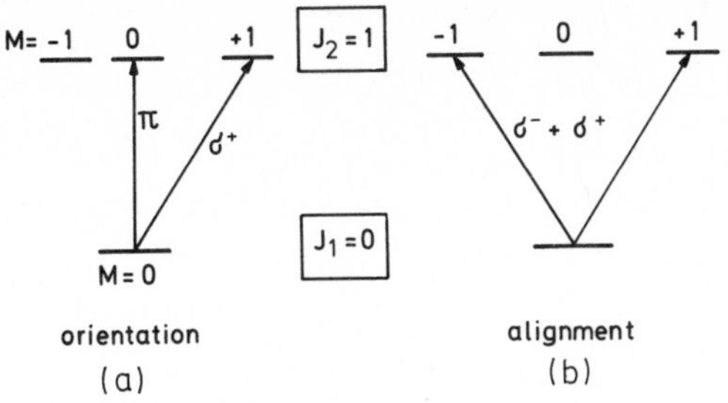

Fig. 8.22. Optical pumping of a transition $J_1=0 \to J_2=1$ by (a) σ^+ light, or π light; (b) pumping of a $J_1=0 \to J_2=1$ transition by directional unpolarized light

(π polarization) induces $\Delta M = 0$ transitions and therefore populates only the $M = 0$ level. Unpolarized directional light (e.g., an unpolarized laser beam) which may be regarded as a random superposition of σ^+ and σ^- light, can induce both $\Delta M = \pm 1$ transitions and therefore populates the two sublevels $M = \pm 1$. While in the first case *orientation* has been generated in the upper state, the last case is an example of *alignment*.

For a more quantitative discussion we start from the rate equations for the optical pumping process. Assume that the laser has been tuned to a transition $E_1 \to E_2$ between two atomic levels $E_1(J_1,M_1)$ and $E_2(J_2,M_2)$. Without external fields the $2J + 1$ sublevels M of the degenerate level J are equally populated before optical pumping starts.

$$N^0(J,M) = N^0(J)/(2J+1) \quad . \tag{8.24}$$

The population changes $\Delta N(J,M)$ induced by the optical pumping process depend on the intensity I_p of the pump laser, its polarization characteristics, its spectral line profile, and also on possible relaxation processes, such as fluorescence or collision-induced transitions which try to restore the equilibrium population. The rate equation for the depopulation $dN_1(M)/dt$ of a lower sublevel M is

$$\frac{d}{dt} N_1(J_1M_1) = \sum_{M_2} P(J_1,M_1,J_2,M_2)(N_2-N_1) + \sum_k R_{k1}N_k - \sum_k R_{1k}N_1 \quad , \tag{8.25}$$

where the optical pumping rate $P(N_2-N_1)$ is determined by the square of the corresponding matrix element (see Sect.2.9)

$$P_{12}(J_1,M_1,J_2,M_2) = |\langle J_1M_1|\boldsymbol{\mu}_{12}\cdot\boldsymbol{E}|J_2M_2\rangle|^2 \tag{8.26}$$

and the sum extends over all possible transitions between the lower level M_1 to all upper levels M_2 which can be induced by the pump field. The second and third terms in (8.25) represent the total net repopulation rate due to relaxation between level E_1 and all other levels E_k.

Since the pumping probability P_{12} depends on the scalar product

$$\underline{\mu}\cdot\underline{E} = |\underline{\mu}_{12}| \cdot |\underline{E}| \cdot \cos\alpha$$

of the transition moment $\underline{\mu}_{12}$ with the electric vector $\underline{E}$ of the pump wave, the depletion of the lower level depends on the orientation of $\underline{\mu}_{12}$ with respect to $\underline{E}$.

Under stationary conditions $dN_1/dt = 0$ we obtain from (8.25) under the assumption $N_2 \ll N_1$ (which is always true if N_2 is depopulated sufficiently quickly by relaxation into other levels E_k)

$$N_1^S = N_1^0\left[1 - \frac{\sum\limits_{M_2} P(J_1M_1J_2M_2)}{\sum\limits_{M_2} P(J_1M_1J_2M_2) + \sum\limits_k R_{1k}}\right] , \tag{8.27}$$

where $N_1^0 = \sum_k R_{k1}N_k/\sum_k R_{1k}$ is the unsaturated population of E_1 without optical pumping.

Introducing the cross section $\sigma(J_1M_1J_2M_2)$ for the optical transition $E_1(J_1M_1) \rightarrow E_2(J_2M_2)$ we can write (8.27) as

$$N_1^S(J_1M_1) = \frac{N_J^0}{2J+1} \sum_{M2} \left\{[1 - a\sigma(J_1M_1J_2M_2)I_p]\right\} , \tag{8.28}$$

where the factor a stands for the ratio P/(P+R).

The population of the upper level (J_2M_2) is obtained in an analogous way from

$$dN_2/dt = \sum_{M_1} P(J_1M_1J_2M_2)(N_1^S-N_2) + \sum_k R_{k2}N_k - \sum_k R_{2k}N_2 . \tag{8.29a}$$

With (8.28) we obtain under stationary conditions

$$N_2^S(J_2M_2) = C_1 + N_1^0C_2\Big[\sum_{M_1} I_p\sigma(J_1M_1J_2M_2) - C_3\sum_{M_1}\sum_{M_2}\sigma^2 I_p^2\Big] \tag{8.29b}$$

which shows that the upper state population increases less than linearly with increasing pump intensity, due to increasing depletion of the lower levels. This can be observed when the fluorescence intensity $I_{F1}(I_p)$ is monitored as a function of pump intensity (Fig.8.23).

If the pump laser is linearly polarized, the transition probability is largest for those molecules with $\underline{\mu}_{12} || \underline{E}$ which are therefore preferentially

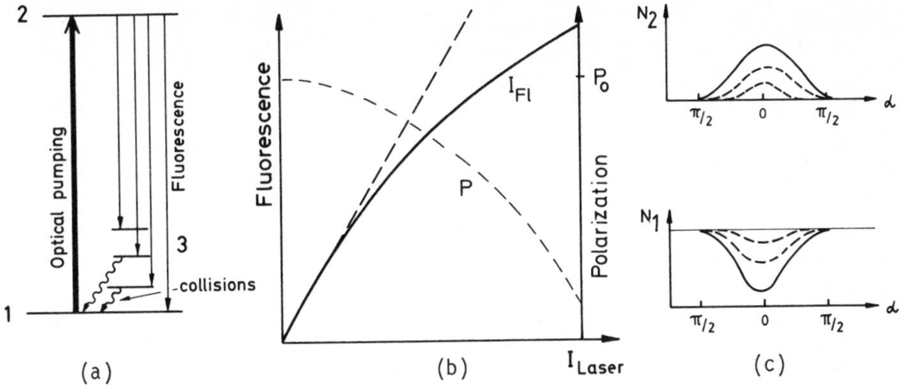

Fig. 8.23a-c. Optical pumping. (a) Level scheme, (b) fluorescence intensity and polarization as a function of intensity of the polarized pump wave, (c) schematic diagram illustrating orientation-dependent saturation for different pump intensities

depleted in the absorbing level. This implies that the upper state becomes less oriented and the degree of polarization of the emitted fluorescence

$$DP = \frac{I_{\parallel} - I_{\perp}}{I_{\parallel} + I_{\perp}}$$

decreases with increasing pump intensity (Fig.8.23).

The *third aspect* of optical pumping is concerned with the *coherent preparation* of states. This means that certain phase relations are established between the phases of the sublevel functions by the optical pumping process. Altering these phase relations by external fields manifests itself in a change of the spatial distribution or the polarization of the emitted fluorescence. These effects can therefore be monitored by observing the polarization of the fluorescence under coherent excitation as a function of an external field (level crossing spectroscopy; see Sect.10.7).

If several nondegenerate sublevels E_{kj} (e.g., different hfs levels) are simultaneously excited by a short pulse, the phase relations between the wave functions develop in time because of the slightly different frequencies $\omega_{kj} = E_{kj}/\hbar$. This results in an intensity modulation of the emitted fluorescence (quantum beats; see Sect.11.4) with a modulation period $T = (h/\Delta E)$ which is determined by the energy separation ΔE of the sublevels.

These coherence effects allow Doppler-free spectroscopy of ground states and excited states in atoms or molecules. While level crossing spectroscopy can be performed with both cw and pulsed lasers, the quantum beat technique

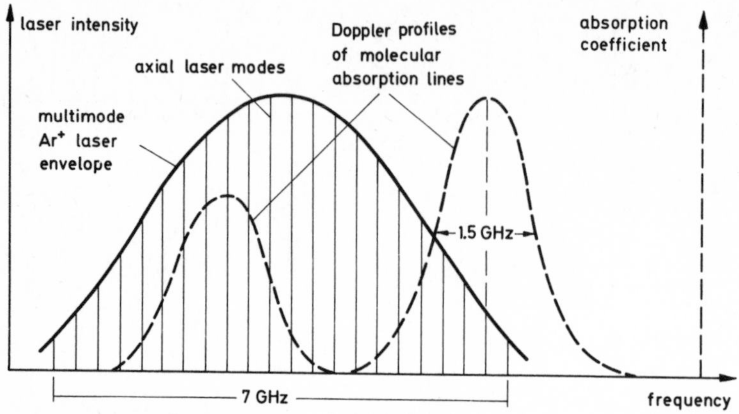

Fig. 8.24. Overlap between a multimode laser profile and two Doppler-broadened molecular absorption lines

demands short excitation pulses and time-resolved detection. It will be therefore discussed in Chap.11.

While optical pumping of *atoms* generally requires *tunable* lasers, most of the experiments on *molecules* up to now have been performed with fixed frequency lasers (e.g., the He-Ne or argon and krypton laser lines). Because most molecules have such dense absorption spectra with numerous lines corresponding to transitions (v",J") → (v',J') between a lower vibrational-rotational level (v",J") and an upper level (v',J') there is nearly always a fortuitous overlap between a fixed laser line and one or even several molecular lines within a molecular absorption band. The number of molecular transitions excited simultaneously depends on the line density and on the laser linewidth (see Fig.8.24).

Using a single-mode laser, which can be tuned within its gain profile onto the center of a molecular absorption line allows one in many cases to select a single molecular transition. However, in this case not all of the molecules in the absorbing ground state can be pumped into the upper level but only those with velocity components $v_z = 0 \pm \Delta v_z$ in a small interval Δv_z which corresponds to the homogeneous linewidth of the transition (see Sects.3.6 and 10.2).

A thorough theoretical treatment of optical pumping can be found in the review of HAPPER [8.53]. The specific aspects of optical pumping by lasers with particular attention to the problems arising from the spectral intensity distribution of the pump laser and from saturation effects are treated in [8.54,55]. Applications of optical pumping methods to the investigation of small molecules are discussed in [8.56].

8.7 Laser-Induced Fluorescence

The application range of laser-induced fluorescence (LIF) extends from the assignment of molecular spectra and the measurements of molecular constants, transition probabilities, and Franck-Condon factors to the study of collision processes or the determination of internal state populations in the reaction products of chemical reactions. Another aspect of LIF concerning sensitive detection of small concentrations of absorbing molecular components was discussed in Sects.8.2-5. Let us first briefly consider the relevance of LIF to molecular spectroscopy.

8.7.1 Molecular Spectroscopy by Laser-Induced Fluorescence

Assume a rovibronic level (v_k',J_k') in an excited electronic state of a diatomic molecule has been selectively populated by optical pumping. With a mean lifetime $\tau_k = 1/\sum_m A_{km}$ the excited molecules undergo spontaneous transitions to lower levels $E_m(v_m'',J_m'')$. At a stationary population density $N_k(v_k',J_k')$ the intensity I_{km} of a fluorescence line with frequency $\nu_{km} = (E_k-E_m)/h$ is given by (see Sect.2.6)

$$I_{km} \propto N_k A_{km} h\nu_{km} \quad . \tag{8.30}$$

The spontaneous transition probability A_{km} is proportional to the square of the matrix element (see Sect.2.9)

$$A_{km} \propto \left|\int \psi_k^* \cdot \underline{r} \cdot \psi_m d\tau_n d\tau_{el}\right|^2 \quad , \tag{8.31}$$

where the integration extends over all nuclear and electronic coordinates. As has been shown in Sect.2.9, the total wave function can be separated into a product

$$\psi = \psi_{el}\psi_{vib}\psi_{rot} \tag{8.32}$$

of electronic, vibrational, and rotational factors if the Born-Oppenheimer approximation holds [8.57]. The total transition probability is then proportional to the product of three factors

$$A_{km} \propto |M_{el}|^2 |M_{vib}|^2 |M_{rot}|^2 \quad , \tag{8.33}$$

where the first factor represents the electronic matrix element which depends on the coupling of the two electronic states, the second the Franck-Condon factor, and the third the Hönl-London factor.

Only those transitions for which all three factors are nonzero appear as lines in the fluorescence spectrum. The Hönl-London factor is always zero unless

$$\Delta J = J_k' - J_m'' = 0, \pm 1 \quad . \tag{8.34}$$

If a *single* upper level (v_k', J_k') has been selectively excited, each vibrational band $v_k' \to v_m''$ consists of at most *three* lines: a *P line* ($J'-J''=-1$), a *Q line* ($J'-J''=0$), and an *R line* ($J'-J''=+1$). For diatomic homonuclear molecules additional symmetry selection rules may further reduce the number of possible transitions. A selectively excited level (v_k', J_k') in a Π state, for example, emits on a $\Pi \to \Sigma$ transition either only Q lines or P and R lines, while on a $\Sigma_u \to \Sigma_g$ transition only P and R lines are allowed [8.58].

The fluorescence spectrum emitted from selectively excited molecular levels of a diatomic molecule is therefore very simple compared with a spectrum obtained under broad-band excitation. Figure 8.25 illustrates this by two fluorescence spectra of the Na_2-molecule, excited by two different argon laser lines. While the $\lambda = 488$ nm line excites a positive Λ component in the ($v' = 6$, $J' = 43$) level which emits only Q lines, the $\lambda = 476.5$ nm line populates the negative Λ component in the $v' = 6$, $J' = 27$ level of the $^1\Pi_u$ state, resulting in P and R lines.

The advantages of LIF spectroscopy for the determination of molecular parameters may be summarized as follows.

a) The relatively simple structure of the spectra allows ready assignment. The fluorescence lines can be resolved with medium sized spectrometers. The demands of experimental equipment are much less stringent than those necessary for complete resolution and analysis of absorption spectra of the same molecule. This advantage still remains if a few upper levels are simultaneously populated under Doppler-limited excitation [8.59].

b) The large intensities of many laser lines allow achievement of large population densities N_k in the excited level. This yields, according to (8.30), correspondingly high intensities of the fluorescence lines and enables detection of even transitions with small Franck-Condon factors. A fluorescence progression $v_k' \to v_m''$ may therefore be measured with sufficiently good signal-to-noise ratio up to very high vibrational quantum numbers v_m''. The potential curve of a diatomic molecule can be determined very accurately from the mea-

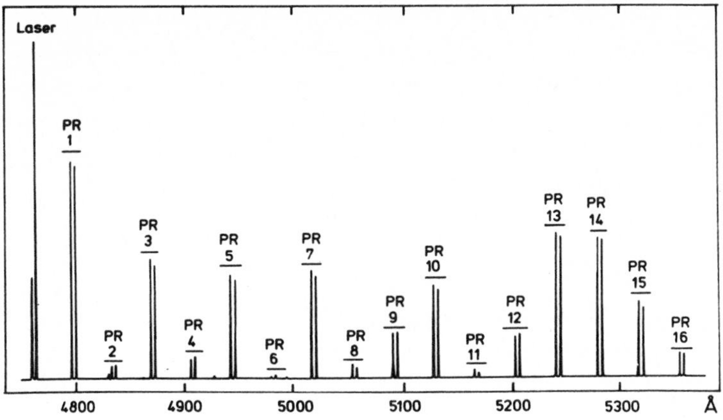

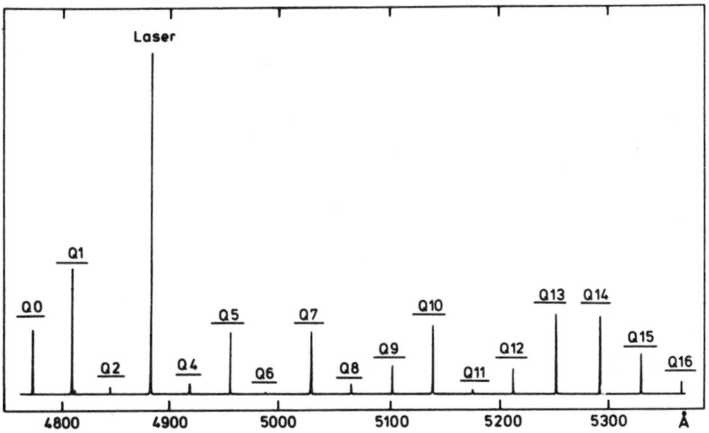

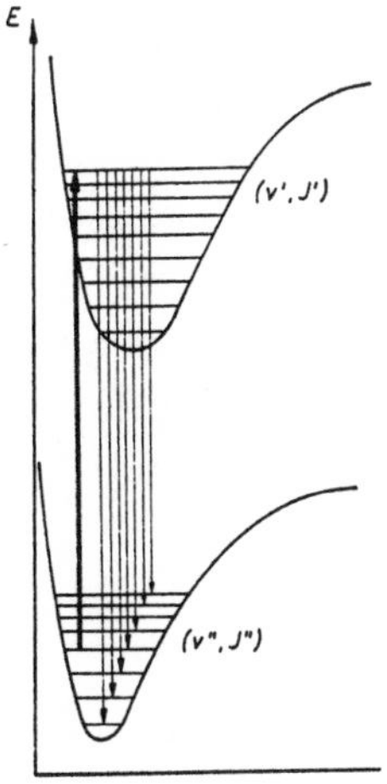

Fig. 8.25. Fluorescence spectra of the Na_2 molecule with Q lines emitted from the upper level v' = 6, J' = 43 under excitation by the argon laser line λ = 488 nm (lower spectrum) and with P and R lines emitted from the (v' = 6, J' = 27) level at λ = 476.5 nm excitation (upper spectrum)

sured term energies $E(v''_m, J''_m)$ using the Rydberg-Klein-Rees (RKR) method which is based on a modified WKB procedure [8.60]. Since the term values $E(v''_m, J''_m)$ can be immediately determined from the wave numbers of the fluorescence lines, the RKR potential can be constructed up to the highest measured level v''_{max}. In some cases fluorescence progressions are found up to levels v" closely below the dissociation limit [8.61]. This allows the spectroscopic determination of the dissociation energy by an extrapolation of the decreasing vibrational spacings $\Delta E_{vib} = E(v''_{m+1}) - E(v''_m)$ to $\Delta E_{vib} = 0$ (Birge-Sponer plot) [8.62].

c) The relative intensities of the fluorescence lines $(v'_k, J'_k \rightarrow v''_m, J''_m)$ are proportional to the Franck-Condon factors. The comparison between calculated FCF obtained with the RKR potential from the Schrödinger equation and the measured relative intensities allows a very sensitive test for the accuracy of the potential. In combination with lifetime measurements these intensity measurements yield absolute values of the electronic transition moment $M_{el}(R)$ and its dependence on the internuclear distance R [8.63].

d) In several cases discrete molecular levels have been excited which emit continuous fluorescence spectra terminating on repulsive potentials of dissociating states [8.64]. The overlap integral between the vibrational eigenfunctions ψ'_{vib} of the upper discrete level and the continuous function $\psi_{cont}(R)$ of the dissociating lower state often shows an intensity modulation of the continuous fluorescence spectrum which reflects the square $|\psi'_{vib}(R)|^2$ of the upper state wave function (see Fig.2.14). If the upper potential is known, the repulsive part of the lower potential can be accurately determined [8.65,65a]. This is of particular relevance for excimer spectroscopy (see Sect.7.4 and [8.66]).

e) For transitions between high vibrational levels of two bound states the main contribution to the transition probability comes from the internuclear distances R close to the classical turning points R_{min}, R_{max} of the vibrating oscillator. There is, however, a nonvanishing contribution from positions R between R_{min} and R_{max}, where the vibrating molecule has kinetic energy $E_{kin} = E(v,J) - V(R)$. During a radiative transition this kinetic energy has to be preserved. If the total energy $E'' = E(v',J') - h\nu = V''(R) + E_{kin} = U(R)$ in the lower state is above the dissociation limit of the potential $V''(R)$, the fluorescence terminates in the dissociation continuum (Fig.8.26). The intensity distribution of these "Condon internal diffraction bands" [8.68] is very sensitive to the difference potential $V''(R) - V'(R)$ and therefore allows an accurate determination of one of the potential curves if the other is known [8.69].

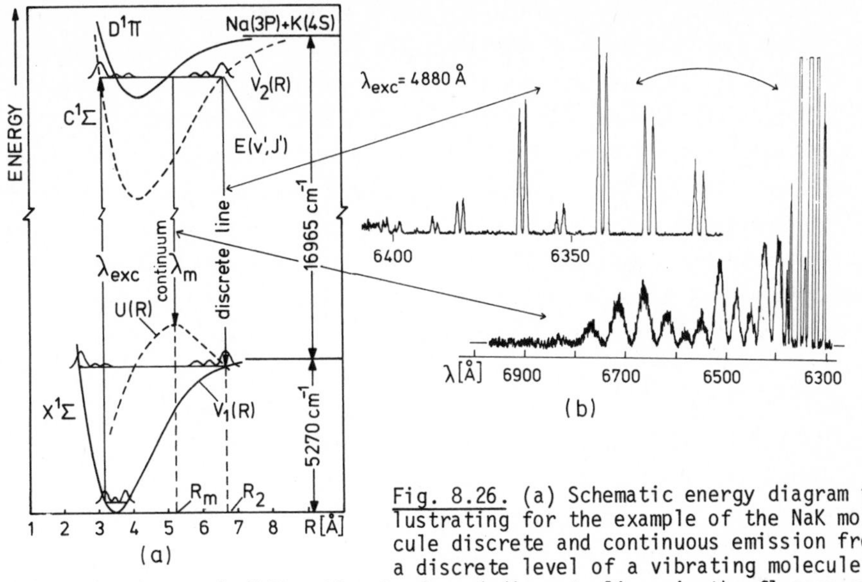

Fig. 8.26. (a) Schematic energy diagram illustrating for the example of the NaK molecule discrete and continuous emission from a discrete level of a vibrating molecule. (b) Condon internal diffraction bands and discrete lines in the fluorescence spectrum of NaK excited by the argon laser at λ = 488 nm under collision-free conditions [8.67]

The technique of LIF is, of course, not restricted to diatomic molecules but has been applied meanwhile to the investigation of triatomic molecules, such as NO_2, SO_2, BO_2, NH_2, and many other polyatomic molecules. In combination with excitation spectroscopy it allows the assignment of transitions and the identification of complex spectra. Examples of such measurements are cited in [1.12,8.70].

8.7.2 Measurements of Internal State Distributions

A very interesting application of LIF is the measurement of population distributions $N(v_i,J_i)$ in rovibronic molecular levels under situations which differ from thermal equilibrium. Examples are chemical reactions of the type $AB + C \rightarrow AC^* + B$, where a reaction product AC^* with internal energy is formed in a reactive collision between the partners AB and C. The measurement of the internal state distribution $N_{AC}(v,J)$ can often provide useful information on the reaction paths and on the potential surfaces of the collision complex (AB)C. The fact that for some of these reactions *population inversion* has been observed allowing the operation of chemical lasers [8.71] may elucidate the importance of such studies. A better knowledge of the reaction mechanisms can help to optimize the conditions for maximum inversion.

Another example is the internal cooling of molecules in supersonic molecular beams (see Sect.10.1.2) resulting in a concentration of the population distribution N(v",J") in the lowest rotational-vibrational levels [8.72]. During the adiabatic expansion through a nozzle from a high-pressure region into the vacuum, the internal energy E(v,J) of the molecules is partly transferred via collisions into directional expansion energy. Since the cross sections for collision-induced rotational-translational transitions (R→T transfer) are larger than that for vibrational-translational (V→T transfer) the rotational energy decreases more than the vibrational energy. Although the different degrees of freedom of translation, rotation, and vibration are no longer at mutual thermal equilibrium, in many cases it is still possible to describe the population distribution by a translational temperature T_{tr}, a rotational temperature T_{rot}, and a vibrational temperature T_{vib}. If the pressure difference is sufficiently large (seeded beam technique [8.73]) "temperatures" of $T_{trans} < 1K$, $T_{rot} < 10K$ and $T_{vib} < 100K$ can be achieved [8.74].

When the laser wavelength λ is tuned to a molecular transition $E_i \to E_k$, the LIF collected from a solid angle $\delta/4\pi$ on the photomultiplier cathode yields, according to (8.10), a photon counting rate

$$n_{PE} = N_i n_L \sigma_{ik} \Delta x \, \eta_k \eta_{Ph} \delta \qquad (8.35)$$

which is proportional to the number density N_i of absorbing molecules, the number n_L of incident laser photons, the absorption path length Δx viewed by the detector, the quantum yield η_k of the excited molecular level E_k, and the quantum efficiency η_{Ph} of the multiplier cathode. At lower laser intensities where saturation effects can be neglected, the LIF signal is therefore proportional to the *density* $N_i(v_i,J_i)$ of molecules in the absorbing level.

When the laser is tuned successively to two different absorbing transitions $1 \to k$ and $2 \to m$, the ratio of the LIF signals

$$\frac{n_{1PE}}{n_{2PE}} = \frac{N_1}{N_2} \frac{\sigma_{1k}}{\sigma_{2m}} \qquad (8.36)$$

is proportional to the ratio N_1/N_2 of the population densities times the ratio of the absorption cross sections, provided the laser intensities, the quantum yields η_k, and the quantum efficiencies η_{Ph} are identical for both cases. The relative population N_1/N_2 can then be directly obtained from the measured signal rates if the absorption cross sections are known.

This technique has been first applied by ZARE and his group [8.75] to determine the internal state distribution of molecular products formed in chem-

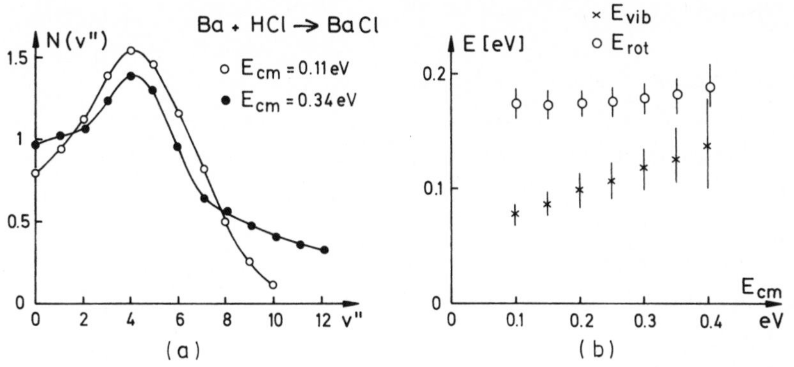

Fig. 8.27. (a) Vibrational population distribution N(v") of BaCl for two different collision energies of the reactants Ba + HCl. (b) Mean vibrational and rotational energy of reactively scattered BaCl as function of the collision energy [8.76b]

ical reactions. One example is the formation of BaCl in the reaction of barium with halogens

$$Ba + HCl \rightarrow BaCl^*(X^2\Sigma^+, v'', J'') + H \quad . \qquad (8.37)$$

Figure 8.27a shows the vibrational population distribution of BaCl for two different collision energies of the reactants Ba and HCl. Figure 8.27b illustrates that the total *rotational* energy of BaCl barely depends on the collision energy in the center of mass system of Ba + HCl, while the *vibrational* energy increases with increasing collision energy.

The interesting question, how the internal state distribution of the products is determined by the internal energy of the reacting molecules, can be answered experimentally with a second laser which pumps the reacting molecule into excited levels (v",J"). The internal state distribution of the product is measured with and without the pump laser. An example which has been studied [8.76a] is the reaction

$$Ba + HF(v''=1) \rightarrow BaF^* + H \quad ,$$

where a chemical HF laser has been used to excite the first vibrational level of HF.

A survey on several experiments where LIF has been applied to measurements of internal state distributions in various chemical reactions or in photodissociation products can be found in the review by KINSEY [8.70].

8.8 Spectroscopy of Excited States

The large population density N_k which can be achieved in selectively excited levels E_k by optical pumping with lasers allows one to perform high-resolution spectroscopy of *excited* molecules with sufficient sensitivity. Many of the techniques applicable to molecules in their ground states (e.g., absorption spectroscopy, optical pumping, LIF, microwave spectroscopy) can be now transferred to transitions between excited states. In the following sections we briefly discuss some of these methods.

8.8.1 Stepwise Excitation

Assume that an excited level E_k has been selectively populated by optical pumping with a laser L_1 (Fig.8.28). If the sample is irradiated with the spectral continuum of a broad-band source, the total absorption spectrum of molecules in all levels is obtained by measuring the transmitted intensity dispersed by a spectrometer (see Sect.8.1). When the intensity I_1 of the pump laser is chopped, the *specific* absorption of molecules in level E_k can be selected by a lock-in detector tuned to the chopping frequency. The spectrometer may be spared if the continuum source is replaced by a tunable laser L_2. The difference in the absorption $dI_2(\omega_2) = \alpha(\omega_2)\Delta x I_2$ with and without the pump laser gives the absorption spectrum of molecules in the excited level E_k directly. More sensitive is the excitation spectroscopy (Sect.8.2) where the fluorescence intensity $I_{fl}(\omega_2)$ induced by L_2 is monitored as a function of the frequency ω_2 of the tunable laser L_2 (Fig.8.28b).

This "two-step excitation" may be regarded as a resonant case of the more general two-photon absorption (see Sects.8.10 and 10.6). Here the intermediate "virtual level" coincides with a real molecular level. Since the upper level of the two-photon absorption must have the same parity as the initial ground state, it cannot be reached by single-photon absorption.

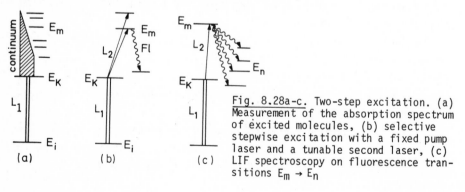

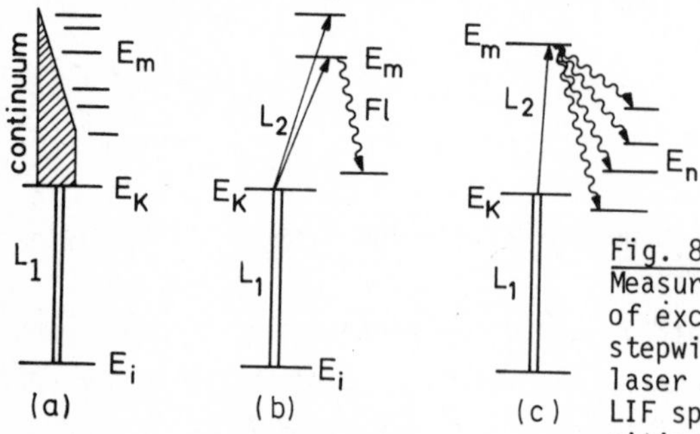

Fig. 8.28a-c. Two-step excitation. (a) Measurement of the absorption spectrum of excited molecules, (b) selective stepwise excitation with a fixed pump laser and a tunable second laser, (c) LIF spectroscopy on fluorescence transitions $E_m \rightarrow E_n$

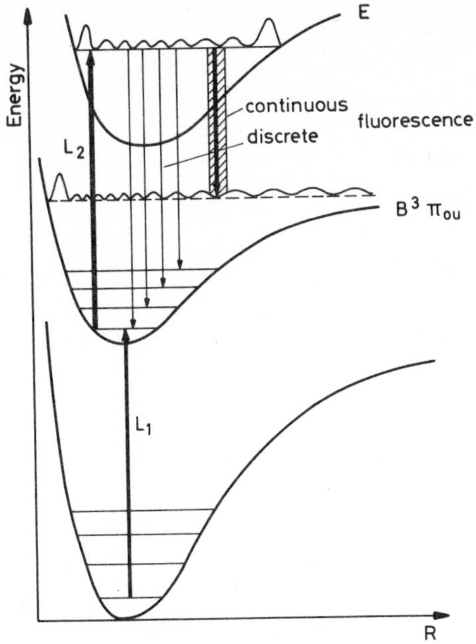

Fig. 8.29. Two-step excitation of the I_2 molecule with a dye laser and a krypton laser [8.77]

The fluorescence induced by the second laser allows the accurate determination of the molecular parameters for those *excited* states on which the fluorescence transitions from E_m are terminating. The LIF method can therefore be extended by stepwise excitation to the investigation of many molecular states which may not even have been found before. Of particular interest are dissociating excited states with repulsive potential curves below bound states E_m. These continuous states often cannot be studied by direct absorption from the ground state because the Franck-Condon factors for the transitions may be quite small. As an example of such investigations we mention the two-step excitation of the iodine molecule I_2 (Fig.8.29). Selected (v_k', J_k') levels in the $B^3\Pi_{0u}$ state are populated by optical pumping with a cw dye laser. Starting from these levels a krypton laser excites further levels (v_m', J_m') in a higher E state. The fluorescence spectrum from these levels (v_m', J_m') consists of discrete lines and continuous bands. The lines terminate on bound levels of the B state while the continuous fluorescence terminates above the dissociation limit of the B state [8.77,78].

8.8.2 Spectroscopy of Rydberg States

The spectroscopy of Rydberg states with large principal quantum numbers n has recently gained increasing interest [8.78a]. This subject is a good example

Table 8.1. Properties of Rydberg atoms [8.78a,b]

Quantity	n dependence	Numerical examples for	
		Na(10d)	H(n=50)
Binding energy	$-Rn^{-2}$	0.14 eV	0.0054 eV
Orbital radius	a_0n^2	147 a_0	2500 a_0
Geometrical cross section	$\pi a_0^2 n^4$	$7\cdot10^4 a_0^2$	$6\cdot10^6 a_0^2$
Dipole moment	$\propto n^2$	143 a_0	
Polarizability	$\propto n^7$	210 $KHzV^{-2}cm^{-2}$	
Radiative life-time	$\propto n^3$	10^{-6}s	10^{-3}s
Stark splitting in electric field E = 1kV/cm	$\Delta W \propto n(n-1)E$	$\sim$15 cm^{-1}	$\sim 10^2$ cm^{-1}
Critical field strength E_C for field ionization	$E_c = \pi\varepsilon_0 R^2 e^{-3} n^{-4}$	3.10^6 V/m	$5\cdot10^3$ V/m

R = Rydberg constant, a_0 = Bohrradius, n = principal quantum number.

to illustrate the application of several experimental techniques discussed in the foregoing sections.

Table 8.1 demonstrates that atoms in highly excited Rydberg states are quite different from our normal conceptions of atomic sizes. The large mean Bohr radius and the small ionization energy of an electron in a level with $n > 30$ make it a very sensitive probe for the interaction of the atom with external fields. Many atomic parameters, such as the core polarizability, configuration interactions, or anomalies in the fine structure splittings, can be calculated for Rydberg states with good approximations. The experimental test of these theories by accurate measurements of energies, lifetimes, or ionization probabilities is therefore of fundamental interest.

The alkali atoms have played a principal role because of their experimental simplicity. Figure 8.30 shows as an example the stepwise excitation of Rydberg levels in the Na atom by two dye lasers which can be pumped by the same N_2 laser [8.79]. The first dye laser is tuned to the Na-D_2 line at λ = 5889 Å and the second dye laser is scanned over the ns and nd Rydberg series. While the excitation of the $3s\,^2S_{1/2} \rightarrow 3p\,^2P_{3/2}$ transition is monitored by the laser-induced fluorescence, the second excitation step $3p\,^2P_{3/2} \rightarrow$ ns,nd is detected utilizing field ionization of the Rydberg states (see Sect.8.2.5).

A possible experimental setup is shown in Fig.8.14b, where the laser beams cross the atomic beam inside a parallel plate capacitor, providing a homogeneous electric field. Electrons escaping through the mesh in the capacitor plate

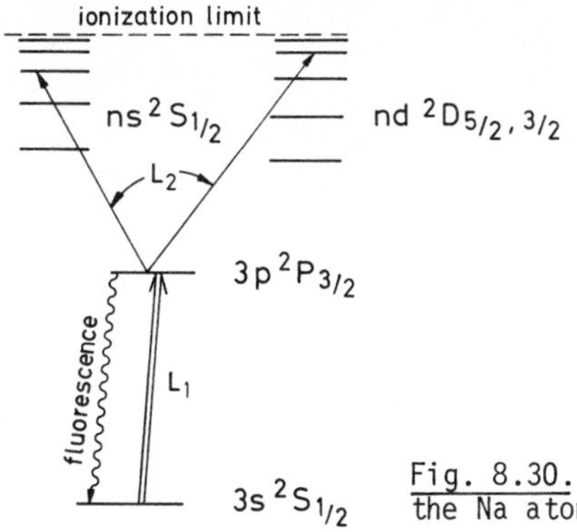

Fig. 8.30. Stepwise excitation of Rydberg levels in the Na atom [8.79]

are detected by an electron multiplier. Since the electric field will perturb the energy levels, the absorption spectrum of the unperturbed atom is obtained by applying a pulsed electric field with a time delay of some hundred ns in respect to the laser pulse.

The study of Rydberg spectra and ionization thresholds of ten lanthanides and actinides has been reported by PAISNER et al. [8.80]. In these experiments, high-lying states were accessed by time-resolved stepwise excitation using pulsed dye lasers tuned to resonant transitions. Atoms excited to levels within 1000 cm^{-1} of the ionization limit were then photoionized by 10.6 μm radiation from a pulsed CO_2 laser. The measurements allowed the accurate determination of ionization thresholds from Rydberg convergences to within 0.0005 eV.

The advantages of infrared laser Stark spectroscopy (Sect.8.3.2) in combination with optical excitation have been utilized by DUCAS et al. [8.81]. The basic technique employs a tunable dye laser for stepwise excitation into a specific Rydberg state (a) with binding energy W_a. The radiation from a cw single-line CO_2 laser is then made incident on these Rydberg atoms. This fixed frequency radiation can induce transitions from the prepared level (a) to still higher levels (b) when the atomic energy difference $W_b - W_a$ is Stark tuned into resonance with the CO_2 laser. Because of the very large Stark shifts in Rydberg states a large number of such resonances can be observed by scanning the electric field E. The resonances are monitored by selective field ionization of the upper Rydberg levels (b). The extrapolation of the Stark splittings to $E = 0$ not only provides the accurate determination of the level energies but also allows one to measure the electric dipole moment and the polarizability of atoms in Rydberg states [8.81a]. Very detailed studies of Rydberg states have been performed using double-resonance methods [8.82] or the quantum beat technique (see Sect.11.4).

8.9 Double Resonance Methods

The stepwise excitation can be regarded as optical-optical double resonance, because two optical fields are simultaneously in resonance with two molecular transitions sharing a common level. This is a special case of more general double-resonance techniques where two E.M. fields with frequencies ω_1 and ω_2, in the radio frequency, microwave, infrared, or optical range, simultaneously interact with a molecule. While optical-rf double resonance has been used for a long time in *atomic* spectroscopy [8.83], the extension to optical-microwave or optical-optical double resonance has been greatly facilitated by optical pumping with lasers. In *molecular* physics these techniques have greatly enlarged the application range of these high-resolution methods. In this section we briefly discuss the basic principles of the various double-resonance techniques, where the main aspect will be the change of level population by optical pumping. Although the spectral resolution of the optical pumping process is in general limited by the Doppler width, some of these double-resonance techniques allow a Doppler-free resolution if the frequency of the second E.M. field is so low that the Doppler width of the double-resonance signal becomes smaller than the natural linewidth of the optical transition. Finer details of the double resonance, such as line profiles or saturation effects or coherent phenomena, will be treated in Chap.10.

8.9.1 Optical-Radio-Frequency Double Resonance

Figure 8.31 illustrates the concept of optical-rf double resonance. The sample is placed in a radio frequency field produced by an rf current through a coil.

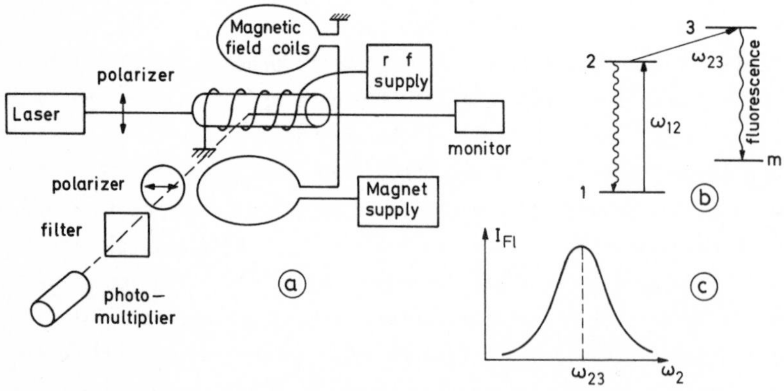

Fig. 8.31a-c. Schematic diagram of optical-rf double resonance. (a) Experimental setup, (b) level diagram, (c) double resonance signal

The molecules are optically pumped into the excited level E_2 by a laser, stabilized to the frequency ω_{12}. When the rf field is tuned into resonance with the transition $2 \rightarrow 3$, some molecules may absorb rf quanta before they return to the initial state 1 by spontaneous emission. The transition $2 \rightarrow 3$ can be monitored either by the decrease of the fluorescence intensity I_{fl} $(2 \rightarrow 1)$ or by the corresponding increase of $I_{fl}(3 \rightarrow m)$. If the orientation of level 3 differs from that of 2 (e.g., when 2 and 3 are different Zeeman sublevels) the change in polarization of the fluorescence can serve as the monitor for the rf transition $2 \rightarrow 3$. Measuring the transition rate $2 \rightarrow 3$ as a function of the radio frequency ω_2 yields a double-resonance signal $I_{fl}(\omega_2)$ with a maximum at $\omega_{23} = (E_3-E_2)/\hbar$ (Fig.8.31c).

This technique monitors the absorption of an rf quantum through the emission of an optical photon which yields an inherent energy amplification of ω_{3m}/ω_{23}. With typical figures of $\nu_{3m} = 5 \cdot 10^{14} s^{-1}$ and $\nu_{23} = 10^6 s^{-1}$, this gives an amplification factor of 5×10^8. In the experiment this energy amplification manifests itself in a much larger detection efficiency for visible photons than for rf quanta.

For *magnetic* dipole transitions $2 \rightarrow 3$ the rf field configuration should have a maximum of the *magnetic* field amplitude at the location of the sample while for *electric* dipole transitions the electric field vector should be maximum. For transitions between Zeeman sublevels it is often more convenient to keep the radio frequency fixed and to tune the level splitting $\Delta E = \hbar\omega_{23}$ by the external dc magnetic field.

The fundamental advantage of the rf double-resonance technique is the high spectral resolution which is not limited by the *optical* Doppler width, see (3.30d)

$$\Delta\nu_D = 7 \cdot 10^{-7} \nu (T/M)^{\frac{1}{2}} \ .$$

The Doppler width of the rf transition is smaller than that of the optical transition by a factor ν_{23}/ν_{12}. This makes the residual Doppler width of the double-resonance signal completely negligible compared with other broadening mechanisms, such as collisional or saturation broadening. In the absence of these additional line-broadening effects the halfwidth of the double-resonance signal

$$\Delta\omega_{23} = (\Delta E_2 + \Delta E_3)/\hbar \tag{8.38}$$

is essentially determined by the energy level widths ΔE_n of the corresponding levels 2 and 3 which are connected to the spontaneous lifetime T by

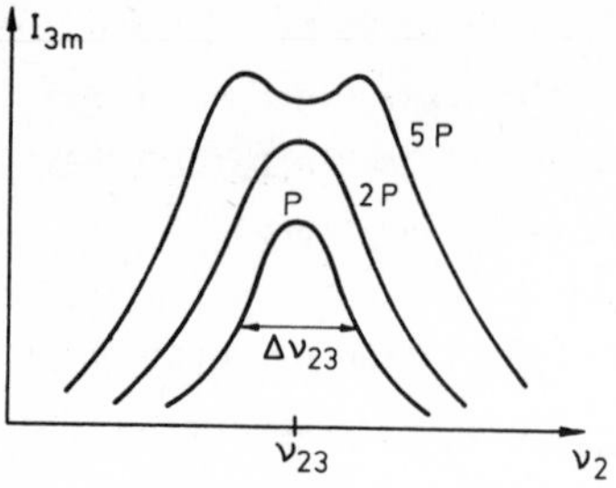

Fig. 8.32. Saturation broadening and splitting of the double resonance signal with increasing rf power P

$$\Delta E_n = \hbar/T \quad . \tag{8.39}$$

With increasing rf intensity, however, saturation broadening is observed (see Sect.3.6) and the double-resonance signal may even exhibit a minimum at the center frequency ω_{23} (Fig.8.32). This can be readily understood from the semiclassical model of Sect.2.9.4. For large rf field amplitudes E_{rf} the Rabi flopping frequency (2.127)

$$\mu = \sqrt{(\omega_{23}-\omega)^2 + e^2R_{ab}^2E_{rf}^2/\hbar} \quad ,$$

at which the population probabilities $|\psi(2)|^2$ and $|\psi(3)|^2$ oscillate, becomes noticeably large. The resultant modulation of the time-dependent populations $N_2(t)$ and $N_3(t)$ causes a splitting $\Delta E = \hbar\mu$ of the energy levels $E_{2,3} = E_{2,3}^0 \pm \hbar\mu$. If the modulation frequency μ exceeds the spontaneous depopulation rate $A_2 \approx A_3 = 1/T$ the splitting exceeds the natural linewidth of the two levels and the side bands can be seen in the double-resonance signal as two separated maxima.

If the halfwidth $\Delta\omega_{23}$ of the double-resonance signal is plotted against the rf power P_{rf} the extrapolation towards $P_{rf} = 0$ yields the linewidth (8.39) and allows one to determine the natural lifetime T. This double-resonance technique therefore allows one to measure splittings of levels which are smaller than the optical Doppler width. It has been mainly applied to measurements of hfs splittings for the determination of nuclear moments [8.83] or of Zeeman splittings to determine Landé factors of excited levels in atoms and molecules [8.84]. Another example is given by the hyperfine structure measurement in many Rydberg states of alkali atoms which had been selectively populated by stepwise excitation with dye lasers [8.82,85].

The achievable accuracy is mainly determined by the signal-to-noise ratio of the double-resonance signal which limits the accuracy of the exact determination of the center frequency ω_{23}. However, the *absolute* accuracy in the determination of the rf frequency ω_{23} is generally several orders of magnitude higher than in conventional optical spectroscopy where the level splitting $\Delta E = \hbar\omega_{23} = \hbar(\omega_{31}-\omega_{21})$ is indirectly deduced from a small difference $\Delta\lambda$ between two directly measured wavelengths $\lambda_{31} = c/\nu_{31}$ and $\lambda_{21} = c/\nu_{21}$.

8.9.2 Microwave-Infrared Double Resonance

Microwave spectroscopy has contributed in an outstanding way to the precise determination of molecular parameters, such as bond lengths, bond angles, and nuclear configurations in polyatomic molecules or fine and hfs splittings in molecular ground states [8.86]. The net absorption $dI = I - I_0$ of the incident microwave passing through a sample with length Δx is given by

$$dI = -\sigma_{ik}(N_i - N_k)\Delta x I_0 \quad . \tag{8.40}$$

At thermal equilibrium the population densities N_i, N_k follow a Boltzmann distribution

$$N_k/N_i = (g_k/g_i)e^{-h\nu_{ik}/kT} \quad . \tag{8.41}$$

For typical microwave frequencies $\nu_{ik} \approx 10^{10}$Hz ($\hat{=}$ 0.3 cm^{-1}) the exponent in (8.41) is very small at room temperatures ($kT \approx 250$ cm^{-1} $\sim h\nu/kT \ll 1$) and we obtain from (8.40) for the relative absorption

$$dI/I_0 = N_i\sigma_{ik}\Delta x[1-(g_i/g_k)(1-h\nu/kT)] = N_i\sigma_{ik}\Delta x h\nu/kT \quad \text{for} \quad g_i = g_k \quad . \tag{8.42}$$

Equation (8.42) illustrates that the net absorption of microwaves is generally much smaller than that of optical light for the following reasons:

1) the absorption coefficient σ_{ik} which is proportional to ν (see Sect.2.6) is at microwave frequencies by many orders of magnitude smaller than at optical frequencies,

2) the small factor $h\nu/kT$ decreases the population difference and therefore further diminishes the obtainable signal.

This implies that conventional absorption spectroscopy at microwave frequencies is restricted to levels E_i, which have a sufficiently large thermal population N_i.

Double-resonance techniques can greatly improve the situation and allow one to extend the applications of microwave spectroscopy to *excited* vibrational or electronic states. They furthermore may enhance the sensitivity of

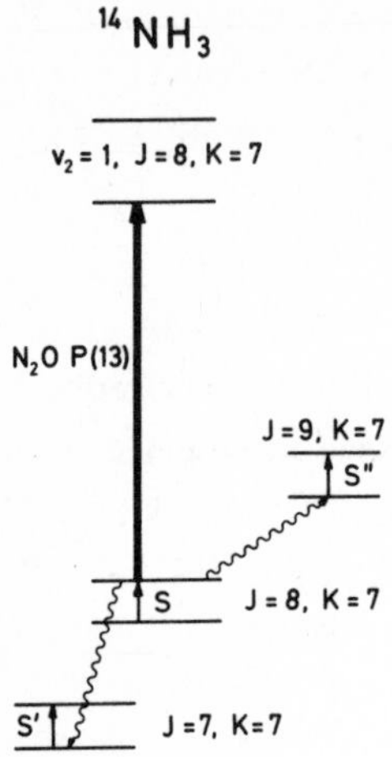

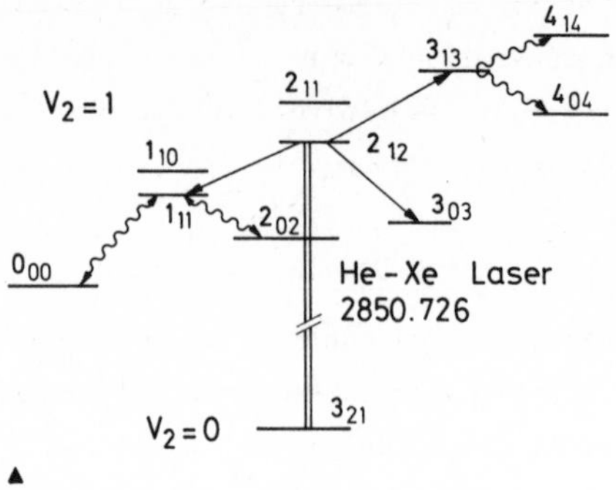

▲ Fig.8.34

◄ Fig.8.33

Fig. 8.33. Infrared-microwave double resonance in NH_3. S, S', and S" are microwave transitions between the inversion sublevels. The wavy arrows indicate collision induced transitions [8.88]

Fig. 8.34. Infrared-microwave double resonance in the (v_2=1) vibrational level of DCCCHO. The solid arrows indicate the double-resonance transitions, the wavy arrows the triple-resonance transitions [8.87]

ground state microwave spectroscopy by increasing the population difference (N_i-N_k) through selective depopulation in one of the levels E_i or E_k.

Many lines of intense infrared lasers (e.g., CO_2, N_2O, CO, HF, DF lasers) coincide with rotational-vibrational transitions of polyatomic molecules. Even for lines which are only close to a molecular transition the molecular lines may be tuned into resonance by external magnetic or electric fields (see Sect.8.3). Figure 8.33 illustrates the enhancement of microwave signals on transitions between the inversion doublets of NH_3 in rotational levels of the vibrational ground state, due to selective depletion of the upper inversion component in the (J=8,K=7) level by optical pumping with an N_2O laser line. The selective depopulation may be partly transferred by collisions to neighboring rotational levels which also causes an enhancement of the corresponding microwave signals S', S".

Figure 8.34 gives an example of microwave spectroscopy in an excited vibrational state, where the ($N_{k_-k_+} = 2_{12}$) rotational level of DCCCHO has been selectively populated by optical pumping with a He-Xe laser. The solid arrows represent the direct microwave transitions from the optically pumped level to adjacent rotational levels, while the wavy arrows correspond to "triple-resonance" transitions from levels which have been populated either by the first microwave or by collisions from the optically pumped level [8.87]. For a detailed discussion of infrared microwave double resonance see [8.87,88]. where many examples are given to illustrate the advantages of this technique.

8.9.3 Optical Microwave Double Resonance

The electronically excited states of most molecules are by far less investigated than their ground states and often the nuclear configuration of polyatomic molecules in excited states is even not known. It would therefore be most desirable to transfer the methods of microwave spectroscopy to excited electronic states. This is just what optical-microwave double resonance allows one to do. One example illustrates the technique.

In the crossing volume of an atomic barium beam and a molecular oxygen beam, BaO molecules in various vibronic levels (v",J") of the electronic ground state $X^1\Sigma$ are formed by the reaction $Ba + O_2 \rightarrow BaO + O$.

With a cw dye laser tuned to transitions $A^1\Sigma(v',J') \leftarrow X^1\Sigma(v'',J'')$, different levels (v',J') can be selectively populated (Fig.8.35). The quantum numbers v and J can be determined from the LIF spectrum. When the crossing volume is irradiated with microwaves from a klystron, transitions between adjacent rotational levels $J' \leftrightarrow J'\pm1$ or $J'' \leftrightarrow J''\pm1$ in both electronic states can be induced by tuning the microwave to the corresponding frequencies. Since optical pumping by the laser has *decreased* the population N(v",J") below its thermal equilibrium value, microwave transitions $J'' \leftrightarrow J''\pm1$ *increase* N(v",J") resulting in a corresponding increase of the total LIF. Transitions $J' \leftrightarrow J'\pm1$ on the other hand *decrease* the population of the optically pumped level (v',J'). They therefore decrease the intensity of fluorescence lines emitted from (v',J') and generate new lines in the fluorescence spectrum originating from the levels (J'±1). This method allows measurement of rotational spacings in

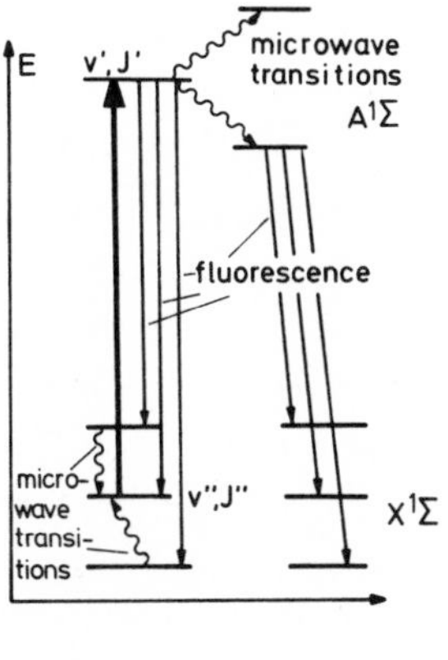

Fig.8.35. Optical-microwave double resonance in diatomic molecules. The lower levels at right are generally not identical with those at left

the excited state and it furthermore enhances the signal-to-noise ratio of microwave spectroscopy in the lower state by increasing the population differences $N(J'') - N(J''\pm 1)$ [8.89].

8.9.4 Optical-Optical Double Resonance

The optical-optical double-resonance method is based on the simultaneous interaction of two optical fields with the molecule. Assume that the pump laser L_1 is tuned to the transition $1 \to 2$ (Fig.8.36). If its intensity I_1 is chopped at a frequency f, the population densities N_1 and N_2 will be modulated.

$$N_1 = N_1^0(1 - aI_1 \sin 2\pi ft) \quad ,$$

$$N_2 = N_2^0(1 + bI_1 \sin 2\pi ft) \quad .$$

The modulation phase is opposite for upper and lower levels. The modulation amplitudes a and b depend on various relaxation processes such as collisions or spontaneous transitions which tend to thermalize the population distribution. When the frequency ω_2 of the second laser is tuned across the molecular absorption spectrum, the intensity $I_2(\omega_2)$ of the fluorescence excited by L_2 will be also modulated if L_2 excites transitions starting from levels 1 or 2. Phase-sensitive detection of $I_2(\omega_2)$ yields positive signals for all transitions $2 \to a$ (see Fig.8.36) and negative signals for transitions $1 \to m$. The frequency difference $\Delta\omega = \omega_m - \omega_{m+1}$ between two signals $1 \to m$ and $1 \to m+1$ directly reflects the energy separation between the corresponding levels in the excited state.

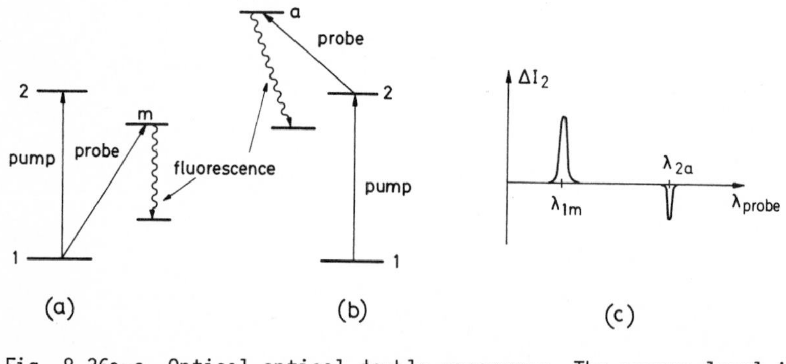

Fig. 8.36a-c. Optical-optical double resonance. The common level is the lower pump level in (a) and the upper pump level in (b). (c) illustrates the different phases of the double-resonance signals in case (a) or (b) monitored with a lock-in amplifier tuned to the chopping frequency f of the pump laser

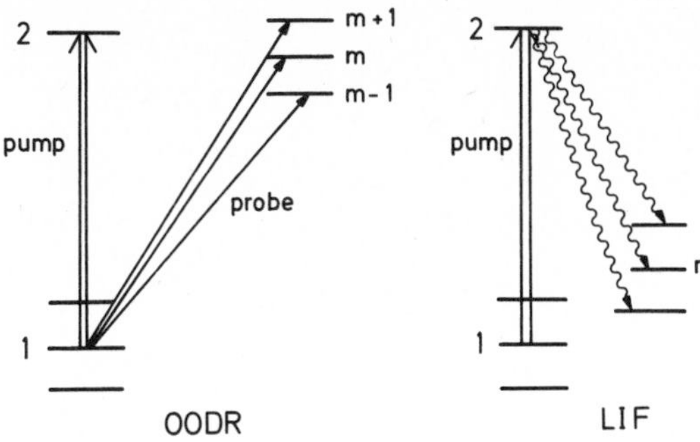

Fig. 8.37. Comparison between OODR and LIF methods

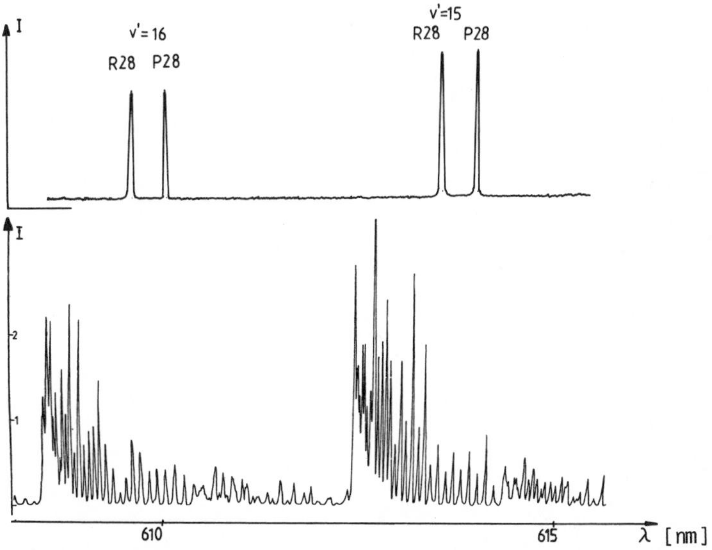

Fig. 8.38. Optical-optical double-resonance signals in the Na_2 spectrum. The pump (argon laser at λ = 476.5 nm) was stabilized onto the (0,28→6,27))X → B transition, while the cw dye laser was tuned across the 0 → 15 and 0 → 16 bands of the X → A system. Note the considerable reduction in line density compared with the linear absorption spectrum shown in the lower part

This optical-optical double resonance may be regarded as the inversion of the LIF method. While in the latter a single *upper* level E_k is selectively populated and all fluorescence lines $E_k \rightarrow E_m$ terminating on lower levels E_m are measured, here a *lower* level 1 or 2 is marked by selective depletion and all absorbing transitions from this level to *higher* levels are monitored (Fig.8.37). The advantage of this method over conventional absorption spectroscopy is the easier and unambiguous assignment of lines. This is of particular relevance if the upper states are perturbed, which considerably impedes the

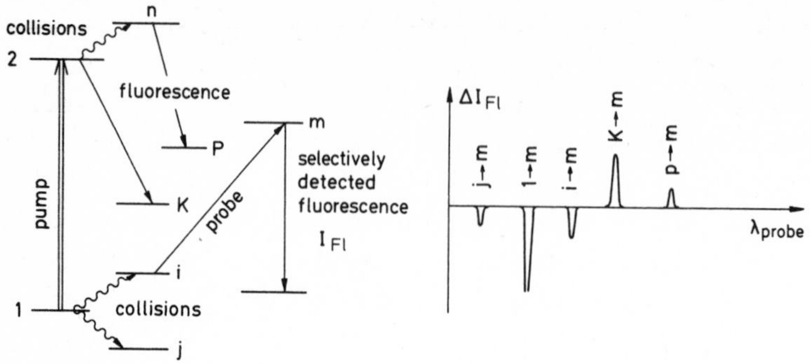

Fig. 8.39. Generation of double-resonance signals by transfer of population modulation via fluorescence or collisions

assignment of conventional absorption spectra. The $X^1\Sigma \rightarrow A^1\Sigma$ transitions of Na_2 are an example which has been studied this way. Figure 8.38 shows the OODR spectrum [8.90a]. Note the significant simplification of the complex linear absorption spectrum (see also [8.90b]).

Since the intensity $I_{fl}(2)$ of the fluorescence emitted from level 2 has the same modulation phase as ΔN_2, the population modulation ΔN_2 is partly transferred to all terminating levels k (Fig.8.39). The modulation ΔN_k of level k is

$$\Delta N_k = \Delta N_2(A_{2k}/\sum_f A_{2f}) \quad .$$

At higher pressures in the sample cell the population modulations ΔN_1 or ΔN_2 may be transferred by collisions to neighboring levels and additional lines appear in the double-resonance spectrum which are excited by L_2 from this collisional modulated levels i (see Sect.12.2). Both effects increase the number of lines in the double-resonance spectrum and may reduce the simplicity of the method. Measuring the pressure dependence of the OODR signals allows one, however, to distinguish between direct and collisional-induced lines.

8.10 Multiphoton Spectroscopy

The stepwise excitation, discussed in Sect.8.8, can be interpreted as two successive one-photon excitations. In this section we consider the *simultaneous* absorption of two or more photons by a molecule which undergoes a transition $E_i \rightarrow E_f$ with $(E_f - E_i) = \hbar\sum_i \omega_i$. The photons may either come from a single laser beam passing through the absorbing sample or they may be provided by two or more beams emitted from one or several lasers.

The first detailed theoretical treatment of two-photon processes was given in 1929 by GÖPPERT-MAYER [8.91], whereas the experimental realization had to wait for sufficiently intense light sources provided by pulsed lasers [8.92]. Multiphoton spectroscopy has several definite advantages for the investigation of atomic and molecular spectra which certainly have contributed to the impressive development of this interesting technique.

1) Excited levels can be reached via two-photon transitions that are parity forbidden by single-photon dipole radiation.

2) The accessible spectral range $\omega = \sum \omega_i$ of multiphoton spectroscopy can be extended into the vacuum ultraviolet region if the participating photons $\hbar\omega_i$ come from visible or uv lasers. By combining tunable lasers or by using combinations of fixed frequency lasers with a tunable laser, continuous tuning ranges in the uv and vuv ranges are possible (see Sect.7.8).

3) A proper combination of the participating photons with momentum $\hbar \sum \underline{k}_i = 0$ allows Doppler-free multiphoton spectroscopy (see Sect.10.6) which opens the way for the investigations of highly excited states with extremely good resolution.

4) Ionizing states can often be reached by multiphoton transitions. This allows one to utilize the extremely high sensitivity of ion detection (see Sect.8.2.5) for the investigation of autoionizing states and opens a new field of molecular ion spectroscopy.

Following a brief discussion of the basic physics of two-photon transitions, we illustrate the relevance of multiphoton spectroscopy by several examples. The Doppler-free multiphoton spectroscopy will be presented in Sect.10.6.

8.10.1 Transition Probability of Two-Photon Transitions

The probability A_{if} for a two-photon transition between the ground state E_i and an excited state E_f of a molecule, induced by photons $\hbar\omega_1$ and $\hbar\omega_2$ from two light waves with wave vectors $\underline{k}_1$, $\underline{k}_2$, polarization vectors $\underline{e}_1$, $\underline{e}_2$, and intensities I_1, I_2, can be written as a product of two factors [8.93]

$$A_{if} \propto \frac{\gamma_{if}}{[\omega_{if} - \omega_1 - \omega_2 - \underline{v} \cdot (\underline{k}_1+\underline{k}_2)]^2 + (\gamma_{if}/2)^2}$$

$$\cdot \left| \sum_k \frac{\underline{R}_{ik} \cdot \underline{e}_1 \cdot \underline{R}_{kf} \cdot \underline{e}_2}{(\omega_{ki} - \omega_1 - \underline{k}_1 \cdot \underline{v})} + \frac{\underline{R}_{ik} \cdot \underline{e}_2 \cdot \underline{R}_{kf} \cdot \underline{e}_1}{(\omega_{ki} - \omega_2 - \underline{k}_2 \cdot \underline{v})} \right|^2 \cdot I_1 I_2 \quad . \tag{8.43}$$

The first factor gives the spectral line profile of the two-photon transition. It corresponds exactly to that of a single-photon transition of a moving molecule at a center frequency $\omega_{if} = \omega_1 + \omega_2 + \underline{v}(\underline{k}_1+\underline{k}_2)$ with a homogeneous linewidth γ_{if} (see Sects.3.1,6). Integration over all molecular velocities v gives a *Voigt profile* with a halfwidth which depends on the relative orientation of $\underline{k}_1$ and $\underline{k}_2$. If both light waves are parallel, the Doppler width which is proportional to $|\underline{k}_1 + \underline{k}_2|$ becomes maximum and is in general large compared to the homogeneous width γ_{if}. For $\underline{k}_1 = -\underline{k}_2$ the Doppler broadening vanishes and we obtain a pure Lorentzian line profile with a homogeneous linewidth γ_{if} provided that the laser linewidth is small compared to γ_{if}. This "Doppler-free two-photon spectroscopy" will be discussed in Sect.10.6.

Because the transition probability (8.43) is proportional to the product of the intensities I_1I_2 (which has to be replaced by I^2 in case of a single laser beam), *pulsed* lasers are generally used which deliver sufficiently large peak powers. The spectral linewidth of these lasers if often comparable to or even larger than the Doppler width and the denominators $(\omega_{ki}-\omega-\underline{k}\cdot\underline{v})$ can then be approximated by $(\omega_{ki}-\omega)$.

The second factor describes the transition probability for the two-photon transition. It can be derived quantum mechanically by second-order perturbation theory (see for example [8.94]). This factor contains a sum of products of matrix elements $R_{ik}R_{kf}$ for transitions between the initial state i and intermediate molecular levels k or between these levels k and the final state f [see (2.96)]. The summation extends over all molecular levels k. The denominator shows, however, that only those levels k which are not too far off resonance with one of the Doppler-shifted laser frequencies $\omega_n' = \omega_n - \underline{k}_n \cdot \underline{v}$ (n=1,2) will mainly contribute.

Often a fictitious "virtual level" is introduced to describe the two-photon transition by a symbolic two-step transition $E_i \to E_v \to E_f$ (Fig.8.40). Since the two possibilities

a) $E_i + \hbar\omega_1 \to E_v$, $E_v + \hbar\omega_2 \to E_f$ (first term)

b) $E_i + \hbar\omega_2 \to E_v$, $E_v + \hbar\omega_1 \to E_f$ (second term)

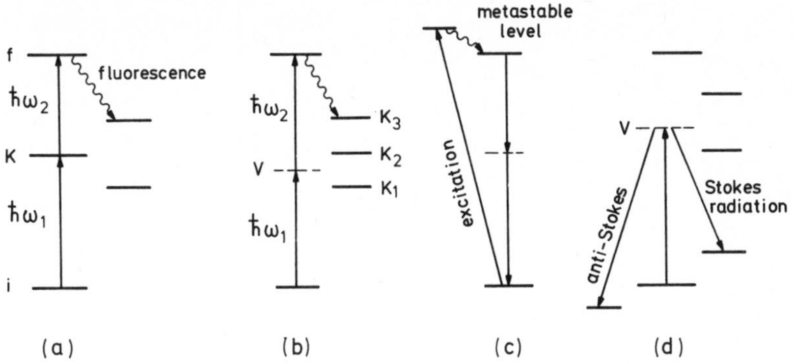

Fig. 8.40a-d. Two-photon processes. (a) Stepwise excitation with a real intermediate level; (b) two-photon absorption, described by a virtual intermediate level; (c) two-photon emission; (d) Raman scattering

lead to the same observable result, namely the excitation of the real level E_f, the total transition probability for $E_i \to E_f$ is the square of the sum over both probability amplitudes, as indicated by the sum of the two terms in (8.43).

Often the frequencies ω_1 and ω_2 can be selected in such a way that the virtual level is close to a real molecular eigenstate. This greatly enhances the transition probability and it is therefore generally advantageous to excite the final level E_f by two different photons with $\omega_1 + \omega_2 = (E_f - E_i)/\hbar$ rather than by two photons out of the same laser with $2\omega = (E_f - E_i)/\hbar$.

The second factor in (8.43) describes quite generally the transition probability for all possible two-photon transitions such as the Raman scattering or the two-photon absorption and emission. Figure 8.40 illustrates schematically these three two-photon processes. The *important point is that the same selection rules are* valid for all two-photon processes. Equation (8.43) shows that both matrix elements R_{ik} and R_{kf} must be nonzero to give a nonvanishing transition probability A_{if}. This means that two-photon transitions can only be observed between two states i and f which are both connected to intermediate levels k by allowed single-photon optical transitions. Because the selection rule for single-photon transitions demands that the levels i and k or k and f have opposite parity, *the two levels* i *and* f, *connected by a two-photon transition, must have the same parity*. In atomic two-photon spectroscopy s → s or s → d transitions are allowed, and also $\Sigma_g \to \Sigma_g$ transitions, for example, in diatomic homonuclear molecules.

It is therefore possible to reach molecular states which cannot be populated by single-photon transitions from the ground state. In this regard two-

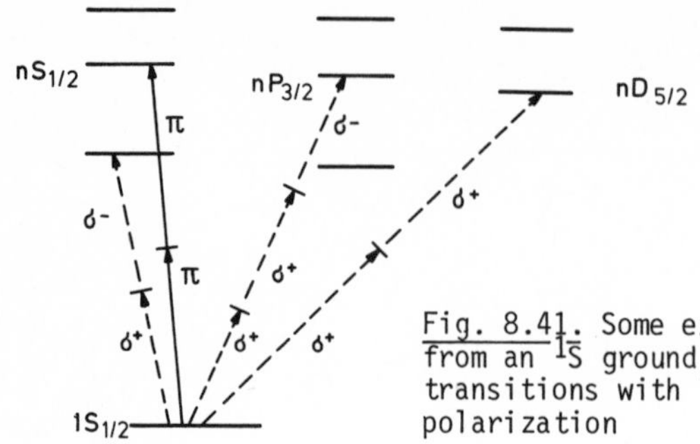

Fig. 8.41. Some excited atomic states, accessible from an ^1S ground state by two- or three-photon transitions with linear (——) and circular (---) polarization

photon absorption spectroscopy is complementary to linear absorption spectroscopy, and its results are of particular interest, because they yield information about states, which often had not been found before. It is not seldom that excited molecular states are perturbed by nearby states of opposite parity and it is generally difficult to deduce the structure of these perturbing states from the degree of perturbations, while two-photon spectroscopy allows direct access to such states. Since the matrix elements $\vec{R}_{ik} \cdot \hat{e}_1$ and $\vec{R}_{kf} \cdot \hat{e}_2$ depend on the polarization characteristics of the incident radiation, it is possible to select the accessible upper states by a proper choice of the polarization. While for single-photon transitions the total transition probability (summed over all M sublevels) is *independent* of the polarization of the incident radiation [see (2.158)], there is a distinct polarization effect in multiphoton transitions, which can be understood by applying successively known selection rules to the two matrix elements in (8.43). For example, two parallel laser beams, which both have right-hand circular polarization, induce two-photon transitions in atoms with $\Delta L = 2$. This allows for instance $s \rightarrow d$ transitions but not $s \rightarrow s$ transitions. When a circularly polarized wave is reflected back on itself, the right-hand circular polarization changes into a left-hand one and if a two-photon transition is induced by one photon from each wave, only $\Delta L = 0$ transitions are selected. Figure 8.41 illustrates the different atomic transitions which are possible by multiphoton transitions with linearly polarized light and with right or left circularly polarized light.

Different upper states can be therefore selected by a proper choice of the polarization. In many cases it is possible to gain information about the symmetry properties of the upper states from the known symmetry of the ground state and the polarization of the two light waves [8.95]. Since the selection rules of two-photon absorption and Raman transitions are identical, one can utilize the group theoretical techniques originally developed for Raman scattering to analyze the symmetry properties of the excited states.

8.10.2 Applications of Multiphoton Absorption to Atomic and Molecular Spectroscopy

The first two-photon absorption in the molecular gas phase was observed in benzene by HOCHSTRASSER et al. [8.96]. An extensive two-photon excitation spectrum of the $^1A_{1g} \rightarrow {}^1B_{2u}$ transition in benzene including the hot bands [8.97] allowed the first assignment of new molecular states. The two-photon absorption is monitored by the fluorescence emitted from the upper excited level. Two-photon spectroscopy in the collisionless low-pressure gas phase together with polarization studies enables one to study in detail the spectroscopy and the dynamics of two-photon prepared states not accessible by one-photon absorption [8.98]. However, polarization criteria as a method of assigning two-photon molecular spectra are more ambiguous than in atomic spectra and must take explicit cognizance of the rotational variations in the polarization ratios, because no unique polarization dependence can be expected for the different rotational lines of a vibrational band [8.99].

These studies of two-photon excitation spectra of benzene, naphtalene, and other organic molecules have been performed with nitrogen laser-pumped dye lasers oscillating around λ = 500 nm which yields two-photon excitations at about 40 000 cm^{-1}. Still higher states can be reached by using frequency-doubled pulsed dye lasers. With a primary photon energy of 2.2 eV ($\hat{=}$ 564 nm) the excitation energy of 8.8 eV is sufficiently high to excite states which emit vacuum ultraviolet fluorescence. While the exciting photons (4.4 eV) can still pass through quartz windows, the emitted fluorescence has to be monitored through windows which transmit in the vuv range, such as MgF_2 windows. With this technique the highly excited states $A^1\Pi$ in CO and $a^1\Pi_g$ in N_2 could be reached and high-resolution spectra for the S branch heads of the (9,0) band of CO and the (5,0) band of N_2 have been recorded [8.100].

The rotational selection rules, which are $\Delta J = 0, \pm 1$ for single-photon transitions, are $\Delta J = 0, \pm 1, \pm 2$ for two-photon transitions, where the value of ΔJ depends on the polarization of the incident radiation (see Fig.8.41).

While for *two*-photon transitions the parity of the excited state is the same as that of the absorbing ground state, *three*-photon transitions excite states with the same parity as single-photon transitions.

Both methods therefore complement each other in high-resolution laser spectroscopy in the vacuum ultraviolet region, where single-photon spectroscopy is impeded by lack of tunable lasers. With intense light of a pulsed tunable dye laser, the $^3P_1$ resonance state of the Xe atom and the P, Q, R branches of the (2-0) band of CO in the fourth positive system ($A^1\Pi \leftarrow X^1\Sigma^+$) have been studied with three-photon spectroscopy [8.101]. The excitation was monitored by detecting the fluorescent decay in the vacuum ultraviolet.

With frequency-doubled tunable dye laser radiation the wavelength range down to 2176 Å can be covered (see Sect.7.6), which allows three-photon spectroscopy at wavelengths as short as 723 Å.

8.10.3 Multiphoton-Ionization Spectroscopy

In the technique of multiphoton-ionization spectroscopy, two or more photons excite atoms or molecules from the ground state to an excited state which may be ionized by several methods (see Sect.8.2.5), e.g., field ionization, photoionization, collisional, or surface ionization. If the laser is tuned to multiphoton resonances, ionization signals are obtained if the upper level is ionized which can be, for instance, monitored with the setup shown in Fig.8.42. The ionization probe is a thin wire inserted into a pipe containing the atomic vapor. If the probe is negatively biased relative to the walls of the pipe, thermionic emission will lead to space-charge-limited current. Ions produced by the laser excitation partly neutralize the space charge, thereby allowing an increased electron current to flow (see Sect.8.2.4).

With this technique a series of even and odd parity states in alkali atoms [8.102] and in the alkaline earth atoms Ca, Sr, and Ba [8.103] has been studied. These experiments have proved the sensitivity of the methods, and the range of possible applications will certainly be greatly increased in the near future.

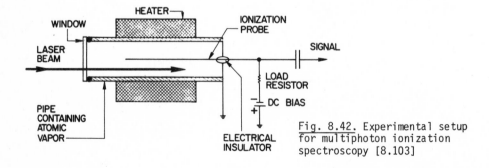

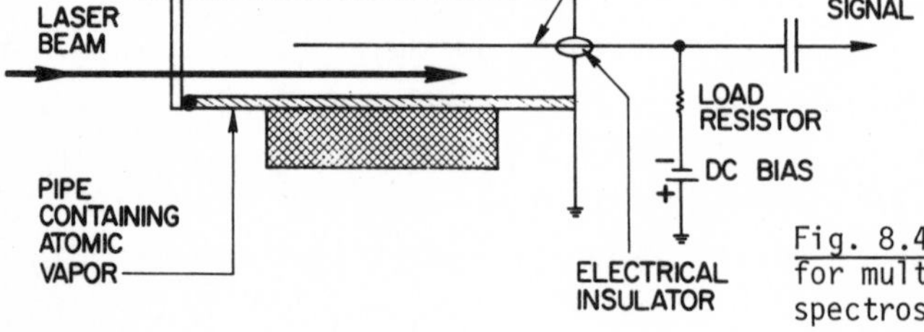

Fig. 8.42. Experimental setup for multiphoton ionization spectroscopy [8.103]

9. Laser Raman Spectroscopy

For many years Raman spectroscopy has been a powerful tool for the investigation of molecular vibrations and rotations. In the pre-laser era, however, its main drawback had been lack of sufficiently intense radiation sources. The introduction of lasers therefore has indeed revolutionized this classical field of spectroscopy. Lasers have not only greatly enhanced the sensitivity of spontaneous Raman spectroscopy but they have furthermore initiated new spectroscopic techniques, based on the stimulated Raman effect, such as coherent anti-Stokes Raman scattering (CARS) or hyper-Raman spectroscopy. The research activities in laser Raman spectroscopy have recently shown an impressive expansion and a vast literature on this field is available. In this chapter we summarize only briefly the basic background of the Raman effect and present some experimental techniques which have been developed. For more thorough studies of this interesting field the textbooks and reviews given in [9.1-4] are recommended.

9.1 Basic Considerations

Raman scattering may be regarded as an inelastic collision of an incident photon $\hbar\omega_i$ with a molecule in the initial energy level E_i. Following the collision a photon $\hbar\omega_s$ with lower energy is detected and the molecule is found in a higher energy level E_f

$$\hbar\omega_i + M(E_i) \rightarrow M^*(E_f) + \hbar\omega_s \quad . \tag{9.1}$$

The energy difference $E_f - E_i = \hbar(\omega_i - \omega_s)$ may appear as vibrational, rotational, or electronic energy of the molecule.

In the energy level scheme (Fig.9.1b), the intermediate state $E_v = E_i + \hbar\omega_i$ of the system "during" the scattering process is often formally described as a "virtual" level which, however, is not necessarily a "real" stationary

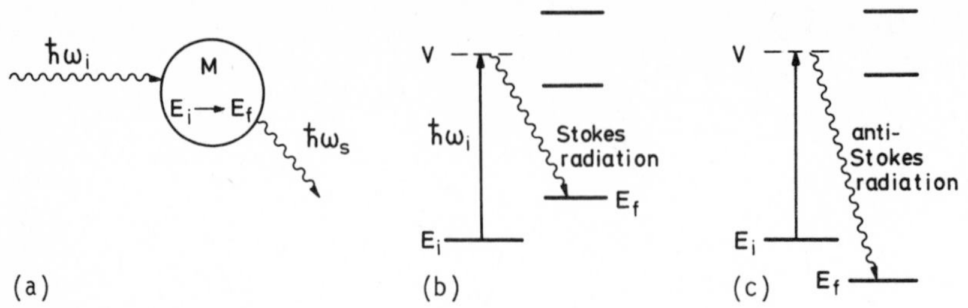

Fig. 9.1. Schematic level diagrams of Raman scattering

eigenstate of the molecule. If the virtual level coincides with one of the molecular eigenstates, one speaks of "resonance Raman effect".

A classical description of the *vibrational* Raman effect (which had been the main process studied before the introduction of lasers) has been developed by PLACEK [9.5]. It starts from the relation

$$\underline{p} = \underline{\mu} + \alpha\underline{E} \tag{9.2}$$

between the electric field amplitude $\underline{E} = \underline{E}_0 \cos\omega t$ of the incident wave and the dipole moment $\vec{p}$ of a molecule. The first term μ represents a possible *permanent* dipole moment while $\alpha\vec{E}$ is the *induced* dipole moment. The polarizability is generally expressed by a tensor (α_{ij}) of rank two, which depends on the molecular symmetry. Dipole moment and polarizability are functions of the coordinates of nuclei and electrons. However, as long as the frequency of the incident radiation is far off resonance with electronic or vibrational transitions, the nuclear displacements, induced by the polarization of the electron cloud, are sufficiently small. Since the electronic charge distribution is determined by the nuclear positions and adjusts "instantaneously" to changes in these positions, we can expand dipole moment and polarizability into Taylor series in the normal coordinates q_n of the nuclear displacements

$$\underline{\mu} = \underline{\mu}(0) + \sum_{n=1}^{3Q} \left(\frac{\partial\mu}{\partial q_n}\right)_0 q_n ; \quad \alpha_{ij}(q) = \alpha_{ij}(0) + \sum_{n=1}^{3Q} \left(\frac{\partial\alpha_{ij}}{\partial q_n}\right)_0 q_n \quad , \tag{9.3}$$

where Q gives the number of nuclei and $\mu(0)$, $\alpha_{ij}(0)$ are the dipole moment and polarizability at the equilibrium configuration $q_n = 0$. For small vibrational amplitudes the normal coordinates $q_n(t)$ of the vibrating molecule can be approximated by

$$q_n(t) = q_{n0} \cos\omega_n t \quad , \tag{9.4}$$

where q_{n0} gives the amplitude and ω_n the vibrational frequency of the nth normal vibration. Inserting (9.4) and (9.3) into (9.2) yields the total dipole moment

$$\underline{p} = \underline{\mu}_0 + \sum_{n=1}^{3Q} \left(\frac{\partial\mu}{\partial q_n}\right)_0 q_{n0} \cos\omega_n t + \alpha_{ij}(0)E_0 \cos\omega t$$

$$+ \frac{1}{2} E_0 \sum_{n=1}^{3Q} \left(\frac{\partial\alpha_{ij}}{\partial q_n}\right)_0 q_{n0} \cos(\omega\pm\omega_n)t \quad . \tag{9.5}$$

The second term describes the infrared spectrum, the third term the Rayleigh scattering, and the last term represents the Raman scattering.

Since an oscillating dipole moment is a source of new waves generated at each molecule, (9.5) shows that an elastically scattered wave at the frequency ω is produced (Rayleigh scattering) but also inelastically scattered components with frequencies $(\omega-\omega_n)$ (*Stokes waves*) and superelastically scattered waves with frequencies $(\omega+\omega_n)$ (*anti-Stokes components*). The microscopic contributions from each molecule add up to macroscopic waves with intensities which depend on the population $N(E_i)$ of molecules in the initial level E_i, on the intensity of the incident radiation and on the expression $(\partial\alpha_{ij}/\partial q_n)q_n$ which describes the dependence of the polarizability components on the nuclear displacements.

Although the classical theory correctly describes the frequencies $\omega \pm \omega_n$ of the Raman lines, it fails to give the correct intensities and a quantum mechanical treatment is demanded. The expectation value of the component α_{ij} of the polarizability tensor is given by

$$<\alpha_{ij}>_{ab} = \int u_b^*(q)\alpha_{ij}u_a(q)dq \quad , \tag{9.6}$$

where the functions $u(q)$ represent the molecular eigenfunctions in the initial level a and the final level b. The integration extends over all nuclear coordinates. This shows that a computation of the intensities of Raman lines is based on the knowledge of the molecular wave functions of initial and final states. In case of vibrational-rotational Raman scattering these are the rotational-vibrational eigenfunctions of the electronic ground state.

For small displacements q_n the molecular potential can be approximated by a harmonic potential, where the coupling between the different normal vibrational modes can be neglected. The functions $u(q)$ are then separable into a product

$$u(q) = \prod_{n=1}^{3N} w_n(q_n,v_n) \tag{9.7}$$

of vibrational eigenfunctions of the nth normal mode with v_n vibrational quanta. Using the orthogonality relation

$$\int w_n \ldots w_m dq = \delta_{nm} \tag{9.8}$$

of the functions $w_n(q_n)$, one obtains from (9.6)

$$\langle \alpha_{ij} \rangle_{ab} = (\alpha_{ij})_0 + \sum_{n=1}^{3N} \left(\frac{\partial \alpha_{ij}}{\partial q}\right)_0 \int w_n(q_n, v_a) q_n w_n(q_n, v_b) dq \quad . \tag{9.9a}$$

The first term is a constant and responsible for the Rayleigh scattering. For nondegenerate vibrations the integrals in the second term vanish unless $v_a = v_b \pm 1$. In these cases it has the value $[(v_a + 1)/2]^{\frac{1}{2}}$. The basic intensity parameter of vibrational Raman spectroscopy is the derivative $(\partial\alpha_{ij}/\partial q)$, which can be determined from the Raman spectra.

The intensity of a Raman line at the Stokes or anti-Stokes frequency $\omega_s = \omega \pm \omega_n$ is determined by the population density $N_i(E_i)$ in the initial level $E_i(v,J)$, by the intensity I_L of the incident pump laser and by the Raman scattering cross section $\sigma_R(i \to f)$ for the Raman transition $E_i \to E_f$.

$$I_s = N_i(E_i)\sigma_R(i \to f) I_L \quad . \tag{9.9b}$$

At thermal equilibrium the population density $N_i(E_i)$ follows a Boltzmann distribution

$$N_i(E_i) = (N/Z)(2J+1)e^{-E_i/kT} \quad , \tag{9.10}$$

where $N = \sum_{v,J} N_i(v,J)$ is the total molecular density obtained by a sum over all rotational-vibrational levels. The partition function

$$Z = \sum g_v(2J+1) \exp[-E(v,J)/kT] \tag{9.11}$$

is a normalization factor which makes $\sum N_i(v,J) = N$, as can be verified by inserting (9.11) into (9.10).

In case of Stokes radiation the initial state of the molecules may be the vibrational ground state, while for the emission of anti-Stokes lines the molecules must have initial excitation energy. Due to the lower population density in these excited levels the intensity of the anti-Stokes lines is down by a factor $\exp(-\hbar\omega_v/kT)$. Typical values are $\hbar\omega_v \approx 1000\ cm^{-1}$; $kT \approx 200\ cm^{-1}$ at $T = 300$ K. The intensity of anti-Stokes lines is therefore down by several orders of magnitude compared with the Stokes intensity.

The scattering cross section depends on the matrix element (9.9) of the polarizability tensor and contains furthermore the ω^4 frequency dependence derived from the classical theory of light scattering. One obtains [9.6] analogous to the two-photon cross section (see Sect.8.10)

$$\sigma_R(i \to f) = \frac{8\pi\omega_s^4}{9\hbar c^4} \left| \sum_j \frac{<\alpha_{ij}> \hat{e}_L <\alpha_{jf}> \hat{e}_s}{(\omega_{ij} - \omega_L - i\Gamma_j)} + \frac{<\alpha_{ji}> \hat{e}_L <\alpha_{jf}> \hat{e}_s}{(\omega_{jf} - \omega_L - i\Gamma_j)} \right|^2 , \qquad (9.12)$$

where $\hat{e}_L$ and $\hat{e}_s$ are unit vectors representing the polarization of the incident laser and the scattered light. The sum extends over all molecular levels j, with homogeneous width Γ_j accessible by single-photon transitions from the initial state i.

We see from (9.12) that initial and final states are connected by *two-photon* transitions which implies that both states have the same parity. For example, the vibrational transitions in homonuclear diatomic molecules, which are forbidden for single-photon infrared transitions, are accessible to Raman transitions. The matrix elements $<\alpha_{ij}>$ depend on the symmetry characteristics of the molecular states. While the theoretical evaluation of the magnitude of $<\alpha_{ij}>$ demands a knowledge of the corresponding wave functions, the question whether $<\alpha_{ij}>$ is zero or not can be answered by group theory. For a more detailed representation with illustrating examples see for instance [9.3,7].

9.2 Stimulated Raman Scattering

If the incident laser intensity I_L becomes very large, an appreciable fraction of the molecules in the initial state E_i is excited into the final state E_f and the intensity of the Raman scattered light is correspondingly large. Under these conditions we have to consider the simultaneous interaction of the molecules with *two* E.M. waves: the laser wave at frequency ω_L and the Stokes wave at frequency $\omega_S = \omega_L - \omega_V$. Both waves are coupled by the molecules vibrating with frequencies ω_V. This parametric interaction leads to an energy exchange between pump wave and Stokes or anti-Stokes waves which may result in a strong, directional output of the Raman waves. This phenomenon of *stimulated* Raman scattering which has been first observed by WOODBURY et al. [9.8] and ECKHARDT et al. [9.9] can be described in a classical picture [9.10].

The Raman medium is taken as consisting of N harmonic oscillators per unit volume, which are independent of each other. Due to the combined action of the incident laser wave and the Stokes wave the oscillators experience a driving force $\underline{F}$ which depends on the total field amplitude $\underline{E}$

$$\underline{E}(z,t) = \underline{E}_L \cos(\omega_L t - k_L z) + \underline{E}_S \cos(\omega_S t - k_S z) \quad , \qquad (9.13)$$

where we have assumed plane waves travelling in the z direction. The potential energy ε_{pot} of a molecule with induced dipole moment $\underline{p} = \alpha \underline{E}$ in an E.M. field with amplitude $\underline{E}$ is

$$\varepsilon_{pot} = -\underline{p} \cdot \underline{E} = -\alpha(q)E^2 \quad . \qquad (9.14)$$

The force acting on the molecule is $\underline{F} = -\mathrm{grad}\varepsilon_{pot}$ which gives, according to (9.3),

$$\vec{F}(z,t) = + \frac{\partial}{\partial q}\left\{[\alpha(q)]E^2\right\} = \left(\frac{\partial\alpha}{\partial q}\right)_0 E^2(z,t) \quad . \qquad (9.15)$$

The equation of motion for an oscillator with mass m and vibrational eigenfrequency ω_v is then

$$\frac{\partial^2 q}{\partial t^2} - \gamma \frac{\partial q}{\partial t} + \omega_v^2 q = \left(\frac{\partial\alpha}{\partial q}\right)_0 E^2/m \quad , \qquad (9.16)$$

where γ is the damping constant which is responsible for the linewidth $\Delta\omega = \gamma$ of spontaneous Raman scattering. With the total field amplitude (9.13) written in the complex notation

$$E(z,t) = \frac{1}{2} E_L \, e^{i(\omega_L t - k_L z)} + \frac{1}{2} E_S \, e^{i(\omega_S t - k_S t)} + \mathrm{c.c.} \qquad (9.13a)$$

we obtain with the trial solution $q = (q_v/2)\, e^{i\omega t}$ + c.c. from (9.16)

$$\left(\omega_v^2 - \omega^2 + i\gamma\omega\right) q_v \, e^{i\omega t} = \frac{1}{2m}\left(\frac{\partial\alpha}{\partial q}\right)_0 E_L E_S \, e^{i[(\omega_L - \omega_S)t - (k_L - k_S)z]} \quad . \qquad (9.17)$$

Comparison of the time-dependent terms on both sides of (9.17) shows that $\omega = \omega_L - \omega_S$. The molecular vibrations

$$q(z,t) = q_v/2)\, e^{i\omega t} + \mathrm{c.c.} \qquad (9.18)$$

are therefore driven at the difference frequency $\omega_L - \omega_S$. Solving (9.17) for q_v yields

$$q_v = \frac{(\partial\alpha/\partial q)_0 E_L E_S}{2m\left[\omega_v^2 - (\omega_L - \omega_S)^2 + i(\omega_L - \omega_S)\gamma\right]} \, e^{-i(k_L - k_S)z} \quad . \qquad (9.19)$$

The nonlinear part of the molecular polarization $\underline{P} = N\underline{p}$, which is responsible for Raman scattering is, according to (9.5) and (9.13a),

$$P_{NL} = N\left(\frac{\partial\alpha}{\partial q}\right)_0 q\left(E_L\, e^{i(\omega_L t-k_L z)} + E_S\, e^{i(\omega_S - k_S z)}\right) \quad . \tag{9.20}$$

This polarization which depends on z and t is the source of new waves at frequencies ω_L, ω_S, and $\omega_L \pm \omega_S$ which superimpose the incident laser wave and the Stokes wave and may amplify or attenuate these waves. The contribution to the Stokes waves is obtained from (9.18) and (9.20) as

$$P_{NL}(\omega_S) = \frac{1}{2} N\left(\frac{\partial\alpha}{\partial q}\right)_0 q_v E_L^* \, e^{-i(\omega_S t-k_S z)} + \text{c.c.} \quad , \tag{9.21}$$

which shows that a "polarization wave" travels through the Raman medium with the same wave vector k_S as the Stokes wave and which therefore can always amplify the Stokes wave. From the general wave equation [9.10]

$$\Delta E = \mu_0 \sigma \frac{\partial E}{\partial t} + \mu_0 \varepsilon \frac{\partial^2 E}{\partial t^2} + \mu_0 \frac{\partial^2}{\partial t^2}(P_{NL}) \tag{9.22}$$

of waves in a medium with conductivity σ we obtain for the one-dimensional problem ($\partial/\partial y = \partial/\partial x = 0$) with the approximation $d^2E/dz^2 \ll k dE/dz$ and with (9.21) the equation for the Stokes wave

$$\frac{dE_S^*}{dz} = -\frac{\sigma}{2}\sqrt{\frac{\mu_0}{\varepsilon}}\, E_S^* + i\, S_S E_L^* q_v \quad , \tag{9.23}$$

where the abbreviation

$$S_S = \frac{1}{2}\frac{k_S}{\varepsilon} N\left(\frac{\partial\alpha}{\partial q}\right)_0$$

has been used. Substituting (9.19) we obtain the final result

$$\frac{dE_S^*}{dz} = \left[-\frac{\sigma}{2}\sqrt{\frac{\mu_0}{\varepsilon}} + \frac{N(\partial\alpha/\partial q)^2 E_L^2}{4m\varepsilon\left[\omega_v^2 - (\omega_L-\omega_S)^2 + i(\omega_L-\omega_S)\gamma\right]}\right] E_S^* = (-f+g)E_S^* \quad . \tag{9.24}$$

Integration of (9.24) yields

$$E_S^* = E_S(0)\, e^{(g-f)z} \quad . \tag{9.25}$$

The Stokes wave is amplified if g exceeds f. The amplification factor g depends on the square of the laser amplitude E_L and on the term $(\partial\alpha/\partial q)^2$. Stimulated Raman scattering is therefore observed only if the incident laser intensity exceeds a threshold value which is determined by the nonlinear term $(\partial\alpha_{ij}/\partial q)_0$ in the polarization tensor of the Raman active normal vibration.

While in spontaneous Raman scattering the intensity of anti-Stokes radiation is very small due to the low thermal population density in excited molecular levels (see Sect.9.1), this is not necessarily true in stimulated Raman scattering. Due to the strong incident pump wave a large fraction of all interacting molecules is excited into higher vibrational levels, and strong anti-Stokes radiation at frequencies $\omega_L + \omega_V$ has been found.

The driving term in the wave equation (9.22) for an anti-Stokes wave at $\omega_a = \omega_L + \omega_V$ is, similar to (9.21), given by

$$P_{NL}^{(\omega_a)} = \frac{1}{2} N \left(\frac{\partial\alpha}{\partial q}\right)_0 q_v E_L \, e^{i[(\omega_L+\omega_V)t - k_L z]} + \text{c.c.} \quad . \tag{9.26}$$

For small amplitudes $E_a << E_L$ of the anti-Stokes waves we can assume that the molecular vibrations are independent of E_a and we can replace q_v by its solution (9.19). This yields for the amplification of E_a the equation

$$\frac{dE_a}{dz} = -\frac{f}{2} E_a \, e^{i(\omega_a - k_a z)} + i \, \frac{\omega_a \sqrt{\mu_0/\varepsilon} \, N(\partial\alpha/\partial q)_0^2}{8m_v} E_L^2 E_S^* \, e^{i(2k_L - k_S - k_a)z} \quad . \tag{9.27}$$

This shows that, analogous to sum or difference frequency generation (Sect. 7.5), a macroscopic wave can build up only if the phase-matching condition

$$k_a = 2k_L - k_S \tag{9.28a}$$

can be satisfied. In a normally dispersive medium this condition cannot be met for collinear waves. From a three-dimensional analysis, however, one obtains the vector equation

$$2\underline{k}_L = \underline{k}_S + \underline{k}_a \tag{9.28b}$$

which shows that the anti-Stokes radiation is emitted into a cone whose axis is parallel to the laser propagation direction (Fig.9.2). This is exactly what has been observed [9.9].

Let us briefly summarize the differences between the linear (spontaneous) and the nonlinear (induced) Raman effect

1) While the intensity of spontaneous Raman lines is down by several orders of magnitude compared with the pump intensity, the stimulated Stokes or anti-Stokes radiations have intensities comparable to that of the pump wave.

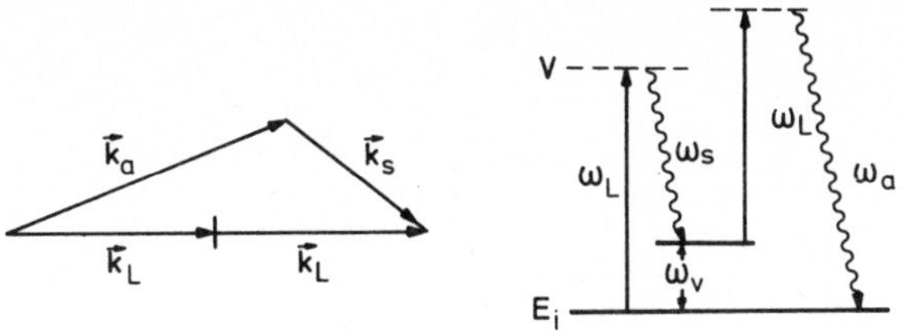

Fig. 9.2. Generation of stimulated anti-Stokes radiation

2) The stimulated Raman effect is observed only above a threshold pump intensity which depends on the gain of the Raman medium and the length of the pump region.

3) Most Raman active substances show only one or two Stokes lines at frequencies $\omega_S = \omega_L - \omega_V$ in stimulated emission. Besides these Stokes lines, however, lines at frequencies $\omega = \omega_L - n\omega_V$ $(n = 1,2,3)$ have been observed which do not correspond to overtones of vibrational frequencies. Because of the anharmonicity of molecular vibrations the spontaneous Raman lines due to vibrational overtones are shiften against ω_L by $\Delta\omega < n\omega_V$.

4) The linewidths of spontaneous and stimulated Raman lines depend on the linewidth of the pump laser. For narrow laser linewidths, however, the width of the stimulated Raman lines becomes smaller than that of the spontaneous lines.

The main merit of the stimulated Raman effect for molecular spectroscopy may be seen in the fact that it allows one to build high-power "Raman lasers" which can be tuned over wide spectral ranges in combination with tunable pump lasers (see Sect.7.6). More details on the stimulated Raman effect and reference to experiments in this field can be found in [9.11-13].

9.3 Coherent Anti-Stokes Raman Spectroscopy (CARS)

In the previous section we discussed the fact that a sufficiently strong incident pump wave at a frequency ω_L can generate an intense Stokes wave at $\omega_S = \omega_L - \omega_V$. Under the combined action of both waves a nonlinear polarization P_{NL} of the medium is generated which contains contributions at the frequencies $\omega_V = \omega_L - \omega_S$, $\omega_S = \omega_L - \omega_V$, and $\omega_a = \omega_L + \omega_V$. These contributions act as driving terms for the generation of new waves. Provided that the phase-matching condition $2\underline{k}_L = \underline{k}_S + \underline{k}_a$ can be satisfied, a strong anti-Stokes wave $E_a \cos(\omega_a t - \underline{k}_a \cdot \underline{r})$ is observed in the direction of $\underline{k}_a$.

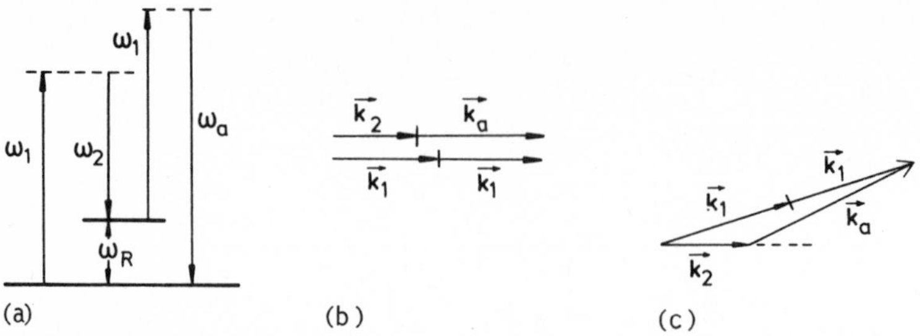

Fig. 9.3. Energy level diagram for CARS (a) and wave vector diagram for phase-matching (b) in gaseous samples neglecting dispersion, (c) in liquid samples

Despite the enormous intensities of stimulated Stokes and anti-Stokes waves, stimulated Raman spectroscopy has been of little use in molecular spectroscopy. The high threshold, which depends, according to (9.24), on the molecular density N, the incident intensity $I = E_L^2$, and the polarizability term $(\partial\alpha_{ij}/\partial q)$, limits stimulated emission to only the strongest Raman lines of materials of high densities N.

The recently developed technique of CARS, however, combines the advantages of signal strength obtained in stimulated Raman spectroscopy with the general applicability of spontaneous Raman spectroscopy [9.14]. In this technique Stokes and anti-Stokes waves are directly generated from *two* incident waves. Assume that two intense collinear laser beams with frequencies ω_1 and ω_2 $(\omega_1 > \omega_2)$ are passing through a Raman sample. If the frequency difference $\omega_1 - \omega_2 = \omega_R$ equals the frequency ω_R of a Raman active molecular transition (see Fig.9.3a), Stokes and anti-Stokes waves are generated in the same way as discussed above. Two waves at the "pump frequency" ω_1 and one at the Stokes frequency ω_2 are mixed by the nonlinear polarization of the medium and a new anti-Stokes wave at the frequency $\omega_a = 2\omega_1 - \omega_2$ is generated in a four-wave parametric mixing process. Similarly a new Stokes wave at $\omega_S = 2\omega_2 - \omega_1$ can be produced.

If the incident waves are at *optical* frequencies, the difference frequency $\omega_R = \omega_1 - \omega_2$ is small compared with ω_1 in case of rotational-vibrational frequencies ω_R. In gaseous Raman samples the dispersion is generally negligible over small ranges $\Delta\omega = \omega_1 - \omega_2$ and satisfactory phase matching is obtained for *collinear* beams. The Stokes wave at $\omega_S = 2\omega_2 - \omega_1$ and the anti-Stokes wave at $\omega_a = 2\omega_1 - \omega_2$ are then generated in the same direction as the incoming beams. In liquids, dispersion effects are more severe and the phase-matching condition can only be satisfied over a sufficiently long coherence length, if the two incoming beams are crossed at the phase-matching angle (see Fig.9.3).

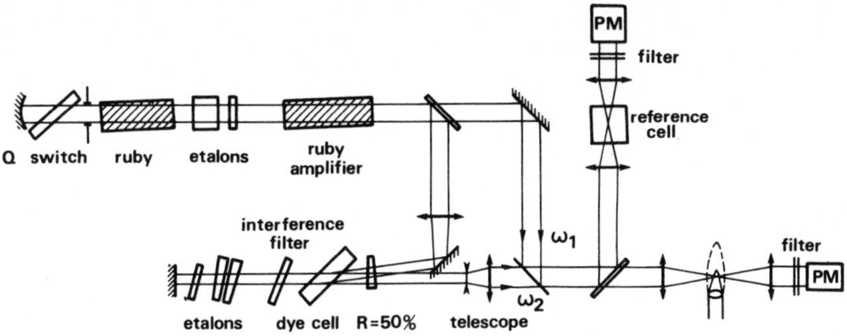

Fig. 9.4. Experimental setup for CARS with a ruby laser and a dye laser pumped by the ruby laser [9.15]

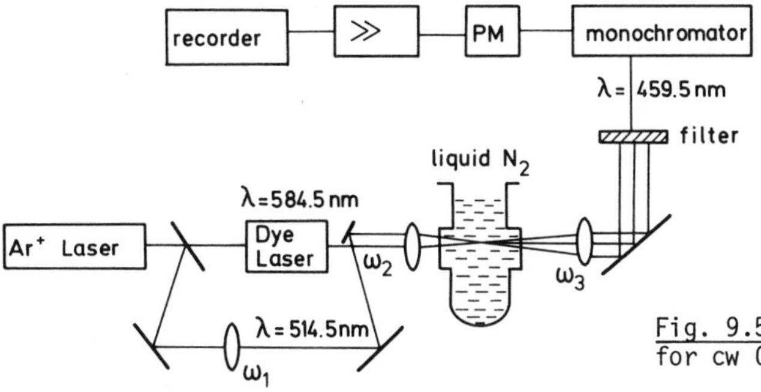

Fig. 9.5. Experimental setup for cw CARS in liquids [9.17]

The anti-Stokes wave at $\omega_a = 2\omega_1 - \omega_2$ ($\omega_a > \omega_1$!) is detected through filters which reject both incident laser beams and also the fluorescence which may be generated in the sample or the filters. Figure 9.4 shows a typical experimental setup used for rotational-vibrational spectroscopy of gases by CARS [9.15]. The two incident laser beams are provided by a Q-switched ruby laser and a tunable dye laser, pumped by this ruby laser. Because the gain of the anti-Stokes wave depends quadratically on the molecular density N [see (9.24)] power levels of the incident beams are required which are in the megawatt range for gaseous samples while kilowatt powers are sufficient for liquid samples. Two dye lasers pumped by a common nitrogen laser provide convenient sources for the radiations at ω_1 and ω_2. Frequency drifts in the lasers and instabilities of the spatial intensity distributions in the beams cause severe fluctuations of the signals measured in CARS. Using stable and compact designs these fluctuations could be reduced below 10% [9.16].

In liquid samples CARS can be performed even with cw lasers [9.17]. Figure 9.5 shows the experimental setup for cw CARS of liquid nitrogen, where the two incident collinear waves are provided by the 514.5 nm argon laser line and a cw dye laser pumped by the same argon laser.

The advantages of CARS may be summarized as follows [9.14]:

1) The signal levels in CARS may exceed that obtained in spontaneous Raman spectroscopy by a factor of 10^4-10^5.

2) The higher frequency $\omega_3 > \omega_1, \omega_2$ of the anti-Stokes waves allows one to use filters which reject the incident light as well as fluorescence light.

3) The small beam divergence allows excellent spatial discrimination against fluorescence or thermal luminous background such as occurs in flames, discharges, or chemiluminescent samples.

4) The main contribution to the anti-Stokes generation comes from a small volume around the focus of the two incident beams. Therefore very small sample quantities (microliters for liquid samples or millitorr pressures for gaseous samples) are required. Furthermore a high spatial resolution is possible, which allows one to probe the spatial distribution of molecules in definite rotational-vibrational levels. The measurements of local temperature variations in flames from the intensity of anti-Stokes lines in CARS is an example where this advantage is utilized.

5) A high spectral resolution can be achieved without using a monochromator. The Doppler width $\Delta\omega_D$, which represents a principal limitation in 90° spontaneous Raman scattering, is reduced to $[(\omega_2 - \omega_1)/\omega_1]\Delta\omega_D$ for the collinear arrangement used in CARS. While a resolution of 0.3 to 0.03 cm^{-1} is readily obtained in CARS with pulsed lasers, even linewidths down to 0.001 cm^{-1} have been achieved with single-mode lasers.

The main disadvantages of CARS are the expensive equipment and strong fluctuations of the signals caused by instabilities of intensities and alignments of the incident laser beams. The sensitivity of detecting low relative concentrations of specified sample molecules is mainly limited by interference with the nonresonant background from the other molecules in the sample.

9.4 Hyper-Raman Effect

For large electric field amplitudes E, such as can be achieved with pulsed lasers, the linear relation (9.2) between induced dipole moment p and electric field E is no longer a satisfactory approximation and higher order terms have to be included. We then obtain instead of (9.2)

$$\underline{p} = \underline{\mu} + \alpha\underline{E} + \beta\underline{E}\cdot\underline{E} + \gamma\underline{E}\cdot\underline{E}\cdot\underline{E} + \dots \quad . \tag{9.29}$$

The coefficients β, γ, ... which are tensors of rank 3,4,... are called hyperpolarizabilities. Analogous to (9.3) we can expand β in a Taylor series in the normal coordinates $q_n = q_{n0} \cos\omega_n t$

$$\beta = \beta_0 + \sum_{n=1}^{3Q} \left(\frac{\partial\beta}{\partial q_n}\right)_0 q_n + \dots \quad . \tag{9.30}$$

Assume that two laser waves $E_1 = E_{01} \cos(\omega_1 t - k_1 z)$ and $E_2 = E_{02} \cos(\omega_2 t - k_2 z)$ are incident on the Raman sample. From the third term in (9.29) we then obtain with (9.30) contributions due to β_0

$$\beta_0 E_{01}^2 \cos 2\omega_1 t \quad , \text{and} \quad \beta_0 E_{02}^2 \cos 2\omega_2 t \quad ,$$

which give rise to *hyper-Rayleigh scattering*. The term $(\partial\beta/\partial q_n) q_{n0} \cos\omega_n t$ of (9.30) inserted into (9.29) yields contributions

$$\left(\frac{\partial\beta}{\partial q}\right)_0 q_{n0} [\cos(2\omega_1 \pm \omega_n) + \cos(2\omega_2 \pm \omega_n)t] \quad ,$$

which are responsible for *hyper-Raman scattering*.

Since the coefficients $(\partial\beta/\partial q)_0$ are very small, one needs large incident intensities to observe hyper-Raman scattering. Similar to second harmonic generation (see Sect.7.7) hyper-Rayleigh scattering is forbidden for molecules with a center of inversion. The hyper-Raman effect obeys selection rules which differ from those of the linear Raman effect. It is therefore very attractive to molecular spectroscopists since molecular vibrations can be observed in the hyper-Raman spectrum which are forbidden for infrared as well as for linear Raman transitions. For example spherical molecules such as CH_4 have no pure rotational Raman spectrum but a hyper-Raman spectrum which has been found by MAKER [9.18]. A general theory for rotational and rotational-vibrational hyper-Raman scattering has been worked out by ALTMANN and STREY [9.19].

9.5 Experimental Techniques of Laser Raman Spectroscopy

The scattering cross sections in spontaneous Raman spectroscopy are very low, typically of the order of 10^{-30} cm^2. The experimental problems of detecting weak signals in the presence of intense background radiation are by no means

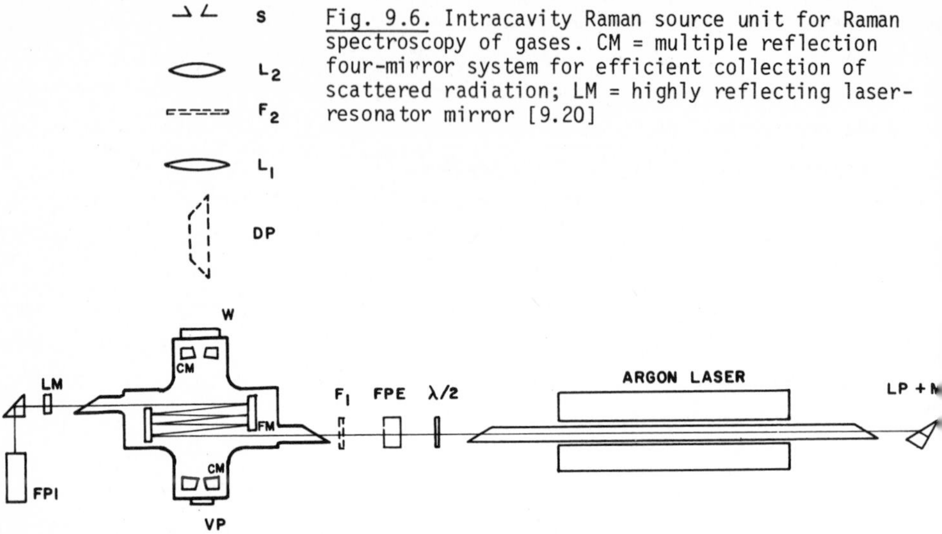

Fig. 9.6. Intracavity Raman source unit for Raman spectroscopy of gases. CM = multiple reflection four-mirror system for efficient collection of scattered radiation; LM = highly reflecting laser-resonator mirror [9.20]

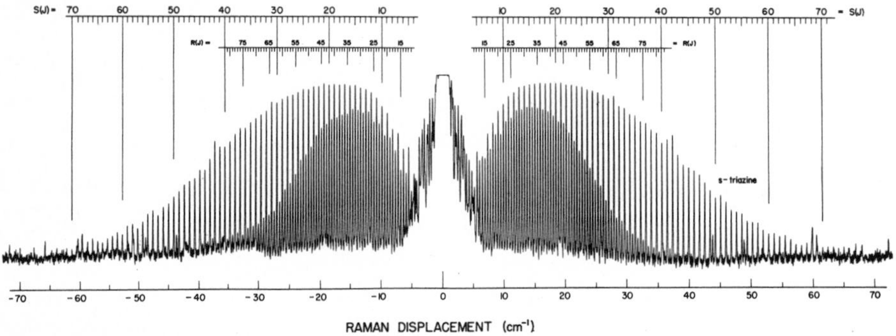

Fig. 9.7. Pure rotational Raman spectrum of $C_3H_3N_3$ excited by the argon laser at λ = 488 nm. The Stokes side lies to the right. Pressure 4 Torr, exposure time 6 h [9.20]

trivial. The achievable signal-to-noise ratio depends both on the pump intensity and on the sensitivity of the detector. The last years have brought remarkable progress on the source as well as on the detector side. The incident light intensity can be greatly enhanced by using multiple reflection cells, intracavity techniques (see Sect.8.2), or a combination of both. Figure 9.6 shows as an example a multiple reflection Raman cell inside the resonator of an argon laser. The laser can be tuned by the prism LP to the different laser lines [9.20]. A sophisticated system of mirrors CM collects the scattered light which is further imaged by the lens L_1 onto the entrance slit of the spectrometer. A Dove prism [9.21] turns the image of the line source by 90°

to make it parallel to the entrance slit. Figure 9.7 shows for illustration of the achieved sensitivity the pure rotational Raman spectrum of s-triazine $C_3H_3N_3$, obtained with this set up.

In earlier days of Raman spectroscopy the photographic plate was the only detector used to record the Raman spectra. The introduction of sensitive photomultipliers and in particular the development of image intensifiers and optical multichannel analyzers with cooled photocathodes (see Sect.4.5) have greatly enhanced the detection sensitivity. Image intensifiers and OMA (or OSA) instrumentation allow simultaneous recording of extended spectral ranges with sensitivities comparable to those of photomultipliers. The third experimental component which has contributed to the further improvement of the quality of Raman spectra is the introduction of digital computers to control the experimental procedure and to read out and to analyze the data. This has greatly reduced the time spent for interpretation of the results [9.21a].

A great increase of sensitivity in linear Raman spectroscopy of liquids has been achieved with the optical fiber Raman spectroscopy. This technique

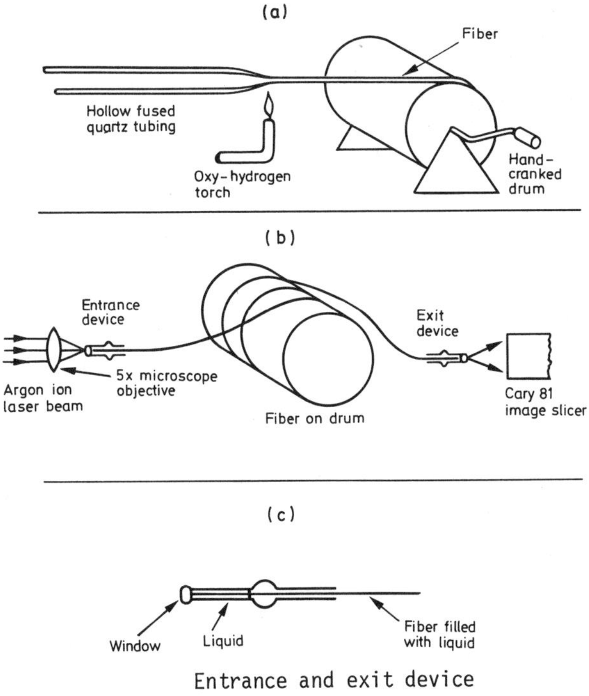

Fig.9.8. Raman spectroscopy of liquid samples in an optical fiber [9.24]. The laser light is coupled into the fiber by a microscope objective

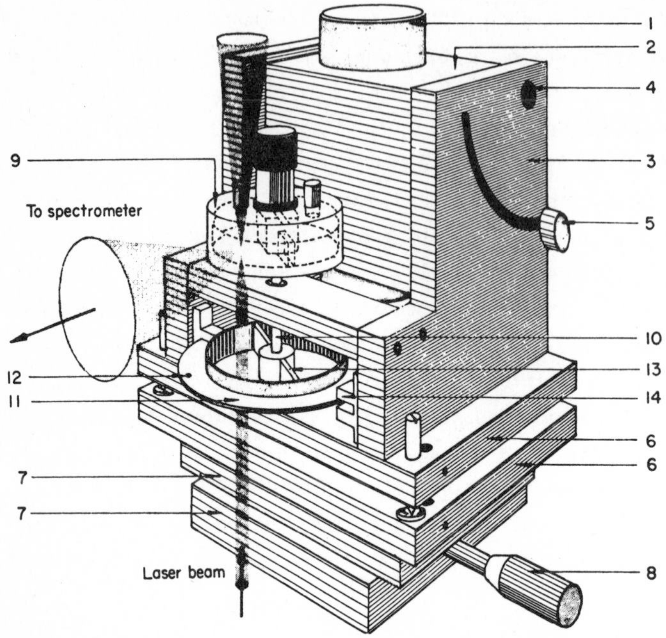

Fig. 9.9. Rotating system for difference Raman spectroscopy of liquids. 1) motor; 2) motor block; 3) side parts; 4) motor block axis; 5) set screw; 6) kinematic mount; 7) X-Y precision ball movement; 8) adjustment screw; 9) liquid cell for difference Raman spectroscopy; 10) axis for trigger wheel and for holders of liquid cells; 11) trigger wheel; 12) trigger hole; 13) bar; 14) optoelectronic array consisting of a luminescence diode and phototransistor [9.24]

uses a capillary optical fiber with refractive index n_f, filled with a liquid of refractive index $n_e > n_f$. If the incident laser beam is focused into the fiber, the laser light as well as the Raman light is trapped in the core due to internal reflection and therefore travels inside the capillary. With sufficiently long capillaries (10-30 m) and low losses, very high spontaneous Raman intensities are achieved which may exceed those of conventional techniques by factors of 10^3 [9.22]. Figure 9.8 shows schematically the experimental arrangement where the fiber is wound on a drum. Because of the increased sensitivity this fiber technique allows one to record also second- and third-order Raman bands which facilitates complete assignments of vibrational spectra [9.23].

Just as in absorption spectroscopy, the sensitivity may be enhanced by difference laser Raman spectroscopy, where the pump laser passes alternately through a cell containing the sample molecules solved in some liquid and through a cell containing only the liquid. The basic advantages of this difference technique are the cancellation of unwanted Raman bands of the solvent in the spectrum of the solution and the accurate determination of small frequency shifts due to interactions with the solvent molecules.

In the case of strongly absorbing Raman samples the heat production at the focus of the incident laser may become so large that the sample molecules may thermally decompose. A solution to this problem is the rotating sample technique [9.24] where the sample is rotated with an angular velocity Ω. If the interaction with the laser beam is R cm away from the axis, the time T spent by the molecules within the focal region with diameter d (m) is $T = d/(R\Omega)$. This technique, which allows much higher input powers and therefore better signal-to-noise ratios, can be combined with the difference technique by mounting a cylindrical cell onto the rotation axis. One half of the cell is filled with the liquid Raman sample in solution, the other half is filled with only the solvent (Fig.9.9).

More details of recent techniques in laser Raman spectroscopy can be found in the review of KIEFER [9.24].

9.6 Applications of Laser Raman Spectroscopy

The primary object of Raman spectroscopy is the determination of molecular energy levels and transition probabilities connected with molecular transitions which are not accessible to infrared spectroscopy. Linear laser Raman spectroscopy, CARS, and hyper-Raman scattering have very successfully collected many spectroscopic data which could not have been obtained with other techniques. Besides these basic applications to molecular spectroscopy there are, however, a number of scientific and technical applications of Raman spectroscopy to other fields which have become feasible with the new methods discussed in the previous sections. We can give only a few examples.

Since the intensity of spontaneous Raman lines is proportional to the density $N(v_i,J_i)$ of molecules in the initial state (v_i,J_i), Raman spectroscopy can provide information on the population distribution $N(v_i,J_i)$, its local variation, and on concentrations of molecular constituents in samples. This allows one, for instance, to probe the temperature in flames or hot gases from the rotational Raman spectra [9.25,26] and to detect deviations from thermal equilibrium.

Using CARS the spatial resolution is greatly increased and the focal volume from which the signal radiation is generated can be made smaller than

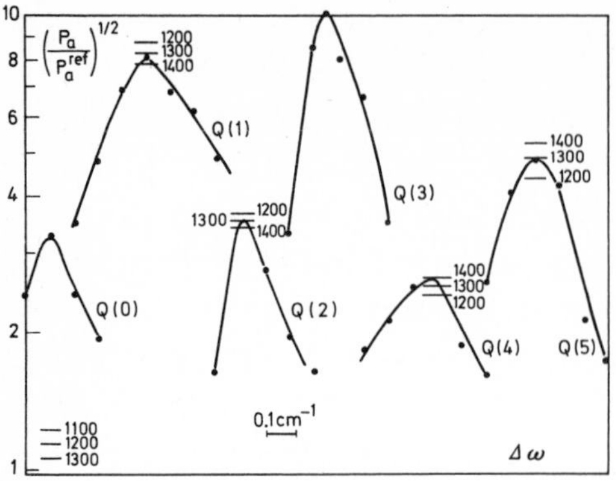

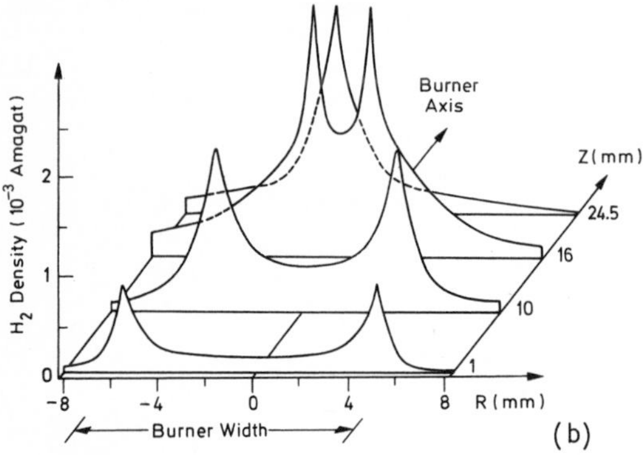

Fig. 9.10. (a) Q lines of H_2, obtained by CARS in a flame near their maxima. The points at line centers represent averages over 10 laser shots. This and the flame fluctuations explain the irregular shapes and widths. (b) H_2 distribution in a horizontal gas flame. R is the distance from the burner axis, z the distance along this axis. The results are deduced from the Q(1) intensity [9.15]

0.1 mm^3 [9.15]. The local density profiles of reaction products formed in flames or discharges can therefore be accurately probed without disturbing the sample conditions. The intensity of the stimulated anti-Stokes radiation is proportional to N^2 [see (9.27)]. Figure 9.10 shows for illustration the H_2 distribution in a horizontal Bunsen flame, measured from the CARS spectrum of the Q branch in H_2. The H_2 molecules are formed by the pyrolysis of hydrocarbon molecules [9.15]. Another example is the measurement of CARS spectra of water vapor in flames which allowed one to probe the temperature in the postflame region of a premixed CH_4 air flame [9.26].

With a detection sensitivity of 10 to 100 ppm, CARS is not as good as some other techniques in monitoring pollutant gases at low concentrations (see Sect.8.2) but its advantage is the capability of examining a large number of species quickly by tuning the dye lasers. The good background rejection allows the use of this technique under conditions of bright background radiation where other methods may fail [9.27].

Nondestructive analysis of various materials, such as rocks, composite materials, phases and inclusions in solids, can be performed with a laser molecular microprobe [9.28], which is based on a combination of an optical microscope with a Raman spectrometer. The laser beam is focused into the sample and the Raman spectrum, emitted from the small focal spot, is monitored through the microscope. The principal modes of operation were illustrated by studying the different phases of "fluid inclusions" contained in a quartz crystal from the Swiss Alps. The Raman spectra obtained permit determination of CO_2 as a gaseous inclusion, water as liquid inclusion, and the fact that the mineral, thought to be $CaSO_4$, is in fact $CaCO_3$.

Raman spectroscopy has been extensively used in biology, biophysics, and medicine to investigate the molecular structure of biomolecules and to study the transient behavior of chemical reactions in biological processes. Some examples are given in Chap.14. For further applications, see [9.29,30].

10. High-Resolution Sub-Doppler Laser Spectroscopy

The spectral resolution of all methods discussed in Chap.8 was in principle limited by the Doppler width of lines in the molecular absorption—or emission—spectra, although the laser linewidth might have been much smaller. In this chapter we present several techniques which overcome this resolution limit and which allow one to resolve the natural linewidth even in the presence of a much larger Doppler width. These techniques, which have been developed recently, have already stimulated experimental and theoretical atomic physics in an outstanding way. These various Doppler-free methods certainly represent an important step towards a more detailed knowledge of molecular structure and of deeper details regarding the interaction of E.M. radiation and matter.

While for most experiments in Doppler-limited spectroscopy—discussed in Chaps.8 and 9—*multimode* lasers can be used (e.g., for optical pumping experiments, laser-induced fluorescence of atoms and simple molecules, or for Doppler-limited absorption spectroscopy) only some of the sub-Doppler methods, treated in this chapter, may be performed with pulsed or cw multimode lasers. Most of these techniques demand narrow-band tunable *single-mode* lasers with a bandwidth which should be smaller than the desired spectral resolution. If the natural linewidth $\delta\nu_n$ has to be resolved, the laser frequency jitter should be smaller than $\delta\nu_n$. This demands frequency stabilization techniques (see Sect.6.5) and there are many examples in this branch of high-resolution laser spectroscopy where the achieved resolution is indeed limited by the stability of the laser.

The basic principle of most Doppler-free techniques relies on a proper selection of a subgroup of molecules with velocity components v_z in the direction of the incident monochromatic wave, which fall into a small interval Δv_z around $v_z = 0$. This selection can be achieved either by mechanical apertures which select a collimated molecular beam, or by selective saturation within the velocity distribution of absorbing molecules caused by a strong pump wave, and a successive probing of this selective "hole burning" by a monochromatic tunable probe wave (see Sect.3.6).

Another class of Doppler-free techniques is based on a coherent preparation of an atomic or molecular state and a detection scheme which is capable of detecting the phase relations of the wave functions in this state.

Most of these "coherent excitation methods" rely on optical pumping by pulsed or cw lasers, which may not even be single-mode lasers as long as the selective population of the levels under study is guaranteed.

We will now discuss these different methods in more detail.

10.1 Spectroscopy in Collimated Molecular Beams

Let us assume molecules effusing into a vacuum tank from a small hole A in an oven which is filled with a gas or vapor at pressure p. The molecular density behind A and the background pressure in the vacuum tank shall be sufficiently low to assure a large mean free path of the effusing molecules, such that collisions can be neglected. The number $N(\vartheta)$ of molecules which travel into the cone $\vartheta \pm d\vartheta$ around the direction ϑ against the symmetry axis (which we choose to be the z axis) is proportional to $\cos\vartheta$. A slit B with width b selects at a distance d from the point source A a small angular interval $-\varepsilon \leq \vartheta \leq +\varepsilon$ around $\vartheta = 0$ (see Fig.10.1). The molecules passing through the slit B, which is parallel to the y axis, form a molecular beam in the z direction, collimated with respect to the x direction. The collimation ratio is defined by (see Fig.10.1b)

$$v_x/v_z = \tan\varepsilon = b/(2d) \quad . \tag{10.1}$$

If the source diameter is small compared with the slit width b and if $b \ll d$ (which means $\varepsilon \ll 1$), the flux density behind the slit B is approximately constant across the beam diameter, since $\cos\vartheta \approx 1$ for $\vartheta \ll 1$. For this case the density profile of the molecular beam is illustrated in Fig.10.1b.

The density $n(v)dv$ of molecules with velocities $v = |\underline{v}|$ inside the interval v to v + dv in a molecular beam at thermal equilibrium which effuses with a most probable velocity $v_p = (2kT/m)^{\frac{1}{2}}$ into the z direction can be described at a distance $r = (z^2 + x^2)^{\frac{1}{2}}$ from the source A as

$$n(v,r)dv = C \frac{\cos\vartheta}{r^2} Nv^2 e^{-(v/v_p)^2} dv \quad , \tag{10.2}$$

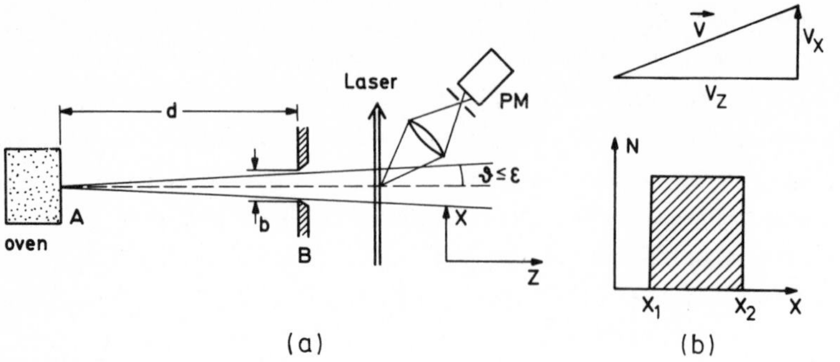

Fig.10.1. Schematic illustration of laser excitation spectroscopy with reduced Doppler width in a collimated molecular beam (a). Collimation ratio and density profile n(x) in a collimated beam effusing from a point source (b)

where the normalization factor $C = (4/\sqrt{\pi})v_p^{-3}$ assures that the total density N of the molecules is $N = \int n(v)dv$.

10.1.1 Reduction of Doppler Width

If the collimated molecular beam is crossed perpendicularly with a monochromatic laser beam with frequency ω propagating into the x direction, the absorption probability for each molecule depends on its velocity component v_x. In Sect.3.2 it was shown that the center frequency of a molecular transition, which is ω_0 in the rest frame of the moving molecule, is Doppler shifted to a frequency ω_0' according to

$$\omega_0' = \omega_0 - \underline{k}\underline{v} = \omega_0 - kv_x \quad ; \quad k = |\underline{k}| \quad . \tag{10.3}$$

Only those molecules with velocity components v_x in the interval $dv_x = \delta\omega_n/k$ around $v_x = (\omega - \omega_0)/k$ essentially contribute to the absorption of the monochromatic laser wave, because these molecules are shifted into resonance with the laser frequency ω within the natural linewidth $\delta\omega_n$ of the absorbing transition.

When the laser beam in the x-z plane passes along the x direction through the molecular beam at y = 0, its intensity decreases as

$$I(\omega) = I_0 \exp\left[-\int_{x_1}^{x_2} \alpha(\omega,x)dx\right] \ . \tag{10.4}$$

For small absorptions the spectral profile of the absorbed intensity $dI = I(x_1) - I(x_2)$ is with $I(x_1) = I_0$

$$dI(\omega) = I_0 \int_{-v\sin\varepsilon}^{+v\sin\varepsilon} \left[\int_{x_1}^{x_2} n(v_x,x)\sigma(\omega,v_x)dx\right]dv_x \quad . \tag{10.5}$$

The absorption cross section $\sigma(\omega,v_x)$ describes the absorption of a monochromatic wave of frequency ω by a molecule with a velocity component v_x. Its spectral profile is represented by a Lorentzian (see Sect.3.6)

$$\sigma(\omega,v_x) = \sigma_0 \frac{(\gamma/2)^2}{(\omega - \omega_0 - kv_x)^2 + (\gamma/2)^2} = \sigma_0 \cdot L(\omega - \omega_0,\gamma) \quad . \tag{10.6}$$

The density $n(v_x,x)dv_x$ of molecules with velocity components v_x in the interval dv_x at a point (x,z) in the molecular beam can be derived from (10.2). Using the relations $r^2 = x^2 + z^2$; $v_x/v = x/r \Rightarrow dv_x = (x/r)dv$ we obtain with $\cos\vartheta = z/r$ (note that dv_x changes sign with x since $dv > 0$)

$$n(v_x,x)dv_x = CN \frac{z}{x^3} v_x^2 \, e^{-(rv_x/xv_p)^2} dv_x \quad . \tag{10.7}$$

Equation (10.3) gives the relation between the velocity component v_x and the Doppler shift $\Delta\omega_0 = \omega_0 - \omega_0' = kv_x = v_x\omega_0/c$ of the center frequency ω_0.

Inserting (10.3,6,7) into (10.5) yields the absorption profile

$$dI(\omega) = a \int_{-\infty}^{+\infty} \left\{\int_{x_1}^{x_2} \frac{\Delta\omega_0^2}{x^3} L(\omega - \omega_0,\gamma) \exp\left[-\Delta\omega_0^2 c^2/\omega_0^2 v_p^2\left(1 + \frac{z^2}{x^2}\right)\right] dx\right\} d(\Delta\omega_0) \tag{10.8}$$

with $a = I_0 cN\sigma_0(c^3/\,_0^3)z$. The integration over $\Delta\omega_0$ extends from $-\infty$ to $+\infty$ since the velocities v in (10.5) range from 0 to ∞. The integration over x can be carried out and gives with $x_1 = r\sin\varepsilon$, $r^2 = x^2 + z^2$

$$dI(\omega) = I_0 b \int_{-\infty}^{+\infty} L(\omega - \omega_0,\gamma) \exp-\left[(\omega - \omega')c/\omega_0 vp \sin\varepsilon\right]^2 d\omega' \tag{10.9}$$

with

$$b = a(zc/\omega_0)^2 \quad .$$

This represents a *Voigt profile*, i.e., a convolution product of a Lorentzian function with halfwidth γ and a Doppler function. A comparison with (3.33) shows, however, *that the Doppler width is reduced by a factor* $\sin\varepsilon$, which equals the collimation ratio of the beam. The collimation of the molecular beam therefore reduces the Doppler width $\Delta\omega_0$ of the absorption lines to a width

$$\boxed{\Delta\omega_D^* = \Delta\omega_D \sin\varepsilon} \qquad \text{with} \qquad \Delta\omega_D = 2\omega_0(v_p/c)(\ln 2)^{\frac{1}{2}} \ , \qquad (10.10)$$

where $\Delta\omega_D$ is the corresponding Doppler width in a gas at thermal equilibrium. Note that for larger diameters of the oven hole A the density profile n(x) of the molecular beam is no longer rectangular but decreases gradually beyond the limiting angles $\vartheta = \pm\varepsilon$. For $\Delta\omega_D^* > \gamma$ the absorption profile is altered compared to that in (10.9), while for $\Delta\omega_D^* \ll \gamma$ the difference is negligible because the Lorentzian profile is dominant in the latter case [10.1].

The technique of reducing the Doppler width by the collimation of molecular beams was used before the invention of lasers to produce light sources with narrow emission lines (10.2). Atoms in a collimated beam had been excited by electron impact. The fluorescence lines emitted by the excited atoms show a reduced Doppler width if observed in a direction perpendicular to the atomic beam. However, the intensity of these atomic beam light sources was very weak and only the application of intense monochromatic tunable lasers has allowed one to take full advantage of this method of Doppler-free spectroscopy.

Figure 10.2 shows a typical experimental arrangement for sub-Doppler spectroscopy in molecular beams. The photomultiplier PM 1 monitors the total fluorescence $I_{Fl}(\lambda_L)$ as a function of the laser wavelength λ_L (excitation spectrum, see Sect.8.2), whereas PM 2 records the dispersed fluorescence spectrum excited at a fixed laser wavelength, where the laser is stabilized

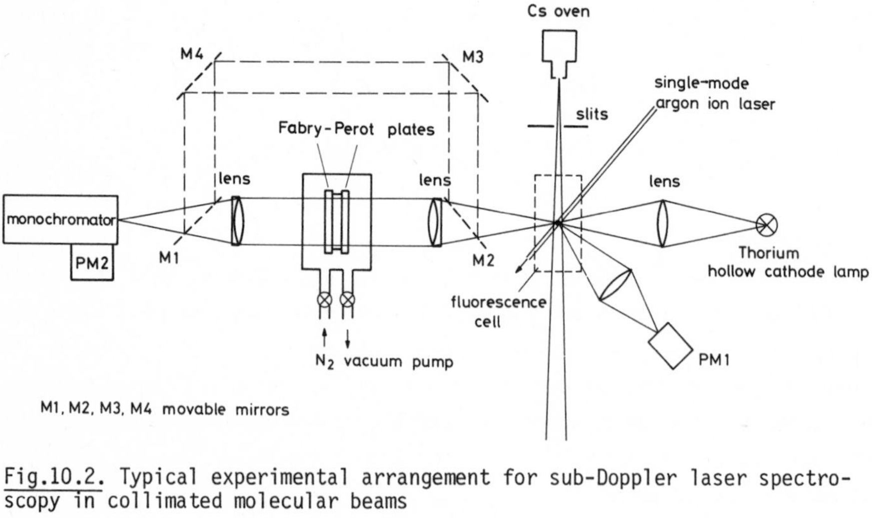

Fig.10.2. Typical experimental arrangement for sub-Doppler laser spectroscopy in collimated molecular beams

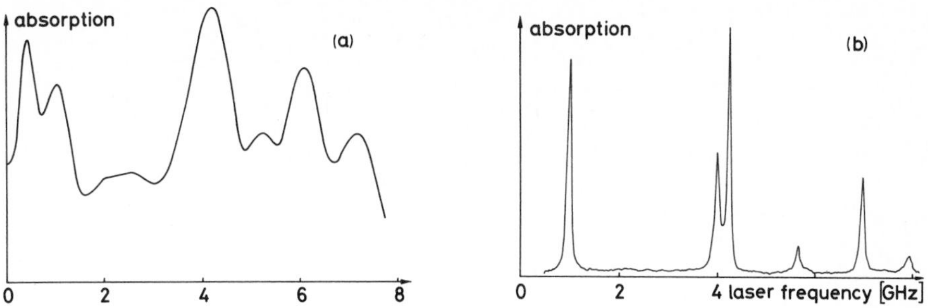

Fig.10.3a,b. Section of the excitation spectrum of the Cs_2 molecule in the spectral range around λ = 476.5 nm. (a) Excited in a cesium vapor cell, (b) excited in a collimated cesium beam

onto a molecular absorption line. A Fabry-Perot can be inserted to increase the spectral resolution of the fluorescence spectrum. A thorium hollow cathode spectral lamp serves as the wavelength calibration standard.

Figure 10.3 illustrates the achievable spectral resolution by comparing a section of the Cs_2 absorption spectrum obtained in a cesium vapor cell (left curve) and in a collimated cesium beam (right spectrum). Both spectra were taken with a single-mode argon laser which could be continuously tuned around λ = 476.5 nm. The absorption was monitored through the total fluorescence intensity $I_{Fl}(\lambda_L)$ as a function of laser wavelength λ_L (excitation spectroscopy, see Sect.8.2.1). The different lines correspond to rotational-vibrational transitions (v",J") → (v',J') between two different electronic states of the Cs_2 molecule [10.3].

Particularly for polyatomic molecules with their complex visible absorption spectra, the reduction of the Doppler width is essential for the resolution of single lines [10.4]. This is illustrated by a section from the excitation spectrum of the NO_2 molecule, excited with a single-mode tunable argon laser around λ = 488 nm (Fig.10.4). For comparison the same section of the spectrum as obtained with Doppler-limited laser spectroscopy in an NO_2 cell is shown in the upper trace [10.5].

Limiting the collimation angle of the molecular beam to 2×10^{-3} rad, the residual Doppler width can be reduced to 500 KHz. High-resolution iodine spectra with linewidths of less than 150 KHz could be achieved in a molecular iodine beam [10.6]. At such small linewidths the time of flight broadening due to the finite interaction time of the molecules with a focussed laser beam is no longer negligible, since the spontaneous lifetime already exceeds the transit time.

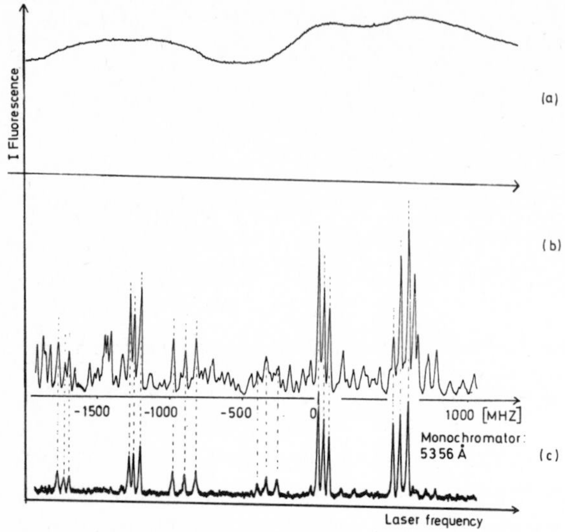

Fig.10.4a-c. Section of the excitation spectrum of NO_2 obtained with a single-mode argon laser, tunable around $\lambda = 488$ nm. (a) In an NO_2 cell ($p = 0.01$ torr), (b) in a collimated NO_2 beam with a collimation ratio of 1 : 80, (c) filtered excitation spectrum. Instead of the total fluorescence as in (b) only the (0,1,0) fluorescence band was monitored

Besides these three examples a large number of atoms and molecules have been studied in molecular beams with high spectral resolution. Mainly hyperfine structure splittings, isotope shifts, and Zeeman splittings have been investigated by this technique because these splittings are generally so small that they may be completely masked in Doppler-limited spectroscopy. An impressive illustration of the sensitivity of this technique is the measurement of nuclear charge radii and nuclear moments of stable and radioactive instable isotopes through the resolution of optical hfs splittings and isotope shifts performed by several groups [10.7-11]. Even spurious concentrations of short-lived radioactive isotopes could be measured in combination with an on-line mass separator.

More examples of sub-Doppler spectroscopy in atomic or molecular beams can be found in reviews on this field by JACQUINOT [10.11] and LANGE et al. [10.12].

10.1.2 Laser Spectroscopy in Supersonic Beams

An interesting development in the high-resolution spectroscopy of complex molecules in molecular beams takes advantage of the internal cooling of molecules in supersonic beams [10.13]. When a gas expands freely from a high-pressure region into the vacuum, adiabatic cooling of the internal energy occurs. This means that the thermal energy of the molecules in the source, which is composed of translational, rotational, and vibrational energy, is partly transferred into expansion energy. This transfer occurs during the expansion in the orifice at densities where the collision probability is high. The degree of cooling depends on the number of collisions during the expansion which is proportional to the product nd of density and orifice diameter [10.14].

Since the cross sections for elastic collisions are larger than those for collision-induced rotational transitions which are still larger than those for vibrational transitions, the translational cooling [which means a narrowing of the velocity distribution $n(v_z)$] will be more effective than the rotational or vibrational cooling. While at thermodynamic equilibrium in the oven all degrees of freedom have the same mean energy $(1/2)kT$, which implies that the translational temperature T_t equals the rotational temperature T_r as well as the vibrational temperature T_v, after the adiabatic expansion the translational energy T_t has decreased more than T_r and T_v, and we obtain at the intersection region with the laser beam

$$T_t < T_r < T_v \quad .$$

Typical values achieved for these temperatures in supersonic beams at oven pressures of some atmospheres are

$$T_t \approx 0.5\text{-}20\ \mathrm{K} \quad , \quad T_r \approx 2\text{-}50\ \mathrm{K} \quad , \quad T_v \approx 10\text{-}100\ \mathrm{K} \quad .$$

This internal cooling has two definite advantages for high-resolution laser spectroscopy.

a) Because of the low values of T_r and T_v, only the lowest rotational-vibrational levels in the electronic ground state are populated. Assuming thermodynamic equilibrium within each degree of freedom, the population in a level (v,J) with vibrational quantum number v and rotational number J is given by

$$n(v,J) = \frac{2(J+1)}{N}\, e^{-(v+\frac{1}{2})\hbar\omega_e/kT_v}\, e^{-B_eJ(J+1)/kT_r} \quad , \qquad (10.11)$$

where $N = \sum n(v,J)$ is the total number of molecules, $v\hbar\omega_e$ the vibrational energy in level v, and $B_eJ(J+1)$ the rotational energy.

Since the intensities of absorption lines are proportional to the population densities n(v,J) of the absorbing levels, this reduction of n(v,J) to a few populated levels implies a considerable reduction of the number of

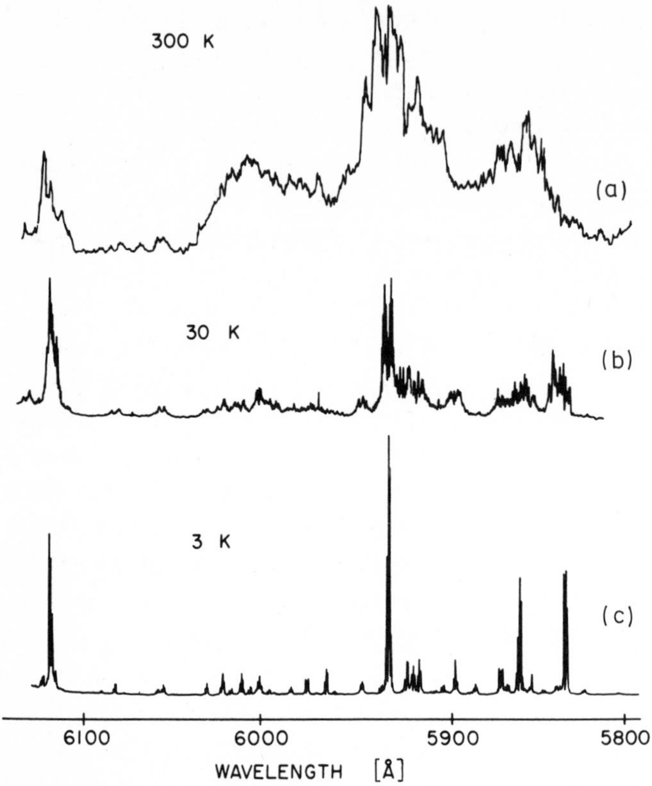

Fig.10.5a-c. Reduction of the complexity of the NO_2 excitation spectrum by internal cooling in a supersonic beam. (a) Conventional room temperature sample of pure NO_2 at 0.04 torr, (b) supersonic beam of pure NO_2, (c) supersonic beam of 5% NO_2 in Ar. The excitation source was a cw dye laser with 0.5 Å bandwidth [10.13]

absorption lines and leads to an appreciable simplification of the spectrum. This is illustrated by Fig.10.5, which shows three times the same section of the NO_2 excitation spectrum, obtained in a conventional absorption cell at room temperature, in a pure NO_2 supersonic beam with medium cooling (rotational temperature 30 K) and in a beam of argon, seeded with 5% of NO_2. Due to many collisions between the argon atoms at high density and the NO_2 molecules, the rotational temperature drops to 3 K.

b) Because of these low temperatures, loosely bound molecules with small dissociation energies D_e can be formed in supersonic beams, which would immediately dissociate at room temperature if $kT \gg D_e$. The spectroscopy of these "van der Waals molecules" opens a new area for the investigation of interaction potentials which before could only be studied through scattering experiments. The much higher accuracy of spectroscopic measurements will improve the exact knowledge of long-range interactions, as for instance experienced by molecules in $^3\Sigma$ ground states (see below) [10.15].

This is not only important to understand the chemistry of loosely bound complexes but also contributes to elucidating molecular structure in the transition region between free molecules and solids. One example of this is the formation of clusters in supersonic alkali beams where molecules Na_x have been observed from $x = 2$ to $x = 12$ [10.16]. The spectroscopy of such multimers (clusters) yields dissociation energies, ionization energies, and vibrational structure as a function of the number x of atoms in the cluster. The comparison of these figures with the values in the solid allows the proof of theoretical models, which explain the transition from molecular orbitals to the band structure of solids.

Besides its merits for high-resolution spectroscopy the absorption spectrum in a molecular beam, generated by a tunable monochromatic laser, can be also used to measure the velocity distribution $n(v)$ of molecules in a defined level E_i at arbitrary locations in the beam. For this purpose the incident laser beam is split into two parts (see Fig.10.6a). The first beam 1 crosses the molecular beam perpendicularly while beam 2 intersects the common crossing point at an angle β with the z axis. When the laser wavelength λ is continuously tuned across a molecular absorption line, an absorption profile around the unshifted center frequency ω_0 with a reduced Doppler width is obtained from beam 1 while beam 2 produces a Doppler-shifted and Doppler-broadened absorption profile. Since each molecule which travels at a velocity $v \approx v_z$ absorbs at the Doppler-shifted frequency $\omega_0' = \omega_0 - kv\cos\beta$, the absorption profile $\alpha(\omega) = n(v\cos\beta)\sigma(\omega - \omega_0)$ of beam 2 reflects the velocity distribution $n(v)$.

From the maximum of the shifted absorption profile the most probable velocity v_p can be deduced [10.17]. Figure 10.6b illustrates the measurement of velocity distributions of Na_2 molecules in specified quantum states at different oven temperatures. The narrowing of the velocity distribution with increasing oven temperature and the increase of the most probable velocity v_p characterize the transition region from a thermal effusive molecular beam to a supersonic beam. The velocity distribution in the supersonic beam can be described by

$$n(v)dv = C\,\frac{\cos\vartheta}{r^2}\,Nv^2\,e^{-(v-u)^2/v_p^2}\,dv \quad , \tag{10.12}$$

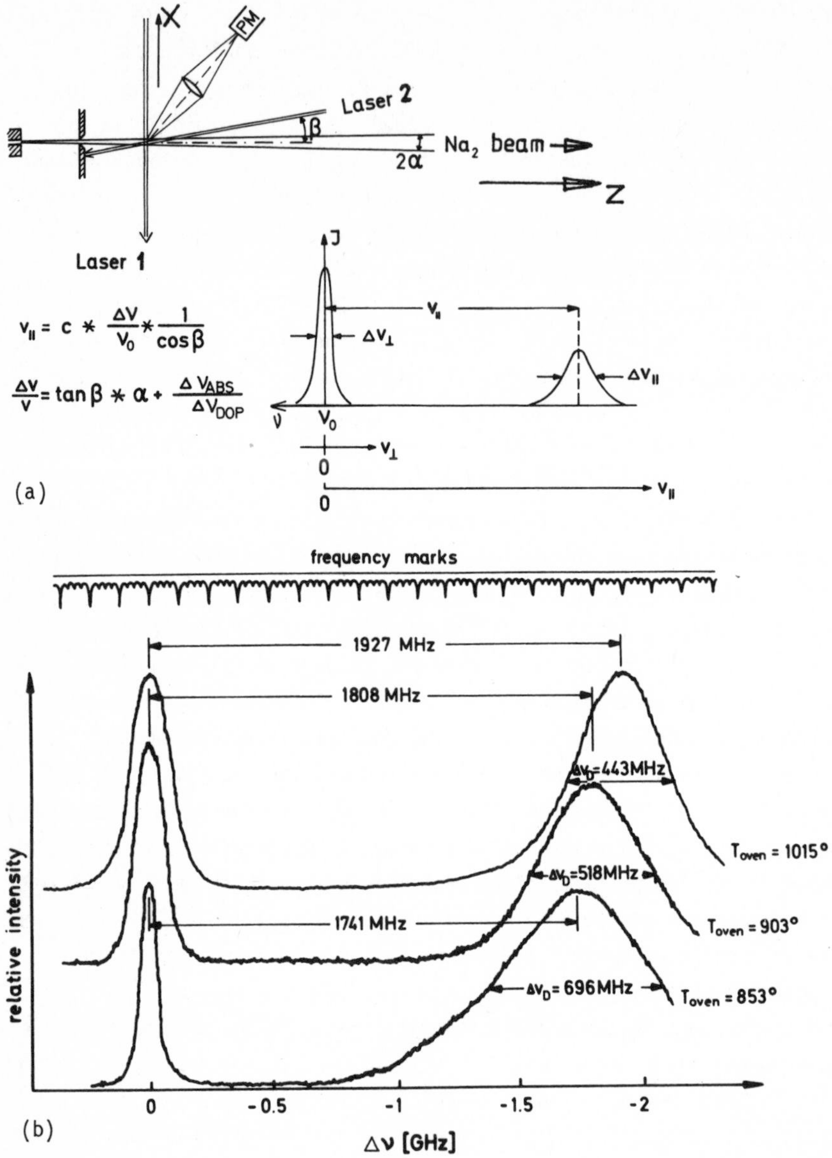

Fig.10.6a,b. Doppler shift laser spectroscopy of the velocity distributions of Na_2 molecules and their dependence on the vapor pressure p(T) in the oven. (a) Experimental arrangement, (b) velocity distributions n(v,v"=0,J"=28) of molecules in the vibrational level v"=0 and rotational level J"=28 at three different oven temperatures [10.17]

where $v_p = (2kT_t/m)^{\frac{1}{2}}$ is much smaller because of the smaller translational temperature T_t. The parameter u characterizes the deviation of the supersonic distribution from a thermal distribution.

Compared with conventional methods of measuring velocity distributions with mechanical selectors or with conventional time-of-flight tubes, this spectroscopic detection method has the following advantages:

a) The accuracy of spectroscopic techniques is generally higher than that obtained with mechanical velocity selectors.

b) The velocity distribution of molecules *in selected vibrational-rotational levels* can be measured. This is of particular importance in crossed beam experiments where inelastic collision processes are studied and the transfer of energy from translational to internal energy can be detected in detail.

c) The velocity and its distribution can be measured at localized points in the beam which allows one to detect local variations of $n(v,E_i)$ along and across the beam. This gives information about the formation of molecules during the expansion and in the region where collisions still occur.

Note, that at low laser intensities the densities n(v)dv are measured. However, at higher laser intensities, where complete saturation occurs, the *flux densities* vn(v)dv are detected.

The relative populations $n(v_i,J_i)/(n(v_k,J_k)$ in the supersonic beam can be measured in the following way. Part of the laser beam is split into a glass cell containing Na-Na_2 vapor at thermal equilibrium. When the laser frequency is tuned through the Na_2 absorption lines, the laser-induced fluorescence both from the cell and from the crossing point of laser beam 1 with the molecular beam is monitored simultaneously. Since the fluorescence intensity at nonsaturating laser intensities is proportional to the population densities in the absorbing levels, the relative populations $n_B(v_i,J_i)/n_B(v_k,J_k)$ in the molecular beam can be obtained from the ratio of the fluorescence intensities

$$\frac{I_c(v_i,J_i)/I_c(v_k,J_k)}{I_B(v_i,J_i)/I_B(v_k,J_k)} = \frac{n_c(v_i,J_i)/n_c(v_k,J_k)}{n_B(v_i,J_i)/n_B(v_k,J_k)} \quad .$$

The population ratio in the cell at thermal equilibrium

$$\frac{n_c(v_i,J_i)}{n_c(v_k,J_k)} = \frac{g_i}{g_k}\, e^{-(E_i-E_k)/kT}$$

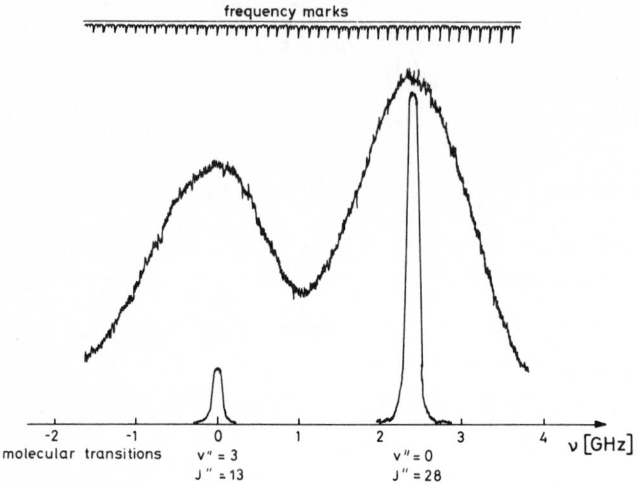

Fig.10.7. Relative population distribution of Na_2 molecules in two levels (v" = 0, J" = 28, v" = 3, J" = 43) in a thermal vapor cell at T = 500 K and in a supersonic beam

follows a Boltzmann distribution. Figure 10.7 shows as an example the Doppler-broadened line profiles from the cell and the corresponding lines with reduced Doppler width in the supersonic beam. One clearly sees that the vibrational level at v" = 3 is much less populated in the supersonic beam than in the cell, which had a temperature of about 500 K. This indicates the vibrational cooling in the beam [10.17a].

10.1.3 Laser Spectroscopy in Fast Ion Beams

In the examples considered so far, the laser beam was crossed *perpendicularly* with the molecular beam, and the reduction of the Doppler width was achieved through the limitation of the maximum velocity components V_x by geometrical apertures. One therefore often calls this reduction of the transverse velocity components "*geometrical cooling*". KAUFMANN [10.18a] and WING et al. [10.18b] have independently proposed another arrangement, where the laser beam travels *collinear* with a fast ion or atom beam and the reduction of the *longitudinal* velocity distribution $n(v_z)$ is achieved by an acceleration voltage ("acceleration cooling").

Assume that two ions start from the ion source (Fig.10.8) with different thermal velocities $v_1(0) = 0$ and $v_2(0) > 0$. After being accelerated by the voltage U their kinetic energies are

$$E_1 = e\,U = \frac{m}{2}\,v_1^2$$

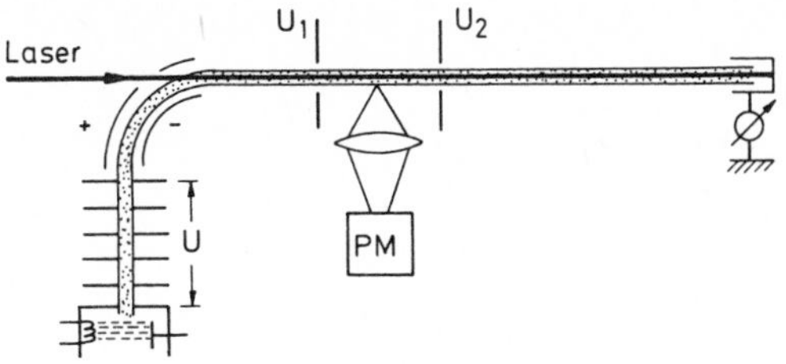

Fig.10.8. Laser spectroscopy in fast ion beams. The laser beam is *collinear* with the ion beam

$$E_2 = \frac{m}{2} v_2^2(0) + eU = \frac{m}{2} v_2^2 \quad .$$

Subtracting the first from the second equation yields

$$v_2^2 - v_1^2 = v_2^2(0) \Rightarrow \Delta v = v_2 - v_1 = v_2^2(0)/2v \quad \text{with} \quad v = (v_1 + v_2)/2 \quad .$$

This means that the initial velocity spread $v_2(0)$ has decreased to

$$\Delta v = v_2(0)\left(\frac{v_2^2(0)m}{8eU}\right)^{\frac{1}{2}} = v_2(0)\left(\frac{\Delta E_{th}}{4eU}\right)^{\frac{1}{2}} \quad . \tag{10.13}$$

Example

$\Delta E_{th} = 0.1$ eV; $eU = 10$ KeV $\Rightarrow \Delta v = 1{,}5 \times 10^{-3}\, v_2(0)$.

This reduction of the velocity spread results from the fact that *energies* rather than velocities are added. If the energy $eU \gg E_{th}$ the velocity change is mainly determined by U but hardly by the fluctuations of the initial thermal velocity. This implies, however, that the acceleration voltage has to be extremely well stabilized to take advantage from this "acceleration cooling".

A definite advantage of this parallel arrangement is the longer interaction zone between the two beams because the laser-induced fluorescence can be collected by a lens from a path length ΔZ, of several cm, compared to a few mm in the perpendicular arrangement. A further advantage is the possibility of Doppler tuning. The absorption spectrum of the ions can be scanned across a *fixed* laser frequency ω simply by tuning the acceleration voltage U. An absorption line at ω_0 is Doppler tuned into resonance with the laser field, at frequency ω, if

$$\boxed{\omega = \omega_0 - kv_z = \omega_0\left(1 - \sqrt{\frac{2eU}{mc^2}}\right)} \quad . \tag{10.14}$$

This allows one to use high-intensity fixed frequency lasers, such as the argon laser, which have a high gain and even allow one to place the interaction zone inside the laser cavity.

Instead of tuning the acceleration voltage U (which influences the beam collimation), the velocity of the ions in the interaction zone with the laser beam is tuned by retarding or accelerating potentials U_1 and U_2 (see Fig.10.8).

This superimposed ion-laser beam geometry is quite suitable for high-sensitivity spectroscopy with high resolution, as has been demonstrated by MEIER et al. [10.19] who resolved the hfs pattern of odd xenon isotopes and by OTTEN et al. [10.20] for Hg isotopes. Such a laser-ion coaxial beam spectrometer has been described in detail by HUBER et al. [10.21]. If the laser photodissociates the molecular ions, the photo-fragments can be analyzed with great efficiency. In this case the coaxial arrangement is clearly superior to the crossed beam arrangement because the interaction time of the molecular ions with the laser field can be made several orders of magnitude longer.

Figure 10.9 shows a velocity-tuned spectrum of O^+ photofragments from the process

$$O_2^+ + h\nu \rightarrow O_2^{+*} \rightarrow O^+ + O \quad .$$

The laser wavelength was kept at $\lambda = 5815$ Å and the velocity of the O_2^+ ions was continuously tuned by controlling the acceleration voltage. More examples, including saturation spectroscopy in fast ion beams, can be found in the review by DUFAY and GAILLARD [10.22].

The fast ions can be converted to neutral atoms by charge exchange collisions while passing through a target gas chamber. Since charge exchange occurs preferentially at large impact parameters the velocity change during the collisions is nearly negligible and the neutral atoms still have a rather narrow velocity distribution $n(v_z)$ [10.23]. The merits of the collinear arrangement can therefore be extended to neutral atoms and molecules.

The charge exchange collisions may result in highly excited states of the neutral species. Often the population of these excited states accumulates in metastable states with long lifetimes. This opens the possibility of performing laser spectroscopy on transitions between *excited* states. While the resonance lines, corresponding to transitions from the ground state, are for most atoms in the ultraviolet region, the transitions between *excited* states are generally in the visible to infrared range and are therefore more readily accessible to the tuning range of dye lasers.

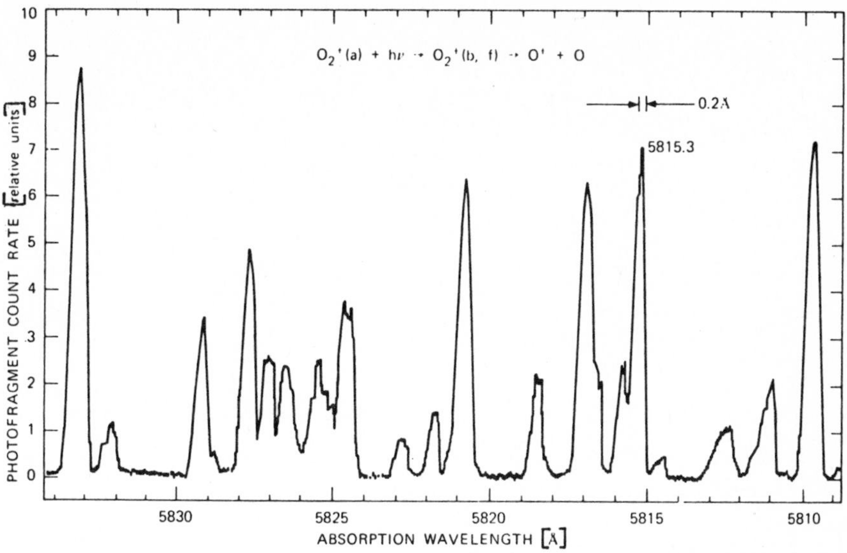

Fig.10.9. Velocity tuned spectrum of O^+-photo fragments obtained by photo-dissociation of O_2^+ in a "laser ion coaxial beam spectrometer" [10.22]

10.1.4 Optical Pumping in Molecular Beams

If the laser frequency ω is tuned to a molecular transition $(v_i'',J_i'' \to v_k',J_k')$, the molecules are excited into the upper level (v_k',J_k') where they can undergo spontaneous transitions $(v_k',J_k') \to (v_m'',J_m'')$ to many rotational-vibrational levels in the electronic ground state. Contrary to optical pumping in *atoms*, only the small fraction $\eta = (A_{ki}/\sum_m A_{km})$ of all excited molecules, determined by the spontaneous transition probabilities A_{km}, returns into the initial state (v_i'',J_i'').

In the intersection region of laser beam and molecular beam the pressure is already sufficiently low to exclude collisional transitions from $(v_m'',J_m'') \to (v_k'',J_k'')$. The radiative transition $(v_k',J_k') \to (v_i'',J_i'')$ is therefore the only relaxation process in diatomic homonuclear molecules which can repopulate the level (v_i'',J_i'') depleted by optical pumping. The mean recycling time for this repopulation process depends on the absorption rate $B_{ik}\rho_{ik}$ and on the spontaneous lifetime τ_k of the excited level. For $B_{ik}\rho_{ik} \gg A_k = 1/\tau_k$ one pump cycle $(i \to k \to i)$ takes a time of about τ_k. The number of pump cycles is determined by the ratio T_f/τ_k of time-of-flight T_f through the laser beam to spontaneous lifetime τ_k. Assuming a diameter of the focussed laser beam of about 0.1 mm and a mean molecular velocity $\underline{v} = 5\times10^4$ cm/s, we obtain

$T_f \approx 2\times10^{-7}$ s. At a spontaneous lifetime $\tau_k = 10^{-8}$ s this implies that about 20 pump cycles are possible. With a value of $\eta = 0.1$ the population of the pumped level (v_i'', J_i'') decreases after each cycle to 10% which means that after 2 pump cycles the population has already dropped to about 1% of its unsaturated value.

This illustrates that optical pumping in molecular beams is a very efficient way to completely deplete a specified molecular level (v_i'', J_i''). This can be used to gain detailed information on velocity distributions in supersonic beams of molecules in defined quantum states and on relaxation processes, molecular formation, and collision processes in crossed molecular beams.

The following is an example of one of the many possible applications of optical pumping to a time-of-flight spectrometer with velocity selection for molecules in defined quantum states in supersonic beams [10.24].

The monochromatic laser beam is split into two beams 1 and 2, both of which cross the molecular beam perpendicularly but at different locations $z_1 = A$ and $z_2 = B$ (see Fig.10.10). When the laser frequency is tuned to a molecular transition $(v_i'', J_i'' \rightarrow v_k', J_k')$, the molecules passing through the pump beam 1 are optically pumped and there are nearly no molecules left in the depleted level (v_i'', J_i''). This means that the fluorescence excited by the probe beam 2 is very low. If the pump beam 1 is interrupted for a short time interval Δt (e.g., by a fast mechanical chopper) the molecules can pass during this interval Δt without being pumped. Because of their different velocities they reach the probe beam 2 at different times $t = L/v$, where $L = z_2 - z_1$. The time-resolved fluorescence intensity induced by the cw probe beam therefore reflects the velocity distribution of molecules in the level (v_i'', J_i''). Figure 10.11 shows for the case of Na_2 molecules in a supersonic beam that these velocities are *different* for different quantum states. This has to do with the fact that during the adiabatic expansion in the nozzle the internal energy of the molecules is partly converted into expansion energy and the expansion velocity depends on the degree of energy transfer a molecule has suffered [10.24a].

10.1.5 Optical-Optical Double-Resonance Spectroscopy in Molecular Beams

This selective depletion of specified rotational-vibrational levels by optical pumping can also be used to facilitate the identification of complex molecular spectra. The basic concept is illustrated in Fig.10.12. Differing from Fig.10.10 the pump beam and the probe beam come from two different

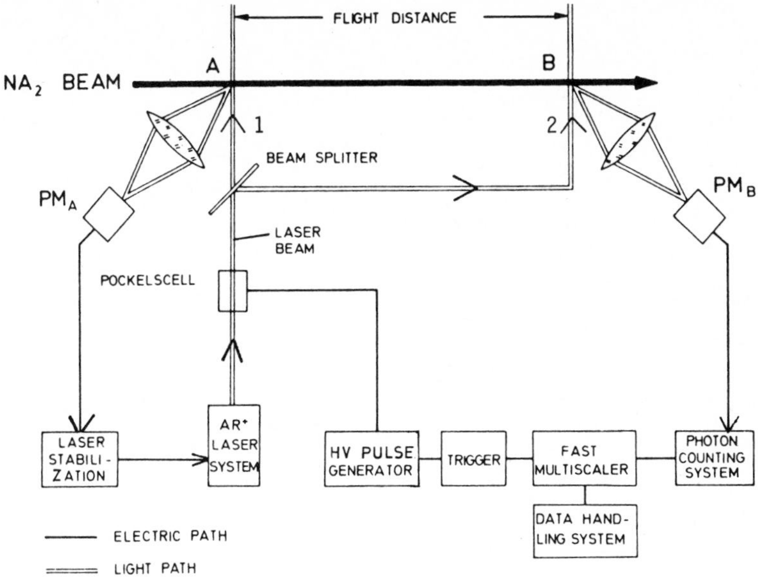

Fig.10.10. Optical time-of-flight spectrometer for the determination of velocity distributions $n_i(v_z)$ of molecules in defined quantum states (v_i, J_i)

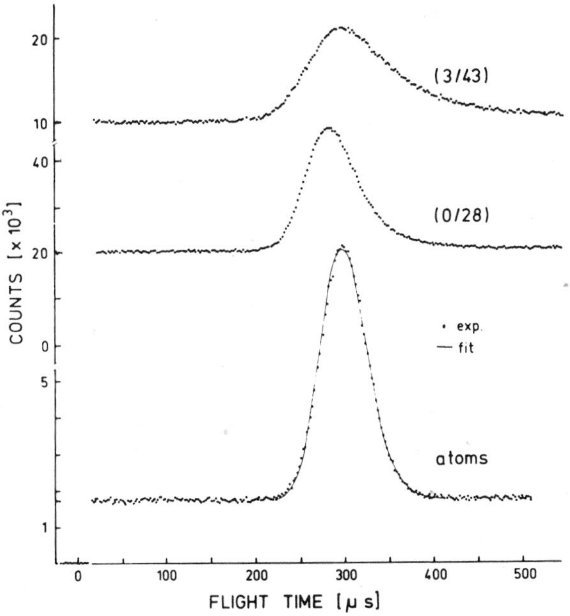

Fig.10.11. Time-of-flight spectrum of Na_2 molecules in two different quantum states ($v'' = 3$, $J'' = 43$) and ($v'' = 0$, $J'' = 28$) in comparison with the Na atom velocity distribution [10.24]

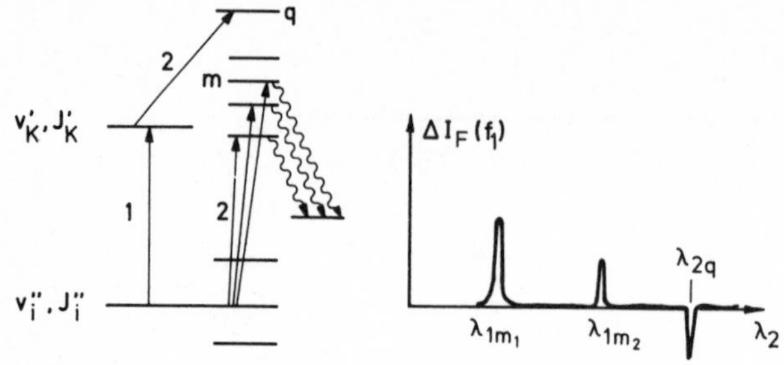

Fig.10.12. Labeling of specified upper or lower levels (v,J) by optical pumping and successive probing (optical-optical double resonance)

lasers. If the pump laser 1 which is stabilized onto the molecular transition $(v_i'',J_i'') \rightarrow (v_k',J_k')$ is chopped at the frequency f_1, the population densities n_i and n_k are also chopped, where the phase of $n_i(f_1)$ is opposite to that of $n_k(f_1)$. If the wavelength λ of the unchopped probe laser 2 is tuned across the absorption spectrum of the molecules, the laser-induced fluorescence intensity I_{Fl} which is proportional to n_iI_2 or n_kI_2, respectively, will always be modulated at f_1 if the probe laser hits a transition which starts from one of the optically pumped levels E_i or E_k. The modulation phase is opposite for the two cases. When the fluorescence is detected through a lock-in amplifier, tuned to the chopping frequency f_1, only those of the many possible transitions induced by the tunable probe laser are detected that are connected to one of the pumped levels E_i or E_k (see Fig.10.12b).

This optical-optical double-resonance technique has already been used for other Doppler-free techniques [10.25], such as polarization spectroscopy (see Sect.10.3). Its applications to molecular beams has, however, the following advantages compared to spectroscopy in gas cells. When the chopped pump laser periodically depletes the level E_i and populates level E_k, there are two relaxation mechanisms in gas cells which may transfer the population modulation to other levels. These are collision processes and laser-induced fluorescence (see Fig.8.39). The neighboring levels therefore also show a modulation and the modulated excitation spectrum induced by the probe laser includes all lines which are excited from those levels. If several absorption lines overlap within their Doppler width, the pump laser simultaneously excites several upper states and also partly depletes several lower levels. The simplification and the unambiguous assignment of the lines is therefore partly lost again. However, in the molecular beam, Doppler-free excitation is possible and collisions can be neglected. Since the fluorescence from the excited level E_k terminates on *many* lower levels (v_m'',J_m''), only a small frac-

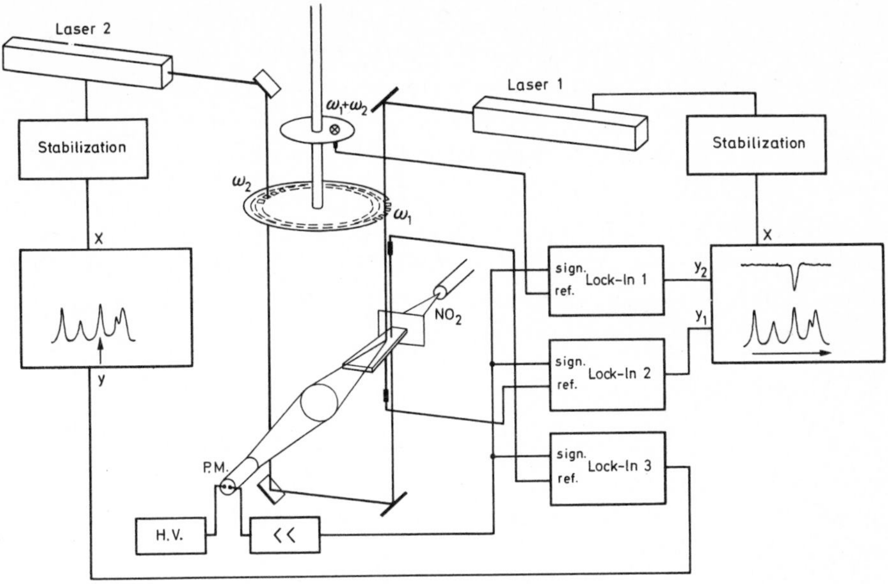

Fig.10.13. Experimental arrangement for optical-optical double resonance in molecular beams

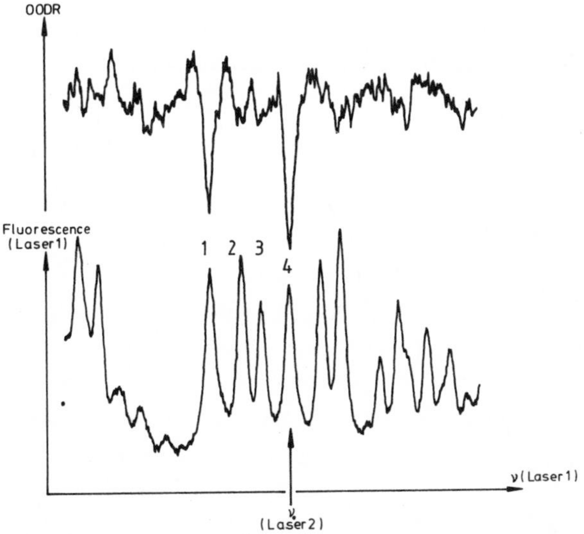

Fig.10.14. Section of linear NO_2 spectrum around λ = 488 nm (lower trace) and double-resonance spectrum (upper trace). The pump laser was stabilized on line No. 4 while the probe laser was tuned [10.26]

tion of the population n_i from the depleted level E_i is pumped into each of the levels E_m which are populated by fluorescence decay from E_k. This means that the modulation amplitude is much smaller if the probe laser is tuned to levels E_m rather than to E_i.

Molecules with long spontaneous lifetimes may fly several centimeters before they radiate and may therefore emit their modulated fluorescence at the location of the probe beam. In order to distinguish this modulated fluorescence from that excited by the probe laser, both lasers are chopped at two different frequencies f_1 and f_2. The fluorescence intensity excited by the probe laser tuned to the transition $(v_i'',J_i'') \rightarrow (v_m,J_m)$ is then

$$I_{Fl}(I_2) \propto N_i I_2 = (N_{i0} - aI_1)I_2$$

$$= [N_{i0} - aI_{01}(1 - \cos 2\pi f_1 t)]I_{02}(1 - \cos 2\pi f_2 t)$$

and therefore contains terms with the sum frequency $(f_1 + f_2)$. Tuning the detection system to $(f_1 + f_2)$ therefore selects those transitions which are due to transitions excited by the probe laser from a level E_i which had been depleted by laser 1. Figure 10.13 shows the experimental arrangement. Figure 10.14 compares a section of the linear excitation spectrum of NO_2 with the double-resonance spectrum detected at $(f_1 + f_2)$ where the pump laser 1 had been stabilized to line No. 4. The double resonance spectrum shows that lines No. 1 and 4 share the same lower level [10.26], information which could not have been deduced easily from the linear spectrum.

10.1.6 Radio-Frequency Spectroscopy in Molecular Beams

The "Rabi technique" of radio-frequency or microwave spectroscopy in atomic and molecular beams [10.27a] has made an outstanding contribution to the accurate determination of ground state parameters, such as the hfs splittings in atoms and molecules or the Coriolis splittings in molecules, etc. Figure 10.15 depicts schematically a conventional Rabi beam apparatus. Molecules with a permanent dipole moment μ effuse from the orifice of the oven and are collimated by the slit S_1. In an inhomogeneous magnetic field (A field) they experience a force $\underline{F} = -\underline{\mu}$ grad $\underline{B}$ and are therefore deflected. In a second reversed inhomogeneous field (B field) they experience an opposite force and are therefore deflected back onto the detector. If an rf field in a homogeneous field region C between the A and the B field induces transitions to other molecular levels the dipole moment $\underline{\mu}$ changes and the B field

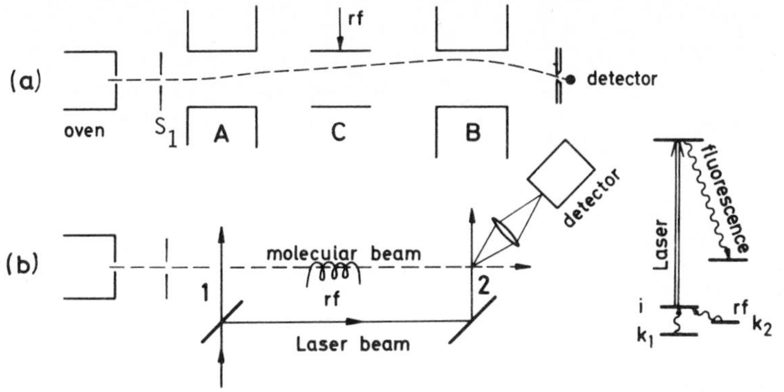

Fig.10.15. Comparison between conventional Rabi method (a) and its laser equivalent (b)

does not deflect the molecules exactly back onto the detector. This means that the detected beam intensity $I(\nu_{rf})$ as a function of the radio frequency ν_{rf} shows a dip in case of resonance. Because of the long spontaneous lifetimes in electronic ground states the linewidth of this dip is mainly determined by the time of flight through the rf field and by saturation broadening [10.27b].

Optical pumping with lasers offers a new and very convenient variation of the conventional Rabi method, which is illustrated in Figs.10.15b and 10.16. The two inhomogeneous magnetic fields A and B are replaced by two parallel laser beams which cross the molecular beam perpendicularly. The first "pump beam" depletes the population N(i) of the lower level i. This is monitored by the resultant decrease of the fluorescence intensity I_{Fl} induced by the probe laser. When the rf field induces transitions $k \to i$ between other levels k and the depleted level i, the population N(i) is increased, resulting in a corresponding increase of I_{Fl}.

The laser version of the Rabi method does not rely on the mechanical deflection of molecules but uses the change of the population density in specified levels due to optical pumping. The conventional technique is restricted to atoms or molecules with magnetic or electric dipole moments and the sensitivity depends on the difference between the dipole moments in the two levels connected by the rf transition. The laser version can be applied to all atoms and molecules which can be optically pumped by existing lasers. The background noise is small because only those molecules which are in the specified level contribute to the probe-induced fluorescence.

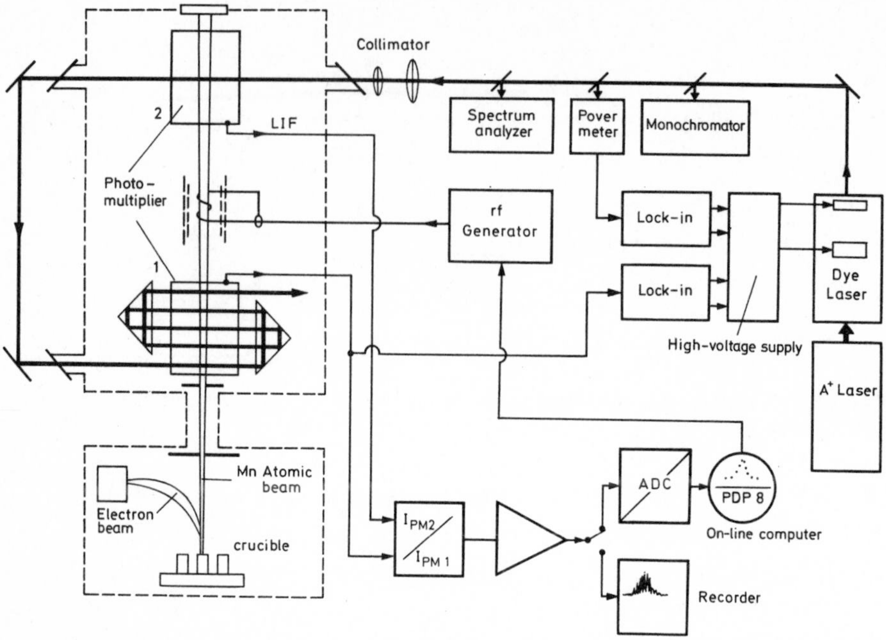

Fig.10.16. Schematic diagram of atomic beam magnetic resonance method using optical pumping and probing with lasers [10.27e]

Some examples illustrate the achievable sensitivity and accuracy: ERTMER and HOFER [10.27c] measured the hyperfine structure of metastable states of scandium which were populated by electron impact. The population achieved in these states was only about 1% of the ground state population. Optical pumping to a higher electronic state with a single-mode dye laser allowed selective depletion of the single hfs sublevels of the metastable F state. The hfs constants could be determined with an accuracy of a few kHz.

ROSSNER et al. [10.27d] measured the very small hfs splitting (< 100 kHz) of a rovibronic level (v" = 0, J" = 28) in the X $^1\Sigma_g$ ground state of the Na_2 molecule. This splitting is much smaller than the natural linewidth of the optical transitions. The quadrupole coupling constant was determined to $e\,q\,Q = 463.7 \pm 0.9$ kHz and the nuclear spin-molecular rotation interaction constant came out to be $a = 0.17 \pm 0.03$ kHz.

Figure 10.16 shows a block diagram of the whole apparatus, used by PENSELIN and his group for atomic beam magnetic resonance spectroscopy detected by laser-induced fluorescence [10.27e]. The pump laser beam crosses the atomic beam several times to assure high pumping efficiency. The rf transitions induce population changes, which are detected by the laser-induced fluorescence in the probe region. For further examples see [10.27f].

10.2 Saturation Spectroscopy

Saturation spectroscopy is based on the selective saturation of an inhomogeneously broadened molecular transition by optical pumping with a monochromatic tunable laser. As has been outlined in Sect.3.6 the population density $n_i(v_z)dv_z$ of molecules in the absorbing state E_i is selectively depleted of molecules with velocity components

$$v_z \pm dv_z = (\omega_0 - \omega \pm \delta\omega)/k$$

in the interval dv_z, because these molecules are Doppler shifted into resonance with the laser frequency ω and are excited from E_i to the higher level E_k ($E_k - E_i = \hbar\omega_0$).

The monochromatic laser therefore "burns a hole" into the population distribution $n_i(v_z)$ of the absorbing state and produces simultaneously a peak at the same velocity component v_z in the upper state distribution $n_k(v_z)$ (Fig.10.17).

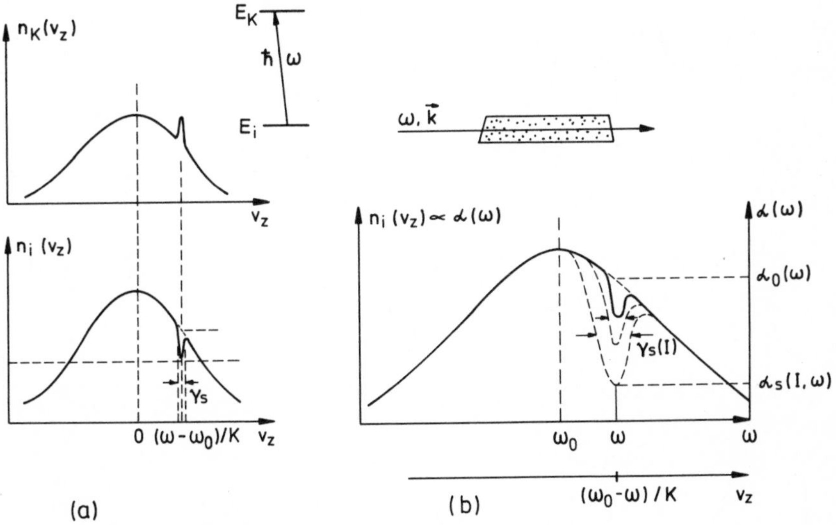

Fig.10.17. (a) "Hole burning" in the lower level population distribution $n_i(v_z)$ of an absorbing transition and generation of a corresponding population peak in the upper level. (b) Increase of Bennet hole width with increasing saturating intensity

Because of this population depletion the absorption coefficient $\alpha(\omega) = \Delta n(v_z)\sigma_{ik}(\omega_0 - \omega - v_z \cdot k)$ *decreases* from its unsaturated value $\alpha_0(\omega)$ to the saturated value

$$\alpha_s(\omega) = \alpha_0(\omega)/\sqrt{1 + S_0} \qquad (10.15)$$

as has been shown in (3.81a).

$\Delta n = n_i - (g_i/g_k)n_k$ gives the population difference between the two levels E_i and E_k.

$S_0 = S(\omega_0)$ is the saturation parameter at the line center (3.70) and $S = B_{ik}\rho_{ik}(\omega)/R$ is the ratio of the depleting absorption rate $B_{ik}\rho(\omega)$ to the sum R of all relaxation processes which refill the depleted level E_i.

Because the intensity absorbed over a path length dz

$$dI(\omega) = \alpha_s(\omega)I_0(\omega)dz = \alpha_0 I_0/\sqrt{1 + aI_0}\, dz \quad \text{with} \quad a = 2B_{ik}/(c\pi\gamma R) \qquad (10.16)$$

is no longer linearly dependent on the incident intensity I_0, spectroscopic methods based on saturation effects are often called *nonlinear spectroscopy*.

We now briefly discuss the basic concepts and experimental arrangements of saturation spectroscopy and show that this technique allows essentially Doppler-free spectral resolution.

10.2.1 Basic Concepts

Assume that a monochromatic laser wave $E = E_0 \cos(\omega t - kz)$ is travelling in the +z direction through a gaseous sample with molecules at thermal equilibrium. The absorption cross section $\sigma_{ik}(\omega)$ of a molecule in level E_i that moves with velocity v is, according to (2.73) and (3.76),

$$\sigma_{ik}(\omega, v_z) = (\hbar\omega/c)B_{ik}g(\omega_0 - \omega - kv_z) \quad , \qquad (10.17)$$

where the function

$$g(\omega_0 - \omega - kv_z) = \frac{\gamma_s/2\pi}{(\omega_0 - \omega - kv_z)^2 + (\gamma_s/2)^2} \qquad (10.18)$$

represents the normalized homogeneous line profile of the molecular transition with $\hbar\omega_0 = (E_k - E_i)$ [see (3.10)].

The homogeneous linewidth

$$\gamma_s = \gamma\sqrt{1 + S_0} \quad \text{with} \quad \gamma = \gamma_n + \gamma_c \qquad (10.19)$$

is determined by the natural linewidth γ_n, the collisional broadening γ_c, and by saturation broadening (see Chap.3). At sufficiently low pressures and small laser intensities, pressure broadening and saturation broadening may be neglected and the homogeneous linewidth γ_s approaches the natural linewidth γ_n, provided the interaction time of the molecules with the radiation field is longer than the spontaneous lifetime, so that time-of-flight broadening can be neglected (see Sect.3.4). In the visible region γ_n is several orders of magnitude smaller than the Doppler width $\Delta\omega_D$.

Since $g(\omega_0 - \omega - kv_z)$ has its maximum for $\omega_0 - \omega = kv_z$, those molecules with velocity components $v_z = (\omega_0 - \omega)/k$ have the largest absorption probability. This implies that a hole centered at $v_z = (\omega_0 - \omega)/k$ appears in the population distribution $n(v_z)$ (Fig.10.17) with a width γ_s and a depth depending on the saturation parameter S_0 at $\omega = \omega_0$. According to (3.77b) the difference $\Delta n_0(v_z) = n_i(v_z) - (g_i/g_k)n_k(v_z)$ decreases to its saturated value

$$\Delta n_s(v_z)dv_z = \Delta n_0(v_z)\left(1 - \frac{(\gamma/2)^2 S_0}{(\omega_0 - \omega - kv_z)^2 + (\gamma_s/2)^2}\right)dv_z \tag{10.20}$$

$$= C\Delta N_0\left(1 - \frac{(\gamma/2)^2 S_0}{(\omega_0 - \omega - kv_z)^2 + (\gamma_s/2)^2}\right)e^{-(v_z/v_p)^2}\, dv_z$$

with $v_p = (2kT/m)^{\frac{1}{2}}$ and $\Delta N_0 = \int \Delta n_0(v_z)dv_z$.

The spectral width γ_s of this "*Bennett hole*" [10.28] represents the homogeneous linewidth of the molecular transition and is, as mentioned above, in the visible region for $\gamma_s \to \gamma_n$ very much smaller than the Doppler width (see the examples in Chap.3).

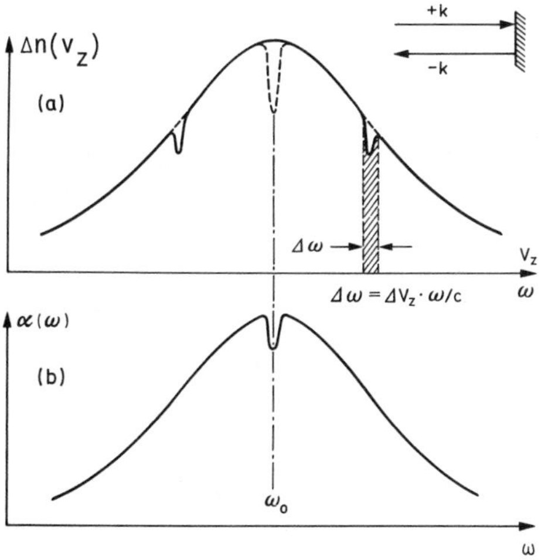

Fig.10.18. (a) Bennet hole burned symmetrically into the velocity distribution $n(v_z)$ by two counterpropagating waves of frequency $\omega \neq \omega_0$. (b) Lamb dip at the center $\omega = \omega_0$ of a Doppler-broadened absorption line $\alpha(\omega)$

In order to detect this Bennet hole, which has been burned into the population distribution $n_i(v_z)$ by the so-called *pump wave*, a second light wave, called the *probe wave* has to be sent through the sample to probe the population depletion in the hole. This probe wave may be either split from the same laser which provides the pump wave, or it may also come from another laser.

In a simple arrangement the pump wave may be reflected by a mirror back into the sample (Fig.10.18a). Since the reflected wave $E = E_0 \cos(\omega t + kz)$ has the opposite wave vector $-\underline{k}$, it interacts with another group of molecules centered around $-v_z$. As long as $\omega \neq \omega_0$, therefore two different holes at $v_z = \pm(\omega_0 - \omega)/k$ are burned into the population distribution $n_i(v_z)$ (Fig.10.18), which merge together at the line center $v_z = 0$ for $\omega \to \omega_0$ (dotted curve). The superposition of the two counterpropagating waves $E = E_0 \cos(\omega t - kz) + E_0 \cos(\omega t + kz) = 2E_0 \cos\omega t \cos kz$ represents a standing wave field. The crucial point is now that the absorption in this field (which means the total absorption of both counterrunning waves) *has a minimum for* $\omega = \omega_0$. In this case namely both waves interact with the *same molecules*, which are therefore exposed to twice the intensity. This means that the saturation parameter S becomes twice as large and the depletion of $\Delta n(v_z = 0)$ therefore is larger for $v_z = 0$ than for $v_z \neq 0$ [see (10.20) and Fig.10.18b].

To be more quantitative let us calculate the absorption coefficient $\alpha(\omega)$ in the standing wave field. According to (2.68) we can use the relation $\alpha(\omega) = \Delta N\sigma(\omega)$ and obtain

$$\alpha_s(\omega) = \int \Delta n_s(v_z)[\sigma(\omega_0 - \omega - kv_z) + \sigma(\omega_0 - \omega + kv_z)]dv_z \quad , \tag{10.21}$$

where the population difference $\Delta n(v_z)$, saturated in the standing wave, is according to (10.20) given by

$$\Delta n_s(v_z) = \Delta n_0(v_z)\left[1 - \frac{(\gamma/2)^2 S_0}{(\omega_0 - \omega - kv_z)^2 + (\gamma_s/2)^2} - \frac{(\gamma/2)^2 S_0}{(\omega_0 - \omega + kv_z)^2 + (\gamma_s/2)^2}\right]. \tag{10.22}$$

Equation (10.21) can be solved in the weak field approximation for $S_0 \ll 1$. After some elaborate calculations (see [10.53]) one obtains

$$\alpha_s(\omega) = \alpha_0(\omega)\left[1 - \frac{S_0}{2}\left(1 + \frac{(\gamma_s/2)^2}{(\omega - \omega_0)^2 + (\gamma_s/2)^2}\right)\right] \quad . \tag{10.23}$$

This is a Doppler profile $\alpha_0(\omega) = CN_0 \exp-[(\ln 2)(\omega - \omega_0)^2/\delta\omega_D^2]$ modified by the expression in brackets, which represents a small dip at the center (Fig.10.18b) which is called *Lamb dip*, after W.E. Lamb, who first described this effect in the approximation of weak saturation [10.29].

The Lamb dip profile is Lorentzian with a halfwidth γ_S(FWHM). At the line center $\alpha_s(\omega_0)$ decreases to $\alpha_0(\omega_0)(1 - S_0)$, far off resonance to $\alpha_0(\omega)(1 - S_0/2)$. This is because for $\omega = \omega_0$ the molecules are saturated by twice the intensity, while for $\omega - \omega_0 > \gamma$ both fields interact with different molecules.

The Lamb dip in the distribution $\Delta n(v_z)$ of the population difference $\Delta n = n_1 - (g_1/g_2)n_2$ appears not only in the inhomogeneously broadened *absorption* profile $\alpha(\nu)$ but in case of inversion ($\Delta n < 0$) also in the gain profile $-\alpha(\omega)$ of an amplifying medium with an inhomogeneous linewidth. If the frequency of a single-mode gas laser is tuned over the Doppler-broadened gain profile, the output intensity shows a dip around the center frequency ω_0 [10.30]. This Lamb dip in the laser output can be used to stabilize the laser frequency to the center of the gain profile (see below).

10.2.2 Doppler-Free Saturation Spectroscopy

The narrow Lamb dips in the Doppler-broadened absorption coefficient $\alpha(\omega)$, as seen by a monochromatic standing wave, can be used to resolve closely spaced absorption lines, which would be completely masked in Doppler-limited spectroscopy. Figure 10.19 gives an example for two transitions, between hyperfine components of two molecular states, which are separated by less than their Doppler width. Although the Doppler-broadened line profiles completely overlap, their Lamb dips can be clearly resolved. The detection of these narrow resonances can be realized with different experimental arrangements, which will be discussed in the following.

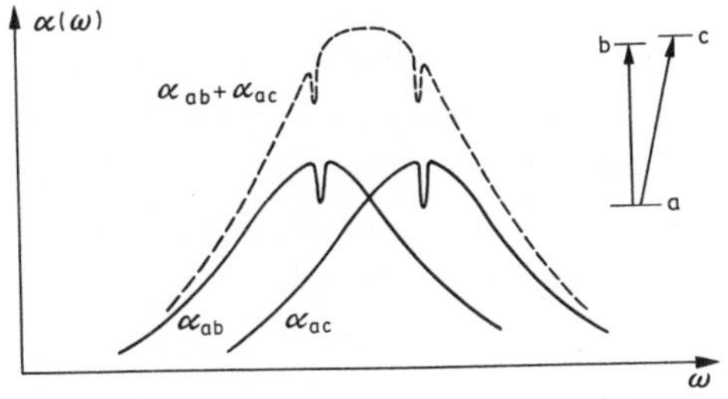

Fig.10.19. Resolution of the Lamb dips of two closely spaced transitions with overlapping Doppler profiles

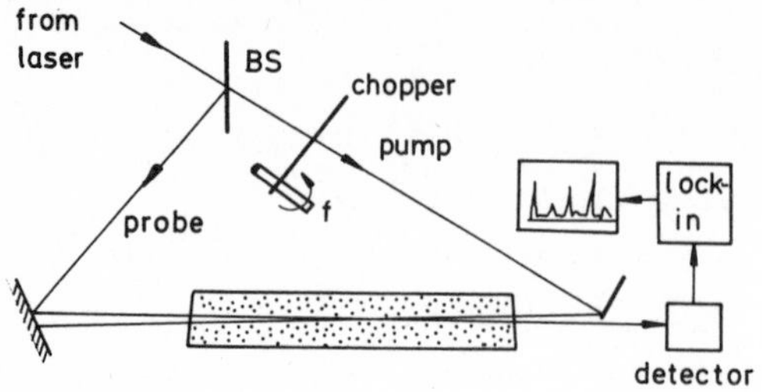

Fig.10.20. Possible experimental arrangement for saturation spectroscopy outside the laser resonator

A typical example for a possible experimental setup is shown in Fig.10.20. The output beam from a tunable laser is split by the beam splitter BS into a strong pump beam and a weak probe beam, which pass through the absorbing sample in opposite directions. The attenuation of the probe beam is measured as a function of the laser frequency ω. To enhance the sensitivity, the probe beam can be again split into two parts. One beam passes the region of the sample which is saturated by the pump beam the other passes the sample cell at an unsaturated region (see Fig.10.46). The difference of the two probe beam outputs yields the saturation signal. Figure 10.21 illustrates a saturation spectrum of the H_α line in atomic hydrogen [10.31] obtained in a hydrogen discharge with a pulsed narrow-band dye laser with a linewidth of about 7 MHz (see Sect.7.3). While the fine structure of the H_α line ($2S \rightarrow 3P$) is masked in Doppler-limited spectroscopy, a spectral resolution of 30 MHz could be achieved with saturation spectroscopy. Absolute wavelength measurements of the strong $2P_{3/2} - 3D_{5/2}$ component provided a new tenfold improved value of the Rydberg constant [10.31].

The two transitions $2S_{1/2} \rightarrow 3P_{1/2}$ at ω_1 and $2S_{1/2} \rightarrow 3P_{3/2}$ at ω_2 share a common lower level. In such cases a "crossover resonance" is observed at a laser frequency $\omega_c = (\omega_1 + \omega_2)/2$, if the Doppler width $\delta\omega_D$ is larger than $(\omega_1 - \omega_2)$. At the frequency ω_c the pump beam interacts with atoms which are Doppler-shifted into resonance on transition 1. The same atoms are then in resonance on transition 2 with the counterpropagating probe beam. Such crossover signals, which appear exactly at the mean frequency of two coupled transitions, can help to assign transitions with a common level and to separate the upper level splittings from the lower ones.

Figure 10.22 shows an example of the saturation spectrum of a mixture of different cesium isotopes contained in a glass cell heated to about 100°C [10.32]. The hyperfine structure and the isotope shifts of the different isotopes can be derived from these measurements with high accuracy.

Instead of measuring the *attenuation* of the probe beam, the absorption can also be monitored by the laser-induced *fluorescence intensity*, which is

3 refers to n=3.

2 refers to n=2

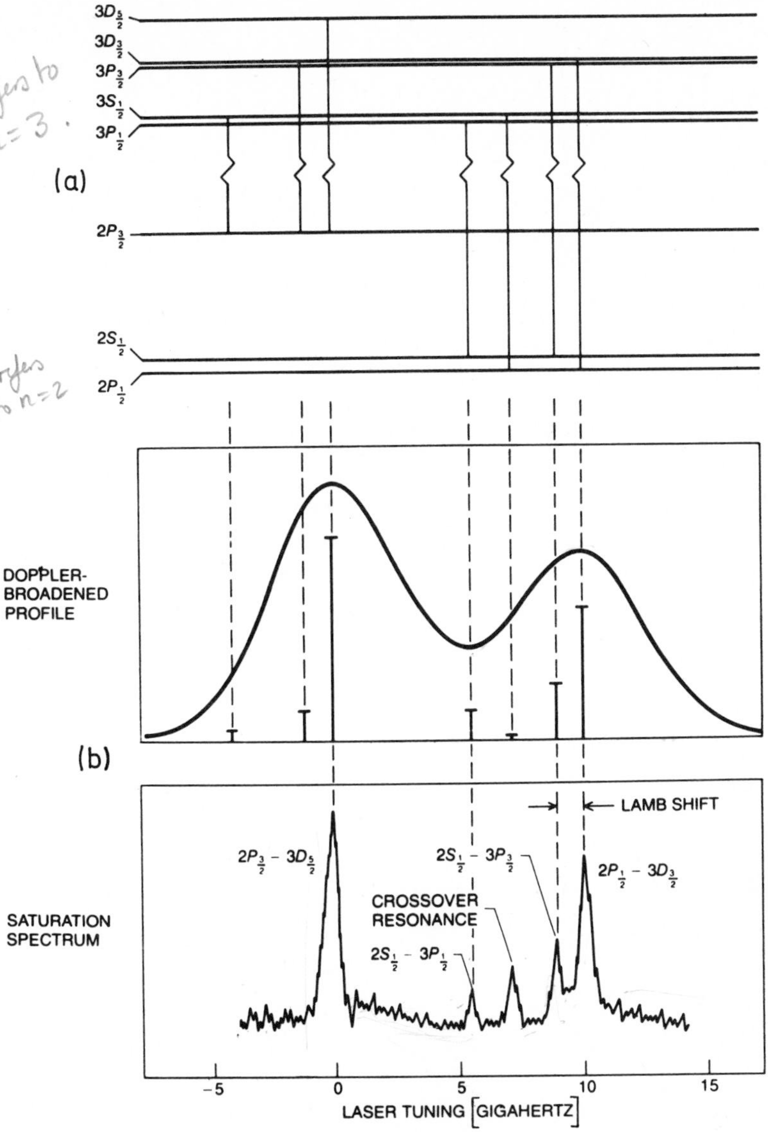

Fig.10.21a,b. Measurement of the Rydberg constant by saturation spectroscopy of the Hydrogen Balmer α transition. (a) Level scheme, (b) Doppler profiles and saturation spectrum of the Balmer α line in a hydrogen discharge [10.31a]

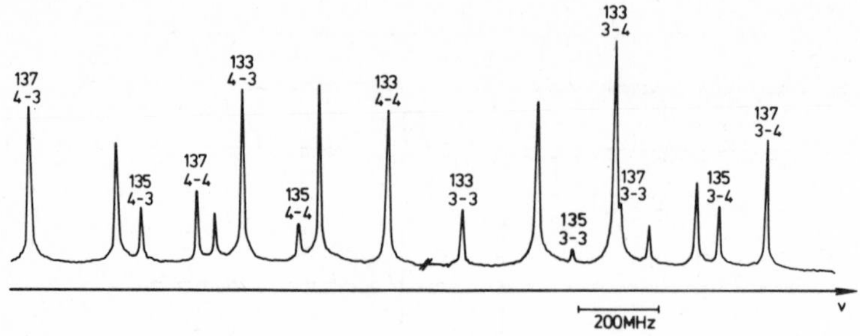

Fig.10.22. Survey scan of the saturation spectrum of all hfs components of the $6^2S_{1/2} \rightarrow 7^2P$ - transition at $\lambda = 459.3$ nm in a vapor mixture of Cs 133, 135, and 137 [10.32]

proportional to the absorbed laser intensity. In cases where the saturation is very small, the change in the attenuation of the probe beam is difficult to detect and the small Lamb dips may be nearly buried under the noise of the Doppler-broadened background. SOREM and SCHAWLOW [10.33] have demonstrated a very sensitive "intermodulated fluorescence technique", where pump beam and probe beam are chopped at two different frequencies f_1 and f_2. Assume the intensities of the two beams to be $I_1 = I_0(1 + \cos 2\pi f_1 t)$ and $I_2 = I_0(1 + \cos 2\pi f_2 t)$. The intensity of the laser-induced fluorescence is then

$$I_{Fl} = Cn_s(I_1 + I_2) \quad , \tag{10.24}$$

where n_s is the saturated population density of the absorbing state and the constant C includes the transition probabilities and the collection efficiency of the fluorescence detector. According to (10.22) the saturated population density at the center of an absorption line is $n_s = n_0(1 - S_0) = n_0[1 - a(I_1 + I_2)]$.

Inserting this into (10.24) gives

$$I_{Fl} = C[n_0(I_1 + I_2) - an_0(I_1 + I_2)^2] \quad , \tag{10.25}$$

which shows that the fluorescence intensity contains linear terms, modulated at the chopping frequencies f_1 and f_2, respectively, and quadratic terms with modulation frequencies $(f_1 + f_2)$ and $(f_1 - f_2)$, respectively. While the linear terms represent the normal laser-induced fluorescence with a Doppler-broadened line profile, the quadratic terms describe the saturation effect, because they depend on the decrease of the population density $n_i(v_z = 0)$ due to the simultaneous interaction of the molecules with both fields. When the fluorescence is monitored through a lock-in amplifier,

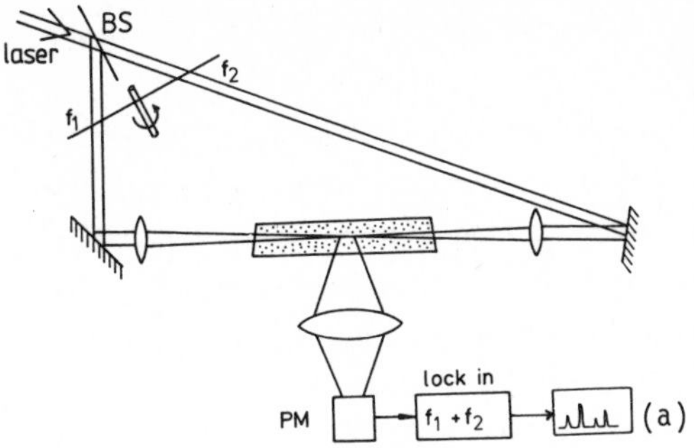

Fig.10.23a,b. Saturation spectroscopy using intermodulated fluorescence technique. (a) Experimental arrangement, (b) hyperfine spectrum of the (v" = 1, J" = 98) → (v' = 58, J' = 99) line in the $X^1\Sigma \rightarrow {}^3\Pi_{u0}$ system of I_2 at λ = 514.5 nm, monitored at the chopping frequency f_1 of the pump beam (upper trace) and at the sum frequency (lower spectrum) [10.34]

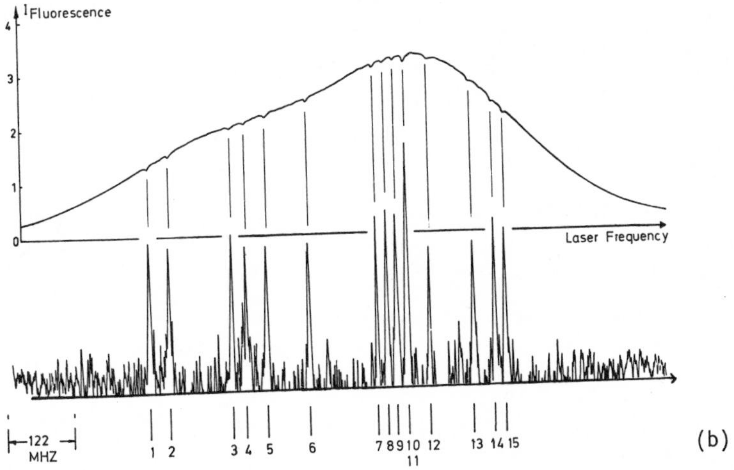

tuned to the sum frequency $f_1 + f_2$, the linear background is suppressed and only the saturation signals are detected. This is demonstrated by Fig.10.23, which shows the 15 hyperfine components of the rotational line (v" = 1, J" = 98) (v' = 58, J' = 99) in the $X^1\Sigma_g^+ \rightarrow B^3\Pi_{u0}$ transition of the iodine molecule I_2 [10.34]. The two laser beams were chopped by a rotating disc with two rows of a different number of holes which interrupted the beams at $f_1 = 600\ s^{-1}$ and $f_2 = 900\ s^{-1}$. The upper spectrum was monitored at the frequency f_1 at which the pump beam was chopped while the probe was not modulated. The Doppler-broadened background caused by the linear terms in (10.25) and the Lamb-dips both show a modulation at the frequency f_1 and are therefore recorded simultaneously. The center frequencies of the hfs components, however, can be obtained more accurately from the intermodulated fluorescence spectrum (lower spectrum) which was monitored at the sum frequency $(f_1 + f_2)$ where the linear background is suppressed.

This transition is very weak and the signal to noise ratio is therefore not very high.

Although the iodine molecule has always served as a standard example to demonstrate new sub-Doppler techniques and many papers have been published about saturation spectroscopy of I_2 [10.35], meanwhile the intermodulated fluorescence version has been applied to a number of other atoms and molecules. One example is the BO_2 molecule [10.36] where the hfs components of 37 R branch transitions involving several vibronic bands in the ground ($X^2\pi_g$) and excited ($A^2\pi_u$) electronic states have been resolved. For many molecules the hfs splittings are caused by two effects: an electric interaction with the electric quadrupole of a nucleus and a magnetic interaction between the nuclear spin moment and the magnetic field produced by the rotation of the molecule (spin-rotation interaction). Measuring the hfs splittings for different rotational levels allows one in many cases to separate the two contributions [10.37a,b].

Another very sensitive method of monitoring saturation signals is based on intracavity absorption. If the sample is placed inside the laser resonator, the Lamb dips in the absorption profiles represent minima of the intracavity losses. Because the laser output depends sensitively on the internal losses it will show sharp peaks when the laser frequency is tuned across the Lamb dips of the absorption profile (see Fig.10.24b).

Due to the enhancement of sensitivity in intracavity absorption (see Sect.8.2.3) these peaks in the laser output generally have a much better signal-to-noise ratio than the Lamb dips obtained with saturation spectro-

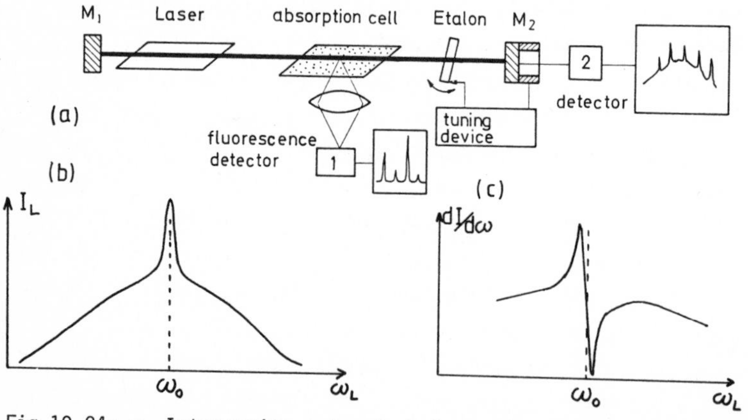

Fig.10.24a-c. Intracavity saturated absorption spectroscopy. (a) Experimental setup, (b) "Lamb peak" in the laser output, (c) derivative of a Lamb peak obtained by modulation of the laser frequency

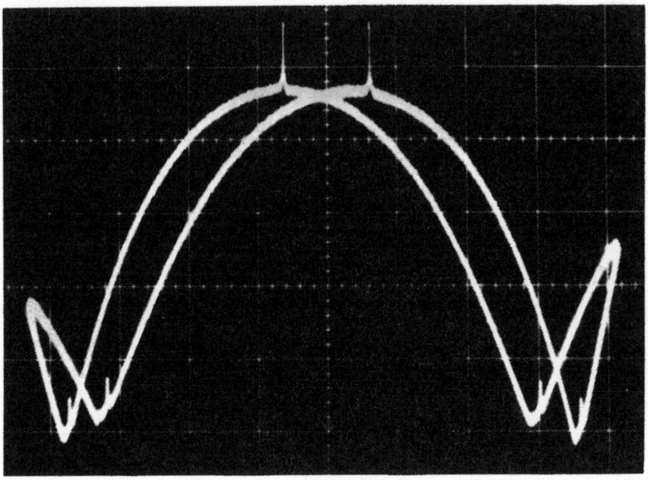

Fig.10.25. Saturation peak in the output of a He-Ne laser at λ = 3.39 nm caused by the Lamb dip of a CH_4 absorption line [10.43]. The laser frequency was swept twice over the gain profile

scopy outside the laser resonator. Furthermore, because of the nonlinear dependence of the laser output on the losses, the halfwidth of the peaks may be smaller than γ_n, particularly when the laser is operated close above threshold. Figure 10.25 illustrates such a "Lamb peak", monitored in the output of a He-Ne laser at $\lambda = 3.39$ μm with an internal methane cell [10.43]. By fortuitous coincidence an absorption line of the CH_4 molecule falls into the tuning range of the 3.39 μm laser transition. This line corresponds to a rotational line in a vibrational transition of CH_4. Because of the long lifetime of the excited vibrational level in the electronic ground state ($\tau \approx 20$ ms) the natural width γ_n of the transition is very small and the width of the Lamb dip is mainly determined by time-of-flight broadening and pressure broadening (see Chap.3). Enlarging the laser beam diameter and reducing the CH_4 pressure allows one to achieve extremely narrow resonances [10.38,39], with linewidths in the kilohertz range.

The sensitivity may be further enhanced by modulation of the laser frequency. This allows sensitive lock-in detection and yields signals which represent the derivative of the Lorentzian line profiles (see Sect.8.1). For illustration of the resolution achieved with extremely well-stabilized lasers, large magnifications of the laser beam diameter, and long absorption cells at low pressures, Fig.10.26 shows the modulated saturation spectrum of the $CH_3{}^{35}Cl$ molecule at $\underline{\nu}$ = 2947.821 cm^{-1} ($\hat{=}$ λ = 3.39 μm) [10.40].

Because of the chlorine nuclear spin I = 3/2, each rotational level with $J \geq 2$ splits into 4 hfs components with total angular momentum $\underline{F} = \underline{J} + \underline{I}$. The four lines in Fig.10.26 correspond to the four hfs transitions.

If the center frequency ω_0 of the molecular absorption line is different from the center frequency ω_1 of the gain profile $G(\omega - \omega_1)$ of the amplifying laser medium, the absorbing sample inside the laser resonator causes a Lamb peak at ω_0 in the laser output I_L, which lies at the *slope* of the gain profile (see Fig.10.27). This not only causes a slight shift of the Lamb peak

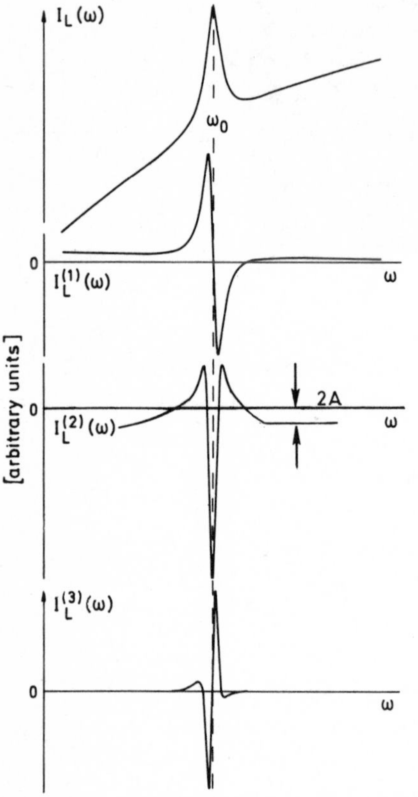

Fig.10.27. Lamb peak and its first, second, and third derivatives in the output $I_L(\omega)$ of a gas laser with intracavity absorber, when the absorption frequency ω_0 is placed at the slope of the laser gain profile

center frequency, due to the sloped background, but also makes it unsafe to stabilize the laser frequency directly onto this peak. Any slight perturbation which drives the laser frequency away from the Lamb peak by more than the peak width, may induce the feedback control to drive the frequency further away to the center frequency ω_1 of the gain profile because here it also finds a zero point of the derivative $dI_L/d\omega$.

In such cases it is often advantageous to use the "*third derivative technique*" which is based on the following considerations. The frequency dependence $I_L(\omega)$ of the laser intensity is determined by the superposition of the

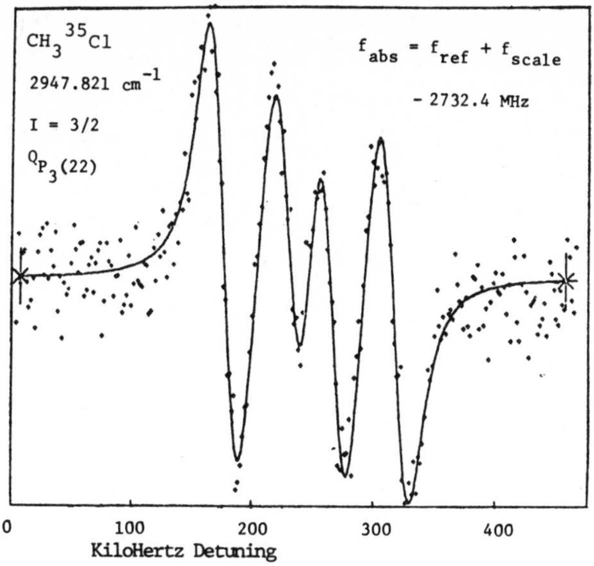

Fig.10.26. Saturated absorption spectrum of $CH_3{}^{35}Cl$ at $\lambda = 2947.821\ cm^{-1}$. The spectrum represents the derivative of four hfs components of the $^QP_3(22)$ rotational line [10.40]

10.2.3 Lamb Dip Stabilization of Gas Lasers

The extremely small linewidths, achieved with saturation spectroscopy of vibrational transitions, are not only essential for sub-Doppler resolution of closely spaced molecular lines but also allow some interesting applications in other fields of physics. One aspect is concerned with the realization of superstabilized lasers, which may serve as frequency standards in the visible or near infrared region. If the laser frequency is locked to the center of a narrow Lamb peak, its stability depends on the widths of the peak, on the signal-to-noise ratio, and on possible frequency shifts of the molecular transition caused by external perturbations, such as magnetic or electric fields, collisions, or light fields. Transitions with a small homogeneous linewidth in molecules without permanent dipole moments are therefore preferable candidates for laser stabilization. Locking the He-Ne laser at $\lambda = 0.63\ \mu m$ to a hyperfine component of a rotational line in an electronic transition of I_2 allows a frequency stability of a few kHz [10.41]. The He-Ne laser transition at $\lambda = 3.39\ \mu m$ can be locked to the vibrational transition of the CH_4 molecule even with a stability of $\Delta\nu/\nu \leqq 10^{-13}$ [10.42].

gain profile $G(\omega - \omega_1)$, and the Doppler-broadened absorption profile of the absorbing probe inside the cavity with the Lamb dip at ω_0.

$$I_L(\omega) \propto \left[G(\omega - \omega_1) - \alpha_0(\omega)\left(1 - \frac{\gamma S_0}{(\omega - \omega_0)^2 + (\gamma/2)^2}\right)\right] \ . \tag{10.26}$$

In a small interval around ω_0 we may approximate the Doppler profiles of $G(\omega - \omega_1)$ and $\alpha_0(\omega)$ by a quadratic function and obtain for (10.26) the approximation

$$I_L(\omega) = A\omega^2 + B\omega + C + \frac{D}{(\omega - \omega_0)^2 + (\gamma/2)^2} \ , \tag{10.27}$$

where the constants A, B, C, D depend on ω_0, ω_1, γ, and S. The derivatives $I_L^{(n)}(\omega) = [(d^n I_L)(\omega)/d\omega^n]$ are

$$\begin{aligned} I_L^{(1)}(\omega) &= 2A\omega + B - \frac{2D(\omega - \omega_0)}{[(\omega - \omega_0)^2 + (\gamma/2)^2]^2} \\ I_L^{(2)}(\omega) &= 2A + \frac{6D(\omega - \omega_0)^2 - D\gamma^2/2}{[(\omega - \omega_0)^2 + (\gamma/2)^2]^3} \\ I_L^{(3)}(\omega) &= + \frac{4D\gamma^2(\omega - \omega_0)}{[(\omega - \omega_0)^2 + (\gamma/2)^2]^4} \ . \end{aligned} \tag{10.28}$$

These derivatives are shown in Fig.10.27 which demonstrates that the Doppler-broadened background disappears for the higher derivatives. It is therefore advantageous to stabilize the laser frequency ω_L to the zero point ω_0 of the third derivative. This can be performed as follows.

If the laser frequency ω_0 is modulated at a frequency Ω, the laser intensity

$$I_L(\omega) = I_L(\omega_0 + a \sin\Omega t)$$

can be expanded into a Taylor series around ω_0

$$\begin{aligned} I_L(\omega) &= I_L(\omega_0) \\ &\quad + a \sin\Omega t I_L^{(1)}(\omega_0) \\ &\quad + \frac{a^2}{2} \sin^2\Omega t \ I_L^{(2)}(\omega_0) \end{aligned}$$

$$+ \frac{a^3}{6} \sin^3\Omega t \; I_L^{(3)}(\omega_0) + \dots \quad . \tag{10.29}$$

Rearrangement of (10.29) yields, after applying some trigonometric relations,

$$\begin{aligned} I_L(\omega) = {} & I_L(\omega_0) + \frac{a^2}{4} I_L^{(2)}(\omega_0) + \frac{a^4}{64} I_L^{(4)}(\omega_0) + \dots \\ & + \left[a\, I_L^{(1)}(\omega_0) + \frac{a^3}{8} I_L^{(3)}(\omega_0) + \dots\right] \sin\Omega t \\ & + \left[-\frac{a^2}{4} I_L^{(2)}(\omega_0) - \frac{a^4}{48} I_L^{(4)}(\omega_0) + \dots\right] \cos 2\Omega t \\ & + \left[-\frac{a^3}{24} I_L^{(3)}(\omega_0) - \frac{a^5}{384} I_L^{(5)}(\omega_0) + \dots\right] \sin 3\Omega t \\ & + \dots \quad . \end{aligned} \tag{10.30}$$

The expressions in brackets yield the laser intensity at the nth harmonic of the modulation frequency Ω. The signal at 3Ω is therefore essentially determined by the third derivative $I_L^{(3)}(\omega_0)$ since at a sufficiently small modulation amplitude a, the second terms in the brackets are small compared to the first ones.

Figure 10.28 shows the experimental performance. The modulation frequency Ω is tripled by forming rectangular pulses, where the third harmonic is filtered and is fed into the reference input of a lock-in amplifier, which is tuned to 3Ω. The lock-in output is used for the stabilization feedback. Figure 10.29 illustrates a third derivative spectrum of the hyperfine structure components of I_2, obtained in Fig.10.23 with the intermodulated fluorescence technique.

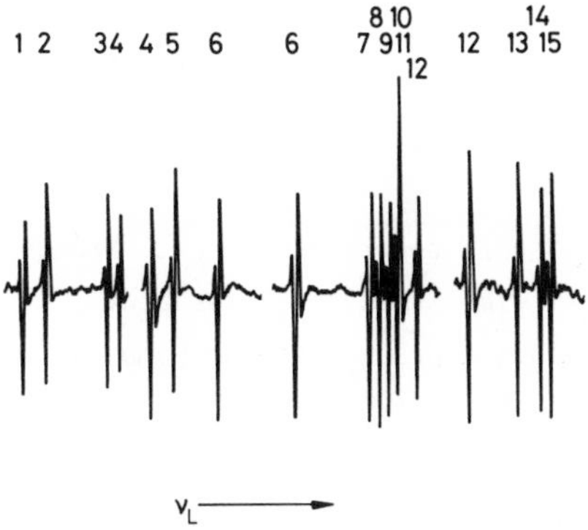

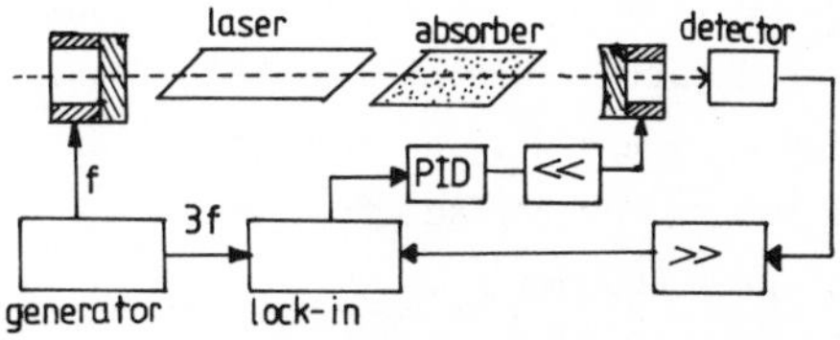

Fig.10.28. Schematic arrangement for third derivative frequency stabilization

Fig.10.29. Third derivative intracavity absorption spectrum of I_2 around $\lambda = 514.5$ nm, showing the same hfs components as Fig.10.23 [10.37a]

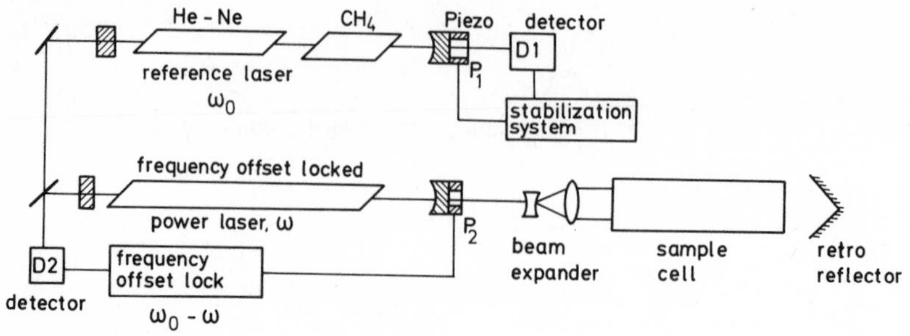

Fig.10.30. Schematic diagram of a frequency offset-locked spectrometer

Lamb dip stabilization onto very narrow saturation peaks allows one to achieve a very good frequency stability, especially when pressure or power broadening or transit broadening can be minimized, as in the case of the 3.39 μm He-Ne laser stabilized onto a CH_4 transition.

Using a double servo loop for fast stabilization of the laser frequency onto the transmission peak of a Fabry-Perot interferometer and a slow loop to stabilize the F.P.I. onto the first derivative of a forbidden narrow calcium transition BARGER et al. [10.44] constructed an ultrastable cw dye laser with a short-term linewidth of approximately 800 Hz and a long-term drift of less than 2 kHz/hr [10.45].

This extremely high stability can be transferred to tunable lasers by a special "frequency offset locking" technique [10.43] (see also Sect.6.9). Its basic principle is illustrated in Fig.10.30. A reference laser is frequency stabilized onto the Lamb dip of a molecular transition. The output from a second, more powerful laser at frequency ω is mixed in detector D 2 with the output from the reference laser at frequency ω_0. An electronic device compares the difference frequency $\omega_0 - \omega$ with the frequency ω' of a stable but tunable rf oscillator, and controls the piezo P_2 such that always $\omega_0 - \omega = \omega'$. The frequency ω of the power laser is therefore always locked to the "offset frequency" $\omega_0 - \omega'$ which can be controlled by tuning the rf frequency ω'.

The output beam of the power laser is expanded before it is sent through the sample cell in order to minimize transit time broadening (see Sect.3.4). A retroreflector provides the counter propagating probe wave for Lamb dip spectroscopy.

The spectrum in Fig.10.26 has been obtained with such a frequency offset locked laser spectrometer. The real experimental setup is somewhat more complicated. A third laser is used to eliminate the troublesome region near zero offset frequency. Furthermore optical decoupling elements have to be inserted to avoid optical feedback between the three lasers. A detailed description of the whole system can be found in [10.38].

10.2.4 Saturation Spectroscopy of Coupled Transitions

Assume that two laser fields are interacting simultaneously with a molecular system. If the two laser frequencies ω_1 and ω_2 are tuned to two molecular transitions which share a common level (a), coupling phenomena occur due to the nonlinear interaction between the fields and the molecular system. The absorption of one of the laser waves is influenced by the presence of the other wave. This coupling is caused by several different effects.

The first coupling effect is due to the selective saturation of the level population by each wave. When, for example, the frequency ω_1 is tuned close to the molecular transition $a \rightarrow b$ with $\omega_{ab} = (E_b - E_a)/\hbar$, the wave can be absorbed by molecules with velocity components $v_z \pm \Delta v_z = (\omega_{ab} - \omega_1 \pm \delta\omega)/k_1$. This causes a Bennet hole in the population distribution $n_a(v_z)$ (see Fig. 10.17). When the second laser is tuned over the absorption profile of the transition $a \rightarrow c$, the Bennet hole causes a decrease in absorption of this laser. This can be monitored for instance by the corresponding decrease of the fluorescence intensity on the transition $c \rightarrow m$ (Fig.10.31).

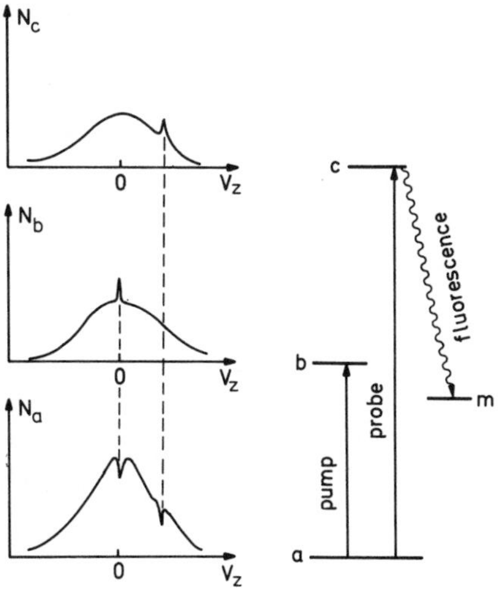

Fig.10.31. Coupling of two transitions $a \rightarrow b$ and $a \rightarrow c$ by saturation of the common level (a)

This optical-optical double resonance has been already discussed in Sect.8.9 as a method of labelling molecular levels and identifying molecular transitions in complex spectra. There we had not, however, considered the line profiles of the double resonance limits, which will be the subject of this section.

Since the two transitions are coupled only by those molecules within the velocity range Δv_z which have been pumped by one of the lasers, the double-resonance signals show similarly small homogeneous linewidths as in saturation spectroscopy with a single laser. However, for precise spectroscopy the common Bennet hole should be *exactly at the center* of the population distribution and not anywhere around $v_z = 0$. This can be achieved for instance by using the Lamb dip, produced in the *standing wave* of the pump field, to stabilize the pump laser frequency ω to the center frequency ω_{ab}.

The general case can be described as follows.

The absorption coefficient for the weak probe wave with frequency ω_2 is, according to (10.21),

$$\alpha(\omega_2) = \int [n_a(v_z) - n_c(v_z)] \frac{\sigma_0 \gamma_{ac}^2}{4(\omega_{ac} - \omega_2 - k_2 v_z)^2 + \gamma_{ac}^2} dv_z \quad . \tag{10.31}$$

The population density $n_a(v)$ is altered by the saturating pump transition $a \rightarrow b$. With a pump laser frequency ω_1 we obtain, according to (10.20),

$$n_a(v_z) - n_c(v_z) = n_a^0(v_z) - n_c^0(v_z) - \frac{[n_a^0(v_z) - n_b^0(v_z)](\gamma_{ab}/2)^2 S}{(\omega_{ab} - \omega_1 - k_1 v_z)^2 + (\gamma_{ab}/2)^2} \tag{10.32}$$

where the saturation parameter $S = B_{ab}\rho(\omega_1)/\gamma_{ab}$ depends on the intensity of the pump wave and γ_{ab} is the homogeneous width of transition $a \rightarrow b$.

Inserting (10.32) into (10.31) yields the absorption coefficient $\alpha(\omega_2)$ of the probe wave in the presence of the saturating pump wave at frequency ω_1

$$\alpha(\omega_2) = \alpha_0 \exp -[(\ln 2)(\omega_{ac} - \omega_2)^2/\Delta\omega_D^2] \tag{10.33}$$

$$\times \left\{ 1 - \frac{N_a^0 - N_b^0}{N_a^0 - N_c^0} \frac{k_2}{2k_1} \frac{S}{\sqrt{1+S}} \frac{\gamma^*}{[(\omega_{ac} - \omega_2) \pm (k_2/k_1)(\omega_{ab} - \omega_1)]^2 + \gamma^{*2}} \right\}$$

where $\gamma^* = \gamma_{ac} + (k_2/k_1)\gamma_{ab}\sqrt{1+S}$ and $N = \int n(v_z) dv_z$.

The + or - signs describe the cases where both waves travel in opposite directions, or in the same direction.

If the pump wave is stabilized to the line center ω_{ab}, the second term in the denominator vanishes, since $\omega_1 - \omega_{ab} = 0$ and the line contour of the saturation dip in the absorption profile $\alpha(\omega_2)$ is described by a Lorentzian-shaped profile

$$L^*(\omega) = \frac{\gamma^*}{(\omega_{ac} - \omega_2)^2 + \gamma^{*2}}$$

with a linewidth which is the sum of the homogeneous width γ_{ac} and the saturated Lamb dip width $(k_2/k_1)\gamma_{ab}\sqrt{1 + S}$ of the pump wave times the ratio of the two laser frequencies ω_1 and ω_2.

One of many examples for the application of this optical-optical double-resonance technique is the precise determination of the CH_3F dipole moment values both for the ground and excited vibrational states [10.46]. The beams from two independent CO_2 lasers propagate collinearly through the sample in an electric field. Both lasers oscillate on the same CO_2 transitions, and their frequency difference $\omega_1 - \omega_2$ is set within the range 0-50 MHz by piezoelectric tuning of the optical cavity lenghts. Measurements of the beat frequency indicate a 20-30 KHz acoustic jitter over a 1 s interval. If the beat frequency is kept fixed but the voltage for the electric field in the sample is swept, the Stark splittings can be tuned into resonance with the beat frequency.

The two monochromatic laser waves can also be provided by two simultaneously oscillating modes of a multimode laser. If the absorbing sample is placed inside the laser resonator in a magnetic field, the molecular levels are split into the (2J + 1) Zeeman sublevels with a separation

$$\Delta\omega_z = \mu_B gB/\hbar \quad ,$$

where μ_B is the Bohr magneton, g the Landé factor, and B the magnetic field strength. If the Zeeman splitting $\Delta\omega_z$ equals the mode separation $\Delta\nu = c/2d$ of the laser resonator, two laser modes share a common level (see Fig.10.32).

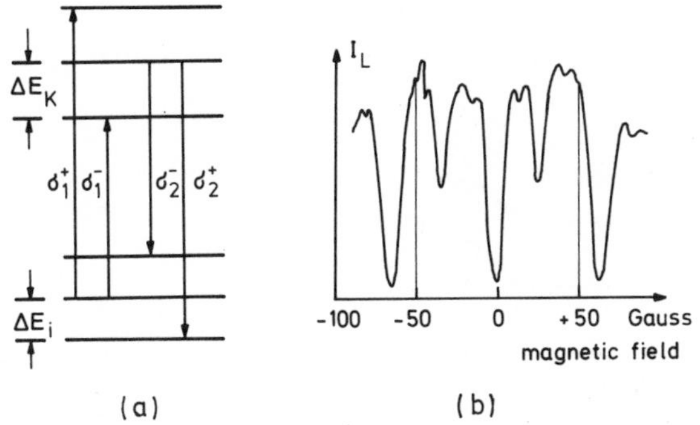

Fig.10.32a,b. Three-level laser spectroscopy with a multimode laser and Zeeman scanning. (a) Level scheme, (b) resonances in the laser output as a function of the magnetic field [10.47]

If both transitions share a common *lower* level, the absorption *decreases* for the resonance case due to the saturation of the lower level by both modes. The laser intensity therefore *increases*. If the two transitions share a common *upper* level, the laser output *decreases* at resonance, because the population of the *upper* level decreases, which diminishes the gain. This allows one to distinguish between the Zeeman splittings in the upper or the lower level and the Landé factors of both levels can be accurately measured [10.47,48].

An experimental advantage of this method is the fact that only the *difference* frequency between two modes and not the absolute frequency is essential for the spectral resolution. Such experiments can therefore be performed with unstabilized lasers.

Examples of two coupled transitions which have been thoroughly studied by several groups [10.49-51] are given by the two neon transitions at λ = 0.63 μm and λ = 1.15 μm which share the $^2P_4$ state as a common lower level, or the 0.63 μm and the 3.39 μm lines which share a common upper level (see Fig.6.7). Both coupled transitions show inversion in a He-Ne discharge and laser oscillation can be simultaneously achieved on both lines. The interaction of the two transitions can therefore be studied in a resonator arrangement which allows one to detect both laser lines separately. Figure 10.33 shows the concept and experimental arrangement. A single mode of the

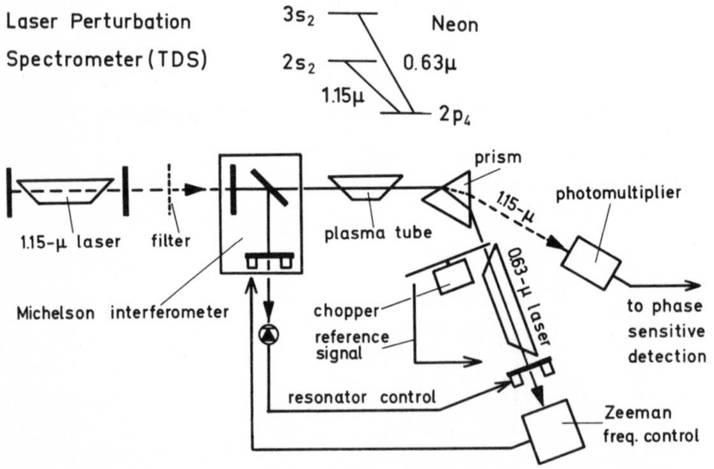

Fig.10.33. Optical-optical double resonance of two coupled Ne laser transitions studied in a tuned differential laser spectrometer [10.50]

He-Ne laser oscillating at $\lambda = 0.63\ \mu m$ is selected by a Fox-Smith cavity (see Sect.6.5) with a frequency ω_1 which can be continuously tuned through the gain profile of the laser transition. The output from a 1.15 µm He-Ne laser is sent collinear with the 0.63 µm beam through the gas discharge and is separated from the 0.63 µm by a prism. The attenuation or amplification of the 1.15 µm beam is detected as a function of the laser frequency ω_1 and of the inversion $n(3S_2) - n(2P_4)$, which can be altered by controlling the discharge conditions. The authors call this spectroscopic technique "tuned differential sepctroscopy".

The detailed studies show that the saturation of the common level, as discussed above, is not the only coupling mechanism. The interaction of the atom with a light wave generates an induced dipole moment which is at small intensities proportional to the field amplitude. At higher intensities the nonlinear terms in the induced polarization become important. If two waves with frequencies ω_1 and ω_2 simultaneously act in resonance with the atom, these nonlinear terms produce sum and difference frequencies $\omega_1 \pm \omega_2$. For two transitions which share a common level the difference frequency $\omega_1 - \omega_2$ is in resonance with the atomic transition $b \leftrightarrow c$ $(2S_2 - 3S_2)$ and will therefore modulate the atomic polarization at the frequency $\omega_1 - \omega_2$. The phenomenon can be regarded as a *resonant Raman process* where the 0.63 µm transition generates the Stokes line at $\lambda = 1.15\ \mu m$ and both waves force the electronic polarization to oscillate at the difference frequency (see Sect.9.4).

The study of these parametric interactions allows one to gain detailed information on the nonlinear polarization of the atom, and on the energy shift of the participating levels under the influence of the two fields for resonance and off-resonance conditions. The interaction is different for the two cases where the waves at 1.15 and 0.63 nm propagate in the same or opposite directions [10.52], see (10.33).

This section on saturation spectroscopy could only give a brief survey on this field and tried to illustrate the method by a few examples. A much more detailed discussion with more examples and extensive references can be found in [10.53-55].

10.3 Polarization Spectroscopy

While saturation spectroscopy monitors the decrease of *absorption* of a probe beam caused by a pump wave which has selectively depleted the absorbing level, the signals in polarization spectroscopy come mainly from the change of *refractive index* induced by a polarized pump wave [10.56]. This very sensitive Doppler-free spectroscopic technique has many advantages over conventional saturation spectroscopy and will certainly gain increasing attention [10.56a-c]. We therefore discuss the basic principle and some of its experimental modifications in more detail.

10.3.1 Basic Principle

The basic idea of polarization spectroscopy can be understood in a simple way (Fig.10.34).

The output from a monochromatic tunable laser is split into a weak probe beam with intensity I_1 and a stronger pump beam with intensity I_2. The probe beam passes through a linear polarizer P_1, the sample cell, and a second linear polarizer P_2 which is nearly crossed with P_1. At a crossing angle $(\pi/2 - \theta)$ with $\theta \ll 1$, only the component $E_t = E_0 \sin\theta \approx E_0\theta$ can pass through P_2 and reaches the detector D.

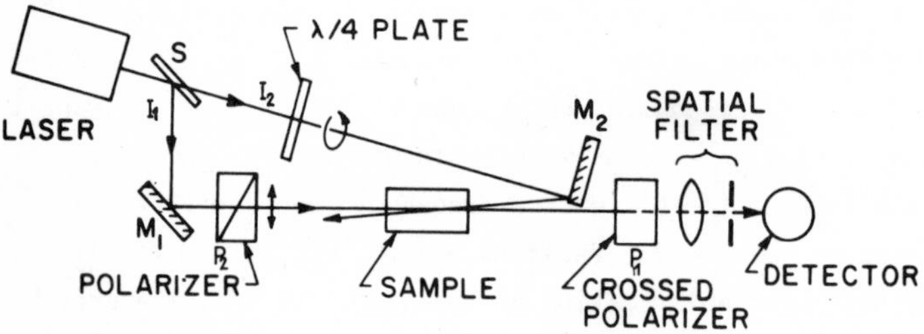

Fig.10.34. Schematic experimental arrangement for polarization spectroscopy [10.57]

After having passed a $\lambda/4$ plate which produces a circular polarization, the pump beam travels *in the opposite direction* through the sample cell. When the laser frequency ω is tuned to a molecular transition $(J'', M'') \to (J', M')$, molecules in the lower level (J'', M'') can absorb the pump wave. The quantum number M, which describes the projection of J onto the direction of light propagation, follows the selection rule $\Delta M = +1$ for transitions $M'' \to M'$ induced by left-hand circularly polarized light $(M'' \to M' = M'' + 1)$. Due to saturation the degenerate M sublevels of the rotational level J" become partially or completely depleted. The degree of depletion depends on the pump intensity I_2, the absorption cross section $\sigma(J'', M'' \to J', M')$, and on possible relaxation processes which may repopulate the level (J'', M''). The cross

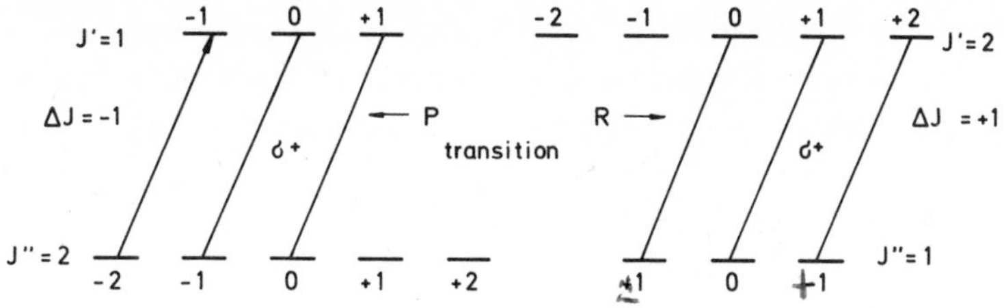

Fig.10.35. Selective depopulation of different M" sublevels and selective population of different M' sublevels by optical pumping with circularly polarized σ^+ light

section σ depends on J", M", J', and M'. From Fig.10.35 it can be seen that in case of P or R transitions ($\Delta J = +1$ or -1), not all of the M sublevels are pumped. For example from levels with $M'' = +1, +2$, no P transitions with $\Delta M = +1$ are possible while for R-transitions the levels $M' = -1, -2$, are not populated. This implies that *the pumping process produces an unequal saturation and with it a nonuniform population of the M sublevels*, which is equivalent to an anisotropic distribution for the orientations of the angular momentum vector J.

Such an anisotropic sample becomes birefringent for the incident linearly polarized probe beam, and the plane of polarization is slightly rotated. This effect is quite analogous to the Faraday effect where the nonisotropic orientiation of J is caused by an external magnetic field. For polarization spectroscopy no magnetic field is needed. Contrary to the Faraday effect where all molecules are oriented, here only those molecules which interact with the monochromatic pump wave show this nonisotropic orientation. As has already been discussed in Sect.10.1, this is the subgroup of molecules with velocity components

$$v_z \pm \Delta v_z = (\omega_0 - \omega)/k \pm \delta\omega/k \quad ,$$

where Δv_z is determined by the homogeneous linewidth $\delta\omega = \gamma$.

For $\omega \neq \omega_0$ the probe wave which passes in the opposite direction through the sample interacts with a *different* group of molecules in the velocity interval $v_z \pm \Delta v_z = -(\omega_0 - \omega \pm \delta\omega)/k$, and will therefore not be influenced by the pump wave. If, however, the laser frequency ω coincides with the center frequency ω_0 of the molecular transition within its homogeneous linewidth $\delta\omega$ (i.e., $\omega = \omega_0 \pm \delta\omega \Rightarrow v_z = 0 \pm \Delta v_z$), both waves can be absorbed by the same molecules and the probe wave experiences a birefringence due to the nonisotropic M distribution of the absorbing molecules.

Only in this case will the plane of polarization of the probe wave be slightly rotated by $\Delta\theta$ and the detector D will receive a Doppler-free signal every time the laser frequency ω is tuned across the center of a molecular absorption line.

10.3.2 Line Profiles of Polarization Signals

Let us now discuss the generation of this signal in a more quantitative way following the presentation in [10.57]. The linearly polarized probe wave

$$\underline{E} = \underline{E}_0\, e^{i(\omega t-kz)} \quad , \quad \underline{E}_0 = \{E_{0x}\ , 0\ , 0\}$$

can be always composed of a right and a left circularly polarized component (Fig.10.36). While passing through the sample the two components experience different absorption coefficients α^+ and α^- and different refractive indices n^+ and n^- due to the nonisotropic saturation caused by the left circularly polarized pump wave. After a path length L through the pumped region of the sample the two components are

$$E^+ = E_0^+\, e^{i[\omega t-k^+L\, +\, i(\alpha^+/2)L]} \quad ; \quad 2E_0^+ = E_{0x} + iE_{0y}$$

$$E^- = E_0^-\, e^{i[\omega t-k^-L\, +\, i(\alpha^-/2)L]} \quad ; \quad 2E_0^- = E_{0x} - iE_{0y} \quad . \tag{10.34}$$

Due to the differences $\Delta n = n^+ - n^-$ and $\Delta\alpha = \alpha^+ - \alpha^-$ caused by the nonisotropic saturation, a phase difference

$$\Delta\varphi = (k^+ - k^-)L = (\omega L/c)(n^+ - n^-)$$

has developed between the two components and also a small amplitude difference

$$\Delta E = (E_0/2)\left[e^{-(\alpha^+/2)L} - e^{-(\alpha^-/2)L}\right] \ .$$

If both components are again superimposed at $z = L$ after having passed through the sample cell, an elliptically polarized wave comes out with a major axis which is slightly rotated against the x axis. The y component of this elliptical wave is

$$E_y = -i(E_0/2)\, e^{i[k^+-k^-+i(\alpha^+-\alpha^-)/2]L+ib}\, e^{i(\omega t+\varphi)} \quad , \tag{10.35}$$

where the term ib with $b \ll 1$ in the exponent takes into account that the windows of the sample cell may have a small birefringence which introduces an additional ellipticity. In all practical cases the differences $\Delta\alpha$ and Δk are small,

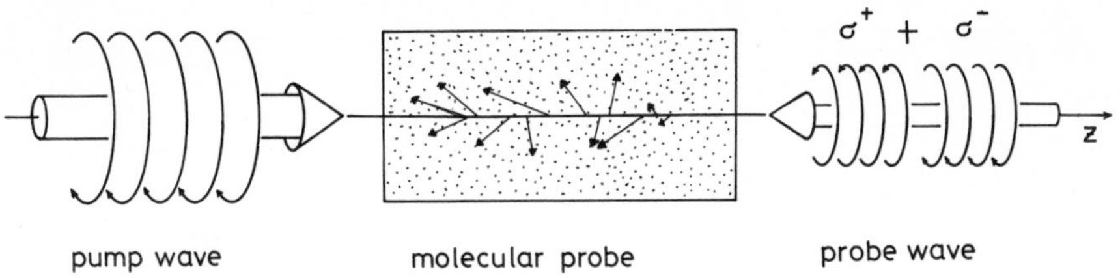

Fig.10.36. Schematic illustration of partial alignment of angular momenta by the left-hand circularly polarized (σ^+) pump wave, which results in different absorption coefficients for σ^+ and σ^- probe light

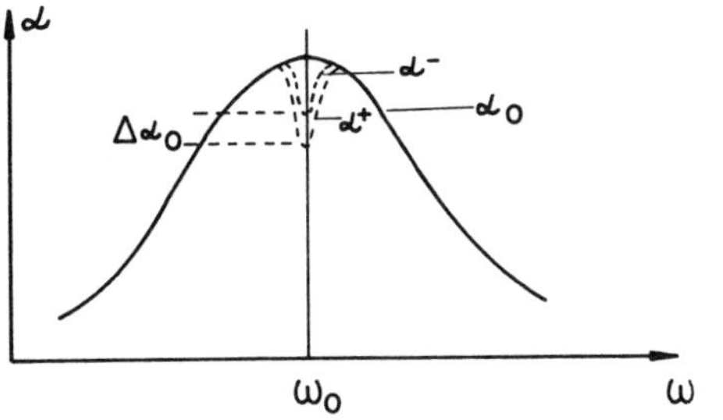

Fig.10.37. Spectral profiles of the absorption coefficients α^+ and α^- for a weak probe, in the presence of a saturating counterpropagating pump

$$(\alpha^+ - \alpha^-)L \ll 1 \quad \text{and} \quad (k^+ - k^-)L \ll 1 \quad ,$$

and we can expand the first exponential factor. If the transmission axis of the second polarizer P_2 is close to the y axis ($\theta \ll 1$), we obtain for the transmitted amplitude

$$E_t = E_0[\theta + ib + (\omega L/2c)(n^+ - n^-) + i(\alpha^+ - \alpha^-)L/4]e^{i(\omega t+\varphi)} \quad . \qquad (10.36)$$

The differences $\Delta\alpha = \alpha^+ - \alpha^-$ in absorption coefficients and $\Delta n = n^+ - n^-$ in refractive indices are due to the different degrees of M sublevel depopulations experienced by the right or the left circularly polarized probe component. Although each coefficient α^+ and α^- itself shows a Doppler-broadened spectral profile with a Lamb dip at the center, the *difference* $\Delta\alpha$ just exhibits the small difference between these Lamb dips (see Fig.10.37) The spectral profile of $\Delta\alpha$ is therefore Lorentzian

$$\Delta\alpha = \frac{\Delta\alpha_0}{1 + x^2} \quad \text{with} \quad x = (\omega_0 - \omega)/\gamma \quad , \qquad (10.37)$$

where $\Delta\alpha_0$ is the maximum difference at the center $\omega = \omega_0$. Since absorption and dispersion are related by the *Kramers-Kronig dispersion relations* (see Sect.2.6), we obtain a dispersion shaped profile for Δn

$$\Delta n = \Delta\alpha_0 \frac{c}{\omega_0} \frac{x}{1 + x^2} \quad . \tag{10.38}$$

The transmitted intensity is $I_T = E_T E_T^*$. Taking into account that even perfectly crossed polarizers show a residual transmission $I_0\xi$ with $\xi \ll 1$ due to imperfect extinction, we obtain the spectral line profile of the transmitted intensity I_T from (10.35-38).

$$I_T = I_0\left[\xi + \theta^2 + b^2 - \frac{1}{2}\,\theta\Delta\alpha_0 L \frac{x}{1 + x^2} + \frac{b}{2}\,\Delta\alpha_0\,L \frac{1}{1 + x^2} + \frac{1}{4}\,(\Delta\alpha_0 L)^2 \frac{1}{1 + x^2}\right] \quad . \tag{10.39}$$

The transmitted intensity contains a constant background term $\xi + \theta^2 + b^2$ which is caused: 1) by the finite transmission $I_0\xi$ of the crossed polarizers at $\theta = 0$; 2) by the birefringence of the cell windows, described by the term $I_0 b^2$ and 3) by the finite uncrossing angle θ of P_2. The birefringence of the windows can be controlled, for example, by squeezing the windows slightly to compensate for the birefringence which is mainly caused by the pressure difference of 1 atm at both sides of the windows. A proper selection of the uncrossing angle θ and the window birefringence b allows one to make either the first Lorentzian term dominant ($\theta = 0$, b large) or the dispersion term ($\theta \neq 0$, b small). If for the latter choice $\theta \gg \Delta\alpha_0 \cdot L$, the last Lorentzian term becomes negligible and one obtains nearly pure dispersion profiles, while for the former choice pure Lorentzian profiles are monitored.

Figure (10.38) shows an example of a section from the polarization spectrum of the cesium molecule Cs_2, recorded with $\theta = 0$ (lower part) and with $\theta = 2.5' \mathrel{\hat{=}} \theta = 7 \times 10^{-4}$ rad (upper part). In this experiment the term $(\xi + b^2)$ was smaller than 10^{-6}.

Note that the main contribution to the signal comes from the *rotation* of the plane of polarization of the linearly polarized probe wave, and only to a minor extent from the change of absorption $\Delta\alpha$. In (10.39) $\Delta\alpha$ appears because $\Delta n = n^+ - n^-$ has been replaced by $\Delta\alpha$ using the dispersion relations. It is the difference in *phase shifts* for the two probe components E^+ and E' rather than the slightly different amplitudes which gives the major part of the signal. The reason why this method is much more sensitive than the saturation method, is due to the crossed polarizers which suppress the background.

A linearly, rather than circularly, polarized pump wave can also be used with the plane of polarization inclined by 45° against that of the probe. Assume that the pump wave is polarized in the x direction. The probe wave then

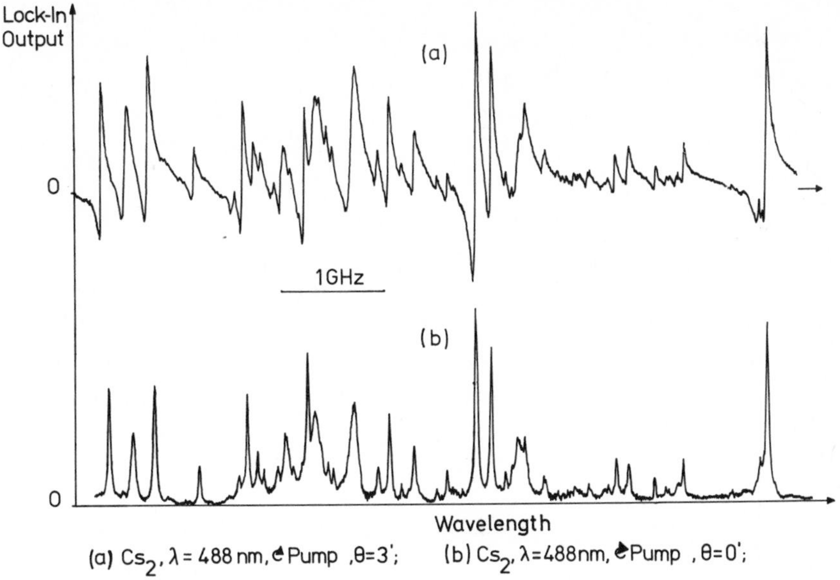

Fig.10.38a,b. Two identical sections of Cs_2 polarization spectra obtained with circular pump polarization and linear probe polarization, but different uncrossing angles θ of the two polarizers [10.59]

can be composed of two linearly polarized components E_x and E_y and the saturation by the pump will cause a difference $\alpha_x - \alpha_y$ and $n_x - n_y$. Analogous to the derivation above we obtain in this case for the probe intensity, transmitted through the polarizer P_2,

$$I_T = I_0\left[\xi + \theta^2 + b^2 + \frac{1}{2}\,\theta\Delta\alpha L\,\frac{1}{1+x^2} + \frac{1}{2}\,b\Delta\alpha L\,\frac{x}{1+x^2} + \frac{1}{4}\,(\Delta\alpha L)^2\,\frac{1}{1+x^2}\right] , \tag{10.40}$$

which shows that the dispersion term and the Lorentzian term are just interchanged compared to (10.39). For b = 0 (no birefringence of the cell windows) and $\theta \neq 0$ (polarizers not completely crossed) we obtain pure Lorentzian profiles.

10.3.3 Magnitude of Polarization Signals

In order to estimate the expected magnitudes of the polarization signals in (10.39) let us consider the difference $\Delta\alpha = \alpha^+ - \alpha^-$ in the absorption coefficients for the right- and left-hand circularly polarized probe wave components. The absorption of a circularly polarized wave tuned to a rotational transition $J \to J_1$ is due to the sum of all allowed transitions $M_J \to M_{J_1}$ with $\Delta M = \pm 1$ between the $(2J+1)$ degenerate sublevels M in the lower level J and the $(2J_1+1)$ sublevels in the upper level J_1.

$$\alpha^+ - \alpha^- = \sum_M n_M(\sigma^+_{JJ_1M} - \sigma^-_{JJ_1M}) \quad , \tag{10.41}$$

where $\sigma^{\pm}_{JJ_1M}$ is the absorption cross section for the transition $(J, M) \to (J_1, M+1)$ or $(J, M) \to (J_1, M-1)$.

The M dependence of the cross sections σ_{JJ_1M} can be expressed in terms of Clebsch-Gordan coefficients $C(J, J_1, M, M_1)$ and a reduced matrix element $\tilde{\sigma}_{JJ_1}$ which is independent of M and describes the total rotational transition $J \to J_1$ [10.58]. The explicit evaluation yields for a circularly polarized pump wave

$$\sigma^{\pm}_{JJ_1M} = \begin{cases} \tilde{\sigma}_{J,J+1}(J \pm M + 1)\,(J \pm M + 2) & \text{for} \quad J_1 = J + 1 \\ \tilde{\sigma}_{J,J}(J \pm M)\,(J \pm M + 1) & \text{for} \quad J_1 = J \\ \sigma_{J,J-1}(J \mp M)\,(J \pm M + 1) & \text{for} \quad J_1 = J - 1 \ . \end{cases} \tag{10.42}$$

For a linearly polarized pump wave the analogous calculation yields

$$\sigma_{JJ_1M} = \begin{cases} \tilde{\sigma}_{J,J+1}(J + 1)^2 - M^2 & \text{for R lines} \\ \tilde{\sigma}_{J,J}\ M^2 & \text{for Q lines} \\ \tilde{\sigma}_{J,J-1}(J^2 - M^2) & \text{for P lines} \ . \end{cases} \tag{10.42a}$$

The total cross section

$$\sigma_{JJ_1} = \frac{1}{2J+1} \sum_M \sigma_{J,J_1M}$$

for the transition $J \to J_1$ is independent of the kind of polarization. Inserting (10.42) and evaluating the sum over M yields

$$\sigma_{JJ_1} = \frac{1}{3}\,\tilde{\sigma}_{JJ_1} \begin{cases} (J + 1)(2J + 3) & \Delta J = +1 \\ J(J + 1) & \text{for} \quad \Delta J = 0 \\ J(2J - 1) & \Delta J = -1 \ . \end{cases} \tag{10.43}$$

The unsaturated level population of a sublevel M is

$$n^0_M = N_0/(2J + 1) \ . \tag{10.44}$$

where N_0 is the unsaturated total population of the rotational level J. Without the saturating pump wave, we obtain by inserting (10.43) and (10.44) into (10.41)

$$\Delta\alpha^0 = \alpha^+ - \alpha^- = 0 \quad .$$

Due to saturation by the pump wave with intensity I_2 the population of the M sublevels decreases according to (2.167) to

$$n_M^{(s)} = \frac{N_0}{2J+1} \cdot \frac{1}{1+S_M} \tag{10.44a}$$

and the absorption of the probe wave by molecules in sublevel M decreases according to (10.23) to

$$\alpha_M^s = \alpha_M^0(1 - S_M) \quad .$$

The saturation parameter S_M at the line center ω_0 [see (2.73) and (3.70)],

$$S_{0M} = 2S_M/(\pi\gamma_s) = \frac{8I_2\sigma_{JJ_1M}}{c\gamma_s R\hbar\omega} \quad , \tag{10.45}$$

depends on the absorption cross section σ_{JJ_1M} of the pump wave, on the number $n_p = I_2/\hbar\omega$ of pump photons incident on the sample per cm^2 and s, on the saturation-broadened homogeneous linewidth γ_s, and on the collision induced relaxation rate R $[s^{-1}]$ which tries to refill the depleted level and depopulate the upper level. In Fig.10.39 the M dependence of σ_{JJ_1M} is plotted for right-hand and left-hand circular polarization and for linear polarization (second column). These diagrams illustrate that saturation by a *circularly* polarized pump wave results in larger differences $(\alpha^+ - \alpha^-)$ for P and R lines than for Q lines, while a linearly polarized pump wave favors Q lines in the detected probe transmission.

Putting the relations (10.41-45) together, we can express the difference $\alpha^+ - \alpha^-$ at the line center ω_0 by the unsaturated absorption coefficient $\alpha_0 = N_J^0\sigma_{JJ_1}$, the saturation parameter $S_0(\omega_0)$, and a numerical factor $C^*_{JJ_1}$ which stands for the sum over the Clebsch-Gordan coefficients and which is tabulated in [10.57]. The final result is

$$\alpha^+ - \alpha^- = \alpha^0 S_0 C^*_{JJ_1} \quad . \tag{10.46a}$$

This is often written in the form

$$\boxed{\alpha^+ - \alpha^- = \alpha^0 C^*_{JJ_1}\ (I_2/I_s)} \quad , \tag{10.46b}$$

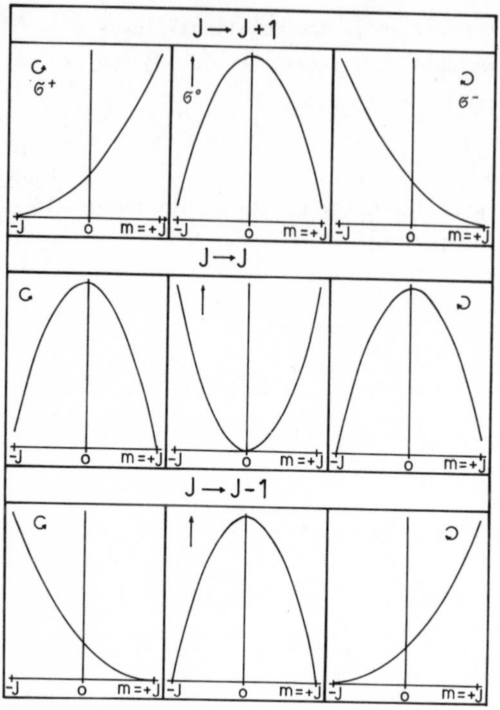

Fig.10.39. M dependence of the cross section σ_{JJ_1M} for P, Q, and R lines for σ^+, σ^-, and $\sigma^0 = \pi$-polarized light

where $I_S = I_2/S_0$ represents that pump intensity which causes a saturation parameter $S_0 = 1$ at the line center.

10.3.4 Sensitivity of Polarization Spectroscopy

In the following we briefly discuss the sensitivity and the signal-to-noise ratio achievable with polarization spectroscopy. The amplitude of the dispersion signal in (10.39) is approximately the difference $\Delta I_T = I_T(x = +1) - I_T(x = -1)$ between the maximum and the mimimum of the dispersion curve. From (10.39) we obtain

$$\Delta I_T = \frac{1}{2}\,\theta\Delta\alpha L \quad . \tag{10.47}$$

Under general laboratory conditions the main contribution to the noise comes from fluctuations of the probe laser intensity, while the principal limit set by shot noise (see Chap.4) is seldom reached. The noise level is therefore essentially proportional to the transmitted intensity, which is given by the background term in (10.39).

Because the crossed polarizers greatly reduce the background level, we can expect a better signal-to-noise ratio than in saturation spectroscopy, where the full intensity of the probe beam is detected.

In the absence of window birefringence (i.e., b = 0) the signal-to-noise ratio (S/N) which is, besides a constant factor a, equal to the signal-to-background ratio, becomes

$$S/N = \frac{\theta\alpha_0 L(I_p/I_s)}{a(\xi + \theta^2)} C^*_{JJ_1} \quad . \tag{10.48}$$

This ratio has a maximum for $d(S/N)/d\theta = 0$ which yields

$$(S/N)_{max} = \frac{\alpha_0 L(I_p/I_s)}{a\xi^{\frac{1}{2}}} C^*_{JJ_1} \tag{10.49}$$

for $\theta^2 = \xi$.

According to (10.23) the signal-to-background ratio is for saturation spectroscopy outside the laser resonator

$$(S/N)_{sat} = \alpha_0 ILS_0/I = \alpha_0 LI_p/I_s \quad ,$$

while a value of I_p/I_s may be reached with the intermodulated fluorescence technique [10.33]. Equation (10.49) shows that a factor of $\xi^{-\frac{1}{2}}$ is gained in polarization spectroscopy. Since good polarizers allow an extinction ratio of $\xi = 10^{-6} - 10^{-7}$, the enhancement of the signal-to-background ratio may become three orders of magnitude. Only for $\alpha_0 L < 10^{-3}$ can the intermodulated fluorescence method compete with polarization spectroscopy. The sensitivity of polarization spectroscopy is demonstrated by Fig.10.40 which shows a section of the polarization spectrum of NO_2 where the three hfs components of each transition are resolved. The transition probabilities of NO_2 transitions are very small. This impedes saturation spectroscopy but still allows polarization spectroscopy because of its higher sensitivity.

With a circularly polarized pump wave the P and R lines show dispersion profiles (Fig.10.40a) while for a linearly polarized pump wave the Q lines are monitored with Lorentzian profiles. Note that the whole range between the arrows in Fig.10.40a covers only 0.01 cm^{-1} = 0.0024 Å. The linewidth of about 5 MHz is limited mainly by the finite crossing angle between pump beam and probe beam. It can be further reduced by enlarging the length L of the sample cell.

For illustration of the achievable signal-to-noise ratio, Fig.10.41 shows the polarization spectrum of the same R(98) hfs components of I_2 as obtained

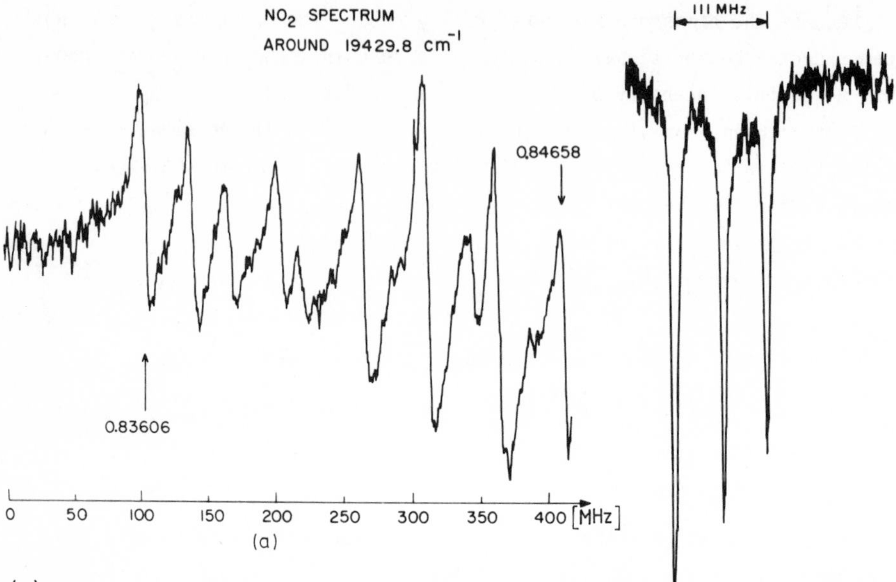

(a) (b)

Fig.10.40a,b. Section of a polarization spectrum of NO_2 around λ = 488 nm. (a) Circular pump polarization, (b) linear pump polarization. The linewidth of 5 MHz is mainly determined by the residual Doppler width, due to the crossing angle between pump and probe beam

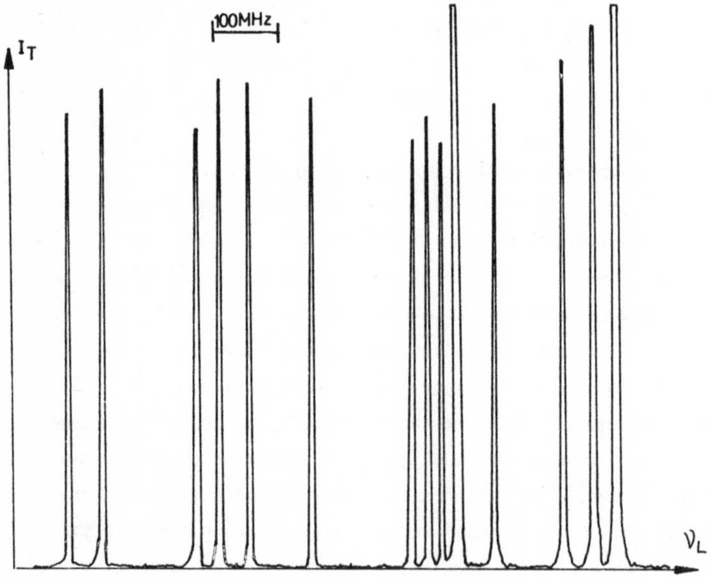

Fig.10.41. Polarization spectrum of the hfs components of the R(98), 58 - 1 line of I_2 at 514.5 nm (compare with Fig.10.23)

in Fig.10.23 with intermodulated fluorescence saturation spectroscopy. The polarization spectrum shows a signal-to-noise ratio which is better by two orders of magnitude.

10.3.5 Polarization Labelling Spectroscopy

A very powerful method for the assignment of complex molecular spectra is the "polarization labelling technique", which is based on a combination of polarization spectroscopy and optical double resonance. This method employs two different lasers (see Fig.10.42). The output beam from the first laser is

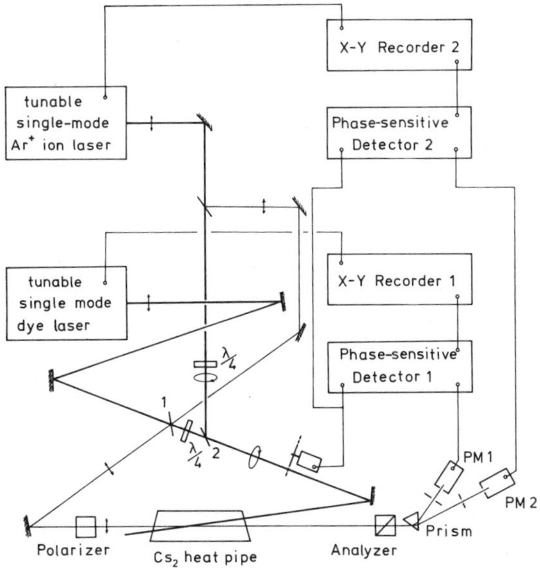

Fig.10.42. Experimental arrangement for polarization labelling spectroscopy

again split into a pump beam and a probe beam and the laser frequency is stabilized onto the center frequency ω_0 of a polarization signal. The probe beam from the second laser passes collinearly with the probe beam from the first laser through the sample cell. Both beams are separated by a prism and are separately detected. When the frequency of the "probe laser" is tuned across the molecular absorption spectrum, a polarization signal is obtained every time the probe laser hits a transition which shares a common level with the pump transition (see Fig.8.36).

Evaluation of the cross sections (see [10.57]) shows that the form, sign, and magnitude of the signals depend on the kind of polarization of the pump wave and on the value of ΔJ_1 and ΔJ_2 for pump and probe transition. This gives

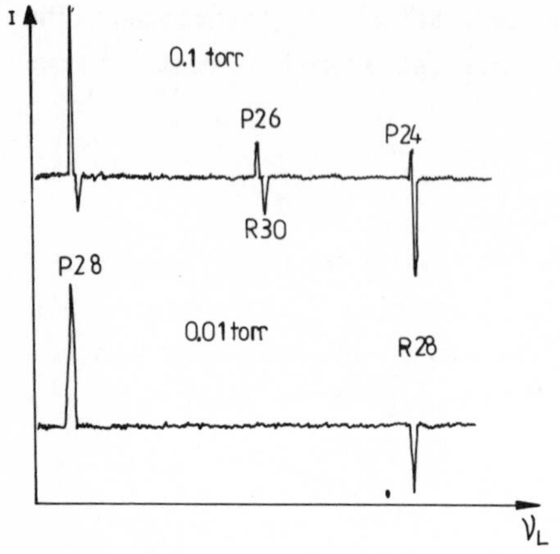

Fig.10.43. Optical-optical double-resonance lines in the polarization spectrum of Na_2. The pump laser was a single-mode argon laser, stabilized onto the transition $X^1\Sigma$ ($v'' = 0$, $J'' = 28$) → B^1($v' = 6$, $J' = 27$). The broad-band dye laser was scanned across the X→A-spectrum. At low pressures only the transitions ($v'' = 0$, $J'' = 28 \rightarrow J' = 28$ 1) appear as double resonance signals while at 0.1 torr collision-induced satellites are generated

the possibility of distinguishing even P lines from R lines in the double-resonance spectrum, because the two signals have opposite signs if the pump laser is stabilized on an R or P transition [10.59]. Figure 10.43 illustrates such a double-resonance spectrum of Na_2 where the opposite sign of P and R lines can be seen.

This technique is particularly useful if the upper state is perturbed and the assignment by conventional spectroscopy is impeded [10.61].

The second probe beam may be a broadband laser. If the polarization signal is detected on a photoplate behind a spectrograph, many transitions from the labelled level can then be detected simultaneously [10.61a].

The orientation of the molecules due to saturation by a polarized pump wave can be partly transferred to neighbouring levels by inelastic collisions of the oriented molecules with other atoms or molecules. This produces "satellite lines" in the double-resonance spectrum which may partly diminish the unambiguity of line assignment. Some of the weaker lines in Fig.10.43 are such collision-induced polarization signals.

10.3.6 Advantages of Polarization Spectroscopy

Let us briefly summarize the advantages of polarization spectroscopy, discussed in the previous sections.

1) With the other sub-Doppler techniques it shares the advantage of high spectral resolution which is mainly limited by the residual Doppler width due to the finite angle between pump beam and probe beam. This limitation corresponds to that imposed to linear spectroscopy in collimated molecular beams by the divergence angle of the molecular beam. The time-of-flight broadening can be reduced if pump and probe beam are less tightly focussed.

2) The sensitivity is 2-3 orders of magnitude larger than that of saturation spectroscopy. It is surpassed only at very low sample pressures by that of the intermodulated fluorescence technique (Sect.10.2.3).

3) The possibility of distinguishing between P, R, and Q lines is a particular advantage for the assignment of complex molecular spectra.

4) The dispersion profile of the polarization signals allows a stabilization of the laser frequency to the line center without any frequency modulation. The large achievable signal-to-noise ratio assures an excellent frequency stability.

5) The combination of optical-optical double resonance techniques and polarization spectroscopy opens the way to detailed studies of perturbed excited molecular states.

10.3.7 Doppler-Free Laser-Induced Dichroism and Birefringence

A slight modification of the experimental arrangement used for polarization spectroscopy allows the simultaneous observation of saturated absorption *and* dispersion [10.62]. While in the setup of Fig.10.34 the probe beam had linear polarization, here a circularly polarized probe and a linearly polarized pump beam are used (Fig.10.44). The probe beam can be composed of two components with linear polarization parallel and perpendicular to the pump beam polarization. Due to anisotropic saturation by the pump, the absorption coefficients $\alpha_{\parallel}$ and $\alpha_{\perp}$ and the refractive indices $n_{\parallel}$ and $n_{\perp}$ experienced by the probe beam are different for the parallel and the perpendicular polarizations. This causes a change of the probe beam polarization which is monitored behind a linear analyzer rotated through an angle β from the reference direction π. Analogous to the derivation in Sect.10.3.2, one can show that the transmitted intensity of a circularly polarized probe wave with incident intensity I is for $\alpha L \ll 1$ and $\Delta n(L/\lambda) \ll 1$

$$I_t(\beta) = \frac{I}{2}\left(1 - \frac{\alpha_{\parallel} + \alpha_{\perp}}{2} L - \frac{L}{2}\Delta\alpha\cos 2\beta - \frac{\omega L}{c}\Delta n \sin 2\beta\right) \tag{10.50}$$

with $\Delta\alpha = \alpha_{\parallel} - \alpha_{\perp}$ and $\Delta n = n_{\parallel} - n_{\perp}$.

The difference of the two transmitted intensities

$$\Delta_1 = I_t(\beta = 0^\circ) - I_t(\beta = 90^\circ) = IL\Delta\alpha/2 \tag{10.51}$$

gives the pure dichroism signal (anisotropic saturated absorption) while the difference

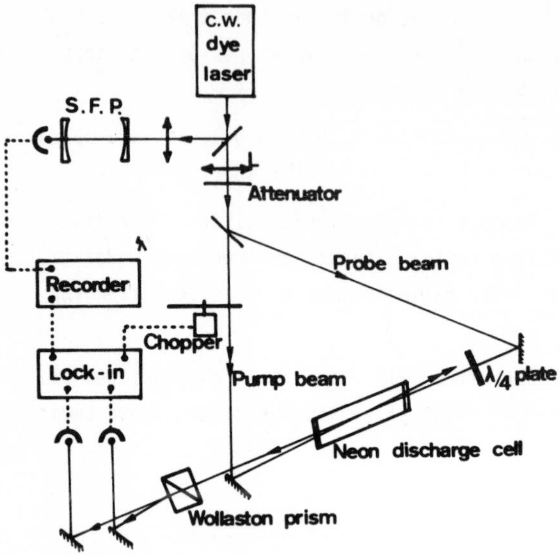

Fig.10.44. Experimental arrangement for observation of Doppler-free laser-induced dichroism and birefringence [10.62]

$$\Delta_2 = I_t(45^0) - I_t(-45^0) = I(\omega L/c)\Delta n \qquad (10.52)$$

yields the pure birefringence signal (saturated dispersion). A birefringent Wollaston prism after the interaction region allows the spatial separation of the two probe beam components with mutual orthogonal polarizations. The two beams are monitored by two identical photodiodes and after a correct balance of the output signals a differential amplifier records directly the desired differences Δ_1 and Δ_2 if the axes of the birefringent prism have suitable orientations.

Figure 10.45 illustrate the advantages of this technique. The upper spectrum represents a "Lamb peak" in the intracavity saturation spectrum of the neon line (1s → 2p) at λ = 588.2 nm (see Sect.10.2). Due to the collisional redistribution of the atomic velocities a broad and rather intense background appears in addition to the narrow peak. This broad structure is not present in the dichroism and birefringent curves (b) and (c). This improves the signal-to-noise ratio and the spectral resolution. Analogous to the technique of "polarization labelling", this method can be extended to optical-optical double-resonance techniques using two different lasers for pump and probe. Signals appear if both lasers are tuned to transitions which share a common level.

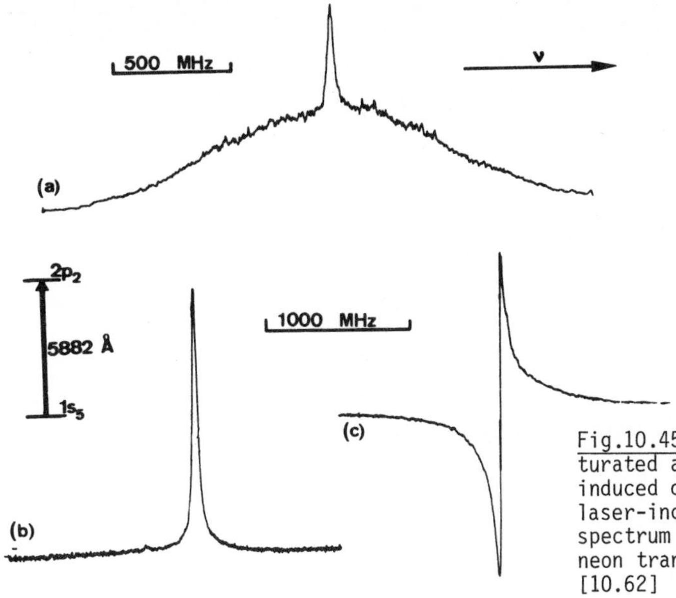

Fig.10.45. Comparison of saturated absorption (a), laser-induced dichroism (b), and laser-induced birefringence spectrum (c) of the $1s_2$-$2p_2$ neon transition at λ = 588.2 nm [10.62]

10.4 Saturated Interference Spectroscopy

The higher sensitivity of polarization spectroscopy compared with conventional saturation spectroscopy results from the detection of *phase differences* rather than amplitude differences. This advantage is also used in a method which monitors the interference between two probe beams where one of the beams suffers saturation-induced phase shifts. This saturated interference spectroscopy was independently developed in different laboratories [10.63,64]. The basic principle can be easily understood from Fig.10.46. We follow here the presentation in [10.63].

The probe beam is split by the plane parallel plate Pl_1 into two beams. One beam passes through that region of the absorbing sample which is saturated by the pump beam; the other passes through an unsaturated region of the same sample cell. The two beams are recombined by a second plane parallel plate Pl_2. The two carefully aligned parallel plates form a Jamin interferometer [4.12] which can be adjusted by a piezoelement in such a way that without the saturating pump beam the two probe waves with intensities I_1 and I_2 interfere destructively.

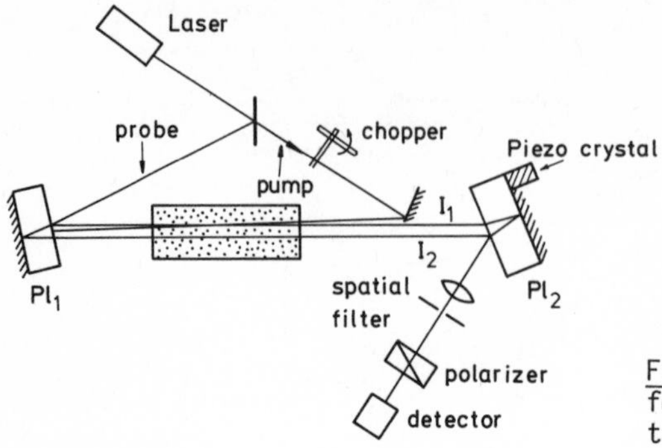

Fig.10.46. Experimental arrangement for saturated interference spectroscopy [10.63]

If the saturation by the pump wave introduces a phase shift ϑ, the resultant intensity at the detector becomes

$$I = I_1 + I_2 - 2\sqrt{I_1 I_2}\cos\vartheta \quad . \tag{10.53}$$

The intensities I_1 and I_2 of the two interfering probe waves can be made equal by placing a polarizer P_1 into one of the beams and a second polarizer P_2 in front of the detector. Due to a slight difference δ in the absorptions of the two beams by the sample molecules, their intensities at the detector are related by

$$I_1 = I_2(1 + \delta) \quad \text{with} \quad \delta \ll 1 \quad .$$

For small phase shifts $\vartheta(\vartheta \ll 1)$ we can approximate (10.53) by

$$I = \left(\frac{1}{4}\delta^2 + \vartheta^2\right) I_2 \quad . \tag{10.54}$$

The amplitude difference δ and the phase shift are both caused by selective saturation of the sample through the monochromatic pump wave which travels in the opposite direction. Analogous to the situation in polarization spectroscopy we therefore obtain Lorentzian and dispersion profiles for the frequency dependence of both quantities

$$\delta(\omega) = \frac{\Delta\delta_0}{1 + x^2} \quad ; \quad \vartheta(\omega) = \frac{1}{2}\Delta\delta_0 \frac{x}{1 + x^2} \quad , \tag{10.55}$$

where $x = (\omega - \omega_0)/\gamma$ and γ is the homogeneous linewidth.

Inserting (10.55) into (10.54) yields, for the total intensity I at the minimum of the interference patterns the Lorentzian profile,

$$I = \frac{1}{4} I_2 \frac{(\Delta\delta_0)^2}{1 + x^2} \quad . \tag{10.56}$$

According to (10.55) the phase differences $\vartheta(\omega)$ depends on the laser frequency ω. However, it can be always adjusted to zero while the laser frequency is scanned. This can be accomplished by a sine wave voltage at the piezoelement which causes a modulation

$$\vartheta(\omega) = \vartheta_0(\omega) + a\,\sin(2\pi f_1 t) \quad .$$

When the detector signal is fed to a lock-in amplifier which is tuned to the modulation frequency f_1, the lock-in output can drive a servo loop to bring the phase difference ϑ_0 back to zero. For $\vartheta(\omega) \equiv 0$ we obtain from (10.54,55)

$$I(\omega) = \frac{1}{4}\,\delta(\omega)^2 I_2 = \frac{1}{4}\,\frac{\Delta\delta_0^2}{(1 + x^2)^2} \quad . \qquad (10.57)$$

The halfwidth of this signal is reduced from γ to $(\sqrt{2} - 1)^{\frac{1}{2}}\gamma \approx 0.62\gamma$.

Contrary to the situation in polarization spectroscopy, where for slightly uncrossed polarizers the line shape of the polarization signal is a superposition of Lorentzian and dispersion profiles, here a *pure dispersion* line profile can be obtained without distortion by a Lorentzian term. To achieve this, the output of the lock-in amplifier that controls the phase is fed into another lock-in, tuned to a frequency f_2 ($f_2 \ll f_1$) at which the saturating pump beam is chopped. The output of the first lock-in is always proportional to $\vartheta(\omega)$ since it drives the servo loop to bring $\vartheta(\omega)$ back to zero. The second lock-in filters the desired signal caused by the saturating pump beam out of all other background which may cause a phase shift.

The method has been applied so far to the spectroscopy of Na_2 [10.63] and I_2 [10.64]. Figure 10.47a shows saturated absorption signals in I_2 obtained with a dye laser at λ = 600 nm with 10 mW pump power and 1 mW probe power. Figure 10.47b displays the first derivative of the spectrum in a) and Fig.10.47c, the first derivative of the saturated dispersion signal.

The sensitivity of the saturated interference technique is comparable to that of polarization spectroscopy. While the latter can be applied only to transitions from levels with a rotational quantum number $J \geq 1$, the former works also for $J = 0$. An experimental drawback may be the critical alignment of the Jamin interferometer and its stability during the measurements.

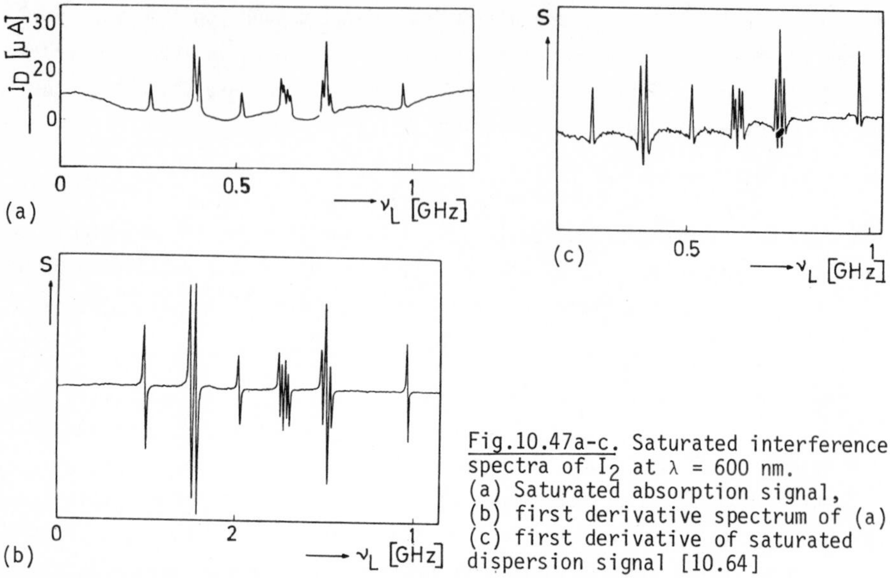

Fig.10.47a-c. Saturated interference spectra of I_2 at $\lambda = 600$ nm. (a) Saturated absorption signal, (b) first derivative spectrum of (a), (c) first derivative of saturated dispersion signal [10.64]

10.5 Heterodyne Spectroscopy

In most of the methods discussed so far in high-resolution laser spectroscopy, the laser frequency was tuned across the molecular absorption spectrum and the frequency separation of different lines was determined either by interpolation between frequency marks provided from a long F.P.I., or by absolute wavelength measurements, using one of the methods explained in Sect.4.4. For many problems in spectroscopy, however, it is important to know accurately the *splittings* of closely separated lines rather than their absolute wavelengths. Examples are the hfs, Zeeman, or Stark splittings.

Up to now the heterodyne technique is the most accurate method to determine such line splittings. Its accuracy is comparable with the optical-rf double-resonance method but its application range is more general. Two independent lasers are stabilized onto the line centers of two different molecular transitions (Fig.10.48). The output of the two lasers is superimposed on a nonlinear detector, such as a photomultiplier in the visible range or a semiconductor diode in the infrared.

The electrical output signal

$$S \propto |E_1\, e^{i\omega_1 t} + E_2\, e^{i\omega_2 t}|^2$$

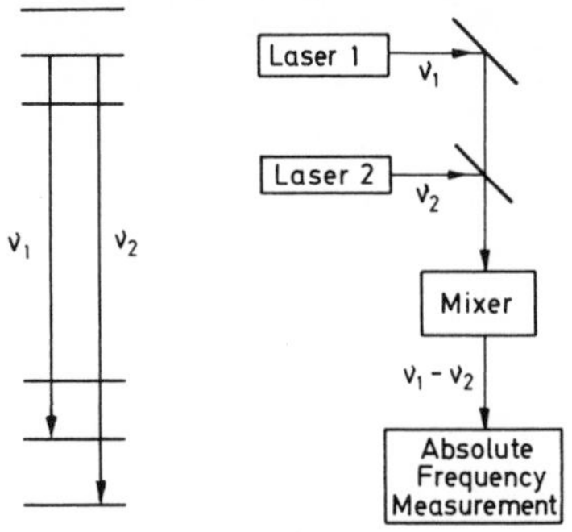

Fig.10.48. Schematic diagram of heterodyne spectroscopy using two stabilized lasers

of this nonlinear detector is proportional to the square of the sum of the two amplitudes and therefore contains terms with the difference frequency $(\omega_1 - \omega_2)$ and the sum frequency $(\omega_1 + \omega_2)$. The difference frequency is filtered through the amplifier and is absolutely counted. The accuracy depends on the frequency stability of both lasers and on the attainable signal-to-noise ratio. It is therefore essential to have small linewidths for the laser stabilization. Either the narrow Lamb dips of Doppler-broadened transitions can be used (see Sect.10.2.5) or the reduction of the Doppler width in collimated molecular beams can be utilized (see Sect.10.1.1). Figures of $\Delta\omega/\omega \leqq 10^{-10}$ have been achieved already.

If both transitions share a common level, the difference frequency immediately gives the separation of the two other levels. We illustrate this technique by several examples.

a) BRIDGES and CHAN [10.65] stabilized two CO_2 lasers onto different rotational lines in the vibrational bands $(00^01) - (10^00)$ at 10.4 μm and $(00^01) - (02^00)$ at 9.4 μm of the CO_2 molecule (see Sect.6.3). A GaAs crystal served as nonlinear mixer to generate the difference frequencies $(\omega_1 - \omega_2)$ of 37 line pair with a common level for each pair. These frequencies fall into the 50-80 GHz range and could be measured with an accuracy of better than 1 MHz. The molecular constants deduced from these measurements are 25-200 times more accurate than those previously determined from infrared spectroscopy.

The range of difference frequencies which is experimentally accessible can be enlarged by mixing these frequencies further with harmonics of a microwave generator. The difference frequency of this second mixing is then directly counted [10.66].

b) In the visible region the heterodyne technique has been used to measure the hfs splittings of rotational lines in an electronic transition of the I_2

molecule [10.67]. The output beams of two single-mode argon lasers were crossed perpendicularly with a highly collimated I_2 beam and the laser frequencies were stabilized onto two different I_2 hfs components. The difference frequency could be measured with an accuracy of about 5 kHz. The direct measurement of the optical lines with frequencies of $6 \times 10^{14}\ s^{-1}$ would demand a relative accuracy of 10^{-11} to achieve the same result. Instead of using a molecular beam, the argon lasers can also be stabilized onto the Lamb dips obtained with intracavity I_2 cells [10.68]. Because of the higher molecular density in cells, this method allows one to measure very weak transitions between levels with high rotational quantum numbers [10.37a]. These levels are not populated in supersonic molecular beams because of internal cooling (see Sect.10.1).

c) A particularly interesting and promising field is the application of laser heterodyne techniques to astronomical spectroscopy [10.69]. The radiation from the extraterrestrical source is collected by a telescope and is mixed with the coherent output of a cw infrared laser which serves as a local oscillator. The mixing element is a high-speed infrared photodiode which produces a difference spectrum at radio to microwave frequencies. The spectral analysis is performed electronically in a multichannel filter system with the multiplex advantage of simultaneous detection of many spectral intervals.

This technique has been applied for example to study infrared line profiles in planetary atmospheres. For further applications and more details see [10.69,70].

10.6 Doppler-Free Multiphoton Spectroscopy

In all methods discussed in the previous sections the Doppler width had been reduced or even completely eliminated by proper selection of a *subgroup* of molecules with velocity components $v_z = 0 \pm \Delta v_z$, either through geometrical apertures or by selective saturation. The recently developed technique of Doppler-free multiphoton spectroscopy does not need such a velocity selection because *all* molecules in the absorbing state, regardless of their velocities, can contribute to the Doppler-free transition.

While the general concepts and the transition probability of multiphoton transitions have been already discussed in Sect.8.10, we concentrate in this section on the aspect of *Doppler-free* multiphoton spectroscopy [10.71-75a].

10.6.1 Basic Principle

When a molecule at rest is exposed to two light waves $E_1 = A_1\, e^{i(\omega_1 t - \underline{k}_1 \underline{r})}$ and $E_2 = A_2\, e^{i(\omega_2 t - \underline{k}_2 \underline{r})}$ it may simultaneously absorb two photons, one out of each wave, if the resonance condition

$$E_f - E_i = \hbar(\omega_1 + \omega_2) \tag{10.58}$$

for a molecular transition $E_i \rightarrow E_f$ between two levels E_i, E_f is fulfilled. Assume the molecule moves with a velocity $\underline{v}$ in the laboratory frame. In the reference frame of the moving molecule the frequency ω of an E.M. wave with wave vector $\underline{k}$ is Doppler shifted to (see Sect.3.2)

$$\omega' = \omega - \underline{k} \cdot \underline{v} \quad . \tag{10.59}$$

The resonance condition (10.58) changes now to

$$(E_f - E_i)/\hbar = (\omega_1' + \omega_2') = \omega_1 + \omega_2 - \underline{v} \cdot (\underline{k}_1 + \underline{k}_2) \quad . \tag{10.60}$$

For two light waves with equal frequencies $\omega_1 = \omega_2 = \omega$ which travel in opposite directions we obtain $\underline{k}_1 = -\underline{k}_2$ and (10.60) shows that the Doppler shift of the two-photon transition becomes zero. This means that *all* molecules, independent of their velocities, absorb at the same sum frequency $\omega_1 + \omega_2 = 2\omega$.

Figure 10.49 shows a possible experimental arrangement for the observation of Doppler-free two-photon absorption. The two oppositely travelling waves are formed by reflection of the output beam from a single-mode tunable dye laser. The Faraday rotator prevents feedback into the laser. The two-photon absorption is monitored by the fluorescence emitted from the final state E_f into the initial state E_i or into other states E_m. We show in the next section that the probability of two-photon absorption is proportional to the square of the power density. Therefore the two beams are focussed into the sample cell by the lens L and a spherical mirror M.

Although the probability of a two-photon transition is generally much lower than that of a single-photon transition, the fact that *all* molecules in the absorbing state can contribute to the signal may outweight the lower transition probability and the signal amplitude may even become in favorable cases larger than that of saturation signals.

The considerations above can be generalized to many photons. When the moving molecule is simultaneously interacting with several plane waves with wave vectors $\underline{k}_i$, the total Doppler shift is $\underline{v} \cdot \sum_i \underline{k}_i$ and becomes zero if the condition

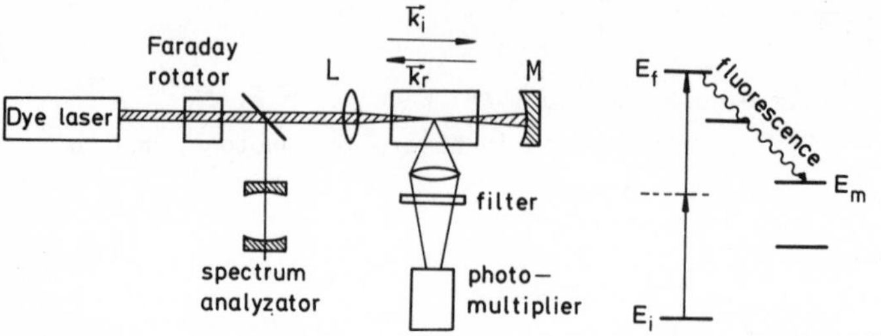

Fig.10.49. Experimental arrangement for the observation of Doppler-free two-photon absorption

$$\sum_i \underline{k}_i = 0$$

is fulfilled.

Another way of interpreting this result is based on momentum conservation. If the total momentum $\sum \hbar \underline{k}_i$ of all photons absorbed by the molecule is equal to zero, the velocity of the molecule does not change during the absorption. This means that its kinetic energy remains constant and all of the photon energy is transferred into internal energy. The corresponding equation

$$E_f - E_i = \sum_i c\hbar|\underline{k}_1| = \sum_1 \hbar\omega_i \tag{10.61}$$

is equivalent to the elimination of Doppler broadening.

Since the total photon momentum is zero, there is no recoil of the absorbing atom. It should be noted, however, that only the first-order Doppler effect is eliminated, not the second-order shift. Since the magnitude of this second-order effect is $(E_f - E_i)(v^2/2c^2)$, it can be neglected in most cases.

Figure 10.50 shows a possible optical arrangement for Doppler-free three-photon spectroscopy. The three beams, which are generated from the dye laser beam by a beam splitter and reflectors, form in the crossing point angles of 120° between each other.

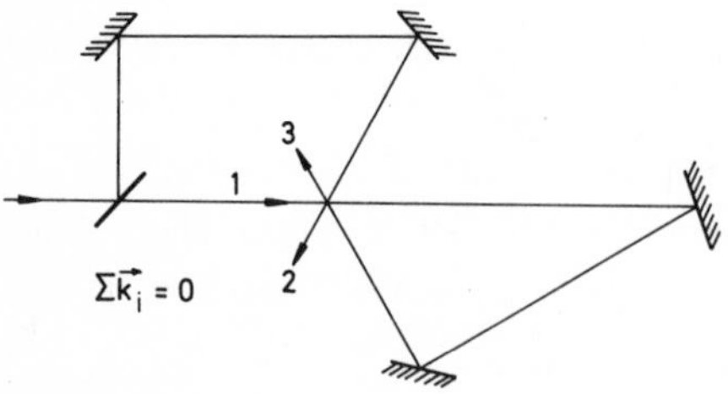

Fig.10.50. Possible arrangement for Doppler-free three-photon spectroscopy

10.6.2 Line Profiles of Two-Photon Transitions

As has been discussed in Sect.8.10, the line profile $g(\omega_{if} - \omega)$ of a two-photon transition with $\omega = \omega_1 + \omega_2$ is mainly determined by the first factor in (8.43),

$$g(\omega_{if} - \omega) \propto \frac{\gamma_{if}}{[\omega_{if} - \omega_1 - \omega_2 - \underline{v}(\underline{k}_1 + \underline{k}_2)]^2 + (\gamma_{if}/2)^2} \ . \qquad (10.62)$$

The linewidth depends on the relative orientation of the two wave vectors $\underline{k}_1$ and $\underline{k}_2$. In the experimental arrangement of Fig.10.49 the two vectors are either parallel ($\underline{k}_1 = \underline{k}_2$ if the two photons come from the same wave) or anti-parallel ($\underline{k}_1 = -\underline{k}_2$) if one photon comes from the incident wave, the other from the reflected wave). While in the second case the Doppler shift becomes $\underline{v}(\underline{k}_1 + \underline{k}_2) = 0$, it is maximum in the first case. Integrating over all molecular velocities leads to a Doppler-broadened line profile for the two-photon transition with parallel wave vectors which equals that of a single-photon transition at the frequency 2ω.

If the reflected beam $(I, \omega, -\underline{k}, \underline{e})$ has the same intensity I and the same polarization $\underline{e}$ as the incident beam $(I, \omega, +\underline{k}, \underline{e})$, the two numerators of the second factor in (8.43) become identical and (8.43) reduces to

$$A_{if} \propto \left[\frac{4\gamma_{if}}{(\omega_{if} - 2\omega)^2 + (\gamma_{if}/2)^2} + \frac{\gamma_{if}}{(\omega_{if} - 2\omega - 2\underline{k}\cdot\underline{v})^2 + (\gamma_{if}/2)^2} + \frac{\gamma_{if}}{(\omega_{if} - 2\omega + 2\underline{k}\cdot\underline{v})^2 + (\gamma_i/2)^2}\right] \cdot \left|\sum_k \frac{\underline{R}_{ik}\,\underline{e}\,\underline{R}_{kf}\,\underline{e}}{\omega - \omega_{ik}}\right|^2 I^2 \ . \qquad (10.63)$$

Here the first term describes the cases where the two photons come from different waves, in the second they both come from the incident wave, and in the third term, from the reflected wave. Integration over all velocities yields the Lorentzian Doppler-free profile from the first term and the Doppler-broadened background from the second and third terms.

When both waves have the same polarization, the probability that the two photons come from different waves, is just twice as large as the probability that both come from the same wave. This can be understood as follows: In the latter case the two photons can come either from the incident (a a) or from the reflected wave (b b) and the total probability is the sum $(a\ a)^2 + (b\ b)^2$ of two independent events. The total intensity of the Doppler-broadened background is therefore twice the intensity of a two-photon absorption in a travelling wave. For the Doppler-free signal, however, the two possible processes (ab) or (ba) are indistinguishable and the total probability $(ab + ba)^2$ is the square over the sum of probability amplitudes.

The total intensity of the Doppler-free signal is therefore four times as large as a two-photon absorption in a travelling wave. This means that the area under the Doppler-broadened background is one half of that under the Doppler-free two-photon signal (see Fig.10.51). When the homogeneous width of the Doppler-free Lorentzian profile is γ, its magnitude therefore exceeds that of the background by a factor $2\delta\omega_D/\gamma$. With $\gamma = 10$ MHz and $\delta\omega_D = 1000$ MHz this factor becomes 200! The background can be often completely eliminated by choosing the right kind of polarization (see below). In most cases we can therefore regard the first term in (10.63) to be the dominant one and can neglect the two other contributions. For the resonance case $\omega_{if} = 2\omega$ the first term is equal to 1 and we obtain for the total two-photon transition probability.

$$\boxed{A_{if} \propto \left| \sum_k \frac{\underline{R}_{ik}\,\underline{e}\;\underline{R}_{kf}\,\underline{e}}{\omega - \omega_{ik}} \right|^2 I^2} \quad . \qquad (10.64)$$

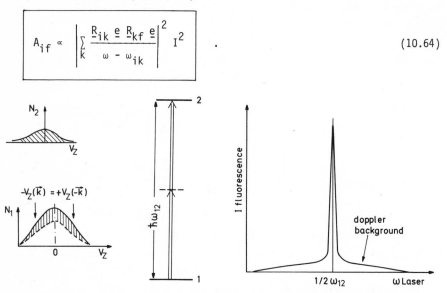

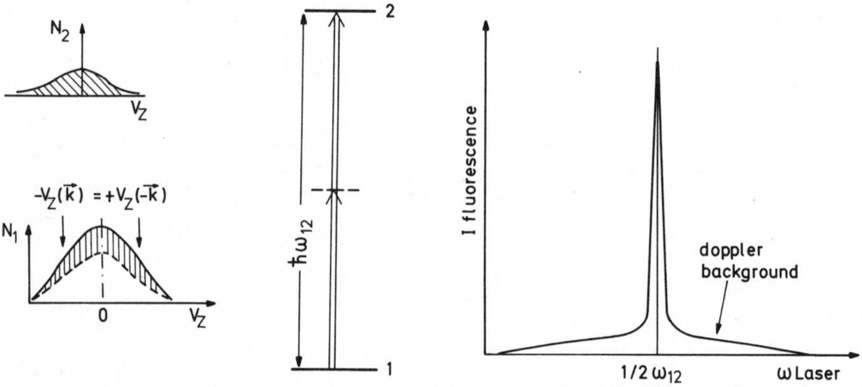

Fig.10.51. Schematic illustration of Doppler-free two-photon signal with Doppler-broadened background. The background amplitude is greatly exaggerated

For two-photon transitions with $\Delta M = 0$ the Doppler-broadened background can be eliminated by a proper choice of the laser polarization. If the incident laser wave has a σ^+ circular polarization, the wave reflected at M_2 in Fig.10.49 has σ^- polarization. Transitions with $\Delta M = 0$ (e.g., $s \rightarrow s$ transitions) can only be induced by two photons from different beams since two photons from the same beam always induce transitions with $\Delta M = +2$.

Let us briefly consider how focussing of the two laser beams into the sample affects the magnitude of the two-photon absorption signal. Assume that the beams are travelling into the $\pm z$ direction. The two-photon absorption is monitored through the fluorescence emitted from the excited level E_f, which can be collected from a total length $\Delta z = 2L$. The two-photon absorption signal S_{if} is then (Fig.10.52)

$$S_{if} = \int_{z=-L}^{+L} \int_{r=0}^{\infty} A_{if}\, n(E_i) 2\pi r\, dr\, dz \tag{10.65}$$

$$\propto \left| \sum_k \frac{\underline{R}_{ik}\, \underline{e}\, \underline{R}_{kf}\, \underline{e}}{\omega - \omega_{ik}} \right|^2 n(E_i) \int_{z=-L}^{+L} \int_{r=0}^{\infty} r I^2(z) dr\, dz \quad ,$$

where $n(E_i)$ is the density of molecules in the absorbing level E_i, $w(z)$ is the radius of the Gaussian beam in the vicinity of the focus, and $I(z)$ is the intensity [W/cm^2] which of course depends on the beam waist. The beam waist of a Gaussian beam (see Sect.5.11)

$$w(z) = w_0 [1 + (\lambda z/\pi w_0^2)^2]^{1/2}$$

is approximately constant over the Rayleigh length $\Delta z = \pi w_0^2/\lambda$. With the radial intensity distribution

$$I(r) = \frac{2I_0}{\pi w^2} e^{-(2r^2/w^2)} \tag{5.14}$$

we see that an optimum signal is achieved, when the integral

$$\int_{-L}^{+L} \frac{dz}{1 + (\lambda z/\pi w_0^2)^2}$$

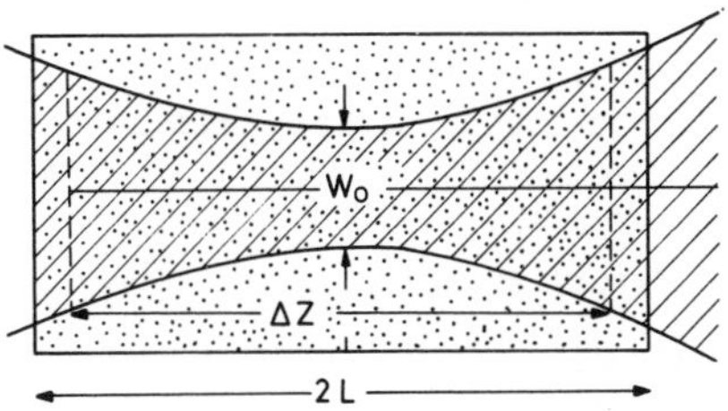

Fig.10.52. Optimum focussing with $\Delta z = 2L$ gives maximum signals in two-photon spectroscopy

becomes maximum. Although w_0 and w decreases with decreasing focal length f of the focussing lens, it is not worthwhile to decrease w_0 below the value which gives a Rayleigh length Δz equal to the observation region L. For tighter focussing the value of the integral increases only slightly but time-of-flight broadening, power broadening, and light shifts may deteriorate the line shape of the two-photon absorption signal. We may therefore regard a focal length f which gives a Rayleigh length $\Delta z = L$ as an optimum choice.

10.6.3 Examples

The first experiments on Doppler-free two-photon absorption have been performed on the 3S → 5S and 3S → 4D transitions in sodium atoms [10.74-76a]. The energy defect $\hbar(\omega - \omega_{ik})$ to the intermediate 3P level is small and therefore the transition probability fairly large. Figure 10.53 illustrates the hyperfine components of the 3S-5S transition and Fig.10.54 the Zeeman splitting of the hfs lines in the 3S-4D transition, measured by BIRABEN et al. [10.74].

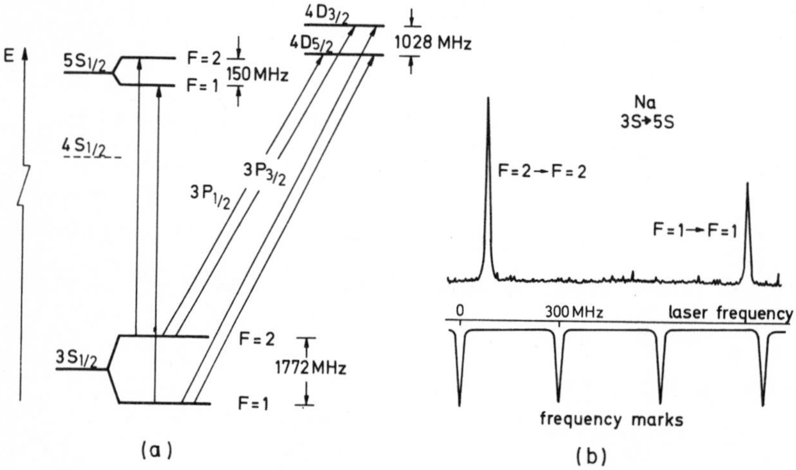

Fig.10.53. (a) Level scheme of the Na atom. The hfs splitting is not to scale. (b) Two-photon transitions 3s → 5s with frequency markers from an F.P.I. with d = 25 cm [10.74]

The S → S and S → D transitions of the different alkali atoms have been most thoroughly investigated. Fine structure splittings of high-lying D states, hfs splittings of S states, isotope shifts, and Stark splittings have been measured to a high precision. Of particular interest are the high-lying Rydberg levels which can be excited by two- or three-photon

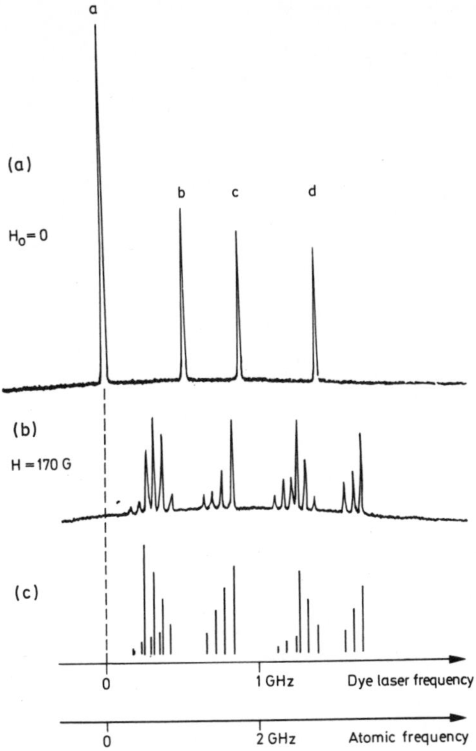

Fig.10.54a-c. Zeemann effect of the 3S-4D transition in Na. (a) hfs with zero magnetic field, (b) Zeeman pattern at 170 Gauss, (c) calculated positions and intensities of the Zeeman components for $\Delta M = 2$ transitions [10.74]

transitions. For example the rubidium levels were measured up to n = 50 with better than 1×10^{-7} absolute accuracy [10.77]. Because of the relatively long lifetimes of the high levels the linewidth of the two-photon transition is very small. One obtains, for example, for the $30^2S_{1/2}$ level in rubidium a lifetime of $\sim 25\ \mu s$ and a linewidth of 6 kHz. Such narrow lines may become precise reference wavelengths for dye lasers.

A very interesting application of Doppler-free two-photon spectroscopy was demonstrated by HÄNSCH et al. [10.78]. The frequency ω of a nitrogen laser pumped dye laser at λ = 486 nm was frequency doubled in a lithium formiate monohydrate crystal. The fundamental dye laser output was used to perform saturation spectroscopy of the Balmer β line in a hydrogen discharge cell (see Fig.10.55), while the frequency-doubled output at λ = 243 nm simultaneously excited a two-photon transition 1S $\rightarrow$ 2S in a second flow cell where H atoms are produced in a gas discharge. The two-photon transition was observed through the collision-induced 2P-1S fluorescence at λ = 121.5 nm. Without the Lamb shift the frequency of the 1S-2S transition should be four

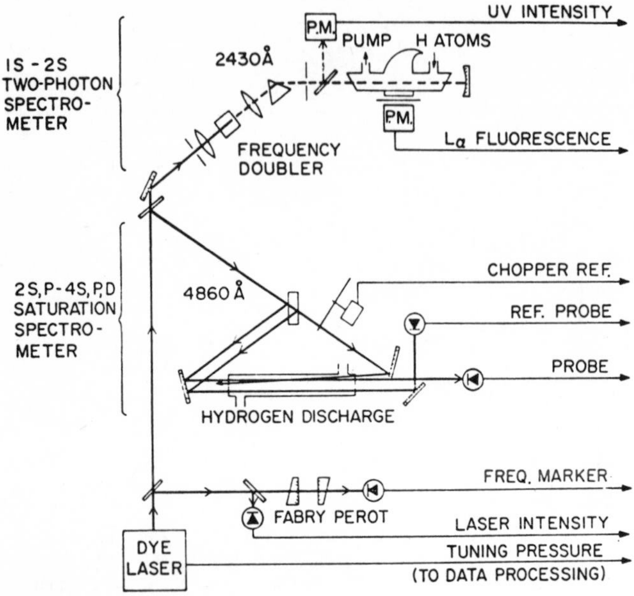

Fig.10.55. Experimental setup for the determination of hydrogen 1S Lamb shift by combination of saturation and two-photon spectroscopy [10.78]

times that of the Balmer β line. From the frequency difference $\Delta\omega = \left[\frac{1}{4}\,\omega(1S_{1/2} \rightarrow 2S_{1/2}) - \omega(2S_{1/2} \rightarrow 4P_{1/2})\right]$ the Lamb shift of the 1S state could be determined (Fig.10.56a). Figure 10.56b shows the resolution of the hyperfine splitting of the 1S-2S transition in hydrogen while the hfs components of the deuterium line D(1S-2S) could not be resolved. The frequency markers from a Fabry-Perot allow one to measure the isotope shift of the D(1S-2S) transitions against the H(1S-2S) line. This yields a very precise determination of the Rydberg constant [10.78a].

Two-photon spectroscopy has already been applied to a number of molecules, such as C_6H_6, NO, NH_3, CH_3F, and Na_2. For the larger molecules the selective excitation of excited states with definite symmetry determined by the selection of the polarization of the two laser beams is helpful for the assignment of these states [10.80].

With three-photon spectroscopy, states with the same parity are reached as in single-photon transitions. The advantage is, however, that transitions in the vacuum-ultraviolet region can be excited with visible light [10.79]. Since single-photon laser spectroscopy in this region is impeded by the lack of suitable and easy to handle laser systems, the application of nitrogen laser-pumped dye lasers to Doppler-free three-photon spectroscopy will cer-

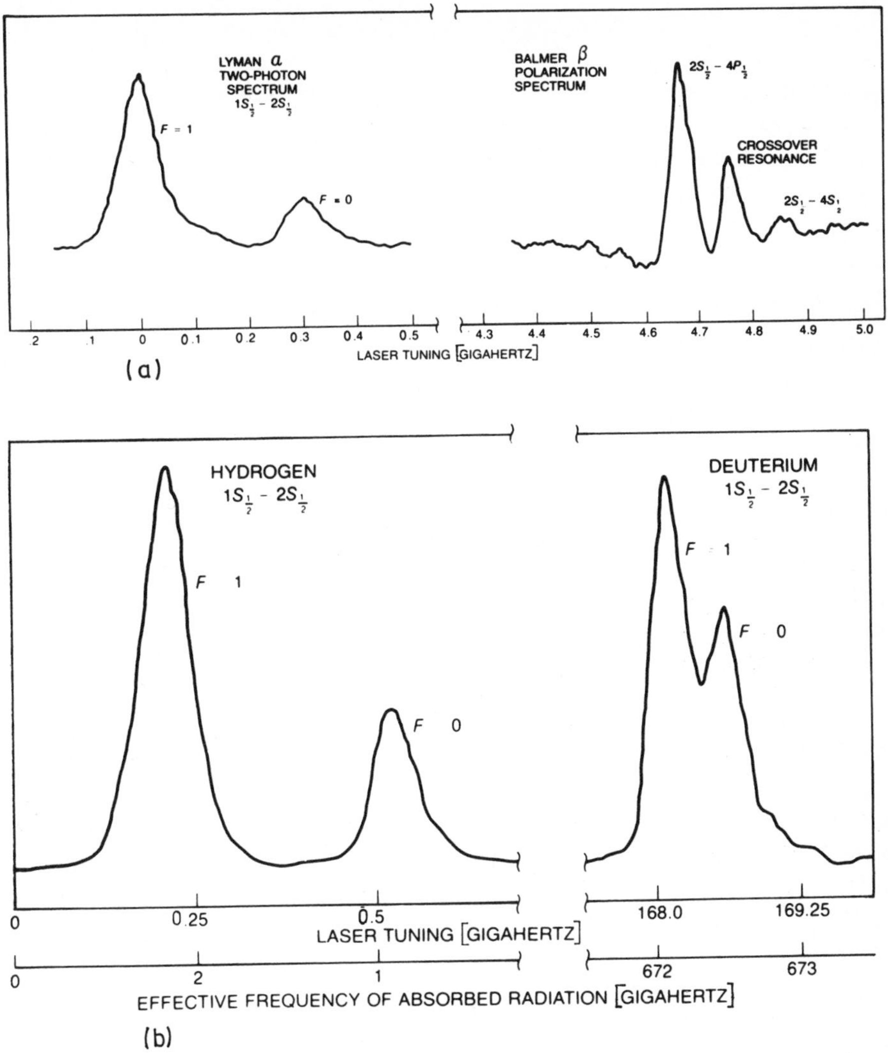

Fig.10.56. (a) Simultaneous measurement of the 1S → 2S transition and the Balmer β line ($2S - 4P_{1/2}$) of the H atom. (b) Hfs of the $1S_{1/2} \rightarrow 2S_{1/2}$ transition in hydrogen and deuterium, resolved with two-photon spectroscopy [10.31a]

tainly open a whole new area of investigations in the far ultraviolet range. More examples can be found in the recent excellent review by GRYNBERG and CAGNAC [10.74,75a] or in [10.70,80].

10.7 Level-Crossing Spectroscopy with Lasers

Level-crossing spectroscopy is a well-known Doppler-free technique which has been used for many years in atomic spectroscopy, even before the invention of lasers [10.81]. However, in the pre-laser era most investigations had been restricted to atomic resonance transitions which can be excited by strong atomic resonance lamps. Only a few molecules have been studied with this technique, using fortuitous coincidences between molecular transitions and atomic resonance lines [10.82]. Optical pumping with tunable lasers or with one of the numerous lines of fixed frequency lasers has greatly enlarged the possibilities of level-crossing spectroscopy particularly in *molecular* physics. Lasers furthermore allow new variations of this technique, such as stimulated level crossing.

In this section we briefly summarize the fundamentals of level-crossing spectroscopy and illustrate its relevance for the investigation of angular momentum coupling schemes in excited molecular states. A more detailed presentation of the theory can be found in [8.53] and a survey on the pre-laser work in [8.52].

10.7.1 Basic Principle

The experimental arrangement of level-crossing spectroscopy is shown in Fig.10.57. The sample of atoms or molecules is placed inside a uniform magnetic field B and is illuminated by polarized light. The fluorescence emitted from the excited levels is monitored through a polarizer as a function of the magnetic field. From the shape of the measured signal $I_{Fl}(B)$ the product $g\tau$ of lifetime τ and Landé factor g of the upper state can be derived. For an explanation of the effect, let us start with a simple classical picture which gives a clear physical insight into the basic phenomenon. We regard the special case of zero-field level crossing (Hanle effect [10.83]). Assume that an atomic electron has been excited from a lower level 1 into an upper level 2 at the time $t = t_0$, by absorption of polarized light. We choose the x axis as the propagation of the incident light with its electric vector pointing into the y direction. The excited electron oscillates at the angular frequency $\omega_{12} = (E_2 - E_1)/\hbar$ in a direction specified by the polarization direction of the incident radiation. The oscillating electron looses its energy by radiation damping (see Sect.3.1) and produces the well-known dipole field pattern (Fig.10.58) with a time-dependent field amplitude

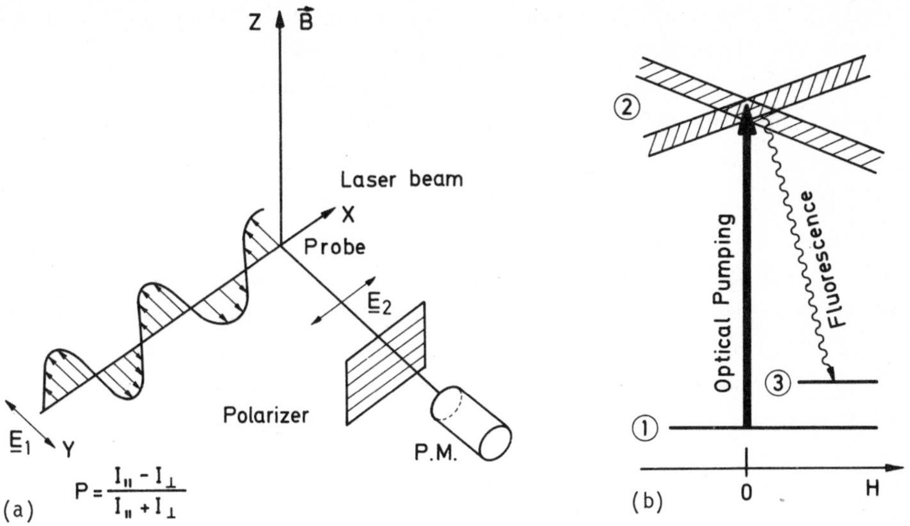

Fig.10.57. (a) Schematic experimental arrangement for level-crossing spectroscopy; (b) energy level scheme

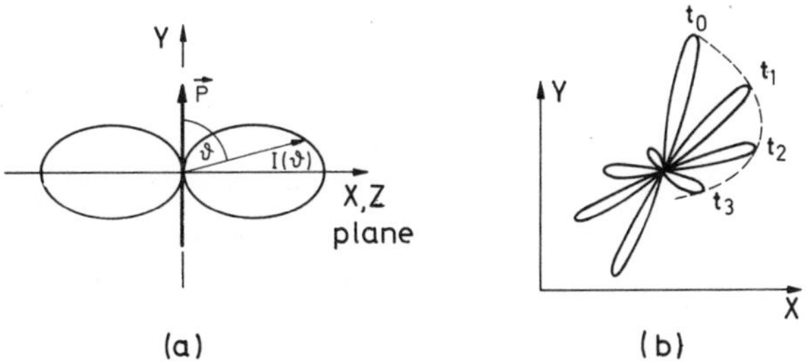

Fig.10.58. (a) Spatial distribution of the radiation emitted by a classical dipole oscillator with dipole moment p; (b) intensity distribution at different times t, due to the precession of p in an external magnetic field parallel to the z direction

$$E(t) = E(0)\, e^{-(i\omega_{12}-\gamma/2)(t-t_0)}\hat{y} \quad . \tag{10.66}$$

When an external magnetic field B is applied in the z direction, the oscillating electron experiences a Lorentz force $e(\underline{v} \times \underline{B})$ which causes the plane of oscillation to precess about the field direction with the Larmor angular frequency

$$\omega_L = g_J \mu_0 B/\hbar \quad , \tag{10.67}$$

where g_J is the Landé factor and μ_0 the Bohr magneton. The axis of the emitted dipole radiation pattern of course precesses in the same way.

An observer, looking through a linear polarizer with a transmission axis at an angle α against the y direction, receives the intensity $I \propto E E^*$

$$I(B,\alpha,t) = I_0 \, e^{-\gamma(t-t_0)} \cos^2[\omega_L(t - t_0) - \alpha] \quad , \tag{10.68}$$

where I_0 is the intensity emitted at $t = t_0$. The detector averages over the rapid light oscillations ω_{12} (see Sect.2.4). This intensity may be observed either time resolved, following pulsed excitation, or under steady-state conditions where the sample is irradiated at a constant rate R and the intensity at time t is due to the excitation during the whole time from $t_0 = -\infty$ to the time of observation t. In the latter case we obtain

$$I(B,\alpha) = R \int_{t_0=-\infty}^{t} I_0 \, e^{-\gamma(t-t_0)} \cos^2[\omega_L(t - t_0) - \alpha] dt_0 \quad . \tag{10.69}$$

Using the relation $2\cos^2 x = 1 + \cos 2x$, we can solve the integral and obtain

$$I(B,\alpha) = \frac{I_0 R}{2}\left(\frac{1}{\gamma} + \frac{\gamma \cos 2\alpha}{\gamma^2 + 4\omega_L^2} + \frac{2\omega_L \sin 2\alpha}{\gamma^2 + 4\omega_L^2}\right) \quad . \tag{10.70}$$

The shape of the Hanle signal (10.70) depends on the orientation of the polarizer. For $\alpha = \pi/4$, for example, the signal shows a dispersion profile, while for $\alpha = 0$ or $\alpha = \pi/2$ (polarizer in the y or x direction) the signal has a Lorentzian profile. With (10.67) we obtain from (10.70) with $\alpha = \pi/2$ (polarizer in the x direction)

$$I(B,\alpha = \pi/2) = \frac{I_0 R}{2\gamma}\left[1 - \frac{1}{1 + \left(\frac{2g\mu_0 B}{\hbar\gamma}\right)^2}\right] \quad . \tag{10.71}$$

The full halfwidth of the Lorentzian signal is

$$\Delta B_{1/2} = \frac{\hbar\gamma}{g\mu_0} = \frac{\hbar}{g\tau\mu_0} \quad \text{with} \quad \gamma = 1/\tau \quad . \tag{10.72}$$

From the measured full halfwidth $\Delta B_{1/2}$ the product of lifetime τ times Landé factor g can be determined. If the Landé factor g is known, the level-crossing method gives the effective lifetime of the excited level. For many excited *molecular* levels the angular momentum coupling scheme is not known, in particular if hyperfine splittings complicate the situation and the total

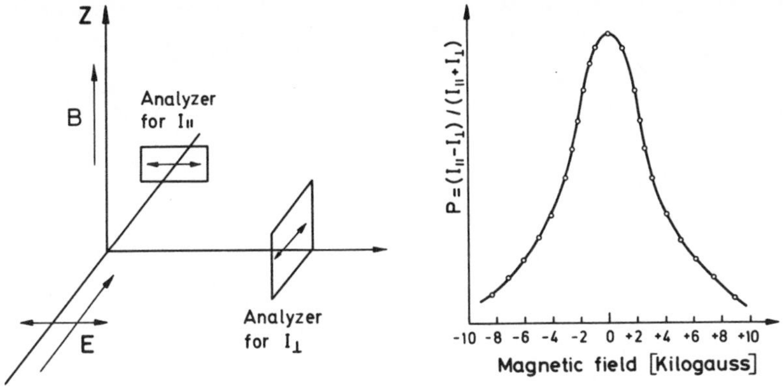

Fig.10.59. Hanle signal in the fluorescence from the laser-excited ($v' = 10$, $J' = 12$) level in the $B^1\Pi_u$ state of Na_2 [10.84]

angular momentum F is the sum of nuclear spin I, and molecular angular momentum J, composed of molecular rotation N, electronic spin S, and electronic angular momentum L. If the lifetime τ has been measured with other methods, the level-crossing signal allows determination of the Landé factor and therefore gives information about the coupling scheme. Figure 10.59 illustrates a Hanle signal obtained from the laser excited level ($v' = 10$, $J' = 12$) in the $B^1\Pi_u$ state of the Na_2 molecule. The Landé factor of this level is, according to Hund's coupling case a [10.84],

$$g_f = \frac{1}{J(J+1)} \; \frac{F(F+1) + J(J+1) - I(I+1)}{2F(F+1)} \quad .$$

This shows that the Landé factor decreases with increasing rotational quantum number J, demanding higher magnetic fields.

In a quantum mechanical treatment the intensity of the fluorescence emitted on a transition $2 \to 3$ with polarization vector $\underline{E}_2$ following excitation on a transition $1 \to 2$ by optical pumping with linearly polarized light (polarization vector $\underline{E}_1$) is described by the product of the two corresponding matrix elements

$$I_{F1}(2 \to 3) \propto |\langle 1|\underline{\mu}_{12} \cdot \underline{E}_1|2\rangle|^2 \cdot |\langle 2|\underline{\mu}_{23} \cdot \underline{E}_2|3\rangle|^2 \quad . \tag{10.73}$$

The spatial distribution of the fluorescence intensity is determined by the x, y, z components of the matrix elements which depend on the polarization $\underline{E}_1$ of the pump wave and on the orientations of the transition moments $\underline{\mu}_{12}$ and $\underline{\mu}_{23}$ with respect to $\underline{E}_1$ [10.85].

An excited level with angular momentum quantum number J is composed of (2J + 1) magnetic sublevels with quantum number M, which are degenerate in the absence of an external field. The wave function

$$\psi_2 = \sum_M c_M \psi_{2,M} \exp(-i\omega_{2M}t) \tag{10.74}$$

can be represented by a linear combination of the sublevel functions $\psi_{2,M}$. The matrix elements in (10.73) therefore contain interference terms $c_{M_1} c^*_{M_2} \psi_{2M_1} \psi^*_{2M_2} \exp[-i(\omega_{2M_1} - \omega_{2M_2})t]$.

Without external fields the frequencies ω_{2M} are all equal which means that the interference terms become *time independent*. *These interference terms determine the spatial distribution of the fluorescence*. If the degeneracy is removed by an external field, the sublevels split, the frequencies ω_{2M} become different, and the phase factors are time dependent. Even if all sublevels have been excited coherently (which means *in phase*) their phases develop at a different rate and the interference pattern, resulting from the superposition of the different contributions, is washed out, due to the rapidly varying phase relations. The spatial distribution of the emitted fluorescence becomes isotropic for $2g_J\mu_0 B/\hbar \gg \gamma$. Although the total fluorescence intensity is not altered by the magnetic field, its spatial distribution and its polarization characteristics are changed.

In the case of Zeeman sublevels of a degenerate J level, the M levels cross each other at B = 0 when the field is tuned from negative to positive field values. This is called *zero field level crossing* or *Hanle effect*. In the general case Zeeman sublevels of different fine structure or hyperfine structure levels may cross at $B \neq 0$.

The quantum mechanical description may be illustrated by a specific example, taken from [10.86], where optical pumping on a transition (J",M") → (J',M') is achieved by linearly polarized light with a polarization vector in the x, y plane, perpendicular to the field direction along the z axis. The linearly polarized light may be regarded as a superposition of left- and right-hand circularly polarized components σ^+ and σ^-. The excited state wave function for excitation by light polarized in the x direction is (Fig.10.60)

$$|2\rangle_x = (-1/\sqrt{2})[a_{M+1}|M+1\rangle + a_{M-1}|M-1\rangle] \tag{10.75}$$

and for excitation with y polarization

$$|2\rangle_y = (-i/\sqrt{2})[a_{M+1}|M+1\rangle - a_{M-1}|M-1\rangle] \quad . \tag{10.76}$$

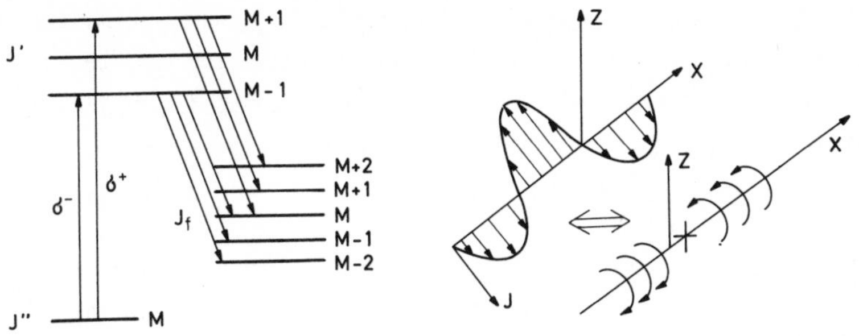

Fig.10.60. Level scheme for optical pumping of Zeeman levels by light linearly polarized in y direction. Each of the two excited Zeeman levels can decay into three final state sublevels. Superposition of the two different routes $(J'',M) \rightarrow (J_f,M)$ generates interference effects

The coefficients $a_{M'}$ are proportional to the matrix element of the transition $(J'',M'') \rightarrow (J',M')$. The optical pumping process generates a coherent superposition of eigenstates $(M' + 1)$ and $(M' - 1)$ as long as the spectral width of the pump radiation is broader than the level splitting.

The time development of the excited state wave function is described by the time-dependent Schrödinger equation

$$(-\hbar/i)\,\frac{\partial\psi_2}{\partial t} = H\psi_2 \tag{10.77}$$

where the operator H has the eigenvalues $E_\mu = E_0 + \mu_0 gMB$. Including the spontaneous emission in a semiclassical way (see Sect.2.9), we may write the solution of (10.77) as

$$\psi_2(t) = e^{-(\gamma/2t)}\, e^{-iHt}\psi_2(0) \tag{10.78}$$

where the operator $\exp(-iHt)$ is defined by its power series expansion. From (10.75-78) we find for excitation with y-polarization

$$\begin{aligned}\psi_2(t) &= \frac{-i}{\sqrt{2}}\, e^{-(\gamma/2)t}\, e^{-iHt/\hbar}\left[a_{M+1}|M+1\rangle - a_{M-1}|M-1\rangle\right] \\ &= \frac{-i}{\sqrt{2}}\, e^{-(\gamma/2)t}\, e^{-i(E_0+\mu_0 gMBt/\hbar)}\left[a_{M+1}\, e^{-i\mu_0 gBt/\hbar}|M+1\rangle\right. \\ &\qquad \left. - a_{M-1}\, e^{i\mu_0 gBt/\hbar}|M-1\rangle\right] \qquad (10.79) \\ &= e^{-(\gamma/2)t}\, e^{-i(E_0+\mu_0 gMBt/\hbar)}\left[|\psi_2\rangle_x \sin\mu_0 gBt/\hbar + |\psi_2\rangle_y \cos\mu_0 gBt/\hbar\right] \;.\end{aligned}$$

This shows that the excited state wave function under the influence of the magnetic field B changes continuously from $|\psi_2\rangle_x$ to $|\psi_2\rangle_y$ and back. If the fluorescence is detected through a polarizer with transmission axis parallel to the x axis, only light from the $|\psi_2\rangle_x$ component is detected. The intensity of this light is

$$I(E_x,t) = C\, e^{-\gamma t} \sin^2(\mu_0 gBt/\hbar) \tag{10.80}$$

which gives after integration over all time intervals from $t = -\infty$ to the time t of observation, the Lorentzian intensity profile observed in the y direction behind a polarizer in the x direction (Fig.10.59)

$$I_y(E_x) = \frac{C\tau}{2} \frac{(2\mu_0 g\tau B/\hbar)^2}{1 + (2\mu_0 g\tau B/\hbar)^2} \,, \tag{10.81}$$

which turns out to be identical with the classical result (10.71).

10.7.2 Experimental Realization and Examples of Level-Crossing Spectroscopy with Lasers

Level-crossing spectroscopy has some definite experimental advantages. Compared with other Doppler-free techniques it demands a relatively simple experimental arrangement. Neither single-mode lasers and frequency stabilization techniques nor collimated molecular beams are required. The experiments can be performed in simple vapor cells and the experimental expenditure is modest. In many cases no monochromator is needed since sufficient selectivity in the excitation process can be achieved to avoid simultaneous excitation of different molecular levels with a resulting overlap of several level-crossing signals. The width $\Delta B_{1/2}$ of the Hanle curve is determined by the effective lifetime τ_{eff} of the excited level [see (10.72)]. At a pressure p we obtain from $1/\tau_{eff} = 1/\tau_{sp} + (8/\pi\mu kT)^{\frac{1}{2}} p\sigma$ (see Sect.2.7.1)

$$\Delta B_{\frac{1}{2}} = \left[\frac{1}{\tau_{sp}} + \frac{4}{\sqrt{\pi M kT}}\, p\sigma\right] \frac{1}{\mu_0 g_J} \,, \quad M = \text{molecular mass} \,, \tag{10.82}$$

where σ is the cross section for destruction of the coherence between the Zeeman sublevels. The sign of the dispersion-shaped Hanle curve gives the sign of the Landé factor (10.70).

There are of course also some disadvantages. One major problem is the change of absorption profile with magnetic field. The laser bandwidth must be sufficiently large in order to assure that all Zeeman components can absorb the laser radiation independent of the field strength B. On the other hand, the laser bandwidth should not be too large, to avoid simultaneous excitation of different closely spaced transitions. This problem arises particularly in *molecular* level crossing where often several molecular lines overlap within their Doppler widths. In such cases a compromise has to be found for an intermediate laser bandwidth, and the fluorescence has to be monitored through a monochromator to discriminate against other transitions. Because of the high magnetic fields required for Hanle signals from short-lived molecular levels with small Landé factors, careful magnetic shielding of the photomultiplier is essential to avoid a variation of the multiplier gain factor with magnetic field strength.

The level-crossing signal may be only a few percent of the field-independent background intensity. In order to improve the signal-to-noise ratio, either the field is modulated or the polarizer in front of the detector is made to rotate and the signal is recovered by lock-in detection.

Instead of magnetic fields electric fields can also be used to achieve crossings of Stark sublevels. From the width of the electric level-crossing signal DALBY et al. [10.87] derived the anisotropic polarizability of the iodine molecule which had been excited by an argon laser into a (v',J') level of the $B^3\Pi_0^+$ state.

The iodine molecule has been very thoroughly studied with electric and magnetic level-crossing spectroscopy. The hyperfine structure of the rotational levels affects the profile of the level-crossing curves [10.88]. A computer fit to the non-Lorentzian superposition of all Hanle curves from the different hfs levels allows simultaneous determination of Landé factor g and lifetime τ [10.89]. Because of different predissociation rates the effective lifetimes of different hfs levels differ considerably.

In larger molecules the phase coherence time of excited levels may be shorter than the population lifetime because of perturbations between closely spaced levels of different electronic states, which cause a dephasing of the excited level wave functions. One example is the NO_2 molecule where the width of the Hanle signal turns out to be more than one order of magnitude larger than expected from independent measurements of population-lifetime and Landé factors [10.90].

A large number of atoms and molecules have been meanwhile investigated by level-crossing spectroscopy using laser excitation. Because of the available high laser intensity highly excited states can be studied, which have been populated by stepwise excitation (see Sect.8.8). Often resonance lamps are used to excite the first resonance level and a dye laser pumps the next step. Many experiments have been performed with two different dye lasers, either in a pulsed or a cw mode [10.91]. These techniques allow measurement of natural linewidth, fine structure, and hfs parameters of high-lying Rydberg states. Most experiments have been performed on alkali atoms. A compilation of measurements up to 1975 can be found in the review of WALTHER [1.12].

Level-crossing experiments with time-resolved detection following pulsed excitation may even allow a spectral resolution within the natural linewidth. If only those fluorescence photons are detected which have been emitted at times $t > a\tau$ after the excitation process ($a \gg 1$) the spectral profile of the signal is narrowed [10.92]. This technique allows one to reach a spectral resolution beyond the natural linewidth (see Sect.13.5).

10.7.3 Stimulated Level-Crossing Spectroscopy

So far we have considered level crossing monitored through the spontaneous emission. A level-crossing resonance can manifest itself also as a change in absorption of an intense monochromatic laser wave tuned to the molecular transition. The physical origin of this stimulated level-crossing spectroscopy is based on saturation effects and may be illustrated by a simple example [10.93].

Consider a molecular transition between two levels a and b with angular momentum J = 1 and J = 0 (Fig.10.61). We denote the center frequencies of

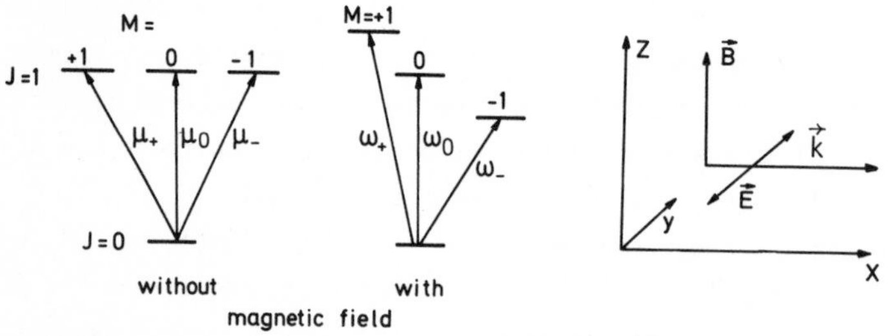

Fig.10.61. Stimulated level crossing

the $\Delta M = +1, 0, -1$ transitions by ω_+, ω_0, and ω_- and the corresponding matrix elements by μ_+, μ_0, and μ_-. Without external fields the M sublevels are degenerate and $\omega_+ = \omega_- = \omega_0$. A monochromatic laser $E = E_0 \cos(\omega t - kx)$, linearly polarized in the y direction, induces transitions with $\Delta M = 0$ without external field. The saturated absorption of the laser wave is then, according to (3.81b),

$$\alpha_s^0(\omega) = \frac{\alpha_0^0(\omega_0)}{\sqrt{1 + S_0^0}} \; e^{-[(\omega-\omega_0)/\Delta\omega_D]^2} \quad , \tag{10.83}$$

where $\alpha_0 = (N_a - N_b)|\mu|^2\omega/(\hbar c\gamma)$ is the unsaturated absorption coefficient and $S_0 = E_0^2|\mu|^2/(\hbar\gamma^2)$ is the saturation parameter (see Sect.3.6).

If an external electric or magnetic field is applied in the z direction, the laser wave, polarized in the y direction, induces transitions $\Delta M = \pm 1$ because it can be composed of $\sigma^+ + \sigma^-$ contributions (see previous section). If the level splitting $\hbar(\omega_+ - \omega_-) >> \gamma$, the absorption coefficient is now the sum of two contributions,

$$\alpha_s = \frac{\alpha^+}{\sqrt{1 + S_0^+}} e^{-[(\omega-\omega_+)/\Delta\omega_D]^2} + \frac{\alpha_0^-}{\sqrt{1 + S_0^-}} e^{-[(\omega-\omega^-)/\Delta\omega_D]^2} . \qquad (10.84)$$

For a $J = 1 \rightarrow 0$ transition $|\mu_+|^2 = |\mu_-|^2 = |\mu_0|^2/2$. Neglecting the difference $(\omega_+ - \omega_-) << \Delta\omega_D$ we may approximate (10.84) by

$$\alpha_s^+ = \frac{\alpha_0^0}{\sqrt{1 + \frac{1}{2} S_0}} e^{-[(\omega-\omega_0)/\Delta\omega_D]^2} , \qquad (10.85)$$

which differs from (10.83) by the factor 1/2 in the denominator. This demonstrates that the effect of the level splitting on the absorption appears only in the saturated absorption and disappears for $S \rightarrow 0$.

The advantage of stimulated versus spontaneous level crossing is the larger signal-to-noise ratio and the fact, that also level crossings in the ground state can be detected. Most experiments on stimulated level crossing are based on intracavity absorption techniques, because of their increased sensitivity (see Sect.8.2.3). LUNTZ and BREWER [10.94] demonstrated that even such small Zeeman splittings as occur in molecular $^1\Sigma$ ground states can be precisely measured. They used a single-mode He-Ne laser oscillating on the 3.39 μm line which coincides with a vibration-rotation transition in the $^1\Sigma$ ground state of CH_4. Level crossings were detected as resonances in the laser output when the CH_4 transition was tuned by an external magnetic field. The rotational magnetic moment of the $^1\Sigma$ state of CH_4 was measured to be $0.36 \pm 0.07\ \mu_N$. Also Stark-tuned level-crossing resonances in the excited vibrational level of CH_4 have been detected with this method [10.95].

A number of stimulated level-crossing experiments have been performed on the active medium of gas lasers where the gain of the laser transition is changed when sublevels of the upper or lower laser level cross each other. The whole gain tube is for instance placed in a longitudinal magnetic field and the laser output is observed as a function of the magnetic field. One example is the observation of stimulated hyperfine level crossings in a Xe laser [10.96], where accurate hyperfine splittings could be determined.

Although saturation effects may influence the line shape of the level-crossing signal, for small saturation it may still be essentially Lorentzian. Stimulated level-crossing spectroscopy has been used to measure Landé factors of atomic laser levels with high precision. One example is the determination of $g(^2P_4) = 1.3005 \pm 0.1\%$ in neon by HERMANN et al. [10.97].

There is one important point to note. The width $\Delta B_{1/2}$ of the level-crossing signal reflects the average width $\gamma = (\gamma_1 + \gamma_2)/2$ of the two crossing levels. If these levels have a smaller width than the other level of the optical transition, level-crossing spectroscopy allows a higher spectral resolution than, for example, saturation spectroscopy where the limiting linewidth $\gamma = \gamma_a + \gamma_b$ is given by the sum of upper and lower level widths. Examples are all cw laser transitions where the upper level always has a longer spontaneous lifetime than the lower level (otherwise inversion could not be maintained). Level-crossing spectroscopy of the upper level then yields a higher spectral resolution than the natural linewidth of the fluorescence between both levels. This is in particular true for level-crossing spectroscopy in electronic ground states where the spontaneous lifetimes are infinite and other broadening effects, such as time-of-flight broadening or the finite linewidth of the laser, limit the resolution.

11. Time-Resolved Laser Spectroscopy

While the previous chapter emphasized the high *spectral* resolution achievable with different sub-Doppler techniques, this chapter concentrates on some methods which allow high *time* resolution. The generation of extremely short and intense laser pulses has opened the way for the study of fast transient phenomena, such as molecular relaxation processes in gases or liquids due to spontaneous or collision-induced transitions. A new field of laser spectroscopy is the time-resolved detection of coherence and interference effects such as quantum beats or coherent transients monitored with *pulse Fourier transform spectroscopy*.

Recent progress in the generation of picosecond and sub-picosecond light pulses makes investigations of ultrafast processes occurring during excitation and deactivation of molecular states in solvents possible. The population decay times and the phase decay times of coherently excited molecules can be studied with extremely high time resolution.

The *spectral* resolution $\Delta\nu$ of most time-resolved techniques is in principle limited by the Fourier limit $\Delta\nu = a/\Delta T$ where ΔT is the duration of the short light pulse and the factor a depends on the profile I(t) of the pulse. Generally the spectral bandwidth $\Delta\nu$ of these pulses is still much narrower than, that of light pulses from incoherent light sources, such as flashlamps or sparks. Some time-resolved methods based on regular trains of short pulses even circumvent the Fourier limit $\Delta\nu$ of a single pulse and simultaneously reach extremely high spectral and time resolutions (see Sect.11.4).

We discuss first some techniques of generating short laser pulses and then illustrate different applications. Methods for lifetime measurements, the quantum beat technique, pulse Fourier transform spectroscopy, and multiple coherent interactions are some of the recently developed methods which demonstrate the capabilities of pulsed lasers for high time-resolution studies. For a more extensive representation of this field some monographs and reviews are recommended [11.1-3a].

11.1 Generation of Short Laser Pulses

In active laser media pumped by a pulsed source (e.g., flashlamps, electron pulses, or pulsed lasers) the population inversion necessary for oscillation threshold can be maintained only over a time interval ΔT that depends on duration and power of the pump pulse. A schematic time diagram of pump pulse, population inversion, and laser output is shown in Fig.11.1. As soon as threshold is reached, the laser emission starts. If the pump power is still increasing, the gain becomes high and the laser power rises faster than the inversion, until the increasing induced emission reduces the inversion to the threshold value.

The behavior of the laser output strongly depends on the time scale of relaxation processes. If they are slow, the induced emission drives the population inversion even below threshold and the laser output discontinues until the pump has reproduced sufficient population inversion to reach threshold, where induced emission starts again. The laser output consists in such cases of more or less irregular "spikes", with typical pulsewidths of a few μs, which appear during the whole time interval ΔT while the pump power is above threshold (Fig.11.2). The flashlamp-pumped ruby laser is a typical example of such a spiking laser.

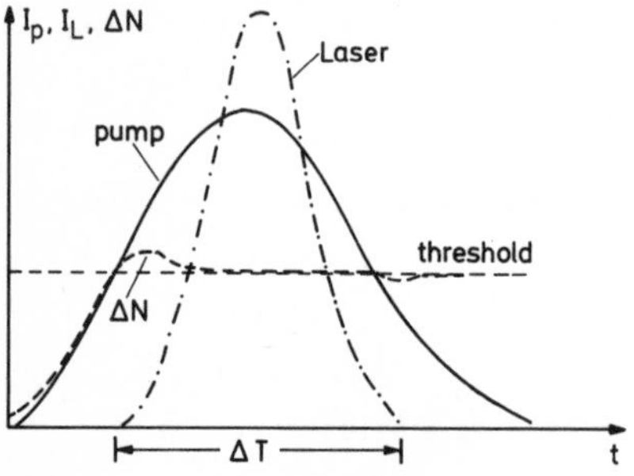

Fig.11.1. Schematic diagram of pump pulse, time-dependent inversion ΔN, and laser intensity $I_L(t)$

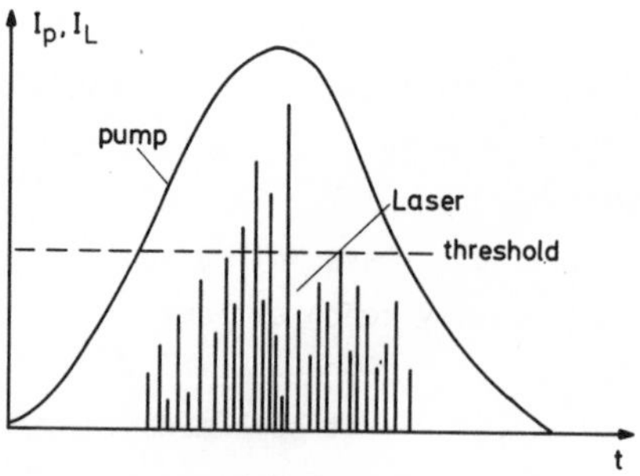

Fig.11.2. Spiking laser emission for laser media with long relaxation times

If the relaxation processes are sufficiently fast, they rapidly damp the oscillatory fluctuations of the population inversion and a steady-state inversion is reached which equals the threshold inversion ΔN_{thr} (see Sect.5.7). In these cases the laser output follows the pump pulse (see Sect.5.9) and ceases when the pump power drops below threshold.

Because of the short radiative lifetimes of most electronically excited free atoms or molecules, the output power of many gas lasers does not show spikes, but follows quasistationarily the time dependence of the pump pulse. Since the relaxation rate of excited dye molecules is in the sub-nanosecond range, dye lasers also generally have no spiking behavior and Fig.11.1 applies to flashlamp-pumped and laser-pumped dye lasers.

In some active media of pulsed lasers the lower laser level has a longer lifetime than the upper level. After a time T the increasing induced emission populates the lower level faster than it is depopulated by spontaneous emission and the inversion drops below threshold. In such "self-terminating" lasers the duration of the laser pulse is limited by the slow depopulation rate of the lower level and the laser emission may stop even before the pulsed pump drops below the initial threshold. One example of this laser type is the N_2 laser [11.4] which is pumped by a fast transverse gas discharge ($\approx$ 100 ns duration) and delivers output pulses of a few ns (1-10 ns) with typical peak powers of 0.1-1 MW.

Dye lasers, pumped with these short N_2 laser pulses (see Sect.7.3) yield light pulses of tunable wavelength with a pulse width of a few ns, peak powers of 1-100 KW, and repetition rates up to 1 KHz. With different dyes the whole spectral region between 0.36 and 1.2 μm can be covered. Optical frequency-doubling or nonlinear mixing techniques (see Sect.7.5) even allow extension of the spectral range into the ultraviolet region. Such lasers therefore represent very useful light sources for measuring lifetimes of selectively excited atomic or molecular states [11.5].

For experiments demanding high time resolution in the vacuum-ultraviolet range, excimer lasers pulsed by short electron pulses (see Sect.7.4) offer vuv pulses with tunable wavelengths, ns widths, and MW peak powers.

The subnanosecond range in time resolution can be reached by mode-locked lasers [11.6]. The basic principle may be understood as follows. If the intensity of a monochromatic light wave of frequency ν_0 is modulated at a frequency f, the Fourier analysis of the modulated light shows that, besides the carrier frequency ν_0, sidebands at frequencies $\nu_0 \pm mf$ (m = 1,2,3...) are generated. The intensity of the sidebands depends on the modulation function and the modulation factor (Fig.11.3).

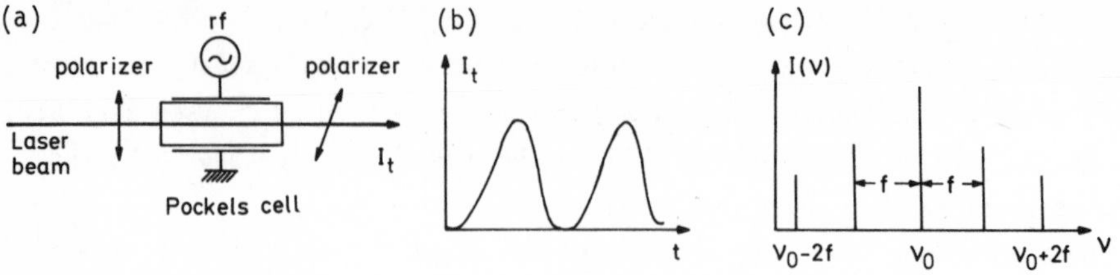

Fig.11.3. Intensity modulation using a Pockels cells. a) Schematic experimental arrangement. b) Transmitted intensity. c) Frequency spectrum of transmitted intensity

The modulation may be performed with Pockels cells [11.7] based on the Kerr effect, or by optoacoustic light modulators [11.8], which use the diffraction of light by ultrasonic waves for the modulation. When the modulator is placed inside the laser resonator (Fig.11.4) and the modulation frequency f is tuned to the mode separation $\Delta\nu = c/2d$, the carrier and the generated sidebands correspond to possible resonator modes, where laser oscillation is possible within the gain profile of the active medium. Since the sidebands are phase coupled by the modulation function, the oscillating laser modes are also coupled, as can be seen from the following consideration. Assume a sinusoidal modulation of the modulator transmission $T = (1+\delta\cos\Omega t)/2$ with modulation factor $\delta < 1$ and modulation frequency $f = \Omega/2\pi$. The amplitude of the mth mode is then

$$\begin{aligned} A_m(t) &= TA_0 \cos\omega_m t \\ &= (A_0/2)(1 + \delta\cos\Omega t)\cos\omega_m t \quad . \end{aligned} \tag{11.1}$$

This can be written as

$$A_m(t) = A_0/2 \cos\omega_m t + (A_0\delta/4)[\cos(\omega_m + \Omega)t + \cos(\omega_m - \Omega)t] \quad . \tag{11.2}$$

If the modulation frequency $\Omega/2\pi$ equals the mode spacing $\Delta\nu = c/2d$, sideband amplitudes

$$A_{m\pm1} = (A_0\delta/4)\cos\omega_{m\pm1}t \tag{11.3}$$

are generated in the adjacent resonator modes which are further amplified by stimulated emission. *The phases of these sidebands are determined by that of the carrier and by the modulation phase*. All three waves are in phase at times $T_k = 2\pi k/\Omega$ $(k = 0,1,2,3...)$. The modulation of these waves generates new sidebands at $\omega_2 = \omega_0 \pm 2\Omega$, etc., until all modes within the gain profile of the active medium oscillate with mutually coupled phases.

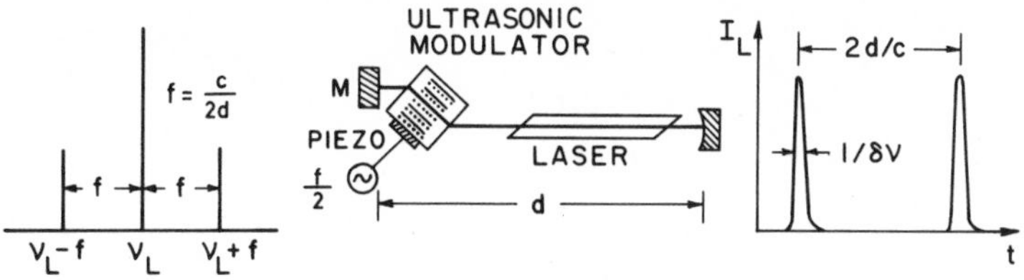

Fig.11.4. Active mode locking via intracavity amplitude modulation using diffraction by a standing ultrasonic wave

Within the bandwidth $\delta\nu$ of the gain profile 2p + 1 modes may oscillate. The superposition of these phase-coupled modes results in a total amplitude

$$A(t) = \sum_{m=-p}^{+p} A_m \cos(\omega_0 + m\Omega)t \quad , \quad \text{with} \quad 2p + 1 = \delta\nu/\Delta\nu \quad . \tag{11.4}$$

The resultant total laser intensity $I(t) \propto A(t)A^*(t)$ becomes then

$$I(t) \propto A_0^2 \frac{\sin^2[(2p + 1)(\Omega/2)t]}{\sin^2[(\Omega/2)t]} \cos^2\omega_0 t \quad . \tag{11.5}$$

For cw lasers the amplitude A_0 is constant and (11.5) represents a periodic function with a period $T = 1/\Delta\nu = 2\pi/\Omega$, which depends on the mode spacing $\Delta\nu = c/(2nd)$ (n = refractive index). Figure 11.5 illustrates this function for different values of the number $2p+1$ of simultaneously oscillating modes. For large numbers p the output of the mode-locked laser consists of a regular train of short pulses with a repetition rate $\Delta\nu$ which equals the mode separation. The pulse width $\Delta T = a/\delta\nu$ depends on the spectral width $\delta\nu = p\Delta\nu$ of the gain profile at the threshold line (see Fig.6.12). The exact value of the constant $a \approx 1$ depends on the shape of the gain profile.

Note that (11.5) is completely analogous to the interference of coherent waves, diffracted by the grooves of a grating (see Sect.4.1). Replacing the product Ωt by the optical path difference δ between waves from adjacent grooves of the grating yields (4.26). While (4.26) describes the interference of *stationary* coherent waves *in space*, (11.5) represents the interference of *phase-coupled* waves of different frequencies on a *time axis*. Figure 11.5 therefore is identical to Fig.4.18 if the time axis is replaced by the path difference.

Another way to look at the generation of mode-locked pulses is the following. The time separation

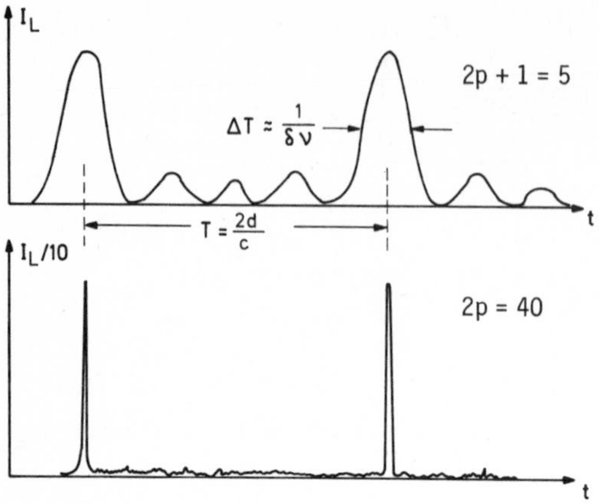

Fig.11.5. Schematic diagram of the output from a mode-locked laser when 5 or 41 longitudinal modes are locked

$$T = 1/\Delta\nu = 2nd/c \tag{11.6}$$

between subsequent pulses is equal to the transit time through the resonator. Laser photons travelling through the modulator at those times where its transmission T has a maximum, experience a minimum loss and will therefore have a maximum gain per round trip. Light passing always at the right time through the modulator will therefore be amplified, while light at all other times will be attenuated and cannot reach threshold. This explains why the peak intensity of the pulses is many orders of magnitude larger than the background intensity between the pulses.

Besides this *active mode coupling* by internal modulators, *passive mode-locking* by saturable absorbers can be used to generate ultrashort light pulses [11.9]. This method is preferentially used for mode locking of pulsed lasers with high peak powers, such as the neodymium-glass laser or the ruby laser. The principle of passive mode locking may be understood as follows. Assume that the laser with active medium and saturable absorber within the resonator is operating just at threshold. The intensity I(t) of the laser output fluctuates. Due to the saturation of the absorber the intensity peaks bleach the absorber more than the average intensity does. The saturated absorption coefficient

$$\alpha(I) = \alpha_0(1 - bI) \tag{11.7}$$

decreases with intensity. The peaks of the fluctuating intensity therefore suffer fewer losses and will be amplified more rapidly than the average intensity. After several round trips this leads to the generation of high peak power pulses, if the saturated absorber refills its depleted lower level population fast enough by relaxation processes.

When a cw laser is mode locked its average output power decreases by a factor β which depends on various parameters of the active medium, such as the lifetime of the upper laser level. For mode-locked argon lasers β is about 2-4. The peak power, however, is increased by a factor $\delta\nu/(\beta\Delta\nu)$.

Examples

a) *Argon-ion laser*: The spectral width of the gain profile above threshold for one of the laser transitions is about $\delta\nu \approx 7$ GHz, which implies that pulse widths of about 150 ps can theoretically be expected from a cw mode-locked argon laser. The experimental results yields somewhat larger pulse widths of 200-300 ps. Figure 11.6 shows for illustration two pulses from a regular train emitted at $\lambda = 488$ nm by an argon laser actively mode locked by an acousto-optic modulator. The apparent width of 500 ps is mainly determined by the time constant of the detection system.

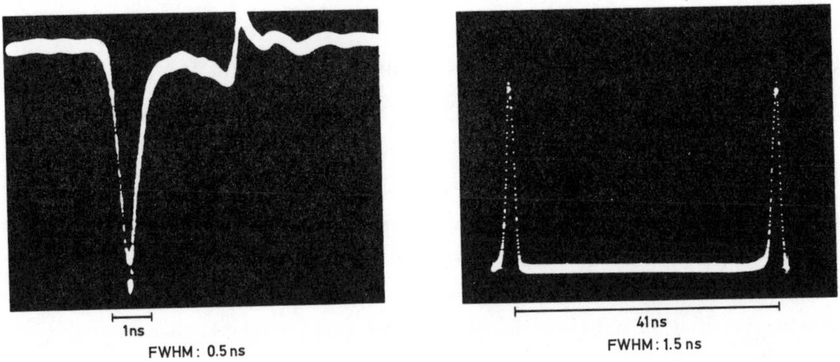

(a) (b)

Fig.11.6a,b. Observed output pulses from an argon laser at $\lambda = 488$ nm, actively mode locked by an acousto-optic modulator. (a) Detected with a fast photodiode and a sampling oscilloscope, (b) detected by single-photon counting technique using a photomultiplier. The oscillations following the optical pulse in (a) are due to cable reflections of the electric output signal from the diode. The pulse width in (b) is limited by electron transit time variations in the photomultiplier [11.10]

b) *Dye laser*: The spectral width $\delta\nu$ of the gain profile is very large ($\delta\nu \approx 3\times10^{13}$ s^{-1} $\triangleq$ $\Delta\lambda = 30$ nm). The bandwidth of the laser oscillation depends on the wavelength-selecting elements inside the laser resonator (see Sect.7.3). With $\delta\nu = 3\times10^{12}$ s^{-1} a pulse width of $\Delta T \approx 3\times10^{-13}$ s should be possible. Such short subpicosecond pulses have actually been observed

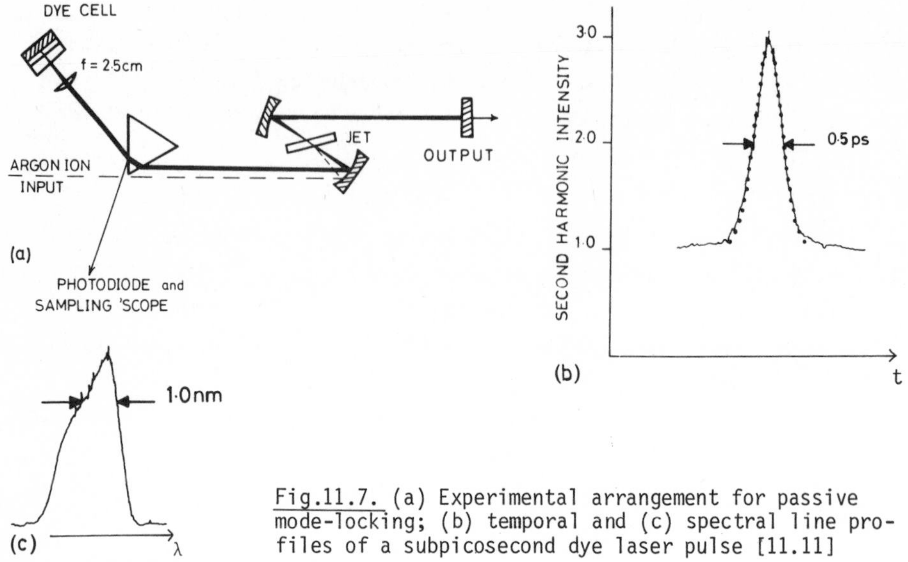

Fig.11.7. (a) Experimental arrangement for passive mode-locking; (b) temporal and (c) spectral line profiles of a subpicosecond dye laser pulse [11.11]

[11.11,11a]. Figure 11.7 shows the experimental arrangement for passive mode locking and spectral and temporal line profiles of a subpicosecond dye laser pulse.

c) *Neodymium-glass laser*: The mode-locked Nd:glass laser [11.12] generates pulses in the range from 5 to 15 ps and is capable of delivering very high peak powers ($\geqq 10^{10}$ W) at $\lambda = 1.06$ μm, which can be readily frequency doubled or tripled to give ultrashort pulses in the ultraviolet region. Mode locking is usually achieved by the use of a saturable absorber (e.g., a dye) inside the laser cavity. Figure 11.8 illustrates the output from a pulsed mode-locked neodymium laser [11.13].

A very efficient way of achieving a regular train of ultrashort pulses from cw dye lasers is *synchronous mode locking*. In this technique a mode-locked ion laser is used as excitation source and the cavity length of the dye laser is matched with that of the pump laser [11.14,15].

For many applications the time interval $\Delta T = 2nd/c$ between two successive pulses should not be too short. This implies that the cavity length d has to be sufficiently long. In order to avoid inconveniently large geometrical extensions of the arrangement, an optical delay line can be used. Figure 11.9 illustrates a synchronously pumped mode-locked cw dye laser system with such an optical delay line.

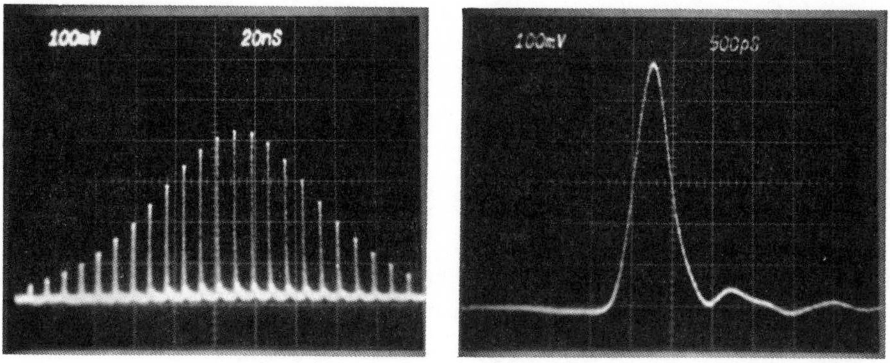

(a) (b)

Fig.11.8. (a) Output pulse train from a pulsed mode-locked neodymium laser. Time scale 20 ns/div. (b) Selection of a single pulse by a Pockels cell. Time scale 500 ps/div [11.13]

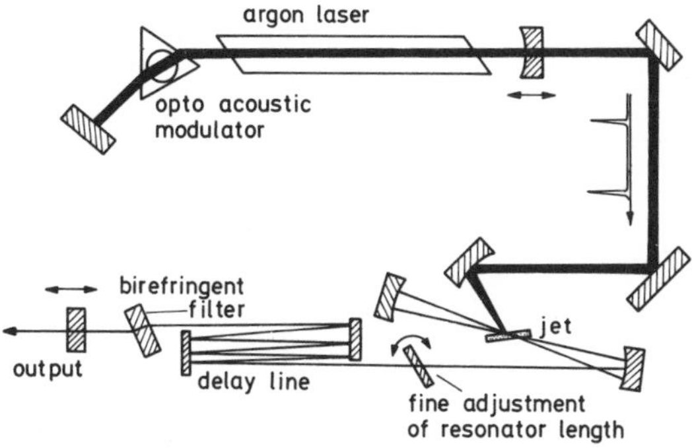

Fig.11.9. Synchronous mode locking of a cw dye laser. In order to match the cacity lengths of argon laser and dye laser (which are not drawn to scale) an optical delay line can be used in the dye laser resonator (courtesy of coherent radiation)

An elegant way to produce mode-locked pulses with variable repetition frequency uses a cavity dumping system (Fig.11.10) [11.16]. Inside a folded laser resonator with high reflectivity mirrors, a quartz cube is inserted where ultrasonic pulses with controllable repetition rate generate a periodic modulation of the refractive index. During the ultrasonic pulse a phase grating is produced in the quartz cube which causes Bragg diffraction of the laser wave. Choosing the optimum ultrasonic wave amplitude, nearly all of the intracavity laser intensity is being diffracted into the first order

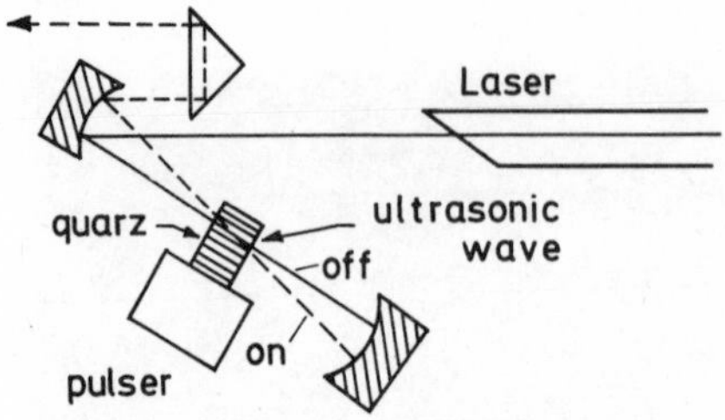

Fig.11.10.
Cavity dumping system

which is coupled out of the resonator by a prism. Combined with a mode-locked laser, such a system allows repetition rates between 1 Hz and 3 MHz and pulse widths between 0.6-2 ns. The cavity dumper allows larger output peak powers because it releases nearly all of the intracavity power P_i while in other systems only the fraction TP_i is transmitted through the output mirror with transmission T (T is, for example, only 4% in cw argon or dye lasers).

Figure 11.11 shows such a combination of cavity dumping and mode locking. The time interval ΔT between successive pulses can be controlled between 1s and 0.3 μs. The cavity dumping technique can be also applied to synchronously pumped mode-locked cw dye lasers.

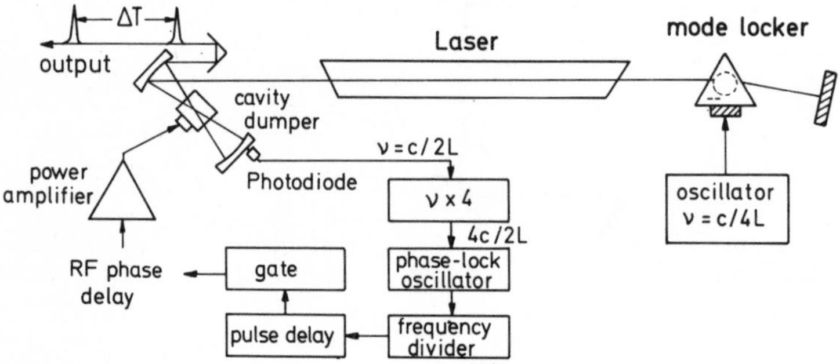

Fig.11.11. Schematic diagram of a "combo laser" which represents a combination of mode locking and cavity dumping (courtesy of Spectra Physics)

11.2 Lifetime Measurements with Lasers

The various techniques of producing short light pulses with variable wavelength, duration, and repetition rate, allow time-resolved measurements of the decay of selectively excited atomic, ionic, or molecular levels. The most commonly used experimental techniques for direct measurements of lifetimes are the phase-shift method, the pulse excitation, the technique of delayed coincidences with single-photon counting, and the time-of-flight method. We briefly describe the basic principles of these techniques. For a more detailed treatment the reader is referred to the literature [11.5].

11.2.1 Phase Shift Method

The intensity I_{exc} of the incident light which excites the molecular level E_i is sinusoidally modulated at a frequency Ω which is assumed to be small compared with the light frequency ω_{ik}

$$I_{exc} = I_0(1 + a \sin\Omega t)\cos\omega_{ik}t \quad . \tag{11.8}$$

The modulation may be achieved for example with a Pockels cell or an ultrasonic light modulator (Fig.11.12).

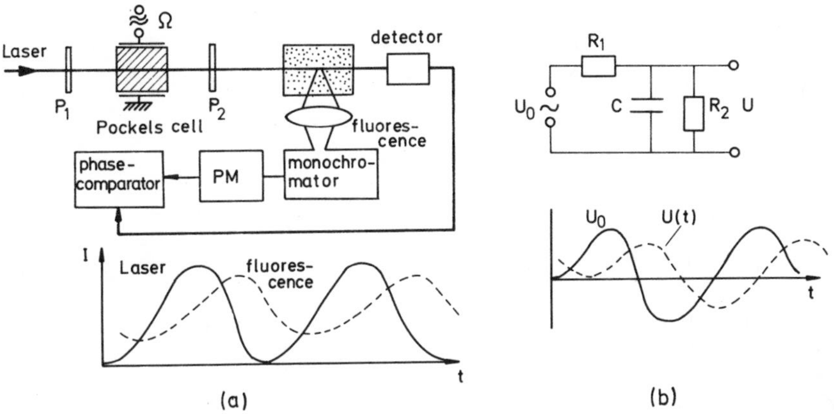

Fig.11.12a,b. Lifetime measurement with the phase shift method. (a) Experimental arrangement and modulation of laser light and fluorescence, (b) analogous electrical circuit

The differential equation for the time-dependent density $N_i(t)$ of molecules in level E_i with $g_i = g_k = 1$ and $I_{exc} = \rho c$ is

$$dN_i/dt = B_{ik}\rho(\omega_{ik})(N_k - N_i) - N_iA_{ik} \quad . \tag{11.9}$$

The solution of (11.9) is straightforward and yields $N_i(t)$. The fluorescence intensity $I_{Fl} = N_i(t)A_{ik}$ is observed perpendicular to the incident light beam where no induced emission can be seen. One obtains

$$I_{Fl} = bI_0\left[1 + \frac{a}{\sqrt{1+\Omega^2\tau^2}}\sin(\Omega t + \varphi)\right]\cos\omega_{ik}t \quad , \tag{11.10}$$

where b is a proportionality constant.
Equation (11.10) shows that the fluorescence intensity is modulated at the same frequency Ω as the exciting light, but the modulation amplitude has decreased and its phase is shifted by φ against the modulation phase of I_{exc}. For $\rho B_{ik} \ll A_{ik}$ the effective lifetime

$$\tau = 1/(\rho B_{ik} + A_{ik}) \tag{11.11}$$

is related to this phase shift φ and to the modulation frequency Ω by

$$\boxed{\tan\varphi = \Omega\tau} \quad . \tag{11.12}$$

Note that this problem is completely equivalent to the well-known electrical problem of charging a capacitor C through a resistor R_1 by an ac source with a voltage $U_0 \sin\Omega t$ (see Fig.11.12). The time-dependent voltage across C and the resistor R_2 parallel to C can be obtained from the equation

$$C\frac{dU}{dt} = (U_0 - U)/R_1 - U/R_2 \quad , \tag{11.13}$$

which is equivalent to (11.9) and has the solution

$$U = U_2 \sin(\Omega t - \varphi) \text{ with } \tan\varphi = R_1R_2C\Omega/(R_1+R_2) \text{ and } U_2 = \frac{R_2U_0}{[(R_1+R_2)^2 + \Omega^2C^2R_1^2R_2^2]^{1/2}} \quad .$$

The lifetime τ corresponds to the time constant R_2C and the exciting light intensity to $(U_0 - U)/R_1$.

If $I_{exc} = c\rho_{exc}$ is sufficiently small, the induced emission term may be neglected and the extrapolation to $I_{exc} \to 0$ yields the spontaneous lifetime $\tau_i = 1/\sum_k A_{ik}$ if collisions are negligible. The effect of collisions on τ can be determined by plotting $1/\tau$ against the pressure (Stern-Vollmer plot), which yields at low pressures a straight line (see Sect.12.2). The extrapolation to zero pressure gives the unperturbed spontaneous lifetime of the excited level of the free molecule. Figure 11.13 illustrates such a Stern-Vollmer plot for the excited $^1D\pi_\mu$ state of the NaK molecule.

If the modulation of the exciting light is not sinusoidal, the time dependence of the fluorescence will be more complicated. However, with a Fourier analysis it is always possible to filter out of the modulation frequency spectrum the first harmonics with frequency Ω for which the relation (11.12) holds.

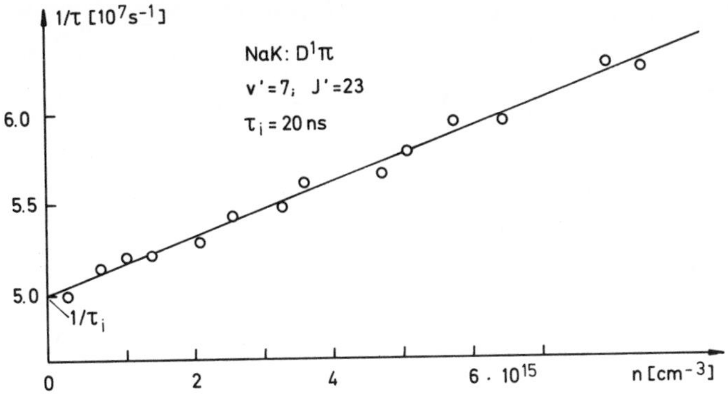

Fig.11.13. Stern-Vollmer plot: Lifetime dependence of a rovibronic excited level of the NaK molecule on the density of collision partners

The phase shift method is not so well suited to the measurement of *non-exponential* decays (if, for example, the fluorescence from several levels with different lifetimes overlap). Although measurements at different modulation frequencies Ω_k enable one to fit the measured phase shifts φ_k to a sum of exponential $\sum C_k\,(\exp(-t/\tau_k)$, the decay curve cannot be viewed directly, and the fit may not be unambiguous.

The main disadvantage of the phase shift method is the influence of induced emission [see (11.11)] which is a handicap especially when using lasers for excitation [11.17]. The necessary extrapolation to zero excitation intensity is time consuming and limits the accuracy because of the low signal-to-noise ratio at low intensities.

11.2.2 Pulse Excitation

The molecules are excited by a light pulse with a trailing edge short compared to the mean lifetime τ of the excited level. The subsequent decay of the level population, monitored by the decay of the fluorescence intensity, is either viewed directly on a scope or is monitored with a boxcar integrator or a transient recorder (see Sect.4.5.10). This method does not suffer from the influence of induced emission since the exciting light is already switched off when the fluorescence is observed. It is especially adapted to the use of pulsed or mode-locked lasers as excitation sources. From the decay curve the mean lifetime can be derived directly. Deviations from exponential decays, caused for instance by cascade effects, can be seen immediately. The accuracy is comparable to that of the phase shift method, if an absolute time calibration has been accurately performed. Generally the signals are detected in an analog form and many fluorescence photons per excitation pulse are required to obtain a sufficiently large signal-to-noise ratio.

11.2.3 Delayed Coincidence Technique

This method also uses short excitation pulses. Contrary to the previous method, however, the detection probability is kept below one fluorescence photon per excitation pulse and the repetition rate of the excitation pulse is chosen as high as possible. Single photon counting techniques are used (see Sect.4.5.4), which measure the time distribution of the probability $P_{ik}(t)dt$ that a fluorescence photon $\hbar\omega_{ik}$ is emitted within a time interval dt between t and t + dt after the excitation pulse at t = 0 [11.18a].

The experimental realization is performed as follows (see Fig.11.14). The excitation pulse starts a voltage ramp U = at, which is stopped by the first fluorescence photon detected after the excitation pulse. The output pulse with voltage U of the ramp generator (time to pulse height converter TPC) is stored in a multichannel analyzer. If the detection probability P_{ik} of a fluorescence photon per excitation pulse is small compared with one, the probability that two photons arrive at the detector during the on cycle is negligible and this method (which would suppress the second photon) can be used safely. The voltage distribution on the MCA displays the decay curve directly. This method is especially useful if mode-locked cw lasers are used for excitation because the repetition rate R is high, allowing a sufficiently large number of fluorescence photons

$$N(t) = P_{ik}(t)R\Delta T \tag{11.14}$$

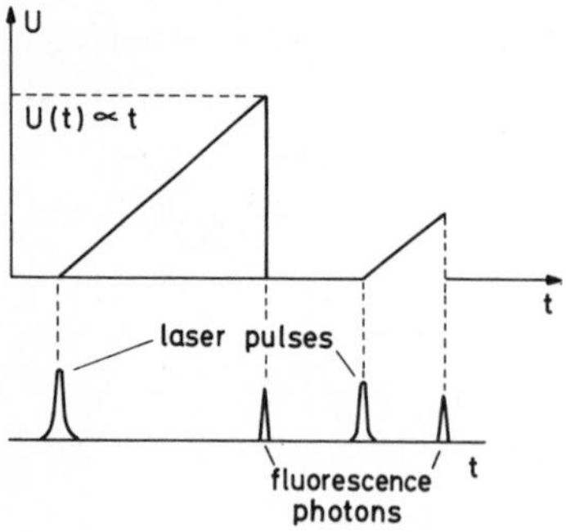

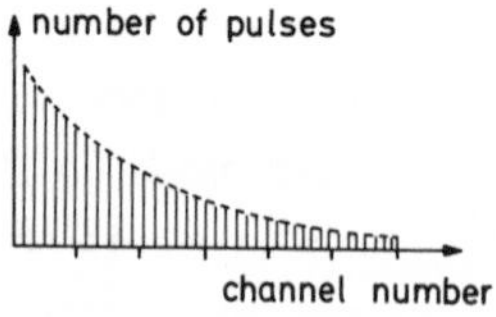

Fig.11.14. Schematic illustration of time-to-pulse-height conversion used for lifetime measurements with delayed coincidences

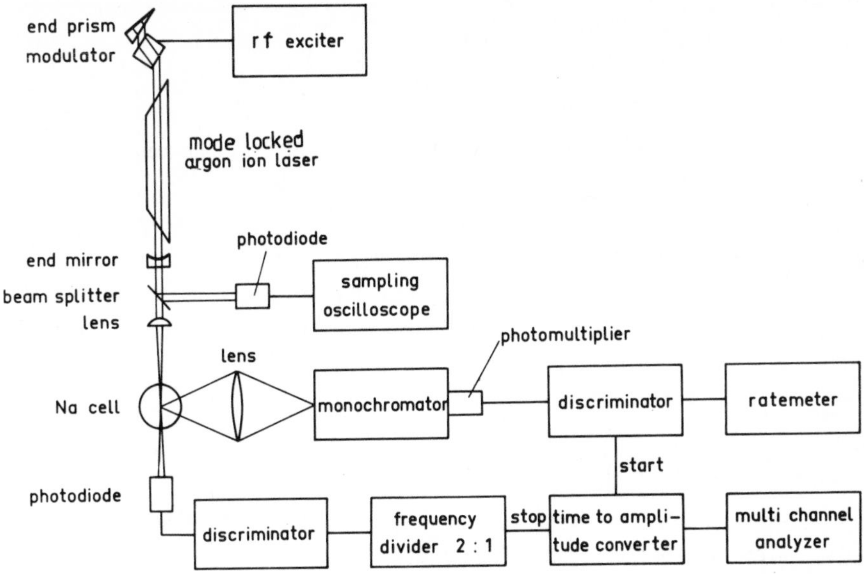

Fig.11.15. Schematic diagram of lifetime apparatus with mode-locked laser and single-photon counting techniques

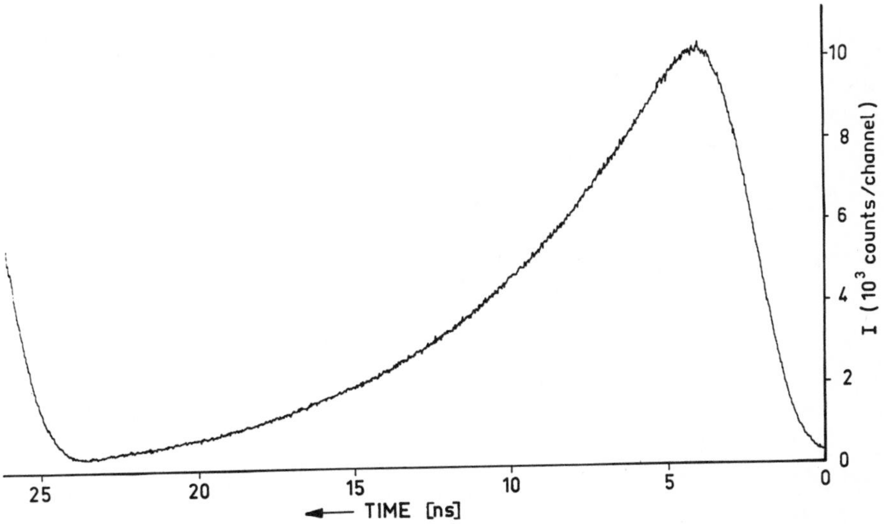

Fig.11.16. Fluorescence decay from the (v' = 6, J' = 43) level in the excited $B^1\pi_u$ state of Na_2 [11.10]

to be counted during the counting time ΔT. Figure 11.15 illustrates the experimental arrangement for measuring lifetimes of selectively excited molecular levels, using the delayed coincidence technique [11.10]. A typical decay curve of the fluorescence from the excited ($v' = 6$, $J' = 43$) level in the $B^1\pi_\mu$ state of the Na_2 molecule is shown in Fig.11.16. The repetition rate of the mode-locked argon laser was 42 MHz [11.10].

Such high pulse frequencies cannot be used to start the TPC which has a dead time of at least 100 ns after a start pulse. Therefore the fluorescence photons (counting rate about 10-100 KHz) are used to start the ramp and the subsequent laser pulse stops it. This corresponds to a time reversal and yields the probability distribution $P_{ik}(T-t)$ if T is the constant time interval between successive mode-locked pulses.

11.2.4 Lifetime Measurements in Fast Atomic Beams

This method, which is a modernized version of the old Wien method, uses fast atomic, molecular, or ion beams with kinetic energies in the keV to MeV range. The atoms which move into the x direction are excited at a well-defined small interval Δx around $x = 0$. The excitation source may be a laser or collisions with other atoms in foils or gas chambers. The subsequent fluorescence $I_{Fl}(x)$ is measured as a function of the distance x from the point of excitation (see Fig.11.17). The transformation to a time scale $I_{Fl}(t)$ uses the relation $x = vt$ where the velocity $v = (2e\ U/m)^{\frac{1}{2}}$ of the ions is determined by the acceleration voltage U and the mass m of the ions. Neutral atoms or molecules can be produced from ions by charge exchange. The energy loss during charge exchange is negligible in most cases.

The accuracy of the method is in principle limited only by the accuracy of distance measurement, and many lifetimes of highly excited atoms or ions have been measured this way (beam foil spectroscopy) [11.18b]. However, a severe drawback of this method with nonselective simultaneous excitation of many upper levels results from cascade effects. The level population

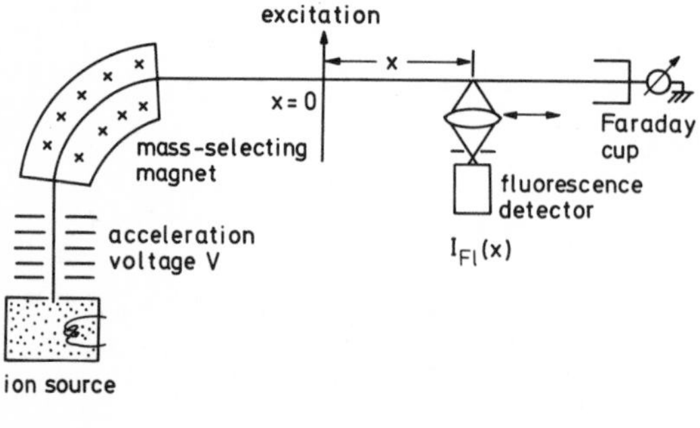

Fig.11.17. Schematic experimental arrangement for lifetime measurements in fast atomic or ion beams

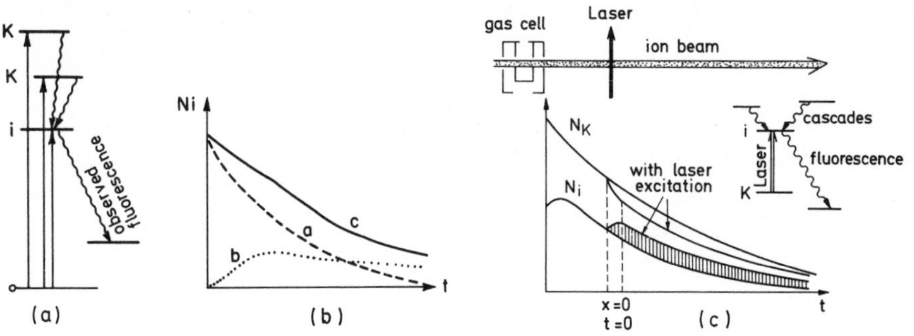

Fig.11.18a-c. Influence of cascade transitions on measured decay curves. (a) Level scheme for broad band excitation with resultant cascade transitions to level i. (b) Observed fluorescence decay including cascade effects (curve c) due to the overlap of pure exponential decay of level i (curve a) and population of i by cascades (curve b). (c) Combined gas cell and laser excitation with resultant population changes of levels

$N_i(t)$ decays by fluorescence to lower levels but is simultaneously fed by radiative transitions from upper levels (Fig.11.18a). We obtain the rate equation

$$dN_i/dt = -N_i \sum_k A_{ik} + \sum_m N_m A_{mi} \quad , \tag{11.15}$$

where $E_m > E_i > E_k$. Generally the populations N_m are not known and the observed fluorescence intensity $I_{Fl} = N_i(t)A_{ik}$ is nonexponential due to the superposition of cascades (Fig.11.18b).

Selective excitation with a laser tuned to a transition $E_0 \to E_i$ solves the cascade problem [11.19]. In this variant many excited levels of the atoms or ions are populated by broad-band excitation via collisions with target gas atoms in a differentialy pumped gas cell (Fig.11.18c). A few cm behind the exit aperture of the gas cell a laser crosses the ion beam. If the laser frequency is tuned to transition $E_k \to E_i$ the populations of both levels are changed, due to optical pumping, by an amount ΔN which depends on the laser intensity, the transition probability A_{ik}, and the initial population of the two levels. The laser intensity can be chopped and the fluorescence intensity $I_{Fl}(E_i \to E_k)$ is observed as a function of x alternatively with (I_1) or without (I_2) laser excitation. The difference

$$\Delta I_{Fl}(x) = I_1(x) - I_2(x)$$

gives that part of the fluorescence signal which has been induced by selective laser excitation. The difference signal therefore yields a pure exponential decay even if cascading is present (curve a in Fig.11.18b).

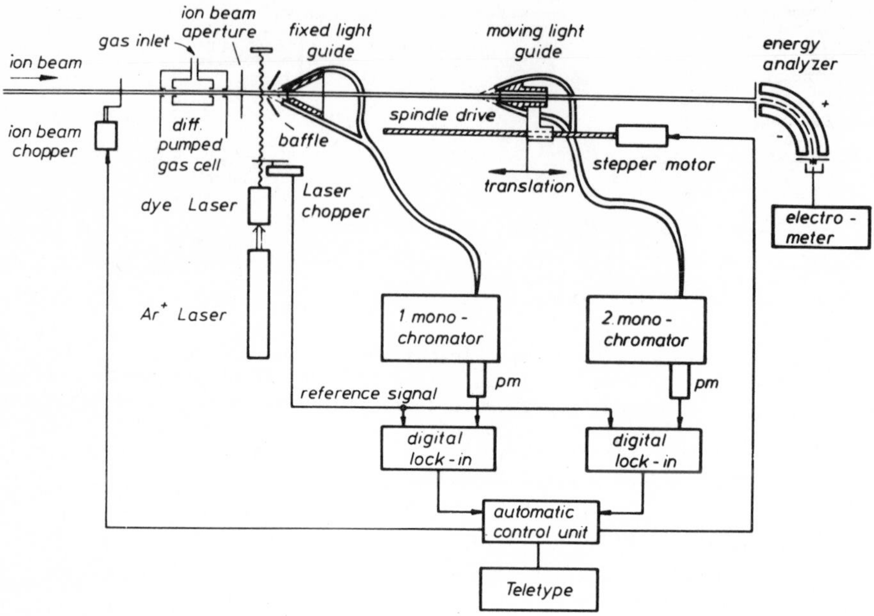

Fig.11.19. Experimental arrangement for accurate lifetime measurements in fast atomic or ion beams using a combined gas cell laser excitation [11.20]

Figure 11.19 shows a possible experimental realization. The fluorescence is detected by a cone of fiber bundles, centered around the ion beam. The other end of the fiber bundle forms a thin rectangle which is imaged onto the monochromator slit. For intensity normalization a second fixed fiber detector is placed close behind the crossing point with the laser [11.20].

11.3 Picosecond Spectroscopy

Optoelectronic detection systems such as fast photodiodes and sampling oscilloscopes have reached a time resolution of 10^{-10} s. However, this is still not sufficient to resolve many fast transient events on a picosecond time scale. In picosecond spectroscopy, therefore, new techniques had to be invented to measure durations and profiles of picosecond pulses and to probe ultrafast relaxation processes.

For many applications in the picosecond range the streak camera [11.21] can be used which may reach time resolutions of a few picoseconds. It consists essentially of a fast image intensifier (see Sect.4.5.5) where the electron beam is deflected by a fast voltage ramp applied to the deflection plates. This causes the intensified image of a spatially fixed point to be swept over the phosphor screen. Writing speeds of 2×10^{10} cm/s can be achieved with a jitter of less than 50 ps. The phosphor screen is photographed and a microdensitometer trace of the photoplate yields the time resolution of a few ps.

Using powerful polarized light pulses instead of electrical pulses to induce birefringence in traditional Kerr cell liquids, an optical Kerr shutter has been developed with wide ranging applications to picosecond measurements [11.22].

Most of the methods used to measure picosecond phenomena are based on optical delay lines where the picosecond laser pulse is divided by a beam splitter and the two replica pulses travel different path lengths before they are recombined. The measurement of a time interval Δt is thus transferred to that of a path difference $\Delta x = c\Delta t$ where c is the velocity of light.

Figure 11.20 illustrates a possible arrangement which uses the second harmonic generation (SHG) in KDP crystal [11.23]. The two light pulses with intensities $I_1(\omega,t) = E_1(\omega,t)E_1(\omega,t)$ and $I_2(\omega, t + \Delta t)$ and a variable time separation Δt travel collinearly through the nonlinear KDP crystal. Since the intensity of the SH wave is proportional to the square of the total fundamental intensity (see Sect.7.5), we obtain

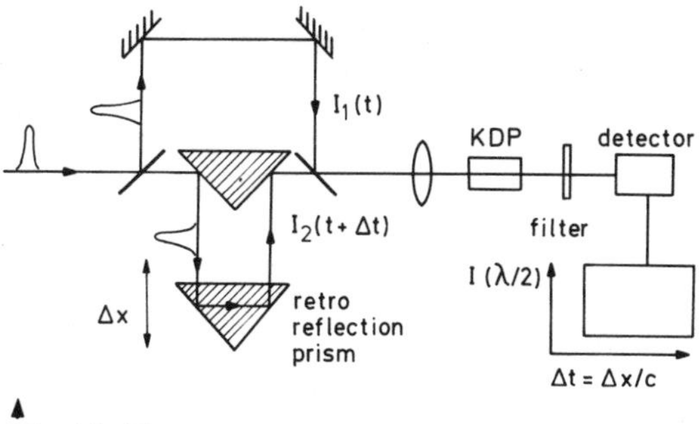

Fig.11.20

Fig.11.21

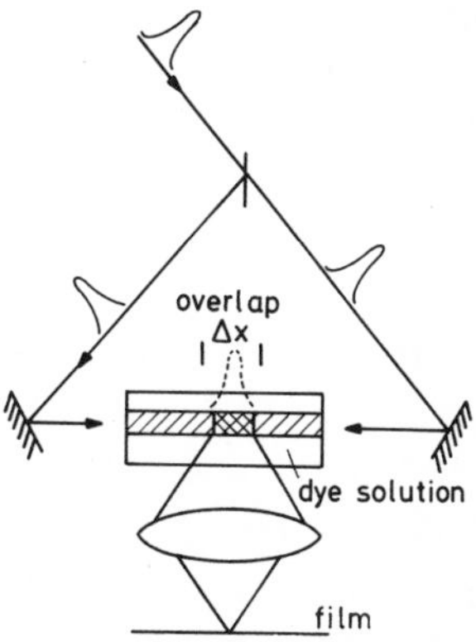

Fig.11.20. Schematic arrangement for measuring the pulse width of a picosecond pulse based on second harmonic generation with autocorrelation measurement

Fig.11.21. Picosecond pulse width measurement using two-photon induced fluorescence

$$I(2\omega,t,\Delta t) = A\left\{[E_1(\omega,t) + E_2(\omega,t+\Delta t)][E_1^*(\omega,t) + E_2^*(\omega,t+\Delta t)]\right\}^2 \quad . \tag{11.16}$$

The detector averages over the rapid oscillations of the light field ($\omega \approx 10^{15}\ s^{-1}$). The detected signal is therefore better described by correlation functions [11.24]. The normalized second-order autocorrelation function of the time-dependent intensity $I(t) = E(t)E^*(t)$ (see Sect.2.10) is defined by

$$G^{(2)}(\tau) = \frac{\int_{-\infty}^{+\infty} I(t)I(t+\tau)dt}{\int_{-\infty}^{+\infty} I^2(t)\,dt} = \frac{\langle I(t)I(t+\tau)\rangle}{\langle I^2(t)\rangle} \quad . \tag{11.17}$$

The measured averaged intensity $\langle I(2\omega,\tau)\rangle$ of the SH wave displayed as a function of the delay time $\Delta t = \tau$ between the two pulses can be expressed by the correlation function $G^2(\tau)$. For equal intensities $I_1 = I_2$ one obtains

$$\begin{aligned} I(2\omega,\tau) &= A[G^{(2)}(0) + 2G^{(2)}(\tau)] \\ &= A[1 + 2G^2(\tau)] \quad , \end{aligned} \tag{11.18}$$

where the relation $G^{(2)}(0) = 1$ has been used. The other cross terms in (11.16) average to zero. The SH intensity is maximum for $\tau = 0$ where $I(2\omega,0) = 3A$ and approaches a constant background value $I(2\omega,\infty) = A$ for $\tau \to \infty$. The signal-to-background ratio is therefore

$$R = \frac{I(2\omega,0)}{I(2\omega,\tau)} = \frac{3G^2(0)}{G^2(0) + 2G^{(2)}(\tau)} = \frac{3}{1 + 2G^{(2)}(\tau)} \quad . \tag{11.19}$$

Instead of SHG in nonlinear crystals, often two-photon induced fluorescence (see Sect.8.5) is used to monitor and to visualize the pulse profiles of picosecond pulses [11.25]. The experimental arrangement is shown in Fig. 11.21. The liquid dye solution has no absorption at the fundamental frequency ω but shows fluorescence at high light intensity due to two-photon absorption. For two identical optical pulses moving into opposite direction through the dye solution the observed fluorescence intensity as a function of the delay time τ is again [11.25] given by

$$I_{Fl}(\tau) = [1 + 2G^2(\tau)] \quad . \tag{11.20}$$

Since $\tau = \Delta x/c$, the spatial distribution $I_{Fl}(\Delta x)$ yields the autocorrelation function $G(\tau)$ directly and with it the pulse profile [11.26].

A few examples may serve to illustrate the applications of picosecond pulses to the study of fast relaxation processes.

a) Many dye molecules, solved in organic solutions, have very short lifetimes, below 1 ns. The dye molecule rose bengal dissolved in methanol has for instance a mean lifetime of $\tau = 597$ ps [11.27]. With mode-locked excitation pulses the limiting factor for time resolution is no longer the pulse duration but the spread of transit times in the photomultiplier used to detect the fluorescence (see Sect.4.5).

b) The limitations of the detection system can be overcome by a pump and probe technique. A single picosecond pulse, selected by an electro-optic shutter out of a train of pulses from a mode-locked laser, is divided into two replica pulses, a pump and a probe pulse. The delay Δt between the two pulses can be controlled by varying the length of the optical path difference. The pump pulse is, for instance, used to excite molecular vibrations of molecules in liquids. The probe pulse probes the population in this level as a function of the delay time. This allows determination of the depopulation time due to fast relaxation of the excited molecules with the liquid. Relaxation times as fast as a few ps have been measured [11.28]. With this technique it is also possible to measure the dephasing time of Raman active molecular vibrations [11.29].

c) With consecutive excitation by two picosecond pulses with variable delay, followed by fluorescence detection, it is possible to make direct observations of phenomena originating from short-lived upper states of molecules in solutions [11.30]. The intensity of the fluorescence induced by the probe pulse (Fig.11.22) which is measured as a function of the delay time τ is directly proportional to the decaying population $N_1(t)$.

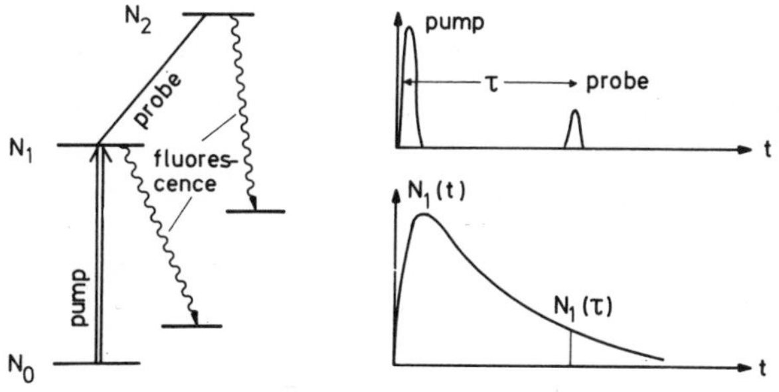

Fig.11.22. Pump and probe technique to measure decay times of short-lived excited states

Such measurements also allow measurement of quantum yields in large molecules. Since radiationless transitions may contribute considerably to the deactivation of excited states, the quantum yield

$$\eta = n_{Fl}/n_{ab} \quad ,$$

which gives the ratio of fluorescence photons to absorbed ones, is generally smaller than unity.

There are numerous examples where picosecond and subpicosecond pulses have been used to study the kinetics of molecular collisions, of electron transfer and photosynthesis, and of coherent transients in molecules. For more thorough information see [11.1,2,29-32].

11.4 Coherent Transients and Pulse Fourier Transform Spectroscopy

In Chap.10 we discussed several techniques of sub-Doppler laser spectroscopy which overcome the Doppler width of molecular lines by selecting a subgroup of molecules within a narrow velocity interval Δv_x to interact with the laser field. The spectral resolution of these techniques cannot exceed the bandwidth of the laser. Single-frequency stabilized lasers are therefore required to take full advantage of the spectral resolution, in principle achievable with these methods. Another class of Doppler-free techniques presented in Chap.10 was based on the coherent preparation of molecular states (level crossing spectroscopy). Here broad-band lasers could be used because the spectral resolution was achieved by utilizing phase relations between the wave functions of coherently excited sublevels. The detection scheme did not employ time resolution but was based on time-integrated signals.

In this section we discuss some methods of time-resolved laser spectroscopy which are based on the time evolution of coherently excited states. Such techniques have been successfully used for about 25 years in radio frequency spectroscopy and are well known under the slogans "spin-echo" and time-resolved nuclear magnetic resonance spectroscopy (NMR).

For the last 10 years these methods have been extended into the optical region, taking advantage of sufficiently intense coherent radiation sources provided by pulsed lasers. Time-resolved coherent optical spectroscopy has rapidly developed into an important branch of high-resolution laser spectroscopy [11.32a-c].

11.4.1 Quantum Beat Spectroscopy

If two or more closely spaced molecular levels are simultaneously excited by a short laser pulse, the time-resolved total fluorescence intensity emitted from these coherently prepared levels shows a modulated exponential decay. The modulation pattern, known as *quantum beats*, is due to interference between the fluorescence amplitudes emitted from these coherently excited levels. Although a more thorough discussion of quantum beats demands the theoretical framework of quantum electrodynamics [11.33], it is possible to understand the basic principle by using more simple argumentation.

Assume that two closely spaced levels 1 and 2 of an atom or molecule are simultaneously populated by optical pumping with a short laser pulse from a common initial lower level i (Fig.11.23a). In order to achieve coherent excitation of both levels by a laser pulse with duration ΔT, the Fourier limited spectral bandwidth $\Delta\nu = a/\Delta T$ [a is a constant of the order of unity which depends on the pulse profile $I_p(t)$] must be larger than the frequency separation $(E_2 - E_1)/h$.

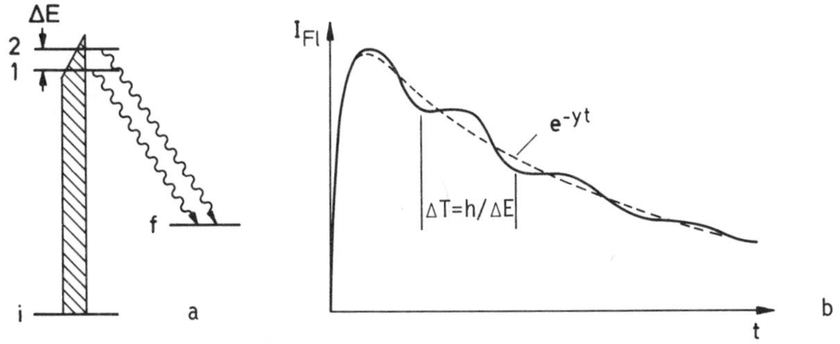

Fig.11.23a,b. Quantum beat spectroscopy. (a) Level scheme illustrating coherent excitation of levels 1 and 2 with a short broad-band pulse. (b) Fluorescence intensity showing a modulation of the exponential decay

If the pulse excitation occurs at t = 0, the wave function of the excited states at this time can be written as a linear superposition of the sublevel functions $\varphi_k(k = 1,2,)$

$$\psi(0) = \sum_k a_k\varphi_k(0) \quad , \tag{11.21}$$

where the coefficients a_k represent the probability amplitudes that the light pulse has prepared the atom in level E_k. Because of spontaneous decay into a final state f the excited state evolves as a function of t as

$$|\psi(t)\rangle = \sum_k a_k|\varphi_k(0)\rangle e^{-(iE_{kf}/\hbar + \gamma_k/2)t} \quad , \quad \hbar\omega_{kf} = E_{kf} \qquad (11.22)$$

where γ_k represents the decay constant of level k. The time-dependent intensity of the fluorescence emitted from the excited levels is determined by the transition matrix element

$$I(t) = C|\langle\varphi_f|\hat{\varepsilon}\underline{r}|\psi(t)\rangle|^2 \quad , \qquad (11.23)$$

where $\hat{\varepsilon}$ is the polarization vector of the emitted fluorescence photon, $e\underline{r}$ is the dipole operator, and C is a proportionality factor depending on the experimental parameters. Inserting (11.22) into (11.23) yields, with $\gamma_1 = \gamma_2 = \gamma$,

$$I(t) = C\, e^{-\gamma t}(A + B\cos\omega_{21}t) \quad , \qquad (11.24)$$

where

$$A = a_1^2|\langle\varphi_f|\hat{\varepsilon}\underline{r}|\varphi_1\rangle|^2 + a_2^2|\langle\varphi_f|\hat{\varepsilon}\underline{r}|\varphi_2\rangle|^2$$

$$B = 2a_1a_2|\langle\varphi_f|\hat{\varepsilon}\underline{r}|\varphi_1\rangle||\langle\varphi_f|\hat{\varepsilon}\underline{r}|\varphi_2\rangle|$$

$$\omega_{21} = (E_2 - E_1)/\hbar \quad .$$

This shows that a modulation of the exponential decay is observed if both matrix elements for the transitions 1 → f and 2 → f are nonzero (Fig.11.23b). The measurement of the modulation frequency ω_{21} allows determination of the energy separation of the two levels, even if their splitting is less than the Doppler width. *Quantum beat spectroscopy therefore allows Doppler-free resolution.*

The physical interpretation of the quantum beats is based on the following fact. When the molecule has re-emitted a photon, there is no way to distinguish between transitions 1 → f or 2 → f if the total fluorescence is monitored. As a general rule in quantum mechanics the total probability amplitude of two indistinguishable processes is the sum of the two corresponding amplitudes and the observed signal is the square of this sum. This quantum beat interference effect is analogous to Young's double slit interference experiment.

The experimental realization uses either short-pulse lasers, such as the N_2-laser pumped dye laser (see Sect.11.1) or mode-locked lasers. The time response of the detection system has to be fast enough to resolve time intervals $\Delta t < h/(E_2 - E_1)$. Fast transient digitizers or boxcar detection systems (see Sect.4.5) meet this requirement.

When atoms, ions, or molecules in a fast beam are excited and the fluorescence intensity is monitored as a function of the distance x downstream from the excitation point, the time resolution $\Delta t = \Delta x/v$ is determined by the particle velocity v and the resolvable spatial interval Δx from which the fluorescence is collected [11.19,34]. In this case detection systems can be used which integrate over the intensity and measure the quantity $I(x) = \int I(t,x)dt$. The excitation can even be performed with cw lasers since the bandwidth necessary for coherent excitation of the two levels is assured by the short interaction time $\Delta t = d/v$ of a molecule with velocity v passing through a laser beam with diameter d. With $d = 0.1$ cm and $v = 10^8$ cm/s $\Rightarrow t = 10^{-9}$ s which allows coherent excitation of two levels with a separation up to 1000 MHz.

The Fourier transform of the time-resolved fluorescence intensity I(t) yields its spectral distribution $I(\omega)$ with sub-Doppler resolution. Figure 11.24 illustrates as an example quantum beats measured by ANDRÄ et al. [11.19] in the fluorescence following the excitation of three hfs levels in the $6p\,^2P_{3/2}$ state of the $^{137}Ba^+$ ion. Either a tunable dye laser crossed perpendicularly with the ion beam, or a fixed frequency laser crossed under a tilting angle θ with the ion beam, can be used for excitation. In the latter case Doppler tuning of the ion transitions can be achieved by tuning the ion velocity (see Sect.10.1.3). The lower spectrum in Fig.11.24 is the Fourier transform of the quantum beats, which yields the hfs lines depicted in the energy level diagram.

A very promising novel technique of measuring quantum beats in transmission rather than in fluorescence has been demonstrated by LANGE et al. [11.35]. A short pump pulse excites coherently different upper levels. The time evolution of the superposition of states following the coherent excitation causes time-dependent changes of the complex susceptibility χ of the sample. Similar to the quantum beats in the fluorescence intensity the susceptibility $\chi(t)$ is found to contain oscillating nonisotropic contributions which can be readily detected by placing the sample between crossed polarizers and transmitting a probe pulse with variable delay (see also Sect.10.3 on polarization spectroscopy). Even a cw broadband dye laser can be used for probing if the probe intensity transmitted by the polarizer is monitored with sufficient time resolution.

This method relies on the dependence of stimulated emission and absorption on the coherence between molecular states. Its spectral resolution is

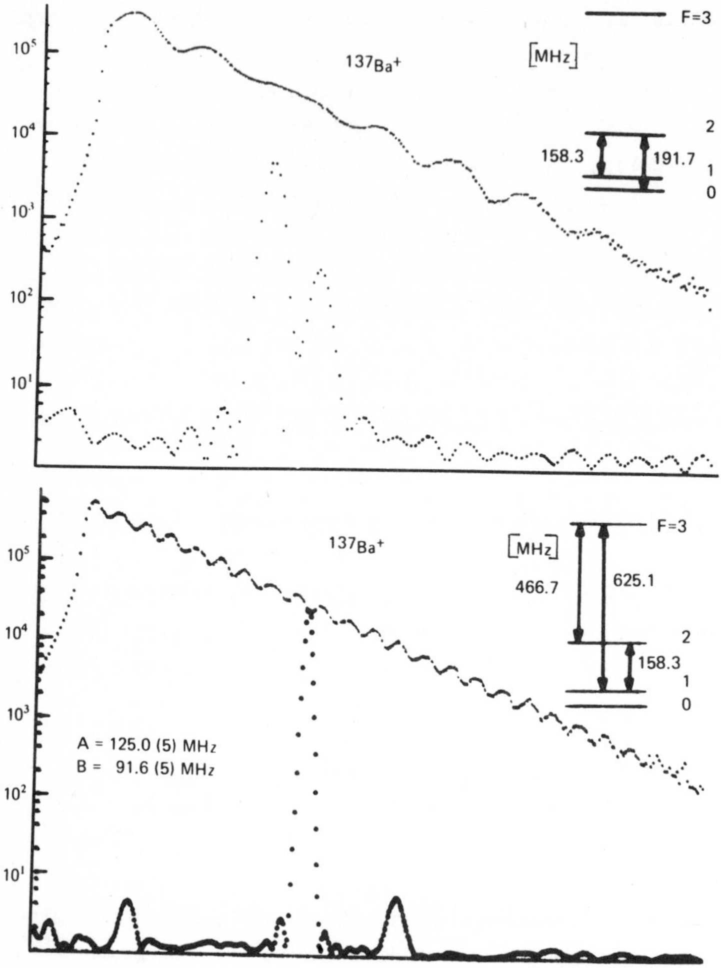

Fig.11.24. Observed quantum beats in the fluorescence of $^{137}Ba^+$ following excitation at λ = 455.4 nm, and corresponding Fourier transform spectrum. The inset level schemes give the hfs of the emitting $6p\,^2P_{3/2}$ state with the observed beat frequencies [11.19]

Doppler free and orders of magnitude better than the spectral width of the laser in use. Its time resolution is comparable to that of the fast beam technique and for a pulsed probe it does not depend on the timing characteristics of the detector. Figure 11.25 shows the experimental arrangement for measuring Zeeman splittings in the Na-groundstate [11.35a]. The pump pulse is provided by a N_2 laser-pumped dye laser and the probe beam by a cw argon laser-pumped dye laser. Both beams are sent nearly collinearly through the sample, which is placed between two crossed polarizers. The

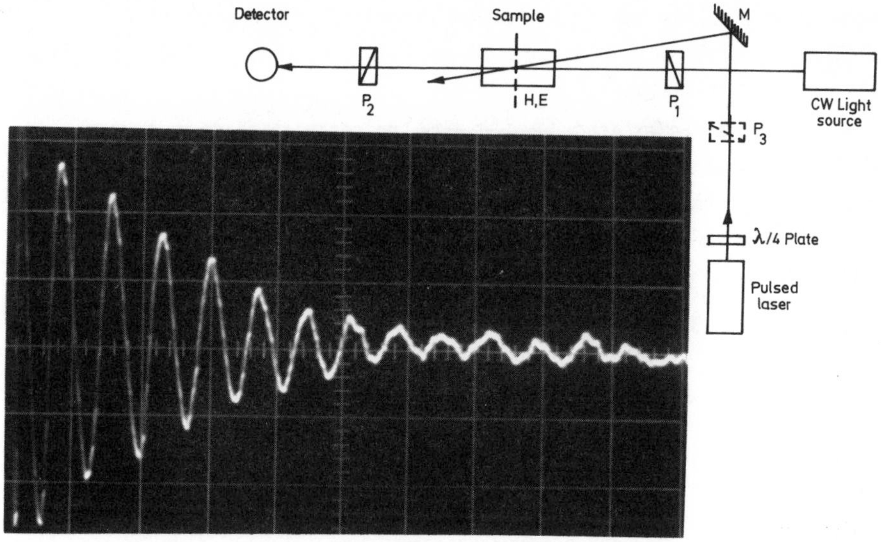

Fig.11.25. Quantum beat spectroscopy in atomic or molecular ground states. The oscilloscope trace shows a Zeeman quantum beat signal for the Na $3^2S_{1/2}$ ground state, recorded with a transient digitizer (time resolution 100 ns) for a single pulse of the pump laser (time scale: 1 μs/div, magnetic field 130 A/m) [11.35a]

probe is polarized either circularly (λ/4 plate) or linearly (without λ/4 plate) at an angle of 45° with respect to the pump pulse. Because of the time-dependent anisotropy induced by the pump pulse, the probe polarization becomes elliptical and the detector receives a signal which is monitored as a function of the delay time.

This time-resolved polarization spectroscopy has, quite similar to its cw counterpart, the advantage of a zero transmission method, avoiding the problem of finding a small signal against a large background. In contrast with cw polarization spectroscopy, no narrow-band single-frequency lasers are required and a broad-band laser source can be utilized which facilitates the experimental setup considerably.

11.4.2 Photon Echoes

Assume that N atoms have been simultaneously excited by a short laser pulse from a lower level E_1 into an upper state E_2. The total fluorescence intensity emitted on a transition $E_2 \rightarrow E_1$ is given by (see Sect.2.9.5)

$$I_{Fl} = \sum_N \hbar\omega A_{21} = \frac{\omega^4}{3\pi\varepsilon_0 c^3}\,\frac{g_1}{g_2}\left|\sum_N \langle\mu_{21}\rangle\right|^2 \quad , \qquad (11.25)$$

where $\mu_{21} = e\, R_{21}$ is the dipole matrix element of the transition $E_2 \rightarrow E_1$ and g_1, g_2 are the weight factors of levels E_1, E_2 [see (2.22,109)]. The sum extends over all N atoms.

If the atoms are excited *incoherently*, no definite phase relations exist between the wave functions of the N excited atoms. The cross terms in the square of the sum (11.25) average to zero and we obtain in the case of identical atoms

$$\left|\sum_N \langle\mu_{12}\rangle\right|^2 = \sum_N |\langle\mu_{12}\rangle|^2 = N|\langle\mu_{12}\rangle|^2 \quad . \tag{11.26}$$

The total fluorescence intensity is therefore

$$I_{F1} = N\hbar\omega A_{12} \quad . \tag{11.27}$$

The situation is drastically changed, however, under *coherent* excitation where definite phase relations are established between the N excited atoms at the time $t = 0$ of excitation. If all N excited atomic states are *in phase* we obtain

$$\left|\sum_N \langle\mu_{12}\rangle\right|^2 = N\sum_N |\langle\mu_{12}\rangle|^2 = N^2|\langle\mu_{12}\rangle|^2 \quad . \tag{11.28}$$

This implies that the fluorescence intensity $I_{F1}(t)$ at a time where all excited atoms oscillate *in phase* is N times larger than in the incoherent case (Dicke superradiance) [11.36]. This phenomenon of superradiance is used in the photon-echo technique for high-resolution spectroscopy to measure population and phase decay times, expressed by the "longitudinal" and "transversel" relaxation times T_1 and T_2. This technique is analogous to the spin-echo method in nuclear magnetic resonance [11.37]. Its basic principle may be understood in a simple model, transferred from NMR to the optical region.

We introduce a pseudo polarization vector $\underline{P} = (P_x, P_y, \mu_{12}\Delta N)$ of the sample, which replaces the magnetization vector in NMR. The components P_x and P_y are the vector sums of the x- and y-components of the molecular transition dipole moments μ_{12}, while the z-component of P is proportional to the population difference $\Delta N = N_1 - N_2$. It can be shown (see for example [1.4]) that the coherent excitation of a sample can be described by an equation for $\underline{P}$, which is analogous to the Bloch equation in NMR

$$d\underline{P}/dt = \underline{P} \times \underline{\omega} - \{P_x/T_2;\ P_y/T_2;\ P_z/T_1\}$$

where the angular frequency $\underline{\omega} = \{(\mu_{12}/2\hbar)E_1;\ 0;\ \omega_{12}\}$ is determined by the amplitude E_1 of the light field and the energy separation $\Delta E = \hbar\omega_{12}$ of the two levels. The time development of $\underline{P}$ describes the time dependent polarization of the system.

At $t = 0$ the atoms in the sample are coherently excited into the state E_2 by a sufficiently intense short laser pulse. If intensity and duration of the pulse are chosen properly, the probabilities $|a_1|^2$ and $|a_2|^2$ of finding an atom in level E_1 or E_2 can change from $|a_1|^2 = 1$, $|a_2|^2 = 0$ before the pulse to $|a_1|^2 = |a_2|^2 = 0.5$ after the pulse. Because such a properly matched pulse changes the phase of the induced polarization by 90°, it is called a $\pi/2$ pulse. Just after the $\pi/2$ pulse, all induced dipoles representing the N atoms are *in phase* (see Sect.2.9.4). This is illustrated in Fig.11.26a by the pseudopolarization vector $\underline{P}$ pointing in the y direction.

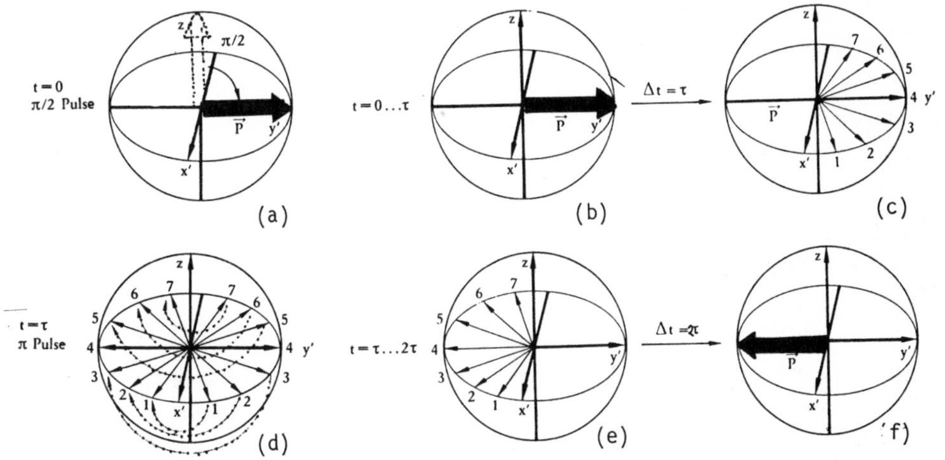

Fig.11.26. Development of the pseudo polarization vector and generation of a photon echo at $t = 2\tau$ after applying a $\pi/2$ pulse at $t = 0$ and a π pulse at $t = \tau$

Because of the finite linewidth $\Delta\omega$ of the transition $E_1 \rightarrow E_2$ (e.g., the Doppler width in a gaseous sample) the frequencies $\omega_{12} = (E_1 - E_2)/\hbar$ of the atomic transitions of our N dipoles are randomly distributed within the interval $\Delta\omega$. This causes the phases of the N atoms to develop in time at a different rate after the end of the $\pi/2$ pulse at $t = 0$. After a time $\tau >> T_2$ which is large compared to the phase relaxation time T_2, the phases are again randomly distributed (Fig.11.26c).

If a second laser pulse which has the proper intensity and duration to invert the phase of the induced polarization (π pulse) is applied to the sample at a time $\tau < T_1$, it causes a reversal of the phase development for each atom (Fig.11.26d,e). This means that after a time $t = 2\tau$ all atoms are again in phase (Fig.11.26f). As discussed above, these excited atoms, while being in phase, emit a superradiant signal which is $N_2(2\tau)$ times larger than the incoherent fluorescence intensity, where $N_2(2\tau)$ is the number of excited atoms at the time $t = 2\tau$. This signal is called a *photon echo*.

There are, however, two relaxation processes which prevent the original state, as prepared just after the first $\pi/2$ pulse at $t = 0$, from being completely re-established at the echo time $t = 2\tau$. Because of spontaneous or collision-induced decay, the population of the upper state decreases to

$$N_2(2\tau) = N_2(0)e^{-2\tau/T_1} \quad . \tag{11.29}$$

This means that also the echo amplitude decreases with the time constant T_1 of the population decay.

A second, generally more rapid relaxation is caused by phase perturbing collisions (see Sect.3.3) which change the phase development of the atoms and therefore prevent all atoms from being again in phase at $t = 2\tau$. Because such phase perturbing collisions give rise to homogeneous line broadening (see Sect.3.5), the phase relaxation time due to these collisions is called T_2^{hom} in contrast to the inhomogeneous phase relaxation, caused for instance by the different Doppler shifts of moving atoms in a gas (Doppler broadening).

The important point, however, is that the *inhomogeneous* phase relaxation which occurs between $t = 0$ and $t = 2$ because of the random Doppler-shifted frequencies of atoms with different velocities does *not* prevent a complete restoration of the initial phases by the π pulse. If the velocity of a single atom does not change within the time 2τ, the different phase development of each atom between $t = 0$ and $t = \tau$ is exactly reversed by the π pulse. This means that even in the presence of inhomogeneous line broadening the homogeneous relaxation processes (i.e., the homogeneous part of the broadening) can be measured with the photon-echo method. *This technique therefore allows Doppler-free spectroscopy.*

The production of the coherent state by the first pulse must of course be faster than these homogeneous relaxation processes. This implies that the laser pulse has to be sufficiently intense. From (2.136) we obtain the condition

$$e\,R_{12}\,E_0 > \hbar(1/T_1 + 1/T_2^{hom}) \tag{11.30}$$

for the product of laser field amplitude E_0 and transition matrix element $e\,R_{12}$. With typical relaxation times of 10^{-6} to 10^{-9} s the condition (11.30) requires laser powers in the KW to MW range which can be readily achieved with pulsed or mode-locked lasers.

With increasing delay time τ between the first $\pi/2$ pulse and the second π pulse the echo intensity I_e decreases exponentially as

$$I_e(2\tau) = I_e(0)E^{-2\tau/T} \quad \text{with} \quad 1/T = (1/T_1 + 1/T_2^{hom}) \quad . \tag{11.31}$$

From the slope of a logarithmic plot of $I_e(2\tau)$ versus the delay time the homogeneous relaxation times can be obtained.

The qualitative presentation of photon echoes, discussed above, may be put on a more quantitative base by using time-dependent perturbation theory. We outline briefly the basic considerations which can be understood from the treatment in Sect.2.9. For a more detailed discussion see [11.38].

A two-level system can be represented by the time-dependent wave function (2.89)

$$\psi(t) = \sum_{n=1}^{2} a_n(t)u_n \, e^{-iE_nt/\hbar} \quad . \tag{11.32}$$

Before the first light pulse is applied, the system is in the lower level E_1 which means $|a_1| = 1$ and $|a_2| = 0$. A harmonic perturbation

$$V = -\underline{p} \cdot \underline{A}_0 \cos\omega t \quad \text{with} \quad \hbar\omega = E_2 - E_1 \tag{11.33}$$

produces a linear superposition

$$\psi(t) = \cos\left(\left|\frac{pA_0}{2\hbar}\right| t\right) u_1 \, e^{-iE_1t/\hbar} + \sin\left(\left|\frac{pA_0}{2\hbar}\right| t\right) u_2 \, e^{-iE_2t/\hbar} \quad . \tag{11.34}$$

If this perturbation consists of a short intense light pulse of duration T such that

$$|(\underline{p}\,\underline{A}/\hbar)|T = \pi/2 \quad , \tag{11.35}$$

the total wave function becomes, with $\cos\pi/4 = \sin\pi/4 = 1/\sqrt{2}$,

$$\psi(t) = \frac{1}{\sqrt{2}} \left[u_1 \, e^{-iE_1t/\hbar} + u_2 \, e^{-iE_2t/\hbar} \right] \quad . \tag{11.36}$$

After a time τ, the phases have developed to $E_n\tau/\hbar$. If now a second π pulse with

$$|(pA_0/\hbar)|T = \pi$$

is applied, the wave functions of the upper and lower states are just interchanged, so that for time t

$$(e^{-iE_1\tau/\hbar})u_1 \rightarrow e^{-iE_1\tau/\hbar}u_2\ e^{-iE_2(t-\tau)/\hbar} \tag{11.37}$$

$$e^{-iE_2\tau/\hbar}u_2 \rightarrow e^{-iE_2\tau/\hbar}u_1\ e^{-iE_1(t-\tau)/\hbar} \quad .$$

The total wave function therefore becomes

$$\psi = \frac{1}{\sqrt{2}}\left[u_2\ e^{-\frac{1}{2}i\omega_k(t-2\tau)} - u_1\ e^{+\frac{1}{2}i\omega_k(t-2\tau)}\right] \tag{11.38}$$

and the dipole moment for each atom is

$$|\psi^*|e\underline{r}|\psi>_k = -<u_2^*|e\underline{r}|u_1>_k\ e^{-i\omega_k(t-2\tau)} \quad . \tag{11.39}$$

If the different atoms have slightly different absorption frequencies $\omega_k = (E_2 - E_1)/\hbar$ (e.g., because of the different velocities in a gas), the phase factors for different atoms are different for $t \neq 2\tau$ and the macroscopic fluorescence intensity is the incoherent superposition (11.26) from all atomic contributions. However, for $t = 2\tau$ the phase factor is zero for all atoms, which implies that all atomic dipole moments are in phase and superradiance is observed.

Photon echoes were first observed in ruby crystals using two ruby laser pulses with variable delay [11.39]. The application of this technique to gases started with CO_2 laser pulses incident on a SF_6 sample. From the time decay of the echo amplitude the collision-induced homogeneous relaxation time T_2^{hom} has been measured. Figure 11.27 is an oscilloscope trace of the $\pi/2$ and π pulses and the echo obtained from a SF_6 cell at a pressure of 0.015 torr [11.40].

Instead of short laser pulses cw lasers can also be used if the molecules are tuned for a short time interval into resonance with the laser frequency. Experimentally this can be achieved either by switching the laser frequency or by Stark tuning of the molecular absorption lines with a pulsed electric field. The laser frequency switching may be performed for instance by acousto-optical modulators outside the laser resonator or by electro-optic crystals inside the resonator. Figure 11.28 is an example of an experimental arrangement [11.42] where a stable tunable cw dye laser is used and frequency switching is achieved with an ammonium dihydrogen phosphate (ADP) crystal driven by a sequence of low-voltage pulses causing a variation of the refractive index n and therefore a shift of the laser wavelength $\lambda = c/(2nd)$.

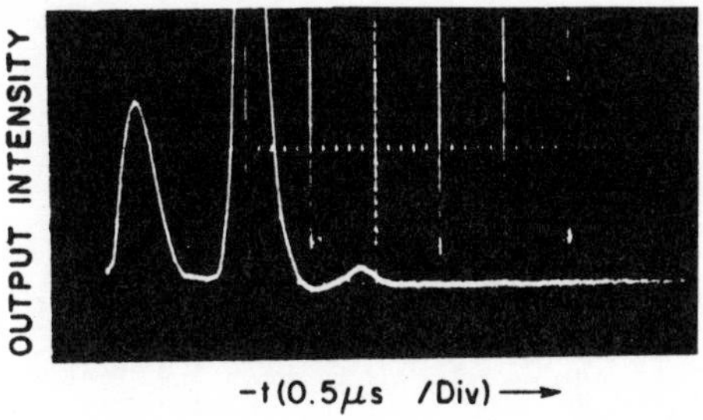

Fig.11.27. Oscilloscope trace of output pulses from an SF_6 cell. The first two pulses are the transmitted CO_2 laser $\pi/2$ and π pulses, the third is the photon echo [11.40]

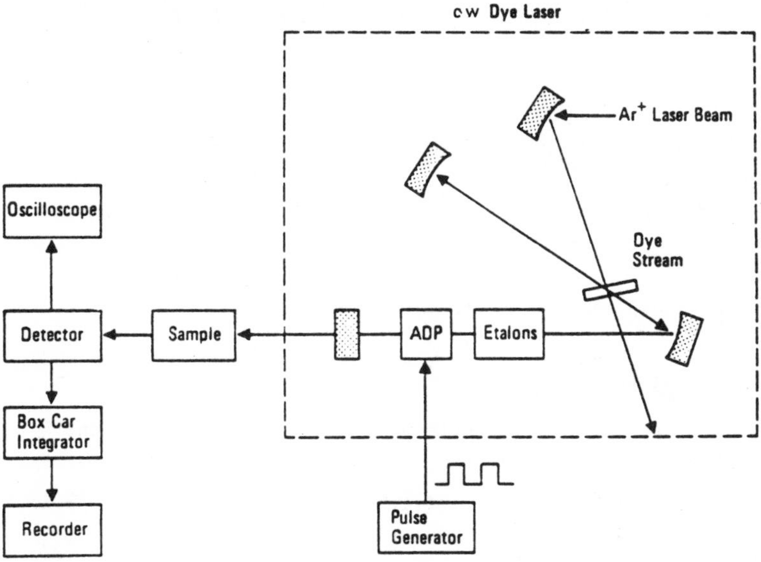

Fig.11.28. Schematic of the laser frequency switching apparatus for observing coherent optical transients [11.42]

Figure 11.29 illustrates the Stark switching technique, which can be applied to all those molecules that show a sufficiently large Stark shift [11.43]. In the case of Doppler-broadened absorption lines the laser of fixed frequency initially excites molecules of velocity v_z. A Stark pulse which abruptly shifts the molecular absorption profile from the solid to the dashed curve causes the velocity group v_z' to come into resonance with the laser frequency Ω. We have assumed that the Stark shift of the molecular eigenfrequencies is larger than the homogeneous linewidth but smaller than the Doppler width. With two Stark pulses the group v_z' emits an echo. This is shown in Fig.11.29b where the CH_3F molecules are switched twice into resonance with a cw CO_2 laser by 60 V/cm Stark pulses.

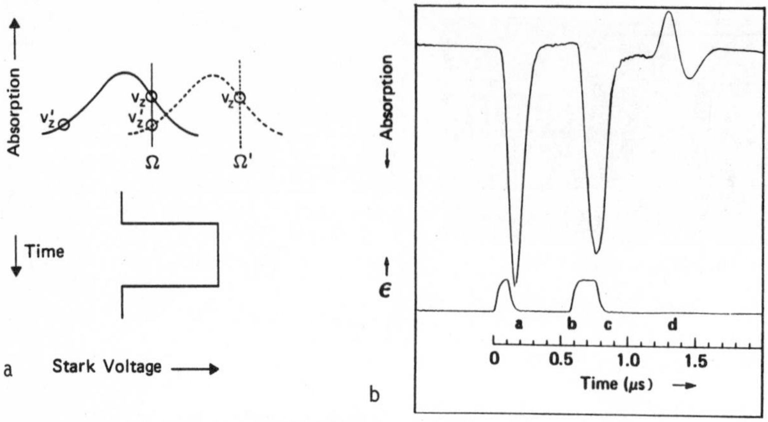

Fig.11.29. (a) Stark switching principle for the case of a Doppler-broadened transition. (b) Infrared photon echo for a $^{13}CH_3F$ vibration rotation transition is shown as the third pulse in the upper trace. The gas sample is switched twice into resonance with a cw CO_2-laser by the two Stark pulses (lower trace). (From ref. [11.43])

For a more detailled discussion of photon echoes see [11.38,43,44].

11.4.3 Optical Nutation and Free Induction Decay

If the laser pulse applied to the sample molecules is sufficiently long and intense, a molecule (represented by a two-level system) will be driven back and forth between the two levels at the Rabi flopping frequency (2.134). The time-dependent probability amplitudes $a_1(t)$ and $a_2(t)$ are now periodic functions of time and we have the situation depicted in Fig.2.30. Since the laser beam is alternately absorbed (induced absorption $E_1 \rightarrow E_2$) and amplified (induced emission $E_2 \rightarrow E_1$), the intensity of the transmitted beam will display an oscillation. Because of relaxation effects this oscillation is damped and the transmitted intensity reaches a steady state determined by the ratio of induced to relaxation transitions. According to (2.133) the Rabi frequency depends on the laser intensity and on the detuning $(\omega_{12} - \omega)$ of the molecular eigenfrequency ω_{12} from the laser frequency ω. This detuning can be performed either by tuning of the laser frequency ω (Fig.2.28) or by Stark tuning of the molecular eigenfrequencies ω_{12}.

We consider the case of a gaseous sample with Doppler-broadened transitions [11.45], where the molecular levels are shifted by a pulsed electric field. The Stark shift of the molecular eigenfrequencies is larger than the homogeneous linewidth but smaller than the Doppler width (Fig.11.29a). During the steady-state absorption preceding the Stark pulse, the single-mode cw

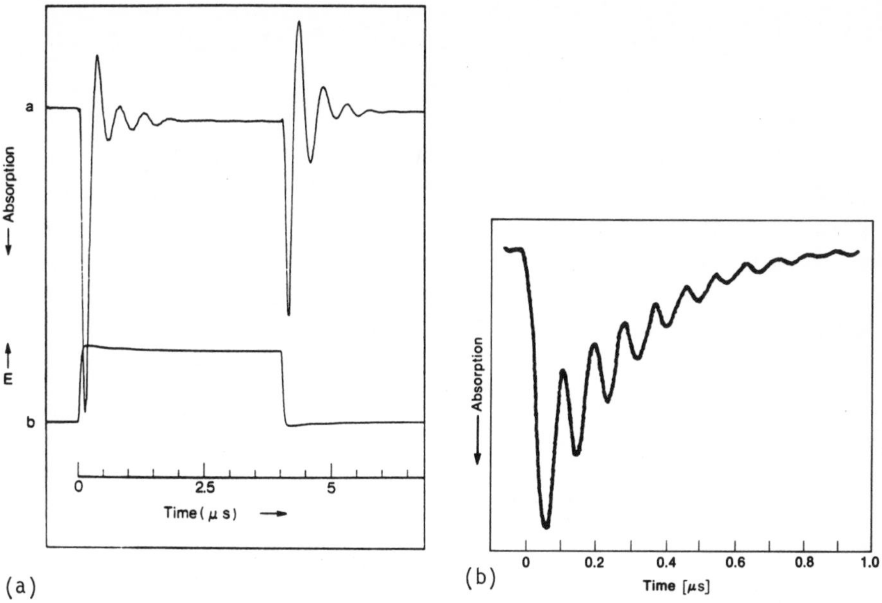

Fig.11.30. (a) Optical nutation in $^{13}CH_3F$ with cw CO_2-laser excitation at $\lambda = 9.7$ μm. The Rabi oscillations appear because the Stark pulse is longer than in Fig.11.29b. (b) Optical free induction decay in H_2D at 10.6 μm following a step function Stark pulse. The beat frequency is the Stark shift and the slowly varying background is a nutation signal similar to that in (a). Note the different time scales in (a) and (b) [11.43]

laser exites only a narrow velocity group of molecules around v_{x1} within the Doppler line shape. Sudden application of a Stark field shifts the transition frequency ω_{12} of this subgroup out of resonance with the laser field and another velocity subgroup around v'_{z2} is shifted into resonance, giving the first optical nutation pattern in Fig.11.30. When the Stark pulse terminates, the initial velocity group is switched back into resonance and generates the second nutation pattern. The amplitude A of this delayed nutation depends on the population of the first subgroup v_{x1} at the time τ where the Stark pulse ends. This population had been partially saturated before $t = 0$, but collisions during the time τ try to refill it. BERMAN et al. [11.46] have shown that

$$A(\tau) \propto N_1(\tau) - N_2(\tau) = \Delta N_0 - [\Delta N_0 - \Delta N(0)]e^{-\tau/T_1} \quad , \tag{11.40}$$

where ΔN_0 is the unsaturated population difference in the absence of radiation and $\Delta N(0)$, $\Delta N(\tau)$ are the saturated population difference at $t = 0$

or the partially refilled difference at $t = \tau$. The dependence of $A(\tau)$ on the length τ of the Stark pulse therefore allows measurement of the relaxation time T_1 for refilling the lower level and depopulating the upper one.

The Stark switching technique also allows observation of the free induction decay of coherently prepared molecules in the optical region [11.47]. During steady-state excitation (in Fig.11.30a for $t < 0$) the molecules of the absorbing velocity subgroup are prepared in a coherent two-level superposition. When a Stark field is suddenly applied at $t = 0$, these molecules are switched out of resonance with the laser field. Due to the coherent preparation during the time $t < 0$ these molecules generate a macroscopic dipole moment oscillating on the molecular eigenfrequency ω_{12} with a phase depending on the phase of the total wave function ψ in (11.39) at $t = 0$. This results in a coherent emission in the direction of the laser beam.

This coherent emission of the N dipoles in phase is the optical analogue of the free induction decay first observed in NMR spectroscopy. The intensity is proportional to N_2 (see Sect.11.4.1). Due to the Stark shift the molecular eigenfrequency ω_{12} differs from the laser frequency ω. The superposition of laser beam and coherent molecular emission therefore results in a beat signal at the difference frequency $(\omega_{12} - \omega)$. Figure 11.30b shows the optical free induction decay in NH_2D following a step function Stark pulse. The beat frequency $\omega_{12} - \omega$ is the Stark shift and the slowly varying background is the optical nutation signal. Note the different time scales in Figs.11.30a,b. The free induction decay is of course also present in Fig.11.30a but cannot be resolved because it is too short lived.

The difference between optical nutation and free induction decay should be clear. While the optical nutation occurs at the Rabi frequency which depends on the product of laser field intensity and transition moment, the free induction decay is monitored as a heterodyne signal at the beat frequency $\omega_{12} - \omega$ which depends on the Stark shift. The importance of these coherent transient phenomena for time-resolved sub-Doppler spectroscopy is discussed in the next section. Its application to the study of collision processes is treated in Chap.12. For more detailed information the excellent reviews of BREWER [11.43,48] are recommended.

11.4.4 Pulse Fourier Transform Spectroscopy

In the previous sections we have assumed that the molecules under investigation had only a single transition between two levels E_1 and E_2, so that a single velocity group could be selected by excitation with a single-mode laser. For the case of several closely spaced transitions that overlap within their Doppler line profiles, a corrresponding number of molecular velocity groups may be prepared simultaneously. After coherent preparation, the different velocity groups radiate at slightly different frequencies, producing an interference pattern in time. These transient signals obtained by free induction decay or by photon echoes can be processed by a computer performing a Fourier transformation. This yields the frequency distribution of the spectrum with a spectral resolution limited only by the homogeneous linewidths.

Using the Stark switching technique with varying pulse delay times in an echo experiment, BREWER et al. [11.49] could map the decay behavior of each line in a three-dimensional diagram of signal amplitude versus frequency and elapsed time. This is illustrated in Fig.11.31 which shows such a time-resolved spectrum with sub-Doppler spectral resolution. It has been obtained in the following way. A frequency-stabilized CO_2 laser excites the $(J,K) = (4,3) \rightarrow (5,3)$ transition in the fundamental ν_3 mode of the $^{13}CH_3F$ molecule. Due to the Stark shift, the M sublevels are split and eight lines appear symmetrically about the laser frequency in the emission spectrum which obey the selection rule $\Delta M = 0$. Since the sample is prepared with the

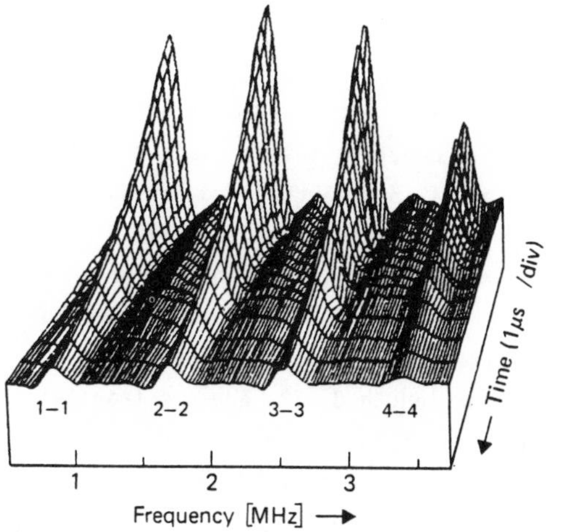

Fig.11.31. Fourier transform heterodyne beat spectrum of $^{13}CH_3F$ derived from Stark switched two-pulse photon echoes [11.49]

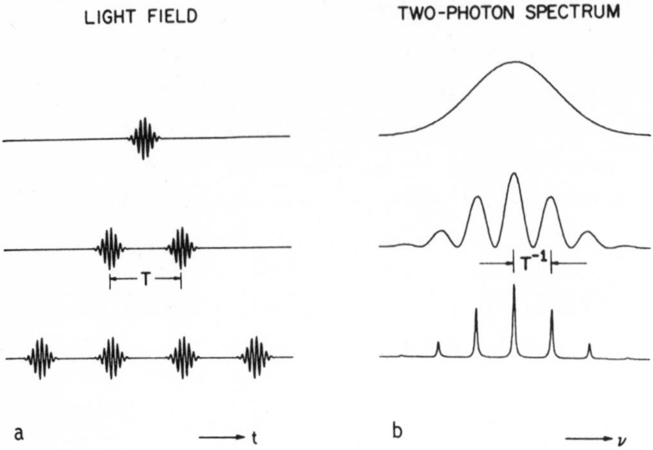

Fig.11.32. Fourier spectrum of a single pulse, two pulses, and of a regular pulse train. a) Field amplitude as a function of time. b) Spectral profile as monitored e.g., by two-photon spectroscopy (after [11.52])

Stark pulse on, but radiates when it is off, the emission is Stark shifted from the laser and four heterodyne beat frequencies are produced at the detector, each beat being due to two transitions $\pm M \rightarrow \pm M$. The four lines are 170 KHz wide and are spaced at 0.83 MHz [11.50]. The exponential decay of the echo signal with increasing delay time τ, which is plotted in Fig.11.31, samples the homogeneous dipole dephasing rate due to elastic and inelastic collisions (see Sect.11.4.1). This technique therefore allows measurement of collision rates separately for different M levels. Since different velocity groups of molecules can be sampled, depending on the Stark shift, information on the velocity dependence of the collision rates can also be obtained (see Chap.12).

Another experimental technique which may also be regarded as pulse Fourier transform spectroscopy is based on multiple coherent interactions of a sample with a regular train of short light pulses [11.51]. Such a regular pulse train may be produced either by cw mode-locked lasers (see Sect.11.1) or by injecting a single laser pulse into an external optical resonator, where it is reflected back and forth. The pulse separation time in the latter case is equal to the transit time through the resonator and can therefore be easily controlled by changing the resonator length d with a piezo translator. The gas sample is placed near one end mirror of the resonator and the molecules are therefore exposed to a pulsed standing wave field with the pulse duration of the injected laser pulse and a pulse separation $T = 2d/c$.

Figure 11.32 shows the Fourier spectrum of a single pulse, of two pulses, and of a regular train of pulses. In the case of a cw mode-locked laser the frequency spectrum of the pulse train is just the mode spectrum of the laser resonator, because the laser output is the phase-locked superposition of all resonator modes. In case of a single laser pulse injected into an external resonator, the frequency spectrum of the resultant pulse train again equals the axial mode spectrum of this resonator. Another interpretation of the same fact is the following. The resonator acts as a spectral filter which filters its eigenfrequencies out of the broad spectrum of the injected single pulse.

Tuning the resonator length allows tuning of the frequency spectrum of Fig.11.32b over the molecular absorption lines. If Doppler-free techniques such as two-photon spectroscopy or polarization or saturation spectroscopy are used (see Chap.10), a similar spectral resolution is obtained as with cw single mode lasers [11.52]. However, the higher peak powers of pulsed lasers facilitate nonlinear spectroscopy and increase the efficiency of optical frequency doubling. This is for instance important for high-resolution photochemistry where high-lying molecular levels can be selectively excited and the strong signal enhancement increases the sensitivity.

12. Laser Spectroscopy of Collision Processes

The two main sources of information about atomic and molecular structure and interatomic interactions are provided by spectroscopic measurements and by the investigation of elastic, inelastic, or reactive collision processes. For a long time these two branches of experimental research have been developed along separate lines without a strong mutual interaction. The main contributions of classical spectroscopy to the study of collision processes have been the investigations of collision-induced spectral line broadening and line shifts (see Sect.3.3).

The situation has changed considerably since lasers were introduced to this field. In fact, laser spectroscopy has already become a powerful tool for studying various kinds of collision processes in more detail. The different spectroscopic techniques, presented in this chapter, illustrate the wide range of laser applications in collision physics and they demonstrate that detailed information about the collision processes may be obtained. This provides a better knowledge of the interaction potentials which often cannot be adequately obtained from classical scattering experiments without lasers.

The high spectral resolution of various Doppler-free techniques, discussed in Chap.10, has opened a new dimension in the measurement of collisional line broadening. While in Doppler-limited spectroscopy small line-broadening effects at low pressures are completely masked by the much larger Doppler width, Doppler-free spectroscopy is well suited to measure line-broadening effects and line shifts in the kHz range. This allows detection of "soft collisions" at large impact parameters of the collision partners which probe the interaction potential at large internuclear separations and which contribute only a small line broadening.

Some techniques of laser spectroscopy, such as the method of separated fields (optical Ramsey fringes, see Sect.13.1) or coherent transient spectroscopy (see Sect.11.4) allow one to distinguish between phase changing, velocity changing, or orientation changing collisions.

The high *time resolution*, achievable with pulsed or mode-locked lasers (see Chap.11) opens the possibility of studying the dynamics of collision processes and relaxation phenomena. The interesting question how and how fast the excitation energy, selectively pumped into a polyatomic molecule by absorption of laser photons, is redistributed among the various degrees of freedom by intermolecular or intramolecular energy transfer can be attacked by time-resolved laser spectroscopy.

One of the attractive goals of laser spectroscopy of reactive collision processes is the basic understanding of chemical reactions. The fundamental question in laser chemistry of how the excitation energy of the reactants influences the reaction probability and the internal state distribution of the reaction products can be at least partly answered by detailed laser spectroscopic investigations. Section 12.4 treats some experimental techniques in this field.

The new and interesting field of "light-assisted collisions", where absorption of laser photons by a collision pair results in effective excitation of one of the collision partners, is briefly treated in the last section of this chapter.

12.1 High-Resolution Laser Spectroscopy of Collisional Line Broadening and Line Shifts

In Section 3.3 we discussed how elastic and inelastic collisions contribute to the broadening of spectral lines. In a semiclassical model, where the colliding particles travel along definite paths, an impact parameter b can be defined (see Fig.12.1) and the collisions may be classified as *soft* collisions (impact parameter b large compared to the minimum location r_0 of the interaction potential) and *hard* collisions ($b < r_0$). The soft collisions probe the

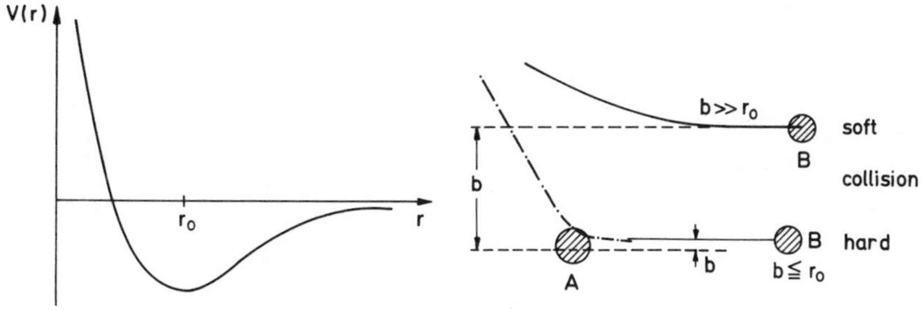

Fig. 12.1. Soft and hard collisions with impact parameters $b \gg r_0$ and $b \ll r_0$, respectively

long range of the interaction potential, result in small scattering angles, and contribute mainly to the *kernel* of the collision-broadened line (i.e., that part of the line profile within the halfwidth). The hard collisions, on the other hand, probe the short-range potential and give rise to the wings of the collision-broadened line.

In Doppler-limited spectroscopy the influence of collisions on the line kernel is generally masked by the much larger Doppler width and the information obtained from pressure broadening is extracted from measurements of the wings, which are described by the convolution of Gaussian and Lorentzian line profiles (Voigt profile, see Sect.3.2). Since the pressure-broadened linewidth is proportional to the pressure, the unambiguous separation of Doppler profile and collisional contributions are possible with sufficient accuracy only at higher pressures.

With increasing pressure, however, many-body collisions are no longer negligible, because the probability that n atoms are simultaneously in a volume $dV \approx r_0^3$ where they interact with each other, increases with the nth power of the density. This implies that not only binary collisions contribute to line broadening but also three- or more body collisions. This makes it much more difficult to extract unambiguous two-body interaction potentials from the pressure-broadened line profiles [3.16].

The Doppler-free techniques, which have been explained in Chap.10, eliminate the bothering Doppler width and therefore already small collision broadening can be sensitively monitored. One example is given by the investigation of broadening or shifts of Lamb dips in saturation spectroscopy (see Sect. 10.2), which can be measured with an accuracy of a few kilohertz if stabilized lasers are used. This high accuracy allows detection of even small interaction forces such as those experienced by atoms or molecules without permanent dipole moments. The interaction potential at large internuclear distances can be described in these cases by a van der Waals potential $V(r) = -ar^{-6}$ where the constant a depends on the polarizability of the collision partners.

Most measurements up to now have been performed with He-Ne lasers on visible transitions [12.1] or on the 3.39 μm line [12.2] with absorption cells inside the laser resonator. In the infrared region CO_2 lasers have also been used [12.3]. The laser output $I_L(\omega)$ is monitored while the laser frequency ω is tuned across the Doppler-broadened absorption line and the Lamb dip at the center of the absorption profile occurs as a peak in the laser output. The line profile and the center frequency ω_0 of this inverse Lamb dip are measured as a function of pressure in the absorption cell. Figure 12.2 shows the half-

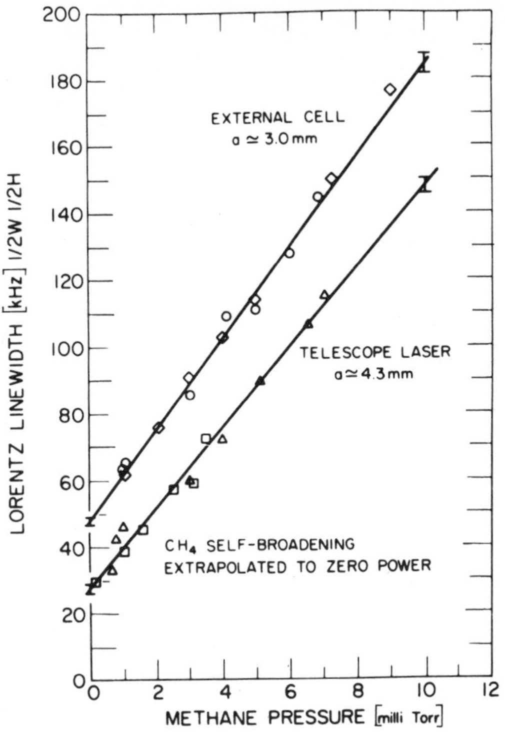

Fig. 12.2. Halfwidth of the "Lamb peak" in the output of an He-Ne laser at $\lambda = 3.39$ µm with an internal methane cell as a function of the CH_4 pressure (lower curve). The upper curve shows pressure broadening in an external CH_4 cell. The two different intersects are mainly due to the different laser beam diameters in the two cells, causing different time of flight broadening [12.4]

width of this inverse Lamb dip at $\lambda = 3.39$ µm as a function of the CH_4 pressure in the intracavity methane cell [12.4].

The halfwidth is determined by pressure and by power broadening (see Sect. 3.6). From the slope of the straight lines in Fig.12.2 the line broadening parameter σ_b in (3.53) can be obtained. Together with line-shift measurements, which yield the second parameter σ_s, (3.55), the interaction potential, dependent on the small polarizability of CH_4, can be deduced.

The understanding of collision broadening of the Lamb dip profile demands some more considerations. As discussed in Sect.10.2, only molecules with velocity components v_x inside a narrow interval $\Delta v_x = \gamma/|\underline{k}|$ around $v_x = 0$, (γ is the homogeneous width of the Lamb dip) can interact resonantly with both travelling waves generating the monochromatic laser field of a standing wave. These are those molecules which move within a cone of angle ε (see Fig.12.3) such that

$$\sin\varepsilon = v_x/|\underline{v}| \leqq \gamma/kv = \gamma\lambda/(2\pi v) \quad ; \quad v = |\underline{v}| \quad . \tag{12.1}$$

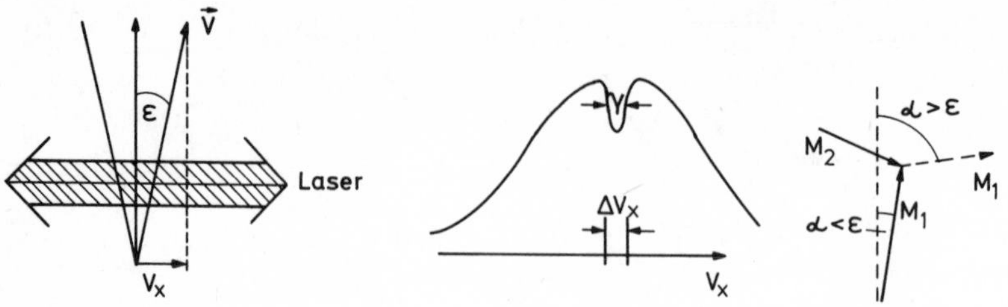

Fig. 12.3. Only those molecules with velocities within a conical slab of angle ε around directions perpendicular to the laser beam axis contribute to the line profile of the Lamb dip. Collisions which change this angle from $\alpha < \varepsilon$ to $\alpha > \varepsilon$ throw the molecule out of resonance with the laser

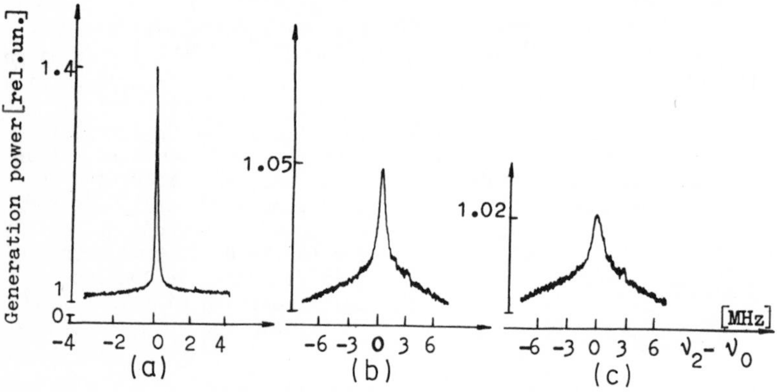

Fig. 12.4a-c. Line profile of the saturation peak of an He-Ne laser at λ = 3.39 µm with internal CH_4 cell. (a) Pure CH_4 at 1 mTorr, (b) addition of 20 mTorr He, (c) 43 mTorr He [12.5]

Since the width γ of the Lamb dip is generally small compared with the Doppler width $\Delta\omega_D$, this implies that $v_x \ll |\underline{v}|$.

During an elastic collision a molecule with velocity $\underline{v}$ may be deflected by an angle ϑ. If

$$v \sin\vartheta < \gamma\lambda/2\pi \quad , \tag{12.2}$$

this molecule may still be in resonance with the laser field. Collisions with scattering angles $\vartheta \ll \varepsilon$ (called "soft" or "weak" collisions) therefore do not appreciably change the probability for a molecule to absorb a laser photon. Because of phase randomization during the collision, weak collisions do, however, broaden the line which still remains Lorentzian in shape (see Sect.3.3). The broadening is proportional to pressure.

Velocity changing collisions with scattering angles $\vartheta > \varepsilon$, on the other hand, do shift the molecules out of resonance with the laser field ("hard" or "strong" collisions). After a strong collision a molecule can only contribute to the absorption at the far wings of the Lorentzian profile of the Lamb dip. Including strong and weak elastic collisions, one observes an inverse Lamb dip with Lorentzian line shape, slightly broadened by weak collisions, with a broad background due to strong collisions. Figure 12.4 illustrates such a line profile obtained as the inverse Lamb dip of a He-Ne laser at $\lambda = 3.39\ \mu m$ with internal methane cell [12.5].

Often the collision broadening of Lamb dip and of the Doppler profile can be measured simultaneously. The comparison of both broadenings leads to a conclusion about the different contributions to line broadening. For collisions with phase randomization there is no difference between the broadenings of the two different line profiles. An additional broadening of the inverse Lamb dip may be due (apart from power broadening) to velocity changing collisions with angular scattering. This contribution may increase nonlinearly with pressure. The case

$$kv\vartheta \ll N\bar{v}\sigma = \gamma_{coll}^{hom} \tag{12.3}$$

corresponds to weak collisions where σ is the elastic cross section and γ_{coll}^{hom} is the line broadening due to phase randomization, while $kv\vartheta \gg N\sigma\bar{v}$ corresponds to strong collisions. This shows that with increasing density the allowed maximum scattering angle for weak collisions becomes larger.

Orientation changing collisions can be sensitively detected with polarization spectroscopy (see Sect.10.3). Since the polarized pump wave produces a partial orientation of the molecular angular momentum, which gives rise to a change in polarization of the linearly polarized probe wave, any collisions which change the orientation alter the probe intensity transmitted by the second polarizer in Fig.10.33. Another way to detect these collisions uses saturation spectroscopy, where the angle β between the planes of polarization of pump and probe wave are varied and the Lamb dip profile is monitored as a function of β [12.6].

12.2 Measurement of Inelastic Collisions by LIF

Assume that a vibrational-rotational level (v_k', J_k') in an excited electronic state of a molecule has been selectively populated by optical pumping. If the excited molecules (density N_A) suffer inelastic collisions with other atoms or molecules (density N_B) before they emit a fluorescence photon, they are

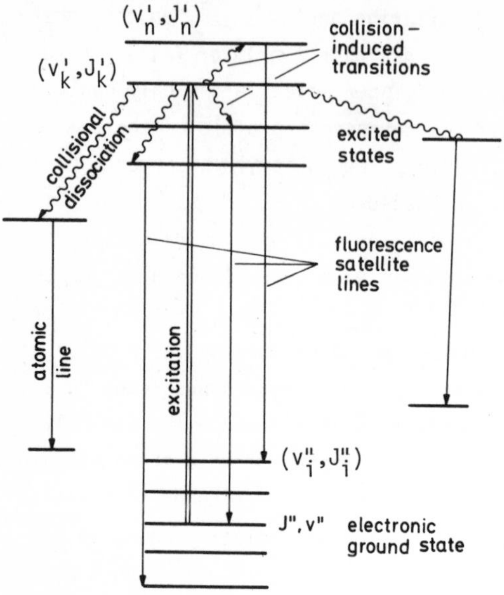

Fig. 12.5. Schematic level diagram for illustration of collision-induced molecular transitions

being transferred into other levels (v'_n, J'_n) either of the same or of another electronic state (Fig.12.5). These collision-induced radiationless transitions, which play an important role in chemical reactions, can be studied in detail by the technique of laser-induced fluorescence (LIF, see Sect.8.7).

The rate equation for the depopulation of level (v'_k, J'_k) after the end of the optical pump pulse is given by

$$dN_k/dt = -\Big(\sum_i A_{ki} + \sum_n R_{kn}\Big)N_k = -P_k N_k \quad , \tag{12.4}$$

where the total depopulation rate $P_k N_k$ is the sum of spontaneous rate $A_k N_k = \sum_i A_{ki} N_k$ for fluorescence transitions into lower levels i and the net relaxation rate $\sum_n R_{kn} N_k$ caused by collision-induced transitions into other levels n. Integration of (12.4) yields

$$N_k = N_k(0)\, e^{-t/\tau_{eff}} \quad , \tag{12.5}$$

where the effective lifetime τ_{eff} is given by

$$1/\tau_{eff} = \sum_i A_{ki} + \sum_n R_{kn} \quad . \tag{12.6}$$

The collision rate R_{kn} is related to the integral cross section σ_{kn} by

$$R_{kn} = \int N_B v_r \sigma_{kn} dv_r \quad , \tag{12.7}$$

where v_r is the relative velocity of the collision partners. It has been assumed that $N_A \ll N_B$, so that collisions between excited molecules can be neglected. At thermal equilibrium $N_B(v_r)$ follows a Maxwellian distribution. If σ_{kn} is not critically dependent on v_r we can approximate (12.7) by

$$R_{kn} = \bar{\sigma}_{kn} \int N_B(v_r) v_r \, d\, v_r = \bar{\sigma}_{kn} \bar{v}_r N_B \quad . \tag{12.8}$$

Here $\bar{\sigma}_{kn}$ is an average cross section and $\bar{v}_r$ is the mean relative velocity

$$\bar{v}_r = (kT/\pi\mu)^{\frac{1}{2}} \quad , \quad \mu = \frac{m_A m_B}{m_A + m_B} \quad . \tag{12.9}$$

At sufficiently low pressures we can use the ideal gas law to relate the density N_B with the pressure p

$$p = N_B kT \quad . \tag{12.10}$$

Inserting (12.8-10) into (12.6) yields with $\sigma_k^{tot} = \sum_n \sigma_{kn}$ the well-known Stern-Vollmer relation (2.82)

$$\frac{1}{\tau_{eff}} = \frac{1}{\tau_{spont}} + \sqrt{\frac{8}{\pi\mu kT}} \, \sigma_k^{tot} p \quad . \tag{12.11}$$

A plot of the inverse measured lifetime $1/\tau_{eff}$ versus the pressure p yields a straight line. From its slope the mean total deactivation cross section can be obtained while the intersection at p = 0 gives the unperturbed spontaneous lifetime. Figure 12.6 shows two such Stern-Vollmer plots where optically excited Na_2 molecules collide either with Na atoms (upper plot) or with He atoms (lower plot). The total deactivation cross section in the upper plot is about four times larger than in the lower plot [12.7]. The reason for this will be discussed below.

The total deactivation cross section σ^{tot} is a sum

$$\sigma^{tot} = \sigma^{rot} + \sigma^{vib} + \sigma^{el} + \sigma^{diss}$$

of cross sections for rotational transitions $(v_k', J_k') \rightarrow (v_k', J_k' + \Delta J)$, for vibrational-rotational transitions $(v_k', J_k') \rightarrow (v_k' + \Delta v_k, J_k' + \Delta J)$, for collision-induced electronic transitions, and for collision-induced dissociation processes. There are several experimental techniques to measure these different contributions separately.

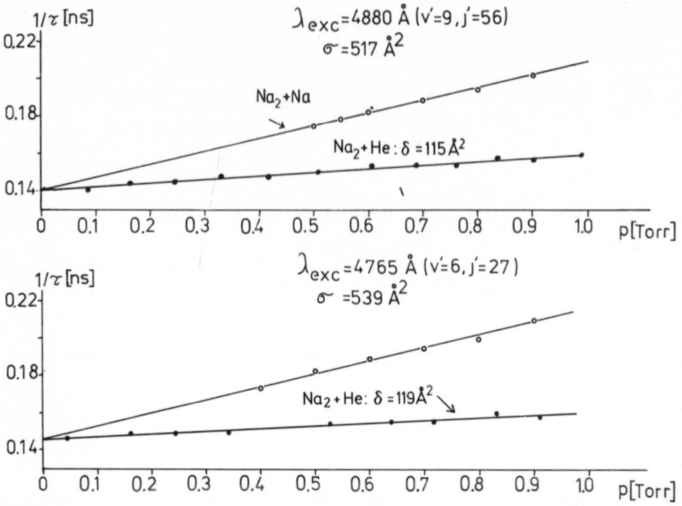

Fig. 12.6. Stern-Vollmer plot showing the dependence of effective lifetime of an excited (v',J') level in the $B^1\Pi_u$ state of Na_2 on the vapor pressure of Na atoms (upper curves) and on the pressure of He added to the vapor cell

If the levels (v_n', J_n') populated by collision-induced transitions from the optically pumped level (v_k', J_k') belong to the same electronic state, the fluorescence from these levels generates "satellite lines" in the fluorescence spectrum around the "parent lines" emitted from (v_k', J_k'). Figure 12.7 illustrates such a collision-induced satellite spectrum emitted from the $B^1\Pi_u$ state of the Na_2 molecule. The spectrum is produced by collision-induced rotational transitions $(v_k', J_k') \rightarrow (v_k', J_k' + \Delta J)$ within the same vibrational level. The number above the lines give the rotational quantum jump ΔJ.

These satellite lines contain the whole information about the inelastic collision process. From the wavelength of a line the emitting level (v_n', J_n') can be generally determined unambiguously, provided that the molecular constants of the excited electronic state and of the ground state are known sufficiently accurately. The intensity ratio $I_s(v_n', J_n')/I_0(v_k', J_k')$ of satellite line to parent line is given by the corresponding population densities N_n, N_k and by the spontaneous transition probabilities A_{nm} and A_{ki}

$$\frac{I_s(v_n', J_n' \rightarrow v_m'', J_m'')}{I_0(v_k', J_k' \rightarrow v_i'', J_i'')} = \frac{N_n A_{nm} h\nu_{nm}}{N_k A_{ki} h\nu_{ki}} \quad . \tag{12.12}$$

If the ratio A_{nm}/A_{ki} of the spontaneous transition probabilities is known from spectroscopic studies, the ratio N_n/N_k of the population densities can be deduced from the measured intensity ratios. Under stationary conditions

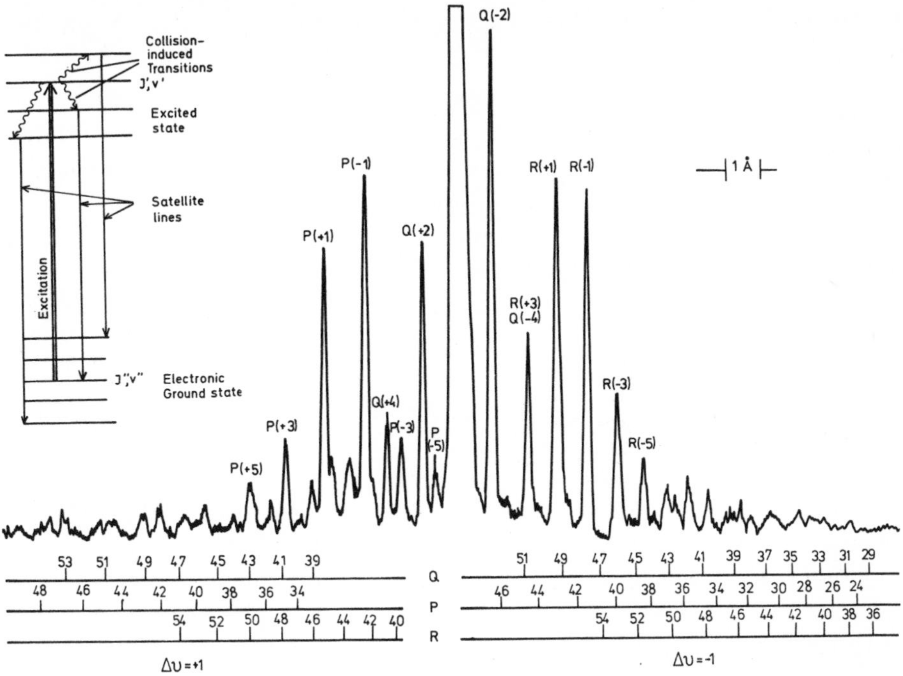

Fig. 12.7. Collision-induced satellite lines in the laser-induced fluorescence spectrum. Example of rotational and vibrational transitions from the (v'=6,J'=43) level in the $B^1\Pi_u$ state of Na_2. The parent line is 20-fold off scale

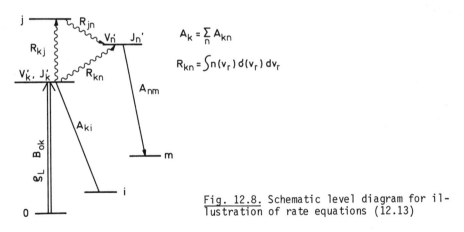

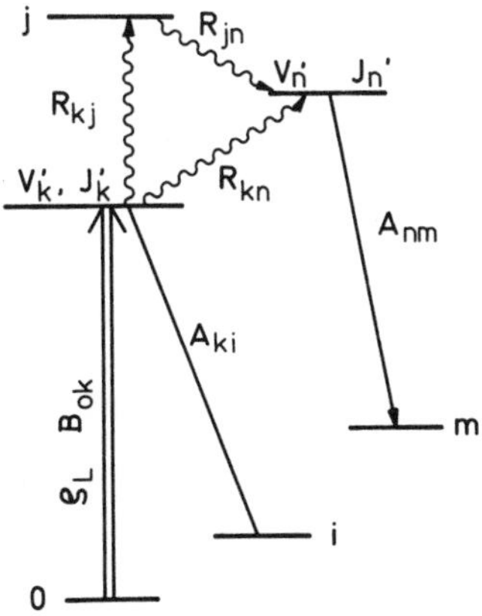

$$A_k = \sum_n A_{kn}$$

$$R_{kn} = \int n(v_r)\,\sigma(v_r)\,dv_r$$

Fig. 12.8. Schematic level diagram for illustration of rate equations (12.13)

with cw laser excitation, we obtain the rate equations for the populations N_k and N_n (Fig.12.8),

$$dN_k/dt = 0 = N_0\rho_L B_{0k} - N_k\left(A_k + \sum_n R_{kn}\right) \quad (12.13a)$$

$$dN_n/dt = 0 = N_k R_{kn} - N_n\Big(A_n + \sum_j R_{nj}\Big) + \sum_j N_j R_{jn} \quad , \tag{12.13b}$$

where the last term in (12.13b) describes the collision-induced transitions from collisional populated levels j to the level n. This term is therefore due to two successive collisions and may be neglected at lower pressures. From (12.13) we obtain the stationary population densities

$$N_k = \frac{N_0 \rho_L B_{0k}}{A_k + \sum\limits_n R_{kn}} \quad , \quad N_n = \frac{N_k R_{kn} + \sum\limits_j N_j R_{jn}}{A_n + \sum\limits_j R_{nj}} \approx N_k \frac{R_{kn}}{A_n} \quad . \tag{12.14}$$

This shows that the intensity ratio of satellite line to parent line

$$\frac{I_{nm}}{I_{ki}} = \frac{N_n A_{nm} h\nu_{nm}}{N_k A_{ki} h\nu_{ki}} = \frac{R_{kn}}{A_n} \frac{A_{nm}}{A_{ki}} \frac{\nu_{nm}}{\nu_{ki}} \tag{12.15}$$

directly yields the collision-induced transition rate R_{kn}, provided the spontaneous lifetime $\tau_n = 1/A_n$ and the ratio A_{nm}/A_{ki} are known.

Such measurements of *integral cross sections* for rotational and vibrational transitions have been performed for a number of molecules, such as I_2 [12.8], Li_2 [12.9], and Na_2 [12.10]. For Na_2, for instance, the cross sections for $\Delta J = \pm 1$ transitions are of the order of 100 $Å^2$ and they rapidly decrease with increasing ΔJ. The cross sections depend on the rotational energy transfer $\Delta E_r = E(J') - E(J'+\Delta J)$ and decrease with increasing ΔE_r [12.11] (Fig.12.9).

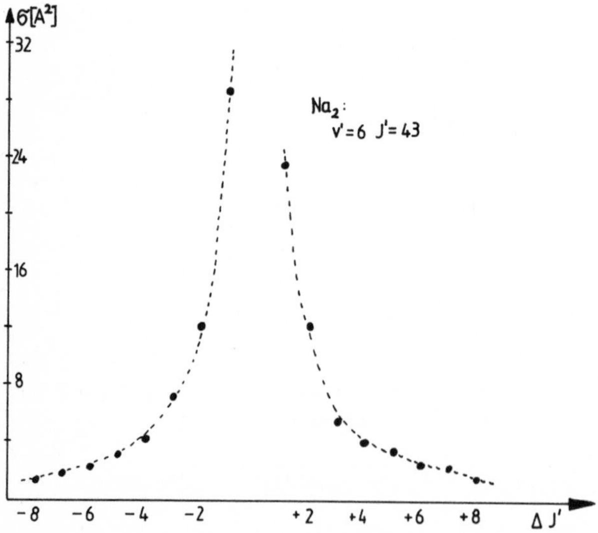

Fig. 12.9. Cross sections for collision-induced rotational transitions in the Na_2 $B^1\Pi_u$ state as a function of quantum jump ΔJ for collisons Na_2 + Ar [12.10]

The interesting question of how *electronic* excitation energy is transferred during inelastic collisions [12.12] has been studied by several groups using laser spectroscopy. One example is the energy transfer process

$$Na^*(3p) + Na_2(X^1\Sigma_g^+) \leftrightarrow Na_2^*(A^1\Sigma_u) + Na(3s) + \Delta E \quad , \tag{12.16}$$

which can be studied in both directions. Either the Na atoms are excited into the 3p state by optical pumping with a dye laser [12.13] and the excitation transfer is observed through the molecular fluorescence, or the *molecule* Na_2 is pumped into an excited state (v',J') and the collision-induced dissociation is monitored via the atomic fluorescence [12.7,14].

If the laser radiation used for optical pumping is polarized, the excited molecules in the upper level (v_k',J_k') are partially oriented. Although this primarily produced orientation is partly washed out by the molecular rotation, a certain degree of polarization of the fluorescence emitted from (v_k',J_k') still remains. If $I_\parallel$ and $I_\perp$ are the fluorescence intensities, polarized parallel or perpendicular to the linear polarization of the laser beam, the degree of polarization can be defined as

$$P = \frac{I_\parallel - I_\perp}{I_\parallel + I_\perp} \quad .$$

The magnitude of P is different for P, Q and R transitions [12.15]. Measuring the polarization ratio $P\Big(I(v_k',J_k')\Big)/P\Big(I(v_k',J_k'+\Delta J)\Big)$ as a function of the rotational quantum jump gives information about the depolarization during the collision-induced transition. These depolarizing collisions change the projection quantum number M and therefore the orientation of the molecule [12.16].

12.3 Study of Energy-Transfer Processes in Electronic Ground States

Most infrared molecular lasers, such as the CO_2 laser, the HCl laser, and many other chemical lasers, are based on collisional energy transfer between vibrational-rotational levels of the laser molecule and other atoms or molecules. Such lasers may therefore be called *energy-transfer lasers* [12.17]. Also in the visible region many molecular cw laser systems, where the laser transition terminates on rotation-vibration levels of the electronic ground state, rely on efficient collisional energy transfer between these levels and neighboring vibration-rotation levels in order to achieve a sufficiently fast depletion of the lower laser levels and to maintain inversion. Examples are the cw dye laser or the molecular dimer lasers (see Ref. [12.17a]). For the development of efficient laser media and for many other applications in chemistry it is therefore of great interest to study collisional energy transfer between vibration-rotation levels in electronic ground states of molecules.

The internal energy $E_{vib} + E_{rot}$ of a molecule M^* may be transferred during a collision with another molecule AB into vibrational energy of AB^* ($V \to V$ transfer), rotational energy ($V \to R$ transfer), electronic energy ($V \to E$ transfer), or translational energy of the collision partners ($V \to T$ transfer). In collisions with atoms $M^* + A \to M + A + \Delta E_{kin}$ only the two latter processes can occur. It turns out that the cross sections for $V \to V$ or $V \to R$ transfer are much larger than for $V \to T$ transfer, particularly if the vibrational energy levels of M and AB are close together (resonance case).

A famous example of a nearly resonant $V \to V$ energy transfer process is the collision-induced energy transfer from vibrationally excited N_2 molecules onto CO_2 molecules

$$CO_2(0,0,0) + N_2(v{=}1) \to CO_2(0,0;1) + N_2(v{=}0) \quad .$$

This is the major excitation process for populating the upper laser level in the CO_2 laser (Fig.12.10).

The $V \to V$ energy transfer

$$M^* + AB \to M + AB^*$$

can be monitored by observing the time-resolved fluorescence from the collisionally populated levels in AB^* after pulsed excitation of M. This is illustrated in an experiment by GREEN et al. [12.18] where HF molecules are excited

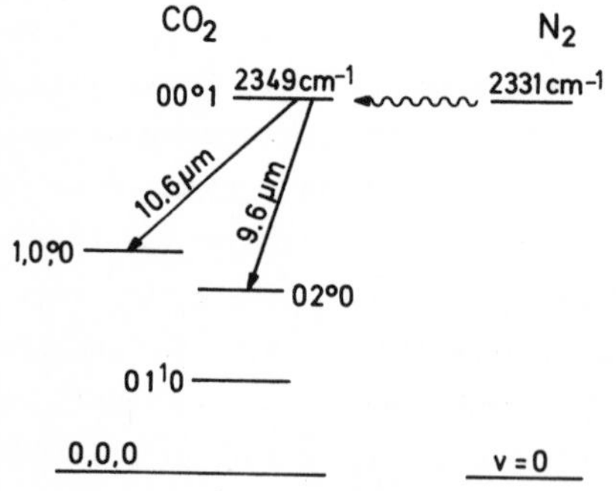

Fig. 12.10. Collisional energy transfer from vibrationally excited N_2 to the upper CO_2 laser level

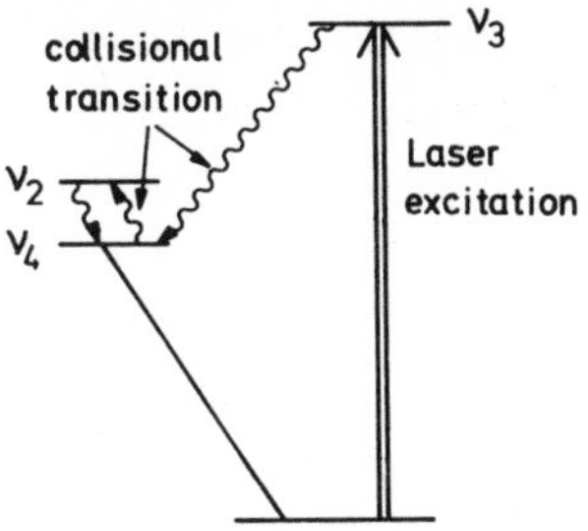

Fig. 12.11. Vibrational relaxation of optically pumped vibrationally excited CH_4 molecules

into the first vibrational level by optical pumping with a HF laser. The transfer of the vibrational energy to other molecules AB

$$HF + h\nu \rightarrow HF^*(v=1) \quad ; \quad HF^*(v=1) + AB(v=0) \rightarrow HF(v=0) + AB^*(v=1) \quad ,$$

is detected through the fluorescence emitted by the AB^* molecule, which is selected by suitable filters from the HF^* fluorescence. As collision partners AB molecules such as NO, CO and others have been chosen.

The two colliding molecules may be of the same kind. Figure 12.11 illustrates the example of CH_4^* relaxation in collisions with CH_4 molecules. The chopped output from a He-Ne laser oscillating at $\lambda = 3.39$ µm excites the ν_3 stretching vibration of CH_4. The excited level is rapidly deactivated by collisions with ground state CH_4 molecules. Because the energy of the ν_3 vibrational level is only slightly larger than twice the ν_4 level energy, the $V \rightarrow V$ transfer

$$CH_4^*(\nu_3) + CH_4(0) \rightarrow CH_4^*(\nu_4) + CH_4^*(\nu_4) + \Delta E_{kin}$$

is very effective. If the phase shift of the modulated fluorescence emitted from the ν_3 level is measured (see Sect.11.2), the collisional deactivation rate can be determined [12.19]. The results show that on the average already 70 gas kinetic collisions between $CH_4(0)$ and $CH_4^*(\nu_3)$ molecules are sufficient to produce molecules in the ν_4 level. At typical pressures of a few Torr the collision-terminated lifetime τ_3^{eff} is much shorter than the radiative lifetime $\tau_R = 37$ ms.

The large cross sections of $V \rightarrow V$ transfer are utilized in "transfer chemical lasers". Here for instance HX molecules (X=F, Cl, or Br) in excited vibrational levels are produced by chemical reactions [12.17]. In a gas mixture of HX and CO_2 molecules the vibrational energy of HX is very effectively transferred to the (001) level of CO_2 and laser action on the 10 µm transitions is achieved.

The time-dependent change in the populations of the various vibrational levels due to collisions can be detected in different ways. In many cases the detection of the fluorescence, emitted from these levels, is a simple and sensitive technique. In some cases absorption methods may be more suitable. Here two lasers irradiate the sample simultaneously: a strong pulsed pump laser which populates the initial vibrational level, and a weak probe laser which probes the time-dependent populations $N_v(t)$. Either the time-resolved absorption of a cw probe laser is monitored or the total time-integrated fluorescence induced by a pulsed probe laser is measured as a function of the delay time between pump and probe [12.20].

Very short vibrational relaxation times of molecules in liquids or in gases at high pressures can be studied with picosecond techniques (see Chap. 11). The excitation of vibrational levels may be performed by stimulated Raman scattering (see Sect.9.3) and the time evolution of the excited state can be monitored by spontaneous Raman scattering. This method allows direct measurements of vibrational relaxation times of large molecules, such as long-chain alcohols [12.21].

In the experiments discussed above, *infrared* lasers had always been used to pump or probe transitions $(v_i'',J_i'') \rightarrow (v_m'',J_m'')$ between rotation-vibration levels in the electronic ground state. The collision-induced rotational-vibrational transitions within electronic ground states of molecules can, however, also be measured by optical pumping of individual levels (v_i'',J_i'') with visible or ultraviolet lasers. This technique, where optical transitions into levels (v_k',J_k') of excited electronic states are involved, is based on the following consideration (Fig.12.12). A short pump pulse from a laser tuned to a mole-

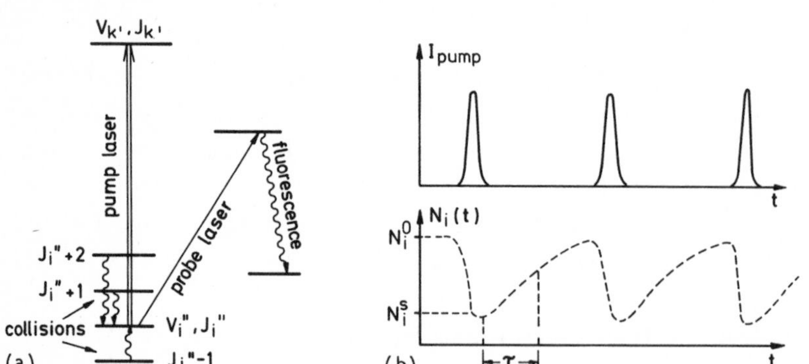

Fig. 12.12. (a) Schematic energy level diagram for detection of collisional relaxation in electronic ground states of molecules by optical-optical double-resonance techniques. (b) Time sequence of pump pulses and lower level population $N_i(t)$

cular transition $(v_i'',J_i'') \rightarrow (v_k',J_k')$ selectively depletes the lower level. The saturated population N_i^S is below the thermal equilibrium population N_i^0 and relaxation processes try to refill N_i^S in order to restore thermal equilibrium. The change $N_i(t)$ after the end of the pump pulse is described by

$$dN_i/dt = [N_i^0 - N_i(t)]N_B K_i \quad , \tag{12.17}$$

where N_B is the buffer gas density (which is assumed to be large compared to the molecular density N_i) and $K_i = \sum_m K_{im}$ is the total rate constant for collision-induced transitions $(v_m,J_m) \rightarrow (v_i,J_i)$. The solution of (12.17),

$$N_i(t) = N_i^0 + [N_i^S(0) - N_i^0]e^{-N_B K_i t} \quad , \tag{12.18}$$

shows that the population $N_i(t)$ exponentially approaches its equilibrium value N_i^0 with a time constant $\tau = 1/(N_B K_i)$.

This time-dependent population $N_i(t)$ can be measured in different ways. One possibility is to pass a weak cw probe laser through the sample and to measure the time-resolved absorption $\Delta I = N_i(t)\alpha L$. Figure 12.12b is a schematic time diagram of pump pulse, lower level relaxation, and probe laser absorption. Another way uses a probe pulse with variable delay T. This probe pulse, which may come either from the same laser as the pump pulse or from another laser, is sent through the pumped region of the sample. The time-integrated fluorescence intensity

$$\int_T^{T+\Delta T} I_{Fl}\, dt = \int_T^{T+\Delta T} N_i(t)\rho_L^{probe} B_{in} A_n\, dt \tag{12.19}$$

induced by the probe pulse with duration ΔT is proportional to the lower level population $N_i(T)$. Measuring I_{Fl} as a function of the delay time T (which can be performed, for example, with a boxcar integrator) yields the rate constant K_i which is the sum of all collision rates repopulating N_i^S.

Such measurements have been performed with polarized laser pulses to measure the orientation relaxation [12.22] and to measure the total rate of inelastic collisions [12.23]. With single-mode lasers even velocity changing elastic collisions can be investigated [12.24].

A more detailed study of the individual collision-induced rotational or vibrational transitions $(v_i'',J_i'') \rightarrow (v_m'',J_m'')$ can be obtained by modulated two-laser spectroscopy. A cw pump laser is tuned to a molecular transition $(v_i'',J_i'') \rightarrow (v_k',J_k')$ (Fig.12.13). If the laser intensity is chopped at a frequency f_1, the population density N_i is, due to saturation, also periodically mo-

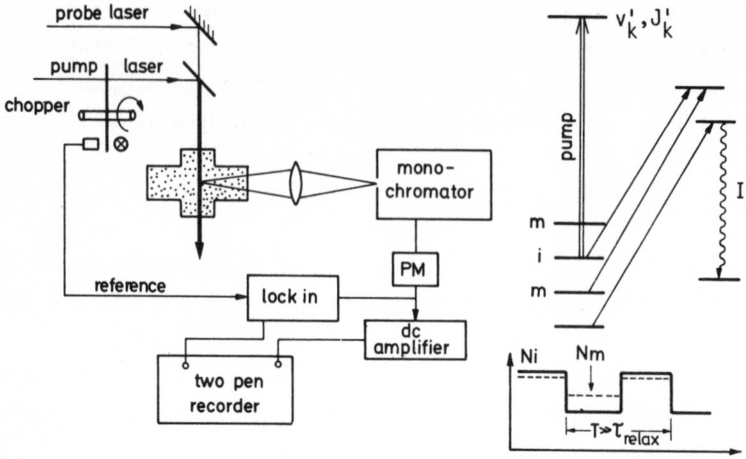

Fig. 12.13. Experimental arrangement and level scheme for measuring collision-induced rotational and vibrational relaxation processes in molecular ground states

dulated between the unsaturated equilibrium value N_i^0 and the saturated value N_i^s. The relaxation processes try to restore thermal equilibrium and partly refill the depleted population N_i^s at the expense of the populations $N_m = N(v_i'' + \Delta v'', J_i'' + \Delta J)$ of neighboring levels. If the collision rate is fast compared with the chopping frequency f_1, we can assume stationary conditions. For the time when the pump laser is off, we have unsaturated populations at thermal equilibrium

$$N_m^0 = (2J_m'' + 1)e^{-E_{vib}/kT} e^{-E_{rot}/kT} \quad , \tag{12.20}$$

while during the period where the pump laser is on, we have stationary saturated populations N_m^s. The probe laser is permanently on but is assumed to be sufficiently weak so that it does not change the populations appreciably.

The quantity measured in such experiments is the relative modulation of the probe laser-induced fluorescence intensity I_{Fl} which reflects the modulation of the lower level population N_m

$$\frac{I^0 - I^s}{I^0} = \frac{N_m^0 - N_m^s}{N_m^0} \quad . \tag{12.21}$$

Under stationary-state conditions the rate equation for the population N_m at a buffer gas density N is given by [12.25]

$$\frac{dN_m}{dt} = 0 = N_B \sum_{j \neq m} (N_j k_{j \to m} - N_m k_{m \to j}) - (N_m - N_m^0) K_m \quad , \tag{12.22}$$

where $k_{j \to m}$ is the rate constant for the collision-induced transition between levels j and m. The collision rate is then $R_{jm} = N_j k_{j \to m}$. The last term in (12.22) describes the general relaxation of the perturbed populations towards their equilibrium value N_m^0. For the equilibrium situation $N_j = N_j^0$ we obtain

$$\frac{dN_m}{dt} = 0 = N_B \sum_{j \neq m} \left(N_j^{(0)} k_{j \to m} - N_m^{(0)} k_{m \to j} \right) \quad . \tag{12.23}$$

Subtracting (12.23) from (12.22), we find an equation for the difference population $\Delta N_m = N_m^0 - N_m(t)$,

$$\frac{d\Delta N_m}{dt} = 0 = N_B \sum_{j \neq m} (\Delta N_j k_{j \to m} - \Delta N_m k_{m \to j}) - \Delta N_m K_m \quad . \tag{12.24}$$

This equation connects the population modulation in level m with that of all other levels, in particular the optically pumped level i. The rate constants $k_{j \to m}$ for collision-induced transitions can be now deduced from the measured modulation ratios $\Delta N_i / \Delta N_m$. The magnitude of the population modulation $\Delta N_j (j \neq i)$ is proportional to the density N_B of the collision partners. At sufficiently low pressures, therefore, all $\Delta N_j (j \neq i)$ are small compared with ΔN_i and we may, in a first approximation, neglect all terms with $j \neq i$ in the sum (12.24). We then obtain

$$\frac{\Delta N_m}{\Delta N_i} = N_B \frac{k_{i \to j}}{K_m} \quad . \tag{12.25}$$

If the unmodulated population densities are simultaneously measured by time-averaged dc signals

$$I_{Dc} = C \, \frac{I^s + I^{(0)}}{2} \propto \frac{N_m^s + N_m^0}{2} \quad ,$$

the normalized population changes $\Delta N_m / N_m^0$ can be obtained and, assuming equal equilibrium relaxation rates K for all levels, we arrive at the equations

$$\eta_m = \frac{\Delta N_m / N_m^0}{\Delta N_i / N_i^0} = N_B \frac{k_{i \to m}}{K_m} \frac{N_i^0}{N_m^0} \quad \text{with} \quad \frac{N_i^0}{N_m^0} = \frac{K_m}{K_i} \quad . \tag{12.26}$$

The unimolecular rate constant K_i which describes the relaxation of the depleted level population N_i^s towards its equilibrium value N_i^0 gives the time scale of the experiment. In the time $\tau = 1/K_i$ a fractional population change

$n_m = N_B k_{i \to m} \tau$ of the satellite level m is produced. Measuring n_m for different levels m therefore allows one to deduce the rate constants $k_{i \to m}$ for collision-induced transitions between different rotational-vibrational levels $(v_i'', J_i'') \to (v_m'', J_m'')$ in the electronic ground state.

Since the detailed study of these different inelastic collision processes is of great interest for the basic understanding of collision physics and chemistry, a large number of experimental and theoretical investigations have been published. For more complete information the reader is referred to some review articles on this field [12.26-29] and to the literature cited therein.

12.4 Measurements of Differential Cross Sections in Crossed Molecular Beams

The various spectroscopic techniques discussed in the previous sections allowed the measurement of absolute rate constants of collision-induced transitions from which mean *integral* cross sections could be deduced, averaged over the thermal velocity distribution and over all directions of the relative velocity of the collision partners. Much more detailed information can be obtained from measurements of *differential* cross sections in crossed molecular beams. In particular nonspherical interaction potentials, which are for instance responsible for rotational transitions in collisions between atoms and molecules and which cannot be obtained from data averaged over all directions, can be deduced from differential cross sections.

Without laser spectroscopy, mainly *elastic* collisions had been measured in crossed beams. Inelastic collisions resulting in vibrational excitation have been studied only for a few molecules and the investigation of rotational transitions with conventional beam techniques with time-of-flight measurements were restricted to hydrogen molecules H_2 because of insufficient energy resolution. Only with Rabi-type spectrometers using electrostatic quadrupole deflection fields have inelastic cross sections for Δm and ΔJ transitions been measured for small rotational quantum numbers J of polar molecules such as CsF [12.30]. Laser spectroscopy can overcome these limitations because its energy resolution is many orders of magnitude larger than that of time-of-flight spectrometers and its application is not restricted to low J levels.

Figure 12.14 shows a possible experimental arrangement for laser spectroscopy in crossed molecular beams, which allows the measurement of differential elastic and inelastic collision cross sections between atoms and molecules, where initial and final states of the molecule are known [12.31]. The supersonic primary beam containing Na atoms and Na_2 molecules is crossed with a

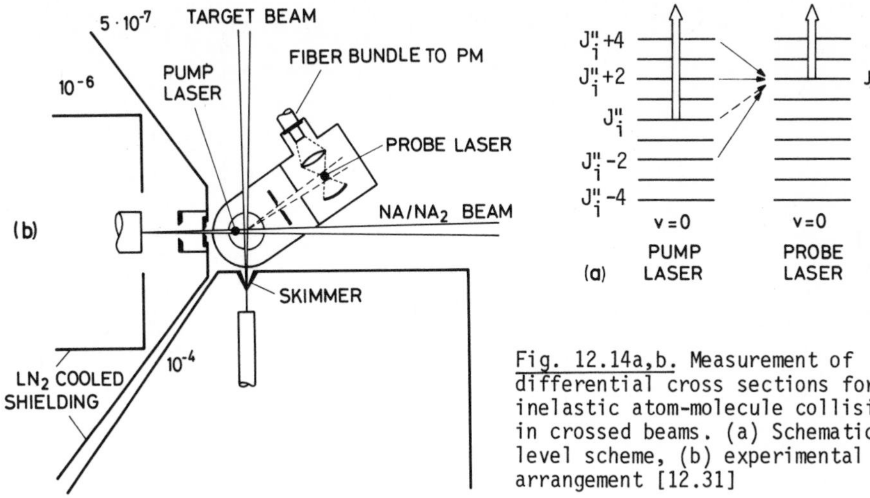

Fig. 12.14a,b. Measurement of differential cross sections for inelastic atom-molecule collisions in crossed beams. (a) Schematic level scheme, (b) experimental arrangement [12.31]

secondary noble gas beam. Molecules in the level (v''_m, J''_m) which have been scattered by an angle ϑ can be monitored through the fluorescence induced by the probe laser which is tuned to a molecular transition $(v''_m, J''_m) \rightarrow (v', J')$. The number $N(v''_m, J''_m)$ of molecules which are scattered into the solid angle $(\vartheta \pm d\vartheta)$ monitored by the probe laser, results from two contributions:

a) the elastically scattered molecules $(v''_m, J''_m) \rightarrow (v''_m, J''_m)$, and
b) the sum of all inelastically scattered molecules $\sum_n [(v''_n, J''_n) \rightarrow (v''_m, J''_m)]$ which have been scattered from all initial levels E_n to the final level E_m.

If the specified level $E_i = (v''_i, J''_i)$ is depleted through optical pumping by the pump laser which intersects the Na_2 beam before the collision region, the signal at the fluorescence detector decreases because all molecules which have undergone collisional transitions $(v''_i, J''_i) \rightarrow (v''_m, J''_m)$ are now missing (see Fig.12.15). If the pump laser is chopped, the difference of the detected signals with and without the pump laser just gives the collision rate for the transitions $(v''_i, J''_i) \rightarrow (v''_m, J''_m)$. Tuning the probe or the pump laser to different transitions allows measurement of the cross sections for different ΔJ transitions. Such a spectroscopic modification of a collision experiment in crossed molecular beams may be called an "ideal scattering experiment", because it provides all necessary information for a full description of the collision process, such as scattering angle, initial and final internal states, and the relative velocity of the collision partners. From these data the probability of a collision-induced rotational transition can be deduced as a function of the impact parameter. This allows one to probe the interaction poten-

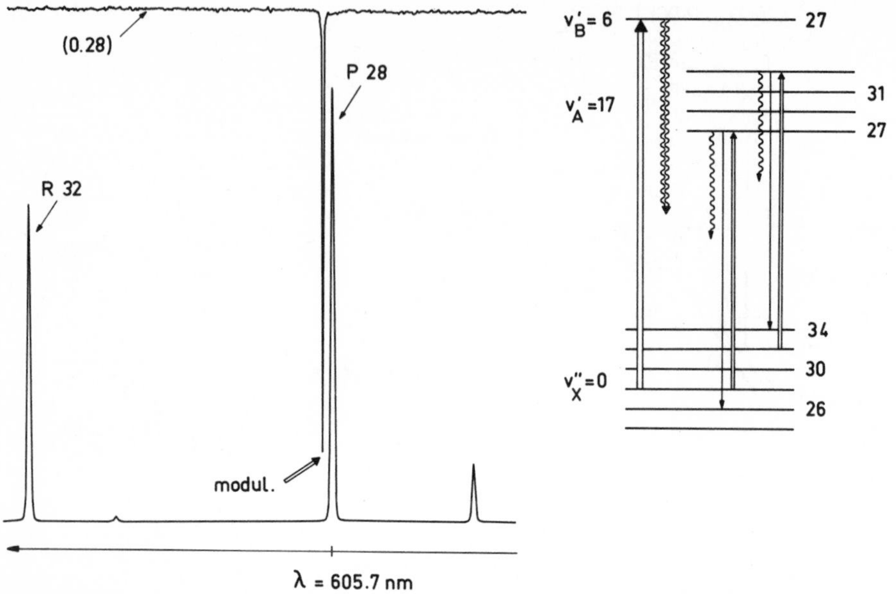

Fig. 12.15. Experimental demonstration of the nearly complete depletion of a molecular level in a collision-free molecular beam. The probe laser is stabilized on the (v"=0,J"=28) → (v'=6,J'=27) transition. The pump laser is tuned. The lower curve is the excitation spectrum of the pump laser, the upper curve shows the fluorescence intensity of the probe laser, which drops nearly to zero when the pump is tuned to the (0,28) → (17,27) transition. The two pens of the recorder were shifted against each other [12.33]

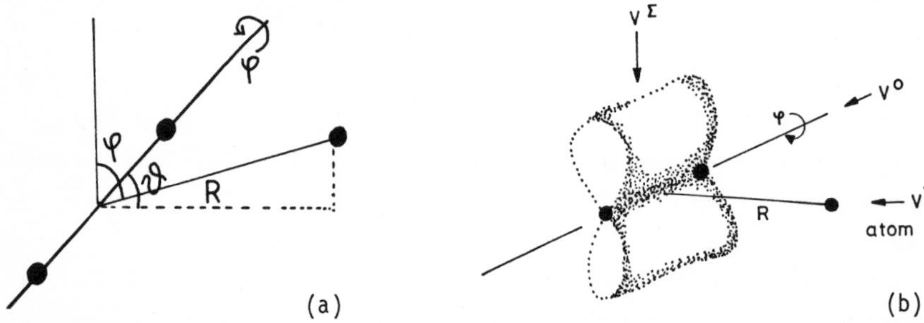

Fig. 12.16a,b. Collision between a diatomic molecule and an atom. (a) Coordinates for describing the interaction potential. (b) Example of a potential which is not symmetrical around the internuclear axis but depends on φ

tial V(R,θ) as a function of the distance R between the two centers of mass and of the angle θ between molecular axis and the line of approach (Fig.12.16). Figure 12.17 shows for illustration some differential cross sections for rotational transitions in Na_2 as a function of the scattering angle [12.32]. From the measured cross sections an interaction potential of the form

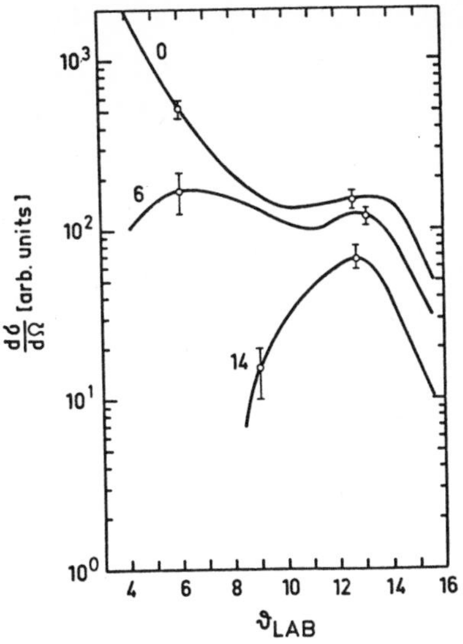

Fig. 12.17. Differential cross sections for rotational transitions $J'' \rightarrow J'' + \Delta J$ for different ΔJ in Na_2-Ne collisions as a function of the scattering angle in the center of mass system. Initial level: $J'' = 28$ [12.32]

$$V(R,\theta) = V_0(R) + V_2(R)P_2(\cos\theta)$$

can be obtained which fits the experimental data very well [12.33].

The process of electronic-to-vibrational energy transfer in collisions between excited atoms and ground state molecules

$$A^* + M(v''=0) \rightarrow A + M(v''>0) + \Delta E_{kin} \quad (12.27)$$

plays an important role in many chemical reactions. While cell experiments generally yield only collisional transition rates and averaged total cross sections, crossed beam experiments allow measurement of differential cross sections. Optical pumping of the atoms by a laser produces a sufficiently large population density of excited atoms.

In the first experiments of this kind, collisions between Na(3p) atoms, excited by a cw dye laser, and N_2 or CO molecules have been studied [12.34]. The vibrational excitation of the molecules could be deduced from the kinematics of the collision because the scattering angle and the velocity of the scattered Na atoms could be simultaneously measured.

Scattering of electrons by laser excited sodium atoms in crossed beams [12.35] can result in elastic ($3p \rightarrow 3p$), inelastic ($3p \rightarrow 3d,4s$), or superelastic ($3p \rightarrow 3s$) collisions. Since the orientation of the excited Na atoms depends on the polarization of the pump laser, orientational effects on the cross section can be studied.

12.5 Light-Induced Collisional Energy Transfer

In inelastic collisions between excited atoms or molecules A^* and ground state atoms B,

$$A^* + B \rightarrow B^* + A + \Delta E_{kin} \quad , \tag{12.28}$$

energy and momentum have to be conserved. The excess energy $\Delta E = E(A^*) - E(B^*)$ between the internal energy $E(A^*)$ before and $E(B^*)$ after the collision has to be converted into translational energy of the collision partners. For $\Delta E_{kin} >> kT$ the cross section for the reaction (12.28) becomes very small, while for "near-resonant" collisions where $\Delta E_{kin} < kT$, the cross section for energy transfer from A to B may exceed the gas kinetic cross section.

In this section we consider energy transfer collisions with $\Delta E >> kT$ where the absorption of photons of energy $\hbar\omega = \Delta E$ during the collision is utilized to conserve energy and to increase the probability for the transfer process. These light-induced collisional energy transfer (LICET) may occur when a collision pair A^*B consisting of an excited atom A^* and a ground state atom B is irradiated by a laser. If a photon of proper energy $\hbar\omega$ is absorbed during the collision, the collision pair may be excited into $(AB)^*$ which separates after the collision into $A + B^*$. Thus the initial excitation energy of A^* plus the photon energy $\hbar\omega$ have been transferred into excitation energy of B^* (Fig.12.18).

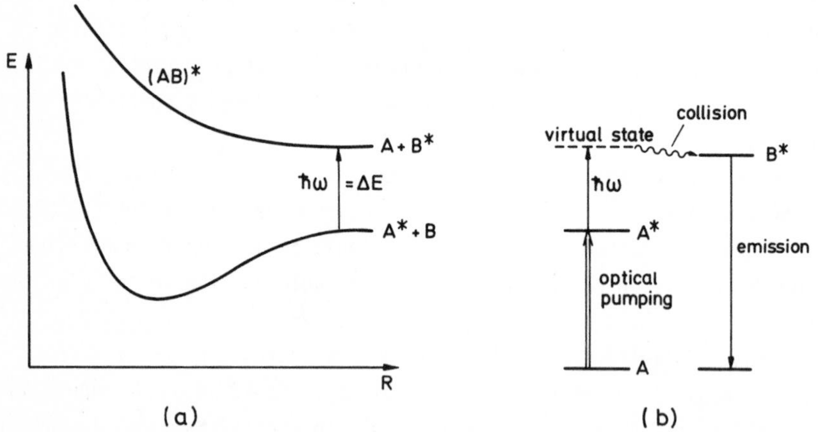

Fig. 12.18a,b. Schematic illustration of LICET processes. (a) Quasi-molecule viewpoint, (b) optical excitation into a virtual state of A with successive collisional energy transfer to B

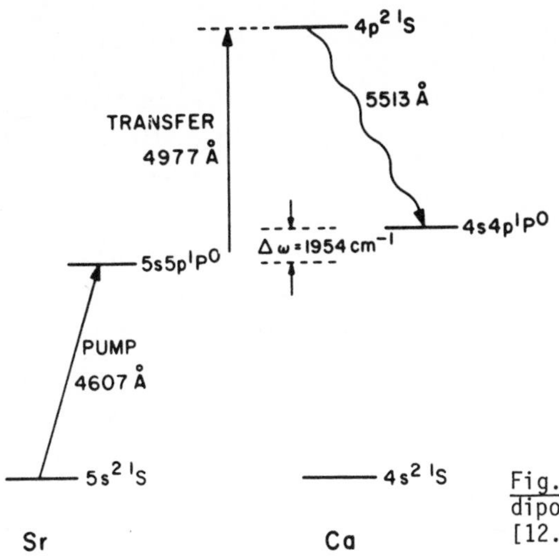

Fig. 12.19. Energy level diagram for dipole-dipole transfer in Sr and Ca [12.37]

The first experimental realization of light-induced collisional energy transfer has been demonstrated for the Sr-Ca system by HARRIS et al. [12.36]. Figure 12.19 illustrates the energy level diagram for dipole-dipole transfer from Sr to Ca [12.37]. The Sr atom is excited into the 5s5p $^1P_1^0$ level by a pump laser. While the excited Sr atom collides with a ground state Ca atom, the collision pair may absorb a photon from another so-called transfer laser. After the atoms have separated, a ground state Sr atom and an excited Ca atom in $4p^2\ ^1S_0$ level are found which can be monitored by the fluorescence at 5513 Å.

This LICET process may be described in the potential curve diagram of the quasi-molecule or collision pair illustrated in Fig.12.18. The lower curve represents the potential curve of the pair A^*B, the upper, that of AB^*. A better description, which also allows calculation of the line shape of the transfer cross section as a function of the wavelength λ_T of the transfer laser, uses the model of virtual transition (Fig.12.18b). The excited atom A^* makes a transition into a virtual state under the influence of the transfer laser field. Excitation is then transferred to the atom B via dipole-dipole coupling. The whole process is thus described as a virtual electromagnetic transition followed by a real collision.

The theoretical treatment [12.38] shows that for dipole-dipole interaction the maximum cross section for light-induced energy transfer occurs at a laser wavelength λ_T such that energy conservation is satisfied for the infinitely separated atoms. This implies for the situation in Fig.12.19 that

$$\hbar(\omega_p + \omega_T) = E(4p^2\ {}^1S_0) \quad .$$

The experiments are performed at high pressures (density of atoms B about $10^{19}\ cm^{-3}$, that of atoms A about $10^{16}\ cm^{-3}$) to increase the collision rate. In order to prevent collisional diffusion of the excitation energy of A^* from the initially pumped level into other excited levels, the excitation and transfer processes have to occur within a short time. A possible experimental arrangement uses picosecond pulses of several MW peak powers with 40 ns width, delivered from two synchronously pumped dye lasers (see Sect.12.1). The delay time between pump pulse and transfer pulse can be controlled and lies around a few ns. This delay is also necessary to separate the systems, where the intermediate level of the atom B is close to the energy-storing level of A^* [12.39].

Both lasers are focused in a metal vapor cell to increase the power density which enhances the cross section for the LICET process. A convenient metal vapor cell is the "heat pipe" [12.40] where the vapor is confined within noble gas zones which prevent the metal vapor from diffusing to the cell windows [12.40a]. The vapor pressure of the metal can be controlled by the noble gas pressure. For mixtures of two metals A and B with different vapor pressures a double heat pipe has been developed [12.41] which allows to operate the heat pipe with p(A) = p(B) = p(noble gas). In [12.41a] a heat pipe is described which allows one to control the vapor pressures p(A) and p(B) indepently within wide ranges. A one dimensional theoretical model of the heat pipe can be found in [12.41b].

The collision cross sections for the different systems investigated so far range from $4 \times 10^{-13}\ cm^{-1}$ at power densities of $3 \times 10^{10}\ W/cm^2$ to $9 \times 10^{-18}\ cm^2$ at $5 \times 10^5\ W/cm^2$ [12.37]. This demonstrates that transfer cross sections can be reached which exceed gas kinetic cross sections by two to three orders of magnitude.

13. The Ultimate Resolution Limit

In Chap.10 several techniques have been presented which allow the Doppler width to be overcome. Provided that all other sources of line broadening could be eliminated, the spectral resolution of these techniques can reach at least in principle the limit imposed by the *natural* linewidth γ_n of a molecular transition. For *allowed* electronic transitions with typical natural linewidths of a few MHz, other broadening effects, such as pressure and power broadening or time-of-flight broadening (see Chap.3) can be indeed made smaller than the natural linewidth by an appropriate experimental arrangement. In such cases the natural linewidth has already been reached experimentally. On the other hand, there is much interest in ultrahigh resolution spectroscopy of lines with extremely small natural linewidths below the kHz range. Examples are visible or uv forbidden transitions between ground states and metastable excited states with long spontaneous lifetimes, or in the infrared region vibrational transitions between long-lived vibrational levels. For such transitions it is not the spontaneous lifetime but the finite interaction time of the molecules with the laser field which limits the spectral resolution. If the time of flight of a molecule passing through the laser beam is small compared with the spontaneous lifetime, the time-of-flight broadening becomes the major broadening mechanism, provided the laser frequency is sufficiently stable.

In this chapter we discuss several techniques which can reduce or even completely avoid time-of-flight broadening. Some of these methods have already been realized experimentally while others are only theoretical proposals which could not be proved up to now. These techniques allow ultrahigh resolution, in some cases even within the natural linewidth. This raises the interesting question about the ultimate resolution limit and the experimental or fundamental factors that determine such a limit.

13.1 Optical Ramsey Fringes

The problem of time-of-flight broadening was recognized many years ago in electric or magnetic resonance spectroscopy in molecular beams [13.1]. In these Rabi-type experiments the natural linewidth of the radio-frequency or microwave transitions is extremely small [because the spontaneous transition probability is according to (2.22) proportional to ω^3]. The spectral widths of the microwave or rf lines are therefore determined mainly by the time of flight $\Delta T = d/\bar{v}$ of molecules with mean velocity $\bar{v}$ through the interaction zone in the C field (see Fig.10.15) with length d.

A considerable reduction of time-of-flight broadening could be achieved by realization of Ramsey's ingenious idea of separated fields [13.2]. The molecules in the beam pass two phase-coherent fields which are spatially separated by a distance $x = L \gg \Delta x = d$ large compared with the extension $\Delta x = d$ of each field (Fig.13.1). The interaction of the molecules with the first field creates a dipole moment of each molecular oscillator with a phase depending on the interaction time $\tau = d/\bar{v}$ and the detuning $\Omega = \omega_0 - \omega$ of the field frequency ω from the center frequency ω_0 of the molecular transition (see Sect.2.8). After having passed the first interaction zone the molecular dipole precesses in the field-free region at its eigenfrequency ω_0. When it enters the second field, it therefore has accumulated a phase angle $\Delta\varphi = \omega_0 T = \omega_0 L/\bar{v}$. In the same time interval the field phase has changed by ωT. The relative phase between dipole and field has therefore changed by $(\omega_0 - \omega)T$ during the flight time through the field-free region.

The interaction between dipole and second field depends on their relative phases. The observed signal is related to the power absorbed by the molecular dipoles in the second field and is therefore proportional to $E_2^2 \cos(\omega - \omega_0)L/\bar{v}$.

When we assume that all N molecules passing the field per second have the same velocity v, we obtain the signal

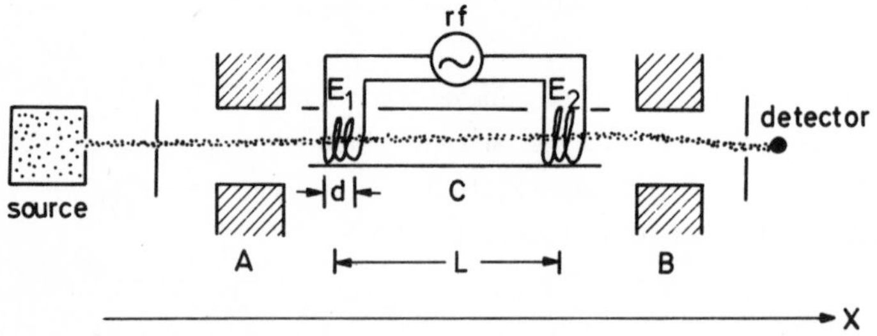

Fig. 13.1. Schematic diagram of Ramsey's method of two separated fields

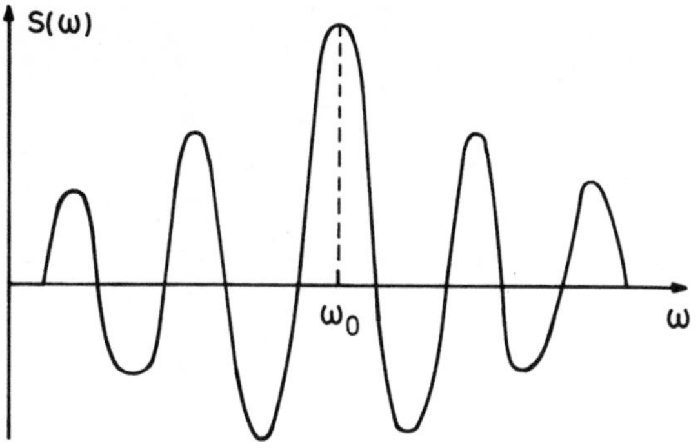

Fig. 13.2. Signal power, absorbed in the second field, as a function of detuning $\Omega = \omega - \omega_0$ (Ramsey fringes for a narrow velocity distribution)

$$S(\omega) = CNE_2^2 \cos[(\omega_0 - \omega)L/v] \quad . \tag{13.1}$$

Measured as a function of the field frequency ω, this signal exhibits an interference pattern called *Ramsey fringes* (Fig.13.2). The full halfwidth of the central fringe, which is $\delta\omega = \pi(v/L)$, decreases with the separation L between the fields.

This interference phenomenon is quite similar to the well-known interference experiment (see Sect.2.10) where two slits are illuminated by coherent light and the superposition of light from both slits is observed as a function of the optical path difference Δs. The number of maxima observed in the two-slit interference pattern depends on the coherence length ℓ_c of the incident light and on the slit separation. The fringes can be observed if $\Delta s \leqq \ell_c$. A similar situation is observed for the Ramsey fringes. Since the velocities of the molecules in the molecular beam are not equal but follow a Maxwellian distribution, the phase differences $(\omega_0 - \omega)L/v$ show a corresponding distribution. The interference pattern is obtained by integrating the contributions to the signal from all molecules n(v) with velocity v

$$S = C\int n(v)E^2 \cos[(\omega_0 - \omega)L/v]dv \quad . \tag{13.2}$$

Similar to Young's interference with partially coherent light the velocity distribution will smear out the interference pattern for the higher order fringes [large $(\omega_0 - \omega)$] but will essentially leave the central fringe [small $(\omega_0 - \omega)$]. For a halfwidth Δv of the velocity distribution n(v) this restricts the maximum field separation to about $L \leqq \pi\Delta v/\omega_0$, since for larger L the higher interference orders for fast molecules overlap with the zeroth order for slow molecules. Using supersonic beams with a narrow velocity distribution (see Sect.10.1), larger separations L can be used. In general, however,

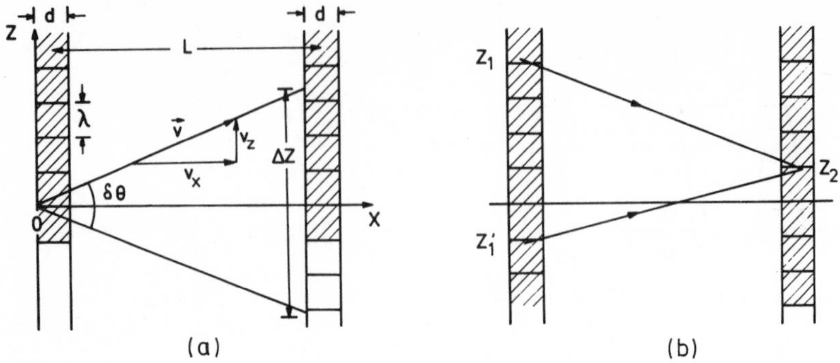

Fig. 13.3. Molecules starting with different v_Z (a) from the same point, (b) from different points in the first zone experience different phases in the second field

only the zeroth order of the Ramsey interference is used in high-resolution spectroscopy and the "velocity averaging" of the higher orders has the advantage that it avoids the overlap of different orders for two closely spaced molecular lines.

The extension of Ramsey's idea to the optical region seems to be quite obvious if the rf fields are replaced by two-phase coherent laser fields. However, the transfer from the rf region, where the wavelength λ is larger than the field extension d, to the optical range where $\lambda \ll d$, meets with some difficulties [13.3]. Molecules with slightly inclined path directions traverse the optical fields at different phases (Fig.13.3). Consider molecules starting from a point $z = 0$, $x = 0$ at the beginning of the first field. Only those molecules with flight directions within a narrow angular slice $\delta\theta \leqq \lambda/2d$ around the x axis experience phases at the end of the first field differing by less than π. These molecules, however, traverse the *second* field at a distance L downstream within an extension $\Delta z = L\delta\theta \leqq L\lambda/2d$ where the phase φ of the optical field has a spatial variation up to $\Delta\varphi \leqq L\pi/d$. If the method of separated fields is to increase the spectral resolution, L has to be large compared to d, which implies that $\Delta\varphi \gg \pi$. Although these molecules have experienced nearly the same phase in the first field, they *do not* generate observable Ramsey fringes when interacting with the second field because their interaction phases are all different. The total signal is obtained by spatial averaging over the different phases, which means that the Ramsey fringes are washed out. The same is true for molecules starting from different points $(0,z_1)$ in the first interaction zone which arrive at the same point (L,z_2) in the second zone (Fig.13.3b). The phases of these molecules are randomly distributed and therefore no macroscopic polarization is observed at (L,z_2).

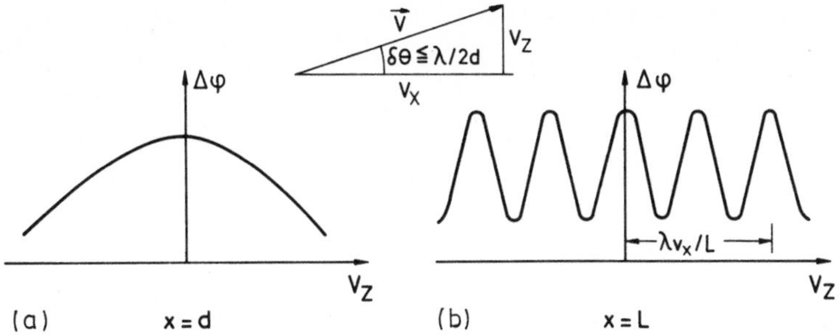

Fig. 13.4. Relative phase $\Delta\varphi(v_Z)$ between the fields at a given point (x,z) and a molecular dipole as a function of the transverse velocity v_Z (a) in the first field, (b) in the second field

Note that the requirement $\delta\theta < \lambda/2d$ for molecules which experience the same phase in the first zone is equivalent to the condition that the residual Doppler width $\delta\omega_D$ in the absorption profile of molecules moving within the angular slice $\delta\theta$ does not exceed the time-of-flight broadening $\delta\omega_{TF} = \pi v_x/d$. This can immediately be seen from the relations

$$\delta\omega_D = \omega v_z/c = \omega\delta\theta v_x/c = \delta\theta v_x 2\pi/\lambda \quad \text{for} \quad \delta\theta < \lambda/2d \rightsquigarrow \delta\omega_D < \pi v_x/d \quad . \tag{13.3}$$

The phase $\varphi(v_z)$ of a molecular dipole starting from a point $(0,z_1)$ in the first zone can be plotted as a function of the transverse velocity v_z, as indicated in Fig.13.4. Although $\varphi(v_z,z_1)$ shows a flat distribution after the first laser beam, it exhibits a modulation with period $\Delta v_z = \lambda/2T = \lambda v_x/2L$ in the second zone. However, this modulation cannot be detected because it is washed out by integrating over all contributions from molecules arriving at a point (L,z_2) with different transverse velocities v_z.

Fortunately several methods have been developed which overcome these difficulties and which allow ultranarrow Ramsey resonances to be obtained. One of these methods is based on Doppler-free two-photon spectroscopy, while another technique uses saturation spectroscopy but introduces a third interaction zone at a distance x = 2L downstream from the first zone to recover the Ramsey fringes. We briefly discuss both methods.

13.1.1 Two-Photon Ramsey Resonances

In Sect.10.6 we have seen that the first-order Doppler effect can be exactly cancelled for a two-photon transition if the two photons $\hbar\omega_1 = \hbar\omega_2$ have opposite wave vectors $\underline{k}_1 = -\underline{k}_2$. A combination of Doppler-free two-photon absorption and Ramsey method therefore avoids the phase dependence $\varphi(v_z)$ on the transverse velocity component. In the first interaction zone the molecular dipoles are excited with a transition amplitude a_1 and precess with their eigenfrequency $\omega_{12} = (E_2 - E_1)/\hbar$. If the two photons come from oppositely travelling waves with frequency ω, the detuning

$$\Omega = \omega + kv_z + \omega - kv_z - \omega_{12} = 2\omega - \omega_{12}$$

of the molecular eigenfrequency ω_{12} from 2ω is independent of v_z. The phase factor $\cos(\Omega T)$, which appears after the transit time $T = L/v_x$ at the entrance of the molecular dipoles into the second field can be composed as $\cos(\varphi_2^- + \varphi_2^+ - \varphi_1^- - \varphi_1^+)$, where each φ comes from one of the four fields (two oppositely travelling waves in each zone). The v_z dependence of each φ cancels in this sum in a folded standing wave geometry [13.4]. If we denote the two-photon transition amplitudes in the first and second field zone by c_1 and c_2, respectively, we obtain the total transition probability as

$$W_{12} = |c_1|^2 + |c_2|^2 + 2|c_1||c_2| \cos\Omega T \quad . \tag{13.4}$$

The first two terms describe the conventional two-photon transitions in the first and second zones while the third term represents the interference leading to the Ramsey resonance. Due to the *longitudinal* thermal velocity distribution $f(v_x)$, only the central maximum of the Ramsey resonance is observed with a theoretical halfwidth (for negligible natural width) of

$$\Delta\Omega = (2/3)\pi/T = 2/3\pi\bar{v}_x/L \quad . \tag{13.5}$$

The higher interference orders are washed out.

A quantitative description starts from the equations of motion (2.98) for the probability amplitudes a_i which lead for the two-photon transitions between levels 1 and 2 to the equations [13.5,5a]

$$\begin{aligned} \dot{a}_1 &= iD_{11}E^2a_1 + iD_{12}E^2\, e^{i\Omega T}\, a_2 \\ \dot{a}_2 &= iD_{12}E^2a_2 + iD_{22}E^2\, e^{-i\Omega T}\, a_1 \quad , \end{aligned} \tag{13.6}$$

where the two-photon transition element is [see (8.43)]

$$D_{ik} \propto \sum_n \frac{R_{in}R_{nk}}{(\omega_{n2} - \omega)} \quad . \tag{13.7}$$

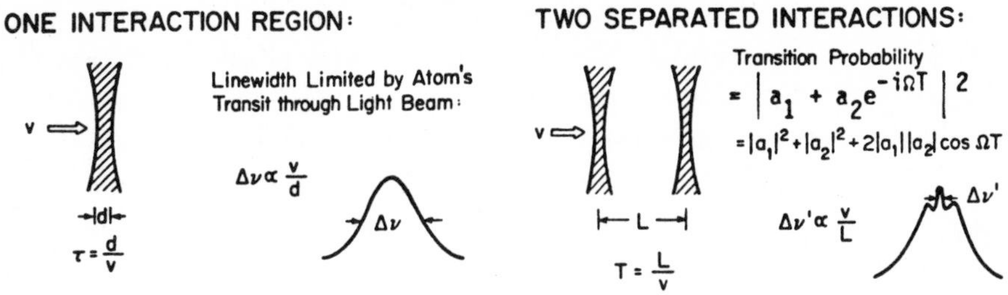

Fig. 13.5. Illustration of two-photon Ramsey resonances [13.6]

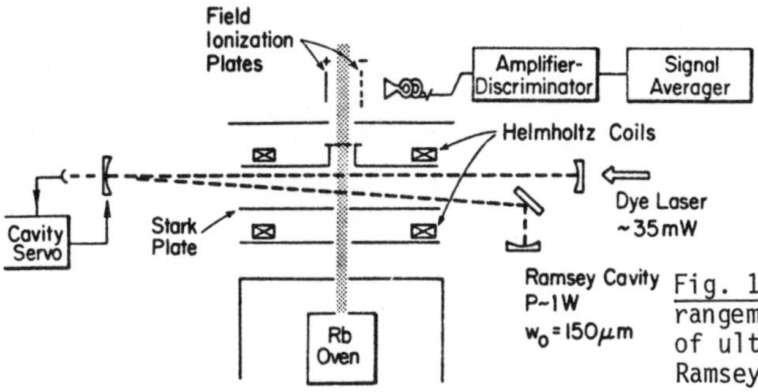

Fig. 13.6. Experimental arrangement for the observation of ultra-narrow two-photon Ramsey fringes [13.6]

Under weak field conditions and for small detuning $\Omega \sim 1/T << 1/\tau$ we obtain with the initial conditions $a_1(0) = 1$, $a_2(0) = 0$ the probability amplitudes for molecules in the upper level (2) in the first and second zones

$$a_2^{(1)} = \frac{i}{2} D_{21} E_1^2 \tau/\hbar^2 \quad ; \quad a_2^{(2)} = \frac{i}{2} D_{21}\, e^{-i\Omega T}\, E_2^2 \tau/\hbar^2 \quad , \tag{13.8}$$

where $\tau = d/v$ is the interaction time with each field. The total probability of detecting an atom in level 2 after having passed through both fields is equal to

$$\begin{aligned} W_{12} &= |a_2^{(1)}|^2 + |a_2^{(2)}|^2 + 2\,\mathrm{Re}\left(a_2^{(1)} a_2^{(2)}\right) \\ &= \frac{|D_{21}|^2 \tau^2}{4\hbar^2}\left(E_1^4 + E_2^4 + 2E_1^2 E_2^2 \cos\Omega T\right) \quad . \end{aligned} \tag{13.9}$$

Figure 13.5 illustrates the narrowing of the two-photon resonance with two separated interaction fields. The technique of two-photon Ramsey resonances has been succesfully applied to the ultrahigh resolution spectroscopy of Rydberg levels in the rubidium atom using a well-stabilized cw dye laser [13.6]. At a separation of 4.2 mm between the two fields, Ramsey fringes with a spectral width of 17 kHz (!) FWHM have been measured while the single zone two-photon resonance width was limited by transit time broadening to about 600 kHz. Figure 13.6 shows the experimental arrangement. The rhodamine 6G dye laser beam is reflected by a folding mirror and crosses the atomic beam twice. The cavity length can be controlled to maintain the resonance condition for a standing wave. The intracavity laser power is about 1 W and the beam waists are $w_0 = 150\ \mu m$. The two-photon transition is detected by field ionization of the Rydberg levels (see Sect.8.2.5).

13.1.2 Nonlinear Ramsey Fringes Using Three Separated Fields

Another solution to restore the Ramsey fringes, which are generally washed out in the second field, is based on the introduction of a third field at a distance 2L downstream from the first field. The idea of this arrangement was first pointed out by BAKLANOV et al. [13.7]. The basic idea may be understood as follows.

In Sect.10.2 it has been discussed in detail that the nonlinear absorption of a molecule in a monochromatic standing wave field leads to the formation of a narrow Lamb dip at the center ω_0 of a Doppler-broadened absorption line (see Fig.10.19). The Lamb-dip formation may be regarded as a two-step process: the depletion of a small subgroup of molecules with velocity components $v_z = 0 \pm \Delta v_z$ by the pump wave (hole burning), and the successive probing of this depletion by a second wave. In the standing wave of the second zone the nonlinear saturation by the pump wave depends on the relative phase between the molecular dipoles and the field. This phase is determined by the starting point $(0,z_1)$ in the first zone and by the transverse velocity component v_z.

Figure 13.7a depicts the collision-free straight path of a molecule with transverse velocity component v_z, starting from a point $(x=0,z=z_1)$ in the first zone, traversing the second field at $z_2 = z_1 + v_zT = z_1 + v_zL/v_x$, and arriving at the third field at $z_3 = z_1 + 2v_zT$. The relative phase between molecule and field at the entrance (L,z_2) into the second field is

$$\Delta\varphi = \varphi_1(z_1) + (\omega_{12} - \omega)T - \varphi_2(z_2) \quad .$$

The macroscopic polarization at (L,z_2) averages to zero because molecules with different velocities v_z arrive at z_2 from different points $(0,z_1)$. Note, however, that the population depletion Δn_a in the second field depends on the relative phase $\Delta\varphi$ and therefore on v_z. If the phases $\varphi(z_1)$ and $\varphi(z_2)$ of the two fields are made equal for $z_1 = z_2$, the phase difference $\varphi(z_1) - \varphi(z_2) = \varphi(z_1 - z_2) = \varphi(v_zT)$ between the two fields at the intersection points depends

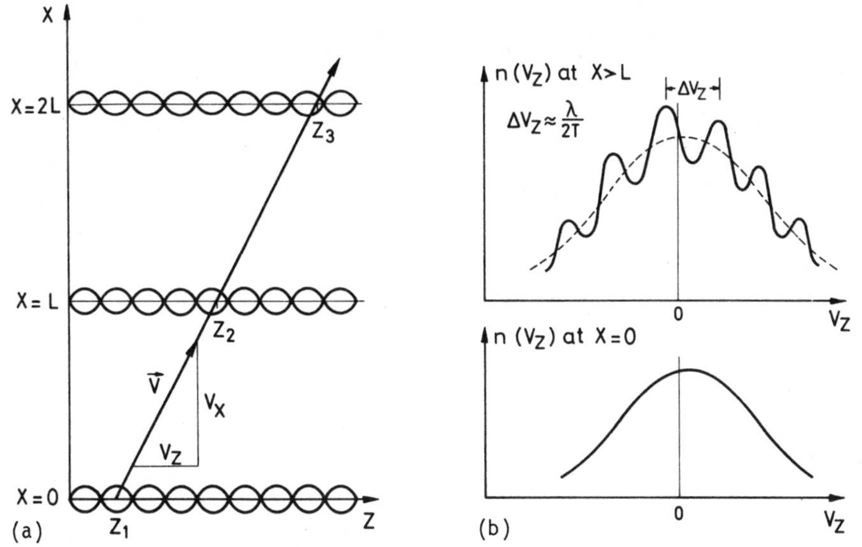

Fig. 13.7. (a) Path of a molecule through the three fields. (b) Modulation of the population $n(v_Z)$ after the nonlinear interaction with the second field (after [13.7])

only on v_z and not on z. *After the nonlinear interaction with the second field the number* $n(v_z)$ *of molecular dipoles shows a characteristic modulation* (Fig. 13.7b). This modulation cannot be detected in the second field because it appears in v_z but not in z, and in the interaction with the probe wave with a phase $\varphi(z_2)$ this modulation is completely washed out. This is, however, not true in the third field. Since the intersection points z_1, z_2, and z_3 are related to each other by the transverse velocity v_z, the modulation $n(v_z)$ in the second beam results in a nonvanishing macroscopic polarization P in the third beam. The energy absorbed in the third field with amplitude E_3 is

$$W = 2\mathrm{Re}\left(E_3 \int\limits_{z=0}^{z_0} \int\limits_{t=2T}^{2T+\tau} \left[P(z,t)\cos(kz+\varphi_3)E^{i\omega t}\right] dz\, dt\right) . \tag{13.10}$$

The detailed calculation, using third-order perturbation theory [13.7], shows that

$$W = \frac{\hbar\omega}{2}\,|G_1G_2^2G_3|\tau^4 \cos^2\Omega T \cos(2\varphi_2 - \varphi_1 - \varphi_3) \quad , \tag{13.11}$$

where $G_n = id_{21}E_n/\hbar$ $(n = 1,2,3)$ and φ_1, φ_2, φ_3 are the spatial phases of the three fields

$$E_n(x,z,t) = 2E_n(x)\cos(kz_n + \varphi_n)\cos\omega t \quad . \tag{13.12}$$

Adjusting the phases φ_n properly, such that $2\varphi_2 = \varphi_1 + \varphi_3$, allows optimization of the signal in the third zone.

The capability of this combination of optical Ramsey fringes with saturation spectroscopy has been impressively demonstrated by BERGQUIST et al. [13.8], who measured the intercombination line $^1S_0 - {}^3P_1$ in calcium at $\lambda = 657$ nm. Linewidths of the central Ramsay fringe as narrow as 3 kHz (!) with the completely resolved photon recoil doublet (see next section) have been obtained, using field separations up to 3.5 cm. The present limitation to the accuracy of this technique is the second-order Doppler effect.

In the previous discussion the third field was used to detect the Ramsay fringes through resonances in the absorbed power. It is also possible to omit the third field. If two standing waves at $x = 0$ and $x = L$ resonantly interact with the molecules, continuous coherent radiation may be observed at $x = 2L$, which is due to polarization transfer. The radiation intensity has a sharp peak at the center frequency ω_{12}. The basic principle of this phenomenon is similar to that of photon echoes (see Sect.11.4.2). Due to a phase jump at the nonlinear interaction with the second field the Doppler phase, caused by the transverse velocity v_z, is exactly cancelled if the transfer times $T_{12} = v_x/(x_2 - x_1)$ and $T_{23} = v_x(x_3 - x_2)$ are equal [13.9].

13.2 Photon Recoil

Assume an atom or molecule with mass M and energy levels E_a, E_b, which moves at a velocity $\underline{v}$, absorbs a photon of energy $\hbar\omega$ and momentum $\hbar\underline{k}$, resulting in a transition $E_a \rightarrow E_b$. Conservation of momentum demands that

$$\underline{P}_a + \hbar\underline{k} = \underline{P}_b$$

where P_a is the momentum of the atom before and P_b that after the absorption of $\hbar\omega$. Conservation of total energy must be expressed by the relativistic relation [13.10]

$$\hbar\omega = \sqrt{P_b^2c^2 + (Mc^2 + E_b)^2} - \sqrt{P_a^2c^2 + (Mc^2 + E_a)^2} \quad . \tag{13.13}$$

Expanding (13.13) in powers of 1/c yields the absorption frequency $\omega_{ab}^{(abs)}$ for the resonance case

$$\omega_{ab}^{(abs)} = \omega_0 + \underline{k}\cdot\underline{v}_a - \omega_0 \frac{v_a^2}{2c^2} + \frac{\hbar\omega_0^2}{2Mc^2} + \dots \quad . \qquad (13.14)$$

The first term $\omega_0 = (E_a - E_b)/\hbar$ represents the eigenfrequency of the atom in rest if recoil is neglected. The second term is the *linear* (first-order) Doppler effect, describing the well-known Doppler shift $\Delta\omega = \underline{k}\cdot\underline{v}$ in the absorption frequency of a moving atom. The third term represents the *second-order* Doppler effect. Note that this term is independent of the direction of $\vec{v}$ and cannot be eliminated by the methods, discussed in Chap.10, which only overcome the first-order Doppler effect. The last term in (13.14) describes the photon recoil effect, where ω_{ab} has been approximated by ω_0.

A similar consideration for the *emission* of a photon by a molecule in level E_b with momentum $\underline{P}_b$ yields with $\underline{P}_b = \underline{P}_a + \hbar\underline{k}$ the emission frequency

$$\omega_{ab}^{(em)} = \omega_0 + \vec{\underline{k}}\cdot\vec{\underline{v}}_b - \frac{\omega_0 v_b^2}{2c^2} - \frac{\hbar\omega_0^2}{2Mc^2} \quad . \qquad (13.15)$$

The frequency difference

$$\Delta\omega = \omega_{ab}^{(abs)} - \omega_{ab}^{(em)} = \hbar\omega_0^2/Mc^2 \qquad (13.16)$$

between absorption and emission frequency of a molecule at rest ($\underline{v}_a = \underline{v}_b = 0$) is due to the recoil. The relative frequency change

$$\frac{\Delta\omega}{\omega} = \frac{\hbar\omega}{Mc^2} \qquad (13.17)$$

equals the ratio of photon energy $\hbar\omega$ to rest mass energy Mc^2.

In the X-ray region the recoil energy can be so large that the frequency of γ quanta emitted by free nuclei is shifted out of resonance with the absorption profile of the same transition in absorbing nuclei of the same kind. The recoil can be avoided by implanting the nuclei into the rigid lattice of a bulk crystal below its Debye temperature. This recoil-free emission and absorption of γ quanta is called the *Mößbauer effect*.

In the optical region the recoil shift is extremely small because of the small ratio of photon energy $\hbar\omega$ to rest mass energy Mc^2. Nevertheless the recoil shift can be observed nowadays in Doppler-free spectroscopy with ultrahigh resolution. This has been demonstrated by HALL et al. [13.11] who observed in high-resolution saturation spectra of CH_4 a splitting of the Lamb dips into two components, called the *recoil doublet*. This splitting may be explained as follows.

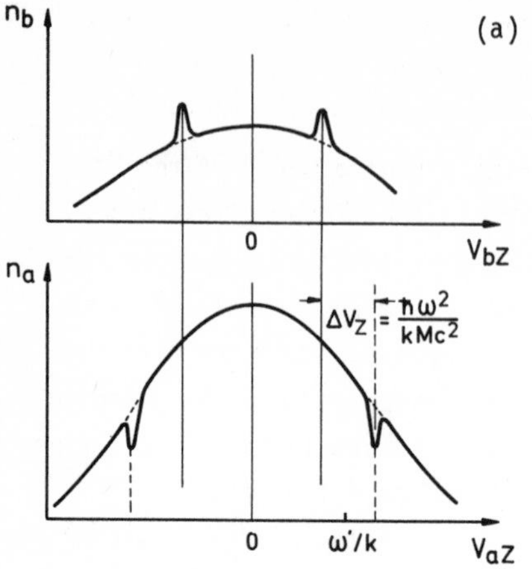

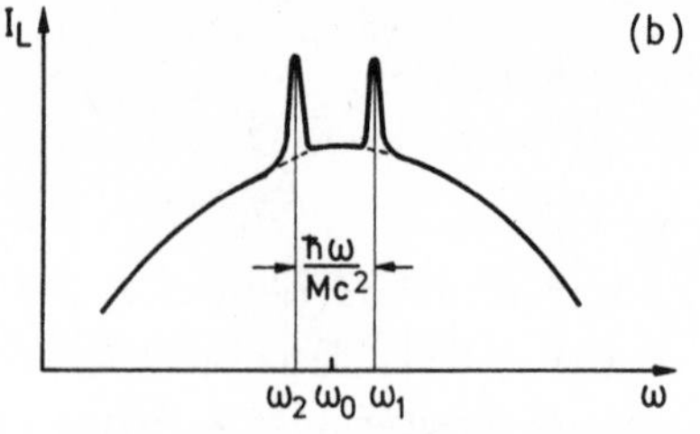

Fig. 13.8. Recoil shift of Bennet holes in the lower state population (b) and of Bennet peaks in the upper state population (a). Recoil doublet in the output of a laser with internal absorption cell, generated by the shift of the Lamb dip against the Lamb peak

When the absorbing probe is placed inside the laser resonator, the standing wave of the monochromatic laser with $\omega \neq \omega_0$ burns two holes into the ground state population $n_a(v_z)$, which appear at the velocity components

$$v_{az} = \pm(\omega' - \hbar\omega^2/2Mc^2)/k \quad ,$$

where $\omega' = (\omega - \omega_0 + \omega_0 v^2/2c^2)$ [see (13.14), Sect.10.2, and Fig.13.8]. The corresponding peaks in the population distribution $n_b(v_z)$ of the upper level are shifted against the holes due to the recoil shift and appear at

$$v_{bz} = \pm(\omega' + \hbar\omega^2/2Mc^2)/k \quad .$$

In Fig.13.8 we have assumed that $\omega < \omega_0$ and $\omega' < 0$. The two population holes in the lower level merge together for a laser frequency $\omega = \omega_1$ where

$$\omega' = \hbar\omega^2/2Mc^2 \sim \omega_1 = \omega_0(1 - v^2/2c^2) + \hbar\omega^2/2Mc^2 \quad ,$$

while the population peaks in the upper level b coincide for

$$\omega_2 = \omega_0(1 - v^2/2c^2) - \hbar\omega^2/2Mc^2 \quad .$$

Since the total absorption of the laser shows minima at the Lamb dip in the ground state population as well as at the Lamp peak in the upper state population (due to stimulated emission), the laser output exhibits *two* peaks which are separated by the recoil splitting $\Delta\omega = \hbar\omega^2/Mc^2$.

For the CH_4 transition at $\lambda = 3.39$ µm the recoil splitting amounts to 2.16 kHz [13.12]. Since such small recoil splittings can be resolved only if the Lamb-dip linewidth becomes smaller than the splitting, time-of-flight broadening and all other broadening effects must be kept extremely small. This may be achieved either by using expanded laser beams or by utilizing the Ramsey method of separated fields [13.8].

13.3 Optical Cooling and Trapping of Atoms

During the last years several proposals have been published concerning the possibilities of optical cooling and trapping of atoms or ions in the gas phase. While some of these ideas could already be realized experimentally, some others are still in the theoretical stage. It turns out that optical cooling techniques may be very efficient to cool atoms or ions down to extremely low temperatures far below 1 K. The corresponding small velocities allow the atoms to stay within the interaction region with the light field for a long time. Ions can even be completely trapped in electrodynamic ion traps while neutral atoms may be trapped in sufficiently strong laser fields. Such ultracold atoms will allow spectroscopy with extremely high resolution because many broadening effects, such as time-of-flight broadening or first- and second-order Doppler effect, can be considerably reduced.

We discuss in this section cooling and trapping of neutral atoms while the next section will deal with spectroscopy in ion traps.

A free atom interacting with a laser field experiences two different kinds of forces. The first is due to the photon recoil during resonance absorption with subsequent spontaneous emission. This force is often called *resonance radiation pressure*. The second force arises from nonresonant stimulated scattering of photons by atoms and occurs only in fields with a nonvanishing field gradient. Let us at first consider how the *spontaneous recoil force* can be utilized to cool atoms [13.13a].

13.3.1 Optical Cooling by Resonance Photon Recoil

Assume a single two-level atom moving through a laser beam with a frequency ω tuned into resonance with the atomic transition $E_a \rightarrow E_b$. During its transit time T through the laser beam the atom may absorb and reradiate a photon many times before it leaves the interaction zone. For sufficiently strong laser intensities the number n of absorption-emission cycles may reach the limit $n = T/\tau$ set by the spontaenous lifetime τ of the upper level E_b.

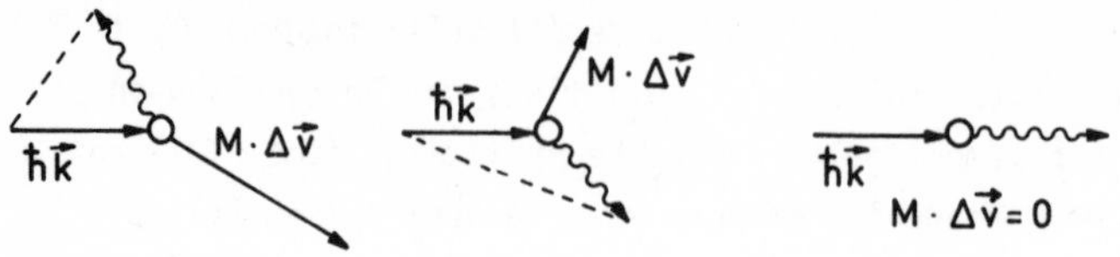

Fig. 13.9. Total recoil of an atom after one absorption-emission cycle for different directions of the emitted photon

Since the emission of spontaneous photons occurs randomly distributed over all directions, the average momentum experienced by the atom due to recoil of the *spontaneously emitted* photons approaches zero for $n \to \infty$. The recoil due to *absorption* of photons, however, does *not* average to zero but accumulates for successive absorptions, since the absorbed photons all come from the same direction (Fig.13.9).

If the initial velocity $\underline{v}$ of the atom is opposite to the light propagation $\underline{k}$ $(\underline{v}\cdot\underline{k} < 0)$, $|\underline{v}|$ decreases for each absorption by an amount

$$\Delta v = \hbar\omega/Mc \quad . \tag{13.18}$$

Example:

1) For a sodium atom with M = 23 AMU which absorbs photons on the $3S \to 3P$ transition with $\hbar\omega \approx 2$ eV, (13.18) yields $\Delta v = 3$ cm/s. This implies that about $n = 2\cdot 10^4$ absorptions are required to decrease the initial velocity from its thermal value of 6×10^4 cm/s at T = 500 K to 2×10^3 cm/s corresponding to $T \approx 0.6$ K. With the spontaneous lifetime $\tau = 16$ ns the minimum cooling time becomes $2\times 10^4 \cdot 1.6\times 10^{-8} \approx 300$ µs. During this time the atom has travelled a mean distance of about $3\times 10^{-4}\cdot 6\times 10^4/2 \approx 9$ cm.

2) For magnesium atoms absorbing photons on the singlet resonance line at $\lambda = 285.2$ nm with an upper state lifetime of 2 ns the situation becomes more favorable. $\Delta v \approx 6$ cm/s, $n \approx 1.3\times 10^4$ and the minimum cooling time becomes 3×10^{-5} s and the path length about 1 cm.

Note that *molecules* cannot be cooled by this mechanism, since the spontaneous emission will only partly repopulate the absorbing level while most of the emission terminates on other rotational-vibrational levels of the electronic ground state. The initial level will therefore be completely depleted after the first few cycles.

Note that increasing the pump intensity beyond the saturation intensity which produces a cycle frequency $n = 1/\tau$ does not help to increase the cooling speed because the induced emitted photons always have the same direction as the absorbed ones.

Since the Doppler shifted absorption frequency $\omega = \omega_0 + \underline{k} \cdot \underline{v}$ changes with $|v|$ during the cooling process, the bandwidth of the laser must be sufficiently broad to cover the *low-frequency side* of the Doppler profile ($\underline{k} \cdot \underline{v} < 0$) to stay always in resonance. If a monochromatic laser is used, its frequency ω has to be tuned during the cooling process according to

$$\omega(t) = \omega_0 + \underline{k} \cdot \underline{v}(t) \pm \delta\omega_n \qquad (13.19)$$

in order to keep it always within the natural linewidth $\delta\omega_n$. The minimum velocity $|v_{min}| \approx \delta\omega_n/k$ that can be reached by optical cooling is determined by the natural linewidth $\delta\omega_n$ even if the atom stays within the radiation field for a sufficiently long time.

So far we have considered a single atom moving towards an unidirectional laser beam. Optical cooling, however, can also be achieved in a gas of atoms with thermal velocity distribution N(v) exposed to isotropic laser radiation [13.13]. If the laser frequency is confined to the lower half of the Doppler-broadened absorption line, the probability of absorption is higher for atoms with $\underline{k} \cdot \underline{v} < 0$ than for atoms with $\underline{k} \cdot \underline{v} > 0$ (Fig.13.10). In other words, more atoms moving towards the light propagation $\underline{k}$ absorb photons than atoms moving in the direction of $\underline{k}$. The *reduction* of $|\underline{v}|$ due to recoil overcompensates the increase of $|\underline{v}|$.

If the resonance transition occurs between a ground state E_a and an excited state E_b with spontaneous decay rate $\gamma = 1/\tau$, the spontaneous force experienced by the atom in a one-dimensional standing wave $E = 2E_0 \cos kz \cos\omega t$ can be derived as [13.14]

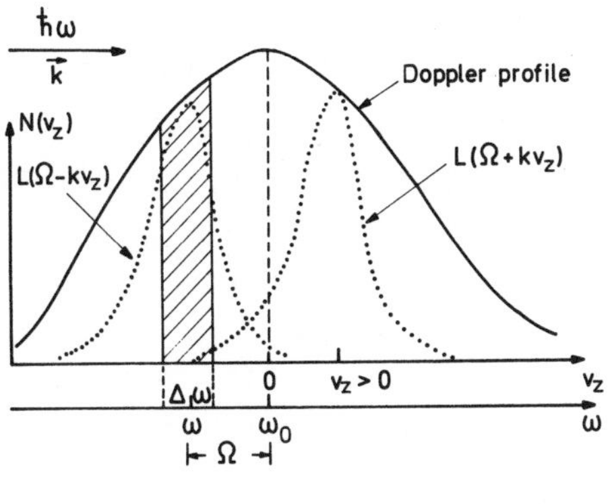

Fig. 13.10. The probability of absorbing photons $\hbar\omega$ within the bandwidth $\Delta\omega$ of the laser with $\omega < \omega_0$ is higher for atoms with $\underline{v} \cdot \underline{k} < 0$ than for atoms with $\underline{v} \cdot \underline{k} > 0$

$$F_{sp} = 2\hbar k\gamma \left(\frac{dE_0}{\gamma}\right)^2 \frac{L(\Omega - kv_z) - L(\Omega + kv_z)\ \sin^2 kt}{1 + 2(dE_0/\gamma)^2[L(\Omega - kv_z) + L(\Omega + kv_z)]} \quad , \tag{13.20}$$

where d is the dipole moment of the transition $E_a \to E_b$ and L represents the Lorentzian profile of width γ with the Doppler-shifted frequencies $\Omega \pm kv_z$ and $\Omega = \omega_{ab} - \omega$. Equation (13.20) expresses the differences in the probabilities of recoil momentum transfer for atoms with $\underline{k} \cdot \underline{v} < 0$ and $\underline{k} \cdot \underline{v} > 0$. The numerator in (13.20) can take positive and negative values depending on ω. If the Doppler shift kv_z becomes smaller than the natural linewidth γ, the positive and negative contributions nearly cancel and the cooling force becomes exceedingly small.

For the example of Na atoms given above, the natural linewidth (10 MHz) is about 100 times smaller than the Doppler width at 500 K. If optical cooling reduces the Doppler width by a factor of 100 this implies a reduction of the temperature by a factor of 10^4, i.e., from 500 to 0.05 K. The real advantage of optical cooling becomes obvious in cases where the natural linewidth is small relative to other broadening effects such as time-of-flight broadening or second-order Doppler effect. Then this technique really becomes superior to other Doppler-free methods.

13.3.2 Induced Dipole Forces

When an atom with polarizability α is placed within an inhomogeneous electric field $\underline{E}$, a dipole moment $\underline{p} = \alpha\underline{E}$ is induced which results in the well-known force on the dipole

$$\underline{F}_D = (-\underline{p}\ \mathrm{grad})\underline{E} \quad . \tag{13.21}$$

A similar relation holds for atoms in optical fields where the field frequency ω is off resonance with the atomic frequency ω_0. In an electromagnetic field the Lorentz force on an atom in a dilute medium (refractive index $n \approx 1$, $n - 1 \ll 1$) is [13.15]

$$\underline{F} = (\underline{p}\cdot\underline{\nabla})\underline{E} + \frac{1}{c}\left(\frac{d\underline{p}}{dt} \times \underline{B}\right) = \alpha(\underline{E}\cdot\underline{\nabla})\underline{E} + \frac{1}{c}\left(\frac{\partial \underline{E}}{\partial t} \times B\right) \quad . \tag{13.22}$$

Using the identity

$$(\underline{E}\cdot\underline{\nabla})\underline{E} = \underline{\nabla}\left(\frac{1}{2}E^2\right) - \underline{E} \times (\underline{\nabla}\times\underline{E})$$

and Maxwell's equation

$$(\underline{\nabla}\times\underline{E}) + \frac{1}{c}\frac{\partial \underline{B}}{\partial t} = 0 \quad ,$$

we obtain for (13.22)

$$\underline{F}_D = \alpha\left[\left(\nabla \frac{1}{2} E^2\right) + \frac{1}{c}\frac{\partial}{\partial t}(\underline{E}\times\underline{B})\right] \quad . \tag{13.23}$$

Averaged over a cycle of the optical field the last term in (13.23) vanishes and we obtain for the average force

$$\overline{\underline{F}_D} = \alpha\nabla\left(\frac{1}{2}\overline{E}^2\right) \quad . \tag{13.24}$$

Since the polarizability α is related to the refractive index $n(\omega)$ of a dilute gas with atom density N by

$$\alpha(\omega) = \frac{1}{2\pi N}[n(\omega) - 1] \quad , \tag{13.25}$$

the frequency dependence of $\alpha(\omega)$ follows like $n(\omega)$ a dispersion profile (see Sect.2.6). Taking into account the saturation broadening $\delta\omega_s = \gamma_s = \delta\omega_n\sqrt{1+S}$ which depends on the saturation parameter $S = I/I_s$ (see Sect.3.6) we obtain for the line profile of $\alpha(\omega)$

$$\alpha(\omega) = \frac{c^3(\gamma/2)}{2\omega^3}\,\frac{\Delta\omega}{\Delta\omega^2 + (\gamma_s/2)^2} \quad , \tag{13.26}$$

where $\Delta\omega = \omega - \omega_0 - \underline{k}\cdot\underline{v}$ is the detuning of the field frequency ω from the Doppler-shifted atomic frequency $(\omega_0 - \underline{k}\cdot\underline{v})$. For sufficiently intense laser fields ($S \gg 1$, $\gamma_s \gg \gamma_n$) the polarizability $|\alpha|$ increases with $|\Delta\omega|$. For $\Delta\omega < \gamma_s$ the induced dipole force therefore increases with $\Delta\omega$ and with the intensity I. It can be shown [13.16] that in the limit $S \gg 1$ the dipole force reduces to

$$\underline{F}_D = \hbar\Delta\omega\cdot\underline{\nabla}I/I \quad . \tag{13.27}$$

For homogeneous fields (e.g., a plane travelling wave) the dipole force vanishes. For a Gaussian beam, travelling into the z direction, the intensity profile in the x-y plane is (see Sect.5.11)

$$I(r) = I_0\, e^{-2r^2/w^2}$$

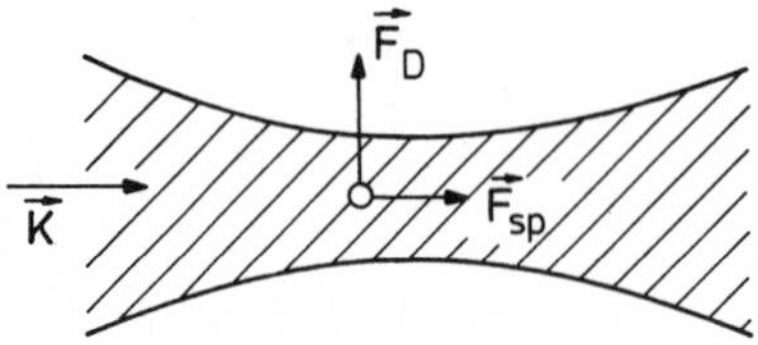

Fig. 13.11. Longitudinal and transverse forces exerted on a neutral atom in a weakly focused Gaussian laser beam [13.16]

and we obtain a dipole force $\underline{F}_D$ in the x-y plane in the direction of the field gradient and a spontaneous force in the propagation direction $\underline{k}$ (Fig.13.11).

13.3.3 Trapping of Atoms in Optical Standing Wave Fields

The induced dipole force may perhaps be used to trap atoms in a standing wave field [13.17,18]. For a one-dimensional plane standing wave

$$E(z,t) = E_0 \sin kz \sin\omega t \quad ,$$

the dipole force according to (13.24) becomes

$$\underline{F}_D = \frac{1}{4}\alpha(\omega)\cdot\underline{k}\,E_0^2 \sin 2kz$$

and the equation of motion $F(z) = M\ddot{z}$ for an atom with mass M is

$$\ddot{z} = \frac{1}{4M}\alpha(\omega)\cdot\underline{k}\,E_0^2 \sin 2kz \quad . \tag{13.28}$$

With the initial conditions $z(0) = z_0$ and $z(0) = v_0$ (13.28) has the solution

$$\dot{z} = \pm E_0(\alpha/2M)^{\frac{1}{2}}\left[(\cos 2kz_0 - \cos 2kz) + \frac{2M}{\alpha E_0^2} v_0^2\right]^{\frac{1}{2}} \quad . \tag{13.29}$$

The sign of $\alpha(\omega)$ depends on the field frequency ω. For $\omega \gtrless \omega_0 + \underline{k}\cdot\underline{v} \sim \alpha(\omega) \gtrless 0$. For $\alpha > 0$ the potential minima are the nodes of the standing wave. If the initial velocity $v_0 = v(z_0)$ in a node at $z = z_0$ is less than a critical value

$$v_{cr} = E_0\left(\frac{\alpha(\omega)}{2M}\right)^{\frac{1}{2}} \quad , \tag{13.30}$$

the velocity in (13.29) becomes zero for certain values of z. This means that the particle becomes trapped in the potential minimum (Fig.13.12). The maximum kinetic energy of the atoms is then less than the potential barrier heights, which in turn depend on the field amplitude E_0. It can be shown that the tunnel effect is negligible, which implies that in a more rigorous quantum mechanical treatment the same results are also obtained.

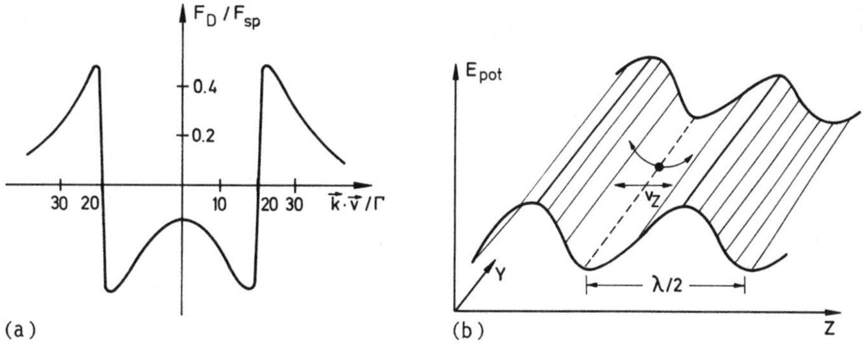

Fig. 13.12a,b. Trapping of neutral atoms in a standing light wave. (a) Induced light pressure force, normalized to the spontaneous force $F_s^0 = 2\hbar K\Gamma$ as a function of the particle velocity $\vec{v}$. (b) One-dimensional oscillation of a trapped particle around the minimum of the potential energy in a plane standing wave

An estimation of this potential barrier height yields, however, the result, that even at field intensities of several kW/cm^2 the maximum thermal energy of atoms to be trapped must be below 1 K. This implies that the atoms must be cooled before they can be efficiently trapped in optical standing wave fields. This cooling can be performed for example by the spontaneous, resonant light pressure discussed in Sect.13.3.1.

The cold atoms may eventually be spatially captured in a three-dimensional standing wave field [13.18]. The motion of the atoms is due to the combined action of spontaneous and induced light pressure. If the light frequency is tuned below resonance, the atoms are cooled by the spontaneous forces until their velocities decrease below the critical value. Due to the induced force they undergo oscillatory motions around the potential minima of the three-dimensional standing wave field. The observation of this motion can be performed with a probe laser, tuned to a transition other than that used for cooling [13.19].

The experimental realization of this idea is certainly not trivial. Besides the necessary frequency stability and spatial stability of the three-dimensional light field, the proper time sequence of cooling and trapping is essential if cooling and trapping are to be performed by the same field. It is therefore not clear whether or not this idea can be realized. Fortunately there are, however, other ways which allow trapping of *charged* particles in Penning traps or in rf quadrupole fields. These trapped ions can then be optically cooled by "sideband cooling". These methods have been already verified experimentally and we discuss them briefly in the next section.

13.4 Trapping and Cooling of Ions

Two different techniques have been developed to store ions within a small volume. In the "rf quadrupole trap" [13.20] the ions are confined within a hyperbolic electric radio-frequency field while in the "Penning trap" [13.21] a dc magnetic field with a superimposed dc electric field of hyperbolic geometry is used to trap the ions.

The electric quadrupole field is formed by applying a voltage U between a ring electrode with hyperbolic surface as one pole and two hyperbolic caps as the other pole (Fig.13.13). The whole system has axial symmetry with respect to the z axis. In a geometrical arrangement where the inner ring radius r_0 is related to the cap separation $2z_0$ by $r_0 = \sqrt{2}\, z_0$ the electric potential ϕ for points within the trap can be written as [13.20]

$$\phi = \frac{U}{2r_0^2}(r^2 - 2z^2) \quad . \tag{13.31}$$

If the applied voltage $U = U_0 - V_0 \cos\omega_0 t$ is a superposition of a dc voltage U_0 and an ac voltage $V_0 \cos\omega_0 t$, the equation of motion $m\ddot{\vec{r}} = \vec{F} = -q \operatorname{grad} \phi$ of a particle with charge q and mass m moving in the potential ϕ can be written for the x direction as

$$\frac{d^2x}{dt^2} + \frac{\omega_0^2}{4}(a - 2b \cos\omega_0 t)x = 0 \quad , \tag{13.32}$$

where the parameters

$$a = \frac{4qU_0}{mr_0^2\omega_0^2} \quad , \quad b = \frac{2qV_0}{mr_0^2\omega_0^2}$$

are determined by the dc voltage U_0 or the ac voltage V_0. Because of the axial symmetry of ϕ the same equation holds for the y component while for the z component r_0^2 has to be replaced by $-z_0^2$.

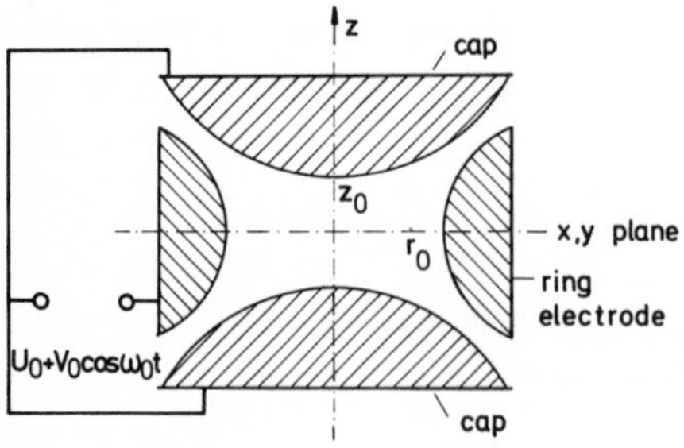

Fig.13.13. Cross section of the rf quadrupole ion trap

The Matthieu differential equation (13.32) has stable oscillatory solutions only for certain ranges of the parameters a and b [13.22] and for certain initial conditions. Charged particles which enter the trap from outside cannot be stabilized. The ions therefore have to be produced inside the electrodes. The general stable solutions of (13.22) can be represented as a superposition of two motions: a "micro motion" with frequency ω_0 of the ion around a "guiding center", which itself performs slower oscillations composed of harmonic oscillations with frequency $\bar{\omega}_r$ in the x and y directions and with frequency $\bar{\omega}_z = 2\bar{\omega}_r$ in the z direction [13.21]. The motion of an ion along the z axis is

$$z(t) = [1 + \sqrt{2}\,(\bar{\omega}_z/\omega_0)\cos\omega_0 t]\bar{z}_0 \cos\bar{\omega}_z t \quad . \tag{13.33}$$

The frequency spectrum of this motion contains the fundamental frequency ω_0 and its harmonics $n\omega_0$ with sidebands at $n\omega_0 \pm \bar{\omega}_z$.

The trapped ions can be detected either by the rf voltage of frequency $\bar{\omega}_z$ they induce in an rf circuit with the cap electrodes serving as capacitance [13.23] or by laser-induced fluorescence [13.24]. The last method is very sensitive. If the laser transition is chosen properly to ensure that all ions pumped out of the initial level E_i return into this level by spontaneous emission, each ion can be recycled $n \leq 1/\tau$ times. This means that at a spontaneous lifetime $\tau = 10^{-8}$ s of the upper level each ion may emit 10^8 fluorescence photons per s which allows detection of even single ions [13.25].

The *Penning trap* consists of a homogeneous magnetic field in the z direction and a dc electric field formed by hyperbolic electrodes resulting in a dc electric potential

$$\emptyset(x,y,z) = A_0(x^2 + y^2 - 2z^2) \quad ; \quad A_0 = \text{const.} \tag{13.34}$$

Stable bound motions of charged particles with mass m and charge e can be superimposed by three simple types of orbits with three characteristic frequencies [13.21].

1) Harmonic oscillations with frequency ω_z along the z axis, where $\omega_z^2 = 4e\,A_0/m$ because the restoring force is

$$e\,E_z = -e\,\text{grad}_z\,\emptyset = +4e\,A_0\,z \quad .$$

2) Circles in the x-y plane around the z axis, which are determined by the balance of the electric force $+e\,E_r$ and the magnetic force evB. The frequency

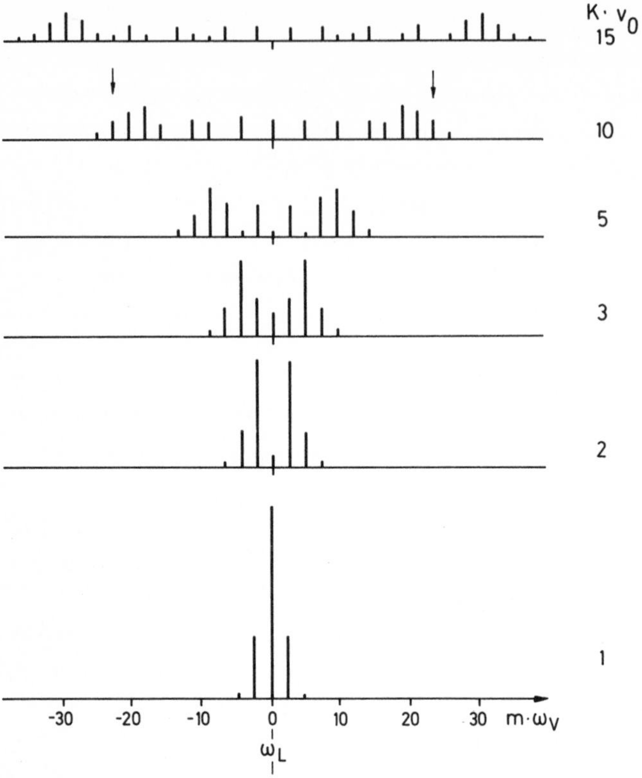

Fig.13.14. Frequency spectrum of an absorber oscillating in the direction of light propagation for different velocity amplitudes v_0

ω_M of this *magnetron motion* is a constant of the trap, $\omega_M = v/r = E_r/(Br) = 2A_0/B$ because $E_r = \text{grad}_r\ \phi = 2A_0 r$.

3) The *cyclotron motion*, determined by the balance of Lorentz force evB and centrifugal force mv^2/r which yields a frequency $\omega_c = e\ B/m$.

Ions confined in either of the two traps can be cooled by optical sideband cooling [13.25,26]. The cooling process can be illustrated by an absorbing atom which oscillates in the x direction with a velocity $v_x = v_0 \cos\omega_v t$, and which is illuminated by light of frequency ω_L propagating in the x direction (Fig.13.14). We assume that the natural linewidth γ_n of the optical transition is smaller than the vibration frequency ω_v. The absorption spectrum of such a vibrating absorber consists of the central resonance frequency ω_0 and sidebands $\omega_m = \omega_0 \pm m\omega_v$ $(m = 1,2,3)$ with relative intensities given by $J_m^2[v_0\omega_0/(c\omega_v)]$, where J_m is the Bessel function of order m. The frequency spectrum of this frequency-modulated harmonic oscillator depends on the amplitude v_0 of the oscillator. In Fig.13.14 the spectrum is plotted for different amplitudes v_0 as expressed in units of $1/k = \lambda/2\pi$.

If the absorber is irradiated with light of frequency $\omega_L = \omega_0 - m\omega_v$ the fluorescence frequency is scattered symmetrically about the central frequency ω_0 because the lifetime τ is large compared to $1/\omega_v$. On the average the oscillator therefore loses more energy than it gains. The energy difference is taken from the kinetic energy of the vibrating ion which therefore decreases on the average by $m\hbar\omega_v$ per scattering event.

This optical sideband cooling is based on the same principle as the optical cooling discussed in Sect.13.3.1. The only difference is the discontinuous absorption spectrum of the vibrating particle compared to the continuous Doppler-broadened spectrum of neutral atoms at thermal velocities. One advantage of sideband cooling is the constant frequency $\omega_0 - m\omega_v$ which does not change during the cooling process, although the amplitude of the mth sideband depends on the oscillation amplitude of the absorber. The laser frequency ω_L therefore does not need to be altered during the cooling process but the cooling rate decreases with decreasing oscillation amplitude.

Optical sideband cooling has been used to cool Mg II ions to temperatures below 0.5 K by scattering frequency-doubled dye laser photons at $\lambda = 560/2$ nm which are nearly resonant with the $3s^2\,S_{1/2} \rightarrow 3p\,^2P_{3/2}$ transition. Probing of the cooled ions can be performed by a second laser which is scanned across the absorption profile. In order to prevent heating of the ions when the probe laser is tuned to the high-frequency side of the sidebands, the power of the probe laser must be lower than that of the cooling laser [13.26]. TOSCHEK and co-workers [13.25] succeeded in confining 10 to 20 barium ions in a miniaturized rf quadrupole trap and observing the effect of optical cooling. A weak electron beam generates ions inside the initially empty trap at a rate of 1 or 2 per minute. The corresponding increments of the resonance fluorescence give rise to steps of the recorded fluorescence signals, thus demonstrating the detection of single ions.

13.5 Resolution Within the Natural Linewidth

Assume that all other line-broadening effects except the natural linewidth have been eliminated by one of the methods discussed in the previous chapters. The question that arises is whether the natural linewidth represents an insurmountable natural limit to spectral resolution. In this section we give some examples of techniques which allow observation of structures *within* the natural linewidth. It is, however, not obvious that such methods may really increase the amount of information about the molecular structure, since the inevitable loss in intensity may outweight the gain in resolution. We discuss under which conditions spectroscopy within the natural linewidth may be a tool which really helps to improve the quality of spectral information. The situation is illustrated by some examples.

If molecules are excited into an upper level with spontaneous lifetime $\tau = 1/\gamma$ by a light pulse ending at $t = 0$, the time resolved fluorescence amplitude is given by

$$A(t) = A(0)\, e^{-(\gamma/2)t} \cos\omega_0 t \quad . \tag{13.35}$$

If a detection method is used which allows to resolve the natural linewidth γ of the transition (e.g., levelcrossing spectroscopy or observation of quantum beats), the observed intensity profile of the spectral line becomes

$$I(\omega) = I_0 \frac{1}{(\omega - \omega_0)^2 + (\gamma/2)^2} \tag{13.36}$$

where $I_0 = (2\pi/\gamma) \int I(\omega)d\omega$ is proportional to the total fluorescence intensity integrated over the spectral line profile. The Lorentzian profile of (13.36) can be obtained from the Fourier transform $A(\omega)$ of the amplitude $A(t)$ by $I(\omega) = A(\omega)A^*(\omega)$, if the integration time (which experimentally means the observation time) extends from $t = 0$ to $t = \infty$ (see Sect.3.1).

If the detection probability for $I(t)$ is not constant, but follows a time dependence $f(t)$, the detected intensity $I_g(t)$ is determined by the "gate function" $f(t)$

$$I_g(t) = I(t)f(t) \quad .$$

The Fourier transform of $I_g(t)$ now depends on the form of $f(t)$ and may no longer be a Lorentzian. Let us consider some specific examples [13.27].

The time resolved fluorescence intensity measured in a quantum beat experiment, can be represented by

$$I(t) = I(0)\, e^{-\gamma t}(1 + a \cos\omega_0 t)$$

where $\hbar\omega_0$ is the energy separation between two levels with equal decay constants $\gamma = 1/\tau$, which had been simultaneously excited by a light pulse ending at $t = 0$ (see Fig.13.15). Assume the detection is gated by a step function

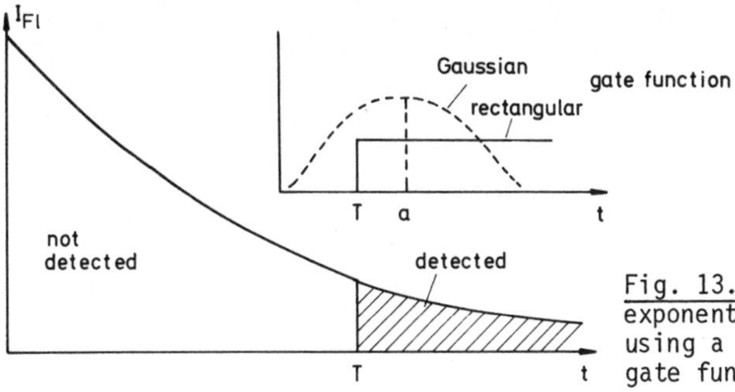

Fig. 13.15. Gated detection of an exponential fluorescence decay, using a rectangular or a Gaussian gate function

$$f(t) = \begin{cases} 0 & \text{for} \quad t < T \\ 1 & \text{for} \quad t \geqq T \end{cases}$$

which may be simply realised by a shutter in front of the detector which opens only for times $t \geqq T$ (see Fig.13.15).

The Fourier-transform of $I_g(t) = I(t)f(t)$ is now

$$I(\omega) = \int_T^\infty I(0)\, e^{-\gamma t}(1 + a \cos\omega_0 t)\, e^{-i\omega t}\, dt \tag{13.37}$$

which yields the real and imaginary parts (cosine and sine Fourier transforms) in the approximation $(\omega - \omega_0) \ll \omega_0$

$$I_c(\omega) = \frac{1}{2}\, \frac{a\gamma\, e^{-\gamma T}}{(\omega_0 - \omega)^2 + \gamma^2}\, [\gamma\cos(\omega_0 - \omega)T - (\omega_0 - \omega)\sin(\omega_0 - \omega)T]$$

$$I_s(\omega) = \frac{1}{2}\, \frac{a\gamma\, e^{-\gamma T}}{(\omega_0 - \omega)^2 + \gamma^2}\, [\gamma\sin(\omega_0 - \omega)T + (\omega_0 - \omega)\cos(\omega_0 - \omega)T] \quad . \tag{13.38}$$

For T = 0 the cosine Fourier-transform represents the normal Lorentzian profile and the sine-transform the dispersion profile. For $T > 0$ the functions (13.38) show an oscillatory structure (see Fig.13.16) where the central peak of the real part can still be approximately described for $|\omega_0 - \omega| < (1/T) < \gamma$ by a Lorentzian with a reduced full halfwidth

$$\Delta\omega_{1/2} = \frac{2\gamma}{\sqrt{1 + \gamma^2 T^2}} \quad . \tag{13.39}$$

This shows that for e.g., $T = 5\tau = 5/\gamma$ the halfwidth $\Delta\omega_{1/2}$ of the central peak has decreased from 2γ to approximately 0.4γ. However, the peak intensity

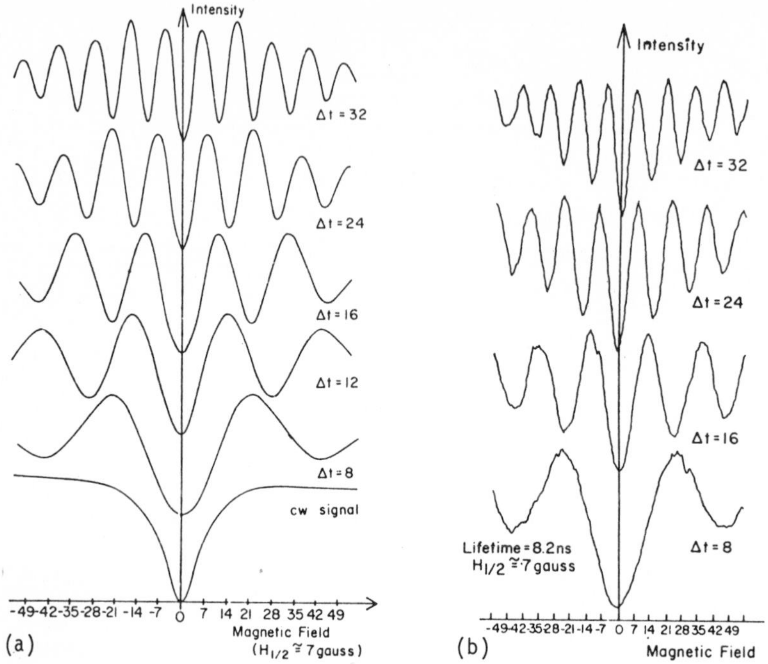

Fig. 13.16. Comparison of calculated (a) and observed (b) Hanle signals, obtained for different time delays T of the gated detector, given in ns [13.28]

of the signal has also drastically decreased by a factor $e^{-\gamma T} \approx 10^{-2}$ to less than 1%. In Fig.3.16 the central peak intensities have been always normalized.

The resultant reduction of the signal to noise ratio leads to a larger uncertainity in determining the line center. It has been shown [13.28a], that the reduction in linewidth still is advantageous in spite of the intensity losses, if the line profiles are not symmetrical. In such cases systematic errors in the determination of the line center, due to insufficient knowledge of the line profile are reduced by the technique of time delayed observation.

It is important to note, that this line narrowing effects only occur if the phase of the decaying signal is preserved. This is always the case when the detection is based on interference effects, such as level crossing, quantum beats or interferometric detection, if the shutter is placed behind the interference device [13.28b]. For these examples the cosine Fourier transform can be measured separately. If, however, the detection only measures the power spectrum of the fluorescence, without phase information, the signal is represented by

$$I(\omega) = |I_c + iI_s| = \left(I_c^2 + I_s^2\right)^{\frac{1}{2}} = \frac{\frac{1}{2}\, a\gamma\, e^{-\gamma T}}{(\omega_0 - \omega)^2 + \gamma^2} \qquad (13.40)$$

which shows that no line narrowing is observed.

The method of line narrowing by selective detection of spontaneous photons from those excited atoms which have survived for times $t \geqq T \gg \tau$ has been used to increase the spectral resolution of Doppler-free level-crossing signals. With $T = 10\tau$ a reduction of the natural linewidth up to a factor of 6 could be achieved [13.29]. The intensity of the central maximum decreases drastically because only 5×10^{-5} of all N atoms which had been excited at $t = 0$ survive for $t \geqq 10\tau$.

Another possibility of obtaining line profiles narrower than the natural linewidth of an optical transition is based on *optical-optical double-resonance* methods (see Sect.10.2.4). Assume that two monochromatic laser waves, a pump wave with frequency ω_p and a weak probe wave with ω_s, interact resonantly with two molecular transitions sharing a common upper level (Fig.13.17). The linewidth γ_s of the probe transition is then given by [13.30;10.51]

$$\gamma_s = \gamma_1 + (1 \pm \omega_s/\omega_p)\gamma_2 + \gamma_3 \quad ,$$

where γ_i is the level width of level i, determined by the population relaxation rate γ_i. The minus sign is for copropagating waves, the plus sign for counterpropagating waves. For $\omega_s/\omega_p \approx 1$, the linewidth of the probe transition in the case of copropagating beams becomes approximately $\gamma_s \approx \gamma_1 + \gamma_3$. If the levels 1 and 3 are long-lived rotational-vibrational levels of the

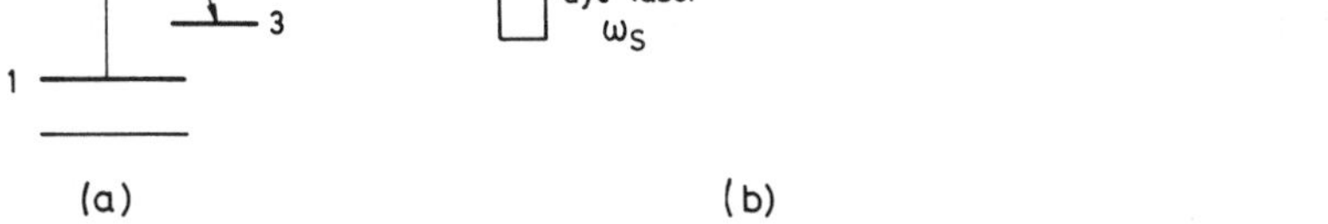

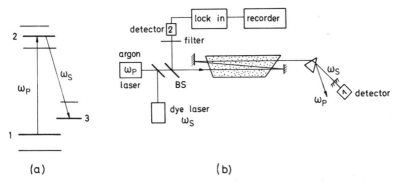

Fig. 13.17. Level scheme (a) and experimental arrangement (b) for observation of subnatural linewidths by optical-optical double-resonance techniques (after [13.31])

electronic ground state, their spontaneous lifetimes are very long (in the case of homonuclear molecules even infinitely long) and $\gamma_1 + \gamma_3 \ll \gamma_2$, which means that the linewidth γ_s of the probe transition becomes much narrower than the natural linewidth of the transition $1 \to 2$.

Figure 13.17b shows the experimental arrangement used by EZEKIEL et al. [13.31] to measure subnatural linewidths in I_2 vapor. An intensity-modulated pump beam and a continuous dye laser probe pass the iodine cell three times. The transmitted probe beam yields the forward scattering signal (detector 1) and the reflected beam the backward scattering. A lock-in amplifier monitors the change in transmission due to the pump beam, which yields the probe signal with width γ_s. The authors achieved a linewidth as narrow as 80 kHz on a transition with a natural linewidth γ_n = 141 kHz. The theoretical limit of γ_s = 16.5 kHz could not be reached because of dye laser frequency jitter.

This extremely high resolution can be used to measure collision broadening at low pressures, caused by long distance collisions (see Sect.12.1) and to

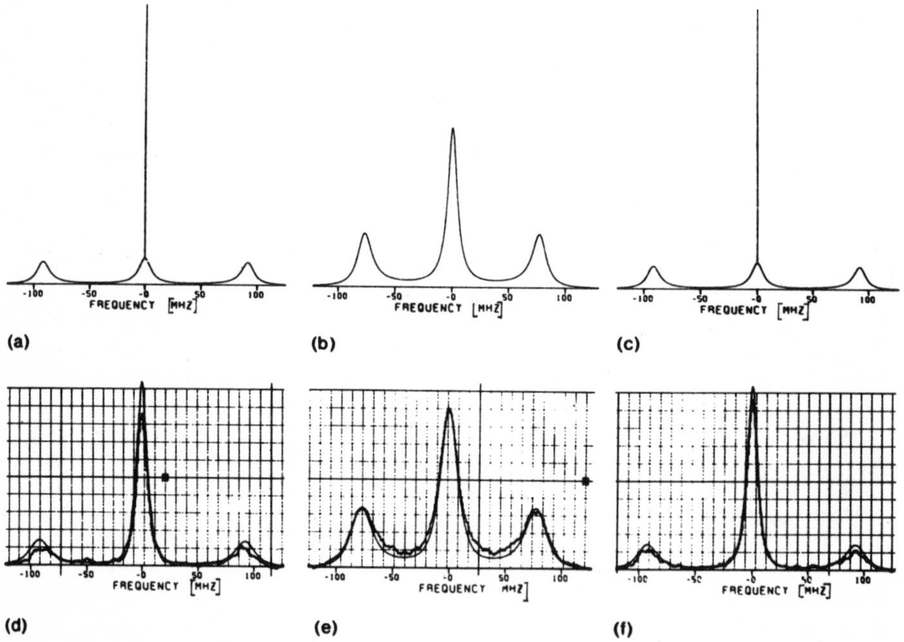

Fig. 13.18a-f. Frequency spectrum of the Na fluorescence excited by a monochromatic dye laser in a collimated Na beam. Natural linewidth γ = 10 MHz, $dE/\hbar$ = 78 MHz, laser detuning $\Delta\nu = (\omega - \omega_0)/2\pi$ from resonance: (a) -50 MHz, (b) 0, (c) +50 MHz. (a-c) Theoretical curves, (d-f) experimental curves [13.34]

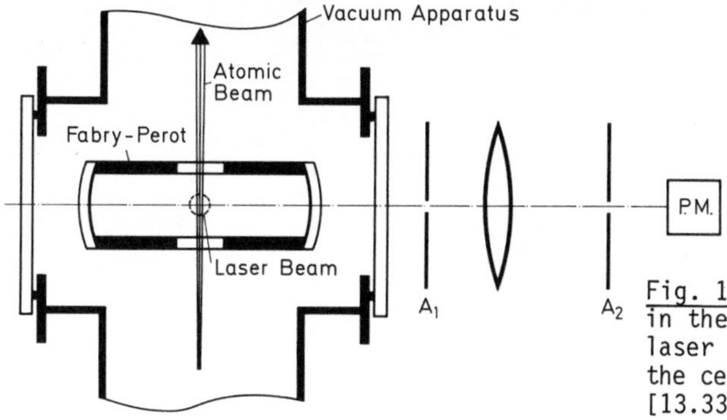

Fig. 13.19. Excitation region in the crossing volume of dye laser beam and atomic beam in the center of a confocal F.P.I. [13.33]

determine the collisional relaxation rates of the levels involved. Another interesting application is the study of power broadening and splitting effects. With increasing pump intensity the probe transition at first broadens and then shows a splitting into two components, due to the ac Stark effect [13.32]. The magnitude of both effects for copropagating waves is different from that of counterpropagating waves.

Of fundamental importance for the understanding of the interaction of atoms with radiation are detailed studies of the resonance fluorescence spectrum emitted by atoms interacting with intense monochromatic radiation. This spectrum depends on the detuning $\Omega = \omega_p - \omega_0$ of the pump frequency ω_p from the line center ω_0 and on the pump intensity $I_p \propto E^2$. The fluorescence spectrum (Fig.13.18) consists of a central peak at ω_0 and two side peaks at $\omega = \omega_0 \pm \omega_R$, where

$$\omega_R = \sqrt{(\omega-\omega_0)^2 + (dE/\hbar)^2}$$

is the Rabi flopping frequency (Sect.2.9.4) and $d = eR_{ab}$ the transition dipole moment.

Such measurements have been performed for Na atoms by several groups [13.33,34]. The experimental arrangement of [13.33] uses a well-collimated Na beam which is crossed perpendicularly by a tunable single-mode laser beam (Fig.13.19). The crossing point of the two beams is in the center of a confocal Fabry-Perot interferometer with a high finesse which serves to resolve the fluorescence spectrum. In another arrangement (see Fig.13.20) [13.34] the fluorescence emitted from the central part of the interaction region is collimated by two apertures and is analyzed by an external F.P.I. with a 2 MHz instrumental width (see Fig.13.18).

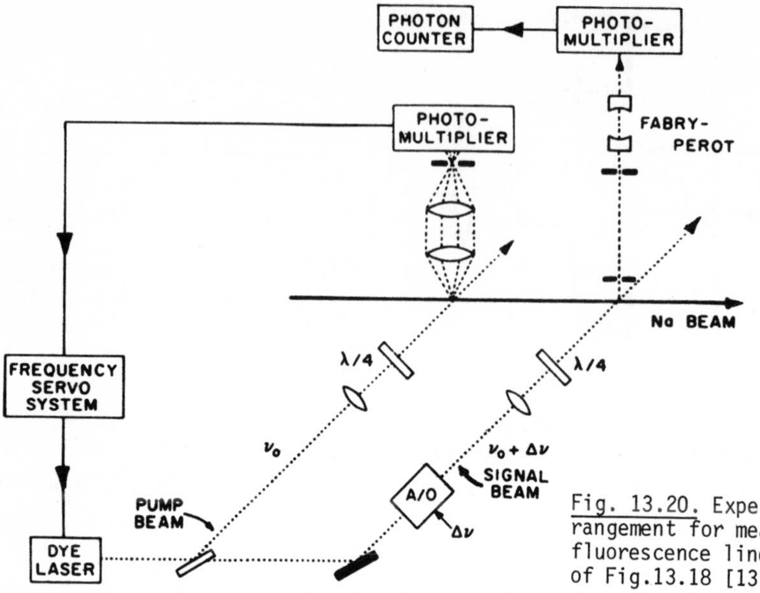

Fig. 13.20. Experimental arrangement for measuring the fluorescence line profiles of Fig.13.18 [13.34]

14. Applications of Laser Spectroscopy

The experimental advantages of laser spectroscopy regarding spectral power density and spectral resolution have brought about a great variety of applications in many scientific and technical fields. The selection of examples presented in this chapter is by far not complete but intends to illustrate the impact of laser spectroscopy on the development of new experimental techniques in chemistry, biology, and environmental sciences. The importance of laser spectroscopic applications is emphasized by the publication of many monographs, review papers, or conference proceedings on this subject. For more detailed information the reader is therefore referred to the cited literature [1.17,19; 14.1-6].

14.1 Laser Photochemistry

There are many ways in which lasers may be utilized in chemistry. Of particular interest, and probably of great economical importance in the near future, is the possibility of enhancing or catalyzing specific chemical reactions by selective excitation of the reactands via optical pumping with lasers. Another area of lasers in chemistry is the study of internal state distributions of reaction products using laser-induced fluorescence techniques (see Sect.8.7). The dependence of this distribution on the internal energy of the reactands or on their translational energy allows far-reaching conclusions about the reaction pathways and the potential surfaces of the intermediate state. A very interesting field comprises spectroscopy investigations of energy transfer processes (see Chap.12) which allow a more detailed insight into the nature of inelastic and reactive collisions [14.6a]. One example is the study of excitation and deactivation processes in chemical lasers or in energy transfer lasers in the infrared and visible regions.

Let us first consider laser-induced chemical reactions. The excitation energy of one or several reactants which initiates and drives the chemical

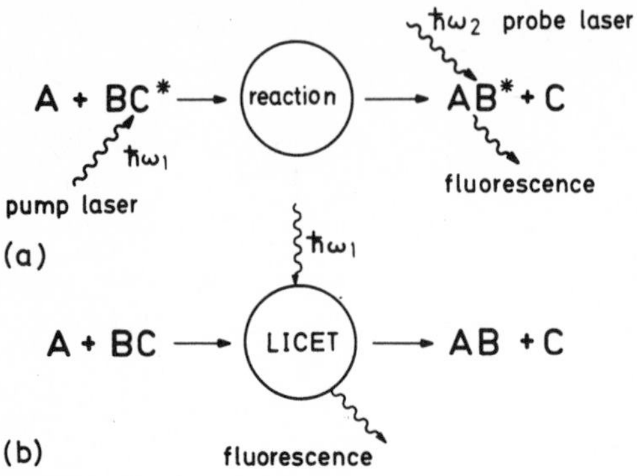

Fig. 14.1a,b. Possible applications of lasers to chemical reactions. (a) Enhancement of reaction rate by laser excitation of a reactant before the collision, and subsequent monitoring of internal state distribution of the reaction product by LIF. (b) Laser-induced collisional energy transfer (LICET)

reaction is supplied by absorption of one or of several laser photons. For these processes the time elapsed between absorption of laser photons and completion of the desired reaction is of crucial importance. The energy pumped into a selected level of a reactant molecule can be dissipated in several ways before the desired reaction takes place. It can be lost by spontaneous emission, it may be redistributed among the numerous degrees of freedom of a large excited molecule, or it may be thermalized and shared with other molecules through collisional deactivation. The time scales of these processes depend on the kind of molecules, the amount of excitation energy, and the pressure in the sample cell. We may distinguish between three different time regimes [14.5].

1) Excitation and reaction takes place in a time regime shorter than all other energy dissipation mechanisms. In the case of highly excited vibrational levels of large molecules, the nonlinear coupling between the different vibrational modes rapidly distributes the energy of the selectively excited level among many other levels on a time scale in the picosecond range and the energy fluctuates statistically between these modes. In order to enhance selectively the desired reaction, the excitation reaction time should be shorter than about 100 ps.

2) On a medium time scale (typically in the ns to μs range, depending on the pressure in the sample) the reaction takes place before the excitation energy of a molecule has been transferred by collisions to other molecules. The excited molecules may react with a larger probability than the ground state molecules.

3) On a still larger time scale (μs to cw operation) the excitation energy is uniformly distributed among all molecules of the sample, resulting in an in-

crease of the temperature. In this case the effect of laser irradiation on the reaction rates is comparable to that of thermally heating the sample.

For the first two time regions mode-locked or pulsed lasers must be used. In most experiments performed so far, pulsed CO_2 lasers or chemical lasers were utilized for vibrational excitation of reactant molecules. As a specific example of a laser-enhanced bimolecular reaction we consider the reaction

$$HCl(v''=1,2) + O(^3p) \rightarrow OH + Cl \quad , \tag{14.1}$$

where the HCl molecule is pumped by a HCl laser into the vibrational levels v"=1 or v"=2 [14.7]. The OH radicals are monitored by laser-induced fluorescence excited by a frequency-doubled dye laser. Tuning the laser to the P(1) line of the (0-0) band in the $^2\Pi - ^2\Sigma$ system of OH at λ = 308.2 nm allows one to probe the concentration of OH radicals in the vibrational ground state v"=0 while tuning to the (1-1) band at 318.2 nm probes the radicals in v"=1. The concentration of OH(v"=0,1) is measured with and without laser excitation of HCl. This yields the influence of vibrational excitation of HCl on the reaction rate.

Another example is the synthesis of SF_5NF_2 by CO_2 laser pulses of 10^{-7} s duration [14.8]. While the standard synthesis requires relatively high pressures of the commencing product S_2F_{10} and heating to 425 K in a sealed reactor for 10-20 hours, a single CO_2 laser pulse already produces substantial amounts of SF_5NF_2, following the reaction scheme

$$\begin{aligned} S_2F_{10} &+ nh\nu \rightarrow 2SF_5 \\ N_2F_4 &+ nh\nu \rightarrow 2NF_2 \\ SF_5 &+ NF_2 \rightarrow SF_5NF_2 \quad . \end{aligned} \tag{14.2}$$

The synthesis can be achieved by multiphoton absorption of 10 μm CO_2 laser photons or by single-photon absorption at λ = 193 nm with the ArF laser.

Of particular importance in many chemical reactions is the catalytic effect of solid surfaces. The possibility that catalytic reactions may be enhanced by laser irradiation of surfaces has stimulated many investigations in this field [14.9]. The laser can excite either molecules adsorbed at a surface or molecules in the gas phase close above the surface. In both cases the desorption-adsorption process may be selectively affected, because the interaction of excited molecules with the surface is different from that of ground state molecules.

With infrared lasers vibrational motions of adsorbed molecules or atom groups in the adsorption potential at the surface can be excited. The surface mobility of these excited vibrational modes is much higher than the mobility of molecules in the ground state. If the vibrational energy reaches the adsorption potential barrier, these atom groups can move nearly freely across the surface. They may collide with other molecules or they may be desorbed. This may allow a selective control of surface chemical reactions by selective excitation of surface potential vibrations. This has been demonstrated by DJIDJOEV et al. [14.10] who studied the stimulation of surface reactions between hydroxyl groups OH and amino groups NH_2 on a silica surface by irradiation with a CO_2 laser.

14.2 Laser Isotope Separation

The classical methods of isotope separation on a large, technical scale, such as thermal diffusion, or gas centrifuge techniques are expensive because they demand costly equipment or consume much energy [14.11]. New techniques based on a combination of laser spectroscopy with photochemistry may considerably reduce the costs. Up to now several methods have been proposed and some of them already proved their feasibility in laboratory experiments. The extension to an industrial scale, however, demands still more efforts and many improvements.

Most methods of laser isotope separation are based on selective excitation of the desired atomic or molecular isotope in the gas phase. Figure 14.2 depicts some possible ways of separating the excited species from the ground state isotopes.

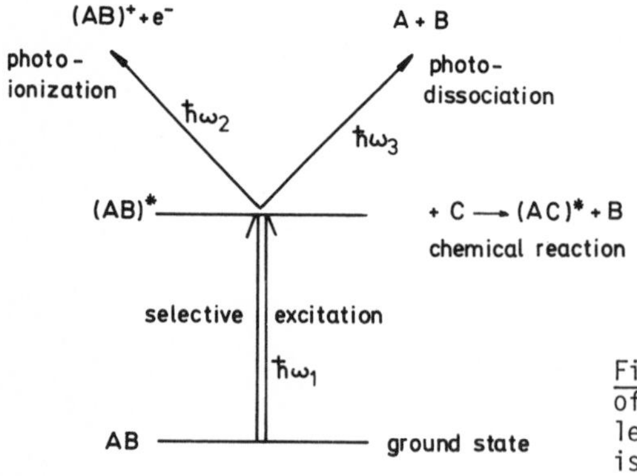

Fig. 14.2. Different possible ways of isotope separation following selective excitation of the wanted isotope

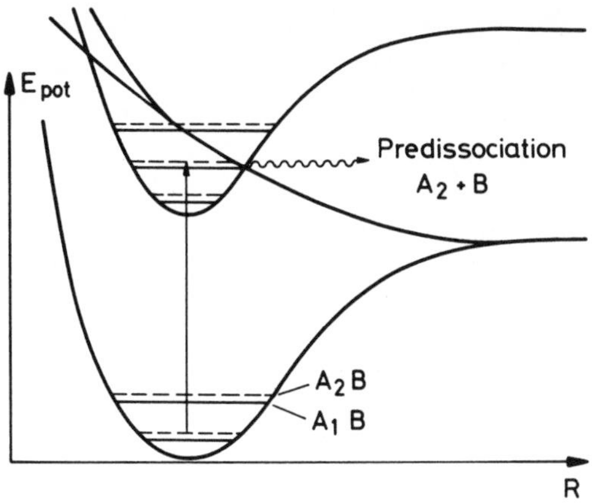

Fig. 14.3. Isotope separation by predissociation of molecules following isotopic selective excitation (after [14.12])

If a selectively excited isotope is irradiated by a second photon during the lifetime of the excited state, photoionization or photodissociation may take place. The ions can be collected and separated from the neutral species by electric fields. The photodissociation fragments may be separated by adding scavenger reactants S which preferably react with the fragments A or B but not with the parent molecules AB.

Another approach to laser isotope separation is offered by predissociation of laser-excited molecular isotopes into stable fragments. If the potential curve of the excited state of AB is intersected by a repulsive potential (Fig.14.3), the molecule may dissociate without absorbing a second photon.

The probably most promising way of laser isotope separation on a large scale is based on the rule that in general excited atoms or molecules have a larger probability of reacting with added reactants than have ground state species of the same kind. If "scavenger" molecules S which are added to the isotopic mixture do not react with ground state isotopes M but do combine irreversibly with *excited* isotopes M^*, these stable compounds SM^* can be separated by chemical means. An example of this chemical scavenging of laser-excited molecules is the separation of chlorine isotopes ^{37}Cl [14.12] following the reaction

$$\begin{aligned} &I^{37}Cl + h\nu \rightarrow I^{37}Cl^* \\ &I^{37}Cl^* + C_6H_5Br \rightarrow {}^{37}ClC_6H_5Br \rightarrow C_6H_5{}^{37}Cl \quad . \end{aligned} \tag{14.3}$$

The iodine monochloride isotope $I^{37}Cl$ was selectively excited by a cw dye laser at λ = 605 nm. The excited $(I^{37}Cl)^*$ molecules reacted in collisions with bromobenzene to form the unstable radical $^{37}ClC_6H_5Br$ which rapidly dissociates into $C_6H_5{}^{37}Cl + Br$. After two hours of irradiation several milligrams of C_6H_5Cl was produced which was enriched sixfold in chlorine 37.

The recent discovery of "multiphoton dissociation" of polyatomic molecules, where molecules, such as SF_6, can be dissociated by multiple absorption of infrared laser photons, has stimulated many theoretical [14.13] and experimental [14.14] investigations about the mechanism of this process. Since the first steps, namely the excitation of lower vibrational levels with moderate level density may be isotope selective, the multiphoton dissociation may turn out to become a cheap and efficient way of laser isotope separation. Infrared lasers, such as the CO_2 laser, have a high conversion efficiency which makes CO_2 laser photons inexpensive. For more detailed discussions of the various aspects of laser isotope separation see [14.15-17].

14.3 Laser Monitoring of the Atmosphere

A detailed understanding of our atmosphere, and of the various photochemical or collisional processes which determine the atmospheric composition, is of fundamental importance for mankind. Since in densely populated industrial areas air pollution has become a serious problem, the study of pollutants and their reactions with natural components in the atmosphere has become an urgent demand. Various techniques of laser spectroscopy can be successfully used in atmospheric and environmental research: direct absorption measurements, laser-induced fluorescence, spontaneous Raman scattering, or coherent anti-Stokes Raman spectroscopy (CARS; see Chap.9) can be utilized either for in situ measurements or for remote sensing.

An obvious way to determine the concentrations n_i of atomic or molecular constituents is the measurement of the total attenuation which a laser beam experiences along a known distance L through the atmosphere. Species with density n_i and absorption cross section $\sigma_i(\omega)$ result in an absorption coefficient

$$\alpha_i(\omega) = n_i(x)\sigma_i(\omega,p,T) \quad , \tag{14.4}$$

where the frequency-dependent absorption cross section $\sigma(\omega,p,T)$ depends on the temperature T (because the Doppler broadening depends on T) and on pressure p (because of pressure broadening). The density $n_i(x)$ may vary along the absorption path length $x = 0$ to $x = L$.

Besides absorption losses the laser light suffers Mie scattering by aerosols and Rayleigh scattering by molecules in the atmosphere. The total attenuation dI of a monochromatic laser beam at frequency ω with intensity $I(0) = I_0$ is then $dI = I_0 - I(L)$ where

$$I = I_0 / \exp\left\{\int_0^L [\alpha^{Mie}(x) + \alpha^{Rayleigh}(x) + \alpha^{abs}(\omega,x)]dx\right\} \quad . \tag{14.5}$$

The total absorption coefficient is

$$\alpha^{abs} = \sum_i n_i \sigma_i(\omega,p,T) \quad ,$$

where the sum extends over all molecular components in the atmosphere which contribute to the absorption at frequency ω.

In order to separate the contribution of a selected molecular species, the laser frequency ω has to be properly chosen to coincide with an absorption line of only one molecular species. If this is not possible, because there may be always overlapping bands from other molecules, the measurement has to be performed at several frequencies ω_{ki}. Since the Rayleigh and Mie scattering cross sections do not differ appreciably in a frequency interval corresponding to the linewidth $\Delta\omega$ of a molecular absorption line, the absorption coefficient $\alpha_i(\omega)$ and with it the density n_i of a specified absorbing molecular component can be determined by measuring the difference in attenuation $dI(\omega) - dI(\omega + \Delta\omega)$ when the laser is alternatively shifted into and out of resonance with the absorbing transition.

Because most polar molecules present in the atmosphere can be identified by their characteristic vibrational-rotational lines in the near infrared (called the "fingerprint region" of molecular spectroscopy) *infrared* lasers can be used. In particular multiline lasers, such as HF, DF, CO_2, or CO lasers with their numerous lines are well suited for simultaneous detection of several atmospheric constituents [14.18].

A definite absorption path length L can be realized with a retroreflector, placed at a distance L/2 from the source. Another possibility is to place the laser in an airplane (or satellite) and the receiver on the ground. Figure 14.4 illustrates a possible experimental setup. The laser output is expanded by a telescope in order to decrease the divergence of the beam. The retroreflector reflects the beam exactly back into a beam splitter and into a polychromator. With a diode array at the spectrometer output (optical multichannel analyzer; see Sect.4.5.9) the whole spectral range of the multiline laser can be recorded simultaneously. This diode array also allows measure-

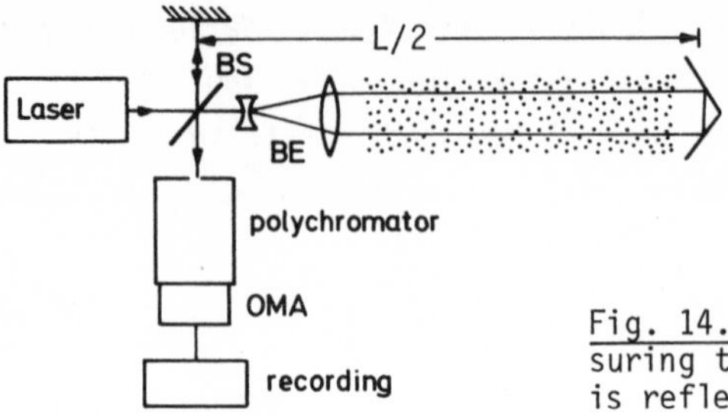

Fig. 14.4. Monitoring air pollution by measuring the attenuation of a laser beam which is reflected by a retroreflector

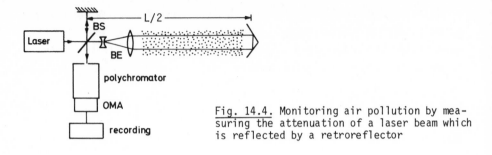

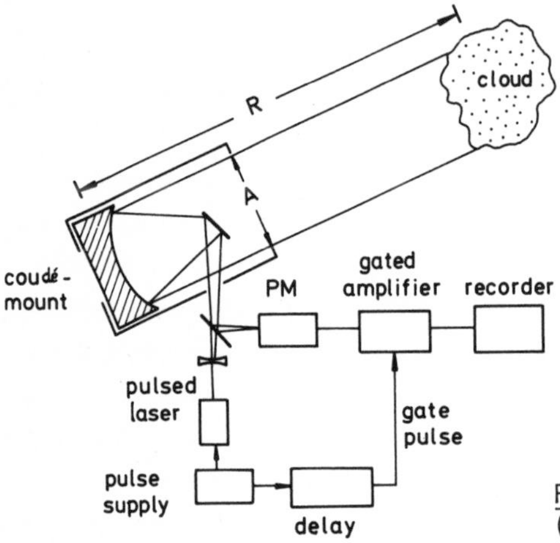

Fig.14.5. Schematic diagram of LIDAR (ligth detection and ranging)

ment of the dispersed laser output on one half of the array while the reflected beam is dispersed on the other half. The difference

$$dI = I_0 - I_R = I_0(1 - e^{-\alpha(\omega)L}) \approx I_0\alpha(\omega)L \quad \text{for} \quad \alpha L \ll 1 \qquad (14.6)$$

between both readings gives directly the absorption coefficient for all recorded lines [14.19].

For remote sensing of the atmosphere often a retroreflector cannot be used. In such cases the Mie scattering of the laser light by aerosols can be utilized to receive a signal. Figure 14.5 illustrates the basic principle of the LIDAR method. The output from a pulsed laser is expanded by a telescope. The light, scattered by particles at distance R from the source, is collected by the telescope and reaches the receiver with a time delay $t = 2R/c$ against the emission of the laser pulse at $t = 0$. Gating of the detector with a time delay $2R/c$ during a time interval Δt allows selection of scattered light from a spatial interval R to $R+c\Delta t$ along the laser path. With laser pulse widths of 10^{-8} s and a gate width of 2×10^{-8} s a spatial interval of $\Delta R = 3$ m can be resolved.

The amplitude S of the signal received by the detector is given by

$$S = aI_0\, e^{[-\alpha(\omega)2R]}\, \sigma^{\mathrm{Mie}} nA/R^2 \quad , \tag{14.7}$$

where $\sigma^{\mathrm{Mie}}n$ is the fraction of the laser intensity $I_0 \exp[-\alpha(\omega)R]$ which is scattered by n aerosols per unit volume element into all directions. Although the angular distribution of the scattered light is not isotropic, the receiver collects a fraction which is proportional to the solid angle accepted by the telescope with aperture A.

The crucial point is that the signal is proportional to $\exp[-\alpha(\omega)2R]$, which means that it measures the attenuation of the laser beam from 0 to R. Gating the receiver alternatively at delay times $t = 2R/c$ and $t+\Delta t = 2(R+\Delta R)/c$ allows measurement of the difference $\Delta S = S(t+\Delta t) - S(t)$ which yields the attenuation between R and $R+\Delta R$.

Tuning the laser frequency ω alternatively into resonance ω_0 with a molecular absorption line and out of resonance $(\omega_0+\Delta\omega)$ allows determination of the absorption $\alpha_i(\omega,R) = n_i(R)\sigma_i(\omega)$ and therefore the density n_i of the absorbing molecules at a definite location. In this way a complete "air pollution map" of industrial areas can be recorded and pollution sources can be localized. With pulsed dye lasers NO_2 concentrations in the ppm range (10^{-6}) at distances up to 5 km could be monitored [14.20].

More details of this LIDAR method (LIDAR is the acronym for *li*ght *d*etection *a*nd *r*anging) can be found in the review of COLLIS and RUSSEL [14.21].

In the higher atmosphere the aerosol density decreases rapidly with altitude and other detection schemes may become more advantageous. Raman spectroscopy or detection of laser-induced fluorescence excited by frequency-doubled pulsed lasers has been utilized [14.22]. Both Raman and fluorescence intensities excited by the laser at a location x are proportional to the density $n_i(x)$ of scattering particles. However, because of the high pressure $(p \leqq 1\,\text{atm})$ the fluorescence is quenched if the collisional deactivation $n\sigma_i^{\text{coll}}\bar{v}$ becomes faster than the spontaneous decay $A_i = 1/\tau_i$ (see Sect.12.2). Transition probabilities and quenching cross sections must therefore be known if quantitative results are to be obtained from measurements of the fluorescence intensity.

The application of these methods to the study of the upper atmosphere has already brought many interesting results on the atomic and molecular composition as a function of altitude and its daily variations. Measurements of air pollution by molecular pollutants such as SO_2, NO_2, NO, etc., are possible up to altitudes of several km. SO_2 concentrations of 100 ppm, for example, can be monitored up to distances of 1 km at night. Because of the daytime background the sensitivity is lower during the day than at night.

A comprehensive presentation of different techniques of laser monitoring the atmosphere and their advantages and limitations can be found in [14.23]; experimental examples are also given in [14.24]. Fundamentals of propagation, scattering and absorption of light in the atmosphere can be found in [14.25].

14.4 Laser Spectroscopy in Biology

The advantages of laser spectroscopy regarding spatial, spectral, and time resolution can be utilized for the spectroscopy of biomolecules and cells. Measurements of fluorescence or excitation spectra, investigations of quantum yields, decay times of excited states, or wavelength shifts of molecular absorption bands in solutions allow one to gain valuable information on structure, kinetics, and interactions of biological molecules. With newly developed techniques, such as fluorescence microscopy, biophysical and biochemical processes in *single* cells can be studied. Resonance Raman spectroscopy of biological systems allows examination of the vibrational and electronic structure of molecules using sample concentrations as low as 10^{-6} M and sample volumes of one mm^3. In this section we give a few examples of laser applications in the spectroscopy of biological systems. For further information see [14.26-28].

14.4.1 Laser Microscope

The output beam of a laser oscillating in the fundamental TEM_{00}-mode shows a Gaussian intensity profile (see Sect.5.11). If the Gaussian beam is properly matched to a focusing system with aperture d and focal length f, a diffraction-limited focal spot of diameter $d_0 \approx 2(f/d)\lambda$ can be reached. With a microscope objective for example, with f/d = 1, focal spot sizes of $w_0 = d_0/2 = 0.5$ μm can be obtained at $\lambda = 500$ nm. This means that the spatial resolution in the focal plane of the microscope allows selective excitation of single biological cells.

The fluorescence emitted from the excited molecules in the cell can be collected by the same microscope and is either imaged onto a TV camera for direct visual observation or is collected onto a photomultiplier cathode for

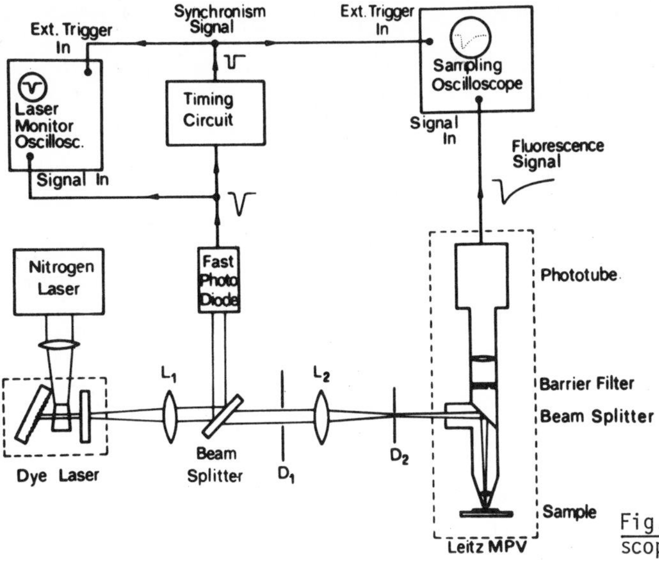

Fig. 14.6. Laser microscope [14.31]

time-resolved detection. Figure 14.6 illustrates a typical experimental design [14.31]. A nitrogen laser pumped dye laser delivers pulses of 0.5 ns duration at a wavelength λ, which can be tuned to the peak of the absorption band of the biological molecules under study. The time-resolved fluorescence is monitored by a fast photomultiplier and signal averaging allows a good signal-to-noise ratio to be obtained even if only a few fluorescence photons per laser pulse are detected.

Several biological molecules have been studied with this design. One example is the investigation of acridine-DNA complexes, where quantum yields, excitation, and fluorescence spectra have been measured [14.31]. The interesting question, at which part of a chromosome the fluorescent dyes are located and how the quantum yield depends on the two base pairs in which the dye is intercalated, can be attacked with the technique of laser fluorimetry.

Pulsed fluorescence spectroscopy can be applied to investigations of dynamics and structure of biopolymers [14.28]. If the fluorescence emission, excited by a linearly polarized laser pulse, is probed with respect to its polarization, the anisotropy $(I_{\parallel}-I_{\perp})/(I_{\parallel}+I_{\perp})$ is related to the orientation of absorbing and emitting molecules. Molecular rotations and orientation changing collisions within the time interval between excitation and fluorescence alter the polarization characteristics of the fluorescence.

14.4.2 Application of Laser Raman Spectroscopy to Biological Problems

Raman spectroscopy is a powerful tool for studying molecular vibrations (see Chap.9). Because vibrational frequencies are sensitive to geometric structure and bonding arrangements of localized groups of atoms in larger molecules, both the frequencies and intensities in a Raman spectrum are affected by structural changes and intermolecular interactions. This means that Raman spectra of biological molecules contain much information about structure and kinetics of these molecules. Conformational changes of chromophoric groups attached to biomolecules, for instance, can be monitored by changes in the Raman spectrum. The other technique of investigating vibrational structure, namely infrared spectroscopy, has the serious drawback that most of the vibrational frequencies of interest fall in a spectral region where water, which is always present in biological cells, has strong absorption bands. This often prevents detailed infrared studies of individual molecular vibrational structure by ir spectroscopy. With excitation wavelengths in the visible or ultraviolet region Raman spectroscopy does not share this drawback because the Raman scattering cross section of water is small. The small sensitivity of ordinary Raman spectroscopy, however, represents a severe experimental disadvantage.

Fortunately several recently developed experimental techniques have greatly enhanced the sensitivity (see Chap.9) and have therefore enlarged the range of possible applications in biology. One example is the *resonance* Raman effect which takes advantage of the enhanced Raman scattering from molecular vibrations coupled to an electronic transition, when the excitation wavelength coincides with an electronic transition of the molecule [14.27].

The obvious way to increase the intensity of the Raman scattered light by increasing the power of the laser used for excitation cannot be used for many biological molecules because they may undergo rapid photochemical changes, such as photolysis, and the sample under examination may soon consist of a mixture of all kinds of molecular conformations. The molecules should therefore stay within the interaction region for only a short time.

Using a rapid flow technique, where a jet of molecules dissolved in a liquid flows through a focused laser beam, it is possible to study also photolabile molecules, such as rhodopsin, by resonance Raman spectroscopy [14.32]. This allows investigation of the interesting problem about the molecular mechanisms and dynamics of visual excitation. Rhodopsin molecules act as photoreceptors in the retina of vertebrates. If the rhodopsin sample is rapidly flowed through the focused laser beam, the fraction of isomerized molecules within the illuminated region stays small. This allows measurement of the resonance Raman spectra of unphotolysed rhodopsin. This technique has been used to study various visual pigments and several of their photolytic intermediates.

The spectra of these intermediates can be obtained by time-resolved resonance Raman spectroscopy. An example is lumirhodopsin spectroscopy. First a high-intensity pump pulse is used to initiate the bleaching of rhodopsin and, after a few ns, a second, low intensity probe pulse generates the Raman scattering from lumirhodopsin [14.32]. Many more examples of resonance Raman spectroscopy of biological molecules can be found in [14.33].

14.5 Medical Applications of Laser Spectroscopy

Numerous books have been meanwhile published on laser applications in medical research and in hospital practice [14.34]. Most of these applications rely on the high laser output power, which can be focused into a small volume. The strong dependence of the absorption coefficient of living tissue on the wavelength allows selection of the penetration depth of the laser beam by choosing the proper laser wavelength. The most spectacular outcomes of laser applications in medicine have been achieved in laser surgery, dermatology, and ophthalmology.

There are, however, also very promising direct applications of laser *spectroscopic* techniques in medicine which can be illustrated by a few examples.

The first example demonstrates the application of laser Raman spectroscopy to in vivo monitoring of respiratory gases [14.35]. The detection sensitivity for Raman light scattered by molecular gases can be greatly enhanced by using a multipass cell (Fig.14.7). The unpolarized laser beam enters the multipass cell through a small hole in one of the resonator mirrors and is reflected back and forth up to 70 times. The power density inside the cell is about the same as inside the laser resonator, but the stability of the whole arrangement can be made much better than for intracavity experiments.

Six detectors are arranged around the cell in a plane perpendicular to the cell axis. The Raman light scattered by the molecules in the cell can be collected within a large acceptance angle. Through interference filters in front of each detector specific Raman lines can be selected. This allows the simultaneous detection of six different lines and therefore of up to six different molecular components in the sample.

The sensitivity of this arrangement is illustrated by Fig.14.8 which shows the time variation of CO_2, N_2, and O_2 concentrations in the exhaled air of a human patient. Note the variation of CO_2 concentrations with changing breath-

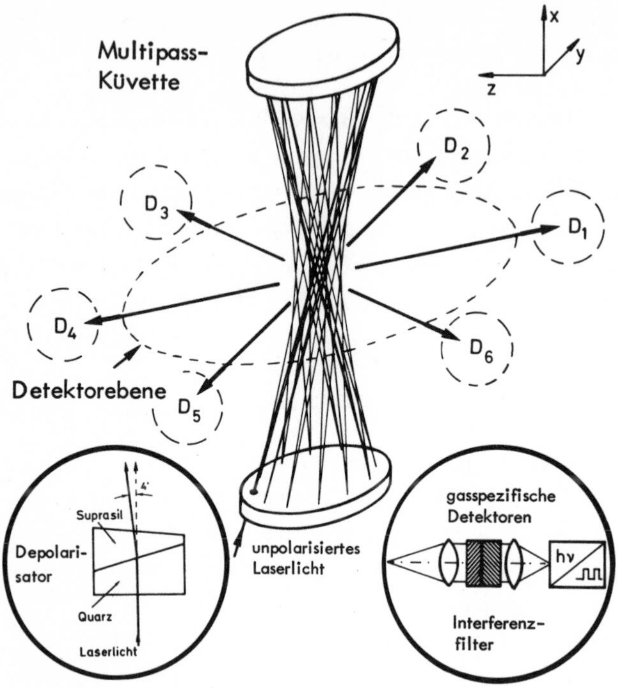

Fig. 14.7. Multipass cell for sensitive Raman diagnostics of molecular gases [14.35]

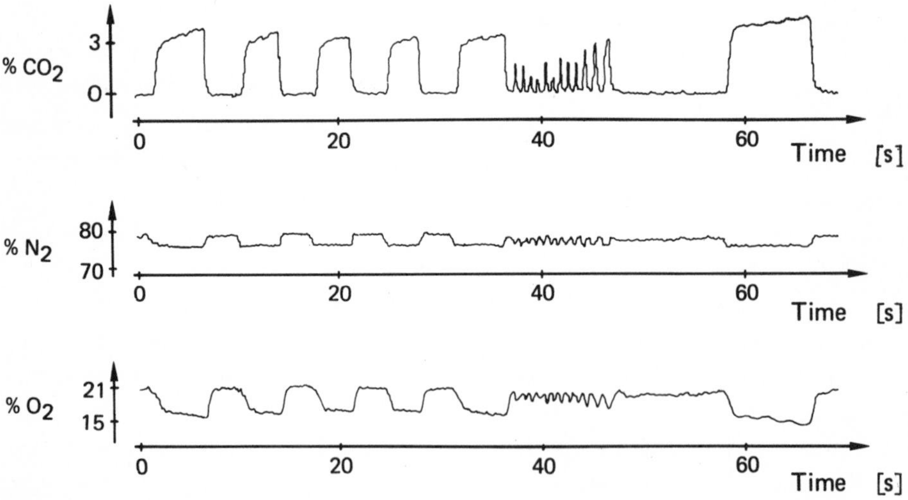

Fig. 14.8. CO_2, N_2, and O_2 concentrations of respiratory gases for varying breath periods, measured in vivo with the multipass cell arrangement of Fig.14.7 [14.35]

ing period. The technique can be routinely used in the clinical practice for anesthetic control during operations and obviously also for alcohol tests of car drivers.

Another example is the application [14.36] of tunable dye lasers in photodermatology. The determination of the spectral erythema effectiveness curve, for instance, gives information on the action spectrum of sunburn. The investigation of light-induced defects in DNA and its spectral dependence is of great importance for photo- and photochemotherapy. Photochemotherapy is a new field in medicine, where the combined action of light and a photosensitizer on living tissue is studied, with regard to applications in dermatology for the treatment of acne, vitiligo, and various eczemas. A relatively narrow wavelength range of optimum healing of psoriatic lesions was found, suggesting a narrow band irradiation of the order of a few nm in the phototherapy of psoriasis.

A wide, but still not very much studied, field is the photodestruction of tumors and its spectral dependence. If the tumors are locally stained with a photosensitizing dye, the interaction between the absorbing dye molecules and the nucleic acids may influence the tumor destruction. Results, obtained from photobiology, can help to clarify the situation.

A very promising technique, useful in medical laboratory practice, is the laser nephelometry [14.37]. This method allows determination of concentrations of different proteins in the blood serum by measuring the Mie scattering of laser light. The output beam of a He-Ne laser passes through the sample cell which contains the serum diluted in a NaCl solution. The scattered light intensity is measured as a function of dilution and as a function of time to control the antigen reactions. Pathological deviation from normal protein concentrations can be safely detected.

The application of laser spectroscopic methods to medicine and biology is only at the beginning of a rapid growth. The development of commercial tunable dye lasers in the ultraviolet region will certainly greatly enhance these applications [14.38,39].

References

Chapter 1

1.1 A.E. Siegman: *An Introduction to Lasers and Masers* (McGraw-Hill, New York 1971)

1.2 O. Svelto: *Principles of Lasers* (Heyden, London, New York 1976)

1.3 A. Levine (ed.): *Lasers, A Series of Advances*, Vol. 1-4 (Marcel Dekker, New York 1966-1971)

1.4 A. Yariv: *Quantum Electronics* (Wiley, New York 1975)

1.5 M. Sargent, III, M.O. Scully, W.E. Lamb: *Laser Physics* (Addison-Wesley, London 1974)

1.6 M.S. Feld, A. Javan, N. Kurnit (eds.): *Fundamental and Applied Laser Physics*: Proceedings of the Esfahan Symposium 1971 (Wiley, London 1973)

1.7 J.C. Lehmann, J.C. Pebay-Peyroula (eds.): *Internat. Colloquium on Doppler-Free Spectroscopic Methods for Simple Molecular Systems*, Aussois, May 1973 (Editions Du Centre National de la Recherche Scientifique 15, Quai Anatole France, Paris, No. 217, 1974)

1.8 R.G. Brewer, A. Mooradian (eds.): *Laser Spectroscopy*, Proceedings of the 1st Intern. Conf. Vale 1973 (Academic Press, New York 1974)

1.9 S. Haroche, J.C. Pebay-Peyroula, T.W. Hänsch, S.E. Harris (eds.): *Laser Spectroscopy*, Proceedings of the 2nd Intern. Conf. Megève 1975, Lecture Notes in Physics, Vol. 43 (Springer, Berlin, Heidelberg, New York 1975)

1.10 A. Mooradian, T. Jaeger, P. Stokseth (eds.): *Tunable Lasers and Applications*, Springer Ser. Opt. Sci., Vol.3 (Springer, Berlin, Heidelberg, New York 1976)

1.11 J.L. Hall, J.L. Carlsten (eds.): *Laser Spectroscopy III*, Proceedings of the 3rd Intern. Conf., Jackson Lake 1977, Springer Series in Optical Sciences, Vol. 7 (Springer, Berlin, Heidelberg, New York 1977)
H. Walther, K.W. Rothe (eds.): *Laser Spectroscopy IV*, Proceedings of the 4th Intern. Conf., Tegernsee 1979, Springer Series in Optical Sciences, Vol. 21 (Springer, Berlin, Heidelberg, New York 1979)
A.R.W. McKellar, T. Oka, B.P. Stoicheff (eds.): *Laser Spectroscopy V*, Proc. 5th. Intern. Conf. Jasper Park Lodge, Alberta, Canada 1981, Springer Ser. Opt. Sci., Vol.30 (Springer, Berlin, Heidelberg, New York 1981)

1.12 H. Walther (ed.): *Laser Spectroscopy of Atoms and Molecules*, Topics in Applied Physics, Vol. 2 (Springer, Berlin, Heidelberg, New York 1976)

1.13 K. Shimoda (ed.): *High-Resolution Laser Spectroscopy*, Topics in Applied Physics, Vol. 13 (Springer, Berlin, Heidelberg, New York 1976)

1.14 R.A. Smith (ed.): *Very High Resolution Spectroscopy* (Academic Press, London, New York 1976)

1.15 St.K. Freeman: *Applications of Laser Raman Spectroscopy* (Wiley, New York 1974)
1.16 *Impact of Lasers in Spectroscopy*, Proc. Soc. Photo-Opt. Instr. Eng., San Diego, CA 1974, Vol. 49 (Society of Photo-Opt. Instrumentation, Palos Verdes Estates, Calif. 1975)
W.O.N. Guimaraes, C.-T. Lin, A. Mooradian (eds.): *Lasers and Applications*, Springer Ser. Opt. Sci., Vol.26 (Springer, Berlin, Heidelberg, New York 1981)
1.17 St.F. Jacobs, M. Sargent, III, J.F. Scott, M.O. Scully (eds.): *Laser Applications to Optics and Spectroscopy*, Lectures of the 1973 Summer School, Crystal Mountain, Washington, Phys. of Quant. Electronics, Vol. 2 (Addison-Wesley, London 1975)
1.18 A. Corney: *Atomic and Laser Spectroscopy* (Clarendon Press, Oxford 1977)
1.19 M.A. West (ed.): *Lasers in Chemistry*, Proceedings of the Conference held at the Royal Inst. London 1977 (Elsevier, Amsterdam 1977)

Chapter 2

2.1 A. Corney: *Atomic and Laser Spectroscopy* (Clarendon Press, Oxford 1977)
2.2 A.P. Thorne: *Spectrophysics* (Chapman & Hall, London 1974)
2.3 E. Hecht, A. Zajac: *Optics* (Addison-Wesley, London 1974)
2.4 H.G. Kuhn: *Atomic Spectra* (Longmans, London 1969)
2.5 M. Born, E. Wolf: *Principles of Optics* (Pergamon Press, Oxford 1970)
2.6 R. Loudon: *The Quantum Theory of Light* (Clarendon Press, Oxford 1973)
2.7 E.S. Steeb, W.E. Forsythe: "Photometry and Illumination", in *Handbook of Physics*, ed. by E.V. Condon, Part 6, Chapter 3 (McGraw-Hill, New York 1958)
2.8 W.J. Smith: *Modern Optical Engineering* (McGraw-Hill, New York 1966)
2.9 A. Stimson: *Photometry and Radiometry for Engineers* (Wiley-Interscience, New York 1974)
2.10 D. Eisel, D. Zevgolis, W. Demtröder: Sub-Doppler Laser Spectroscopy of the NaK-Molecule. J. Chem. Phys. *71*, 2005 (1979)
2.11 N.N. Bogoliubov, P.V. Shirkov: *Introduction to the Theory of Quantized Fields* (Academic Press, New York 1959) W. Heitler: *Quantum Theory of Radiation* (Oxford University Press, New York 1956)
2.12 W.L. Wiese: "Transition Probabilities", in *Methods of Experimental Physics*, Vol. 7A, ed. by B. Bederson, W.L. Fite (Academic Press, New York 1968) p. 117
W.L. Wiese, M.W. Smith, B.M. Glennon: "Atomic Transition Probabilities", National Standard Reference Data Series NBS 4 and NSRDS-NBS 22 (1966-1969)
2.13 R.E. Imhof, F.H. Read: Measurements of lifetimes of atoms, molecules and ions, Rep. Prog. Phys. *40*, 1 (1977)
2.14 A.N. Nesmeyanov: *Vapor Pressure of the Chemical Elements* (Elsevier, Amsterdam 1963)
2.15 G. Yale Eastman: The heat pipe. Sci. Am. *218*, 38 (1968); *Heat Pipe Technology* (Technical Application Center, The University of New Mexico, Albuquerque, N.M. 87131, 1972)
2.16a C.R. Vidal, J. Cooper: Heat pipe oven. A new well defined metal vapor device for spectroscopic measurements. J. Appl. Phys. *40*, 3370 (1969)
2.16b H. Kopfermann, G. Wessel: Die absoluten f-Werte der Fe-I-resonanzlinien. Z. Phys. *130*, 100 (1951)
2.16c I. Meroz (ed.): *Optical Transition Probabilities. A Representative Collection of Russian Articles* (Israel Program for Scientific Translation, Jerusalem 1962)
2.17 W. Marlow: Hakenmethode. Appl. Opt. *6*, 1715 (1967)

2.18a "Bibliography on Atomic Transition Probabilities 1914 Through October 1977"; NBS Publication 505 (US Department of Commerce, 1978)
2.18b E.W. Foster: The measurement of oscillator strength in atomic spectra. Rep. Prog. Phys. *27*, 469 (1964)
2.19 A. Maitland, M.H. Dunn: *Laser Physics* (North-Holland, Amsterdam, London 1969)
2.20 E.U. Condon, G.H. Shortley: *The Theory of Atomic Spectra* (Cambridge University Press, Cambridge 1964)
2.21a A.R. Edmonds: *Angular Momentum in Quantum Mechanics* (Princeton University Press, New York 1960)
2.21b M.Weissbluth: *Atoms and Molecules* (Academic Press, New York 1978)
2.22 G. Herzberg: *Molecular Spectra and Molecular Structure I* (Van Nostrand Reinhold Company, New York 1950)
2.23 C.J.H. Schutte: *The Wave Mechanics of Atoms, Molecules and Ions* (Edward Arnold, London 1968)
2.24 A.C. Hurley: *Introduction to the Electron Theory of Small Molecules* (Academic Press, London 1976)
2.25a G.W. King: *Spectroscopy and Molecular Structure* (Holt Rinehart and Winston, New York 1964)
2.25b J.I. Steinfeld: *Molecules and Radiation* (Harper and Row, New York 1974)
2.26 L. Mandel, E. Wolf: Rev. Mod. Phys. *37*, 231 (1965)
2.27 G.W. Stroke: *An Introduction to Coherent Optics and Holography* (Academic Press, New York 1969)
2.28 J.R. Klauder, E.C.G. Sudarshan: *Fundamentals of Quantum Optics* (Benjamin, New York 1968)

Chapter 3

3.1 I.I. Sobelman: *Atomic Spectra and Radiative Transitions*, Springer Series in Chemical Physics, Vol. 1 (Springer, Berlin, Heidelberg, New York 1979)
3.2 W.R. Hindmarsh, J.M. Farr: "Collision Broadening of Spectral Lines by Neutral Atoms", in *Progress in Quantum Electronics*, Vol. 2, Part 4, ed. by J.H. Sanders, S. Stenholm (Pergamon Press, Oxford 1973)
3.3 R.G. Breen: "Line Width", in *Handbuch der Physik*, Vol. 27, ed. by S. Flügge (Springer, Berlin 1964) p. 1
3.4 S.N. Dobryakov, Ya.S. Lebedev: Analysis of spectral lines whose profile is described by a composition of Gaussian and Lorentz profiles. Sov. Phys. Dokl. *13*, 9 (1969)
3.5 A. Unsöld: *Physik der Sternatmosphären* (Springer, Berlin, Heidelberg, New York 1955)
3.6 E. Lindholm: "Pressure Broadening of Spectral Lines"; Ark. Mat. Astron. Fys. *32A*, Nr. 17 (1945)
3.7 G. Traving: *Über die Theorie der Druckverbreiterung von Spektrallinien* (Verlag Braun, Karlsruhe 1960)
3.8 F. Schuler, W. Behmenburg: Perturbation of spectral lines by atomic interactions. Phys. Rep. *12C*, 274 (1974)
3.9 See, for instance, D. Ter Haar: *Elements of Statistical Mechanics* (Pergamon, New York 1954 and 1977)
3.10 A. Gallagher: "The Spectra of Colliding Atoms", in *Atomic Physics*, Vol. 4, ed. by G. zu Putlitz, E.W. Weber, A. Winnaker (Plenum, New York 1975)
3.11 K. Niemax, G. Pichler: Determination of van der Waals constants from the red wings of self-broadened Cs principal series lines. J. Phys. B, Atom. Mol. Phys. *8*, 2718 (1975)

3.12 R.J. Exton, W.L. Snow: Line shapes for satellites and inversion of the data to obtain interaction potentials. J. Quant. Spectrosc. Radiat. Transfer *20*, 1 (1978)

3.13 H. Griem: *Plasma Spectroscopy* (McGraw-Hill, New York 1964)

3.14 J. Ward, J. Cooper, E.W. Smith: Correlation effects in the theory of combined Doppler and pressure broadening. J. Quant. Spectrosc. Radiat. Transfer *14*, 555 (1974)

3.15 P.R. Berman, W.E. Lamb, Jr.: Influence of resonant and foreign gas collisions on line shapes. Phys. Rev. *187*, 221 (1969)

3.16 J. Hirschfelder, Ch.F. Curtiss, R.B. Bird: *Molecular Theory of Gases and Liquids* (Wiley, New York, 1954)

3.17 C.C. Davis, T.A. King: "Gaseous Ion Lasers", in *Advances in Quantum Electronics*, Vol. 3, ed. by D.W. Godwin (Academic Press, New York 1975)

3.18 R.S. Eng, A.R. Calawa, T.C. Harman, P.L. Kelley: Collisional narrowing of infrared water vapor transitions. Appl. Phys. Lett. *21*, 303 (1972)

3.19 J. Hall: "The Line Shape Problem in Laser Saturated Molecular Absorption", in *Lecture Notes in Theoretical Physics*, Vol. 12A, ed. by K. Mahanthappa, W. Brittin (Gordon and Brach, New York 1971)

3.20 D.S. McClure: "Electronic Spectra of Molecules and Ions in Crystals", in *Solid States Physics*, Vols. 8 and 9, ed. by F. Seitz, and D. Turnbull (Academic Press, New York 1959)

Chapter 4

4.1 K.I. Tarasov: *The Spectroscope* (Adam Hilger, London 1974)

4.2 A.P. Thorne: *Spectrophysics* (Chapman and Hall Science Paperbacks, London 1974)

4.3 P. Bousquet: *Spectroscopy and its Instrumentation* (Adam Hilger, London 1971)

4.4 G.L. Clark: *The Encyclopedia of Spectroscopy* (Reinhold, New York 1960)

4.5 See, for instance, Ref. 2.3 or
W.J. Smith: *Modern Optical Engineering* (McGraw-Hill, New York 1966)

4.5a See Ref. 2.5, p. 333 ff

4.5b D.E. Gray (ed.): *American Institute of Physics Handbook* (McGraw-Hill, New York 1972)

4.6 *Handbook of Diffraction Gratings, Ruled and Holographic* (Jobin Yvon Optical Systems, 20 Highland Ave., Metuchen, N.J. 1970)

4.7 *Bausch & Lomb Diffraction Grating Handbook* (Bausch & Lomb, Rochester, N.Y. 1970)

4.8 G. Schmahl, D. Rudolph: "Holographic Diffraction Gratings", in *Progress in Optics*, Vol. XIV, ed. by E. Wolf (North-Holland, Amsterdam 1977) p. 195 f

4.9 S.P. Davis: *Diffraction Grating Spectrographs* (Holt Rinehard, Winston, New York 1970)

4.10 G.W. Stroke: "Diffraction Gratings", in *Handbuch der Physik*, Vol. 29, ed. by S. Flügge (Springer, Berlin, Heidelberg, New York 1967)

4.11 F. Kneubühl: Diffraction grating spectroscopy. Appl Opt. *8*, 505 (1969)

4.12 Basic treatments of interferometers may be found in general textbooks on optics, more detailed discussions for instance in S. Tolansky: *An Introduction to Interferometry* (Longman, London 1973);
W.H. Steel: *Interferometry* (Cambridge University Press, Cambridge 1967);
J. Dyson: *Interferometry* (Machinery Publ., Brighton 1970);
M. Francon: *Optical Interferometry* (Academic Press, New York 1966).

A recommended review on "New Developments in Interferometry" has been written by H. Polster, J. Pastor, R.M. Scott, R. Crane, P.H. Langenbeck, R. Pilston, G. Steinberg: Appl. Opt. *8*, 521 (1969)
4.13 M. Francon, J. Mallick: *Polarisation Interferometers* (Wiley Interscience, London 1971)
4.14 H. Welling, B. Wellingehausen: High resolution Michelson interferometer for spectral investigations of lasers. Appl. Opt. *11*, 1986 (1972)
4.15 W.W. Rigrod, A.M. Johnson: Resonant prism mode selector for gas lasers. IEEE J. QE-*3*, 644 (1967)
4.16 R.J. Bell: *Introductory Fourier Transform Spectroscopy* (Academic Press, New York 1972)
4.17 M.J.D. Low: Fourier-transform-spectroscopy. Naturwissenschaften *57*, 280 (1970)
4.18 H.A. Gebbie: Fourier transform versus grating spectroscopy. Appl. Opt. *8*, 501 (1969)
4.19 J.B. Bates: Fourier transform infrared spectroscopy. Science *191*, 31 (1976)
4.20 D.C. Champeney: *Fourier Transforms and their Physical Applications* (Academic Press, New York 1973)
4.21 H.R. Chandrasekhar, L. Genzel, J. Kühl: Double beam Fourier spectroscopy with interferometric background compensation. Opt. Commun. *17*, 106 (1976)
4.22 V. Grigull, H. Rottenkolber: Two beam interferometer using a laser. J. Opt. Soc. Am. *57*, 149 (1967)
4.23 W.C. Marlow: Haken-methode. Appl. Opt. *6*, 1715 (1967)
4.24a J.P. Marioge, B. Bonino: Fabry-Perot interferometer surfacing. Opt. Laser Technol. *4*, 228 (1972)
4.24b M. Hercher: Tilted etalons in laser resonators. Appl. Opt. *8*, 1103 (1969)
4.25 V.R. Costich: "Multilayer Dielectric Coatings", in *Handbook of Lasers*, ed. by R.J. Pressley (Chemical Rubber Company, Cleveland, Ohio 1972)
4.26 J. McDonald: *Metal-Dielectric Multilayers* (Adam Hilger, London 1971)
4.27 H. Anders: *Dünne Schichten für die Optik* (Wissenschaft. Verlagsgesellschaft, Stuttgart 1965)
4.28 H.A. Macleod: *Thin Film Optical Filters* (Adam Hilger, London 1969)
4.29 A. Musset, A. Thelen: "Multilayer Antireflection Coatings", in *Progress in Optics*, Vol. III, ed. by E. Wolf (North-Holland, Amsterdam 1970) p. 203 ff
4.30 J.T. Cox, G. Hass: In *Physics of Thin Films*, Vol. 2 (Academic Press, New York 1964)
4.31 E. Delano, R.J. Pegis: "Methods of Synthesis for Dielectric Multilayer Filters", in *Progress in Optics*, Vol. VII, ed. by E. Wolf (North-Holland, Amsterdam 1969) p. 69 ff
4.32 W. Demtröder, M. Stock: Molecular constants and potential curves of Na_2 from laser-induced fluorescence. J. Mol. Spectrosc. *55*, 476 (1975)
4.33 P. Connes: L'etalon de Fabry-Perot spherique. Phys. Radium *19*, 262 (1958), and in *Quantum Electronics and Coherent Light* , ed. by P.H. Miles (Academic Press, New York 1964) p. 198 ff
4.33a J.R. Johnson: A high resolution scanning confocal interferometer. Appl. Opt. *7*, 1061 (1968)
4.34 M. Hercher: The spherical mirror Fabry-Perot interferometer. Appl. Opt. *7*, 951 (1968)
4.35 J. Evans: The birefringent filter. J. Opt. Soc. Am. *39*, 229 (1949)
4.36 H. Walther, J.L. Hall: Tunable dye laser with narrow spectral output. Appl. Phys. Lett. *17*, 239 (1970)
4.36a M. Okada, S. Iliri: Electronic tuning of dye lasers by an electrooptic birefringent Fabry-Perot etalon. Opt. Commun. *14*, 4 (1975)

4.37 B.H. Billings: The electro-optic effect in uniaxial crystals of the type XH_2PO_4. J. Opt. Soc. Am. *39*, 797 (1949)

4.38 B. Zwicker, P. Scherrer: Elektrooptische Eigenschaften der Seignette-elektrischen Kristalle KH_2PO_4 und KD_2PO_4. Helv. Phys. Acta *17*, 346 (1944)

4.38a F. Zernike, J.E. Midwinter: *Applied Nonlinear Optics* (Academic Press, New York 1973)

4.39 A.L. Bloom: Modes of a laser resonator containing tilted birefringent plates. J. Opt. Soc. Am. *64*, 447 (1974)

4.40 J.R. Johnson: A high resolution scanning confocal interferometer. Appl. Opt. *7*, 1061 (1968)

4.41 R.L. Fork, D.R. Herriot, H. Kogelnik: A scanning spherical mirror interferometer for spectral analysis of laser radiation. Appl. Opt. *3*, 1471 (1964)

4.42 V.G. Cooper, B.K. Gupta, A.D. May: Digitally pressure scanned Fabry-Perot interferometer for studying weak spectral lines. Appl. Opt. *11*, 2265 (1972)

4.43 J.M. Telle, C.L. Tang: Direct absorption spectroscopy, using a rapidly tunable cw-dye-laser. Opt. Commun. *11*, 251 (1974)

4.44 P.R. Bevington: *Data Reduction and Error Analysis for the Physical Sciences* (McGraw-Hill, New York 1969)

4.45 Th.W. Hänsch: "A Self-Calibrating Grating", in Ref. 1.11, p. 423 ff

4.46 P. Cerez, S.J. Bennet: New developments in iodine-stabilized He-Ne-lasers. IEEE Trans. IM-*27*, 396 (1978)

4.47 K.M. Evenson, J.S. Wells, F.R. Petersen, B.L. Danielson, G.W. Day, R.L. Barger, J.L. Hall: Speed of light from direct frequency and wavelength measurements of the methane-stabilized laser. Phys. Rev. Lett. *29*, 1346 (1972)

4.48 K.M. Evenson, D.A. Jennings, F.R. Petersen, J.S. Wells: "Laser Frequency Measurements: A Review, Limitations and Extension to 197 THz", in Ref. 1.11, p. 56

4.49 J.J. Snyder: "Fizeau Wavelength Meter", in Ref. 1.11, p. 419 ff

4.50 R.L. Byer, J. Paul, M.D. Duncan: "A Wavelength Meter", in Ref. 1.11, p. 414

4.50a A. Fischer, H. Kullmer, W. Demtröder: Computer-controlled Fabry-Perot-Wavemeter. Opt. Commun. *39*, 277 (1981)

4.51 J.L. Hall, S.A. Lee: Interferometric real time display of cw dye laser wavelength with sub-Doppler accuracy. Appl. Phys. Lett. *29*, 367 (1976)

4.52 F.V. Kowalski, R.E. Teets, W. Demtröder, A.L. Schawlow: An improved wavemeter for cw lasers; J. Opt. Soc. Am. *68*, 1611 (1978)

4.53 R. Best: Theorie und Anwendung des Phase-Locked Loops (AT-Fachverlag, Stuttgart 1976)
F.M. Gardner: *Phase-Lock Techniques* (Wiley, New York 1966)
Phase-Locked Loop Data Book (Motorola Semiconductor Prod., Inc. 1973)

4.54 P. Juncar, J. Pinard: A new method for frequency calibration and control of a laser. Opt. Commun. *14*, 438 (1975)
P. Jacquinot, P. Juncar, J. Pinard: "Motionless Michelson for High Precision Laser Frequency Measurements: The Sigmameter", in Ref. 1.11, p. 417

4.55 G.C. Mönch: *Interferenzlängenmessung und Brechzahlbestimmung* (Pfalz-Verlag, Basel 1966)

4.55a F. Stöckmann: Photodetectors, their performance and limitations. Appl. Phys. *7*, 1 (1975)

4.56 J.J. Keyes (ed.): *Optical and Infrared Detectors*, Topics in Applied Physics, Vol. 19 (Springer, Berlin, Heidelberg, New York 1977)

4.57 E.H. Putley: "Thermal Detectors", in Ref. 4.56, p. 71 ff

4.58 M.J.E. Golay; Rev. Sci. Instr. *18*, 357 (1947)
4.59 W.M. Doyle: Pyroelectric detectors. Laser Focus *6*, 34 (July 1970)
4.60 A.J. Steckl, R.D. Nelson, B.T. French, R.A. Gudmundsen, D. Schecater: Proc. IEEE *63*, 67 (1975)
4.61 C.A. Hamilton, R.J. Phelan, G.W. Day: Pyroelectric radiometers. Opt. Spectra *9*, 37 (October 1975)
4.62 H.R. Zwicker: "Photoemissive Detectors", in Ref. 4.56, p. 149 ff
4.63 R.L. Bell: *Negative Electron Affinity Devices* (Clarendon Press, Oxford 1973)
4.64 G.H. McCall: High speed inexpensive photodiode assembly. Rev. Sci. Instrum. *43*, 865 (1972)
4.65 P.W. Kruse: "The Photon Detection Process", in Ref. 4.56, p. 5
4.66 EMI Electronics, Ltd.: "An Introduction to the Photomultiplier"; Information sheet (1966)
4.67 L.E. Wood, T.K. Grady, M.C. Thompson: Technique for the measurement of photomultiplier transit time variation. Appl. Opt. *8*, 2143 (1969)
4.68 B. Sipp, J.A. Miehe, R. Lopez Delgado: Wavelength dependence of the time resolution of high speed photomultipliers used in single-photon timing experiments. Opt. Commun. *16*, 202 (1976)
4.69 J.D. Rees, M.P. Givens: Variation of time of flight of electrons through a photomultiplier. J. Opt. Soc. Am. *56*, 93 (1966)
4.70 G. Beck: Operation of a 1P28 photomultiplier with subnanosecond response time. Rev. Sci. Instrum. *47*, 537 (1976)
4.71 A.T. Young: Undesirable effects of cooling photomultipliers. Rev. Sci. Instrum. *38*, 1336 (1967)
4.72 J. Sharpe, C. Eng: "Dark Current in Photomultiplier Tubes", EMI Ltd. Information Document. Ref. R/P021Y70
4.73 *Phototubes and Photocells*, RCA Manual
4.74 J.F. James: On the use of a photomultiplier as a photon counter. Mon. Not. R. Astron. Soc. *137*, 15 (1967)
4.75 H.A.W. Tothill: "Measurement of Very Low Spectral Intensities"; EMI, Ltd. Document Ref. R/P029Z70
R.G. Tull: A comparison of photon counting and current measuring techniques in spectrophotometry. Appl. Opt. *7*, 2023 (1968)
4.76 L.M. Biberman, S. Nudelman (eds.): *Photoelectronic Imaging Devices* (Plenum Press, New York 1971)
4.77 G.A. Morton, A.D. Schnitzler: "Cascade Image Intensifiers", in Ref. 4.76, p. 119 ff
4.78 *Catalogue on Image-Intensifier and Image Converter Tubes* (RCA, Harrison, N.J. 1976)
4.79 S. Jeffers, W. Weller: "Image Intensifier Optical Multichannel Analyzer for Astronomical Spectroscopy", in *Advances in Electronics of Electron Physics*, Vol. 40 B (Academic Press, New York 1976) p. 887 ff
4.80 T.S. Moss, G.J. Burrell, B. Ellis: *Semiconductor Opto-Electronics* (Butterworths, London 1973)
4.81 M. Bleicher: *Halbleiter-Optoelektronik* (Hüthig-Verlag, Heidelberg 1976)
4.82 *The Opto-Electronics Data Book* (Texas Instruments, Dallas, Tex. 1978)
4.83 H. Melchior: "Demodulation and Photodetection Techniques", in *Laser Handbook*, Vol. 1, ed. by F.T. Arrecchi, E.O. Schulz-Dubois (North-Holland, Amsterdam 1972) p. 725 ff
4.84 H. Melchior: Sensitive high speed photodetectors for the demodulation of visible and near infrared light. J. Lumin. *7*, 390 (1973)
4.85 H. Melchior, M.B. Fischer, F.R. Arams: Photodetectors for optical communication systems. Proc. IEEE *58*, 1466 (1970)
4.86 D. Long: "Photovoltaic and Photoconductive Infrared Detectors", in Ref. 4.56, p. 101 ff

4.87 E. Sakuma, K.M. Evenson: Characteristics of tungsten-nickel point contact diodes used as laser harmonic generation mixers. IEEE J. QE-*10*, 599 (1974)
4.88 Princeton Applied Research Corp.: Information Catalogue on the OMA system (Princeton, N.J. 1977) B & M-Spektronik Puchheim/München, Information sheet on the OSA-system
4.89a OMA Vidicon Detectors, PAR Information sheet on Optical Multichannel Analysers (Princeton Applied Research, Princeton, N.J. 1978)
4.89b J.L. Weber: "Gated Optical Multichannel Analyzer for Time Resolved Spectroscopy", SPIE, Conf. Proc. Vol. 82 (1976) p. 60 ff (SPIE, Palos Verdes Estates, Calif.)
4.90 L. Perko, J. Haas, D. Osten: "Cooled and Intensified Array Detectors for Optical Spectroscopy", Proceedings of SPIE 21st Int. Technical Symposium and Instr. Display, Vol. 116 (SPIE, Palos Verdes Estates, Calif.)
4.91 Signal Averagers. Princeton Applied Research Information sheet (Princeton, N.J. 1978)
4.92 Biomation, Palo Alto, Calif. Information sheet on transient recorders
4.93 C. Morgan: "Digital signal processing". Laser Focus *13*, 52 (Nov. 1977) Handshake; Information sheets on Waveform Digitizing Instruments (Tektronix, Inc., Beaverton, Ore. 1979)
4.94 E. Wolf (ed.): *Progress in Optics* (North-Holland, Amsterdam 1970-1977)
4.95 R. Kingslake (ed.): *Applied Optics and Optical Engineering* (Academic Press, New York, London 1965)
4.96 A.C.S. van Heel (ed.): *Advanced Optical Techniques* (North-Holland, Amsterdam 1967)

Chapter 5

5.1 B.A. Lengyel: *Lasers*, 2nd ed. (Wiley Interscience, New York 1971)
5.2 H. Weber, G. Herziger: *Laser* (Physik-Verlag, Weinheim 1972)
5.3 F.T. Arrecchi, E.O. Schulz-Dubois (eds.): *Laser Handbook* (North-Holland, Amsterdam 1972)
5.4 W. Brunner, W. Radloff, H. Junge: *Quantenelektronik* (VEB Deutscher Verlag der Wissenschaften, Berlin 1975)
5.5 A. Maitland, M.H. Dunn: *Laser Physics* (North-Holland, Amsterdam 1969)
5.6 J. Vanier: *Basic Theory of Lasers and Masers* (Gordon and Breach, New York 1971)
5.7 R. Loudon: *The Quantum Theory of Light* (Clarendon Press, Oxford 1973)
5.8 G. Koppelmann: "Multiple Beam Interference and Natural Modes in Open Resonators", in *Progress in Optics*, Vol. 7, ed. by E. Wolf (North-Holland, Amsterdam 1969)
5.9 See, for instance, Refs. 2.3 or 2.5
5.10 A.G. Fox, T. Li: "Resonant Modes in a Maser Interferometer", in *Advances in Quantum Electronics*, ed. by J.R. Singer (Columbia University Press, New York, London 1961) p. 308 ff
5.11 G. Boyd, J.P. Gordon: Confocal multimode resonator for millimeter through optical wavelength masers. Bell Syst. Tech. J. *40*, 489 (1961)
5.12 G.D. Boyd, H. Kogelnik: Generalized confocal resonator theory. Bell Syst. Tech. J. *41*, 1347 (1962)
5.13 A.G. Fox, T. Li: Modes in a maser interferometer with curved and tilted mirrors. Proc. IEEE *51*, 80 (1963)
5.14 See, for instance, G. Arfken: *Mathematical Methods for Physicists* (Academic Press, New York 1970)
5.15 H.K. . Lotsch: A scalar resonator theory for optical frequencies. Opt. Acta *12*, 113 (1965)
5.16 W.R. Bennett, Jr.: Appl. Opt., Supplement on Optical Masers 24-61 (1962)

5.17 D. Kühlke, W. Diehl: Mode selection in cw-laser with homogeneously broadened gain. Opt. Quantum Electron. *9*, 305 (1977)
5.18 I.V. Hertel, A. Stamatovič: Spatial hole burning and oligo-mode distance control in CW dye-lasers. IEEE J. QE-*11*, 210 (1975)
5.19 See, for instance, Ref. 1.5, p. 172 ff, or
V.E. Privalov, S.A. Fridrikhov: The ring gas laser. Sov. Phys. Usp. *12*, 153 (1970)
5.20 E.J. Post: Rev. Mod. Phys. *39*, 475 (1967)
5.21 H. Greenstein: Progress on laser gyros stimulates new interest. Laser Focus *14*, 60 (1978)
5.22 H. Kogelnik, T. Li: Laser beams and resonators. Proc. IEEE *54*, 1312 (1966); reprinted in I.P. Kaminov and A.E. Siegman (eds.): *Laser Devices and Applications* (IEEE Press, New York 1973)

Chapter 6

6.1 R. Beck, W. Englisch, K. Gürs: *Table of Laser Lines in Gases and Vapors*, 2nd Ed. Springer Series in Optical Sciences, Vol. 2 (Springer, Berlin, Heidelberg, New York 1978)
6.2 B.J. Orr: A constant deviation laser tuning device. J. Phys. E *6*, 426 (1973)
6.3 L. Allen, D.G.C. Jones: The helium-neon-laser. Adv. Phys. *14*, 479 (1965)
6.3a C.E. Moore: Atomic Energy Levels, Nat. Stand. Ref. Ser. *35*, NBS Circular 467 (U.S. Dept. Commerce, Washington, D.C. 1971).
6.4 K. Bergmann, W. Demtröder: A new cascade laser transition in He-Ne-mixture. Phys. Lett. *29A*, 94 (1969)
6.5 P.W. Smith: On the optimum geometry of a 6328 Å laser oscillator. IEEE J. QE-*2*, 77 (1966)
6.6 W.B. Bridges, A.N. Chester, A.S. Halsted, J.V. Parker: Ion laser plasmas. Proc. IEEE *59*, 724 (1971)
6.7 A. Ferrario, A. Sirone, A. Sona: Interaction mechanisms of laser transitions in argon and krypton ion-lasers. Appl. Phys. Lett. *14*, 174 (1969)
6.8 C.C. Davis, T.A. King: "Gaseous Ion Lasers", in *Advances in Quantum Electronics*, Vol. 3, ed. by D.W. Goodwin (Academic Press, London 1975)
6.9 G. Herzberg: *Molecular Spectra and Molecular Structure*, Vol. II (Van Nostrand Reinhold, New York 1945)
6.9a D.C. Tyle: "Carbon Dioxyde Lasers", in *Advances in Quantum Electronics*, Vol. 1, ed. by D.W. Goodwin (Academic Press, London 1970)
6.9b K. Gürs: Der CO_2-laser. Z. Angew. Phys. *25*, 379 (1968)
6.9c H.W. Mocker: Rotational level competition in CO_2-lasers. IEEE J. QE-*4*, 769 (1968)
6.10 H. Kogelnik, T. Li: Laser beams and resonators. Proc. IEEE *54*, 1312 (1966)
6.11 J. Haisma: Construction and properties of short stable gas lasers. Phillips Res. Rpt., Supplement No. 1 (1967) and Phys. Lett. *2*, 340 (1962)
6.12 M. Hercher: Tunable single mode operation of gas lasers using intracavity tilted etalons. Appl. Opt. *8*, 1103 (1969)
6.13 P.W. Smith: Stabilized single frequency output from a long laser cavity. IEEE J. QE-*1*, 343 (1965)
6.14 P. Zory: Single frequency operation of argon ion lasers. IEEE J. QE-*3*, 390 (1967)
6.15 V.P. Belayev, V.A. Burmakin, A.N. Evtyunin, F.A. Korolyov, V.V. Lebedeva, A.I. Odintzov: High power single-frequency argon ion laser. IEEE J. QE-*5*, 589 (1969)

6.16 P.W. Smith: Mode selection in lasers. Proc. IEEE *60*, 422 (1972)
6.17 W.W. Rigrod, A.M. Johnson: Resonant prism mode selector for gas lasers. IEEE J. QE-*3*, 644 (1967)
6.18 R.E. Grove, E.Y. Wu, L.A. Hackel, D.G. Youmans, S. Ezekiel: Jet stream cw-dye laser for high resolution spectroscopy. Appl. Phys. Lett. *23*, 442 (1973)
6.19 T.W. Hänsch: Repetitively pulsed tunable dye laser for high resolution spectroscopy. Appl. Opt. *11*, 895 (1973)
6.20 J.P. Goldsborough: "Design of Gas Lasers", in *Laser Handbook I*, ed. by F.T. Arrecchi, E.O. Schulz-Dubois (North-Holland, Amsterdam 1972) p. 597 ff
6.21 Schott-Information Sheet (Jenaer Glaserk Schott & Gen., Hattenbergstraße 10, 65 Mainz, W. Germany, 1972)
6.22 I.S. Chelvdew: *Elektrische Kristalle* (Akademie-Verlag, Berlin 1975)
6.23 K. Bystron: *Technische Elektronik* (Hanser Verlag, München 1974)
6.24 F. Paech, R. Schmiedl, W. Demtröder: Collision free lifetimes of excited NO_2 under very high resolution. J. Chem. Phys. *63*, 4369 (1975)
6.25 K.M. Baird, G.R. Hanes: Stabilisation of wavelengths from gas lasers. Rep. Prog. Phys. *37*, 927 (1974)
6.26 W.J. Tomlinson, R.L. Fork: Frequency stabilisation of a gas laser. Appl. Opt. *8*, 121 (1969)
6.27 H. Hellwig, H.E. Bell, P. Kartaschoff, J.C. Bergquist: Frequency stability of methane-stabilized He-Ne-lasers. J. Appl. Phys. *43*, 450 (1972)
6.28 D.G. Youmans, L.A. Hackel, S. Ezekiel: High-resolution spectroscopy of I_2 using laser-molecular-beam techniques. J. Appl. Phys. *44*, 2319 (1973)
6.29 D.W. Allen: Proc. IEEE *54*, 221 (1966)
6.29a P. Cerez, S.J. Bennet: New developments in iodine-stabilised He-Ne-lasers. IEEE Trans. IM-*27*, 396 (1978)
F. Spieweck: "Wavelength Stabilization of the AR^+-Laser Line at λ = 514,5 nm for Length Measurements of Highest Precision", in *Laser 77, Opto-electronics*, ed. by W. Waidelich (IPC Science and Technology Press, Guildford, Surrey 1977)
6.30 J.L. Hall: "Saturated Absorption Spectroscopy", in *Atomic Physics*, Vol. 3, ed. by S.J. Smith, G.W. Walters (Plenum Press, New York 1973) p. 615 ff
6.31 R. Wallenstein, T.W. Hänsch: Linear pressure tuning of a multielement dye laser spectrometer. Appl. Opt. *13*, 1625 (1974)
6.32 W. Jitschin, G. Meisel: "Precise Frequency Tuning of a Single Mode Dye Laser", in *Laser 77, Opto-Electronics*, Conf. Proc. ed. by W. Waidelich (IPC Science and Technology Press, Guildford, Surrey 1977)
6.33 J.L. Hall: "Sub-Doppler-Spectroscopy, Methane Hyperfine Spectroscopy and the Ultimate Resolution Limits", in Ref. 1.7, p. 105 ff
6.34 K.M. Evenson, D.A. Jennings, F.R. Peterson, J.S. Wells: "Laser Frequency Measurements: A Review, Limitations, Extension to 197 THz (1,5 μm)", in Ref. 1.11, p. 56 ff
6.35 W.R. Rowley, B.W. Jolliffe, K.C. Schotton, A.J. Wallard, P.T. Woods: Laser wavelength measurements and the speed of light. Opt. Quantum Electron. *8*, 1 (1976);
D.J.E. Knight, P.T. Woods: Application of nonlinear devices to optical frequency measurements. J. Phys. E *9*, 898 (1976)
6.36 J. Terrien: International agreement on the value of the speed of light. Metrologia *10*, 9 (1974)
6.37 See, for example, Ref. 1.5, p. 287 ff

6.38 W. Brunner, W. Radloff, K. Junge: *Quantenelektronik* (VEB Deutscher Verlag der Wissenschaften, Berlin 1975) p. 212 ff
6.39 A.L. Schawlow, C.H. Townes: Infrared and optical masers. Phys. Rev. *112*, 1940 (1958)
6.40 C.J. Bordé, J.L. Hall: "Ultrahigh Resolution Saturated Absorption Spectroscopy", in *Laser Spectroscopy*, ed. by R.G. Brewer, H. Mooradian (Plenum Press, New York 1974) pp. 125-142
6.41 S.N. Bagayev, V.P. Chebotajev: Frequency stability and reproducibility of the 3,39 μm He-Ne-laser stabilized on the methane line. Appl. Phys. *7*, 71 (1975)

Chapter 7

7.1 M.J. Colles, C.R. Pidgeon: Tunable lasers. Rep. Prog. Phys. *38*, 329 (1975)
7.1a R.S. McDowell: "High Resolution Infrared Spectroscopy with Tunable Lasers", in *Advances in Infrared and Raman Spectroscopy*, Vol. 5, ed. by R.J.H. Clark, R.E. Hester (Heyden, London 1978)
7.2 E.D. Hinkley, K.W. Nill, F.A. Blum: "Infrared Spectroscopy with Tunable Lasers", in Ref. 1.12, p. 127
7.3 R.W. Campbell, F.M. Mims, III: *Semiconductor Lasers* (Howard W. Sams, Indianapolis 1972)
7.4 A. Mooradian: "High Resolution Tunable Infrared Lasers", in *Very High Resolution Spectroscopy*, ed. by R.A. Smith (Academic Press, London 1976)
7.5 A. Mooradian: "Raman Spectroscopy of Solids", in *Laser Handbook*, ed. by F.T. Arrecchi, E.O. Schulz-Dubois (North-Holland, Amsterdam 1972) p. 1409
7.5a C. Vourmard: External-cavity controlled 32 MHz narrow band cw GaAs-diode laser. Opt. Lett. *1*, 61 (1977)
7.6 H.C. Lasey, M.B. Panisch: *Heterostructure Lasers I and II* (Academic Press, New York 1978)
7.7 J.J. Hsieh, J.A. Rossi, J.P. Donnelly: Room-temperature cw operation of Ga In As P/In P double heterostructure diode lasers emitting at 1.1 μm. Appl. Phys. Lett. *28*, 709 (1976)
7.8 I. Melngailis, A. Mooradian: "Tunable Semiconductor Diode Lasers and Applications", in *Laser Applications in Optics and Spectroscopy*, ed. by S. Jacobs, M. Sargent, M. Scully, J. Scott (Addison-Wesley, New York 1975) p. 1 ff
7.9 J.F. Scott: "Spin-Flip Light Scattering and Spin-Flip Lasers", in *Laser Applications in Optics and Spectroscopy*, ed. by S. Jacobs, M. Sargent, M. Scully, J.F. Scott (Addison-Wesley, New York 1975) p. 123 ff
7.9a S.D. Smith, R.B. Dennis, R.G. Harrison: The spin flip Raman laser. Prog. Quantum Electron. *5*, 205 (1977)
7.10 H.G. Häfele: Spin flip Raman laser. Appl. Phys. *5*, 97 (1974)
7.11 S.R. Brueck, A. Mooradian: Frequency stabilization and fine-tuning characteristics of a cw InSb spin-flip laser. IEEE J. QE-*10*, 634 (1974)
7.12 S.D. Smith: "High Resolution Infrared Spectroscopy: The Spin-Flip Raman Lasers", in Ref. 1.14, p. 13 ff
7.13 M.A. Guerra, S.R.J. Brueck, A. Mooradian: Gradient-field permanent-magnet spin-flip laser. IEEE J. QE-*9*, 1157 (1973)
7.14 R.J. Butcher, R.B. Dennis, S.D. Smith: The tunable spin-flip Raman laser. II. Continuous wave molecular spectroscopy. Proc. Roy. Soc., London *A344*, 541 (1975)

7.15a H.J. Gerritsen, M.E. Heller: High resolution tuned-laser spectroscope. Appl. Opt., Suppl. on Chem. Lasers *73* (1965)

7.15b H.J. Gerritsen: "Tuned Laser Spectroscopy of Organic Vapors", in *Physics of Quantum Electronics*, ed. by P.L. Kelley, B. Lax, P.E. Tannenwald (McGraw-Hill, New York 1966) p. 581

7.16 T. Kasuya: Infrared absorption spectrometer with a broad-band tunable He-Xe-laser. Appl. Phys. *3*, 223 (1974)

7.17 F. O'Neill, W.T. Whitney: Continuously tunable multiatmosphere N_2O and CO_2 lasers. Appl. Phys. Lett. *28*, 539 (1976)

7.18 V.N. Bagratashvili, I.N. Knyazev, V.S. Letokhov, V.V. Lobko: Resonance excitation of C_2H_4-molecule. Luminescence by pulsed high pressure continuously tunable CO_2-laser. Opt. Commun. *14*, 426 (1975)

7.19 N.W. Harris, F. O'Neill, W.T. Whitney: Wide-band interferometric tuning of a multiatmosphere CO_2-laser. Opt. Commun. *16*, 57 (1976)

7.20 F. O'Neill, W.T. Whitney: A high power tunable laser for the 9-12.5 μm spectral range. Appl. Phys. Lett. *31*, 270 (1977)

7.21 P.W. Smith: "High Pressure Waveguide Gas Lasers", in Ref. 1.8, p. 247

7.22 R.L. Abrams: "Wide-band Waveguides CO_2-Lasers", in Ref. 1.8, p. 263

7.23 W.B. Fowler: "Electronic States and Optical Transitions of Color Centers", in *Physics of Color Centers*, ed. by W.B. Fowler (Academic Press, New York 1968)

7.24 F. Lüty: "F_A-Centers in Alkali Halide Crystals", in *Physics of Color Centers*, ed. by W.B. Fowler (Academic Press, New York 1968)

7.25 L.F. Mollenhauer, D.H. Olsen: Broadly tunable lasers using color centers. J. Appl. Phys. *46*, 3109 (1975) and in Ref. 1.9, p. 227

7.26 G. Litfin: "Color Center Lasers". Intern. Conf. on Lasers 78, Orlando, Dec. 1978 and J. Phys. E Sci. Instrum. *11*, 984 (1978)

7.27 H.W. Kogelnik, E.P. Ippen, A. Dienes, Ch.V. Shank: Astigmatically compensated cavities for cw dye lasers. IEEE J. QE-*8*, 373 (1972)

7.28 R. Beigang, G. Litfin, H. Welling: Frequency behavior and linewidth of cw single mode color center lasers. Opt. Commun. *22*, 269 (1977)

7.29 H. Welling, G. Litfin, R. Beigang: "Tunable Infrared Lasers Using Color Centers", in Ref. 1.11, p. 370 ff

7.30 G. Litfin, R. Beigang: Design of tunable cw color center laser. J. Phys. *E11*, 984 (1978)

7.30a L.F. Mollenhauer, D.M. Bloom, A.M. DelGaudio: Broadly tunable cw lasers using F_2^+-centers for the 1.26-1.48 μm and 0.82-1.07 μm bands. Opt. Lett. *3*, 48 (1978)

7.30b L.F. Mollenhauer: Room-temperature stable F_2^+-like center yields cw laser tunable over the 0.99-1.22 μm range. Opt. Lett. *5*, 188 (1980)

7.30c W. Gellermann, K.B. Koch, F. Lüty: Recent progress in color center lasers. Laser Focus *18*, 71 (April 1982)

7.30d P.P. Sorokin: Organic Lasers. Scient. Am. *220* (2), 36 (1969)

7.31 F.P. Schäfer (ed.): *Dye Lasers*, 2nd ed., Topics in Applied Physics, Vol.1 (Springer, Berlin, Heidelberg, New York 1978)

7.31a A. Müller, J. Schulz-Henning, H. Tashiro: Excited state absorption of 1,3,3,1'3',3' hexamethylindotricarbocyanine iodide. Appl. Phys. *12*, 333 (1977)

7.32a G. Marowsky, R. Cordray, F.K. Tittel, W.L. Wilson, J.W. Keto: Energy transfer processes in electron beam excited mixtures of laser dye vapors with rare gases. J. Chem. Phys. *67*, 4845 (1977)

7.32b J.G. Small: "The Dye Laser", in *Physics of Quantum Electronics*, Vol. 4, ed. by A.F. Jacobs, M. Sargent, III, M.O. Scully, Ch.T. Walker (Addison-Wesley, London 1976) p. 343 ff

7.32c B. Steyer, F.P. Schäfer: Stimulated and spontaneous emission from laser dyes in the vapor phase. Appl. Phys. *7*, 113 (1975)

7.33 F.B. Dunning, R.F. Stebbings: The efficient generation of tunable near UV radiation using a N_2-pumped dye laser. Opt. Commun. *11*, 112 (1974)

7.34 T.W. Hänsch: Repetitively pulsed tunable dye laser for high resolution spectroscopy. Appl. Opt. *11*, 895 (1972)

7.34a R. Wallenstein: Pulsed narrow band dye lasers. Opt. Acta *23*, 887 (1976)

7.35 I. Soshan, N.N. Danon, V.P. Oppenheim: Narrowband operation of a pulsed dye laser without intracavity beam expansion. J. Appl. Phys. *48*, 4495 (1977)

7.36 S. Saikan: Nitrogen-laser-pumped single-mode dye laser. Appl. Phys. *17*, 41 (1978)

7.37 M.G. Littman: Single-mode operation of grazing-incidence pulsed dye laser. Opt. Lett. *3*, 138 (1978)

7.38 R. Wallenstein, T.W. Hänsch: Linear pressure tuning of a multielement dye laser spectrometer. Appl. Opt. *13*, 1625 (1974)

7.39 S.M. Curry, R. Cubeddu, T.W. Hänsch: Intensity stabilization of dye laser radiation by saturated amplification. Appl. Phys. *1*, 153 (1973)

7.40 R. Wallenstein, T.W. Hänsch: Powerful dye laser oscillator-amplifier system for high resolution spectroscopy. Opt. Commun. *14*, 353 (1975)

7.41 W. Schmidt: Farbstofflaser. Laser *2*, No. 4, 47 (1970);
G.H. Atkinson, M.W. Schuyler: A simple pulsed laser system, tunable in the ultraviolet. Appl. Phys. Lett. *27*, 285 (1975)

7.42 A. Hirth, H. Fagot: High average power from long pulse dye laser. Opt. Commun. *21*, 318 (1977)

7.43 J. Jethwa, F.P. Schäfer, J. Jasny: A reliable high average power dye laser. IEEE J. QE-*14*, 119 (1978)

7.44 J. Kuhl, G. Marowsky, P. Kunstmann, W. Schmidt: A simple and reliable dye laser system for spectroscopic investigations. Z. Naturforsch. *27a*, 601 (1972)

7.45 H. Walther, J.L. Hall: Tunable dye laser with narrow spectral output. Appl. Phys. Lett. *6*, 239 (1970)

7.46 P.J. Bradley, W.G.I. Caugbey, J.I. Vukusic: High efficiency interferometric tuning of flashlamp-pumped dye-lasers. Opt. Commun. *4*, 150 (1971)

7.47 M. Okada, K. Takizawa, S. Ieiri: Tilted birefringent Fabry-Perot etalon for tuning of dye lasers. Appl. Opt. *15*, 472 (1976)

7.48 M. Okada, S. Ieiri: Electronic tuning of dye-lasers by an electro-optic birefringent Fabry-Perot etalon. Opt. Commun. *14*, 4 (1975)

7.49 J. Kopainsky: Laser scattering with a rapidly tuned dye laser. Appl. Phys. *8*, 229 (1975)

7.50 J.J. Turner, E.I. Moses, C.L. Tang: Spectral narrowing and electro-optical tuning of a pulsed dye-laser by injection-locking to a cw-dye laser. Appl. Phys. Lett. *27*, 441 (1975)

7.50a G.M. Gale: A single mode flashlamp-pumped dye laser. Opt. Commun. *7*, 86 (1973)

7.51 S. Leutwyler, E. Schumacher, L. Wöste: Extending the solvent palette for cw jet-stream dye lasers. Opt. Commun. *19*, 197 (1976)

7.52 P. Anliker, H.R. Lüthi, W. Seelig, J. Steinger, H.P. Weber: 33 watt cw dye laser. IEEE J. QE-*13*, 548 (1977)

7.53 H.W. Kogelnik, E.P. Ippen, A. Dienes, Ch.V. Shank: Astigmatically compensated cavities for cw dye lasers. IEEE J. QE-*8*, 373 (1972)

7.54 H.W. Schröder, H. Dux, H. Welling: Single mode operation of cw dye-lasers. Appl. Phys. *7*, 21 (1975)

7.55 H. Gerhardt, A. Timmermann: High resolution dye-laser spectrometer for measurements of isotope and isomer shifts and hyperfine structure. Opt. Commun. *21*, 343 (1977)

7.56 K. Winkler, J. Kowalski: A magnetic tuning system for dye lasers. Appl. Phys. *14*, 25 (1977)

7.57 G. Marowsky: A comparative study of dye prism ring lasers. IEEE J. QE-*10*, 832 (1974)

7.58 G. Marowsky: A tunable flashlamp-pumped dye ring laser of extremely narrow bandwidth. IEEE J. QE-*9*, 245 (1973)

7.58a G. Marowsky: A single mode dye ring laser without output coupler using frustrated total internal reflection. Z. Naturforsch. A*29a*, 536 (1974)

7.58b T.F. Johnston: Design and performance of a broadband optical diode to enforce one-direction traveling wave operation of a ring laser. IEEE J. QE-*16*, 483 (1980); Focus on Science *3*, No.1, Febr. 1980 (Coherent)

7.59 H.W. Schröder, L. Stein, D. Fröhlich, F. Fugger, H. Welling: A high power single mode cw dye ring-laser. Appl. Phys. *14*, 377 (1978)

7.60 D. Kühlke, W. Diehl: Mode selection in cw laser with homogeneously broadened gain. Opt. Quantum Electron. *9*, 305 (1977)

7.61 Such ring dye lasers are commercially available from Coherent Radiation and from Spectra Physics.

7.62 D. Fröhlich, L. Stein, H.W. Schröder, H. Welling: Efficient frequency doubling of cw dye laser radiation. Appl. Phys. *11*, 97 (1976)

7.62a S.J. Bastow, M.H. Dunn: The generation of tunable UV radiation from 238-249 nm, by intracavity frequency doubling of a coumarin 102 dye laser. Opt. Commun. *35*, 259 (1980)

7.63 J.D. Birks: Excimers. Rep. Prog. Phys. *38*, 903 (1977)

7.64 H. Scheingraber, C.R. Vidal: Discrete and continuous Franck-Condon factors of the Mg_2 $A^1\Sigma_\mu^+ - X^1\Sigma_g^+$ system. J. Chem. Phys. *66*, 3694 (1977)

7.65 H.H. Fleischmann: High current electron beams. Phys. Today *28*, 34 (1975)

7.66 C.P. Wang: Performance of XeF/KrF lasers pumped by fast discharges. Appl. Phys. Lett. *29*, 103 (1976)

7.67 M. Rokni: Rare gas fluoride lasers. IEEE J. QE-*14*, 464 (1978)

7.68 M. Rokni, J.H. Jacob, J.A. Mangano, J. Hsia, A.M. Hawryluk: "Dominant Formation and Quenching Processes in E-Beam Pumped ArF* and KrF* -lasers", in *High-Power Lasers and Applications*, Springer Series in Optical Sciences, Vol. 9, ed. by K.L. Kompa, H. Walther (Springer, Berlin, Heidelberg, New York 1978) p.19-31

7.69 M.L. Bhaumik, R.S. Bradford, E.R. Ault: High efficiency KrF excimer laser. Appl. Phys. Lett. *28*, 23 (1976)

7.70 D.J. Bradley: "Coherent Radiation Generation at Short Wavelengths", in *High-Power Lasers and Applications*, Springer Series in Optical Sciences, Vol. 9, ed. by K.L. Kompa, H. Walther (Springer, Berlin, Heidelberg, New York 1978) p.9-18

7.71 Ch.A. Brau: "Excimer Lasers", in *High-Power Lasers and Applications*, Springer Series in Optical Sciences, Vol. 9, ed. by K.L. Kompa, H. Walther (Springer, Berlin, Heidelberg, New York 1978)

7.71a M.H.R. Hutchinson: Excimers and excimer lasers. Appl. Phys. *21*, 15 (1980)

7.72 C.K. Rhodes (ed.): *Excimer Lasers*, Topics in Applied Physics, Vol. 30 (Springer, Berlin, Heidelberg, New York 1979)

7.72a C.R. Vidal: Coherent VUV sources for high resolution spectroscopy. Appl. Opt. *19*, 3897 (1980)

7.73 F. Zernike, J.E. Midwinter: *Applied Nonlinear Optics* (Academic Press, New York 1973)

7.74 P.G. Harper, B.S. Wherrett (eds.): *Nonlinear Optics* (Academic Press, London 1977)

7.75 D.A. Kleinman, A. Ashkin, G.D. Boyd: Second harmonic generation of light by focussed laser beams. Phys. Rev. *145*, 338 (1966)

7.76 G.C. Baldwin: *An Introduction to Nonlinear Optics* (Plenum Press, New York 1969)

7.77 R.L. Byer: "Parametric Oscillators and Nonlinear Materials", in *Nonlinear Optics*, ed. by P.G. Harper, B.S. Wherrett (Academic Press, London 1977)

7.78 F.B. Dunnings, F.K. Tittel, R.F. Stebbings: The generation of tunable coherent radiation in the wavelength range 2300 to 3000 Å using lithium formate monohydride. Opt. Commun. *7*, 181 (1973)

7.79 H. Dewey: Second Harmonic Generation in $KB_5OH \cdot 4H20$ from 217 to 315 nm. IEEE J. QE-*12*, 303 (1976)
ium formate monohydride. Opt. Commun. *7*, 181 (1973)
7.80 F.B. Dunnings: Tunable ultraviolet generation by sum-frequency mixing. Laser Focus *14*, No. 5, 72 (May 1978)

7.81 S. Blit, E.G. Weaver, F.B. Dunnings, F.K. Tittel: Generation of tunable continuous wave ultraviolet radiation from 257 to 320 nm. Opt. Lett. *1*, 58 (1977)

7.82 G.A. Massey, J.C. Johnson: Wavelength-tunable optical mixing experiments between 208 and 259 nm. IEEE J. QU-*12*, 721 (1976)

7.83 C.R. Vidal: Third harmonic generation of modelocked Nd:glass laser pulses in phase matched Rb-Xe-mixtures. Phys. Rev. A*14*, 2240 (1976)

7.83a R. Hilbig, R. Wallenstein: Enhanced production of tunable VUV radiation by phase matched frequency tripling in krypton and xenon. Appl. Opt. *21*, 913 (1982)

7.83b R. Hilbig, R. Wallenstein: Generation of narrow band tunable VUV radiation. Appl. Phys. B*28*, 202 (1982)

7.84 P.P. Sorokin, J.A. Armstrong, R.W. Dreyfus, R.T. Hodgson, J.R. Lankard, L.H. Manganaro, J.J. Wynne: "Generation of Vacuum Ultraviolet Radiation by Nonlinear Mixing in Atomic and Ionic Vapors", in Ref. 1.9, p. 46

7.85 S.E. Harris, J.F. Young, A.H. Kung, D.M. Bloom, G.C. Bjorklund: "Generation of Ultraviolet and VUV-Radiation", in Ref. 1.8, p. 59

7.86 B.P. Stoicheff, S.C. Wallace: "Tunable Coherent VUV-Radiation", in Ref. 1.10, p. 1

7.86a A.H. Kung, J.F. Young, G.C. Bjorklund, S.E. Harris: Phys. Rev. Lett. *29*, 985 (1972)

7.86b D.M. Bloom: "Optical Frequency Conversion in Metal Vapors", in *Physics of Quantum Electronics*, Vol. 3, ed. by St.F. Jacobs, M.O. Scully, M. Sargent, III, C.D. Cantrell, III (Addison-Wesley, London 1976)

7.87 A.S. Pine: "IR-Spectroscopy Via Difference-Frequency Generation", in Ref. 1.11a, p. 376

7.88 A.S. Pine: High-resolution methane V_3-band spectra using a stabilized tunable difference frequence laser system. J. Opt. Soc. Am. *66*, 97 (1976); *64*, 1683 (1974)

7.89 C.F. Dewey, Jr., L.O. Hocker: Infrared difference frequency generation using a tunable dye laser. Appl. Phys. Lett. *18*, 58 (1971)

7.90 R.Y. Shen (ed.): *Nonlinear Infrared Generation*, Topics in Applied Physics, Vol. 16 (Springer, Berlin, Heidelberg, New York 1977)

7.91 R.G. Byer, R.L. Herbst, R.N. Fleming: "Broadly Tunable IR-Source", in Ref. 1.9, p. 207

7.92 S.E. Harris: Tunable optical parametric oscillators. Proc. IEEE *57*, 2096 (1969)

7.93 A. Yariv: "Parametric Processes", in *Progress in Quantum Electronics*, Vol. 1, Part 1, ed. by J.H. Sanders, S. Stenholm (Pergamon Press, Oxford 1969)

7.94 J. Pinard, J.F. Young: Interferometric stabilization of an optical parametric oscillator. Opt. Commun. *4*, 425 (1972)

7.95 V. Wilke, W. Schmidt: Tunable coherent radiation source covering a spectral range from 185 to 880 nm. Appl. Phys. *18*, 177 (1979)
7.96 W. Hartig, W. Schmidt: A broadly tunable IR waveguide raman laser pumped by a dye laser. Appl. Phys. *18*, 235 (1979)
7.96a Ch. Lin, R.H. Stolen, W.G. French, T.G. Malone: A cw tunable near-infrared (1.085-1.175 μm) Raman oscillator. Opt. Lett. *1*, 96 (1977)
7.97 A.Z. Grasiuk, I.G. Zubarev: High power tunable IR raman lasers. Appl. Phys. *17*, 211 (1978)
7.98 H. Rabin, C.L. Tang (eds.): *Quantum Electronics*, Vol. I, Nonlinear Optics (Academic Press, New York 1975)
7.99 N. Bloembergen: *Nonlinear Optics* (Benjamin, New York 1965)
7.100 M. Schubert, B. Wilhelmi: *Einführung in die Nichtlineare Optik*. (Teubner, Leipzig 1978)

Chapter 8

8.1 W.M. Fairbanks, T.W. Hänsch, A.L. Schawlow: Absolute measurement of very low sodium-vapor densities using laser resonance fluorescence. J. Opt. Soc. Am. *65*, 199 (1975)
8.2 K.H. Becker, D. Haaks, T. Tartarczyk: Measurements of C_2-radicals in flames with a tunable dye-laser. Z. Naturforsch. *29a*, 829 (1974)
8.3 P.J. Dagdigian, H.W. Cruse, R.N. Zare: Laser fluorescence study of AlO formed in the reaction $Al + O_2$: Product state distribution, dissociation energy and radiative lifetime. J. Chem. Phys. *62*, 1824 (1975)
8.4 References to the historical development can be found in H.J. Bauer: Son et lumière or the optoacoustic effect in multilevel systems. J. Chem. Phys. *57*, 3130 (1972)
8.5 Yoh-Han Pao (ed.): *Optoacoustic Spectroscopy and Detection* (Academic Press, New York 1977)
8.6 C. Forbes Dewey, Jr.: "Opto-Acoustic Spectroscopy", in *Impact of Lasers on Spectroscopy*. Proc. Soc. Photo Opt. Instrum. Eng., Vol. 49 (1974) p. 13
8.7 C.K.N. Patel: Spectroscopic measurements of stratospheric nitric oxide and water vapor. Science *184*, 1173 (1974)
8.8 A. Rosenwaig: The Spectraphone; Anal. Chem. *47*, 592A (1975)
8.8a S.O. Kanstadt, P.E. Nordal: Photoacoustic and photothermal spectroscopy. Phys. Technol. *11*, 142 (1980)
8.9 L.B. Kreutzer: Laser optoacoustic spectroscopy. A new technique of gas analysis. Anal. Chem. *46*, 239A (1974)
8.10 W. Schnell, G. Fischer: Spectraphone measurements of isotopes of watervapor and nitricoxyde and of phosgene at selected wavelengths in the CO- and CO_2-laser region. Opt. Lett. *2*, 67 (1978)
8.11 S.D. Smith: "High Resolution Infrared Spectroscopy", in Ref. 1.14, p. 13
8.12 C.K.N. Patel: Use of vibrational energy transfer for excited-state opto-acoustic spectroscopy of molecules. Phys. Rev. Lett. *40*, 535 (1978)
8.13 G. Stella, J. Gelfand, W.H. Smith: Photoacoustic detection spectroscopy with dye laser excitation. The 6190 Å CH_4 and the 6450 NH_3-bands. Chem. Phys. Lett. *39*, 146 (1976)
8.13a A.M. Angus, E.E. Marinero, M.J. Colles: Opto-acoustic spectroscopy with a visible cw dye laser. Opt. Commun. *14*, 223 (1975)
8.14 E.E. Marinero, M. Stuke: Quartz optoacoustic apparatus for highly corrosive gases. Rev. Sci. Instrum. *50*, 31 (1979)
8.15 W. Brunner, H. Paul: On the theory of intracavity absorption. Opt. Commun. *12*, 252 (1974)

8.16 K. Tohama: A simple model for intracavity absorption. Opt. Commun. *15*, 17 (1975)

8.17 G.H. Atkinson, A. Laufer, M. Kurylo: Detection of free radicals by an intracavity dye laser technique. J. Chem. Phys. *59*, 350 (1973)

8.18 W. Brunner, H. Paul: Theory of intracavity absorption spectroscopy. Opt. Quantum Electron. *10*, 139 (1978)

8.19 E.M. Belenov, M.V. Danileiko, V.R. Kozubovskii, A.P. Nedavnii, M.T. Shpak: Ultrahigh resolution spectroscopy based on wave competition in a ring laser. Sov. Phys. JETP *44*, 40 (1976)

8.20 E.A. Sviridenko, M.P. Frolov: Possible investigation of absorption line profiles by intracavity laser spectroscopy. Sov. J. Quantum Electron. *7*, 576 (1977)

8.20a V.M. Baev, T.B. Belikova, E.A. Sviridenko, A.F. Suchkov: Intracavity laser spectroscopy with continuous and quasicontinuous lasers. Sov. Phys. JETP *47*, 21 (1978)

8.21 T.W. Hänsch, A.L. Schawlow, P. Toschek: Ultrasensitive response of a cw-dye laser to selective extinction. IEEE QE-*8*, 802 (1972)

8.22 R.N. Zare: Laser separation of isotopes. Sci. Am., Feb. 1977, p. 86

8.23 R.G. Bray, W. Henke, S.K. Liu, R.V. Reddy, M.J. Berry: Measurement of highly forbidden optical transitions by intracavity dye laser spectroscopy. Chem. Phys. Lett. *47*, 213 (1977)

8.24 E.N. Antonov, V.G. Koloshnikov, V.R. Mironenko: Quantitative measurement of small absorption coefficients in intracavity absorption spectroscopy using a cw-dye laser. Opt. Commun. *15*, 99 (1975)

8.25 K.C. Smith, P.K. Schenck: Opto galvanic spectroscopy of a neon discharge. Chem. Phys. Lett. *55*, 466 (1978)

8.26 D.S. King, P.K. Schenck: Opto galvanic spectroscopy. Laser Focus *14*, 50 (March 1978)

8.27 D. King, P. Schenck, K. Smyth, J. Travis: Direct calibration of laser wavelength and bandwidth using the opto galvanic effect in hollow cathode lamps. Appl. Opt. *16*, 2617 (1977)

8.27a V. Kaufman, B. Edlen: Reference wavelength from atomic spectra in the range 15 Å to 25 ooo Å. J. Phys. Chem. Ref. Data *3*, 825 (1974)

8.27b A. Giacchetti, R.W. Stanley, R. Zalubas: Proposed secondary standard wavelengths in the spectrum of thorium. J. Opt. Soc. Am. *60*, 474 (1969)

8.28 J.E. Lawler, A.I. Ferguson, J.E.M. Goldsmith, D.J. Jackson, A.L. Schawlow: "Doppler free Opto Galvanic Spectroscopy", in Ref. 1.11b, p. 188

8.29a W. Bridges: Characteristics of an opto-galvanic effect in cesium and other gas discharge plasmas. J. Opt. Soc. Am. *68*, 352 (1978)

8.29b P. Popescu, M.L. Pascu, C.B. Collins, B.W. Johnson, I. Popescu: Use of space charge amplification techniques in the absorption spectroscopy of Cs and Cs_2. Phys. Rev. *A8*, 1666 (1973)

8.30 H. Hotop: "Electron Spectrometric Studies of Ionizing Thermal Energy Collisions Involving Excited States; Electronic and Atomic Collisions", Proc. XI ICPEAC, Kyoto (North-Holland, Amsterdam 1979)

8.31 G.S. Hurst, M.H. Nayfeh, J.P. Young, M.G. Payne, L.W. Grossman: "Selective Single Atom Detection in a 10^{19} Atom Background", in Ref. 1.11a, p. 44

8.31a G.S. Hurst, M.G. Payne, S.P. Kramer, J.P Young: Resonance ionization spectroscopy and one atom detection. Rev. Mod. Phys. *51*, 767 (1979)

8.31b G.S. Hurst, M.G. Payne, S.D. Kramer, C.H. Chen: Counting the atoms. Physics Today *33*, Sept. 1980, p.24-29

8.31c K.J. Button (ed.): *Infrared and Submillimeter Waves* (Academic, New York 1979)

8.32 P.B. Davies, K.M. Evenson: "Laser Magnetic Resonance (LMR) Spectroscopy of Gaseous Free Radicals", in Ref. 1.9, p. 132 ff

8.33 K.M. Evenson, C.J. Howard: "Laser Magnetic Resonance Spectroscopy", in Ref. 1.8, p. 535
8.34 W. Urban, W. Herrmann: Zeeman modulation spectroscopy with spin-flip Raman laser. Appl. Phys. *17*, 325 (1978)
8.35 Y. Ueda, K. Shimoda: "Infrared Laser Stark Spectroscopy", in Ref. 1.9, p. 186 ff
8.36 K. Uehara, T. Shimizu, K. Shimoda: High resolution Stark spectroscopy of molecules by infrared and far infrared masers. IEEE J. QE-*4*, 728 (1968)
8.37 E.D. Hinkley: High-resolution infrared spectroscopy with a tunable diode laser. Appl. Phys. Lett. *16*, 351 (1976)
8.38 E.D. Hinkley, K.W. Nill, F.A. Blum: "Infrared Spectroscopy with Tunable Lasers", in Ref. 1.12, p. 127 ff
8.39 K.W. Nill: Spectroscopy with tunable diode lasers. Laser Focus *13*, 32 (1977)
8.39a R.S. Eng, J.F. Butler, K.J. Linden: Tunable diode laser spectroscopy. Opt. Eng. *19*, 945 (1980)
8.40 F. Allario, C.H. Bair, J.F. Butler: High resolution spectral measurements of SO_2 from 1176.0 to 1265.8 cm^{-1} using a single PbSe laser with magnetic and current tuning. IEEE J. QE-*11*, 205 (1975)
8.41 G.P. Montgomery, J.C. Hill: High-resolution diode laser spectroscopy of the 949.2 cm^{-1} band of ethylene. J. Opt. Soc. Am. *65*, 579 (1975)
8.42 A.S. Pine: High-resolution methane V_3-band spectra using a stabilized tunable difference-frequency laser system. J. Opt. Soc. Am. *66*, 97 (1976)
8.43 A.S. Pine: "IR Spectroscopy Via Difference-Frequency Generation", in Ref.1.11a, p. 376 ff
8.44 R.J. Butcher, R.B. Dennis, S.D. Smith: The tunable spin-flip Raman laser: Continuous wave molecular spectroscopy. Proc. Roy. Soc. London *344*, 541 (1975)
8.45 C.K.N. Patel, R.J. Kerl: High resolution opto-acoustic spectroscopy of ^{15}NO: Λ-doubling measurements. Opt. Commun. *24*, 294 (1978)
8.46 T.J. Bridges, E.G. Burkhardt: Zeeman spectroscopy of NO with the magnetospectraphone. Opt. Commun. *22*, 248 (1977)
8.47 R.W. Field, D.O. Harris, T. Tanaka: Continuous wave dye laser excitation spectroscopy CaF $A^2\Pi_r - X^2\Sigma^{+1}$. J. Mol. Spectrosc. *57*, 107 (1975)
8.48 J.M. Green, J.P. Hohimer, F.K. Tittel: A high-resolution cw dye laser spectrometer. Opt. Commun. *9*, 407 (1973)
8.49 R.A. Beaudet, K.G. Weyer, H. Walther: Photoexcitation spectroscopy of BO_2 with a single frequency dye laser. Chem. Phys. Lett. *60*, 486 (1979)
8.50 R.A. Bernheim: *Optical Pumping, an Introduction* (Benjamin, New York 1965)
8.51 B. Budick: "Optical Pumping Methods in Atomic Spectroscopy", in *Advances in Atomic and Molecular Physics*, Vol. 3, ed. by D. R. Bates, I. Esterman (Academic Press, New York 1967) p. 73
8.52 R.N. Zare: "Optical Pumping of Molecules", in Ref. 1.7, p. 29
8.53 W. Happer: Optical pumping. Rev. Mod. Phys. *44*, 169 (1972)
8.54 C. Cohen-Tannoudji: "Optical Pumping with Lasers", in *Atomic Physics*, Vol. 4, ed. by G. zu Putlitz, E.W. Weber, A. Winnacker (Plenum Press, New York 1975)
8.55 B. Decomps, M. Dumont, M. Ducloy: "Linear and Nonlinear Phenomena in Laser Optical Pumping", in Ref. 1.12, p. 284 ff
8.56 M. Broyer, G. Gouedard, J.C. Lehmann, J. Vigue: "Optical Pumping of Molecules", in *Advances in Atomic and Molecular Physics*, Vol. 12, ed. by D.R. Bates, B. Bederson (Academic Press, New York 1976)
8.57 C. Schütte: *The Theory of Molecular Spectroscopy* (North-Holland, Amsterdam 1976)

8.58 G. Herzberg: *Molecular Spectra and Molecular Structure*, Vol. I (Van Nostrand, New York 1950)

8.59 G. Höning, M. Cjajkowski, M. Stock, W. Demtröder: High resolution laser spectroscopy of Cs_2. J. Chem. Phys. *71*, 2138 (1979)

8.60 R. Rydberg: Graphische Darstellung einiger bandenspektroskopischer Ergebnisse. Z. Phys. *73*, 376 (1932)
O. Klein: Zur Berechnung von Potentialkurven zweiatomiger Moleküle mit Hilfe von Spektraltermen. Z. Phys. *76*, 226 (1938)
A.L.G. Rees: The calculation of potential-energy curves from band spectroscopic data. Proc. Phys. Soc., London *A59*, 998 (1947)
R.N. Zare, A.L. Schmeltzkopf, W.J. Harrop, D.L. Albritton: J. Mol. Spectrosc. *46*, 37 (1973)

8.61 G. Ennen, Ch. Ottinger: Laser fluorescence measurements of the ^{7}LiD $(X^1\Sigma^+)$-potential up to high vibrational quantum numbers. Chem. Phys. Lett. *36*, 16 (1975)

8.62 A.G. Gaydon: *Dissociation Energies and Spectra of Diatomic Molecules*. (Chapman and Hall, London 1968)

8.63 W. Demtröder, W. Stetzenbach, M. Stock, J. Witt: Lifetimes and Franck-Condon factors for the $B^1\Pi_u \to X^1\Sigma_g^+$-system of Na_2. J. Mol. Spectrosc. *61*, 382 (1976)

8.64 E.J. Breford, F. Engelke: Laser induced fluorescence in supersonic nozzle beams: Applications to the NaK $D^1\Pi \to X^1\Sigma$ and $D^1\Pi \to a^3\Sigma$ systems. Chem. Phys. Lett. *53*, 282 (1978); J. Chem. Phys. *71*, 1949 (1979)

8.65 H. Scheingraber, C.R. Vidal: "Discrete and Continuous Franck-Condon Factors of the Mg_2 $A^1\Sigma_u \to X^1\Sigma_g^+$ System and Their J-dependence", Third Summer Colloq. on Electronic Transition Lasers, Snowmass Village, 1976 (MIT Press, London 1977)

8.65a J. Tellinghuisen, G. Pichler, W.L. Snow, M.E. Hillard, R.J. Exton: Analysis of the diffuse bands near 6100 Å in the fluorescence spectrum of Cs_2. Chem. Phys. *50*, 313 (1980)

8.66 C.A. Brau, J.J. Ewing: "Spectroscopy, Kinetics and Performance of Rare Gas Halide Lasers", in *Electronic Transition Lasers*, ed. by J.I. Steinfeld (MIT Press, Cambridge, Mass. 1976)

8.67 D. Eisel, D. Zevgolis, W. Demtröder: Sub-Doppler laser spectroscopy of the NaK-molecule. J. Chem. Phys. *71*, 2005 (1979)

8.68 E.V. Condon: Nuclear motions associated with electronic transitions in diatomic molecules. Phys. Rev. *32*, 858 (1928)

8.69 J. Tellinghuisen: The McLennan bands of I_2: A highly structured continuum. Chem. Phys. Lett. *29*, 359 (1974)

8.70 J.L. Kinsey: Laser-induced fluorescence. Ann. Rev. Phys. Chem. *28*, 349 (1977)

8.71 K.L. Kompa: *Chemical Lasers*, Topics in Current Chemistry, Vol. 37 (Springer, Berlin, Heidelberg, New York 1975)
R.W.F. Gross, J.B. Scott (eds.): *Handbook of Chemical Lasers* (Wiley Interscience, New York 1976)

8.72 D.H. Levy, L. Wharton, R.E. Smalley: "Laser Spectroscopy in Supersonic Jets", in *Chemical and Biochemical Applications of Lasers*, Vol. 2, ed. by C.B. Moore (Academic Press, New York 1977) p. 1

8.73 J.B. Anderson: "Molecular Beams", in *Molecular Beams and Low Density Gas-Dynamics*, ed. by P. Wegener (Dekker, New York 1974)

8.74 R.E. Smalley, L. Wharton, D.H. Levy: The fluorescence excitation spectrum of rotationally cooled NO_2. J. Chem. Phys. *63*, 4977 (1975)

8.75 P.J. Dagdigian, H.W. Cruse, A. Schultz, R.N. Zare: Product state analysis of BaO from the reactions $Ba+CO_2$ and $Ba+O_2$. J. Chem. Phys. *61*, 4450 (1974)

8.76a J.G. Pruett, R.N. Zare: State-to-state reaction rates: Ba+HF(v=0,1) $\to$BaF(v=0-12)+H. J. Chem. Phys. *64*, 1774 (1976)

8.76b A. Schultz, A. Siegel: Intern. Conf. Phys. Electr. At. Coll., Paris 1977, Abstracts of Papers (Comissariat A L'Energie Atomique, Paris 1977)
8.77 D.L. Rousseau, P.F. Williams: Discrete and diffuse emission following two photon excitation of the E-state in molecular iodine. Phys. Rev. Lett. *33*, 1369 (1974)
8.78 J. Tellinghuisen: E→B structured continuum in I_2. Phys. Rev. Lett. *34*, 1137 (1975)
8.78a S.A. Edelstein, T.F. Gallagher: "Rydberg Atoms", in *Advances in Atomic and Molecular Physics*, Vol. 14, ed. by D.R. Bates, B. Bederson (Academic Press, New York 1978)
8.78b C.J. Latimer: Recent experiments involving highly excited atoms. Contemporary Physics *20*, 631 (1979)
8.79 G. Leuchs, H. Walther: "Investigation of the Fine Structure Splitting of Rydberg States", in Ref.1.11a, p.299
8.80 J.A. Paisner, R.W. Solarz, E.F. Worden: "Identification of Rydberg States in the Atomic Lanthanides and Actinides", in Ref.1.11a, p.161
8.81 Th.W. Ducas, M.L. Zimmerman: Infrared Stark spectroscopy of sodium Rydberg states. Phys. Rev. *A15*, 1523 (1977)
8.81a K. Fredrikson, S. Svanberg: Stark interaction for excited states in alkali atoms, investigated by laser spectroscopy. Z. Physik *A281*, 189 (1977)
8.82 J. Farley, P. Tsekeris, R. Gupta: "Hyperfine-structure measurements in the Rydberg S and P-states of rubidium and cesium. Phys. Rev. *A15*, 1530 (1977)
8.83 G. zu Putlitz: "Determination of Nuclear Moments with Optical Double Resonance", in Springer Tracts in Modern Physics, Vol. 37 (Springer, Berlin, Heidelberg, New York 1965) p. 105
8.84 H.G. Weber, Ph. Brucat, W. Demtröder, R.N. Zare: Measurement of NO_2 $^2B_2$-state g-values by optical radiofrequency double-resonance. J. Mol. Spectrosc. *75*, 58 (1979)
8.85 G. Belin, L. Holmgren, S. Svanberg: Hyperfine interaction, Zeeman and Stark effects for excited states in rubidium. Phys. Scr. *13*, 351 (1976)
8.86 C.H. Townes, A.L. Schawlow: *Microwave Spectroscopy* (Dover, New York 1975)
8.87 K. Shimoda: "Infrared-Microwave Double-Resonance", in Ref.1.11a, p.279
8.88 K. Shimoda: "Double Resonance Spectroscopy by Means of a Laser", in Ref. 1.12, p. 197
8.89 R.W. Field, A.D. English, T. Tanaka, D.O. Harris, P.A. Jennings: Microwave optical double resonance spectroscopy with a cw dye laser: BaO $X^1\Sigma$ and $A^1\Sigma$. J. Chem. Phys. *59*, 2191 (1973)
8.90a F.K. Klein: Diplomthesis, Fachbereich Physik, Univ. Kaiserslautern 1977
8.90b M.E. Kaminsky, R.T. Hawkins, F.V. Kowalski, A.L. Schawlow: Identification of absorption lines by modulated lower level population: Spectrum of Na_2. Phys. Rev. Lett. *36*, 671 (1976)
8.91 M. Göppert-Mayer: Über Elementarakte mit zwei Quantensprüngen. Ann. Phys. *9*, 273 (1931)
8.92 W. Kaiser, C.G. Garret: Two photon excitation in Ca F_2:Eu^{2+}. Phys. Rev. Lett. *7*, 229 (1961)
8.93 N. Bloembergen, M.D. Levenson: "Doppler-Free Two Photon Absorption Spectroscopy", in Ref. 1.13, p. 315
8.94 P. Bräunlich: "Multiphoton Spectroscopy", in *Progress in Atomic Spectroscopy*, ed. by W. Hanle, H. Kleinpoppen (Plenum Press, New York 1978)
8.95 J.M. Worlock: "Two Photon Spectroscopy", in *Laser Handbook*, ed. by F.T. Arrecchi, E.O. Schulz-Dubois (North-Holland, Amsterdam 1972)

8.96 R.M. Hochstraßer, J.E. Wessel, H.N. Sung: Two-photon excitation spectrum of benzene in the gas phase and the crystal. J. Chem. Phys. *60*, 317 (1974)
8.97 L. Wunsch, H.J. Neusser, E.W. Schlag: Two photon excitation spectrum of benzene and benzene-d_6 in the gas phase. Chem. Phys. Lett. *31*, 433 (1975); *32*, 210 (1975)
8.98 R.G. Bray, R.M. Hochstraßer, H.N. Sung: Two photon excitation spectra of molecular gases: New results for benzene and nitricoxide. Chem. Phys. Lett. *33*, 1 (1975)
8.99 L. Wunsch, H.J. Neusser, E.W. Schlag: Polarization effects in the rotational structure of two-photon spectra in the gas phase. Chem. Phys. Lett. *38*, 216 (1976)
8.100 S.V. Filseth, R. Wallenstein, H. Zacharias: Two photon excitation of CO ($A^1\Pi$) and N_2 ($a^1\Pi_g$). Opt. Commun. *23*, 231 (1977)
8.101 F.H. Faisal, R. Wallenstein, H. Zacharias: Three photon excitation of xenon and carbon monoxide. Phys. Rev. Lett. *39*, 1138 (1977)
8.102 D. Popescu, C.B. Collins, B.W. Johnson, I. Popescu: Multiphoton excitation and ionization of atomic cesium with a tunable dye laser. Phys. Rev. *A9*, 1182 (1974)
8.103 P. Esherik, J.J. Wynne, J.A. Armstrong: "Multiphoton Ionization Spectroscopy of the Alkaline Earths", in Ref. 1.11a, p. 170

Chapter 9

9.1 A. Anderson: *The Raman Effect*, Vols. 1 and 2 (Dekker, New York 1971 and 1973)
9.2 D.A. Long: *Raman Spectroscopy* (McGraw-Hill, New York 1977)
9.3 M.C. Tobin: *Laser Raman Spectroscopy* (Wiley Interscience, New York 1971)
9.4 A. Weber (ed.): *Raman Spectroscopy of Gases and Liquids*, Topics in Current Physics, Vol. 11 (Springer, Berlin, Heidelberg, New York 1979)
9.5 G. Placzek: "Rayleigh-Streuung und Raman-Effekt", in *Handbuch der Radiologie*, Vol. VI, ed. by E. Marx (Akademische Verlagsgesellschaft, Leipzig 1934)
9.6 D.L. Rousseau: "The Resonance Raman Effect", in Ref. 9.4, p. 203 ff
9.7 H.W. Schrötter, H.W. Klöckner: "Raman Scattering Cross Sections in Gases and Liquids", in Ref. 9.4, p. 123 ff
9.8 E.J. Woodbury, W.K. Ny: Proc. IRE *50*, 2367 (1962)
9.9 G. Eckardt: Selection of Raman laser materials. IEEE J. QE-*2*, 1 (1966)
9.10 A. Yariv: *Quantum Electronics*, (Wiley, New York 1967)
9.11 W. Kaiser, M. Maier: "Stimulated Rayleigh, Brillouin and Raman-Spectroscopy", in *Laser Handbook*, ed. by F.T. Arrecchi, E.O. Schulz-Dubois (North-Holland, Amsterdam 1972) p. 1077 ff
9.12 N. Bloembergen: *Nonlinear Optics*, 3rd printing (Benjamin, New York 1977)
9.13 C.S. Wang: "The Stimulated Raman Process", in *Quantum Electronics: A Treatise*, Vol. 1, ed. by H. Rabin, C.L. Tang (Academic Press, New York 1975) Chap. 7
9.14 J.W. Nibler, G.V. Knighten: "Coherent Anti-Stokes Raman Spectroscopy", in Ref. 9.4, Chap. 7
9.15 F. Moya, S.A.J. Druet, J.P.E. Taran: "Rotation-Vibration Spectroscopy of Gases by CARS", in Ref. 1.9, p. 66 ff
9.16 J.P. Taran: "Coherent Anti-Stokes Raman Spectroscopy", in Ref.1.11a, p. 315
9.17 S.A. Akhmanov, A.F. Bunkin. S.G. Ivanov, N.I. Koroteev, A.I. Kourigin, I.L. Shumay: "Development of CARS for Measurement of Molecular Parameters", in Ref. 1.10, p. 389 ff

9.18 P.D. Maker: "Nonlinear Light Scattering in Methane", in *Physics of Quantum Electronics*, ed. by P.L. Kelley, B. Lax. P.E. Tannenwaldt (McGraw-Hill, New York 1960) p. 60

9.19 K. Altmann, G. Strey: Enhancement of the scattering intensity for the hyper-Raman effect. Z. Naturforsch. *32a*, 307 (1977)

9.20 A. Weber: "High-Resolution Rotational Raman Spectra of Gases", in Ref. 9.4, Chap. 3

9.21 E.B. Brown: *Modern Optics* (R. Krieger, New York 1974) p. 251

9.21a J.R. Downey, G.J. Janz: "Digital Methods in Raman Spectroscopy", in *Advances in Infrared and Raman Spectroscopy*, Vol. 1, ed. by R.J.H. Clark, R.E. Hesters (Heyden, London 1975) pp. 1-34

9.22 G.W. Walrafen, J. Stone: Intensification of spontaneous Raman spectra by use of liquid core optical fibers. Appl. Spectrosc. *26*, 585 (1972)

9.23 H.W. Schrötter, J. Bofilias: On the assignment of the second-order lines in the Raman spectrum of benzene. J. Mol. Struct. *3*, 242 (1969)

9.24 W. Kiefer: "Recent Techniques in Raman Spectroscopy", in *Advances in Infrared and Raman Spectroscopy*, Vol. 3, ed. by R.J.H. Clark, R.E. Hester (Heyden, London 1977)

9.25 L. Beardmore, H.G.M. Edwards, D.A. Long, T.K. Tan: "Raman Spectroscopic Measurements of Temperature in a Natural Gas/Air-Flame", in *Lasers in Chemistry*, ed. by M.A. West (Elsevier, Amsterdam 1977)

9.26 M. Lapp, C.M. Penney: "Raman Measurements on Flames", in *Advances in Infrared and Raman Spectroscopy*, Vol. 3, ed. by R.J. Clark, R.E. Hester (Heyden, London 1977)

9.27 J.P. Taran: "CARS Techniques and Applications", in Ref. 1.10, p. 378

9.28 P. Dhamelincourt: "Laser Molecular Microprobe", in *Lasers in Chemistry*, ed. by M.A. West (Elsevier, Amsterdam 1977)

9.29 St.K. Freeman: *Applications of Laser Raman Spectroscopy* (Wiley-Interscience, New York 1974)

9.30 M. Lapp, C.M. Penney (eds.): *Laser Raman Gas Diagnostics* (Plenum Press, New York 1974)

Chapter 10

10.1 R. Abjean, M. Leriche: On the shapes of absorption lines in a divergent atomic beam. Opt. Commun. *15*, 121 (1975)

10.2 R.W. Stanley: Gaseous atomic beam light source. J. Opt. Soc. Am. *56*, 350 (1966)

10.3 G. Höning, M. Cjaikowski, W. Demtröder: High resolution laser spectroscopy of Cs_2. J. Chem. Phys. *71*, 2138 (1979)

10.4 W. Demtröder, F. Paech, R. Schmiedl: Hyperfine-structure in the visible spectrum of NO_2. Chem. Phys. Lett. *26*, 381 (1974)

10.5 R. Schmiedl, I.R. Bonilla, F. Paech, W. Demtröder: Laser spectroscopy of NO_2 under very high resolution. J. Mol. Spectrosc. *68*, 236 (1977)

10.6 L.A. Hackel, K.H. Casleton, S.G. Kukolich, S. Ezekiel: Observation of magnetic octopole and scalar spin-spin interaction in I_2 using laser spectroscopy. Phys. Rev. Lett. *35*, 568 (1975); J. Opt. Soc. Am. *64*, 1387 (1974)

10.7 C. Duke, H. Fischer, H.J. Kluge, H. Kremmling, Th. Kühl, E.W. Otten: Determination of the isotope shift of ^{190}Hg by on line laser-spectroscopy. Phys. Lett. *60A*, 303 (1977)

10.8 Th. Kühl, P. Dabkiewicz, C. Duke, H. Fischer, H.J. Kluge, H. Kremmling: Nuclear shape staggering in very neutron-deficient Hg-isotopes detected by laser-spectroscopy. Phys. Rev. Lett. *39*, 180 (1977)

10.9 G. Nowicki, K. Bekk, J. Göring, A. Hansen, H. Rebel, G. Schatz: Nuclear charge radii and nuclear moments of neutron deficient Ba-isotopes from high resolution laser spectroscopy. Phys. Rev. *C18*, 2369 (1978)

10.10 H.T. Duong, P. Jacquinot, P. Juncar, I. Lieberman, J. Pinard, J.L. Vialle: "High Resolution Laser Spectroscopy of the D-Lines of On-Line Produced Radioactive Sodium Isotopes", in *Laser Spectroscopy*, ed. by I. Haroche, J.C. Pebay-Peyroula, T.W. Hänsch, S.E. Harris, Lecture Notes in Physics, Vol. 43 (Springer, Berlin, Heidelberg, New York 1975)
10.11 P. Jacquinot: "Atomic Beam Spectroscopy", in *High Resolution Laser Spectroscopy*, ed. by K. Shimoda, Topics in Applied Physics, Vol. 13 (Springer, Berlin, Heidelberg, New York 1976)
10.12 W. Lange, J. Luther, A. Steudel: "Dyelasers in Atomic Spectroscopy", in *Advances in Atomic and Molecular Physics*, Vol. 10, ed. by D.R. Bates, B. Bederson (Academic Press, New York 1974)
10.13 D.H. Levy, L. Wharton, R.E. Smalley: "Laser Spectroscopy in Supersonic Jets", in *Chemical and Biochemical Applications of Lasers*, Vol. II, ed. by C.B. Moore (Academic Press, New York 1977)
10.14 J.B. Anderson: "Molecular Beams from Nozzle Sources", in *Molecular Beams and Low Density Gas Dynamics*, ed. by P.P. Wegener (Dekker, New York 1974)
10.15 D.E. Pritchard, R. Ahmed-Bitar, W.P. Lapatovich: "Laser Spectroscopy of Bound Na Ne and Related Atomic Physics", in Ref.1.11a, p.355
10.16 A. Hermann, S. Leutwyler, E. Schuhmacher, L. Wöste: On metal atom clusters IV: Photoionization thresholds and multiphoton ionization spectra of alkali-metal molecules. Helv. Chim. Acta *61*, 453 (1978)
10.17 K. Bergmann, W. Demtröder, P. Hering: Laser diagnostics in molecular beams. Appl. Phys. *8*, 65 (1975)
10.17a M.P. Sinha, A. Schultz, R.N. Zare: Internal state distribution of alkali dimers in supersonic nozzle beams. J. Chem. Phys. *58*, 549 (1973)
10.18a S.L. Kaufman: High resolution laser spectroscopy in fast beams. Opt. Commun. *17*, 309 (1976)
10.18b W.H. Wing, G.A. Ruff, W.E. Lamb, J.J. Spezeski: Observation of the infrared spectrum of the hydrogen molecular ion HD^+. Phys. Rev. Lett. *36*, 1488 (1976)
10.19 T. Meier, H. Hühnermann, W. Wagner: High-resolution spectroscopy on a fast beam of metastable xenon ions. Opt. Commun. *20*, 397 (1977)
10.20 E.W. Otten: "Hyperfine and Isotope Shift Measurements", in *Atomic Physics*, Vol. 5, ed. by R. Marrus, M. Prior, H. Shugart (Plenum Press, New York 1977) p. 239
10.21 B.A. Huber, T.M. Miller, P.C. Cosby, H.D. Zeman, R.L. Leon, J.T. Moseley, J.R. Peterson: Laser-ion-coaxial beams spectrometer. Rev. Sci. Instrum. *48*, 1306 (1977)
10.22 M. Dufay, M.L. Gaillard: "High Resolution Studies in Fast Ion Beams", in Ref.1.11a, p.231
10.23 R. Neugart, S.L. Kaufman, W. Klempt, G. Moruzzi: "High Resolution Spectroscopy in Fast Atomic Beams", in Ref.1.11a, p.446
10.24 K. Bergmann, U. Hefter, P. Hering: Molecular beam diagnostics with internal state selection. Chem. Phys. *32*, 329 (1978); J. Chem. Phys. *65*, 488 (1976)
10.24a K. Bergmann, U. Hefter, P. Hering: Molecular beam diagnostics with internal state selection: Velocity distribution and dimer formation in a supersonic Na/Na_2-beam. Chem. Phys. *32*, 329 (1978)
10.24b K. Bergmann, R. Engelhardt, U. Hefter, P. Hering: Molecular beam diagnostics with internal state selection II: Intensity distribution of a Na/Na_2 supersonic beam. Chem. Phys. *44*, 23 (1979)
10.25 M.E. Kaminsky, R.T. Hawkins, F.V. Kowalski, L.A. Schawlow: Identification of absorption lines by modulated lower level population: Spectrum of Na_2. Phys. Rev. Lett. *36*, 671 (1976)
10.26 W. Demtröder, D. Eisel, H.J. Foth, G. Höning, M. Raab, H.J. Vedder, D. Zevgolis: Sub-doppler laser spectroscopy of small molecules. J. Mol. Structure *59*, 291 (1980)

10.27a N.F. Ramsay: *Molecular Beams* (Oxford Univ. Press, New York 1956)
10.27b J.C. Zorn, T.C. English: Molecular beam electric resonance spectroscopy. Adv. At. Mol. Phys. *9*, 243 (1973)
10.27c W. Ertmer, B. Hofer: Zerofield hyperfine structure measurements of the metastable states $3d^24s^4F_{3/2\ 9/2}$ of ^{45}Sc using laser-fluorescence-atomic beam magnetic resonance technique. Z. Phys. A *276*, 9 (1976)
10.27d S.D. Rosner, R.A. Holt, T.D. Gaily: Measurement of the zero field hyperfine structure of a single vibrational-rotational level of Na_2 by a laser-fluorescence molecular beam resonance technique. Phys. Rev. Lett. *35*, 785 (1975)
10.27e J. Pembczynski, W. Ertmer, U. Johann, S. Penselin, P. Stinner: Measurement of the hyperfine structure of metastable atomic states of ^{55}Mn using the ABMR-LIRF method. Z. Physik A*291*, 207 (1979); Z. Physik A*294*, 313 (1980)
M. Dubke, W. Jitschin, G. Meisel, W.J. Childs: Laser-RF-double resonance measurement of the quadrupole moments of ^{95}Mo and ^{97}Mo. Phys. Lett. *65A*, 109 (1978)
W. Zeiske, G. Meisel, H. Gebauer, B. Hofer, W. Ertmer: Hyperfine structure of cw dye-laser populated high lying levels of ^{45}Sc by atomic-beam magnetic-resonance. Phys. Lett. *55A*, 405 (1976)
10.27f P. Grundevik, M. Gustavson, I. Lindgren, G. Olson, L. Robertsson, A. Rosén, S. Svanberg: Precision method for hyperfine-structure studies in low-abundance isotopes. Phys. Rev. Lett. *42*, 1528 (1979)
10.28 W.R Bennet, Jr.: Hole-burning effects in a He-Ne-optical maser. Phys. Rev. *126*, 580 (1962)
10.29 W.E. Lamb: Theory of an optical maser. Phys. Rev. *134A*, 1429 (1964)
10.30 A. Szöke, A. Javan: Isotope shift and saturation behavior of the 1.15 - μ transition of Ne. Phys. Rev. Lett. *10*, 521 (1963)
10.31 T.W. Hänsch, M.H. Nayfeh, S.A. Lee, S.M. Curry, I.S. Shahin: Precision measurement of the Rydberg constant by laser saturation spectroscopy of the Balmer α-line in hydrogen and deuterium. Phys. Rev. Lett. *32*, 1336 (1974)
10.31a T.W. Hänsch, A.L. Schawlow, G.W. Series: The spectrum of atomic hydrogen. Scient. Am. *240*, 94 (March 1979)
10.32 H. Gerhardt, E. Matthias, F. Schneider, A. Timmermann: Isotope shifts and hyperfine structure of the 6s-7p-transitions in the cesium isotopes 133, 135 and 137. Z. Phys. A *288*, 327 (1978)
10.33 M.S. Sorem, A.L. Schawlow: Saturation spectroscopy in molecular iodine by intermodulated fluorescence. Opt. Commun. *5*, 148 (1972)
10.34 H.J. Foth: "Sättigungsspektroskopie an Molekülen"; Diplomthesis, Kaiserslautern (1976)
10.35 M.D. Levenson, A.L. Schawlow: Hyperfine interactions in molecular iodine. Phys. Rev. *A6*, 10 (1972)
10.36 R.S. Lowe, H. Gerhardt, W. Dillenschneider, R.F. Curl, Jr., F.K. Tittel: Intermodulated fluorescence spectroscopy of BO_2 using a stabilized dye laser. J. Chem. Phys. *70*, 42 (1979)
10.37a H.J. Foth, F. Spieweck: Hyperfine structure of the R(98), (58-1)-line of I_2 at $\lambda = 514.5$ nm. Chem. Phys. Lett. *65*, 347 (1979)
10.37b M. Broyer, J. Vigué, J.C. Lehmann: Effective hyperfine Hamiltonian in homonuclear diatomic molecules. J. Phys. *39*, 591 (1978)
10.38 J.L. Hall: "Sub-Doppler Spectroscopy, Methane Hyperfine Spectroscopy and the Ultimate Resolution Limits", in Ref. 1.7, p. 105
10.39 S.N. Bagayev, E.V. Baklanov, E. Titov, V.P. Chebotayev: Reproducibility of the frequency of an He-Ne-laser with a methane absorbing cell. JETP Lett. *20*, 130 (1974)
10.40 J.C. Hall, J.A. Magyar: "High Resolution Saturated Absorption Studies of Methane and Some Methyl-Halides", in Ref. 1.13, p. 173

10.41 W.G. Schweitzer, E.G. Kessler, R.D. Deslattes, H.P. Layer, J.R. Whetstone: Description, performance and wavelength of iodine stabilised lasers. Appl. Opt. *12*, 2927 (1973)
10.42 J.L. Hall: "Saturated Absorption Spectroscopy", in *Atomic Physics*, Vol. 3, ed. by St. Smith, G.K. Walters (Plenum Press, New York 1973) p. 615 ff
V.S. Letokhov, V.P. Chebotayer: Quantum optical frequency standards. Sov. Phys. J. Quantum Electron. *4*, 137 (1974)
10.43 J.L. Hall: The laser absolute wavelength standard problem. IEEE J. QE-*4*, 638 (1968)
10.44 R.L. Barger, J.B. West, T.C. English: Frequency stabilization of a cw dye laser. Appl. Phys. Lett. *27*, 31 (1975)
10.45 J.C. Bergquist, R.L. Barger, D.J. Glaze: "High Resolution Spectroscopy of Calcium Atoms", in Ref. 1.11b, p. 120
10.46 R.G. Brewer: Precision determination of CH_3F-dipole moment by nonlinear infrared spectroscopy. Phys. Rev. Lett. *25*, 1639 (1970)
10.47 G. Hermann, A. Scharman: Resonance effects in the output of a He-Ne-laser with an axial magnetic field. Phys. Lett. *24A*, 606 (1967); Z. Phys. *254*, 46 (1972)
10.48 M.S. Feld: "Laser Saturated Spectroscopy in Coupled Doppler-Broadened Transitions", in Ref. 1.6, pp. 369-420
10.49 P. Toschek: "General Survey of Laser Saturated Spectroscopy", in Ref. 1.7, pp. 13-28
10.50 A. Schabert, R. Keil, P.E. Toschek: Dynamic Stark effect of an optical line observed by cross-saturated absorption. Appl. Phys. *6*, 181 (1975)
10.51 I.M. Beterov, V.P. Chebotayev: "Three Level Systems and Their Interaction with Radiation", in *Progress in Quantum Electronics*, Vol. 3 (Pergamon Press 1974)
10.52 M. Sargent, III, P. Toschek, H.G. Danielsmeyer: Unidirectional saturation spectroscopy. Appl. Phys. *11*, 55 (1976)
10.53 V.S. Letokhov, V.P. Chebotayev: *Nonlinear Laser Spectroscopy*, Springer Series in Optical Sciences, Vol. 4, ed. by D.L. MacAdam (Springer, Berlin, Heidelberg, New York 1977)
10.54 N. Bloembergen (ed.): *Nonlinear Spectroscopy*, Proc. Intern. School Enrico Fermi, Course LXIV, Varenna 1977 (North-Holland, Amsterdam 1977)
10.55 J.L. Hall: "The Lineshape Problem in Laser-Saturated Molecular Absorption", in *Lectures on Theoretical Physics*, Boulder Summer School (Gordon and Breach, New York 1975)
10.55a R. Balian, S. Haroche, S. Liebermann (eds.): *Frontiers in Laser Spectroscopy* (North-Holland, Amsterdam 1977)
10.56 C. Wieman, Th.W. Hänsch: Doppler-free laser polarization spectroscopy. Phys. Rev. Lett. *36*, 1170 (1976)
10.56a V. Stert, R. Fischer: Doppler-free polarization spectroscopy using linear polarized light. Appl. Phys. *17*, 151 (1978)
10.56b H. Gerhardt, T. Huhle, J. Neukammer, P.J. West: High resolution polarization spectroscopy of the 557 nm transition of Kr I. Opt. Commun. *26*, 58 (1978)
10.56c M. Raab, G. Höning, R. Castell, W. Demtröder: Doppler-free polarization spectroscopy of the Cs_2 molecule at λ = 6270 Å. Chem. Phys. Lett. *66*, 307 (1979)
10.57 R.E. Teets, F.V. Kowalski, W.T. Hill, N. Carlson, T.W. Hänsch: "Laser Polarization Spectroscopy", in *Advances in Laser Spectroscopy*, Proc. Soc. Phot. Opt. Instr. Eng., Vol. 113, San Diego (1977) p. 80
10.58 M.E. Rose: *Elementary Theory of Angular Momentum* (Wiley, New York 1957)

B. Judd: *Angular Momentum Theory for Diatomic Molecules* (Academic Press, New York 1975)
10.59 M. Raab, G. Höning, W. Demtröder, C.R. Vidal: High resolution studies of the Cs_2-molecule. Chem. Phys. (in preparation)
10.60 R. Castell, K. Wickert, W. Demtröder: Polarization spectroscopy of I_2 with accurate absolute wavelength measurements using a wavemeter. Opt. Commun. (in preparation)
10.61 R. Teets, R. Feinberg, T.W. Hänsch, A.L. Schawlow: Simplification of spectra by polarization labeling. Phys. Rev. Lett. *37*, 683 (1976)
10.61a N.W. Carlson, A.J. Taylor, A.L. Schawlow: Identification of Rydberg states in Na_2 by two-step polarization labeling. Phys. Rev. Lett. *45*, 18 (1980)
10.62 C. Delsart, J.C. Keller: "Doppler-Free Laser Induced Dichroism and Birefringence", in Ref. 1.11, p. 154
10.63 F.V. Kowalski, W.T. Hill, A.L. Schawlow: Saturated-interference spectroscopy. Opt. Lett. *2*, 112 (1978)
10.64 R. Schieder: Interferometric nonlinear spectroscopy. Opt. Commun. *26*, 113 (1978)
10.65 T.J. Bridge, T.K. Chang: Accurate rotational constants of CO_2 from measurements of cw beats in bulk GaAs between CO_2 vibrational-rotational laser lines. Phys. Rev. Lett. *22*, 811 (1969)
10.66 F.R. Petersen, D.G. McDonald, F.D. Cupp, B.L. Danielson: Rotational constants for $^{12}C^{16}O_2$ from beats between Lamb-dip stabilized laser lines. Phys. Rev. Lett. *31*, 573 (1973); in Ref. 1.8, p. 555
C. Freed, D.L. Spears, R.G. O'Donnell: "Precision Heterodyne Calibration", in Ref. 1.8, p. 171
10.67 L.A. Hackel, K.H. Casleton, S.G. Kukolich, S. Ezekiel: Observation of magnetic octupole and scalar spin-spin interactions in I_2 using laser spectroscopy. Phys. Rev. Lett. *35*, 568 (1975)
10.68 F. Spiewek: "Wavelength Stabilisation of the Ar^+-Laser line at λ = 515 nm for length Measurements of Highest Precision", in *Laser 77 Opto-Electronics*, ed. by W. Waidelich (ipc, science and technology press, Guildford, England 1977)
10.69 A.L. Betz, E.C. Sutton, R.A. McLaren: "Infrared Heterodyne Spectroscopy in Astronomy", in Ref. 1.11, p. 31
10.70 Y.R. Shen (ed.): *Nonlinear Infrared Generation*, Topics in Applied Physics, Vol. 16 (Springer, Berlin, Heidelberg, New York 1977)
10.71 N. Bloembergen, M.D. Levenson: "Doppler-Free Two Photon Absorption Spectroscopy", in Ref. 1.13, p. 315
10.72 J.M. Worlock: "Two Photon Spectroscopy", in *Laser Handbook*, ed. by F.T. Arrecchi, E.O. Schulz-Dubois (North-Holland, Amsterdam 1972)
10.73 B. Cagnac, G. Grynberg, F. Biraben: Spectroscopie d'absorption multiphotonique sans effet Doppler. J. Phys. *34*, 845 (1973)
10.74 G. Grynberg, B. Cagnac: Doppler-free multiphoton spectroscopy. Rep. Prog. Phys. *40*, 791-841 (1977)
10.75 F. Biraben, B. Cagnac, G. Grynberg: Experimental evidence of two photon transition without Doppler broadening. Phys. Rev. Lett. *32*, 643 (1974)
10.75a G. Grynberg, B. Cagnac, F. Biraben: "Multiphoton Resonant Processes in Atoms", in *Coherent Nonlinear Optics*, ed. by M.S. Feld, V.S. Letokhov (Springer, Berlin, Heidelberg, New York 1980)
10.76 T.W. Hänsch, K. Harvey, G. Meisel, A.L. Schawlow: Two photon spectroscopy of Na 3s - 4d without Doppler-broadening using cw dye laser. Opt. Commun. *11*, 50 (1974)
10.76a M.D. Levenson, N. Bloembergen: Observation of two-photon absorption without Doppler-broadening on the 3s - 5s transition in sodium vapor. Phys. Rev. Lett. *32*, 645 (1974)

10.77 S.A. Lee, J. Helmcke, J.L. Hall, P. Stoicheff: Doppler-free two photon transitions to Rydberg levels. Opt. Lett. *3*, 141 (1978)

10.78 T.W. Hänsch, S.A. Lee, R. Wallenstein, C. Wieman: Doppler-free two-photon spectroscopy of hydrogen 1S - 2S. Phys. Rev. Lett. *34*, 307 (1975)

10.78a A.I. Ferguson, J.E.M. Goldsmith, T.W. Hänsch, E.W. Weber: "High Resolution Spectroscopy of Atomic Hydrogen", in Ref.1.13b, p.31

10.79 F.H.M. Faisal, R. Wallenstein, H. Zacharias: Three-photon excitation of xenon and carbon-monoxide. Phys. Rev. Lett. *39*, 1138 (1977)

10.80 P. Bräunlich: "Multiphoton Spectroscopy", in *Progress in Atomic Spectroscopy*; Part B, ed. by W. Hanle, H. Kleinpoppen (Plenum Press, New York 1978) p. 777

10.80a J.H. Eberly (ed.): *Multiphoton Processes* (Wiley, New York 1978)

10.81 P. Franken: Interference effects in the resonance fluorescence of "crossed" excited states. Phys. Rev. *121*, 508 (1961)

10.82 R.W. Field, T.H. Bergeman: Radio-frequency spectroscopy and perturbation analysis in CS A^1 (v = 0). J. Chem. Phys. *54*, 2936 (1971)

10.83 W. Hanle: Über magnetische Beeinflussung der Polarisation der Resonanzfluoreszenz. Z. Phys. *30*, 93 (1924)

10.84 M. McClintock, W. Demtröder, R.N. Zare: Level crossing studies of Na_2 using laser induced fluorescence. J. Chem. Phys. *51*, 5509 (1969)

10.85 R.N. Zare: Molecular level crossing spectroscopy. J. Chem. Phys. *45*, 4510 (1966)

10.86 R.N. Zare: Interference effects in molecular fluorescence. Acc. Chem. Res. *4*, 361 (1971)

10.87 F.W. Dalby, M. Broyer, J.C. Lehmann: "Electric Field Level Crossing in Molecular Iodine", in Ref. 1.7, p. 226

10.88 J.C. Lehmann: "Probing Small Molecular with Lasers", in *Frontiers in Laser Spectroscopy*, Vol. I, ed. by R. Balian, S. Haroche, S. Liberman (North-Holland, Amsterdam 1977)

10.89 M. Broyer, J.C. Lehmann, J. Vigue: g-factors and lifetimes in the B-state of molecular iodine. J. Phys. *36*, 235 (1975)

10.90 H. Figger, D.L. Monts, R.N. Zare: Anomalous magnetic depolarization of fluorescence from the NO_2 $^2B_2$-state. J. Mol. Spectrosc. *68*, 388 (1977); J.R. Bonilla, W. Demtröder: Level crossing spectroscopy of NO_2 using Doppler reduced laser excitation in molecular beams. Chem. Phys. Lett. *53*, 223 (1978)

10.91 S. Svanberg: "Measurement and Calculation of Excited Alkali Hyperfine and Stark Parameters", in Ref.1.11a, p.183

10.92 H. Figger, H. Walther: Optical resolution beyond the natural linewidth: A level crossing experiment on the 3 $^2P_{3/2}$-level of sodium using a tunable dye laser. Z. Phys. *267*, 1 (1974)

10.93 M.S. Feld, A. Sanchez, A. Javan: "Theory of Stimulated Level Crossing", in Ref. 1.7, p. 87

10.94 A.C. Luntz, R.G. Brewer: Zeeman-tuned level crossing in $^1\Sigma$ CH_4. J. Chem. Phys. *53*, 3380 (1970)

10.95 A.C. Luntz, R.G. Brewer, K.L. Foster, J.D. Swalen: Level crossing in CH_4 observed by nonlinear absorption. Phys. Rev. Lett. *23*, 951 (1969)

10.96 J.S. Levine, P. Boncyk, A. Javan: Observation of hyperfine level crossing in stimulated emission. Phys. Rev. Lett. *22*, 267 (1969)

10.97 G. Hermann, A. Scharmann: Untersuchungen zur Zeeman-Spektroskopie mit Hilfe nichtlinearer Resonanzen eines Multimoden Lasers. Z. Phys. *254*, 46 (1972)

Chapter 11

11.1 S.L. Shapiro (ed.): *Ultrashort Light Pulses*, Topics in Applied Physics, Vol. 18 (Springer, Berlin, Heidelberg, New York 1977)

11.2 C.V. Shank, E.P. Ippen, S.L. Shapiro (eds.): *Picosecond Phenomena III*, Springer Ser. Chem. Phys., Vol.4 (Springer, Berlin, Heidelberg, New York 1978)
R.M. Hochstrasser, W. Kaiser, C.V. Shank (eds.): *Picosecond Phenomena II*, Springer Ser. Chem. Phys., Vol.14 (Springer, Berlin, Heidelberg, New York 1980)
11.3 K.J. Kaufmann, P.M. Rentzepies: Picosecond spectroscopy in chemistry and biology. Acc. Chem. Res. *8*, 407 (1975)
11.3a M.S. Feld, V.S. Letokhov (eds.): *Coherent Nonlinear Optics*. Springer Topics in Current Physics, Vol.21 (Springer, Berlin, Heidelberg, New York 1980)
11.4 A.W. Ali, A.C. Kolb, A.D. Anderson: Theory of the pulsed molecular nitrogen laser. Appl. Opt. *6*, 2115 (1967)
11.5 R.E. Imhof, F.H. Read: Measurements of lifetimes of atoms, molecules and ions. Rep. Prog. Phys. *40*, 1 (1977)
11.6 P.W. Smith, M.A. Duguay, E.P. Ippen: "Mode-Locking of Lasers", in *Progress in Quantum Electronics*, Vol. 3, ed. by J.H. Sanders, S. Stenholm (Pergamon Press, Oxford 1974)
11.7 E. Hartfield, B.J. Thompson: "Optical Modulators", in *Handbook of Optics*, ed. by W. Driscall, W. Vaugham (McGraw-Hill, New York 1974)
11.8 E.I. Gordon: A review of acousto-optical deflectors and modulator devices. Proc. IEEE *54*, 1391 (1966)
11.9 F.O. Neill: Picosecond pulses from a passively mode-locked cw dye laser. Opt. Commun. *6*, 360 (1972)
11.10 W. Demtröder, W. Stetzenbach, M. Stock, J. Witt: Lifetimes and Franck-Condon factors for the $B^1\Pi_u \to X^1\Sigma_g^+$- system of Na_2. J. Mol. Spectrosc. *61*, 382 (1976)
11.11 C.V. Shank, E.P. Ippen: "Mode-Locking of Dye Lasers", in *Dye Lasers*, ed. by F.P. Schäfer, Topics in Applied Physics, Vol. 1, 2nd ed. (Springer, Berlin, Heidelberg, New York 1977)
11.11a R.K. Jain, C.P. Ausschnitt: Subpicosecond pulse generation in a synchronously mode-locked cw rhodamine 6 G dye laser. Opt. Lett. *2*, 117 (1978)
11.12 N.G. Basov, P. Kryukov, Y.V. Senatskii, S.V. Cherkalin: Production of powerful ultrashort light pulses in a neodynium glass laser. Sov. Phys. JETP *10*, 641 (1970)
11.13 W. Rudolf: Thesis, Fachbereich Physik, Univ. Kaiserslautern (1980)
11.14 J. Kuhl, H. Klingenberg, P. von der Linde: Picosecond and subpicosecond pulse generation in synchronously pumped mode-locked cw dye lasers. Appl. Phys. *18*, 279 (1979)
11.15 Ch.K. Chan: "Synchronously Pumped Dye Lasers"; Laser Technical Bulletin No. 8 (Spectra Physics, February 1978)
11.16 P.B. Carlin, W.R. Bennet, Jr.: Mode-locked cavity dumped laser design considerations. Appl. Opt. *15*, 2020 (1976)
11.17 L. Armstrong, Jr., S. Fenerille: Theoretical analysis of the phase shift measurement of lifetimes using monochromatic light. J. Phys. *B*, Atom. Mol. Phys. *8*, 546 (1975)
11.18a L.J. Cine Love, L.A. Shaver: Time correlated single photon technique; fluorescence lifetimes. Anal. Chem. *48*, 364A (1976)
11.18b I.A. Sellin, D.J. Pegg (eds.): *Beam Foil Spectroscopy*, Vol. 2 (Plenum Press, New York 1976)
11.19 H.J. Andrä: "Quantum Beats and Laser Excitation in Fast Beam Spectroscopy", in *Atomic Physics*, Vol. 4, ed. by G. zu Putlitz, E.W. Weber, A. Winnacker (Plenum Press, New York 1975)
11.20 D. Schulze-Hagenest, H. Harde, W. Brand, W. Demtröder: Fast beam-spectroscopy by combined gas-cell laser excitation for cascade-free measurements of highly excited states. Z. Phys. *A282*, 149 (1977)

11.21 P.J. Bradley, W. Sibbet: Subpicosecond chronoscopy. Appl. Phys. Lett. *27*, 382 (1975)
11.22 M.A. Duguay, J.W. Hansen: An ultrafast light gate. Appl. Phys. Lett. *15*, 192 (1969)
11.23 H.P. Weber: Generation and measurement of ultrashort light pulses. J. Appl. Phys. *39*, 6041 (1968)
11.24 R.J. Glauber: *Quantum Optics and Electronics* (Gordon and Breach, New York 1964)
11.25 K.H. Drexhage: Multiphoton excitation of fluorescence in standing light waves and measurement of picosecond pulses. Appl. Phys. Lett. *14*, 318 (1969)
11.26 H.P. Weber, R. Dändliker: Intensity interferometry by two photon excitation of fluorescence. IEEE J. QE-*4*, 1009 (1968)
11.27 G.R. Fleming, A.E.W. Knight, J.M. Morris, R.J.S. Robinson: Picosecond fluorescence studies of xanthene dyes. Am. Chem. Soc. *99*, 4306 (1977)
11.28 W. Kaiser, A. Seilmeier, A. Lauberau: "Dynamic Spectroscopy of Polyatomic Molecules with Tunable Picosecond Pulses", in Ref. 11.2, p. 2 ff
11.29 A. Lauberau, W. Kaiser: "Picosecond Investigations of Dynamic Processes in Polyatomic Molecules in Liquids", in *Chemical and Biochemical Applications of Lasers*, Vol. II, ed. by C.B. Moore (Academic Press, New York 1977), and "Coherent Picosecond Interaction", in Ref.11.32a, p.271 ff.
11.30 K.J. Choi, H.B. Linn, R. Topp: "Fluorescence Spectroscopy of Subpicosecond States in Liquids", in Ref. 11.2, p. 27 ff
11.31 C.V. Shank, E.P. Ippen: Subnanosecond dye laser pulses. Laser Focus *13*, No. 7, 44 (1977)
11.32 W.W. Parson: "Rapid Reactions in Photobiology", in *Chemical and Biochemical Applications of Lasers*, Vol. V, ed. by C.B. Moore (Academic Press, New York 1974)
11.32a M.S. Feld, V.S. Letokhov (eds.): *Coherent Nonlinear Optics* (Springer, Berlin, Heidelberg, New York 1980)
11.32b J.I. Steinfeld: *Laser and Coherence Spectroscopy* (Plenum Press, New York 1978)
11.33 S. Haroche: "Quantum Beats and Time Resolved Fluorescence Spectroscopy", in Ref. 1.13, p. 253 ff
11.34 H.J. Andrä: Fine structure hyperfine structure and Lamb-shift measurements by the beam foil technique. Phys. Scr. *9*, 257 (1974)
11.35 W. Lange, J. Mlynek: Quantum beats in transmission by time-resolved polarization spectroscopy. Phys. Rev. Lett. *40*, 1373 (1978)
11.35a J. Mlynek, K.H. Drake, W. Lange: "Observation of Transient and Stationary Zeeman Coherence by Polarization Spectroscopy", in Ref.1.11b, p.616
11.36 R.H. Dicke: Coherence in spontaneous radiation processes. Phys. Rev. *93*, 99 (1954)
11.37 E.L. Hahn: Spin echoes. Phys. Rev. *80*, 580 (1950)
11.38 I.D. Abella: "Echoes at Optical Frequencies", in *Progress in Optics*, Vol. VII, ed. by E. Wolf (North-Holland, Amsterdam 1969) p. 140 ff
11.39 S.R. Hartmann: "Photon Echoes", in *Lasers and Light; Readings from Scientific American* (Freeman, San Francisco 1969) p. 303
11.40 C.K.N. Patel, R.E. Slusher: Photon echoes in gases. Phys. Rev. Lett. *20*, 1087 (1968)
11.41 R.G. Brewer: Nonlinear spectroscopy. Science *178*, 247 (1972)
11.42 R.G. Brewer, A.Z. Genack: Optical coherent transients by laser frequency switching. Phys. Rev. Lett. *36*, 1959 (1976)
11.43 R.G. Brewer: Coherent optical transients. Phys. Today *30*, 50 (May 1977)

11.44 A. Schenzle, S. Grossman, R.G. Brewer: Theory of modulated photon echoes. Phys. Rev. *A13*, 1891 (1976)
11.45 R.G. Brewer, R.L. Shoemaker: Photo echo and optical nutation in molecules. Phys. Rev. Lett. *27*, 631 (1971)
11.46 P.R. Berman, J.M. Levy, R.G. Brewer: Coherent optical transient study of molecular collisions. Phys. Rev. *A11*, 1668 (1975)
11.47 R.G. Brewer, R.L. Shoemaker: Optical free induction decay. Phys. Rev. *A6*, 2001 (1972)
11.48 R.G. Brewer: "Coherent Optical Spectroscopy", in Ref. 1.14, p. 127
11.49 R.G. Brewer, A.Z. Genack, S.B. Grossman: "Coherent Transients and Pulse Fourier Transform Spectroscopy", in Ref.1.11a, p.220 ff.
11.50 S.B. Grossman, A. Schenzle, R.S. Brewer: Pulse Fourier-transform optical spectroscopy. Phys. Rev. Lett. *38*, 275 (1977)
11.51 T.W. Hänsch: "Multiple Coherent Interactions", in Ref.1.11a, p.149 ff.
11.52 A.I. Ferguson, J.N. Eckstein, T.W. Hänsch: Polarization spectroscopy with ultrashort light pulses. Appl. Phys. *18*, 257 (1979)

Chapter 12

12.1 Th.W. Hänsch, P. Toschek: On pressure broadening in a He-Ne-laser. IEEE J. QE-*5*, 61 (1969)
12.2 J.L. Hall: "Saturated Absorption Spectroscopy", in *Atomic Physics*, Vol. 3, ed. by S.J. Smith, G.K. Walthers (Plenum Press, New York 1973) p. 615 ff
12.3 L.S. Vasilenko, V.P. Kochanov, V.P. Chebotajev: Nonlinear dependence of optical resonance widths at CO_2-transitions on pressure. Opt. Commun. *20*, 409 (1977)
12.4 J.L. Hall: *The Lineshape Problem in Laser-Saturated Molecular Absorption*, Inst. for Theoretical Physics Boulder Summer School (Gordon & Breach, New York 1969)
12.5 S.N. Bagyev: "Spectroscopic Studies into Elastic Scattering of Excited Particles", in Ref. 1.11b, p. 222
12.6 P. Toschek, Th.W. Hänsch: Laser differential spectrometric measurement on neon depolarisation. Phys. Lett. *22*, 150 (1966)
12.7 D. Zevgolis, W. Demtröder: Inelastic collision cross sections of Na_2-molecules in groundstates and excited states. Chem. Phys. (in preparation)
12.8 R.B. Kurzel, J.I. Steinfeld, D.A. Hatzenbuhler, G.E. Leroi: Energy transfer processes in monochromatically excited iodine molecules. J. Chem. Phys. *55*, 4822 (1971)
12.9 G. Ennen, Ch. Ottinger: Rotation-vibration-translation energy transfer in laser excited $Li_2(B^1\Pi_u)$. Chem. Phys. *3*, 404 (1974)
12.10 K. Bergmann, W. Demtröder: Inelastic cross sections of excited molecules. J. Phys. B. Atom. Mol. Phys. *5*, 1386, 2098 (1972)
12.11 T.A Brunner, R.D. Driver, N. Smith, D.E. Pritchard: Rotational energy transfer in Na_2-Xe collisions. J. Chem. Phys. *70*, 4155 (1979)
12.12 St. Lemont, G.W. Flynn: Vibrational state analysis of electronic-to-vibrational energy transfer processes. Ann. Rev. Phys. Chem. *28*, 261 (1977)
12.13 L.K. Lam, T. Fujiimoto, A.C. Gallagher, M. Hessel: Collisional excitation transfer between Na and Na_2. J. Chem. Phys. *68*, 3553 (1978)
12.14 G. Ennen, Ch. Ottinger: Collision induced dissociation of laser excited $li_2(B^1\Pi_u)$. J. Chem. Phys. *40*, 127 (1979); *41*, 415 (1979)
12.15 R.N. Zare: Optical pumping of molecules. J. Chem. Phys. *45*, 4510 (1966)
12.16 S.R. Jeyes, A.J. McCaffery, M.D. Rowe: Selection rules for collisional energy transfer in homonuclear diatomics. Chem. Phys. Lett. *48*, 91 (1977)
12.17 T.A. Cool: "Transfer Chemical Lasers", in *Handbook of Chemical Lasers*, ed. by R.W.F. Gross, J.F. Bott (Wiley, New York 1976)

12.17a B. Wellegehausen: Optically pumped cw dimer lasers. IEEE J. QE-*15*, 1108 (1979)
12.18 W.H. Green, J.K. Hancock: Laser excited vibrational energy exchange studies of HF, CO and NO. IEEE J. QE-*9*, 50 (1973)
12.19 Y.T. Yardley, C.B. Moore: Vibrational energy transfer in methane. J. Chem. Phys. *49*, 1111 (1968)
12.20 G.W. Flynn: "Energy Flow in Polyatomic Molecules", in *Chemical and Biochemical Applications of Lasers*, Vol. I, ed. by C.B. Moore (Academic Press, New York 1974) p. 163
12.21 I.E. Clark, M.J. French, D.A. Long: "Direct Measurement of Vibrational Relaxation Times of Some Alcohols and Alkyl Halides", in *Lasers in Chemistry*, ed. by M. West (Elsevier, Amsterdam 1977)
12.22 B. Drullinger, R.N. Zare: Optical pumping of molecules I and II. J. Chem. Phys. *51*, 5532 (1969); *59*, 4225 (1973)
12.23 F. König, H.G. Weber: Relaxation studies of groundstate Na_2 by optical pumping transients. Chem. Phys. *45*, 91 (1980)
12.24 S.G. Rautian: "Investigation of Collisions by Nonlinear Spectroscopic Methods", in *Atomic Physics*, Vol. 6, ed. by R. Damburg (Plenum Press, New York 1979) p. 493
12.25 Ch. Ottinger, M. Schröder: Rate constants for collision induced transitions in groundstate Li_2 from two laser spectroscopy. Chem. Phys. *45*, (1980)
12.26 P.R. Berman: "Study of Collisions by Laser Spectroscopy", Advances in Atomic and Molecular Physics, Vol. 13 (Academic Press, New York 1977) pp. 57-112
12.27 I.V. Hertel, W. Stoll: "Collision Experiments with Laser Excited Atoms in Crossed Beams", Advances in Atomic and Molecular Physics, Vol. 13 (Academic Press, New York 1977) pp. 113-228
12.28 G.W. Flynn: "Energy Flow in Polyatomic Molecules", in *Chemical and Biochemical Applications of Lasers*, Vol. I, ed. by C.B. Moore (Academic Press, New York 1974)
12.29 J.I. Steinfeld: "Energy Transfer Processes", in *Chemical Kinetics*, Phys. Chemistry Series One, Vol. 9, ed. by J.C. Polany (Butterworths, London 1972)
J.I. Steinfeld: *Laser and Coherence Spectroscopy* (Plenum Press, New York 1978)
12.30 V. Borkenhagen, H. Malthau, J.P. Toennies: Molecular beam measurements of inelastic cross sections for transitions between defined rotational states of CsF. J. Chem. Phys. *71*, 1722 (1979)
12.31 K. Bergmann, R. Engelhardt, U. Hefter, J. Witt: State-resolved differential cross sections for rotational transitions in Na_2 + Ne-collisions. Phys. Rev. Lett. *40*, 1446 (1978)
12.32 K. Bergmann, R. Engelhardt, U. Hefter, J. Witt: State-to-state differential cross sections for rotational transitions in Na_2 + He-collisions. J. Chem. Phys. *71*, 2726 (1979)
12.33 K. Bergmann, U. Hefter, J. Witt: State-to-state differential cross sections for rotationally inelastic scattering of Na_2 by He. J. Chem. Phys. *72*, 4777 (1980)
12.34 I.V. Hertel, H. Hofmann, K.A. Rost: Electronic to vibrational-rotational energy transfer in collisions of Na(3^2P) with simple molecules. Chem. Phys. Lett. *47*, 163 (1977)
12.35 H.W. Hermann, I.V. Hertel, W. Reiland, A. Stamatovic, W. Stoll: Measurement of the collision-induced alignment in the differential scattering by laser excited atoms. J. Phys. B, Atom. Mol. Phys. *9*, 251 (1977)
12.36 S.E. Harris, R.W. Falcone, W.R. Green, P.B. Lidow, J.C. White, J.F. Young: "Laser Induced Collisions", in Ref. 1.10, p. 193

12.37 S.E. Harris, J.F. Young, W.R. Green, R.W. Falcone, J. Lukasik, J.C. White, J.R. Wilson, M.D. Wright, G.A. Zdasivk: "Laser Induced Collisional and Radiative Energy Transfer", in Ref. 1.11a, p. 349 ff
12.38 Th.F. George, J.M. Yuan, I.H. Zimmerman, J.R. Laing: Radiative transitions for molecular collisions in an intense laser field. Discuss. Faraday Soc. No. 62, 246 (1976)
12.39 Ph. Cahuzak, P.E. Toschek: "Light Assisted Collisional Energy Transfer", in Ref. 1.11, p. 431
12.40 C.R. Vidal, J. Cooper: Heat pipe oven: A new well-defined metal vapor device for spectroscopic measurements. J. Appl. Phys. *40*, 3370 (1969)
12.40a H. Scheingraber, C.R. Vidal: Heat-pipe oven with well defined column density. Rev. Scient. Instrum. *52*, 1010 (1981)
12.41 M.M. Hessel, P. Jankowski: Two metal heat-pipe oven: Operation, dynamics and use in spectroscopic investigations. J. Appl. Phys. *43*, 209 (1972)
12.41a C.R. Vidal, M.M. Hessel: Heat-pipe oven for homogeneous mixtures of saturated and unsaturated vapors. J. Appl. Phys. *43*, 2776 (1972)
12.41b C.R. Vidal: Spectroscopic Observations of subsonic and sonic vapor flow inside an open ended heat-pipe. J. Appl. Phys. *44*, 2225 (1973)

Chapter 13

13.1 Th.C. English, J.C. Zorn: "Molecular Beam Spectroscopy", in *Methods of Experimental Physics*, Vol. 3. ed. by D. Williams (Academic Press, New York 1974)
13.2 N.F. Ramsey: *Molecular Beams* (Clarendon Press, Oxford 1956)
13.3 J.C. Berquist, S.A. Lee, J.L. Hall: "Ramsey Fringes in Saturation Spectroscopy", in Ref.1.11a, p.142 ff.
13.4 Ch. Bordé: Sur les franges de Ramsay en spectroscopie sans elargissement Doppler. C. R. Acad. Sci. Paris *284*, 101 (1977)
13.5 Y.V. Baklanov, V.P. Chebotajev, B.Y. Dubetsky: The resonance of two-photon absorption in separated optical fields. Appl. Phys. *11*, 201 (1976)
13.5a V.P. Chebotayev: "Coherence in High Resolution Spectroscopy", in Ref.11.32a, p.59 ff.
13.6 S.A. Lee, J. Helmcke, J.L. Hall: "High Resolution Two-Photon Spectroscopy of Rb Rydberg Levels", in Ref. 1.11b, p. 130 ff
13.7 Y.V. Baklanov, B.Y. Dubetsky, V.P. Chebotayev: Non-linear Ramsay resonance in the optical region. Appl. Phys. *9*, 171 (1976)
13.8 J.C. Bergquist, R.L. Barger, D.J. Glaze: "High Resolution Spectroscopy of Calcium Atoms", in Ref. 1.11b, p. 120 ff
13.9 V.P. Chebotayev: The method of separated optical fields for two-level atoms. Appl. Phys. *15*, 219 (1978)
13.10 Ch.J. Bordé, J.L. Hall: "Ultrahigh Saturated Absorption Spectroscopy", in Ref. 1.8, p. 125
13.11 J.L. Hall: "Sub-Doppler Spectroscopy; methane hyperfine spectroscopy and the ultimate resolution limits", in Ref. 1.9, p. 105
13.12 Ch.J. Bordé: "Progress in Understanding Sub-Doppler-Line Shapes", in Ref. 1.11, p. 121
13.13 T.W. Hänsch, A.L. Schawlow: Cooling of gases by laser radiation. Opt. Commun. *13*, 68 (1975)
13.13a D.J. Wineland, W.M. Itano: Laser cooling of atoms. Phys. Rev. *A20*, 1521 (1979)
13.14 V.S. Letokhov, V.G. Minogin, B.D. Pavlik: Cooling and trapping of atoms and molecules by a resonant laser field. Opt. Commun. *19*, 72 (1976)

13.15 J.P. Gordon: Radiation forces and momenta in dielectric media. Phys. Rev. *A8*, 14 (1973)

13.16 J.E. Bjorkholm, R.R. Freeman, A. Ashkin, D.B. Pearson: "Transverse Resonance Radiation Pressure on Atomic Beams and the Influence of Fluctuations", in Ref. 1.11b, p. 49 ff

13.17 V.S. Letokhov: "New Possibilities for the Spectroscopy Inside the Doppler Line in the Optical and γ-Ranges", in Ref. 1.7, p. 128 ff

13.18 V.S. Letokhov, V.G. Mignon, B.D. Pavlik: Cooling and capture of atoms and molecules by a resonant light field. Opt. Commun. *19*, 72 (1976)

13.19 V.S. Letokhov, D. Pavlik: Spectral line narrowing in a gas by atoms trapped in a standing light wave. Appl. Phys. *9*, 229 (1976)

13.20 E. Fischer: Die dreidimensionale Stabilisierung von Ladungsträgern in einem Vierpolfeld. Z. Phys. *156*, 1 (1959)

13.21 H.G. Dehmelt: Radiofrequency spectroscopy of stored ions. Adv. At. Mol. Phys. *3*, 53 (1967); *5*, 109 (1969)

13.22 See, for instance, E.T. Whittacker, G.N. Watson: *A Course of Modern Analysis* (Cambridge University Press, Cambridge 1963)

13.23 D.A. Church, H.G. Dehmelt: Radiative cooling of an electrodynamically contained proton gas. J. Appl. Phys. *40*, 3421 (1969)

13.24 W. Neuhauser, M. Hohenstatt, P.E. Toschek: Visual observation and optical cooling of electrodynamically contained ions. Appl. Phys. *17*, 123 (1978)

13.25 W. Neuhauser, M. Hohenstatt, P.E. Toschek, H. Dehmelt: "Preparation, Cooling and Spectroscopy of Single Localized Ions", in Ref. 1.11b, P. 73 ff (1979); Phys. Rev. Lett. *41*, 233 (1978)

13.26 R.E. Drullinger, D.J. Wineland: "Laser Cooling of Ions Bound to a Penning Trap", in Ref. 1.11b, p. 66 ff (1979); Phys. Rev. Lett. *40*, 1639 (1978)

13.27 J.N. Dodd, G.W. Series: "Time-Resolved Fluorescence Spectroscopy", in *Progress in Atomic Spectroscopy A*, ed. by W. Hanle, H. Kleinpoppen (Plenum Press, New York 1978)

13.28 S. Schenk, R.C. Hilborn, H. Metcalf: Time-resolved fluorescence from Ba and Ca, excited by a pulsed tunable dye laser. Phys. Rev. Lett. *31*, 189 (1973)

13.28a H. Metcalf, W. Phillips: Time resolved subnaturalwidth spectroscopy. Opt. Lett. *5*, 540 (1980)

13.28b F. Shimizu, K. Umezu, H. Takuma: Subnatural linewidth spectroscopy by phase switching of the optical field. Appl. Phys. B*28*, 297 (1982)

13.29 H. Figger, H. Walther: Optical resolution beyond the natural line width: A level crossing experiment on the $3^2P_{3/2}$-level of sodium using a tunable dye laser. Z. Phys. *267*, 1 (1974)

13.30 I.M. Beterov, V.P. Chebotayev: in *Progress in Quantum Electronics*, Vol. 3, ed. by J.H. Sanders (Pergamon, Oxford 1974)

13.31 R.P. Hackel, S. Ezekiel: Observation of subnatural linewidths by two-step resonant scattering in I_2-vapor. Phys. Rev. Lett. *42*, 1736 (1979); and in Ref.1.11b, p.88

13.32 C. Delsart, J.C. Keller: The optical Autler-Townes effect in Doppler-broadened three level systems. J. Phys. (Paris) *39*, 350 (1978)

13.33 W. Hartig, W. Rasmussen, R. Schieder, H. Walther: Study of the frequency distribution of the fluorescent light induced by monochromatic radiation. Z. Phys. *A278*, 205 (1977)

13.34 R.E. Grove, F.Y. Wu, S. Ezekiel: Measurement of the spectrum of resonance fluorescence from a two level atom in an intense monochromatic laser field. Phys. Rev. *A15*, 227 (1977)

Chapter 14

14.1 A.H. Zewail (ed.): *Advances in Laser Chemistry*, Springer Series in Chemical Physics, Vol. 3 (Springer, Berlin, Heidelberg, Jew York 1978)
14.2 K.L. Kompa, S.D. Smith (eds.): *Laser-Induced Processes in Molecules*, Springer Series in Chemical Physics, Vol. 6 (Springer, Berlin, Heidelberg, New York 1979)
14.3 C.B. Moore (ed.): *Chemical and Biological Applications of Lasers*, Vols. I-IV (Academic Press, New York 1974-1979)
14.4 S. Kimel, Sh. Speiser: Lasers and chemistry. Chem. Rev. *77*, 437 (1977)
14.5 A.M. Ronn: Laser chemistry. Sci. Am. *240*, 102 (1979)
14.6 L. Goldman: *Applications of the Laser* (CRC Press, Cleveland, Ohio 1973)
14.6a R.N. Zare, R.B. Bernstein: State-to-state reaction dynamics. Phys. Today *33* (Nov. 1980)
14.7 A. Baronarski, J.E. Butler, J.W. Hudgens, M.C. Lin, J.R. McDonald, M.E. Umstedd: "Chemical Applications of Lasers", in Ref. 14.1, p. 62
14.8 J.H. Clark, K.M. Leary, T.R. Loree: "Laser Synthesis Chemistry and Laser Photogeneration of Catalysis", in Ref. 14.1, p. 74 ff
14.9 R.P. van Duyne: "Laser Excitation of Raman Scattering from Adsorbed Molecules on Electrode Surfaces", in Ref. 14.3, p. 101
14.10 M.S. Djidjoev, R.V. Khokhlov, A.V. Kiselev, V.I. Lygin, V.A. Namiot, A.I. Osipov, V.I. Panchenko, B.I. Provotorov: "Laser Chemistry at Surfaces", in Ref. 1.10, p. 100
14.11 W. Spindel: *Isotope Separation Processes*, Am. Chem. Soc. Symp. Ser. No. 11 (1975)
14.12 R.N. Zare: Laser separation of isotopes. Sci. Am. *236*, 86 (1977)
14.13 W.E. Lamb, Jr.: "Classical Model of SF_6 Multiphoton Dissociation", in Ref. 1.11b, p. 296
14.14 R.V. Ambartzumian: "Dissociation of Polyatomic Molecules by an Intense Infrared Laser Field", in Ref. 1.10, p. 150
14.15 V.S. Letokhov, C.B. Moore: "Laser Isotope Separation", in Ref. 14.3, Vol. II
14.15a N.V. Karlov, B.B. Krynetskii, V.A. Mishin, A.M. Prokhorov: Atomic photoionization and its use in isotope separation. Sov. Phys. USPEKHI *22*, 220 (1979)
14.16 R.G. Harrison, S.R. Butcher: Multiple photon intrared processes in polyatomic molecules. Contemp. Phys. *21*, 19 (1980)
14.17 St.F. Jacobs, M. Sargent, III, M.O. Scully, Ch.T. Walker (eds.): *Physics of Quantum Electronics*, Vol. 4 (Addison-Wesley, New York 1976)
14.17a J.P. Aldrige III, J.H. Birely, C.D. Cantrell III, D.C. Cartwright: Experimental and theoretical studies of laser isotope separation, in [14.17]
14.17b C.P. Robinson, J.J. Jensen, C.D. Cantrell: Laser Isotope Separation, in *Laser Chemistry*, ed. by M.A. West (Elsevier, Amsterdam 1977)
14.18 A. Tönnißen, J. Wanner, K.W. Rothe, H. Walther: Application of a cw chemical laser for remote pollution monitoring and process control. Appl. Phys. *18*, 297 (1979)
14.19 K.W. Rothe, H. Walther: "Remote Sensing Using Tunable Lasers", in Ref. 1.10, p. 279
14.19a J.W. Strobehn (ed.): *Laser Beam Propagation in the Atmosphere*, Topics in Applied Physics, Vol. 25 (Springer, Berlin, Heidelberg, New York 1978)
14.19b V.E. Zuev: *Propagation of Visible and Infrared Radiation in the Atmosphere* (Wiley, New York 1974)
14.20 K.W. Rothe, U. Brinkmann, H. Walther: Remote measurement of NO_2-emission from a chemical factory by the differential absorption technique. Appl. Phys. *4*, 181 (1974)

14.21 R.T.H. Collis, P.B. Russel: "Lidar Measurements of Particles and Gases", in Ref. 14.23, p. 71

14.21a D.T. Ggessing: Environmental remote sensing. Phys. Technol. *10*, 266 (1979); Phys. Technol. *11*, 23 (1980)

14.22 H. Inaba: "Detection of Atoms and Molecules by Raman Scattering and Resonance Fluorescence", in Ref. 14.23, p. 151

14.23 E.D. Hinkley (ed.): *Laser Monitoring of the Atmosphere*, Topics in Applied Physics, Vol. 14 (Springer, Berlin, Heidelberg, New York 1976)

14.24 S. Ezekiel, St. M. Clainer (chairmen): *Impact of Lasers in Spectroscopy*, Proc. of the SPIE, Vol. 49 (Palos Verdes Estates, Calif. 1975)

14.25 E.J. McCartney: *Optics of the Atmosphere* (Wiley, New York 1976)

14.26 A. Andreoni, C. Sacchi, O. Svelto: "Structural Studies of Biological Molecules via Laser-Induced Fluorescence", in Ref. 14.3, Vol. IV, p. 1

14.27 R. Mathies: "Biological Applications of Resonance Raman Spectroscopy in the Visible and Ultraviolet", in Ref. 14.3, Vol. IV, p. 55

14.28 M. Ehrenberg, R. Riegler: "Fluorescence Spectroscopy Applied to Dynamics and Structure of Biopolymers", in Ref. 1.10, p. 314

14.29 M.W. Berns: *Biological Microirradiation Classical and Laser Sources* (Prentice Hall, Englewood Cliffs 1974)

14.30 J. Joussot-Dubien (ed.): *Lasers in Physical Chemistry and Biophysics* (Elsevier, Amsterdam 1975)

14.31 A. Andreoni, A. Longoni, C.A. Sacchi, O. Svelto: "Laser-Induced Fluorescence of Biological Molecules", in Ref. 1.10, p. 303 ff

14.32 R. Mathies, A.R. Oseroff, T.B. Freedman, L. Stryer: "Resonance Raman Spectroscopy: Application of Tunable Lasers to the Study of the Molecular Mechanism and Dynamics of Visual Excitation", in Ref. 1.10, p. 295 ff

14.33 Th.G. Spiro: "Raman Spectra of Biological Materials" in Ref. 14.3, p. 29 ff

14.34 L. Goldman (ed.): *Laser Medicine*, Papers Presented at the Third Conference on the Laser, New York 1976 (New York, Academy of Sciences, New York 1976)

14.35 H. Albrecht, G. Müller, M. Schaldach: Entwicklung eines Raman-spektroskopischen Gasanalysesystems. Biomed. Tech. *22*, 361 (1977); Proc. of the VIIth Intern. Summer School on Quantum Optics, Wiezyca, Poland (1979)

14.36 A. Anders, P. Aufmuth: Dye lasers in photodermatology. Laser Elektroopt. *4*, 36 (1979)

14.37 W.G. Dorner: Laser-nephelometry at the practice in laboratory. Laser Elektroopt. *4*, 38 (1979)

14.38 R. Pratesi, C.A. Sacchi (eds).: *Lasers in Photomedicine and Photobiology*, Springer Ser. Opt. Sci., Vol.22 (Springer, Berlin, Heidelberg, New York 1980)

14.39 G. von Bally, P. Greguss (eds.): *Optics in Biomedical Sciences*, Springer Ser. Opt. Sci., Vol.31 (Springer, Berlin, Heidelberg, New York 1982)

Subject Index